Some Physical Constants

Quantity	Symbol	Value[a]
Atomic mass unit	u	$1.660\ 538\ 86\ (28) \times 10^{-27}$ kg $931.494\ 043\ (80)$ MeV/c^2
Avogadro's number	N_A	$6.022\ 141\ 5\ (10) \times 10^{23}$ particles/mol
Bohr magneton	$\mu_B = \frac{e\hbar}{2m_e}$	$9.274\ 009\ 49\ (80) \times 10^{-24}$ J/T
Bohr radius	$a_0 = \frac{\hbar^2}{m_e e^2 k_e}$	$5.291\ 772\ 108\ (18) \times 10^{-11}$ m
Boltzmann's constant	$k_B = \frac{R}{N_A}$	$1.380\ 650\ 5\ (24) \times 10^{-23}$ J/K
Compton wavelength	$\lambda_C = \frac{h}{m_e c}$	$2.426\ 310\ 238\ (16) \times 10^{-12}$ m
Coulomb constant	$k_e = \frac{1}{4\pi\epsilon_0}$	$8.987\ 551\ 788 \ldots \times 10^9$ N·m²/C² (exact)
Deuteron mass	m_d	$3.343\ 583\ 35\ (57) \times 10^{-27}$ kg $2.013\ 553\ 212\ 70\ (35)$ u
Electron mass	m_e	$9.109\ 382\ 6\ (16) \times 10^{-31}$ kg $5.485\ 799\ 094\ 5\ (24) \times 10^{-4}$ u $0.510\ 998\ 918\ (44)$ MeV/c^2
Electron volt	eV	$1.602\ 176\ 53\ (14) \times 10^{-19}$ J
Elementary charge	e	$1.602\ 176\ 53\ (14) \times 10^{-19}$ C
Gas constant	R	$8.314\ 472\ (15)$ J/mol·K
Gravitational constant	G	$6.674\ 2\ (10) \times 10^{-11}$ N·m²/kg²
Josephson frequency–voltage ratio	$\frac{2e}{h}$	$4.835\ 978\ 79\ (41) \times 10^{14}$ Hz/V
Magnetic flux quantum	$\Phi_0 = \frac{h}{2e}$	$2.067\ 833\ 72\ (18) \times 10^{-15}$ T·m²
Neutron mass	m_n	$1.674\ 927\ 28\ (29) \times 10^{-27}$ kg $1.008\ 664\ 915\ 60\ (55)$ u $939.565\ 360\ (81)$ MeV/c^2
Nuclear magneton	$\mu_n = \frac{e\hbar}{2m_p}$	$5.050\ 783\ 43\ (43) \times 10^{-27}$ J/T
Permeability of free space	μ_0	$4\pi \times 10^{-7}$ T·m/A (exact)
Permittivity of free space	$\epsilon_0 = \frac{1}{\mu_0 c^2}$	$8.854\ 187\ 817 \ldots \times 10^{-12}$ C²/N·m² (exact)
Planck's constant	h $\hbar = \frac{h}{2\pi}$	$6.626\ 069\ 3\ (11) \times 10^{-34}$ J·s $1.054\ 571\ 68\ (18) \times 10^{-34}$ J·s
Proton mass	m_p	$1.672\ 621\ 71\ (29) \times 10^{-27}$ kg $1.007\ 276\ 466\ 88\ (13)$ u $938.272\ 029\ (80)$ MeV/c^2
Rydberg constant	R_H	$1.097\ 373\ 156\ 852\ 5\ (73) \times 10^7$ m^{-1}
Speed of light in vacuum	c	$2.997\ 924\ 58 \times 10^8$ m/s (exact)

Note: These constants are the values recommended in 2002 by CODATA, based on a least-squares adjustment of data from different measurements. For a more complete list, see P. J. Mohr and B. N. Taylor, "CODATA Recommended Values of the Fundamental Physical Constants: 2002." *Rev. Mod. Phys.* **77**:1, 2005.

[a] The numbers in parentheses for the values represent the uncertainties of the last two digits.

Solar System Data

Body	Mass (kg)	Mean Radius (m)	Period (s)	Distance from the Sun (m)
Mercury	3.18×10^{23}	2.43×10^{6}	7.60×10^{6}	5.79×10^{10}
Venus	4.88×10^{24}	6.06×10^{6}	1.94×10^{7}	1.08×10^{11}
Earth	5.98×10^{24}	6.37×10^{6}	3.156×10^{7}	1.496×10^{11}
Mars	6.42×10^{23}	3.37×10^{6}	5.94×10^{7}	2.28×10^{11}
Jupiter	1.90×10^{27}	6.99×10^{7}	3.74×10^{8}	7.78×10^{11}
Saturn	5.68×10^{26}	5.85×10^{7}	9.35×10^{8}	1.43×10^{12}
Uranus	8.68×10^{25}	2.33×10^{7}	2.64×10^{9}	2.87×10^{12}
Neptune	1.03×10^{26}	2.21×10^{7}	5.22×10^{9}	4.50×10^{12}
Pluto[a]	$\approx 1.4 \times 10^{22}$	$\approx 1.5 \times 10^{6}$	7.82×10^{9}	5.91×10^{12}
Moon	7.36×10^{22}	1.74×10^{6}	—	—
Sun	1.991×10^{30}	6.96×10^{8}	—	—

[a] In August 2006, the International Astronomical Union adopted a definition of a planet that separates Pluto from the other eight planets. Pluto is now defined as a "dwarf planet" (like the asteroid Ceres).

Physical Data Often Used

Average Earth–Moon distance	3.84×10^{8} m
Average Earth–Sun distance	1.496×10^{11} m
Average radius of the Earth	6.37×10^{6} m
Density of air (20°C and 1 atm)	1.20 kg/m^3
Density of water (20°C and 1 atm)	1.00×10^{3} kg/m^3
Free-fall acceleration	9.80 m/s^2
Mass of the Earth	5.98×10^{24} kg
Mass of the Moon	7.36×10^{22} kg
Mass of the Sun	1.99×10^{30} kg
Standard atmospheric pressure	1.013×10^{5} Pa

Note: These values are the ones used in the text.

Some Prefixes for Powers of Ten

Power	Prefix	Abbreviation	Power	Prefix	Abbreviation
10^{-24}	yocto	y	10^{1}	deka	da
10^{-21}	zepto	z	10^{2}	hecto	h
10^{-18}	atto	a	10^{3}	kilo	k
10^{-15}	femto	f	10^{6}	mega	M
10^{-12}	pico	p	10^{9}	giga	G
10^{-9}	nano	n	10^{12}	tera	T
10^{-6}	micro	μ	10^{15}	peta	P
10^{-3}	milli	m	10^{18}	exa	E
10^{-2}	centi	c	10^{21}	zetta	Z
10^{-1}	deci	d	10^{24}	yotta	Y

PHYSICS

for Scientists and Engineers with Modern Physics

PHYSICS

for Scientists and Engineers with Modern Physics

5 Chapters 39–46

Seventh Edition

Raymond A. Serway
Emeritus, James Madison University

John W. Jewett, Jr.
California State Polytechnic University, Pomona

Australia • Brazil • Canada • Mexico • Singapore • Spain • United Kingdom • United States

Physics for Scientists and Engineers with Modern Physics, Chapters 39–46, Seventh Edition
Raymond A. Serway and John W. Jewett, Jr.

Physics Acquisition Editor: Chris Hall
Publisher: David Harris
Vice President, Editor-in-Chief, Sciences: Michelle Julet
Development Editor: Ed Dodd
Assistant Editor: Brandi Kirksey
Editorial Assistant: Shawn Vasquez
Technology Project Manager: Sam Subity
Marketing Manager: Mark Santee
Marketing Assistant: Melissa Wong
Managing Marketing Communications Manager: Bryan Vann
Project Manager, Editorial Production: Teri Hyde
Creative Director: Rob Hugel
Art Director: Lee Friedman
Print Buyers: Barbara Britton, Karen Hunt
Permissions Editors: Joohee Lee, Bob Kauser
Production Service: Lachina Publishing Services
Text Designer: Patrick Devine Design
Photo Researcher: Jane Sanders Miller
Copy Editor: Kathleen Lafferty
Illustrator: Rolin Graphics, Progressive Information Technologies, Lachina Publishing Services
Cover Designer: Patrick Devine Design
Cover Image: Front: © 2005 Tony Dunn; Back: © 2005 Kurt Hoffmann, Abra Marketing
Cover Printer: R.R. Donnelley/Willard
Compositor: Lachina Publishing Services
Printer: R.R. Donnelley/Willard

Printed in the United States of America
1 2 3 4 5 6 7 11 10 09 08 07

Library of Congress Control Number: 2006936870

Student Edition:
ISBN-13: 978-0-495-11293-8
ISBN-10: 0-495-11293-3

Thomson Higher Education
10 Davis Drive
Belmont, CA 94002-3098
USA

NASA

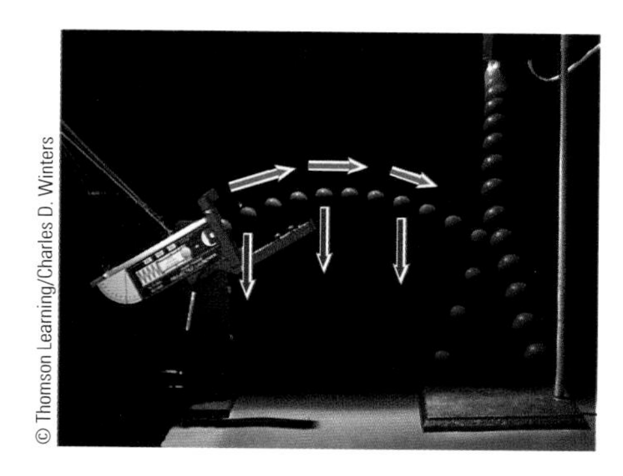

© Thomson Learning/Charles D. Winters

Chapter 7 Energy of a System 163

Chapter 8 Conservation of Energy 195

Chapter 9 Linear Momentum and Collisions 227

John W. Jewett, Jr.

Charles D. Winters

Chapter 10 Rotation of a Rigid Object About a Fixed Axis 269

Chapter 11 Angular Momentum 311

Chapter 12 Static Equilibrium and Elasticity 337

NASA

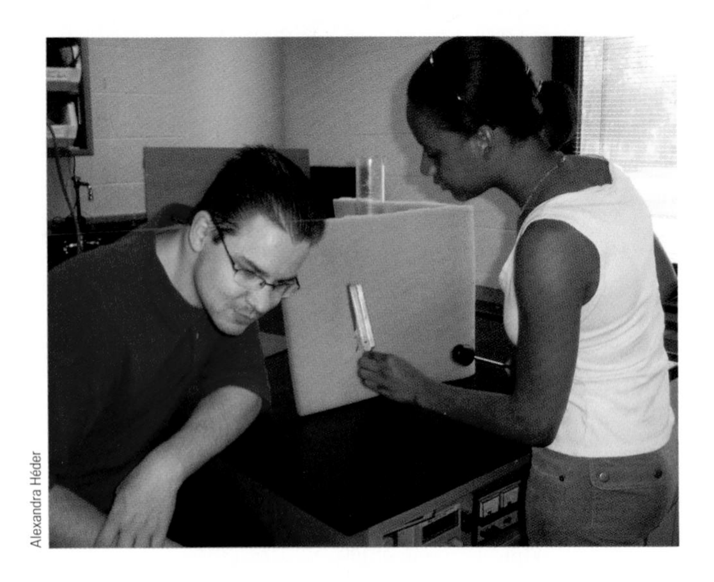
Alexandra Héder

George Semple

© Thomson Learning/Charles D. Winters

PART 4 ELECTRICITY AND MAGNETISM 641

© Thomson Learning/Charles D. Winters

© Thomson Learning/Charles D. Winters

Charles D. Winters

Courtesy of NASA Ames

Raymond A. Serway received his doctorate at Illinois Institute of Technology and is Professor Emeritus at James Madison University. In 1990, he received the Madison Scholar Award at James Madison University, where he taught for 17 years. Dr. Serway began his teaching career at Clarkson University, where he conducted research and taught from 1967 to 1980. He was the recipient of the Distinguished Teaching Award at Clarkson University in 1977 and of the Alumni Achievement Award from Utica College in 1985. As Guest Scientist at the IBM Research Laboratory in Zurich, Switzerland, he worked with K. Alex Müller, 1987 Nobel Prize recipient. Dr. Serway also was a visiting scientist at Argonne National Laboratory, where he collaborated with his mentor and friend, Sam Marshall. In addition to earlier editions of this textbook, Dr. Serway is the coauthor of *Principles of Physics,* fourth edition; *College Physics,* seventh edition; *Essentials of College Physics;* and *Modern Physics,* third edition. He also is the coauthor of the high school textbook *Physics,* published by Holt, Rinehart, & Winston. In addition, Dr. Serway has published more than 40 research papers in the field of condensed matter physics and has given more than 70 presentations at professional meetings. Dr. Serway and his wife, Elizabeth, enjoy traveling, golf, singing in a church choir, and spending quality time with their four children and eight grandchildren.

John W. Jewett, Jr., earned his doctorate at Ohio State University, specializing in optical and magnetic properties of condensed matter. Dr. Jewett began his academic career at Richard Stockton College of New Jersey, where he taught from 1974 to 1984. He is currently Professor of Physics at California State Polytechnic University, Pomona. Throughout his teaching career, Dr. Jewett has been active in promoting science education. In addition to receiving four National Science Foundation grants, he helped found and direct the Southern California Area Modern Physics Institute. He also directed Science IMPACT (Institute for Modern Pedagogy and Creative Teaching), which works with teachers and schools to develop effective science curricula. Dr. Jewett's honors include the Stockton Merit Award at Richard Stockton College in 1980, the Outstanding Professor Award at California State Polytechnic University for 1991–1992, and the Excellence in Undergraduate Physics Teaching Award from the American Association of Physics Teachers in 1998. He has given more than 80 presentations at professional meetings, including presentations at international conferences in China and Japan. In addition to his work on this textbook, he is coauthor of *Principles of Physics,* fourth edition, with Dr. Serway and author of *The World of Physics . . . Mysteries, Magic, and Myth.* Dr. Jewett enjoys playing keyboard with his all-physicist band, traveling, and collecting antiques that can be used as demonstration apparatus in physics lectures. Most importantly, he relishes spending time with his wife, Lisa, and their children and grandchildren.

In writing this seventh edition of *Physics for Scientists and Engineers,* we continue our ongoing efforts to improve the clarity of presentation and include new pedagogical features that help support the learning and teaching processes. Drawing on positive feedback from users of the sixth edition and reviewers' suggestions, we have refined the text to better meet the needs of students and teachers.

This textbook is intended for a course in introductory physics for students majoring in science or engineering. The entire contents of the book in its extended version could be covered in a three-semester course, but it is possible to use the material in shorter sequences with the omission of selected chapters and sections. The mathematical background of the student taking this course should ideally include one semester of calculus. If that is not possible, the student should be enrolled in a concurrent course in introductory calculus.

Objectives

This introductory physics textbook has two main objectives: to provide the student with a clear and logical presentation of the basic concepts and principles of physics and to strengthen an understanding of the concepts and principles through a broad range of interesting applications to the real world. To meet these objectives, we have placed emphasis on sound physical arguments and problem-solving methodology. At the same time, we have attempted to motivate the student through practical examples that demonstrate the role of physics in other disciplines, including engineering, chemistry, and medicine.

Changes in the Seventh Edition

A large number of changes and improvements have been made in preparing the seventh edition of this text. Some of the new features are based on our experiences and on current trends in science education. Other changes have been incorporated in response to comments and suggestions offered by users of the sixth edition and by reviewers of the manuscript. The features listed here represent the major changes in the seventh edition.

QUESTIONS AND PROBLEMS A substantial revision to the end-of-chapter questions and problems was made in an effort to improve their variety, interest, and pedagogical value, while maintaining their clarity and quality. Approximately 23% of the questions and problems are new or substantially changed. Several of the questions for each chapter are in objective format. Several problems in each chapter explicitly ask for qualitative reasoning in some parts as well as for quantitative answers in other parts:

> **19.** ● Assume a parcel of air in a straight tube moves with a constant acceleration of $-4.00\ \text{m/s}^2$ and has a velocity of 13.0 m/s at 10:05:00 a.m. on a certain date. (a) What is its velocity at 10:05:01 a.m.? (b) At 10:05:02 a.m.? (c) At 10:05:02.5 a.m.? (d) At 10:05:04 a.m.? (e) At 10:04:59 a.m.? (f) Describe the shape of a graph of velocity versus time for this parcel of air. (g) Argue for or against the statement, "Knowing the single value of an object's constant acceleration is like knowing a whole list of values for its velocity."

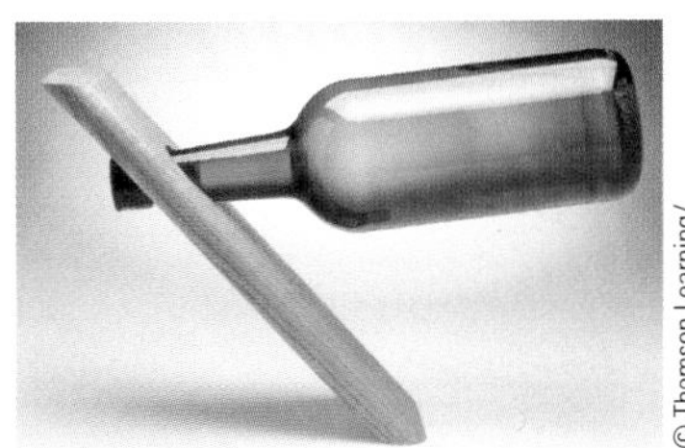

WORKED EXAMPLES All in-text worked examples have been recast and are now presented in a two-column format to better reinforce physical concepts. The left column shows textual information that describes the steps for solving the problem. The right column shows the mathematical manipulations and results of taking these steps. This layout facilitates matching the concept with its mathematical execution and helps students organize their work. These reconstituted examples closely follow a General Problem-Solving Strategy introduced in Chapter 2 to reinforce effective problem-solving habits. A sample of a worked example can be found on the next page.

Each solution has been reconstituted to more closely follow the General Problem-Solving Strategy as outlined in Chapter 2, to reinforce good problem-solving habits.

EXAMPLE 3.2 **A Vacation Trip**

A car travels 20.0 km due north and then 35.0 km in a direction 60.0° west of north as shown in Figure 3.11a. Find the magnitude and direction of the car's resultant displacement.

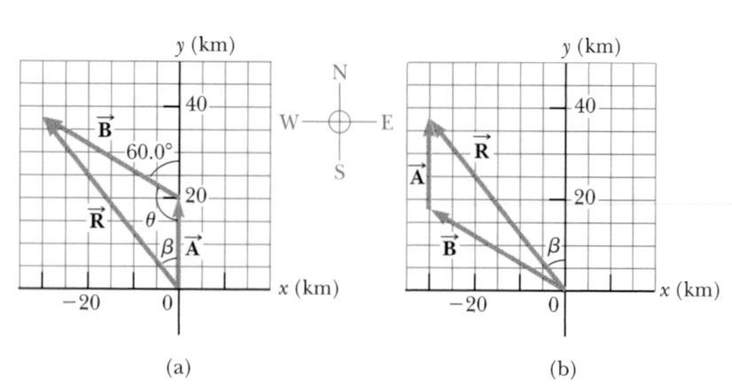

Figure 3.11 (Example 3.2) (a) Graphical method for finding the resultant displacement vector $\vec{R} = \vec{A} + \vec{B}$. (b) Adding the vectors in reverse order ($\vec{B} + \vec{A}$) gives the same result for $\vec{R}$.

SOLUTION

Conceptualize The vectors $\vec{A}$ and $\vec{B}$ drawn in Figure 3.11a help us conceptualize the problem.

Categorize We can categorize this example as a simple analysis problem in vector addition. The displacement $\vec{R}$ is the resultant when the two individual displacements $\vec{A}$ and $\vec{B}$ are added. We can further categorize it as a problem about the analysis of triangles, so we appeal to our expertise in geometry and trigonometry.

Each step of the solution is detailed in a two-column format. The left column provides an explanation for each mathematical step in the right column, to better reinforce the physical concepts.

Analyze In this example, we show two ways to analyze the problem of finding the resultant of two vectors. The first way is to solve the problem geometrically, using graph paper and a protractor to measure the magnitude of $\vec{R}$ and its direction in Figure 3.11a. (In fact, even when you know you are going to be carrying out a calculation, you should sketch the vectors to check your results.) With an ordinary ruler and protractor, a large diagram typically gives answers to two-digit but not to three-digit precision.

The second way to solve the problem is to analyze it algebraically. The magnitude of $\vec{R}$ can be obtained from the law of cosines as applied to the triangle (see Appendix B.4).

Use $R^2 = A^2 + B^2 - 2AB\cos\theta$ from the law of cosines to find R:

$$R = \sqrt{A^2 + B^2 - 2AB\cos\theta}$$

Substitute numerical values, noting that $\theta = 180° - 60° = 120°$:

$$R = \sqrt{(20.0\text{ km})^2 + (35.0\text{ km})^2 - 2(20.0\text{ km})(35.0\text{ km})\cos 120°}$$

$$= 48.2\text{ km}$$

Use the law of sines (Appendix B.4) to find the direction of $\vec{R}$ measured from the northerly direction:

$$\frac{\sin\beta}{B} = \frac{\sin\theta}{R}$$

$$\sin\beta = \frac{B}{R}\sin\theta = \frac{35.0\text{ km}}{48.2\text{ km}}\sin 120° = 0.629$$

$$\beta = 38.9°$$

The resultant displacement of the car is 48.2 km in a direction 38.9° west of north.

Finalize Does the angle β that we calculated agree with an estimate made by looking at Figure 3.11a or with an actual angle measured from the diagram using the graphical method? Is it reasonable that the magnitude of $\vec{R}$ is larger than that of both $\vec{A}$ and $\vec{B}$? Are the units of $\vec{R}$ correct?

Although the graphical method of adding vectors works well, it suffers from two disadvantages. First, some people find using the laws of cosines and sines to be awkward. Second, a triangle only results if you are adding two vectors. If you are adding three or more vectors, the resulting geometric shape is usually not a triangle. In Section 3.4, we explore a new method of adding vectors that will address both of these disadvantages.

What If? Suppose the trip were taken with the two vectors in reverse order: 35.0 km at 60.0° west of north first and then 20.0 km due north. How would the magnitude and the direction of the resultant vector change?

Answer They would not change. The commutative law for vector addition tells us that the order of vectors in an addition is irrelevant. Graphically, Figure 3.11b shows that the vectors added in the reverse order give us the same resultant vector.

What If? statements appear in about 1/3 of the worked examples and offer a variation on the situation posed in the text of the example. For instance, this feature might explore the effects of changing the conditions of the situation, determine what happens when a quantity is taken to a particular limiting value, or question whether additional information can be determined about the problem situation. This feature encourages students to think about the results of the example and assists in conceptual understanding of the principles.

All worked examples are also available to be assigned as interactive examples in the Enhanced WebAssign homework management system (visit **www.pse7.com** for more details).

ONLINE HOMEWORK It is now easier to assign online homework with Serway and Jewett and Enhanced WebAssign. All worked examples, end-of-chapter problems, active figures, quick quizzes, and most questions are available in WebAssign. Most problems include hints and feedback to provide instantaneous reinforcement or direction for that problem. In addition to the text content, we have also added math remediation tools to help students get up to speed in algebra, trigonometry, and calculus.

SUMMARIES Each chapter contains a summary that reviews the important concepts and equations discussed in that chapter. A marginal note next to each chapter summary directs students to additional quizzes, animations, and interactive exercises for that chapter on the book's companion Web site. The format of the end-of-chapter summary has been completely revised for this edition. The summary is divided into three sections: Definitions, Concepts and Principles, and Analysis Models for Problem-Solving. In each section, flashcard-type boxes focus on each separate definition, concept, principle, or analysis model.

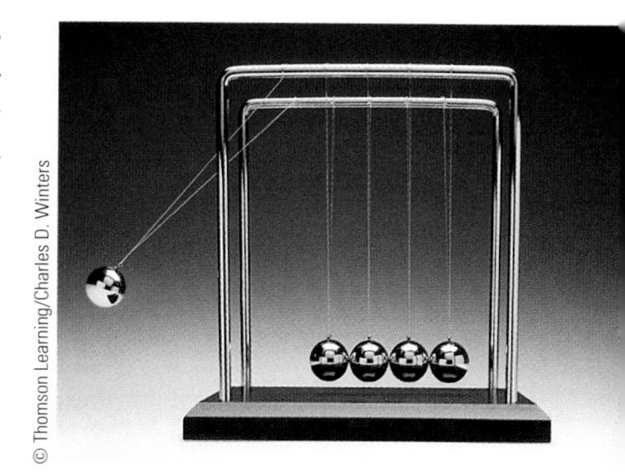

MATH APPENDIX The math appendix, a valuable tool for students, has been updated to show the math tools in a physics context. This resource is ideal for students who need a quick review on topics such as algebra, trigonometry, and calculus.

CONTENT CHANGES The content and organization of the textbook are essentially the same as in the sixth edition. Many sections in various chapters have been streamlined, deleted, or combined with other sections to allow for a more balanced presentation. Vectors are now denoted in boldface with an arrow over them (for example, $\vec{\mathbf{v}}$), making them easier to recognize. Chapters 7 and 8 have been completely reorganized to prepare students for a unified approach to energy that is used throughout the text. A new section in Chapter 9 teaches students how to analyze deformable systems with the conservation of energy equation and the impulse-momentum theorem. Chapter 34 is longer than in the sixth edition because of the movement into that chapter of the material on displacement current from Chapter 30 and Maxwell's equations from Chapter 31. A more detailed list of content changes can be found on the instructor's companion Web site.

Content

The material in this book covers fundamental topics in classical physics and provides an introduction to modern physics. The book is divided into six parts. Part 1 (Chapters 1 to 14) deals with the fundamentals of Newtonian mechanics and the physics of fluids; Part 2 (Chapters 15 to 18) covers oscillations, mechanical waves, and sound; Part 3 (Chapters 19 to 22) addresses heat and thermodynamics; Part 4 (Chapters 23 to 34) treats electricity and magnetism; Part 5 (Chapters 35 to 38) covers light and optics; and Part 6 (Chapters 39 to 46) deals with relativity and modern physics.

Text Features

Most instructors believe that the textbook selected for a course should be the student's primary guide for understanding and learning the subject matter. Furthermore, the textbook should be easily accessible and should be styled and written to facilitate instruction and learning. With these points in mind, we have included many pedagogical features, listed below, that are intended to enhance its usefulness to both students and instructors.

Problem Solving and Conceptual Understanding

GENERAL PROBLEM-SOLVING STRATEGY A general strategy outlined at the end of Chapter 2 provides students with a structured process for solving problems. In all remaining chapters, the strategy is employed explicitly in every example so that students learn how it is applied. Students are encouraged to follow this strategy when working end-of-chapter problems.

MODELING Although students are faced with hundreds of problems during their physics courses, instructors realize that a relatively small number of physical situations form the basis of these problems. When faced with a new problem, a physicist forms a *model* of the problem that can be solved in a simple way by identifying the common physical situation that occurs in the problem. For example, many problems involve particles under constant acceleration, isolated systems, or waves under refraction. Because the physicist has studied these situations extensively and understands the associated behavior, he or she can apply this knowledge as a model for solving a new problem. In certain chapters, this edition identifies Analysis Models, which are physical situations (such as the particle under constant acceleration, the isolated system, or the wave under refraction) that occur so often that they can be used as a model for solving an unfamiliar problem. These models are discussed in the chapter text, and the student is reminded of them in the end-of-chapter summary under the heading "Analysis Models for Problem-Solving."

© Thomson Learning/George Semple

PROBLEMS An extensive set of problems is included at the end of each chapter; in all, the text contains approximately three thousand problems. Answers to odd-numbered problems are provided at the end of the book. For the convenience of both the student and the instructor, about two-thirds of the problems are keyed to specific sections of the chapter. The remaining problems, labeled "Additional Problems," are not keyed to specific sections. The problem numbers for straightforward problems are printed in black, intermediate-level problems are in blue, and challenging problems are in magenta.

- **"Not-just-a-number" problems** Each chapter includes several marked problems that require students to think qualitatively in some parts and quantitatively in others. Instructors can assign such problems to guide students to display deeper understanding, practice good problem-solving techniques, and prepare for exams.
- **Problems for developing symbolic reasoning** Each chapter contains problems that ask for solutions in symbolic form as well as many problems asking for numerical answers. To help students develop skill in symbolic reasoning, each chapter contains a pair of otherwise identical problems, one asking for a numerical solution and one asking for a symbolic derivation. In this edition, each chapter also contains a problem giving a numerical value for every datum but one so that the answer displays how the unknown depends on the datum represented symbolically. The answer to such a problem has the form of a function of one variable. Reasoning about the behavior of this function puts emphasis on the *Finalize* step of the General Problem-Solving Strategy. All problems developing symbolic reasoning are identified by a tan background screen:

> 53. ● A light spring has an unstressed length of 15.5 cm. It is described by Hooke's law with spring constant 4.30 N/m. One end of the horizontal spring is held on a fixed vertical axle, and the other end is attached to a puck of mass m that can move without friction over a horizontal surface. The puck is set into motion in a circle with a period of 1.30 s. (a) Find the extension of the spring x as it depends on m. Evaluate x for (b) $m = 0.070\ 0$ kg, (c) $m = 0.140$ kg, (d) $m = 0.180$ kg, and (e) $m = 0.190$ kg. (f) Describe the pattern of variation of x as it depends on m.

- **Review problems** Many chapters include review problems requiring the student to combine concepts covered in the chapter with those discussed in previous chapters. These problems reflect the cohesive nature of the principles in the text and verify that physics is not a scattered set of ideas. When facing a real-world issue such as global warming or nuclear weapons, it may be necessary to call on ideas in physics from several parts of a textbook such as this one.
- **"Fermi problems"** As in previous editions, at least one problem in each chapter asks the student to reason in order-of-magnitude terms.

- **Design problems** Several chapters contain problems that ask the student to determine design parameters for a practical device so that it can function as required.
- ***"Jeopardy!"* problems** Some chapters give students practice in changing between different representations by stating equations and asking for a description of a situation to which they apply as well as for a numerical answer.
- **Calculus-based problems** Every chapter contains at least one problem applying ideas and methods from differential calculus and one problem using integral calculus.

The instructor's Web site, **www.thomsonedu.com/physics/serway,** provides lists of problems using calculus, problems encouraging or requiring computer use, problems with "**What If?**" parts, problems referred to in the chapter text, problems based on experimental data, order-of-magnitude problems, problems about biological applications, design problems, *Jeopardy!* problems, review problems, problems reflecting historical reasoning about confusing ideas, problems developing symbolic reasoning skill, problems with qualitative parts, ranking questions, and other objective questions.

QUESTIONS The questions section at the end of each chapter has been significantly revised. Multiple-choice, ranking, and true–false questions have been added. The instructor may select items to assign as homework or use in the classroom, possibly with "peer instruction" methods and possibly with "clicker" systems. More than eight hundred questions are included in this edition. Answers to selected questions are included in the *Student Solutions Manual/Study Guide,* and answers to all questions are found in the *Instructor's Solutions Manual.*

19. O (i) Rank the gravitational accelerations you would measure for (a) a 2-kg object 5 cm above the floor, (b) a 2-kg object 120 cm above the floor, (c) a 3-kg object 120 cm above the floor, and (d) a 3-kg object 80 cm above the floor. List the one with the largest-magnitude acceleration first. If two are equal, show their equality in your list. **(ii)** Rank the gravitational forces on the same four objects, largest magnitude first. **(iii)** Rank the gravitational potential energies (of the object–Earth system) for the same four objects, largest first, taking $y = 0$ at the floor.

23. O An ice cube has been given a push and slides without friction on a level table. Which is correct? (a) It is in stable equilibrium. (b) It is in unstable equilibrium. (c) It is in neutral equilibrium (d) It is not in equilibrium.

WORKED EXAMPLES Two types of worked examples are presented to aid student comprehension. All worked examples in the text may be assigned for homework in WebAssign.

The first example type presents a problem and numerical answer. As discussed earlier, solutions to these examples have been altered in this edition to feature a two-column layout to explain the physical concepts and the mathematical steps side by side. Every example follows the explicit steps of the General Problem-Solving Strategy outlined in Chapter 2.

The second type of example is conceptual in nature. To accommodate increased emphasis on understanding physical concepts, the many conceptual examples are labeled as such, set off in boxes, and designed to focus students on the physical situation in the problem.

WHAT IF? Approximately one-third of the worked examples in the text contain a **What If?** feature. At the completion of the example solution, a **What If?** question offers a variation on the situation posed in the text of the example. For instance, this feature might explore the effects of changing the conditions of the situation, determine what happens when a quantity is taken to a particular limiting value, or question whether additional information can be determined about the situation. This feature encourages students to think about the results of the example, and it also assists in conceptual understanding of the principles. **What If?** questions also prepare students to encounter novel problems that may be included on exams. Some of the end-of-chapter problems also include this feature.

QUICK QUIZZES Quick Quizzes provide students an opportunity to test their understanding of the physical concepts presented. The questions require students to make decisions on the basis of sound reasoning, and some of the questions have been written to help students overcome common misconceptions. Quick Quizzes have been cast in an objective format, including multiple-choice, true–false, and ranking. Answers to all Quick Quiz questions are found at the end of each chapter. Additional Quick Quizzes that can be used in classroom teaching are available on the instructor's companion Web site. Many instructors choose to use such questions in a "peer instruction" teaching style or with the use of personal response system "clickers," but they can be used in standard quiz format as well. Quick Quizzes are set off from the text by horizontal lines:

Quick Quiz 7.5 A dart is loaded into a spring-loaded toy dart gun by pushing the spring in by a distance x. For the next loading, the spring is compressed a distance $2x$. How much faster does the second dart leave the gun compared with the first? (a) four times as fast (b) two times as fast (c) the same (d) half as fast (e) one-fourth as fast

---

PITFALL PREVENTION 16.2
Two Kinds of Speed/Velocity

Do not confuse v, the speed of the wave as it propagates along the string, with v_y, the transverse velocity of a point on the string. The speed v is constant for a uniform medium, whereas v_y varies sinusoidally.

PITFALL PREVENTIONS More than two hundred Pitfall Preventions (such as the one to the left) are provided to help students avoid common mistakes and misunderstandings. These features, which are placed in the margins of the text, address both common student misconceptions and situations in which students often follow unproductive paths.

Helpful Features

STYLE To facilitate rapid comprehension, we have written the book in a clear, logical, and engaging style. We have chosen a writing style that is somewhat informal and relaxed so that students will find the text appealing and enjoyable to read. New terms are carefully defined, and we have avoided the use of jargon.

IMPORTANT STATEMENTS AND EQUATIONS Most important statements and definitions are set in **boldface** or are highlighted with a background screen for added emphasis and ease of review. Similarly, important equations are highlighted with a background screen to facilitate location.

MARGINAL NOTES Comments and notes appearing in the margin with a ▶ icon can be used to locate important statements, equations, and concepts in the text.

PEDAGOGICAL USE OF COLOR Readers should consult the **pedagogical color chart** (inside the front cover) for a listing of the color-coded symbols used in the text diagrams. This system is followed consistently throughout the text.

MATHEMATICAL LEVEL We have introduced calculus gradually, keeping in mind that students often take introductory courses in calculus and physics concurrently. Most steps are shown when basic equations are developed, and reference is often made to mathematical appendices near the end of the textbook. Vector products are introduced later in the text, where they are needed in physical applications. The dot product is introduced in Chapter 7, which addresses energy of a system; the cross product is introduced in Chapter 11, which deals with angular momentum.

SIGNIFICANT FIGURES Significant figures in both worked examples and end-of-chapter problems have been handled with care. Most numerical examples are worked to either two or three significant figures, depending on the precision of the data provided. End-of-chapter problems regularly state data and answers to three-digit precision.

UNITS The international system of units (SI) is used throughout the text. The U.S. customary system of units is used only to a limited extent in the chapters on mechanics and thermodynamics.

APPENDICES AND ENDPAPERS Several appendices are provided near the end of the textbook. Most of the appendix material represents a review of mathematical concepts and techniques used in the text, including scientific notation, algebra, geometry, trigonometry, differential calculus, and integral calculus. Reference to these appendices is made throughout the text. Most mathematical review sections in the appendices include worked examples and exercises with answers. In addition to the mathematical reviews, the appendices contain tables of physical data, conversion factors, and the SI units of physical quantities as well as a periodic table of the elements. Other useful information—fundamental constants and physical data, planetary data, a list of standard prefixes, mathematical symbols, the Greek alphabet, and standard abbreviations of units of measure—appears on the endpapers.

Course Solutions That Fit Your Teaching Goals and Your Students' Learning Needs

Recent advances in educational technology have made homework management systems and audience response systems powerful and affordable tools to enhance the way you teach your course. Whether you offer a more traditional text-based course, are interested in using or are currently using an online homework management system such as WebAssign, or are ready to turn your lecture into an interactive learning environment with JoinIn on TurningPoint, you can be confident that the text's proven content provides the foundation for each and every component of our technology and ancillary package.

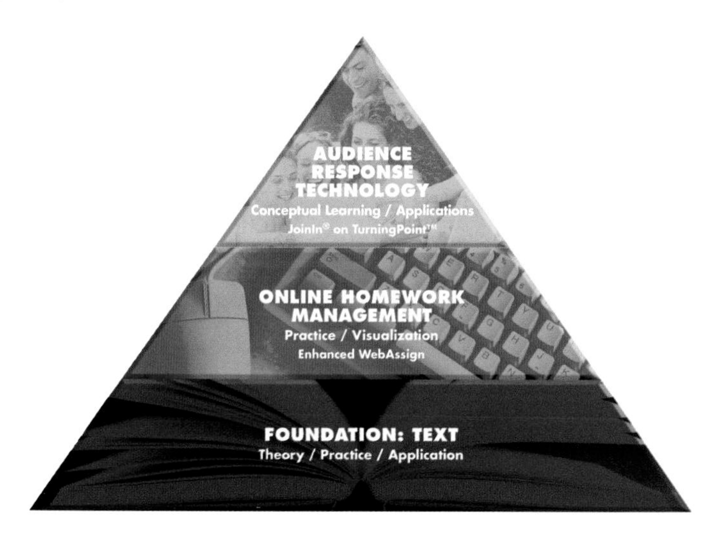

Homework Management Systems

Enhanced WebAssign Whether you're an experienced veteran or a beginner, Enhanced WebAssign is the perfect solution to fit your homework management needs. Designed by physicists for physicists, this system is a reliable and user-friendly teaching companion. Enhanced WebAssign is available for *Physics for Scientists and Engineers*, giving you the freedom to assign

- every end-of-chapter Problem and Question, enhanced with hints and feedback
- every worked example, enhanced with hints and feedback, to help strengthen students' problem-solving skills
- every Quick Quiz, giving your students ample opportunity to test their conceptual understanding.

- animated Active Figures, enhanced with hints and feedback, to help students develop their visualization skills
- a math review to help students brush up on key quantitative concepts

Please visit **www.thomsonedu.com/physics/serway** to view a live demonstration of Enhanced WebAssign.

The text also supports the following Homework Management Systems:

LON-CAPA: A Computer-Assisted Personalized Approach
http://www.lon-capa.org/

The University of Texas Homework Service
contact **moore@physics.utexas.edu**

Personal Response Systems

JoinIn on TurningPoint Pose book-specific questions and display students' answers seamlessly within the Microsoft® PowerPoint slides of your own lecture in conjunction with the "clicker" hardware of your choice. JoinIn on TurningPoint works with most infrared or radio frequency keypad systems, including Responsecard, EduCue, H-ITT, and even laptops. Contact your local sales representative to learn more about our personal response software and hardware.

Personal Response System Content Regardless of the response system you are using, we provide the tested content to support it. Our ready-to-go content includes all the questions from the Quick Quizzes, test questions, and a selection of end-of-chapter questions to provide helpful conceptual checkpoints to drop into your lecture. Our series of Active Figure animations have also been enhanced with multiple-choice questions to help test students' observational skills.

We also feature the Assessing to Learn in the Classroom content from the University of Massachusetts at Amherst. This collection of 250 advanced conceptual questions has been tested in the classroom for more than ten years and takes peer learning to a new level.

Visit **www.thomsonedu.com/physics/serway** to download samples of our personal response system content.

Lecture Presentation Resources

The following resources provide support for your presentations in lecture.

MULTIMEDIA MANAGER INSTRUCTOR'S RESOURCE CD An easy-to-use multimedia lecture tool, the Multimedia Manager Instructor's Resource CD allows you to quickly assemble art, animations, digital video, and database files with notes to create fluid lectures. The two-volume set (Volume 1: Chapters 1–22; Volume 2: Chapters 23–46) includes prebuilt PowerPoint lectures, a database of animations, video clips, and digital art from the text as well as editable electronic files of the *Instructor's Solutions Manual* and *Test Bank.*

TRANSPARENCY ACETATES Each volume contains approximately one hundred transparency acetates featuring art from the text. Volume 1 contains Chapters 1 through 22, and Volume 2 contains Chapters 23 through 46.

Assessment and Course Preparation Resources

A number of resources listed below will assist with your assessment and preparation processes.

INSTRUCTOR'S SOLUTIONS MANUAL by Ralph McGrew. This two-volume manual contains complete worked solutions to all end-of-chapter problems in the textbook as well as answers to the even-numbered problems and all the questions. The solutions to problems new to the seventh edition are marked for easy identification. Volume 1 contains

Chapters 1 through 22, and Volume 2 contains Chapters 23 through 46. Electronic files of the Instructor's Solutions are available on the Multimedia Manager CD as well.

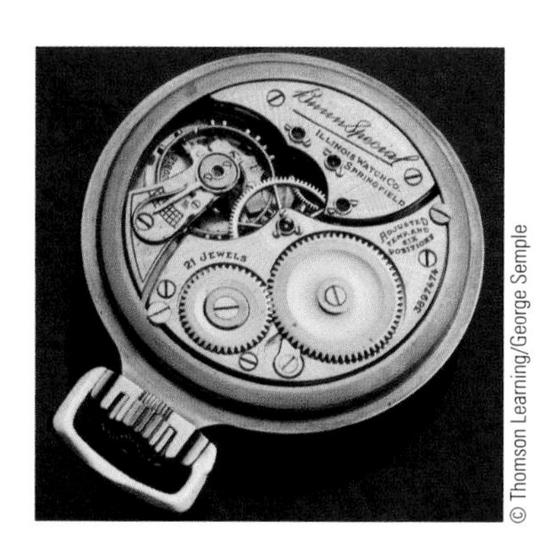

© Thomson Learning/George Semple

PRINTED TEST BANK by Edward Adelson. This two-volume test bank contains approximately 2 200 multiple-choice questions. These questions are also available in electronic format with complete answers and solutions in the ExamView test software and as editable Word® files on the Multimedia Manager CD. Volume 1 contains Chapters 1 through 22, and Volume 2 contains Chapters 23 through 46.

EXAMVIEW This easy-to-use test generator CD features all of the questions from the printed test bank in an editable format.

WEBCT AND BLACKBOARD CONTENT For users of either course management system, we provide our test bank questions in the proper format for easy upload into your online course. In addition, you can integrate the ThomsonNOW for Physics student tutorial content into your WebCT or Blackboard course, providing your students a single sign-on to all their Web-based learning resources. Contact your local sales representative to learn more about our WebCT and Blackboard resources.

INSTRUCTOR'S COMPANION WEB SITE Consult the instructor's site by pointing your browser to **www.thomsonedu.com/physics/serway** for additional Quick Quiz questions, a detailed list of content changes since the sixth edition, a problem correlation guide, images from the text, and sample PowerPoint lectures. Instructors adopting the seventh edition of *Physics for Scientists and Engineers* may download these materials after securing the appropriate password from their local Thomson•Brooks/Cole sales representative.

Student Resources

STUDENT SOLUTIONS MANUAL/STUDY GUIDE by John R. Gordon, Ralph McGrew, Raymond Serway, and John W. Jewett, Jr. This two-volume manual features detailed solutions to 20% of the end-of-chapter problems from the text. The manual also features a list of important equations, concepts, and notes from key sections of the text in addition to answers to selected end-of-chapter questions. Volume 1 contains Chapters 1 through 22, and Volume 2 contains Chapters 23 through 46.

THOMSONNOW PERSONAL STUDY This assessment-based student tutorial system provides students with a personalized learning plan based on their performance on a series of diagnostic pre-tests. Rich interactive content, including Active Figures, Coached Problems, and Interactive Examples, helps students prepare for tests and exams.

Teaching Options

The topics in this textbook are presented in the following sequence: classical mechanics, oscillations and mechanical waves, and heat and thermodynamics followed by electricity and magnetism, electromagnetic waves, optics, relativity, and modern physics. This presentation represents a traditional sequence, with the subject of mechanical waves being presented before electricity and magnetism. Some instructors may prefer to discuss both mechanical and electromagnetic waves together after completing electricity and magnetism. In this case, Chapters 16 through 18 could be covered along with Chapter 34. The chapter on relativity is placed near the end of the text because this topic often is treated as an introduction to the era of "modern physics." If time permits, instructors may choose to cover Chapter 39 after completing Chapter 13 as a conclusion to the material on Newtonian mechanics.

For those instructors teaching a two-semester sequence, some sections and chapters could be deleted without any loss of continuity. The following sections can be considered optional for this purpose:

2.8	Kinematic Equations Derived from Calculus
4.6	Relative Velocity and Relative Acceleration
6.3	Motion in Accelerated Frames
6.4	Motion in the Presence of Resistive Forces
7.9	Energy Diagrams and Equilibrium of a System
9.8	Rocket Propulsion
11.5	The Motion of Gyroscopes and Tops
14.7	Other Applications of Fluid Dynamics
15.6	Damped Oscillations
15.7	Forced Oscillations
17.5	Digital Sound Recording
17.6	Motion Picture Sound
18.6	Standing Waves in Rods and Membranes
18.8	Nonsinusoidal Wave Patterns
22.8	Entropy on a Microscopic Scale
25.7	The Millikan Oil-Drop Experiment
25.8	Applications of Electrostatics
26.7	An Atomic Description of Dielectrics
27.5	Superconductors
28.5	Electrical Meters
28.6	Household Wiring and Electrical Safety
29.3	Applications Involving Charged Particles Moving in a Magnetic Field
29.6	The Hall Effect
30.6	Magnetism in Matter
30.7	The Magnetic Field of the Earth
31.6	Eddy Currents
33.9	Rectifiers and Filters
34.6	Production of Electromagnetic Waves by an Antenna
36.5	Lens Aberrations
36.6	The Camera
36.7	The Eye
36.8	The Simple Magnifier
36.9	The Compound Microscope
36.10	The Telescope
38.5	Diffraction of X-Rays by Crystals
39.10	The General Theory of Relativity
41.6	Applications of Tunneling
42.9	Spontaneous and Stimulated Transitions
42.10	Lasers
43.7	Semiconductor Devices
43.8	Superconductivity
44.8	Nuclear Magnetic Resonance and Magnetic Resonance Imaging
45.5	Radiation Damage
45.6	Radiation Detectors
45.7	Uses of Radiation

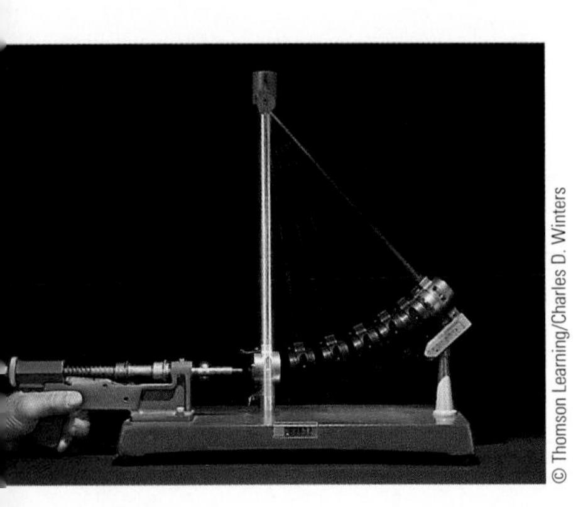

Acknowledgments

This seventh edition of *Physics for Scientists and Engineers* was prepared with the guidance and assistance of many professors who reviewed selections of the manuscript, the prerevision text, or both. We wish to acknowledge the following scholars and express our sincere appreciation for their suggestions, criticisms, and encouragement:

David P. Balogh, *Fresno City College*
Leonard X. Finegold, *Drexel University*
Raymond Hall, *California State University, Fresno*

Bob Jacobsen, *University of California, Berkeley*
Robin Jordan, *Florida Atlantic University*
Rafael Lopez-Mobilia, *University of Texas at San Antonio*
Diana Lininger Markham, *City College of San Francisco*
Steven Morris, *Los Angeles Harbor City College*
Taha Mzoughi, *Kennesaw State University*
Nobel Sanjay Rebello, *Kansas State University*
John Rosendahl, *University of California, Irvine*
Mikolaj Sawicki, *John A. Logan College*
Glenn B. Stracher, *East Georgia College*
Som Tyagi, *Drexel University*
Robert Weidman, *Michigan Technological University*
Edward A. Whittaker, *Stevens Institute of Technology*

This title was carefully checked for accuracy by Zinoviy Akkerman, *City College of New York;* Grant Hart, *Brigham Young University;* Michael Kotlarchyk, *Rochester Institute of Technology;* Andres LaRosa, *Portland State University;* Bruce Mason, *University of Oklahoma at Norman;* Peter Moeck, *Portland State University;* Brian A. Raue, *Florida International University;* James E. Rutledge, *University of California at Irvine;* Bjoern Seipel, *Portland State University;* Z. M. Stadnik, *University of Ottawa;* and Harry W. K. Tom, *University of California at Riverside.* We thank them for their diligent efforts under schedule pressure.

We are grateful to Ralph McGrew for organizing the end-of-chapter problems, writing many new problems, and suggesting improvements in the content of the textbook. Problems and questions new to this edition were written by Duane Deardorff, Thomas Grace, Francisco Izaguirre, John Jewett, Robert Forsythe, Randall Jones, Ralph McGrew, Kurt Vandervoort, and Jerzy Wrobel. Help was very kindly given by Dwight Neuenschwander, Michael Kinney, Amy Smith, Will Mackin, and the Sewer Department of Grand Forks, North Dakota. Daniel Kim, Jennifer Hoffman, Ed Oberhofer, Richard Webb, Wesley Smith, Kevin Kilty, Zinoviy Akkerman, Michael Rudmin, Paul Cox, Robert LaMontagne, Ken Menningen, and Chris Church made corrections to problems taken from previous editions. We are grateful to authors John R. Gordon and Ralph McGrew for preparing the *Student Solutions Manual/Study Guide.* Author Ralph McGrew has prepared an excellent *Instructor's Solutions Manual.* Edward Adelson has carefully edited and improved the test bank. Kurt Vandervoort prepared extra Quick Quiz questions for the instructor's companion Web site.

Special thanks and recognition go to the professional staff at the Brooks/Cole Publishing Company—in particular, Ed Dodd, Brandi Kirksey (who managed the ancillary program and so much more), Shawn Vasquez, Sam Subity, Teri Hyde, Michelle Julet, David Harris, and Chris Hall—for their fine work during the development and production of this textbook. Mark Santee is our seasoned marketing manager, and Bryan Vann coordinates our marketing communications. We recognize the skilled production service and excellent artwork provided by the staff at Lachina Publishing Services, and the dedicated photo research efforts of Jane Sanders Miller.

Finally, we are deeply indebted to our wives, children, and grandchildren for their love, support, and long-term sacrifices.

Raymond A. Serway
St. Petersburg, Florida

John W. Jewett, Jr.
Pomona, California

It is appropriate to offer some words of advice that should be of benefit to you, the student. Before doing so, we assume you have read the Preface, which describes the various features of the text and support materials that will help you through the course.

How to Study

Instructors are often asked, "How should I study physics and prepare for examinations?" There is no simple answer to this question, but we can offer some suggestions based on our own experiences in learning and teaching over the years.

First and foremost, maintain a positive attitude toward the subject matter, keeping in mind that physics is the most fundamental of all natural sciences. Other science courses that follow will use the same physical principles, so it is important that you understand and are able to apply the various concepts and theories discussed in the text.

Concepts and Principles

It is essential that you understand the basic concepts and principles before attempting to solve assigned problems. You can best accomplish this goal by carefully reading the textbook before you attend your lecture on the covered material. When reading the text, you should jot down those points that are not clear to you. Also be sure to make a diligent attempt at answering the questions in the Quick Quizzes as you come to them in your reading. We have worked hard to prepare questions that help you judge for yourself how well you understand the material. Study the **What If?** features that appear in many of the worked examples carefully. They will help you extend your understanding beyond the simple act of arriving at a numerical result. The Pitfall Preventions will also help guide you away from common misunderstandings about physics. During class, take careful notes and ask questions about those ideas that are unclear to you. Keep in mind that few people are able to absorb the full meaning of scientific material after only one reading; several readings of the text and your notes may be necessary. Your lectures and laboratory work supplement the textbook and should clarify some of the more difficult material. You should minimize your memorization of material. Successful memorization of passages from the text, equations, and derivations does not necessarily indicate that you understand the material. Your understanding of the material will be enhanced through a combination of efficient study habits, discussions with other students and with instructors, and your ability to solve the problems presented in the textbook. Ask questions whenever you believe that clarification of a concept is necessary.

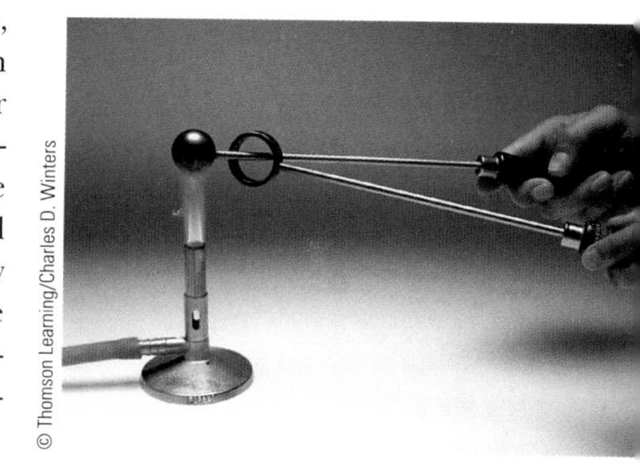

Study Schedule

It is important that you set up a regular study schedule, preferably a daily one. Make sure that you read the syllabus for the course and adhere to the schedule set by your instructor. The lectures will make much more sense if you read the corresponding text material *before* attending them. As a general rule, you should devote about two hours of study time for each hour you are in class. If you are having trouble with the course, seek the advice of the instructor or other students who have taken the course. You may find it necessary to seek further instruction from experienced students. Very often, instructors offer review sessions in addition to regular class periods. Avoid the practice of delaying study until a day or two before an exam. More often than not, this approach has disastrous results. Rather than undertake an all-night study session before a test, briefly review the basic concepts and equations, and then get a good night's rest. If you believe that you need additional help in understanding the concepts, in preparing for exams, or in problem solving, we suggest that you acquire a copy of the *Student Solutions Manual/Study Guide* that accompanies this textbook; this manual should be available at your college bookstore or through the publisher.

Use the Features

You should make full use of the various features of the text discussed in the Preface. For example, marginal notes are useful for locating and describing important equations and concepts, and **boldface** indicates important statements and definitions. Many useful tables are contained in the appendices, but most are incorporated in the text where they are most often referenced. Appendix B is a convenient review of mathematical tools used in the text.

Answers to odd-numbered problems are given at the end of the textbook, answers to Quick Quizzes are located at the end of each chapter, and solutions to selected end-of-chapter questions and problems are provided in the *Student Solutions Manual/Study Guide.* The table of contents provides an overview of the entire text, and the index enables you to locate specific material quickly. Footnotes are sometimes used to supplement the text or to cite other references on the subject discussed.

After reading a chapter, you should be able to define any new quantities introduced in that chapter and discuss the principles and assumptions that were used to arrive at certain key relations. The chapter summaries and the review sections of the *Student Solutions Manual/Study Guide* should help you in this regard. In some cases, you may find it necessary to refer to the textbook's index to locate certain topics. You should be able to associate with each physical quantity the correct symbol used to represent that quantity and the unit in which the quantity is specified. Furthermore, you should be able to express each important equation in concise and accurate prose.

Problem Solving

R. P. Feynman, Nobel laureate in physics, once said, "You do not know anything until you have practiced." In keeping with this statement, we strongly advise you to develop the skills necessary to solve a wide range of problems. Your ability to solve problems will be one of the main tests of your knowledge of physics; therefore, you should try to solve as many problems as possible. It is essential that you understand basic concepts and principles before attempting to solve problems. It is good practice to try to find alternate solutions to the same problem. For example, you can solve problems in mechanics using Newton's laws, but very often an alternative method that draws on energy considerations is more direct. You should not deceive yourself into thinking that you understand a problem merely because you have seen it solved in class. You must be able to solve the problem and similar problems on your own.

The approach to solving problems should be carefully planned. A systematic plan is especially important when a problem involves several concepts. First, read the problem several times until you are confident you understand what is being asked. Look for any key words that will help you interpret the problem and perhaps allow you to make certain assumptions. Your ability to interpret a question properly is an integral part of problem solving. Second, you should acquire the habit of writing down the information given in a problem and those quantities that need to be found; for example, you might construct a table listing both the quantities given and the quantities to be found. This procedure is sometimes used in the worked examples of the textbook. Finally, after you have decided on the method you believe is appropriate for a given problem, proceed with your solution. The General Problem-Solving Strategy will guide you through complex problems. If you follow the steps of this procedure *(Conceptualize, Categorize, Analyze, Finalize),* you will find it easier to come up with a solution and gain more from your efforts. This Strategy, located at the end of Chapter 2, is used in all worked examples in the remaining chapters so that you can learn how to apply it. Specific problem-solving strategies for certain types of situations are included in the text and appear with a blue heading. These specific strategies follow the outline of the General Problem-Solving Strategy.

Often, students fail to recognize the limitations of certain equations or physical laws in a particular situation. It is very important that you understand and remember the assumptions that underlie a particular theory or formalism. For example, certain equations in kinematics apply only to a particle moving with constant acceleration. These equations are not valid for describing motion whose acceleration is not constant such as the motion of an object connected to a spring or the motion of an object through a fluid. Study the Analysis Models for Problem-Solving in the chapter summaries carefully so that you know how each model can be applied to a specific situation.

Experiments

Physics is a science based on experimental observations. Therefore, we recommend that you try to supplement the text by performing various types of "hands-on" experiments either at home or in the laboratory. These experiments can be used to test ideas and models discussed in class or in the textbook. For example, the common Slinky toy is excellent for studying traveling waves, a ball swinging on the end of a long string can be used to investigate pendulum motion, various masses attached to the end of a vertical spring or rubber band can be used to determine their elastic nature, an old pair of Polaroid sunglasses and some discarded lenses and a magnifying glass are the components of various experiments in optics, and an approximate measure of the free-fall acceleration can be determined simply by measuring with a stopwatch the time it takes for a ball to drop from a known height. The list of such experiments is endless. When physical models are not available, be imaginative and try to develop models of your own.

New Media

We strongly encourage you to use the **ThomsonNOW** Web-based learning system that accompanies this textbook. It is far easier to understand physics if you see it in action, and these new materials will enable you to become a part of that action. **ThomsonNOW** media described in the Preface and accessed at **www.thomsonedu.com/physics/serway** feature a three-step learning process consisting of a pre-test, a personalized learning plan, and a post-test.

It is our sincere hope that you will find physics an exciting and enjoyable experience and that you will benefit from this experience, regardless of your chosen profession. Welcome to the exciting world of physics!

The scientist does not study nature because it is useful; he studies it because he delights in it, and he delights in it because it is beautiful. If nature were not beautiful, it would not be worth knowing, and if nature were not worth knowing, life would not be worth living.

—Henri Poincaré

PART 6

Modern Physics

At the end of the 19th century, many scientists believed they had learned most of what there was to know about physics. Newton's laws of motion and theory of universal gravitation, Maxwell's theoretical work in unifying electricity and magnetism, the laws of thermodynamics and kinetic theory, and the principles of optics were highly successful in explaining a variety of phenomena.

At the turn of the 20th century, however, a major revolution shook the world of physics. In 1900, Max Planck provided the basic ideas that led to the formulation of the quantum theory, and in 1905, Albert Einstein formulated his special theory of relativity. The excitement of the times is captured in Einstein's own words: "It was a marvelous time to be alive." Both theories were to have a profound effect on our understanding of nature. Within a few decades, they inspired new developments in the fields of atomic physics, nuclear physics, and condensed-matter physics.

In Chapter 39, we shall introduce the special theory of relativity. The theory provides us with a new and deeper view of physical laws. Although the predictions of this theory often violate our common sense, the theory correctly describes the results of experiments involving speeds near the speed of light. The extended version of this textbook, *Physics for Scientists and Engineers with Modern Physics,* covers the basic concepts of quantum mechanics and their application to atomic and molecular physics, and we introduce solid-state physics, nuclear physics, particle physics, and cosmology.

Even though the physics that was developed during the 20th century has led to a multitude of important technological achievements, the story is still incomplete. Discoveries will continue to evolve during our lifetimes, and many of these discoveries will deepen or refine our understanding of nature and the Universe around us. It is still a "marvelous time to be alive."

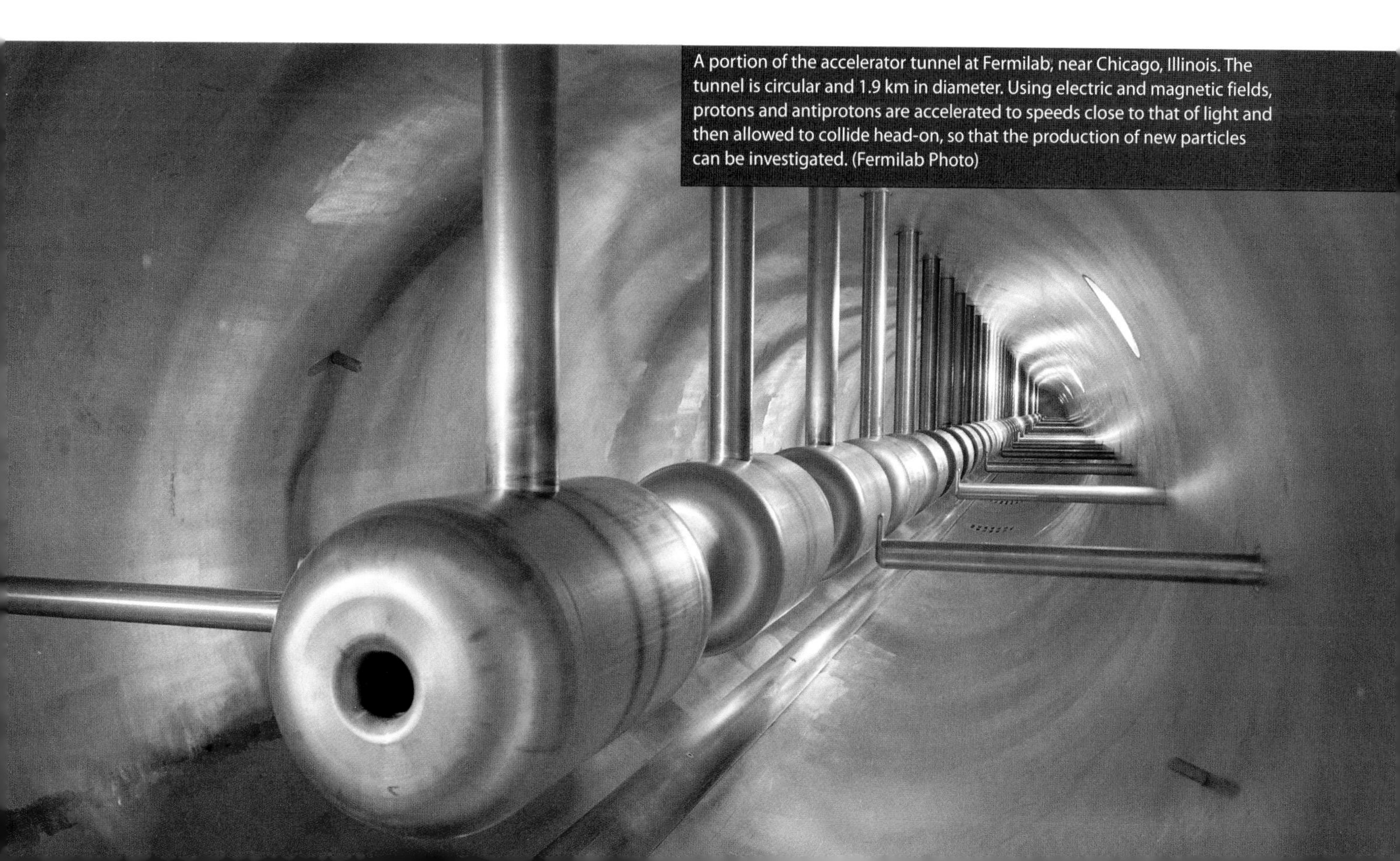

A portion of the accelerator tunnel at Fermilab, near Chicago, Illinois. The tunnel is circular and 1.9 km in diameter. Using electric and magnetic fields, protons and antiprotons are accelerated to speeds close to that of light and then allowed to collide head-on, so that the production of new particles can be investigated. (Fermilab Photo)

Standing on the shoulders of a giant. David Serway, son of one of the authors, watches over two of his children, Nathan and Kaitlyn, as they frolic in the arms of Albert Einstein's statue at the Einstein memorial in Washington, D.C. It is well known that Einstein, the principal architect of relativity, was very fond of children. (Emily Serway)

39 Relativity

Our everyday experiences and observations involve objects that move at speeds much less than the speed of light. Newtonian mechanics was formulated by observing and describing the motion of such objects, and this formalism is very successful in describing a wide range of phenomena that occur at low speeds. Nonetheless, it fails to describe properly the motion of objects whose speeds approach that of light.

Experimentally, the predictions of Newtonian theory can be tested at high speeds by accelerating electrons or other charged particles through a large electric potential difference. For example, it is possible to accelerate an electron to a speed of $0.99c$ (where c is the speed of light) by using a potential difference of several million volts. According to Newtonian mechanics, if the potential difference is increased by a factor of 4, the electron's kinetic energy is four times greater and its speed should double to $1.98c$. Experiments show, however, that the speed of the electron—as well as the speed of any other object in the Universe—always remains less than the speed of light, regardless of the size of the accelerating voltage. Because it places no upper limit on speed, Newtonian mechanics is contrary to modern experimental results and is clearly a limited theory.

In 1905, at the age of only 26, Einstein published his special theory of relativity. Regarding the theory, Einstein wrote:

> The relativity theory arose from necessity, from serious and deep contradictions in the old theory from which there seemed no escape. The strength of the new theory lies in the consistency and simplicity with which it solves all these difficulties.[1]

Although Einstein made many other important contributions to science, the special theory of relativity alone represents one of the greatest intellectual achievements of all time. With this theory, experimental observations can be correctly predicted over the range of speeds from $v = 0$ to speeds approaching the speed of light. At low speeds, Einstein's theory reduces to Newtonian mechanics as a limiting situation. It is important to recognize that Einstein was working on electromagnetism when he developed the special theory of relativity. He was convinced that Maxwell's equations were correct, and to reconcile them with one of his postulates, he was forced into the revolutionary notion of assuming that space and time are not absolute.

This chapter gives an introduction to the special theory of relativity, with emphasis on some of its predictions. In addition to its well-known and essential role in theoretical physics, the special theory of relativity has practical applications, including the design of nuclear power plants and modern global positioning system (GPS) units. These devices do not work if designed in accordance with nonrelativistic principles.

39.1 The Principle of Galilean Relativity

To describe a physical event, we must establish a frame of reference. You should recall from Chapter 5 that an inertial frame of reference is one in which an object is observed to have no acceleration when no forces act on it. Furthermore, any frame moving with constant velocity with respect to an inertial frame must also be an inertial frame.

There is no absolute inertial reference frame. Therefore, the results of an experiment performed in a vehicle moving with uniform velocity must be identical to the results of the same experiment performed in a stationary vehicle. The formal statement of this result is called the **principle of Galilean relativity:**

The laws of mechanics must be the same in all inertial frames of reference.

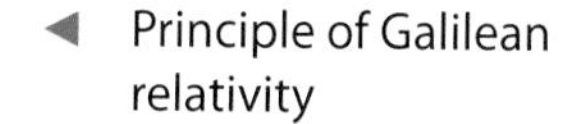

Let's consider an observation that illustrates the equivalence of the laws of mechanics in different inertial frames. A pickup truck moves with a constant velocity as shown in Figure 39.1a. If a passenger in the truck throws a ball straight up

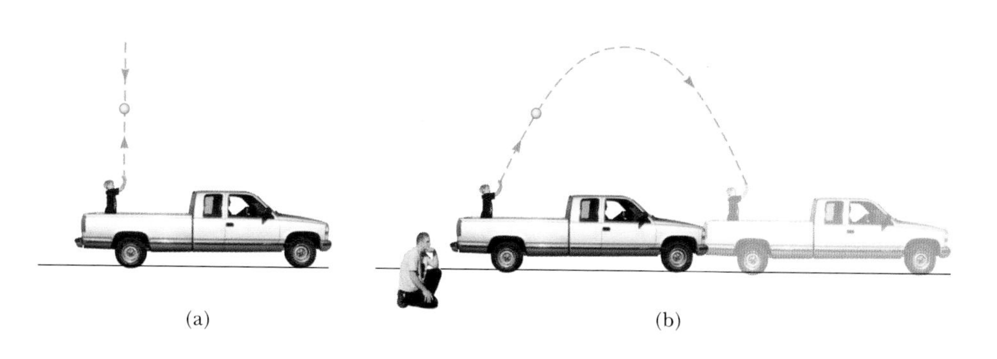

Figure 39.1 (a) The observer in the truck sees the ball move in a vertical path when thrown upward. (b) The Earth-based observer sees the ball's path as a parabola.

[1] A. Einstein and L. Infeld, *The Evolution of Physics* (New York: Simon and Schuster, 1961).

and if air effects are neglected, the passenger observes that the ball moves in a vertical path. The motion of the ball appears to be precisely the same as if the ball were thrown by a person at rest on the Earth. The law of universal gravitation and the equations of motion under constant acceleration are obeyed whether the truck is at rest or in uniform motion.

Both observers agree on the laws of physics: they each throw a ball straight up, and it rises and falls back into their own hand. Do the observers agree on the path of the ball thrown by the observer in the truck? The observer on the ground sees the path of the ball as a parabola as illustrated in Figure 39.1b, while, as mentioned earlier, the observer in the truck sees the ball move in a vertical path. Furthermore, according to the observer on the ground, the ball has a horizontal component of velocity equal to the velocity of the truck. Although the two observers disagree on certain aspects of the situation, they agree on the validity of Newton's laws and on such classical principles as conservation of energy and conservation of linear momentum. This agreement implies that no mechanical experiment can detect any difference between the two inertial frames. The only thing that can be detected is the relative motion of one frame with respect to the other.

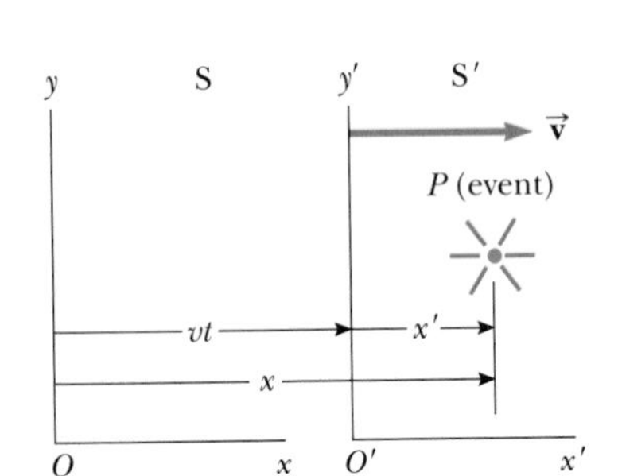

Figure 39.2 An event occurs at a point P. The event is seen by two observers in inertial frames S and S', where S' moves with a velocity $\vec{\mathbf{v}}$ relative to S.

Quick Quiz 39.1 Which observer in Figure 39.1 sees the ball's *correct* path? (a) the observer in the truck (b) the observer on the ground (c) both observers

Suppose some physical phenomenon, which we call an *event,* occurs and is observed by an observer at rest in an inertial reference frame. The wording "in a frame" means that the observer is at rest with respect to the origin of that frame. The event's location and time of occurrence can be specified by the four coordinates (x, y, z, t). We would like to be able to transform these coordinates from those of an observer in one inertial frame to those of another observer in a frame moving with uniform relative velocity compared with the first frame.

Consider two inertial frames S and S' (Fig. 39.2). The S' frame moves with a constant velocity $\vec{\mathbf{v}}$ along the common x and x' axes, where $\vec{\mathbf{v}}$ is measured relative to S. We assume the origins of S and S' coincide at $t = 0$ and an event occurs at point P in space at some instant of time. An observer in S describes the event with space–time coordinates (x, y, z, t), whereas an observer in S' uses the coordinates (x', y', z', t') to describe the same event. As we see from the geometry in Figure 39.2, the relationships among these various coordinates can be written

Galilean transformation equations ▶

$$x' = x - vt \qquad y' = y \qquad z' = z \qquad t' = t \tag{39.1}$$

These equations are the **Galilean space–time transformation equations.** Note that time is assumed to be the same in both inertial frames. That is, within the framework of classical mechanics, all clocks run at the same rate, regardless of their velocity, so the time at which an event occurs for an observer in S is the same as the time for the same event in S'. Consequently, the time interval between two successive events should be the same for both observers. Although this assumption may seem obvious, it turns out to be incorrect in situations where v is comparable to the speed of light.

PITFALL PREVENTION 39.1

The Relationship Between the S and S' Frames

Many of the mathematical representations in this chapter are true *only* for the specified relationship between the S and S' frames. The x and x' axes coincide, except their origins are different. The y and y' axes (and the z and z' axes) are parallel, but they do not coincide due to the displacement of the origin of S' with respect to that of S. We choose the time $t = 0$ to be the instant at which the origins of the two coordinate systems coincide. If the S' frame is moving in the positive x direction relative to S, then v is positive; otherwise, it is negative.

Now suppose a particle moves through a displacement of magnitude dx along the x axis in a time interval dt as measured by an observer in S. It follows from Equations 39.1 that the corresponding displacement dx' measured by an observer in S' is $dx' = dx - v\,dt$, where frame S' is moving with speed v in the x direction relative to frame S. Because $dt = dt'$, we find that

$$\frac{dx'}{dt'} = \frac{dx}{dt} - v$$

or

$$u'_x = u_x - v \tag{39.2}$$

where u_x and u'_x are the x components of the velocity of the particle measured by observers in S and S', respectively. (We use the symbol $\vec{\mathbf{u}}$ rather than $\vec{\mathbf{v}}$ for particle velocity because $\vec{\mathbf{v}}$ is already used for the relative velocity of two reference frames.) Equation 39.2 is the **Galilean velocity transformation equation.** It is consistent with our intuitive notion of time and space as well as with our discussions in Section 4.6. As we shall soon see, however, it leads to serious contradictions when applied to electromagnetic waves.

Quick Quiz 39.2 A baseball pitcher with a 90-mi/h fastball throws a ball while standing on a railroad flatcar moving at 110 mi/h. The ball is thrown in the same direction as that of the velocity of the train. If you apply the Galilean velocity transformation equation to this situation, is the speed of the ball relative to the Earth (a) 90 mi/h, (b) 110 mi/h, (c) 20 mi/h, (d) 200 mi/h, or (e) impossible to determine?

The Speed of Light

It is quite natural to ask whether the principle of Galilean relativity also applies to electricity, magnetism, and optics. Experiments indicate that the answer is no. Recall from Chapter 34 that Maxwell showed that the speed of light in free space is $c = 3.00 \times 10^8$ m/s. Physicists of the late 1800s thought light waves moved through a medium called the *ether* and the speed of light was c only in a special, absolute frame at rest with respect to the ether. The Galilean velocity transformation equation was expected to hold for observations of light made by an observer in any frame moving at speed v relative to the absolute ether frame. That is, if light travels along the x axis and an observer moves with velocity $\vec{\mathbf{v}}$ along the x axis, the observer measures the light to have speed $c \pm v$, depending on the directions of travel of the observer and the light.

Because the existence of a preferred, absolute ether frame would show that light is similar to other classical waves and that Newtonian ideas of an absolute frame are true, considerable importance was attached to establishing the existence of the ether frame. Prior to the late 1800s, experiments involving light traveling in media moving at the highest laboratory speeds attainable at that time were not capable of detecting differences as small as that between c and $c \pm v$. Starting in about 1880, scientists decided to use the Earth as the moving frame in an attempt to improve their chances of detecting these small changes in the speed of light.

Observers fixed on the Earth can take the view that they are stationary and that the absolute ether frame containing the medium for light propagation moves past them with speed v. Determining the speed of light under these circumstances is similar to determining the speed of an aircraft traveling in a moving air current, or wind; consequently, we speak of an "ether wind" blowing through our apparatus fixed to the Earth.

A direct method for detecting an ether wind would use an apparatus fixed to the Earth to measure the ether wind's influence on the speed of light. If v is the speed of the ether relative to the Earth, light should have its maximum speed $c + v$ when propagating downwind as in Figure 39.3a. Likewise, the speed of light should have its minimum value $c - v$ when the light is propagating upwind as in Figure 39.3b and an intermediate value $(c^2 - v^2)^{1/2}$ when the light is directed such that it travels perpendicular to the ether wind as in Figure 39.3c. If the Sun is assumed to be at rest in the ether, the velocity of the ether wind would be equal to the orbital velocity of the Earth around the Sun, which has a magnitude of approximately 30 km/s or 3×10^4 m/s. Because $c = 3 \times 10^8$ m/s, it is necessary to detect a change in speed of approximately 1 part in 10^4 for measurements in the upwind or downwind directions. Although such a change is experimentally measurable, all attempts to detect such changes and establish the existence of the ether wind (and hence the absolute frame) proved futile! We shall discuss the classic experimental search for the ether in Section 39.2.

The principle of Galilean relativity refers only to the laws of mechanics. If it is assumed the laws of electricity and magnetism are the same in all inertial frames, a

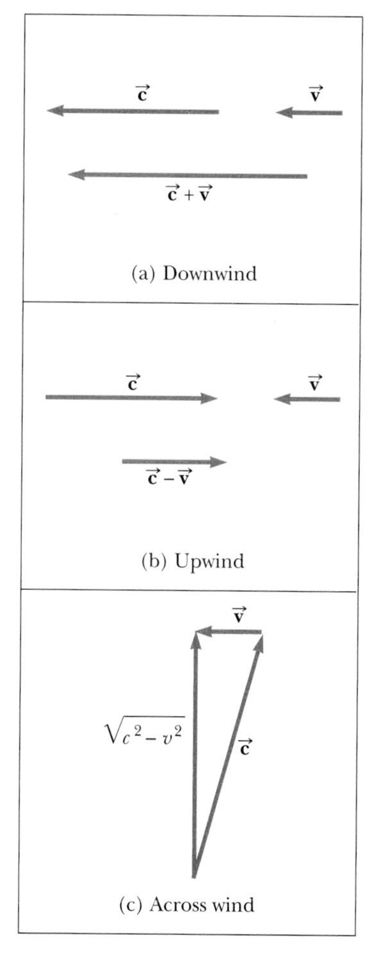

Figure 39.3 If the velocity of the ether wind relative to the Earth is $\vec{\mathbf{v}}$ and the velocity of light relative to the ether is $\vec{\mathbf{c}}$, the speed of light relative to the Earth is (a) $c + v$ in the downwind direction, (b) $c - v$ in the upwind direction, and (c) $(c^2 - v^2)^{1/2}$ in the direction perpendicular to the wind.

paradox concerning the speed of light immediately arises. That can be understood by recognizing that Maxwell's equations imply that the speed of light always has the fixed value 3.00×10^8 m/s in all inertial frames, a result in direct contradiction to what is expected based on the Galilean velocity transformation equation. According to Galilean relativity, the speed of light should *not* be the same in all inertial frames.

To resolve this contradiction in theories, we must conclude that either (1) the laws of electricity and magnetism are not the same in all inertial frames or (2) the Galilean velocity transformation equation is incorrect. If we assume the first alternative, a preferred reference frame in which the speed of light has the value c must exist and the measured speed must be greater or less than this value in any other reference frame, in accordance with the Galilean velocity transformation equation. If we assume the second alternative, we must abandon the notions of absolute time and absolute length that form the basis of the Galilean space–time transformation equations.

39.2 The Michelson–Morley Experiment

The most famous experiment designed to detect small changes in the speed of light was first performed in 1881 by Albert A. Michelson (see Section 37.7) and later repeated under various conditions by Michelson and Edward W. Morley (1838–1923). As we shall see, the outcome of the experiment contradicted the ether hypothesis.

The experiment was designed to determine the velocity of the Earth relative to that of the hypothetical ether. The experimental tool used was the Michelson interferometer, which was discussed in Section 37.7 and is shown again in Active Figure 39.4. Arm 2 is aligned along the direction of the Earth's motion through space. The Earth moving through the ether at speed v is equivalent to the ether flowing past the Earth in the opposite direction with speed v. This ether wind blowing in the direction opposite the direction of the Earth's motion should cause the speed of light measured in the Earth frame to be $c - v$ as the light approaches mirror M_2 and $c + v$ after reflection, where c is the speed of light in the ether frame.

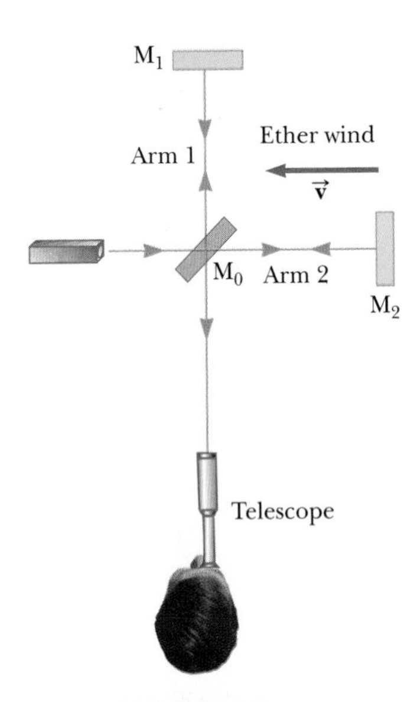

ACTIVE FIGURE 39.4

According to the ether wind theory, the speed of light should be $c - v$ as the beam approaches mirror M_2 and $c + v$ after reflection.

Sign in at www.thomsonedu.com and go to ThomsonNOW to adjust the speed of the ether wind and see the effect on the light beams if there were an ether.

The two light beams reflect from M_1 and M_2 and recombine, and an interference pattern is formed as discussed in Section 37.7. The interference pattern is observed while the interferometer is rotated through an angle of 90°. This rotation interchanges the speed of the ether wind between the arms of the interferometer. The rotation should cause the fringe pattern to shift slightly but measurably. Measurements failed, however, to show any change in the interference pattern! The Michelson–Morley experiment was repeated at different times of the year when the ether wind was expected to change direction and magnitude, but the results were always the same: **no fringe shift of the magnitude required was ever observed.**[2]

The negative results of the Michelson–Morley experiment not only contradicted the ether hypothesis, but also showed that it is impossible to measure the absolute velocity of the Earth with respect to the ether frame. Einstein, however, offered a postulate for his special theory of relativity that places quite a different interpretation on these null results. In later years, when more was known about the nature of light, the idea of an ether that permeates all of space was abandoned. **Light is now understood to be an electromagnetic wave, which requires no medium for its propagation.** As a result, the idea of an ether in which these waves travel became unnecessary.

[2] From an Earth-based observer's point of view, changes in the Earth's speed and direction of motion in the course of a year are viewed as ether wind shifts. Even if the speed of the Earth with respect to the ether were zero at some time, six months later the speed of the Earth would be 60 km/s with respect to the ether and as a result a fringe shift should be noticed. No shift has ever been observed, however.

Details of the Michelson–Morley Experiment

To understand the outcome of the Michelson–Morley experiment, let's assume the two arms of the interferometer in Active Figure 39.4 are of equal length L. We shall analyze the situation as if there were an ether wind because that is what Michelson and Morley expected to find. As noted above, the speed of the light beam along arm 2 should be $c - v$ as the beam approaches M_2 and $c + v$ after the beam is reflected. We model a pulse of light as a particle under constant speed. Therefore, the time interval for travel to the right for the pulse is $\Delta t = L/(c - v)$, and the time interval for travel to the left is $\Delta t = L/(c + v)$. The total time interval for the round trip along arm 2 is

$$\Delta t_{\text{arm 2}} = \frac{L}{c + v} + \frac{L}{c - v} = \frac{2Lc}{c^2 - v^2} = \frac{2L}{c}\left(1 - \frac{v^2}{c^2}\right)^{-1}$$

Now consider the light beam traveling along arm 1, perpendicular to the ether wind. Because the speed of the beam relative to the Earth is $(c^2 - v^2)^{1/2}$ in this case (see Fig. 39.3c), the time interval for travel for each half of the trip is $\Delta t = L/(c^2 - v^2)^{1/2}$ and the total time interval for the round trip is

$$\Delta t_{\text{arm 1}} = \frac{2L}{(c^2 - v^2)^{1/2}} = \frac{2L}{c}\left(1 - \frac{v^2}{c^2}\right)^{-1/2}$$

The time difference Δt between the horizontal round trip (arm 2) and the vertical round trip (arm 1) is

$$\Delta t = \Delta t_{\text{arm 2}} - \Delta t_{\text{arm 1}} = \frac{2L}{c}\left[\left(1 - \frac{v^2}{c^2}\right)^{-1} - \left(1 - \frac{v^2}{c^2}\right)^{-1/2}\right]$$

Because $v^2/c^2 \ll 1$, we can simplify this expression by using the following binomial expansion after dropping all terms higher than second order:

$$(1 - x)^n \approx 1 - nx \quad (\text{for } x \ll 1)$$

In our case, $x = v^2/c^2$, and we find that

$$\Delta t = \Delta t_{\text{arm 2}} - \Delta t_{\text{arm1}} \approx \frac{Lv^2}{c^3} \qquad \textbf{(39.3)}$$

This time difference between the two instants at which the reflected beams arrive at the viewing telescope gives rise to a phase difference between the beams, producing an interference pattern when they combine at the position of the telescope. A shift in the interference pattern should be detected when the interferometer is rotated through 90° in a horizontal plane so that the two beams exchange roles. This rotation results in a time difference twice that given by Equation 39.3. Therefore, the path difference that corresponds to this time difference is

$$\Delta d = c(2\,\Delta t) = \frac{2Lv^2}{c^2}$$

Because a change in path length of one wavelength corresponds to a shift of one fringe, the corresponding fringe shift is equal to this path difference divided by the wavelength of the light:

$$\text{Shift} = \frac{2Lv^2}{\lambda c^2} \qquad \textbf{(39.4)}$$

In the experiments by Michelson and Morley, each light beam was reflected by mirrors many times to give an effective path length L of approximately 11 m. Using this value, taking v to be equal to 3.0×10^4 m/s (the speed of the Earth around the Sun), and using 500 nm for the wavelength of the light, we expect a fringe shift of

$$\text{Shift} = \frac{2(11\text{ m})(3.0 \times 10^4\text{ m/s})^2}{(5.0 \times 10^{-7}\text{ m})(3.0 \times 10^8\text{ m/s})^2} = 0.44$$

The instrument used by Michelson and Morley could detect shifts as small as 0.01 fringe, but **it detected no shift whatsoever in the fringe pattern!** The experiment has been repeated many times since by different scientists under a wide variety of conditions, and no fringe shift has ever been detected. Therefore, it was concluded that the motion of the Earth with respect to the postulated ether cannot be detected.

Many efforts were made to explain the null results of the Michelson–Morley experiment and to save the ether frame concept and the Galilean velocity transformation equation for light. All proposals resulting from these efforts have been shown to be wrong. No experiment in the history of physics received such valiant efforts to explain the absence of an expected result as did the Michelson–Morley experiment. The stage was set for Einstein, who solved the problem in 1905 with his special theory of relativity.

39.3 Einstein's Principle of Relativity

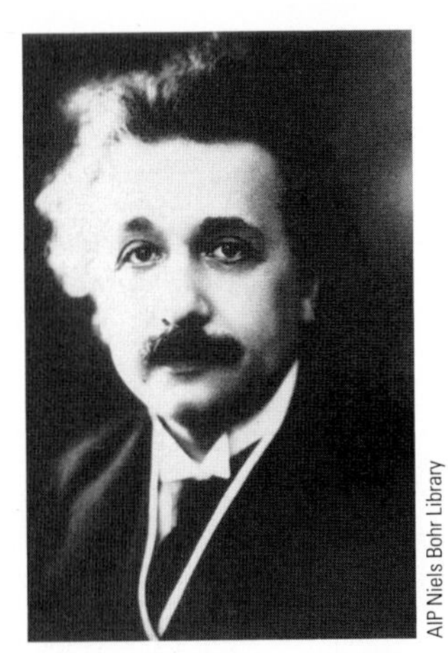

AIP Niels Bohr Library

ALBERT EINSTEIN
German-American Physicist (1879–1955)
Einstein, one of the greatest physicists of all times, was born in Ulm, Germany. In 1905, at age 26, he published four scientific papers that revolutionized physics. Two of these papers were concerned with what is now considered his most important contribution: the special theory of relativity.

In 1916, Einstein published his work on the general theory of relativity. The most dramatic prediction of this theory is the degree to which light is deflected by a gravitational field. Measurements made by astronomers on bright stars in the vicinity of the eclipsed Sun in 1919 confirmed Einstein's prediction, and Einstein became a world celebrity as a result. Einstein was deeply disturbed by the development of quantum mechanics in the 1920s despite his own role as a scientific revolutionary. In particular, he could never accept the probabilistic view of events in nature that is a central feature of quantum theory. The last few decades of his life were devoted to an unsuccessful search for a unified theory that would combine gravitation and electromagnetism.

In the previous section, we noted the impossibility of measuring the speed of the ether with respect to the Earth and the failure of the Galilean velocity transformation equation in the case of light. Einstein proposed a theory that boldly removed these difficulties and at the same time completely altered our notion of space and time.[3] He based his special theory of relativity on two postulates:

1. **The principle of relativity:** The laws of physics must be the same in all inertial reference frames.
2. **The constancy of the speed of light:** The speed of light in vacuum has the same value, $c = 3.00 \times 10^8$ m/s, in all inertial frames, regardless of the velocity of the observer or the velocity of the source emitting the light.

The first postulate asserts that *all* the laws of physics—those dealing with mechanics, electricity and magnetism, optics, thermodynamics, and so on—are the same in all reference frames moving with constant velocity relative to one another. This postulate is a generalization of the principle of Galilean relativity, which refers only to the laws of mechanics. From an experimental point of view, Einstein's principle of relativity means that any kind of experiment (measuring the speed of light, for example) performed in a laboratory at rest must give the same result when performed in a laboratory moving at a constant velocity with respect to the first one. Hence, no preferred inertial reference frame exists, and it is impossible to detect absolute motion.

Note that postulate 2 is required by postulate 1: if the speed of light were not the same in all inertial frames, measurements of different speeds would make it possible to distinguish between inertial frames. As a result, a preferred, absolute frame could be identified, in contradiction to postulate 1.

Although the Michelson–Morley experiment was performed before Einstein published his work on relativity, it is not clear whether or not Einstein was aware of the details of the experiment. Nonetheless, the null result of the experiment can be readily understood within the framework of Einstein's theory. According to his principle of relativity, the premises of the Michelson–Morley experiment were incorrect. In the process of trying to explain the expected results, we stated that when light traveled against the ether wind, its speed was $c - v$, in accordance with the Galilean velocity transformation equation. If the state of motion of the observer or of the source has no influence on the value found for the speed of

[3] A. Einstein, "On the Electrodynamics of Moving Bodies," *Ann. Physik* **17**:891, 1905. For an English translation of this article and other publications by Einstein, see the book by H. Lorentz, A. Einstein, H. Minkowski, and H. Weyl, *The Principle of Relativity* (New York: Dover, 1958).

light, however, one always measures the value to be c. Likewise, the light makes the return trip after reflection from the mirror at speed c, not at speed $c + v$. Therefore, the motion of the Earth does not influence the fringe pattern observed in the Michelson–Morley experiment, and a null result should be expected.

If we accept Einstein's theory of relativity, we must conclude that relative motion is unimportant when measuring the speed of light. At the same time, we must alter our commonsense notion of space and time and be prepared for some surprising consequences. As you read the pages ahead, keep in mind that our commonsense ideas are based on a lifetime of everyday experiences and not on observations of objects moving at hundreds of thousands of kilometers per second. Therefore, these results may seem strange, but that is only because we have no experience with them.

39.4 Consequences of the Special Theory of Relativity

As we examine some of the consequences of relativity in this section, we restrict our discussion to the concepts of simultaneity, time intervals, and lengths, all three of which are quite different in relativistic mechanics from what they are in Newtonian mechanics. In relativistic mechanics, for example, the distance between two points and the time interval between two events depend on the frame of reference in which they are measured.

Simultaneity and the Relativity of Time

A basic premise of Newtonian mechanics is that a universal time scale exists that is the same for all observers. Newton and his followers took simultaneity for granted. In his special theory of relativity, Einstein abandoned this assumption.

Einstein devised the following thought experiment to illustrate this point. A boxcar moves with uniform velocity, and two bolts of lightning strike its ends as illustrated in Figure 39.5a, leaving marks on the boxcar and on the ground. The marks on the boxcar are labeled A' and B', and those on the ground are labeled A and B. An observer O' moving with the boxcar is midway between A' and B', and a ground observer O is midway between A and B. The events recorded by the observers are the striking of the boxcar by the two lightning bolts.

The light signals emitted from A and B at the instant at which the two bolts strike later reach observer O at the same time as indicated in Figure 39.5b. This observer realizes that the signals traveled at the same speed over equal distances and so concludes that the events at A and B occurred simultaneously. Now consider the same events as viewed by observer O'. By the time the signals have reached observer O, observer O' has moved as indicated in Figure 39.5b. Therefore, the signal from B' has already swept past O', but the signal from A' has not yet reached O'. In other words, O' sees the signal from B' before seeing the signal from A'. According to Einstein, *the two observers must find that light travels at the same*

PITFALL PREVENTION 39.2
Who's Right?

You might wonder which observer in Figure 39.5 is correct concerning the two lightning strikes. *Both are correct* because the principle of relativity states that *there is no preferred inertial frame of reference.* Although the two observers reach different conclusions, both are correct in their own reference frame because the concept of simultaneity is not absolute. That, in fact, is the central point of relativity: any uniformly moving frame of reference can be used to describe events and do physics.

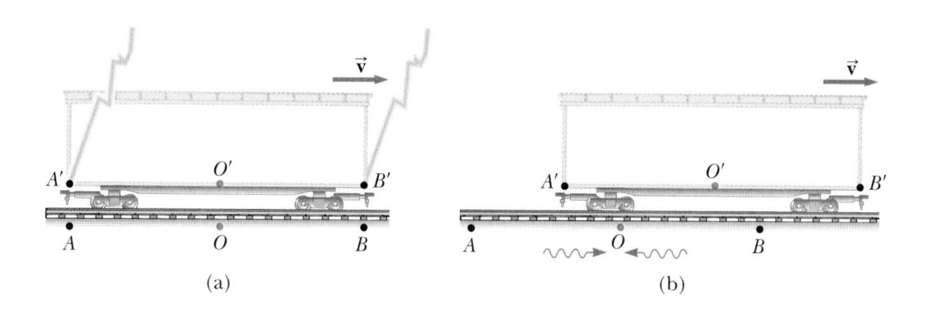

Figure 39.5 (a) Two lightning bolts strike the ends of a moving boxcar. (b) The events appear to be simultaneous to the stationary observer O who is standing midway between A and B. The events do not appear to be simultaneous to observer O', who claims that the front of the car is struck before the rear. Notice in (b) that the leftward-traveling light signal has already passed O', but the rightward-traveling signal has not yet reached O'.

speed. Therefore, observer O' concludes that one lightning bolt strikes the front of the boxcar *before* the other one strikes the back.

This thought experiment clearly demonstrates that the two events that appear to be simultaneous to observer O do *not* appear to be simultaneous to observer O'. In other words,

two events that are simultaneous in one reference frame are in general not simultaneous in a second frame moving relative to the first.

Simultaneity is not an absolute concept but rather one that depends on the state of motion of the observer. Einstein's thought experiment demonstrates that two observers can disagree on the simultaneity of two events. **This disagreement, however, depends on the transit time of light to the observers and therefore does *not* demonstrate the deeper meaning of relativity.** In relativistic analyses of high-speed situations, simultaneity is relative even when the transit time is subtracted out. In fact, in all the relativistic effects that we discuss, we ignore differences caused by the transit time of light to the observers.

Time Dilation

To illustrate that observers in different inertial frames can measure different time intervals between a pair of events, consider a vehicle moving to the right with a speed v such as the boxcar shown in Active Figure 39.6a. A mirror is fixed to the ceiling of the vehicle, and observer O' at rest in the frame attached to the vehicle holds a flashlight a distance d below the mirror. At some instant, the flashlight emits a pulse of light directed toward the mirror (event 1), and at some later time after reflecting from the mirror, the pulse arrives back at the flashlight (event 2). Observer O' carries a clock and uses it to measure the time interval Δt_p between these two events. (The subscript p stands for *proper,* as we shall see in a moment.) We model the pulse of light as a particle under constant speed. Because the light pulse has a speed c, the time interval required for the pulse to travel from O' to the mirror and back is

$$\Delta t_p = \frac{\text{distance traveled}}{\text{speed}} = \frac{2d}{c} \tag{39.5}$$

Now consider the same pair of events as viewed by observer O in a second frame as shown in Active Figure 39.6b. According to this observer, the mirror and the flashlight are moving to the right with a speed v, and as a result, the sequence of

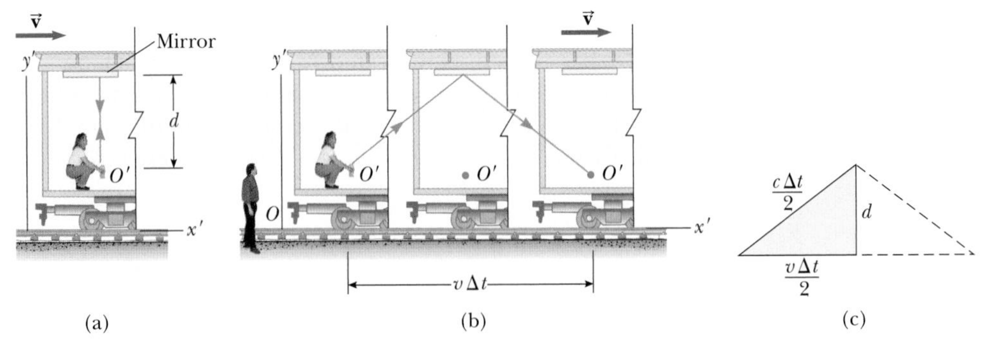

ACTIVE FIGURE 39.6

(a) A mirror is fixed to a moving vehicle, and a light pulse is sent out by observer O' at rest in the vehicle. (b) Relative to a stationary observer O standing alongside the vehicle, the mirror and O' move with a speed v. Notice that what observer O measures for the distance the pulse travels is greater than $2d$. (c) The right triangle for calculating the relationship between Δt and Δt_p.

Sign in at www.thomsonedu.com and go to ThomsonNOW to observe the bouncing of the light pulse for various speeds of the train.

events appears entirely different. By the time the light from the flashlight reaches the mirror, the mirror has moved to the right a distance $v\,\Delta t/2$, where Δt is the time interval required for the light to travel from O' to the mirror and back to O' as measured by O. Observer O concludes that because of the motion of the vehicle, if the light is to hit the mirror, it must leave the flashlight at an angle with respect to the vertical direction. Comparing Active Figure 39.6a with Active Figure 39.6b, we see that the light must travel farther in part (b) than in part (a). (Notice that neither observer "knows" that he or she is moving. Each is at rest in his or her own inertial frame.)

According to the second postulate of the special theory of relativity, both observers must measure c for the speed of light. Because the light travels farther according to O, the time interval Δt measured by O is longer than the time interval Δt_p measured by O'. To obtain a relationship between these two time intervals, let's use the right triangle shown in Active Figure 39.6c. The Pythagorean theorem gives

$$\left(\frac{c\,\Delta t}{2}\right)^2 = \left(\frac{v\,\Delta t}{2}\right)^2 + d^2$$

Solving for Δt gives

$$\Delta t = \frac{2d}{\sqrt{c^2 - v^2}} = \frac{2d}{c\sqrt{1 - \dfrac{v^2}{c^2}}} \quad (39.6)$$

Because $\Delta t_p = 2d/c$, we can express this result as

$$\Delta t = \frac{\Delta t_p}{\sqrt{1 - \dfrac{v^2}{c^2}}} = \gamma \Delta t_p \quad (39.7)$$

◄ Time dilation

where

$$\gamma = \frac{1}{\sqrt{1 - \dfrac{v^2}{c^2}}} \quad (39.8)$$

Because γ is always greater than unity, Equation 39.7 shows that **the time interval Δt measured by an observer moving with respect to a clock is longer than the time interval Δt_p measured by an observer at rest with respect to the clock. This effect is known as time dilation.**

Time dilation is not observed in our everyday lives, which can be understood by considering the factor γ. This factor deviates significantly from a value of 1 only for very high speeds as shown in Figure 39.7 and Table 39.1. For example, for a speed of $0.1c$, the value of γ is 1.005. Therefore, there is a time dilation of only 0.5% at one-tenth the speed of light. Speeds encountered on an everyday basis are far slower than $0.1c$, so we do not experience time dilation in normal situations.

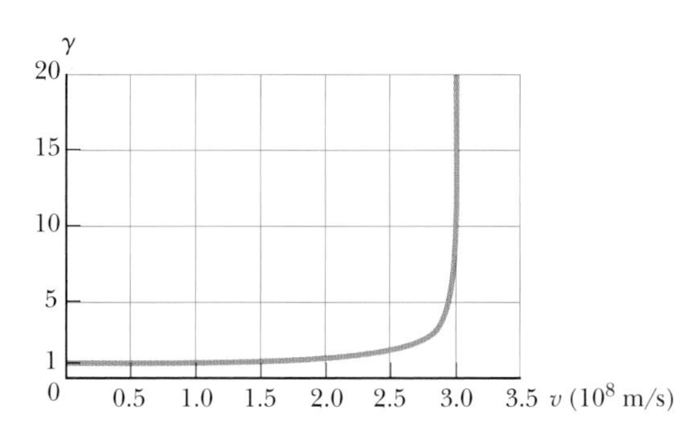

Figure 39.7 Graph of γ versus v. As the speed approaches that of light, γ increases rapidly.

TABLE 39.1

Approximate Values for γ at Various Speeds

v/c	γ
0.001 0	1.000 000 5
0.010	1.000 05
0.10	1.005
0.20	1.021
0.30	1.048
0.40	1.091
0.50	1.155
0.60	1.250
0.70	1.400
0.80	1.667
0.90	2.294
0.92	2.552
0.94	2.931
0.96	3.571
0.98	5.025
0.99	7.089
0.995	10.01
0.999	22.37

PITFALL PREVENTION 39.3

The Proper Time Interval

It is *very* important in relativistic calculations to correctly identify the observer who measures the proper time interval. The proper time interval between two events is always the time interval measured by an observer for whom the two events take place at the same position.

The time interval Δt_p in Equations 39.5 and 39.7 is called the **proper time interval.** (Einstein used the German term *Eigenzeit,* which means "own-time.") In general, **the proper time interval is the time interval between two events measured by an observer who sees the events occur at the same point in space.**

If a clock is moving with respect to you, the time interval between ticks of the moving clock is observed to be longer than the time interval between ticks of an identical clock in your reference frame. Therefore, it is often said that a moving clock is measured to run more slowly than a clock in your reference frame by a factor γ. We can generalize this result by stating that all physical processes, including mechanical, chemical, and biological ones, are measured to slow down when those processes occur in a frame moving with respect to the observer. For example, the heartbeat of an astronaut moving through space keeps time with a clock inside the spacecraft. Both the astronaut's clock and heartbeat are measured to slow down relative to a clock back on the Earth (although the astronaut would have no sensation of life slowing down in the spacecraft).

Quick Quiz 39.3 Suppose the observer O' on the train in Active Figure 39.6 aims her flashlight at the far wall of the boxcar and turns it on and off, sending a pulse of light toward the far wall. Both O' and O measure the time interval between when the pulse leaves the flashlight and when it hits the far wall. Which observer measures the proper time interval between these two events? (a) O' (b) O (c) both observers (d) neither observer

Quick Quiz 39.4 A crew on a spacecraft watches a movie that is two hours long. The spacecraft is moving at high speed through space. Does an Earth-based observer watching the movie screen on the spacecraft through a powerful telescope measure the duration of the movie to be (a) longer than, (b) shorter than, or (c) equal to two hours?

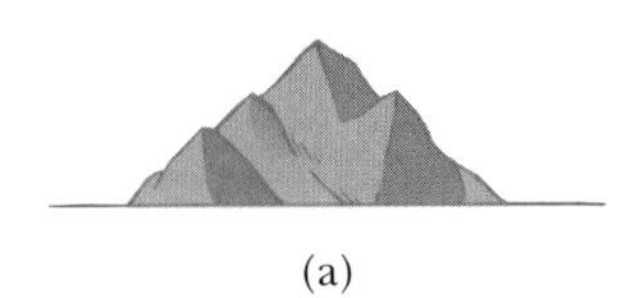

(a)

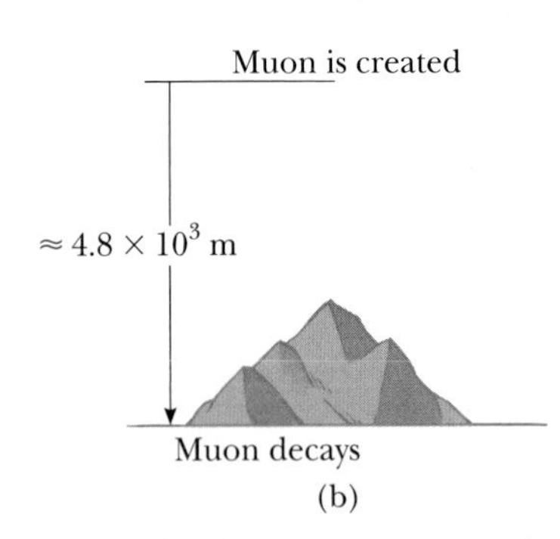

(b)

Figure 39.8 Travel of muons according to an Earth-based observer. (a) Without relativistic considerations, muons created in the atmosphere and traveling downward with a speed of $0.99c$ travel only about 6.6×10^2 m before decaying with an average lifetime of 2.2 μs. Therefore, very few muons reach the surface of the Earth. (b) With relativistic considerations, the muon's lifetime is dilated according to an observer on the Earth. Hence, according to this observer, the muon can travel about 4.8×10^3 m before decaying. The result is many of them arriving at the surface.

Time dilation is a very real phenomenon that has been verified by various experiments involving natural clocks. One experiment reported by J. C. Hafele and R. E. Keating provided direct evidence of time dilation.[4] Time intervals measured with four cesium atomic clocks in jet flight were compared with time intervals measured by Earth-based reference atomic clocks. To compare these results with theory, many factors had to be considered, including periods of speeding up and slowing down relative to the Earth, variations in direction of travel, and the weaker gravitational field experienced by the flying clocks than that experienced by the Earth-based clock. The results were in good agreement with the predictions of the special theory of relativity and were explained in terms of the relative motion between the Earth and the jet aircraft. In their paper, Hafele and Keating stated that "relative to the atomic time scale of the U.S. Naval Observatory, the flying clocks lost 59 ± 10 ns during the eastward trip and gained 273 ± 7 ns during the westward trip."

Another interesting example of time dilation involves the observation of *muons,* unstable elementary particles that have a charge equal to that of the electron and a mass 207 times that of the electron. Muons can be produced by the collision of cosmic radiation with atoms high in the atmosphere. Slow-moving muons in the laboratory have a lifetime that is measured to be the proper time interval $\Delta t_p = 2.2\ \mu$s. If we take 2.2 μs as the average lifetime of a muon and assume their speed is close to the speed of light, we find that these particles can travel a distance of approximately $(3.0 \times 10^8\ \text{m/s})(2.2 \times 10^{-6}\ \text{s}) \approx 6.6 \times 10^2$ m before they decay (Fig. 39.8a). Hence, they are unlikely to reach the surface of the Earth from high in the atmosphere where they are produced. Experiments show, however, that a large number of muons *do* reach the surface. The phenomenon of time dilation explains this effect. As measured by an observer on the Earth, the muons have a dilated lifetime

[4] J. C. Hafele and R. E. Keating, "Around the World Atomic Clocks: Relativistic Time Gains Observed," *Science* **177**:168, 1972.

equal to $\gamma \, \Delta t_p$. For example, for $v = 0.99c$, $\gamma \approx 7.1$, and $\gamma \, \Delta t_p \approx 16 \, \mu s$. Hence, the average distance traveled by the muons in this time interval as measured by an observer on the Earth is approximately $(0.99)(3.0 \times 10^8 \text{ m/s})(16 \times 10^{-6} \text{ s}) \approx 4.8 \times 10^3$ m as indicated in Figure 39.8b.

In 1976, at the laboratory of the European Council for Nuclear Research in Geneva, muons injected into a large storage ring reached speeds of approximately $0.999\,4c$. Electrons produced by the decaying muons were detected by counters around the ring, enabling scientists to measure the decay rate and hence the muon lifetime. The lifetime of the moving muons was measured to be approximately 30 times as long as that of the stationary muon, in agreement with the prediction of relativity to within two parts in a thousand.

EXAMPLE 39.1 What Is the Period of the Pendulum?

The period of a pendulum is measured to be 3.00 s in the reference frame of the pendulum. What is the period when measured by an observer moving at a speed of $0.960c$ relative to the pendulum?

SOLUTION

Conceptualize Let's change frames of reference. Instead of the observer moving at $0.960c$, we can take the equivalent point of view that the observer is at rest and the pendulum is moving at $0.960c$ past the stationary observer. Hence, the pendulum is an example of a clock moving at high speed with respect to an observer.

Categorize Based on the Conceptualize step, we can categorize this problem as one involving time dilation.

Analyze The proper time interval, measured in the rest frame of the pendulum, is $\Delta t_p = 3.00$ s.

Use Equation 39.7 to find the dilated time interval:

$$\Delta t = \gamma \, \Delta t_p = \frac{1}{\sqrt{1 - \dfrac{(0.960c)^2}{c^2}}} \Delta t_p = \frac{1}{\sqrt{1 - 0.921\,6}} \Delta t_p$$

$$= 3.57(3.00 \text{ s}) = 10.7 \text{ s}$$

Finalize This result shows that a moving pendulum is indeed measured to take longer to complete a period than a pendulum at rest does. The period increases by a factor of $\gamma = 3.57$.

What If? What if the speed of the observer increases by 4.00%? Does the dilated time interval increase by 4.00%?

Answer Based on the highly nonlinear behavior of γ as a function of v in Figure 39.7, we would guess that the increase in Δt would be different from 4.00%.

Find the new speed if it increases by 4.00%:

$$v_{\text{new}} = (1.040\,0)(0.960c) = 0.998\,4c$$

Perform the time dilation calculation again:

$$\Delta t = \gamma \, \Delta t_p = \frac{1}{\sqrt{1 - \dfrac{(0.998\,4c)^2}{c^2}}} \Delta t_p = \frac{1}{\sqrt{1 - 0.996\,8}} \Delta t_p$$

$$= 17.68(3.00 \text{ s}) = 53.1 \text{ s}$$

Therefore, the 4.00% increase in speed results in almost a 400% increase in the dilated time!

EXAMPLE 39.2 How Long Was Your Trip?

Suppose you are driving your car on a business trip and are traveling at 30 m/s. Your boss, who is waiting at your destination, expects the trip to take 5.0 h. When you arrive late, your excuse is that clock in your car registered the passage of 5.0 h but that you were driving fast and so your clock ran more slowly than the clock in your boss's office. If your car clock actually did indicate a 5.0-h trip, how much time passed on your boss's clock, which was at rest on the Earth?

SOLUTION

Conceptualize The observer is your boss standing stationary on the Earth. The clock is in your car, moving at 30 m/s with respect to your boss.

Categorize The speed of 30 m/s suggests we might categorize this problem as one in which we use classical concepts and equations. Based on the problem statement that the moving clock runs more slowly than a stationary clock however, we categorize this problem as one involving time dilation.

Analyze The proper time interval, measured in the rest frame of the car, is $\Delta t_p = 5.0$ h.

Use Equation 39.8 to evaluate γ:

$$\gamma = \frac{1}{\sqrt{1 - \frac{v^2}{c^2}}} = \frac{1}{\sqrt{1 - \frac{(3.0 \times 10^1 \text{ m/s})^2}{(3.0 \times 10^8 \text{ m/s})^2}}} = \frac{1}{\sqrt{1 - 10^{-14}}}$$

If you try to determine this value on your calculator, you will probably obtain $\gamma = 1$. Instead, perform a binomial expansion:

$$\gamma = (1 - 10^{-14})^{-1/2} \approx 1 + \tfrac{1}{2}(10^{-14}) = 1 + 5.0 \times 10^{-15}$$

Use Equation 39.7 to find the dilated time interval measured by your boss:

$$\Delta t = \gamma \, \Delta t_p = (1 + 5.0 \times 10^{-15})(5.0 \text{ h})$$

$$= 5.0 \text{ h} + 2.5 \times 10^{-14} \text{ h} = 5.0 \text{ h} + 0.090 \text{ ns}$$

Finalize Your boss's clock would be only 0.090 ns ahead of your car clock. You might want to think of another excuse!

The Twin Paradox

An intriguing consequence of time dilation is the *twin paradox* (Fig. 39.9). Consider an experiment involving a set of twins named Speedo and Goslo. When they are 20 years old, Speedo, the more adventuresome of the two, sets out on an epic journey from the Earth to Planet X, located 20 lightyears away. One lightyear (ly) is the distance light travels through free space in 1 year. Furthermore, Speedo's spacecraft is capable of reaching a speed of $0.95c$ relative to the inertial frame of his twin brother back home on the Earth. After reaching Planet X, Speedo becomes homesick and immediately returns to the Earth at the same speed $0.95c$.

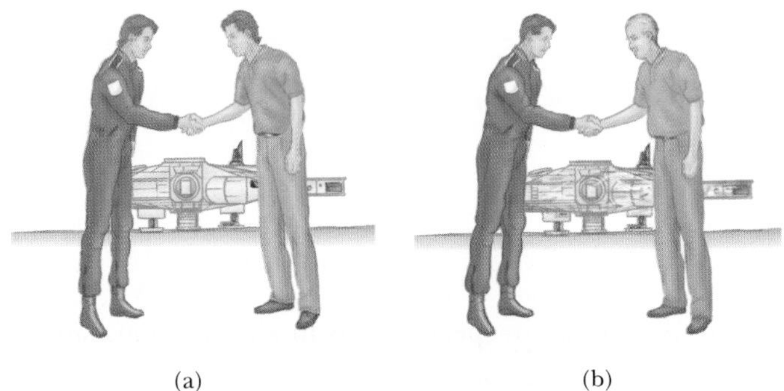

Figure 39.9 (a) As one twin leaves his brother on the Earth, both are the same age. (b) When Speedo returns from his journey to Planet X, he is younger than his twin Goslo.

Upon his return, Speedo is shocked to discover that Goslo has aged 42 years and is now 62 years old. Speedo, on the other hand, has aged only 13 years.

The paradox is *not* that the twins have aged at different rates. Here is the paradox. From Goslo's frame of reference, he was at rest while his brother traveled at a high speed away from him and then came back. According to Speedo, however, he himself remained stationary while Goslo and the Earth raced away from him and then headed back. Therefore, we might expect Speedo to claim that Goslo ages more slowly than himself. The situation appears to be symmetrical from either twin's point of view. Which twin *actually* ages more slowly?

The situation is actually not symmetrical. Consider a third observer moving at a constant speed relative to Goslo. According to the third observer, Goslo never changes inertial frames. Goslo's speed relative to the third observer is always the same. The third observer notes, however, that Speedo accelerates during his journey when he slows down and starts moving back toward the Earth, *changing reference frames in the process.* From the third observer's perspective, there is something very different about the motion of Goslo when compared to Speedo. Therefore, there is no paradox: only Goslo, who is always in a single inertial frame, can make correct predictions based on special relativity. Goslo finds that instead of aging 42 years, Speedo ages only $(1 - v^2/c^2)^{1/2}(42 \text{ years}) = 13$ years. Of these 13 years, Speedo spends 6.5 years traveling to Planet X and 6.5 years returning.

Quick Quiz 39.5 Suppose astronauts are paid according to the amount of time they spend traveling in space. After a long voyage traveling at a speed approaching c, would a crew rather be paid according to (a) an Earth-based clock, (b) their spacecraft's clock, or (c) either clock?

Length Contraction

The measured distance between two points in space also depends on the frame of reference of the observer. **The proper length L_p of an object is the length measured by someone at rest relative to the object.** The length of an object measured by someone in a reference frame that is moving with respect to the object is always less than the proper length. This effect is known as **length contraction.**

To understand length contraction, consider a spacecraft traveling with a speed v from one star to another. There are two observers: one on the Earth and the other in the spacecraft. The observer at rest on the Earth (and also assumed to be at rest with respect to the two stars) measures the distance between the stars to be the proper length L_p. According to this observer, the time interval required for the spacecraft to complete the voyage is $\Delta t = L_p/v$. The passages of the two stars by the spacecraft occur at the same position for the space traveler. Therefore, the space traveler measures the proper time interval Δt_p. Because of time dilation, the proper time interval is related to the Earth-measured time interval by $\Delta t_p = \Delta t/\gamma$. Because the space traveler reaches the second star in the time Δt_p, he or she concludes that the distance L between the stars is

$$L = v\,\Delta t_p = v\frac{\Delta t}{\gamma}$$

Because the proper length is $L_p = v\,\Delta t$, we see that

$$L = \frac{L_p}{\gamma} = L_p\sqrt{1 - \frac{v^2}{c^2}} \qquad (39.9)$$

◀ Length contraction

where $\sqrt{1 - v^2/c^2}$ is a factor less than unity. **If an object has a proper length L_p when it is measured by an observer at rest with respect to the object, its length L when it moves with speed v in a direction parallel to its length is measured to be shorter according to $L = L_p\sqrt{1 - v^2/c^2} = L_p/\gamma$.**

PITFALL PREVENTION 39.4
The Proper Length
As with the proper time interval, it is *very* important in relativistic calculations to correctly identify the observer who measures the proper length. The proper length between two points in space is always the length measured by an observer at rest with respect to the points. Often, the proper time interval and the proper length are *not* measured by the same observer.

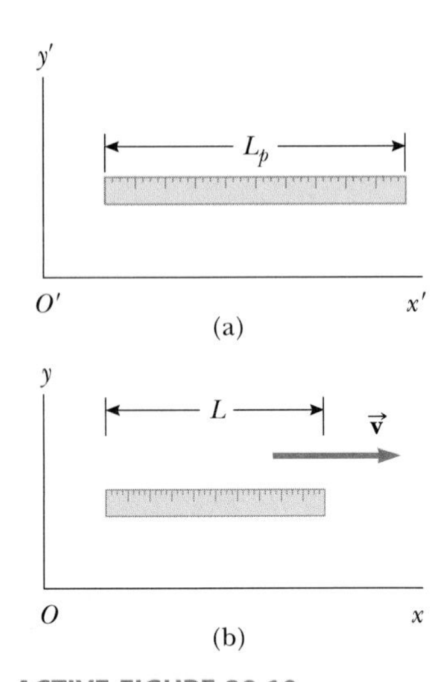

ACTIVE FIGURE 39.10

(a) A meterstick measured by an observer in a frame attached to the stick (that is, both have the same velocity) has its proper length L_p. (b) The meterstick measured by an observer in a frame in which the stick has a velocity $\vec{v}$ relative to the frame is measured to be shorter than its proper length L_p by a factor $(1 - v^2/c^2)^{1/2}$.

Sign in at www.thomsonedu.com and go to ThomsonNOW to view the meterstick from the points of view of two observers and compare the measured length of the stick.

For example, suppose a meterstick moves past a stationary Earth-based observer with speed v as in Active Figure 39.10. The length of the meterstick as measured by an observer in a frame attached to the stick is the proper length L_p shown in Active Figure 39.10a. The length of the stick L measured by the Earth observer is shorter than L_p by the factor $(1 - v^2/c^2)^{1/2}$ as suggested in Active Figure 39.10b. Notice that **length contraction takes place only along the direction of motion.**

The proper length and the proper time interval are defined differently. The proper length is measured by an observer for whom the end points of the length remain fixed in space. The proper time interval is measured by someone for whom the two events take place at the same position in space. As an example of this point, let's return to the decaying muons moving at speeds close to the speed of light. An observer in the muon's reference frame measures the proper lifetime, whereas an Earth-based observer measures the proper length (the distance between the creation point and the decay point in Fig. 39.8b). In the muon's reference frame, there is no time dilation, but the distance of travel to the surface is shorter when measured in this frame. Likewise, in the Earth observer's reference frame, there is time dilation, but the distance of travel is measured to be the proper length. Therefore, when calculations on the muon are performed in both frames, the outcome of the experiment in one frame is the same as the outcome in the other frame: more muons reach the surface than would be predicted without relativistic effects.

Quick Quiz 39.6 You are packing for a trip to another star. During the journey, you will be traveling at $0.99c$. You are trying to decide whether you should buy smaller sizes of your clothing, because you will be thinner on your trip due to length contraction. You also plan to save money by reserving a smaller cabin to sleep in because you will be shorter when you lie down. Should you (a) buy smaller sizes of clothing, (b) reserve a smaller cabin, (c) do neither of these things, or (d) do both of these things?

Quick Quiz 39.7 You are observing a spacecraft moving away from you. You measure it to be shorter than when it was at rest on the ground next to you. You also see a clock through the spacecraft window, and you observe that the passage of time on the clock is measured to be slower than that of the watch on your wrist. Compared with when the spacecraft was on the ground, what do you measure if the spacecraft turns around and comes *toward* you at the same speed? (a) The spacecraft is measured to be longer, and the clock runs faster. (b) The spacecraft is measured to be longer, and the clock runs slower. (c) The spacecraft is measured to be shorter, and the clock runs faster. (d) The spacecraft is measured to be shorter, and the clock runs slower.

Space–Time Graphs

It is sometimes helpful to represent a physical situation with a *space–time graph,* in which ct is the ordinate and position x is the abscissa. The twin paradox is displayed in such a graph in Figure 39.11 from Goslo's point of view. A path through space–time is called a **world-line.** At the origin, the world-lines of Speedo (blue) and Goslo (green) coincide because the twins are in the same location at the same time. After Speedo leaves on his trip, his world-line diverges from that of his brother. Goslo's world-line is vertical because he remains fixed in location. At Goslo and Speedo's reunion, the two world-lines again come together. It would be impossible for Speedo to have a world-line that crossed the path of a light beam that left the Earth when he did. To do so would require him to have a speed greater than c (which, as shown in Sections 39.6 and 39.7, is not possible).

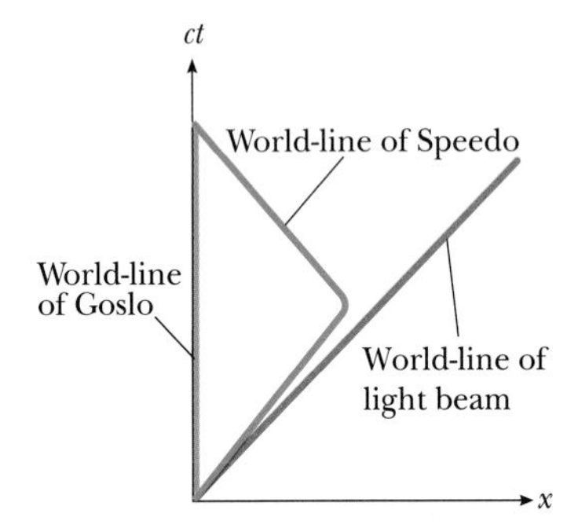

Figure 39.11 The twin paradox on a space–time graph. The twin who stays on the Earth has a world-line along the ct axis (green). The path of the traveling twin through space–time is represented by a world-line that changes direction (blue).

World-lines for light beams are diagonal lines on space–time graphs, typically drawn at 45° to the right or left of vertical (assuming that the x and ct axes have the same scales), depending on whether the light beam is traveling in the direction of increasing or decreasing x. These two world-lines mean that all possible future events for Goslo and Speedo lie within two 45° lines extending from the origin. Either twin's presence at an event outside this "light cone" would require that twin to move at a speed greater than c, which we have said is not possible. Also, the only past events that Goslo and Speedo could have experienced occur between two similar 45° world-lines that approach the origin from below the x axis.

EXAMPLE 39.3 A Voyage to Sirius

An astronaut takes a trip to Sirius, which is located a distance of 8 lightyears from the Earth. The astronaut measures the time of the one-way journey to be 6 years. If the spaceship moves at a constant speed of $0.8c$, how can the 8-ly distance be reconciled with the 6-year trip time measured by the astronaut?

SOLUTION

Conceptualize An observer on the Earth measures light to require 8 years to travel from Earth to Sirius. The astronaut measures a time interval of only 6 years. Is the astronaut traveling faster than light?

Categorize Because the astronaut is measuring a length of space between Earth and Sirius that is in motion with respect to her, we categorize this example as a length contraction problem.

Analyze The distance of 8 ly represents the proper length from the Earth to Sirius measured by an observer on the Earth seeing both objects nearly at rest.

Calculate the contracted length measured by the astronaut using Equation 39.9:

$$L = \frac{8 \text{ ly}}{\gamma} = (8 \text{ ly})\sqrt{1 - \frac{v^2}{c^2}} = (8 \text{ ly})\sqrt{1 - \frac{(0.8c)^2}{c^2}} = 5 \text{ ly}$$

Use the particle under constant speed model to find the travel time measured on the astronaut's clock:

$$\Delta t = \frac{L}{v} = \frac{5 \text{ ly}}{0.8c} = \frac{5 \text{ ly}}{0.8(1 \text{ ly/yr})} = 6 \text{ yr}$$

Finalize Note that we have used the value for the speed of light as $c = 1$ ly/yr. The trip takes a time interval shorter than 8 years for the astronaut because, to her, the distance between the Earth and Sirius is measured to be shorter.

What If? What if this trip is observed with a very powerful telescope by a technician in Mission Control on the Earth? At what time will this technician *see* that the astronaut has arrived at Sirius?

Answer The time interval the technician measures for the astronaut to arrive is

$$\Delta t = \frac{L_p}{v} = \frac{8 \text{ ly}}{0.8c} = 10 \text{ yr}$$

For the technician to *see* the arrival, the light from the scene of the arrival must travel back to the Earth and enter the telescope. This travel requires a time interval of

$$\Delta t = \frac{L_p}{v} = \frac{8 \text{ ly}}{c} = 8 \text{ yr}$$

Therefore, the technician sees the arrival after 10 yr + 8 yr = 18 yr. If the astronaut immediately turns around and comes back home, she arrives, according to the technician, 20 years after leaving, only 2 years *after the technician saw her arrive*! In addition, the astronaut would have aged by only 12 years.

EXAMPLE 39.4 The Pole-in-the-Barn Paradox

The twin paradox, discussed earlier, is a classic "paradox" in relativity. Another classic "paradox" is as follows. Suppose a runner moving at $0.75c$ carries a horizontal pole 15 m long toward a barn that is 10 m long. The barn has front and rear doors that are initially open. An observer on the ground can instantly and simultaneously close and open the two doors by remote control. When the runner and the pole are inside the barn, the ground observer closes and then opens both doors so that the runner and pole are momentarily captured inside the barn and then proceed to exit the barn from the back door. Do both the runner and the ground observer agree that the runner makes it safely through the barn?

SOLUTION

Conceptualize From your everyday experience, you would be surprised to see a 15-m pole fit inside a 10-m barn.

Categorize The pole is in motion with respect to the ground observer so that the observer measures its length to be contracted, whereas the stationary barn has a proper length of 10 m. We categorize this example as a length contraction problem.

Analyze Use Equation 39.9 to find the contracted length of the pole according to the ground observer:

$$L_{\text{pole}} = L_p\sqrt{1 - \frac{v^2}{c^2}} = (15\text{ m})\sqrt{1 - (0.75)^2} = 9.9\text{ m}$$

Therefore, the ground observer measures the pole to be slightly shorter than the barn and there is no problem with momentarily capturing the pole inside it. The "paradox" arises when we consider the runner's point of view.

Use Equation 39.9 to find the contracted length of the barn according to the running observer:

$$L_{\text{barn}} = L_p\sqrt{1 - \frac{v^2}{c^2}} = (10\text{ m})\sqrt{1 - (0.75)^2} = 6.6\text{ m}$$

Because the pole is in the rest frame of the runner, the runner measures it to have its proper length of 15 m. How can a 15-m pole fit inside a 6.6-m barn? Although this question is the classic one that is often asked, it is not the question we have asked because it is not the important one. We asked, *"Does the runner make it safely through the barn?"*

The resolution of the "paradox" lies in the relativity of simultaneity. The closing of the two doors is measured to be simultaneous by the ground observer. Because the doors are at different positions, however, they do not close simultaneously as measured by the runner. The rear door closes and then opens first, allowing the leading end of the pole to exit. The front door of the barn does not close until the trailing end of the pole passes by.

We can analyze this "paradox" using a space–time graph. Figure 39.12a is a space–time graph from the ground observer's point of view. We choose $x = 0$ as the position of the front door of the barn and $t = 0$ as the instant at which the leading end of the pole is located at the front door of the barn. The world-lines for the two doors of the barn are separated by 10 m and are vertical because the barn is not moving relative to this observer. For the pole, we follow two tilted world-lines, one for each end of the moving pole. These world-lines are 9.9 m apart horizontally, which is the con-

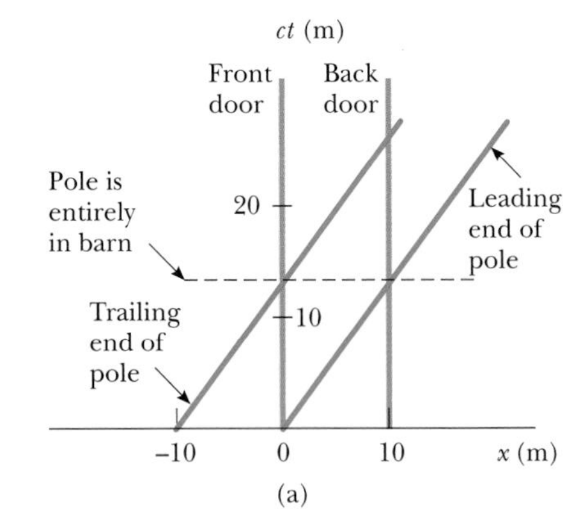

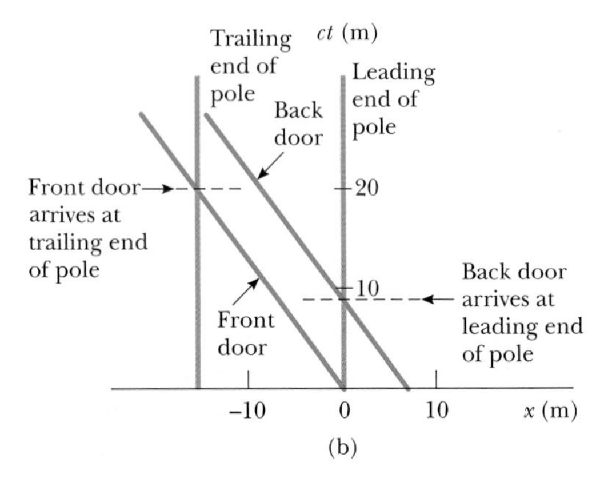

Figure 39.12 (Example 39.4) Space–time graphs for the pole-in-the-barn paradox. (a) From the ground observer's point of view, the world-lines for the front and back doors of the barn are vertical lines. The world-lines for the ends of the pole are tilted and are 9.9 m apart horizontally. The front door of the barn is at $x = 0$, and the leading end of the pole enters the front door at $t = 0$. The entire pole is inside the barn at the time indicated by the dashed line. (b) From the runner's point of view, the world-lines for the ends of the pole are vertical. The barn is moving in the negative direction, so the world-lines for the front and back doors are tilted to the left. The leading end of the pole exits the back door before the trailing end arrives at the front door.

tracted length seen by the ground observer. As seen in Figure 39.12a, the pole is entirely within the barn at one instant.

Figure 39.12b shows the space–time graph according to the runner. Here, the world-lines for the pole are separated by 15 m and are vertical because the pole is at rest in the runner's frame of reference. The barn is hurtling *toward* the runner, so the world-lines for the front and rear doors of the barn are tilted to the left. The world-lines for the barn are separated by 6.6 m, the contracted length as seen by the runner. The leading end of the pole leaves the rear door of the barn long before the trailing end of the pole enters the barn. Therefore, the opening of the rear door occurs before the closing of the front door.

From the ground observer's point of view, use the particle under constant velocity model to find the time after $t = 0$ at which the trailing end of the pole enters the barn:

$$(1) \quad t = \frac{\Delta x}{v} = \frac{9.9 \text{ m}}{0.75c} = \frac{13.2 \text{ m}}{c}$$

From the runner's point of view, use the particle under constant velocity model to find the time at which the leading end of the pole leaves the barn:

$$(2) \quad t = \frac{\Delta x}{v} = \frac{6.6 \text{ m}}{0.75c} = \frac{8.8 \text{ m}}{c}$$

Find the time at which the trailing end of the pole enters the front door of the barn:

$$(3) \quad t = \frac{\Delta x}{v} = \frac{15 \text{ m}}{0.75c} = \frac{20 \text{ m}}{c}$$

Finalize From Equation (1), the pole should be completely inside the barn at a time corresponding to $ct = 13.2$ m. This situation is consistent with the point on the ct axis in Figure 39.12a where the pole is inside the barn. From Equation (2), the leading end of the pole leaves the barn at $ct = 8.8$ m. This situation is consistent with the point on the ct axis in Figure 39.12b where the back door of the barn arrives at the leading end of the pole. Equation (3) gives $ct = 20$ m, which agrees with the instant shown in Figure 39.12b at which the front door of the barn arrives at the trailing end of the pole.

The Relativistic Doppler Effect

Another important consequence of time dilation is the shift in frequency observed for light emitted by atoms in motion as opposed to light emitted by atoms at rest. This phenomenon, known as the Doppler effect, was introduced in Chapter 17 as it pertains to sound waves. In the case of sound, the motion of the source with respect to the medium of propagation can be distinguished from the motion of the observer with respect to the medium. Light waves must be analyzed differently, however, because they require no medium of propagation and no method exists for distinguishing the motion of a light source from the motion of the observer.

If a light source and an observer approach each other with a relative speed v, the frequency f_{obs} measured by the observer is

$$f_{\text{obs}} = \frac{\sqrt{1 + v/c}}{\sqrt{1 - v/c}} f_{\text{source}} \qquad \textbf{(39.10)}$$

where f_{source} is the frequency of the source measured in its rest frame. This relativistic Doppler shift equation, unlike the Doppler shift equation for sound, depends only on the relative speed v of the source and observer and holds for relative speeds as great as c. As you might expect, the equation predicts that $f_{\text{obs}} > f_{\text{source}}$ when the source and observer approach each other. We obtain the expression for the case in which the source and observer recede from each other by substituting negative values for v in Equation 39.10.

The most spectacular and dramatic use of the relativistic Doppler effect is the measurement of shifts in the frequency of light emitted by a moving astronomical object such as a galaxy. Light emitted by atoms and normally found in the extreme violet region of the spectrum is shifted toward the red end of the spectrum for atoms in other galaxies, indicating that these galaxies are *receding* from us. American astronomer Edwin Hubble (1889–1953) performed extensive measurements of this *red shift* to confirm that most galaxies are moving away from us, indicating that the Universe is expanding.

39.5 The Lorentz Transformation Equations

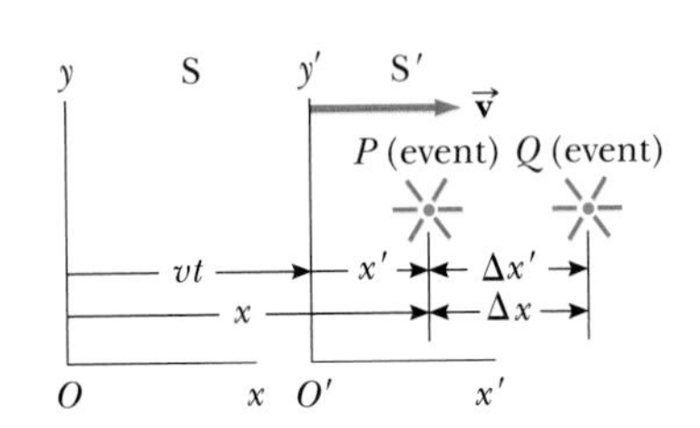

Figure 39.13 Events occur at points P and Q and are observed by an observer at rest in the S frame and another in the S′ frame, which is moving to the right with a speed v.

Suppose two events occur at points P and Q and are reported by two observers, one at rest in a frame S and another in a frame S′ that is moving to the right with speed v as in Figure 39.13. The observer in S reports the events with space–time coordinates (x, y, z, t) and the observer in S′ reports the same events using the coordinates (x', y', z', t'). Equation 39.1 predicts that the distance between the two points in space at which the events occur does not depend on motion of the observer: $\Delta x = \Delta x'$. Because this prediction is contradictory to the notion of length contraction, the Galilean transformation is not valid when v approaches the speed of light. In this section, we present the correct transformation equations that apply for all speeds in the range $0 < v < c$.

The equations that are valid for all speeds and that enable us to transform coordinates from S to S′ are the **Lorentz transformation equations:**

Lorentz transformation for S → S′ ▶

$$x' = \gamma(x - vt) \qquad y' = y \qquad z' = z \qquad t' = \gamma\left(t - \frac{v}{c^2}x\right) \tag{39.11}$$

These transformation equations were developed by Hendrik A. Lorentz (1853–1928) in 1890 in connection with electromagnetism. It was Einstein, however, who recognized their physical significance and took the bold step of interpreting them within the framework of the special theory of relativity.

Notice the difference between the Galilean and Lorentz time equations. In the Galilean case, $t = t'$. In the Lorentz case, however, the value for t' assigned to an event by an observer O' in the S′ frame in Figure 39.13 depends both on the time t and on the coordinate x as measured by an observer O in the S frame, which is consistent with the notion that an event is characterized by four space–time coordinates (x, y, z, t). In other words, in relativity, space and time are *not* separate concepts but rather are closely interwoven with each other.

If you wish to transform coordinates in the S′ frame to coordinates in the S frame, simply replace v by $-v$ and interchange the primed and unprimed coordinates in Equations 39.11:

Inverse Lorentz transformation for S′ → S ▶

$$x = \gamma(x' + vt') \qquad y = y' \qquad z = z' \qquad t = \gamma\left(t' + \frac{v}{c^2}x'\right) \tag{39.12}$$

When $v \ll c$, the Lorentz transformation equations should reduce to the Galilean equations. As v approaches zero, $v/c \ll 1$; therefore, $\gamma \rightarrow 1$ and Equations 39.11 indeed reduce to the Galilean space–time transformation equations in Equation 39.1.

In many situations, we would like to know the difference in coordinates between two events or the time interval between two events as seen by observers O and O'. From Equations 39.11 and 39.12, we can express the differences between the four variables x, x', t, and t' in the form

$$\left.\begin{aligned} \Delta x' &= \gamma(\Delta x - v\,\Delta t) \\ \Delta t' &= \gamma\left(\Delta t - \frac{v}{c^2}\,\Delta x\right) \end{aligned}\right\} \text{S} \rightarrow \text{S}' \tag{39.13}$$

$$\left.\begin{aligned} \Delta x &= \gamma(\Delta x' + v\,\Delta t') \\ \Delta t &= \gamma\left(\Delta t' + \frac{v}{c^2}\,\Delta x'\right) \end{aligned}\right\} \text{S}' \rightarrow \text{S} \tag{39.14}$$

where $\Delta x' = x_2' - x_1'$ and $\Delta t' = t_2' - t_1'$ are the differences measured by observer O' and $\Delta x = x_2 - x_1$ and $\Delta t = t_2 - t_1$ are the differences measured by observer O. (We have not included the expressions for relating the y and z coordinates because they are unaffected by motion along the x direction.[5])

EXAMPLE 39.5 **Simultaneity and Time Dilation Revisited**

(A) Use the Lorentz transformation equations in difference form to show that simultaneity is not an absolute concept.

SOLUTION

Conceptualize Imagine two events that are simultaneous and separated in space such that $\Delta t' = 0$ and $\Delta x' \neq 0$ according to an observer O' who is moving with speed v relative to O.

Categorize The statement of the problem tells us to categorize this example as one involving the Lorentz transformation.

Analyze From the expression for Δt given in Equation 39.14, find the time interval Δt measured by observer O:

$$\Delta t = \gamma\left(\Delta t' + \frac{v}{c^2}\Delta x'\right) = \gamma\left(0 + \frac{v}{c^2}\Delta x'\right) = \gamma\frac{v}{c^2}\Delta x'$$

Finalize The time interval for the same two events as measured by O is nonzero, so the events do not appear to be simultaneous to O.

(B) Use the Lorentz transformation equations in difference form to show that a moving clock is measured to run more slowly than a clock that is at rest with respect to an observer.

SOLUTION

Conceptualize Imagine that observer O' carries a clock that he uses to measure a time interval $\Delta t'$. He finds that two events occur at the same place in his reference frame ($\Delta x' = 0$) but at different times ($\Delta t' \neq 0$). Observer O' is moving with speed v relative to O.

Categorize The statement of the problem tells us to categorize this example as one involving the Lorentz transformation.

Analyze From the expression for Δt given in Equation 39.14, find the time interval Δt measured by observer O:

$$\Delta t = \gamma\left(\Delta t' + \frac{v}{c^2}\Delta x'\right) = \gamma\left(\Delta t' + \frac{v}{c^2}(0)\right) = \gamma\Delta t'$$

Finalize This result is the equation for time dilation found earlier (Eq. 39.7), where $\Delta t' = \Delta t_p$ is the proper time interval measured by the clock carried by observer O'. Therefore, O measures the moving clock to run slow.

39.6 The Lorentz Velocity Transformation Equations

Suppose two observers in relative motion with respect to each other are both observing an object's motion. Previously, we defined an event as occurring at an instant of time. Now let's interpret the "event" as the object's motion. We know that the Galilean velocity transformation (Eq. 39.2) is valid for low speeds. How do the observers' measurements of the velocity of the object relate to each other if the speed of the object is close to that of light? Once again, S′ is our frame moving

[5] Although relative motion of the two frames along the x axis does not change the y and z coordinates of an object, it does change the y and z velocity components of an object moving in either frame as noted in Section 39.6.

at a speed v relative to S. Suppose an object has a velocity component u'_x measured in the S′ frame, where

$$u'_x = \frac{dx'}{dt'} \tag{39.15}$$

Using Equation 39.11, we have

$$dx' = \gamma(dx - v\,dt)$$

$$dt' = \gamma\left(dt - \frac{v}{c^2}\,dx\right)$$

Substituting these values into Equation 39.15 gives

$$u'_x = \frac{dx - v\,dt}{dt - \dfrac{v}{c^2}\,dx} = \frac{\dfrac{dx}{dt} - v}{1 - \dfrac{v}{c^2}\dfrac{dx}{dt}}$$

The term dx/dt, however, is simply the velocity component u_x of the object measured by an observer in S, so this expression becomes

Lorentz velocity transformation for S → S′

$$u'_x = \frac{u_x - v}{1 - \dfrac{u_x v}{c^2}} \tag{39.16}$$

If the object has velocity components along the y and z axes, the components as measured by an observer in S′ are

$$u'_y = \frac{u_y}{\gamma\left(1 - \dfrac{u_x v}{c^2}\right)} \quad \text{and} \quad u'_z = \frac{u_z}{\gamma\left(1 - \dfrac{u_x v}{c^2}\right)} \tag{39.17}$$

Notice that u'_y and u'_z do not contain the parameter v in the numerator because the relative velocity is along the x axis.

When v is much smaller than c (the nonrelativistic case), the denominator of Equation 39.16 approaches unity and so $u'_x \approx u_x - v$, which is the Galilean velocity transformation equation. In another extreme, when $u_x = c$, Equation 39.16 becomes

$$u'_x = \frac{c - v}{1 - \dfrac{cv}{c^2}} = \frac{c\left(1 - \dfrac{v}{c}\right)}{1 - \dfrac{v}{c}} = c$$

This result shows that a speed measured as c by an observer in S is also measured as c by an observer in S′, independent of the relative motion of S and S′. This conclusion is consistent with Einstein's second postulate: the speed of light must be c relative to all inertial reference frames. Furthermore, we find that the speed of an object can never be measured as larger than c. That is, the speed of light is the ultimate speed. We shall return to this point later.

To obtain u_x in terms of u'_x, we replace v by $-v$ in Equation 39.16 and interchange the roles of u_x and u'_x:

$$u_x = \frac{u'_x + v}{1 + \dfrac{u'_x v}{c^2}} \tag{39.18}$$

PITFALL PREVENTION 39.5

What Can the Observers Agree On?

We have seen several measurements that the two observers O and O' do *not* agree on: (1) the time interval between events that take place in the same position in one of the frames, (2) the distance between two points that remain fixed in one of their frames, (3) the velocity components of a moving particle, and (4) whether two events occurring at different locations in both frames are simultaneous or not. The two observers *can* agree on (1) their relative speed of motion v with respect to each other, (2) the speed c of any ray of light, and (3) the simultaneity of two events which take place at the same position *and* time in some frame.

Quick Quiz 39.8 You are driving on a freeway at a relativistic speed. **(i)** Straight ahead of you, a technician standing on the ground turns on a searchlight and a beam of light moves exactly vertically upward as seen by the technician. As you observe the beam of light, do you measure the magnitude of the vertical component of its velocity as (a) equal to c, (b) greater than c, or (c) less than c? **(ii)** If the technician aims the searchlight directly at you instead of upward, do you measure the magnitude of the horizontal component of its velocity as (a) equal to c, (b) greater than c, or (c) less than c?

EXAMPLE 39.6 Relative Velocity of Two Spacecraft

Two spacecraft A and B are moving in opposite directions as shown in Figure 39.14. An observer on the Earth measures the speed of spacecraft A to be $0.750c$ and the speed of spacecraft B to be $0.850c$. Find the velocity of spacecraft B as observed by the crew on spacecraft A.

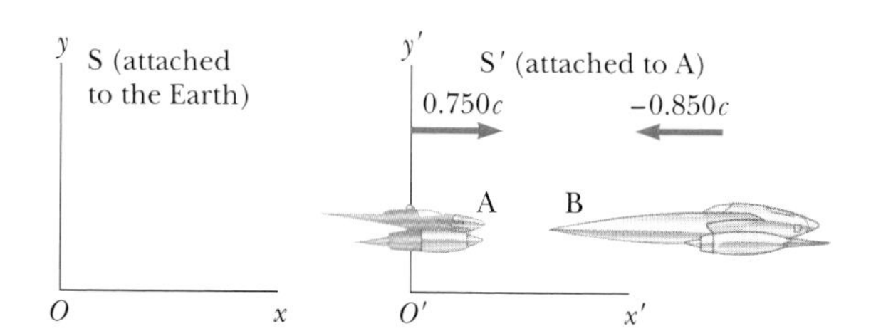

Figure 39.14 (Example 39.6) Two spacecraft A and B move in opposite directions. The speed of spacecraft B relative to spacecraft A is *less* than c and is obtained from the relativistic velocity transformation equation.

SOLUTION

Conceptualize There are two observers, one on the Earth and one on spacecraft A. The event is the motion of spacecraft B.

Categorize Because the problem asks to find an observed velocity, we categorize this example as one requiring the Lorentz velocity transformation.

Analyze The Earth-based observer at rest in the S frame makes two measurements, one of each spacecraft. We want to find the velocity of spacecraft B as measured by the crew on spacecraft A. Therefore, $u_x = -0.850c$. The velocity of spacecraft A is also the velocity of the observer at rest in spacecraft A (the S′ frame) relative to the observer at rest on the Earth. Therefore, $v = 0.750c$.

Obtain the velocity u_x' of spacecraft B relative to spacecraft A using Equation 39.16:

$$u_x' = \frac{u_x - v}{1 - \dfrac{u_x v}{c^2}} = \frac{-0.850c - 0.750c}{1 - \dfrac{(-0.850c)(0.750c)}{c^2}} = -0.977c$$

Finalize The negative sign indicates that spacecraft B is moving in the negative x direction as observed by the crew on spacecraft A. Is that consistent with your expectation from Figure 39.14? Notice that the speed is less than c. That is, an object whose speed is less than c in one frame of reference must have a speed less than c in any other frame. (Had you used the Galilean velocity transformation equation in this example, you would have found that $u_x' = u_x - v = -0.850c - 0.750c = -1.60c$, which is impossible. The Galilean transformation equation does not work in relativistic situations.)

What If? What if the two spacecraft pass each other? What is their relative speed now?

Answer The calculation using Equation 39.16 involves only the velocities of the two spacecraft and does not depend on their locations. After they pass each other, they have the same velocities, so the velocity of spacecraft B as observed by the crew on spacecraft A is the same, $-0.977c$. The only difference after they pass is that spacecraft B is receding from spacecraft A, whereas it was approaching spacecraft A before it passed.

EXAMPLE 39.7 Relativistic Leaders of the Pack

Two motorcycle pack leaders named David and Emily are racing at relativistic speeds along perpendicular paths as shown in Figure 39.15. How fast does Emily recede as seen by David over his right shoulder?

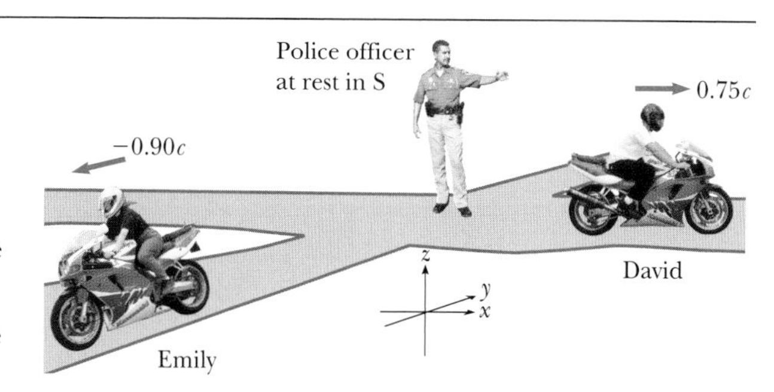

Figure 39.15 (Example 39.7) David moves east with a speed $0.75c$ relative to the police officer, and Emily travels south at a speed $0.90c$ relative to the officer.

SOLUTION

Conceptualize The two observers are David and the police officer in Figure 39.15. The event is the motion of Emily. Figure 39.15 represents the situation as seen by the police officer at rest in frame S. Frame S′ moves along with David.

Categorize Because the problem asks to find an observed velocity, we categorize this problem as one requiring the Lorentz velocity transformation. The motion takes place in two dimensions.

Analyze Identify the velocity components for David and Emily according to the police officer:

$$\text{David: } v_x = v = 0.75c \qquad v_y = 0$$

$$\text{Emily: } u_x = 0 \qquad u_y = -0.90c$$

Using Equations 39.16 and 39.17, calculate u'_x and u'_y for Emily as measured by David:

$$u'_x = \frac{u_x - v}{1 - \dfrac{u_x v}{c^2}} = \frac{0 - 0.75c}{1 - \dfrac{(0)(0.75c)}{c^2}} = -0.75c$$

$$u'_y = \frac{u_y}{\gamma\left(1 - \dfrac{u_x v}{c^2}\right)} = \frac{\sqrt{1 - \dfrac{(0.75c)^2}{c^2}}\,(-0.90c)}{\left(1 - \dfrac{(0)(0.75c)}{c^2}\right)} = -0.60c$$

Using the Pythagorean theorem, find the speed of Emily as measured by David:

$$u' = \sqrt{(u'_x)^2 + (u'_y)^2} = \sqrt{(-0.75c)^2 + (-0.60c)^2} = 0.96c$$

Finalize This speed is less than c, as required by the special theory of relativity.

PITFALL PREVENTION 39.6
Watch Out for "Relativistic Mass"

Some older treatments of relativity maintained the conservation of momentum principle at high speeds by using a model in which a particle's mass increases with speed. You might still encounter this notion of "relativistic mass" in your outside reading, especially in older books. Be aware that this notion is no longer widely accepted; today, mass is considered as *invariant*, independent of speed. The mass of an object in all frames is considered to be the mass as measured by an observer at rest with respect to the object.

39.7 Relativistic Linear Momentum

To describe the motion of particles within the framework of the special theory of relativity properly, you must replace the Galilean transformation equations by the Lorentz transformation equations. Because the laws of physics must remain unchanged under the Lorentz transformation, we must generalize Newton's laws and the definitions of linear momentum and energy to conform to the Lorentz transformation equations and the principle of relativity. These generalized definitions should reduce to the classical (nonrelativistic) definitions for $v \ll c$.

First, recall from the isolated system model that when two particles (or objects that can be modeled as particles) collide, the total momentum of the isolated system of the two particles remains constant. Suppose we observe this collision in a reference frame S and confirm that the momentum of the system is conserved. Now imagine that the momenta of the particles are measured by an observer in a second reference frame S′ moving with velocity $\vec{\mathbf{v}}$ relative to the first frame. Using the Lorentz velocity transformation equation and the classical definition of linear momentum, $\vec{\mathbf{p}} = m\vec{\mathbf{u}}$ (where $\vec{\mathbf{u}}$ is the velocity of a particle), we find that linear momentum is *not* measured to be conserved by the observer in S′. Because the laws of physics are the same in all inertial frames, however, linear momentum of the system must be conserved in all frames. We have a contradiction. In view of this contradiction and assuming the Lorentz velocity transformation equation is correct, we must modify the definition of linear momentum so that the momentum of an isolated system is conserved for all observers. For any particle, the correct relativistic equation for linear momentum that satisfies this condition is

Definition of relativistic linear momentum ▶

$$\vec{\mathbf{p}} \equiv \frac{m\vec{\mathbf{u}}}{\sqrt{1 - \dfrac{u^2}{c^2}}} = \gamma m\vec{\mathbf{u}} \qquad \textbf{(39.19)}$$

where m is the mass of the particle and $\vec{\mathbf{u}}$ is the velocity of the particle. When u is much less than c, $\gamma = (1 - u^2/c^2)^{-1/2}$ approaches unity and $\vec{\mathbf{p}}$ approaches $m\vec{\mathbf{u}}$. Therefore, the relativistic equation for $\vec{\mathbf{p}}$ reduces to the classical expression when u is much smaller than c, as it should.

The relativistic force $\vec{\mathbf{F}}$ acting on a particle whose linear momentum is $\vec{\mathbf{p}}$ is defined as

$$\vec{\mathbf{F}} \equiv \frac{d\vec{\mathbf{p}}}{dt} \tag{39.20}$$

where $\vec{\mathbf{p}}$ is given by Equation 39.19. This expression, which is the relativistic form of Newton's second law, is reasonable because it preserves classical mechanics in the limit of low velocities and is consistent with conservation of linear momentum for an isolated system ($\vec{\mathbf{F}}_{\text{ext}} = 0$) both relativistically and classically.

It is left as an end-of-chapter problem (Problem 61) to show that under relativistic conditions, the acceleration $\vec{\mathbf{a}}$ of a particle decreases under the action of a constant force, in which case $a \propto (1 - u^2/c^2)^{3/2}$. This proportionality shows that as the particle's speed approaches c, the acceleration caused by any finite force approaches zero. Hence, it is impossible to accelerate a particle from rest to a speed $u \geq c$. This argument reinforces that the speed of light is the ultimate speed, the speed limit of the Universe. It is the maximum possible speed for energy transfer and for information transfer. Any object with mass must move at a lower speed.

EXAMPLE 39.8 Linear Momentum of an Electron

An electron, which has a mass of 9.11×10^{-31} kg, moves with a speed of $0.750c$. Find the magnitude of its relativistic momentum and compare this value with the momentum calculated from the classical expression.

SOLUTION

Conceptualize Imagine an electron moving with high speed. The electron carries momentum, but the magnitude of its momentum is not given by $p = mu$ because the speed is relativistic.

Categorize We categorize this example as a substitution problem involving a relativistic equation.

Use Equation 39.19 with $u = 0.750c$ to find the momentum:

$$p = \frac{m_e u}{\sqrt{1 - \frac{u^2}{c^2}}}$$

$$p = \frac{(9.11 \times 10^{-31}\ \text{kg})(0.750)(3.00 \times 10^8\ \text{m/s})}{\sqrt{1 - \frac{(0.750c)^2}{c^2}}}$$

$$= 3.10 \times 10^{-22}\ \text{kg} \cdot \text{m/s}$$

The classical expression (used incorrectly here) gives $p_{\text{classical}} = m_e u = 2.05 \times 10^{-22}$ kg · m/s. Hence, the correct relativistic result is 50% greater than the classical result!

39.8 Relativistic Energy

We have seen that the definition of linear momentum requires generalization to make it compatible with Einstein's postulates. This conclusion implies that the definition of kinetic energy must most likely be modified also.

To derive the relativistic form of the work–kinetic energy theorem, imagine a particle moving in one dimension along the x axis. A force in the x direction causes the momentum of the particle to change according to Equation 39.20. In what follows, we assume the particle is accelerated from rest to some final speed u. The work done by the force F on the particle is

$$W = \int_{x_1}^{x_2} F\,dx = \int_{x_1}^{x_2} \frac{dp}{dt}\,dx \qquad (39.21)$$

To perform this integration and find the work done on the particle and the relativistic kinetic energy as a function of u, we first evaluate dp/dt:

$$\frac{dp}{dt} = \frac{d}{dt}\frac{mu}{\sqrt{1-\frac{u^2}{c^2}}} = \frac{m}{\left(1-\frac{u^2}{c^2}\right)^{3/2}}\frac{du}{dt}$$

Substituting this expression for dp/dt and $dx = u\,dt$ into Equation 39.21 gives

$$W = \int_0^t \frac{m}{\left(1-\frac{u^2}{c^2}\right)^{3/2}}\frac{du}{dt}(u\,dt) = m\int_0^u \frac{u}{\left(1-\frac{u^2}{c^2}\right)^{3/2}}\,du$$

where we use the limits 0 and u in the integral because the integration variable has been changed from t to u. Evaluating the integral gives

$$W = \frac{mc^2}{\sqrt{1-\frac{u^2}{c^2}}} - mc^2 \qquad (39.22)$$

Recall from Chapter 7 that the work done by a force acting on a system consisting of a single particle equals the change in kinetic energy of the particle. Because we assumed the initial speed of the particle is zero, its initial kinetic energy is zero. Therefore, the work W in Equation 39.22 is equivalent to the relativistic kinetic energy K:

Relativistic kinetic energy ▶

$$K = \frac{mc^2}{\sqrt{1-\frac{u^2}{c^2}}} - mc^2 = \gamma mc^2 - mc^2 = (\gamma - 1)mc^2 \qquad (39.23)$$

This equation is routinely confirmed by experiments using high-energy particle accelerators.

At low speeds, where $u/c \ll 1$, Equation 39.23 should reduce to the classical expression $K = \frac{1}{2}mu^2$. We can check that by using the binomial expansion $(1 - \beta^2)^{-1/2} \approx 1 + \frac{1}{2}\beta^2 + \cdots$ for $\beta \ll 1$, where the higher-order powers of β are neglected in the expansion. (In treatments of relativity, β is a common symbol used to represent u/c or v/c.) In our case, $\beta = u/c$, so

$$\gamma = \frac{1}{\sqrt{1-\frac{u^2}{c^2}}} = \left(1-\frac{u^2}{c^2}\right)^{-1/2} \approx 1 + \frac{1}{2}\frac{u^2}{c^2}$$

Substituting this result into Equation 39.23 gives

$$K \approx \left[\left(1 + \frac{1}{2}\frac{u^2}{c^2}\right) - 1\right]mc^2 = \frac{1}{2}mu^2 \quad (\text{for } u/c \ll 1)$$

which is the classical expression for kinetic energy. A graph comparing the relativistic and nonrelativistic expressions is given in Figure 39.16. In the relativistic case, the particle speed never exceeds c, regardless of the kinetic energy. The two curves are in good agreement when $u \ll c$.

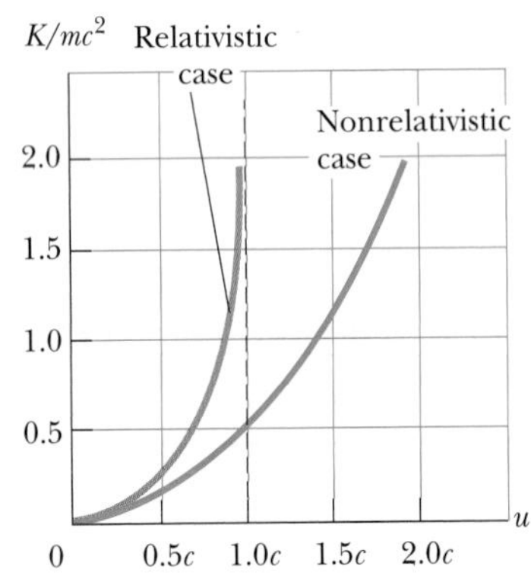

Figure 39.16 A graph comparing relativistic and nonrelativistic kinetic energy of a moving particle. The energies are plotted as a function of particle speed u. In the relativistic case, u is always less than c.

The constant term mc^2 in Equation 39.23, which is independent of the speed of the particle, is called the **rest energy** E_R of the particle:

$$E_R = mc^2 \tag{39.24}$$

◀ Rest energy

Equation 39.24 shows that **mass is a form of energy,** where c^2 is simply a constant conversion factor. This expression also shows that a small mass corresponds to an enormous amount of energy, a concept fundamental to nuclear and elementary-particle physics.

The term γmc^2 in Equation 39.23, which depends on the particle speed, is the sum of the kinetic and rest energies. It is called the **total energy** E:

$$\text{Total energy} = \text{kinetic energy} + \text{rest energy}$$

$$E = K + mc^2 \tag{39.25}$$

or

$$E = \frac{mc^2}{\sqrt{1 - \dfrac{u^2}{c^2}}} = \gamma mc^2 \tag{39.26}$$

◀ Total energy of a relativistic particle

In many situations, the linear momentum or energy of a particle rather than its speed is measured. It is therefore useful to have an expression relating the total energy E to the relativistic linear momentum p, which is accomplished by using the expressions $E = \gamma mc^2$ and $p = \gamma mu$. By squaring these equations and subtracting, we can eliminate u (Problem 37). The result, after some algebra, is[6]

$$E^2 = p^2c^2 + (mc^2)^2 \tag{39.27}$$

◀ Energy–momentum relationship for a relativistic particle

When the particle is at rest, $p = 0$, so $E = E_R = mc^2$.

In Section 35.1, we introduced the concept of a particle of light, called a **photon.** For particles that have zero mass, such as photons, we set $m = 0$ in Equation 39.27 and find that

$$E = pc \tag{39.28}$$

This equation is an exact expression relating total energy and linear momentum for photons, which always travel at the speed of light (in vacuum).

Finally, because the mass m of a particle is independent of its motion, m must have the same value in all reference frames. For this reason, m is often called the **invariant mass.** On the other hand, because the total energy and linear momentum of a particle both depend on velocity, these quantities depend on the reference frame in which they are measured.

When dealing with subatomic particles, it is convenient to express their energy in electron volts (Section 25.1) because the particles are usually given this energy by acceleration through a potential difference. The conversion factor, as you recall from Equation 25.5, is

$$1 \text{ eV} = 1.60 \times 10^{-19} \text{ J}$$

For example, the mass of an electron is 9.11×10^{-31} kg. Hence, the rest energy of the electron is

$$m_ec^2 = (9.109 \times 10^{-31} \text{ kg})(2.998 \times 10^8 \text{ m/s})^2 = 8.187 \times 10^{-14} \text{ J}$$

$$= (8.187 \times 10^{-14} \text{ J})(1 \text{ eV}/1.602 \times 10^{-19} \text{ J}) = 0.511 \text{ MeV}$$

Quick Quiz 39.9 The following *pairs* of energies—particle 1: E, $2E$; particle 2: E, $3E$; particle 3: $2E$, $4E$—represent the rest energy and total energy of three different particles. Rank the particles from greatest to least according to their (a) mass, (b) kinetic energy, and (c) speed.

---

[6] One way to remember this relationship is to draw a right triangle having a hypotenuse of length E and legs of lengths pc and mc^2.

EXAMPLE 39.9 The Energy of a Speedy Proton

(A) Find the rest energy of a proton in units of electron volts.

SOLUTION

Conceptualize Even if the proton is not moving, it has energy associated with its mass. If it moves, the proton possesses more energy, with the total energy being the sum of its rest energy and its kinetic energy.

Categorize The phrase "rest energy" suggests we must take a relativistic rather than a classical approach to this problem.

Analyze Use Equation 39.24 to find the rest energy:

$$E_R = m_pc^2 = (1.673 \times 10^{-27}\text{ kg})(2.998 \times 10^8\text{ m/s})^2$$

$$= (1.504 \times 10^{-10}\text{ J})\left(\frac{1.00\text{ eV}}{1.602 \times 10^{-19}\text{ J}}\right) = \boxed{938\text{ MeV}}$$

(B) If the total energy of a proton is three times its rest energy, what is the speed of the proton?

SOLUTION

Use Equation 39.26 to relate the total energy of the proton to the rest energy:

$$E = 3m_pc^2 = \frac{m_pc^2}{\sqrt{1 - \frac{u^2}{c^2}}} \rightarrow 3 = \frac{1}{\sqrt{1 - \frac{u^2}{c^2}}}$$

Solve for u:

$$\left(1 - \frac{u^2}{c^2}\right) = \frac{1}{9} \rightarrow \frac{u^2}{c^2} = \frac{8}{9}$$

$$u = \frac{\sqrt{8}}{3}c = 0.943c = \boxed{2.83 \times 10^8\text{ m/s}}$$

(C) Determine the kinetic energy of the proton in units of electron volts.

SOLUTION

Use Equation 39.25 to find the kinetic energy of the proton:

$$K = E - m_pc^2 = 3m_pc^2 - m_pc^2 = 2m_pc^2$$

$$= 2(938\text{ MeV}) = \boxed{1.88 \times 10^3\text{ MeV}}$$

(D) What is the proton's momentum?

SOLUTION

Use Equation 39.27 to calculate the momentum:

$$E^2 = p^2c^2 + (m_pc^2)^2 = (3m_pc^2)^2$$

$$p^2c^2 = 9(m_pc^2)^2 - (m_pc^2)^2 = 8(m_pc^2)^2$$

$$p = \sqrt{8}\frac{m_pc^2}{c} = \sqrt{8}\frac{(938\text{ MeV})}{c} = \boxed{2.65 \times 10^3\text{ MeV}/c}$$

Finalize The unit of momentum in part (D) is written MeV/c, which is a common unit in particle physics. For comparison, you might want to solve this example using classical equations.

What If? In classical physics, if the momentum of a particle doubles, the kinetic energy increases by a factor of 4. What happens to the kinetic energy of the proton in this example if its momentum doubles?

Answer Based on what we have seen so far in relativity, it is likely you would predict that its kinetic energy does not increase by a factor of 4.

Find the new doubled momentum:

$$p_{\text{new}} = 2\left(\sqrt{8}\,\frac{m_p c^2}{c}\right) = 4\sqrt{2}\,\frac{m_p c^2}{c}$$

Use this result in Equation 39.27 to find the new total energy:

$$E_{\text{new}}^2 = p_{\text{new}}^2 c^2 + (m_p c^2)^2$$

$$E_{\text{new}}^2 = \left(4\sqrt{2}\,\frac{m_p c^2}{c}\right)^2 c^2 + (m_p c^2)^2 = 33(m_p c^2)^2$$

$$E_{\text{new}} = \sqrt{33}(m_p c^2) = 5.7 m_p c^2$$

Use Equation 39.25 to find the new kinetic energy:

$$K_{\text{new}} = E_{\text{new}} - m_p c^2 = 5.7 m_p c^2 - m_p c^2 = 4.7 m_p c^2$$

This value is a little more than twice the kinetic energy found in part (C), not four times. In general, the factor by which the kinetic energy increases if the momentum doubles depends on the initial momentum, but it approaches 4 as the momentum approaches zero. In this latter situation, classical physics correctly describes the situation.

39.9 Mass and Energy

Equation 39.26, $E = \gamma mc^2$, represents the total energy of a particle. This important equation suggests that even when a particle is at rest ($\gamma = 1$), it still possesses enormous energy through its mass. The clearest experimental proof of the equivalence of mass and energy occurs in nuclear and elementary-particle interactions in which the conversion of mass into kinetic energy takes place. Consequently, we cannot use the principle of conservation of energy in relativistic situations as it was outlined in Chapter 8. We must modify the principle by including rest energy as another form of energy storage.

This concept is important in atomic and nuclear processes, in which the change in mass is a relatively large fraction of the initial mass. In a conventional nuclear reactor, for example, the uranium nucleus undergoes *fission*, a reaction that results in several lighter fragments having considerable kinetic energy. In the case of ^{235}U, which is used as fuel in nuclear power plants, the fragments are two lighter nuclei and a few neutrons. The total mass of the fragments is less than that of the ^{235}U by an amount Δm. The corresponding energy Δmc^2 associated with this mass difference is exactly equal to the total kinetic energy of the fragments. The kinetic energy is absorbed as the fragments move through water, raising the internal energy of the water. This internal energy is used to produce steam for the generation of electricity.

Next, consider a basic *fusion* reaction in which two deuterium atoms combine to form one helium atom. The decrease in mass that results from the creation of one helium atom from two deuterium atoms is $\Delta m = 4.25 \times 10^{-29}$ kg. Hence, the corresponding energy that results from one fusion reaction is $\Delta mc^2 = 3.83 \times 10^{-12}$ J = 23.9 MeV. To appreciate the magnitude of this result, consider that if only 1 g of deuterium were converted to helium, the energy released would be on the order of 10^{12} J! In 2007's cost of electrical energy, this energy would be worth approximately \$30 000. We shall present more details of these nuclear processes in Chapter 45 of the extended version of this textbook.

EXAMPLE 39.10 Mass Change in a Radioactive Decay

The ^{216}Po nucleus is unstable and exhibits radioactivity (Chapter 44). It decays to ^{212}Pb by emitting an alpha particle, which is a helium nucleus, ^{4}He. The relevant masses are $m_i = m(^{216}\text{Po}) = 216.001\,905$ u, and $m_f = m(^{212}\text{Pb}) + m(^{4}\text{He}) = 211.991\,888\text{ u} + 4.002\,603$ u.

(A) Find the mass change of the system in this decay.

SOLUTION

Conceptualize The initial system is the ^{216}Po nucleus. Imagine the mass of the system decreasing during the decay and transforming to kinetic energy of the alpha particle and the ^{212}Pb nucleus after the decay.

Categorize We use concepts discussed in this section, so we categorize this example as a substitution problem.

Calculate the mass change:

$$\Delta m = 216.001\,905 \text{ u} - (211.991\,888 \text{ u} + 4.002\,603 \text{ u})$$
$$= 0.007\,414 \text{ u} = 1.23 \times 10^{-29} \text{ kg}$$

(B) Find the energy this mass change represents.

SOLUTION

Use Equation 39.24 to find the energy associated with this mass change:

$$E = \Delta mc^2 = (1.23 \times 10^{-29} \text{ kg})(3.00 \times 10^8 \text{ m/s})^2$$
$$= 1.11 \times 10^{-12} \text{ J} = 6.92 \text{ MeV}$$

39.10 The General Theory of Relativity

Up to this point, we have sidestepped a curious puzzle. Mass has two seemingly different properties: a *gravitational attraction* for other masses and an *inertial* property that represents a resistance to acceleration. To designate these two attributes, we use the subscripts g and i and write

Gravitational property: $F_g = m_g g$

Inertial property: $\sum F = m_i a$

The value for the gravitational constant G was chosen to make the magnitudes of m_g and m_i numerically equal. Regardless of how G is chosen, however, the strict proportionality of m_g and m_i has been established experimentally to an extremely high degree: a few parts in 10^{12}. Therefore, it appears that gravitational mass and inertial mass may indeed be exactly proportional.

Why, though? They seem to involve two entirely different concepts: a force of mutual gravitational attraction between two masses and the resistance of a single mass to being accelerated. This question, which puzzled Newton and many other physicists over the years, was answered by Einstein in 1916 when he published his theory of gravitation, known as the *general theory of relativity.* Because it is a mathematically complex theory, we offer merely a hint of its elegance and insight.

In Einstein's view, the dual behavior of mass was evidence for a very intimate and basic connection between the two behaviors. He pointed out that no mechanical experiment (such as dropping an object) could distinguish between the two situations illustrated in Figures 39.17a and 39.17b. In Figure 39.17a, a person standing in an elevator on the surface of a planet feels pressed into the floor due to the gravitational force. If he releases his briefcase, he observes it moving toward the floor with acceleration $\vec{\mathbf{g}} = -g\hat{\mathbf{j}}$. In Figure 39.17b, the person is in an elevator in empty space accelerating upward with $\vec{\mathbf{a}}_{el} = +g\hat{\mathbf{j}}$. The person feels pressed into the floor with the same force as in Figure 39.17a. If he releases his briefcase, he observes it moving toward the floor with acceleration g, exactly as in the previous situation. In each situation, an object released by the observer undergoes a downward acceleration of magnitude g relative to the floor. In Figure 39.17a, the person is at rest in an inertial frame in a gravitational field due to the planet. In Figure 39.17b, the person is in a noninertial frame accelerating in gravity-free space. Einstein's claim is that these two situations are completely equivalent.

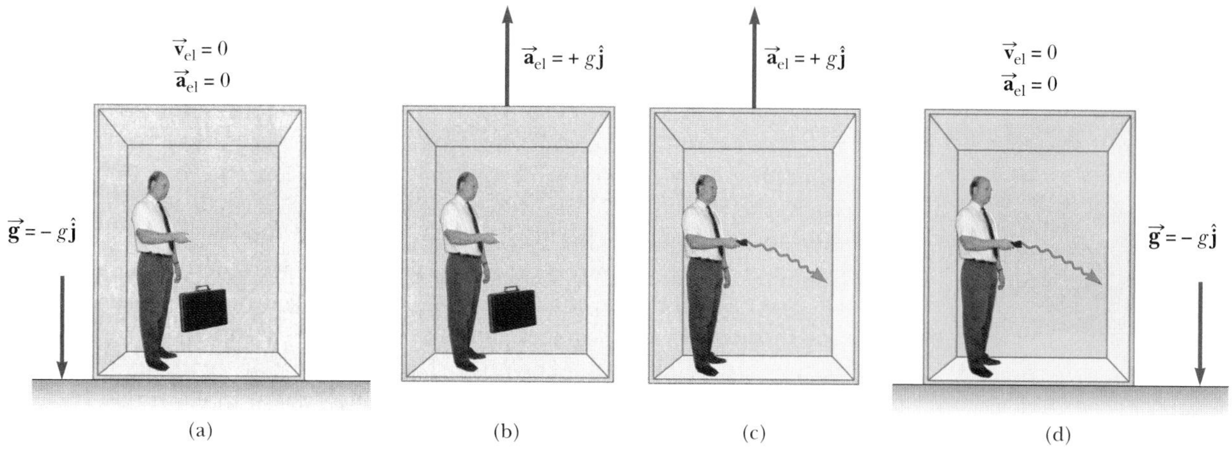

Figure 39.17 (a) The observer is at rest in an elevator in a uniform gravitational field $\vec{\mathbf{g}} = -g\hat{\mathbf{j}}$, directed downward. The observer drops his briefcase, which moves downward with acceleration g. (b) The observer is in a region where gravity is negligible, but the elevator moves upward with an acceleration $\vec{\mathbf{a}}_{el} = +g\hat{\mathbf{j}}$. The observer releases his briefcase, which moves downward (according to the observer) with acceleration g relative to the floor of the elevator. According to Einstein, the frames of reference in (a) and (b) are equivalent in every way. No local experiment can distinguish any difference between the two frames. (c) In the accelerating frame, a ray of light would appear to bend downward due to the acceleration. (d) If (a) and (b) are truly equivalent, as Einstein proposed, (c) suggests that a ray of light would bend downward in a gravitational field.

Einstein carried this idea further and proposed that *no* experiment, mechanical or otherwise, could distinguish between the two situations. This extension to include all phenomena (not just mechanical ones) has interesting consequences. For example, suppose a light pulse is sent horizontally across the elevator as in Figure 39.17c, in which the elevator is accelerating upward in empty space. From the point of view of an observer in an inertial frame outside the elevator, the light travels in a straight line while the floor of the elevator accelerates upward. According to the observer on the elevator, however, the trajectory of the light pulse bends downward as the floor of the elevator (and the observer) accelerates upward. Therefore, based on the equality of parts (a) and (b) of the figure, Einstein proposed that **a beam of light should also be bent downward by a gravitational field** as in Figure 39.17d. Experiments have verified the effect, although the bending is small. A laser aimed at the horizon falls less than 1 cm after traveling 6 000 km. (No such bending is predicted in Newton's theory of gravitation.)

Einstein's **general theory of relativity** has two postulates:

◀ Postulates of the general theory of relativity

- All the laws of nature have the same form for observers in any frame of reference, whether accelerated or not.
- In the vicinity of any point, a gravitational field is equivalent to an accelerated frame of reference in gravity-free space (the **principle of equivalence**).

One interesting effect predicted by the general theory is that time is altered by gravity. A clock in the presence of gravity runs slower than one located where gravity is negligible. Consequently, the frequencies of radiation emitted by atoms in the presence of a strong gravitational field are *redshifted* to lower frequencies when compared with the same emissions in the presence of a weak field. This gravitational redshift has been detected in spectral lines emitted by atoms in massive stars. It has also been verified on the Earth by comparing the frequencies of gamma rays emitted from nuclei separated vertically by about 20 m.

The second postulate suggests a gravitational field may be "transformed away" at any point if we choose an appropriate accelerated frame of reference, a freely falling one. Einstein developed an ingenious method of describing the acceleration necessary to make the gravitational field "disappear." He specified a concept, the *curvature of space–time*, that describes the gravitational effect at every point. In fact, the curvature of space–time completely replaces Newton's gravitational theory. According to Einstein, there is no such thing as a gravitational force. Rather, the presence of a mass causes a curvature of space–time in the vicinity of the mass,

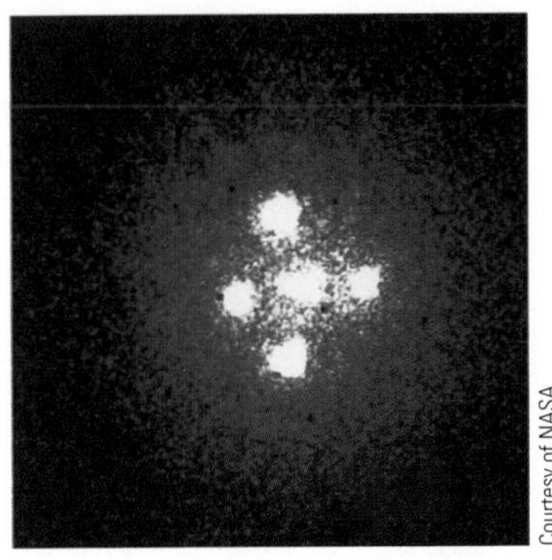

Einstein's cross. The four bright spots are images of the same galaxy that have been bent around a massive object located between the galaxy and the Earth. The massive object acts like a lens, causing the rays of light that were diverging from the distant galaxy to converge on the Earth. (If the intervening massive object had a uniform mass distribution, we would see a bright ring instead of four spots.)

and this curvature dictates the space–time path that all freely moving objects must follow.

As an example of the effects of curved space–time, imagine two travelers moving on parallel paths a few meters apart on the surface of the Earth and maintaining an exact northward heading along two longitude lines. As they observe each other near the equator, they will claim that their paths are exactly parallel. As they approach the North Pole, however, they notice that they are moving closer together and will meet at the North Pole. Therefore, they claim that they moved along parallel paths, but moved toward each other, *as if there were an attractive force between them.* The travelers make this conclusion based on their everyday experience of moving on flat surfaces. From our mental representation, however, we realize they are walking on a curved surface, and it is the geometry of the curved surface, rather than an attractive force, that causes them to converge. In a similar way, general relativity replaces the notion of forces with the movement of objects through curved space–time.

One prediction of the general theory of relativity is that a light ray passing near the Sun should be deflected in the curved space–time created by the Sun's mass. This prediction was confirmed when astronomers detected the bending of starlight near the Sun during a total solar eclipse that occurred shortly after World War I (Fig. 39.18). When this discovery was announced, Einstein became an international celebrity.

If the concentration of mass becomes very great as is believed to occur when a large star exhausts its nuclear fuel and collapses to a very small volume, a **black hole** may form. Here, the curvature of space–time is so extreme that within a certain distance from the center of the black hole all matter and light become trapped as discussed in Section 13.6.

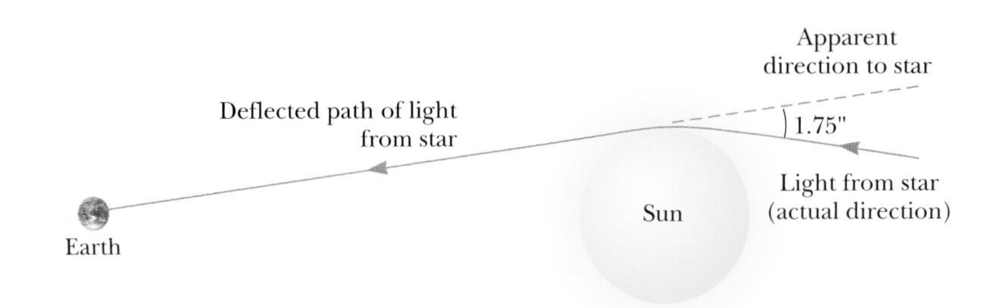

Figure 39.18 Deflection of starlight passing near the Sun. Because of this effect, the Sun or some other remote object can act as a *gravitational lens.* In his general theory of relativity, Einstein calculated that starlight just grazing the Sun's surface should be deflected by an angle of 1.75 s of arc.

Summary

ThomsonNOW™ Sign in at **www.thomsonedu.com** and go to ThomsonNOW to take a practice test for this chapter.

DEFINITIONS

The relativistic expression for the **linear momentum** of a particle moving with a velocity $\vec{\mathbf{u}}$ is

$$\vec{\mathbf{p}} \equiv \frac{m\vec{\mathbf{u}}}{\sqrt{1-\dfrac{u^2}{c^2}}} = \gamma m\vec{\mathbf{u}} \qquad (39.19)$$

The relativistic force $\vec{\mathbf{F}}$ acting on a particle whose linear momentum is $\vec{\mathbf{p}}$ is defined as

$$\vec{\mathbf{F}} \equiv \frac{d\vec{\mathbf{p}}}{dt} \qquad (39.20)$$

(continued)

CONCEPTS AND PRINCIPLES

The two basic postulates of the special theory of relativity are as follows:

- The laws of physics must be the same in all inertial reference frames.
- The speed of light in vacuum has the same value, $c = 3.00 \times 10^8$ m/s, in all inertial frames, regardless of the velocity of the observer or the velocity of the source emitting the light.

Three consequences of the special theory of relativity are as follows:

- Events that are measured to be simultaneous for one observer are not necessarily measured to be simultaneous for another observer who is in motion relative to the first.
- Clocks in motion relative to an observer are measured to run slower by a factor $\gamma = (1 - v^2/c^2)^{-1/2}$. This phenomenon is known as **time dilation.**
- The length of objects in motion are measured to be contracted in the direction of motion by a factor $1/\gamma = (1 - v^2/c^2)^{1/2}$. This phenomenon is known as **length contraction.**

To satisfy the postulates of special relativity, the Galilean transformation equations must be replaced by the **Lorentz transformation equations:**

$$x' = \gamma(x - vt) \qquad y' = y \qquad z' = z \qquad t' = \gamma\left(t - \frac{v}{c^2}x\right) \qquad \textbf{(39.11)}$$

where $\gamma = (1 - v^2/c^2)^{-1/2}$ and the S′ frame moves in the x direction relative to the S frame.

The relativistic form of the **Lorentz velocity transformation equation** is

$$u'_x = \frac{u_x - v}{1 - \dfrac{u_x v}{c^2}} \qquad \textbf{(39.16)}$$

where u'_x is the x component of the velocity of an object as measured in the S′ frame and u_x is its component as measured in the S frame.

The relativistic expression for the **kinetic energy** of a particle is

$$K = \frac{mc^2}{\sqrt{1 - \dfrac{u^2}{c^2}}} - mc^2 = (\gamma - 1)mc^2 \qquad \textbf{(39.23)}$$

The constant term mc^2 in Equation 39.23 is called the **rest energy** E_R of the particle:

$$E_R = mc^2 \qquad \textbf{(39.24)}$$

The total energy E of a particle is given by

$$E = \frac{mc^2}{\sqrt{1 - \dfrac{u^2}{c^2}}} = \gamma mc^2 \qquad \textbf{(39.26)}$$

The relativistic linear momentum of a particle is related to its total energy through the equation

$$E^2 = p^2c^2 + (mc^2)^2 \qquad \textbf{(39.27)}$$

Questions

☐ denotes answer available in *Student Solutions Manual/Study Guide;* **O** denotes objective question

1. The speed of light in water is 230 Mm/s. Suppose an electron is moving through water at 250 Mm/s. Does that motion violate the principle of relativity?

2. **O** You measure the volume of a cube at rest to be V_0. You then measure the volume of the same cube as it passes you in a direction parallel to one side of the cube. The speed of the cube is $0.98c$, so $\gamma \approx 5$. Is the volume you measure close to (a) $V_0/125$, (b) $V_0/25$, (c) $V_0/5$, (d) V_0, (e) $5V_0$, (f) $25V_0$, or (g) $125V_0$?

3. **O** A spacecraft built in the shape of a sphere moves past an observer on the Earth with a speed of $0.5c$. What shape does the observer measure for the spacecraft as it goes by? (a) a sphere (b) a cigar shape, elongated along the direction of motion (c) a round pillow shape, flattened along the direction of motion (d) a conical shape, pointing in the direction of motion

4. **O** A spacecraft zooms past the Earth with a constant velocity. An observer on the Earth measures that an

undamaged clock on the spacecraft is ticking at one-third the rate of an identical clock on the Earth. What does an observer on the spacecraft measure about the Earth clock's ticking rate? (a) It runs more than three times faster than his own clock. (b) It runs three times faster than his own. (c) It runs at the same rate as his own. (d) It runs at approximately half the rate of his own. (e) It runs at one-third the rate of his own. (f) It runs at less than one-third the rate of his own.

5. Explain why, when defining the length of a rod, it is necessary to specify that the positions of the ends of the rod are to be measured simultaneously.

6. **O** Two identical clocks are set side by side and synchronized. One remains on the Earth. The other is put into orbit around the Earth moving toward the east. **(i)** As measured by an observer on the Earth, while in rapid motion does the orbiting clock (a) run faster than the Earth-based clock, (b) run at the same rate, or (c) run slower? **(ii)** The orbiting clock is returned to its original location and brought to rest relative to the Earth. Thereafter, (a) its reading lags farther and farther behind the Earth-based clock, (b) it lags behind the Earth-based clock by a constant amount, (c) it is synchronized with the Earth-based clock, (d) it is ahead of the Earth-based clock by a constant amount, or (e) it gets farther and farther ahead of the Earth-based clock.

7. A train is approaching you at very high speed as you stand next to the tracks. Just as an observer on the train passes you, you both begin to play the same Beethoven symphony on portable CD players. (a) According to you, whose CD player finishes the symphony first? (b) **What If?** According to the observer on the train, whose CD player finishes the symphony first? (c) Whose CD player actually finishes the symphony first?

8. List some ways our day-to-day lives would change if the speed of light were only 50 m/s.

9. How is acceleration indicated on a space–time graph?

10. Explain how the Doppler effect with microwaves is used to determine the speed of an automobile.

11. In several cases, a nearby star has been found to have a large planet orbiting about it, although light from the planet could not be seen separately from the starlight. Using the ideas of a system rotating about its center of mass and of the Doppler shift for light, explain how an astronomer could determine the presence of the invisible planet.

12. A particle is moving at a speed less than $c/2$. If the speed of the particle is doubled, what happens to its momentum?

13. **O** Rank the following particles according to the magnitudes of their momentum from the largest to the smallest. If any have equal amounts of momentum, or zero momentum, display that fact in your ranking. (a) a 1-MeV photon (b) a proton with kinetic energy $K = 1$ MeV (c) an electron with $K = 1$ MeV (d) a grain of dust with $K = 1$ MeV $= 160$ fJ

14. Give a physical argument that shows it is impossible to accelerate an object of mass m to the speed of light, even with a continuous force acting on it.

15. **O** **(i)** Does the speed of an electron have an upper limit? (a) yes, the speed of light c (b) yes, with another value (c) no **(ii)** Does the magnitude of an electron's momentum have an upper limit? (a) yes, m_ec (b) yes, with another value (c) no **(iii)** Does the electron's kinetic energy have an upper limit? (a) yes, m_ec^2 (b) yes, $\frac{1}{2}m_ec^2$ (c) yes, with another value (d) no

16. **O** A distant astronomical object (a quasar) is moving away from us at half the speed of light. What is the speed of the light we receive from this quasar? (a) greater than c (b) c (c) between $c/2$ and c (d) $c/2$ (e) between 0 and $c/2$ (f) 0

17. "Newtonian mechanics correctly describes objects moving at ordinary speeds and relativistic mechanics correctly describes objects moving very fast." "Relativistic mechanics must make a smooth transition as it reduces to Newtonian mechanics in a case in which the speed of an object becomes small compared with the speed of light." Argue for or against each of these two statements.

18. Two cards have straight edges. Suppose the top edge of one card crosses the bottom edge of another card at a small angle as shown in Figure Q39.18a. A person slides the cards together at a moderately high speed. In what direction does the intersection point of the edges move? Show that the intersection point can move at a speed greater than the speed of light.

A small flashlight is suspended in a horizontal plane and set into rapid rotation. Show that the spot of light it produces on a distant screen can move across the screen at a speed greater than the speed of light. (If you use a laser pointer as shown in Figure Q39.18b, make sure the direct laser light cannot enter a person's eyes.) Argue that these experiments do not invalidate the principle that no material, no energy, and no information can move faster than light moves in a vacuum.

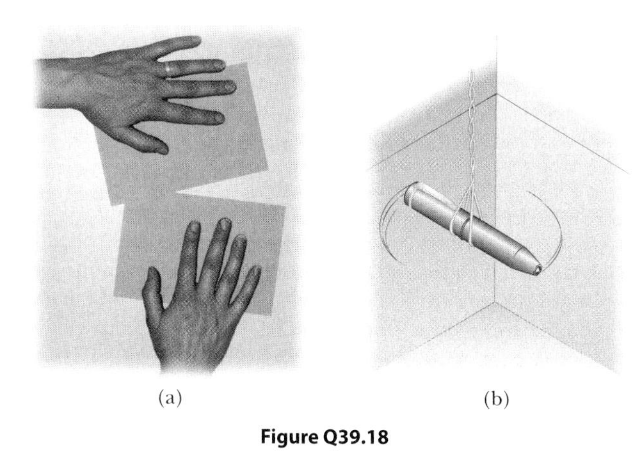

Figure Q39.18

19. With regard to reference frames, how does general relativity differ from special relativity?

20. Two identical clocks are in the same house, one upstairs in a bedroom and the other downstairs in the kitchen. Which clock runs slower? Explain.

Problems

WebAssign The Problems from this chapter may be assigned online in WebAssign.

ThomsonNOW Sign in at **www.thomsonedu.com** and go to ThomsonNOW to assess your understanding of this chapter's topics with additional quizzing and conceptual questions.

1, 2, 3 denotes straightforward, intermediate, challenging; ☐ denotes full solution available in *Student Solutions Manual/Study Guide;* ▲ denotes coached solution with hints available at **www.thomsonedu.com;** denotes developing symbolic reasoning; ● denotes asking for qualitative reasoning; denotes computer useful in solving problem

Section 39.1 The Principle of Galilean Relativity

1. In a laboratory frame of reference, an observer notes that Newton's second law is valid. Show that it is also valid for an observer moving at a constant speed, small compared with the speed of light, relative to the laboratory frame.

2. Show that Newton's second law is *not* valid in a reference frame moving past the laboratory frame of Problem 1 with a constant acceleration.

3. A 2 000-kg car moving at 20.0 m/s collides and locks together with a 1 500-kg car at rest at a stop sign. Show that momentum is conserved in a reference frame moving at 10.0 m/s in the direction of the moving car.

Section 39.2 The Michelson–Morley Experiment

Section 39.3 Einstein's Principle of Relativity

Section 39.4 Consequences of the Special Theory of Relativity

Problem 37 in Chapter 4 can be assigned with this section.

4. How fast must a meterstick be moving if its length is measured to shrink to 0.500 m?

5. At what speed does a clock move if it is measured to run at a rate one-half the rate of a clock at rest with respect to an observer?

6. An astronaut is traveling in a space vehicle moving at $0.500c$ relative to the Earth. The astronaut measures her pulse rate at 75.0 beats per minute. Signals generated by the astronaut's pulse are radioed to the Earth when the vehicle is moving in a direction perpendicular to the line that connects the vehicle with an observer on the Earth. (a) What pulse rate does the Earth-based observer measure? (b) **What If?** What would be the pulse rate if the speed of the space vehicle were increased to $0.990c$?

7. An atomic clock moves at 1 000 km/h for 1.00 h as measured by an identical clock on the Earth. At the end of the 1.00-h interval, how many nanoseconds slow will the moving clock be compared with the Earth clock?

8. A muon formed high in the Earth's atmosphere travels at speed $v = 0.990c$ for a distance of 4.60 km before it decays into an electron, a neutrino, and an antineutrino ($\mu^- \rightarrow e^- + \nu + \bar{\nu}$). (a) For what time interval does the muon live as measured in its reference frame? (b) How far does the Earth travel as measured in the frame of the muon?

9. ▲ A spacecraft with a proper length of 300 m takes 0.750 μs to pass an Earth-based observer. Determine the speed of the spacecraft as measured by the Earth observer.

10. (a) An object of proper length L_p takes a time interval Δt to pass an Earth-based observer. Determine the speed of the object as measured by the Earth observer. (b) A column of tanks, 300 m long, takes 75.0 s to pass a child waiting at a street corner on her way to school. Determine the speed of the armored vehicles. (c) Show that the answer to part (a) includes the answer to Problem 9 as a special case and includes the answer to part (b) as another special case.

11. ● **Review problem.** In 1963, astronaut Gordon Cooper orbited the Earth 22 times. The press stated that for each orbit, he aged 2 millionths of a second less than he would have had he remained on the Earth. (a) Assuming that he was 160 km above the Earth in a circular orbit, determine the difference in elapsed time between someone on the Earth and the orbiting astronaut for the 22 orbits. You may use the approximation

$$\frac{1}{\sqrt{1-x}} \approx 1 + \frac{x}{2}$$

for small x. (b) Did the press report accurate information? Explain.

12. For what value of v does $\gamma = 1.010\,0$? Observe that for speeds lower than this value, time dilation and length contraction are effects amounting to less than 1%.

13. A friend passes by you in a spacecraft traveling at a high speed. He tells you that his spacecraft is 20.0 m long and that the identically constructed spacecraft you are sitting in is 19.0 m long. According to your observations, (a) how long is your spacecraft, (b) how long is your friend's spacecraft, and (c) what is the speed of your friend's spacecraft?

14. The identical twins Speedo and Goslo join a migration from the Earth to Planet X. It is 20.0 ly away in a reference frame in which both planets are at rest. The twins, of the same age, depart at the same moment on different spacecraft. Speedo's spacecraft travels steadily at $0.950c$ and Goslo's at $0.750c$. Calculate the age difference between the twins after Goslo's spacecraft lands on Planet X. Which twin is older?

15. **Review problem.** An alien civilization occupies a brown dwarf, nearly stationary relative to the Sun, several lightyears away. The extraterrestrials have come to love original broadcasts of *I Love Lucy*, on television channel 2, at carrier frequency 57.0 MHz. Their line of sight to us is in the plane of the Earth's orbit. Find the difference between the highest and lowest frequencies they receive due to the Earth's orbital motion around the Sun.

16. Police radar detects the speed of a car (Fig. P39.16) as follows. Microwaves of a precisely known frequency are broadcast toward the car. The moving car reflects the microwaves with a Doppler shift. The reflected waves are received and combined with an attenuated version of the transmitted wave. Beats occur between the two microwave signals. The beat frequency is measured. (a) For an electromagnetic wave reflected back to its source from a mirror approaching at speed v, show that the reflected wave has frequency

$$f = f_{\text{source}}\frac{c+v}{c-v}$$

where f_{source} is the source frequency. (b) When v is much less than c, the beat frequency is much smaller than the transmitted frequency. In this case, use the approximation $f + f_{\text{source}} \approx 2f_{\text{source}}$ and show that the beat frequency can be written as $f_{\text{beat}} = 2v/\lambda$. (c) What beat frequency is measured for a car speed of 30.0 m/s if the microwaves have frequency 10.0 GHz? (d) If the beat frequency measurement is accurate to ±5 Hz, how accurate is the speed measurement?

Figure P39.16

17. *The redshift.* A light source recedes from an observer with a speed v_{source} that is small compared with c. (a) Show that the fractional shift in the measured wavelength is given by the approximate expression

$$\frac{\Delta\lambda}{\lambda} \approx \frac{v_{\text{source}}}{c}$$

This phenomenon is known as the redshift because the visible light is shifted toward the red. (b) Spectroscopic measurements of light at $\lambda = 397$ nm coming from a galaxy in Ursa Major reveal a redshift of 20.0 nm. What is the recessional speed of the galaxy?

18. A physicist drives through a stop light. When he is pulled over, he tells the police officer that the Doppler shift made the red light of wavelength 650 nm appear green to him, with a wavelength of 520 nm. The police officer writes out a traffic citation for speeding. How fast was the physicist traveling, according to his own testimony?

Section 39.5 The Lorentz Transformation Equations

19. Suzanne observes two light pulses to be emitted from the same location, but separated in time by 3.00 μs. Mark observes the emission of the same two pulses to be separated in time by 9.00 μs. (a) How fast is Mark moving relative to Suzanne? (b) According to Mark, what is the separation in space of the two pulses?

20. A moving rod is observed to have a length of 2.00 m and to be oriented at an angle of 30.0° with respect to the direction of motion as shown in Figure P39.20. The rod has a speed of $0.995c$. (a) What is the proper length of the rod? (b) What is the orientation angle in the proper frame?

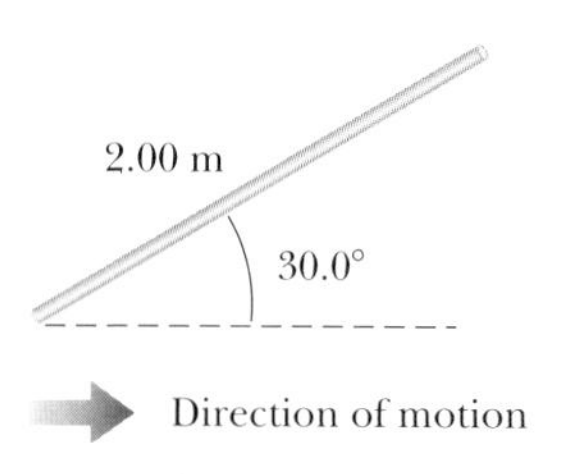

Figure P39.20

21. An observer in reference frame S measures two events to be simultaneous. Event A occurs at the point (50.0 m, 0, 0) at the instant 9:00:00 Universal time on January 15, 2008. Event B occurs at the point (150 m, 0, 0) at the same moment. A second observer, moving past with a velocity of $0.800c\hat{\mathbf{i}}$, also observes the two events. In her reference frame S′, which event occurred first and what time interval elapsed between the events?

22. A red light flashes at position $x_R = 3.00$ m and time $t_R = 1.00 \times 10^{-9}$ s, and a blue light flashes at $x_B = 5.00$ m and $t_B = 9.00 \times 10^{-9}$ s, all measured in the S reference frame. Reference frame S′ moves uniformly to the right and has its origin at the same point as S at $t = t' = 0$. Both flashes are observed to occur at the same place in S′. (a) Find the relative speed between S and S′. (b) Find the location of the two flashes in frame S′. (c) At what time does the red flash occur in the S′ frame?

Section 39.6 The Lorentz Velocity Transformation Equations

23. ▲ Two jets of material from the center of a radio galaxy are ejected in opposite directions. Both jets move at $0.750c$ relative to the galaxy. Determine the speed of one jet relative to the other.

24. A Klingon spacecraft moves away from the Earth at a speed of $0.800c$ (Fig. P39.24). The starship *Enterprise* pursues at a speed of $0.900c$ relative to the Earth. Observers on the Earth measure the *Enterprise* to be overtaking the Klingon craft at a relative speed of $0.100c$. With what speed is the *Enterprise* overtaking the Klingon craft as measured by the crew of the *Enterprise*?

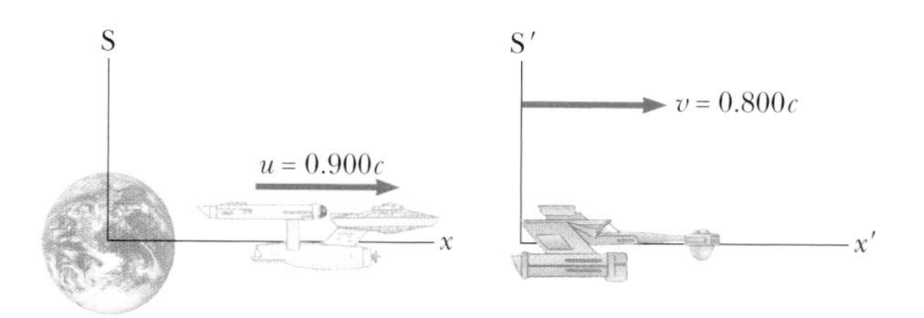

Figure P39.24

Section 39.7 Relativistic Linear Momentum

25. Calculate the momentum of an electron moving with a speed of (a) $0.010\,0c$, (b) $0.500c$, and (c) $0.900c$.

26. The nonrelativistic expression for the momentum of a particle, $p = mu$, agrees with experiment if $u \ll c$. For what speed does the use of this equation give an error in the momentum of (a) 1.00% and (b) 10.0%?

27. A golf ball travels with a speed of 90.0 m/s. By what fraction does its relativistic momentum magnitude p differ from its classical value mu? That is, calculate the ratio $(p - mu)/mu$.

28. The speed limit on a certain roadway is 90.0 km/h. Suppose speeding fines are made proportional to the amount by which a vehicle's momentum exceeds the momentum it would have when traveling at the speed limit. The fine for driving at 190 km/h (that is, 100 km/h over the speed limit) is $80.0. What then will be the fine for traveling (a) at 1 090 km/h? (b) At 1 000 000 090 km/h?

29. ▲ An unstable particle at rest spontaneously breaks into two fragments of unequal mass. The mass of the first fragment is 2.50×10^{-28} kg, and that of the other is 1.67×10^{-27} kg. If the lighter fragment has a speed of $0.893c$ after the breakup, what is the speed of the heavier fragment?

Section 39.8 Relativistic Energy

30. An electron has a kinetic energy five times greater than its rest energy. Find its (a) total energy and (b) speed.

31. A proton in a high-energy accelerator moves with a speed of $c/2$. Use the work–kinetic energy theorem to find the work required to increase its speed to (a) $0.750c$ and (b) $0.995c$.

32. Show that for any object moving at less than one-tenth the speed of light, the relativistic kinetic energy agrees with the result of the classical equation $K = \frac{1}{2}mu^2$ to within less than 1%. Therefore, for most purposes, the classical equation is good enough to describe these objects.

33. Find the momentum of a proton in MeV/c units assuming its total energy is twice its rest energy.

34. ● (a) Find the kinetic energy of a 78.0-kg spacecraft launched out of the solar system with speed 106 km/s by using the classical equation $K = \frac{1}{2}mu^2$. (b) **What If?** Calculate its kinetic energy using the relativistic equation. (c) Explain the result of comparing the results of parts (a) and (b).

35. ▲ A proton moves at $0.950c$. Calculate its (a) rest energy, (b) total energy, and (c) kinetic energy.

36. An unstable particle with a mass of 3.34×10^{-27} kg is initially at rest. The particle decays into two fragments that fly off along the x axis with velocity components $0.987c$ and $-0.868c$. Find the masses of the fragments. *Suggestion:* Use conservation of both energy and momentum.

37. Show that the energy–momentum relationship $E^2 = p^2c^2 + (mc^2)^2$ follows from the expressions $E = \gamma mc^2$ and $p = \gamma mu$.

38. In a typical color television picture tube, the electrons are accelerated from rest through a potential difference of 25 000 V. (a) What speed do the electrons have when they strike the screen? (b) What is their kinetic energy in joules?

39. The rest energy of an electron is 0.511 MeV. The rest energy of a proton is 938 MeV. Assume both particles have kinetic energies of 2.00 MeV. Find the speed of (a) the electron and (b) the proton. (c) By how much does the speed of the electron exceed that of the proton? (d) Repeat the calculations assuming both particles have kinetic energies of 2 000 MeV.

40. Consider electrons accelerated to an energy of 20.0 GeV in the 3.00-km-long Stanford Linear Accelerator. (a) What is the γ factor for the electrons? (b) What is the electrons' speed? (c) How long does the accelerator appear to the electrons?

41. A pion at rest ($m_\pi = 273m_e$) decays to a muon ($m_\mu = 207m_e$) and an antineutrino ($m_{\bar{\nu}} \approx 0$). The reaction is written $\pi^- \rightarrow \mu^- + \bar{\nu}$. Find the kinetic energy of the muon and the energy of the antineutrino in electron volts. *Suggestion:* Use conservation of both energy and momentum for the decay process.

42. Consider a car moving at highway speed u. Is its actual kinetic energy larger or smaller than $\frac{1}{2}mu^2$? Make an order-of-magnitude estimate of the amount by which its actual kinetic energy differs from $\frac{1}{2}mu^2$. In your solution, state the quantities you take as data and the values you measure or estimate for them. You may find Appendix B.5 useful.

Section 39.9 Mass and Energy

43. ● When 1.00 g of hydrogen combines with 8.00 g of oxygen, 9.00 g of water is formed. During this chemical reaction, 2.86×10^5 J of energy is released. Is the mass of the water larger or smaller than the mass of the reactants? What is the difference in mass? Explain whether the change in mass is likely to be detectable.

44. In a nuclear power plant, the fuel rods last 3 yr before they are replaced. If a plant with rated thermal power 1.00 GW operates at 80.0% capacity for 3.00 yr, what is the loss of mass of the fuel?

45. The power output of the Sun is 3.85×10^{26} W. How much mass is converted to energy in the Sun each second?

46. A gamma ray (a high-energy photon) can produce an electron (e^-) and a positron (e^+) when it enters the electric field of a heavy nucleus: $\gamma \rightarrow e^+ + e^-$. What minimum gamma-ray energy is required to accomplish this task? *Note:* The masses of the electron and the positron are equal.

Section 39.10 The General Theory of Relativity

47. An Earth satellite used in the global positioning system (GPS) moves in a circular orbit with period 11 h 58 min. (a) Determine the radius of its orbit. (b) Determine its speed. (c) The satellite contains an oscillator producing the principal nonmilitary GPS signal. Its frequency is 1 575.42 MHz in the reference frame of the satellite. When it is received on the Earth's surface, what is the fractional change in this frequency due to time dilation as described by special relativity? (d) The gravitational "blueshift" of the frequency according to general relativity is a separate effect. It is called a blueshift to indicate a change to a higher frequency. The magnitude of that fractional change is given by

$$\frac{\Delta f}{f} = \frac{\Delta U_g}{mc^2}$$

where ΔU_g is the change in gravitational potential energy of an object–Earth system when the object of mass m is moved between the two points where the signal is observed. Calculate this fractional change in frequency. (e) What is the overall fractional change in frequency? Superposed on both of these relativistic effects is a Doppler shift that is generally much larger. It can be a redshift or a blueshift, depending on the motion of a particular satellite relative to a GPS receiver (Fig. P39.47).

Figure P39.47

Additional Problems

48. *Houston, we've got a problem.* An astronaut wishes to visit the Andromeda galaxy, making a one-way trip that will take 30.0 yr in the spacecraft's frame of reference. Assume the galaxy is 2.00×10^6 ly away and the astronaut's speed is constant. (a) How fast must he travel relative to the Earth? (b) What will be the kinetic energy of his 1 000-metric-ton spacecraft? (c) What is the cost of this energy if it is purchased at a typical consumer price for electric energy of \$0.130/kWh?

49. ▲ The cosmic rays of highest energy are protons that have kinetic energy on the order of 10^{13} MeV. (a) How long would it take a proton of this energy to travel across the Milky Way galaxy, having a diameter $\sim 10^5$ ly, as measured in the proton's frame? (b) From the point of view of the proton, how many kilometers across is the galaxy?

50. An electron has a speed of $0.750c$. (a) Find the speed of a proton that has the same kinetic energy as the electron. (b) **What If?** Find the speed of a proton that has the same momentum as the electron.

51. ● The equation

$$K = \left(\frac{1}{\sqrt{1 - u^2/c^2}} - 1\right)mc^2$$

gives the kinetic energy of a particle moving at speed u. (a) Solve the equation for u. (b) From the equation for u, identify the minimum possible value of speed and the corresponding kinetic energy. (c) Identify the maximum possible speed and the corresponding kinetic energy. (d) Differentiate the equation for u with respect to time to obtain an equation describing the acceleration of a particle as a function of its kinetic energy and the power input to the particle. (e) Observe that for a nonrelativistic particle we have $u = (2K/m)^{1/2}$ and that differentiating this equation with respect to time gives $a = \mathcal{P}/(2mK)^{1/2}$. State the limiting form of the expression in part (d) at low energy. State how it compares with the nonrelativistic expression. (f) State the limiting form of the expression in part (d) at high energy. (g) Consider a particle with constant input power. Explain how the answer to part (f) helps account for the answer to part (c).

52. Ted and Mary are playing a game of catch in frame S′, which is moving at $0.600c$ with respect to frame S, while Jim, at rest in frame S, watches the action (Fig. P39.52). Ted throws the ball to Mary at $0.800c$ (according to Ted), and their separation (measured in S′) is 1.80×10^{12} m. (a) According to Mary, how fast is the ball moving? (b) According to Mary, what time interval is required for the ball to reach her? (c) According to Jim, how far apart are Ted and Mary and how fast is the ball moving? (d) According to Jim, what time interval is required for the ball to reach Mary?

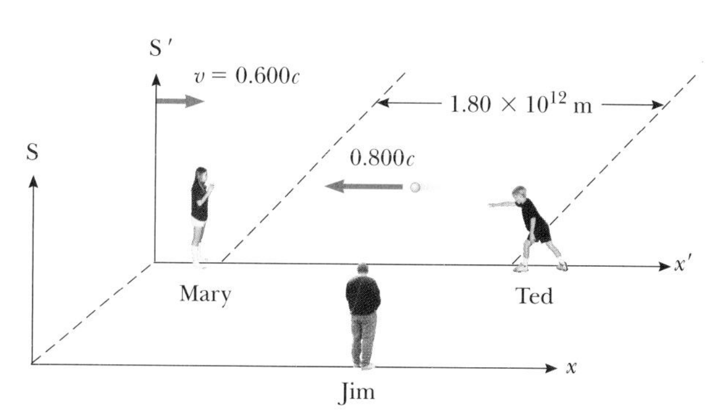

Figure P39.52

53. The net nuclear fusion reaction inside the Sun can be written as $4\,^1H \rightarrow {}^4He + E$. The rest energy of each hydrogen atom is 938.78 MeV, and the rest energy of the helium-4 atom is 3 728.4 MeV. Calculate the percentage of the starting mass that is transformed to other forms of energy.

54. An object disintegrates into two fragments. One fragment has mass 1.00 MeV/c^2 and momentum 1.75 MeV/c in the positive x direction. The other fragment has mass 1.50 MeV/c^2 and momentum 2.00 MeV/c in the positive y direction. Find the (a) mass and (b) speed of the original object.

55. Spacecraft I, containing students taking a physics exam, approaches the Earth with a speed of 0.600c (relative to the Earth), while spacecraft II, containing professors proctoring the exam, moves at 0.280c (relative to the Earth) directly toward the students. If the professors stop the exam after 50.0 min have passed on their clock, for what time interval does the exam last as measured by (a) the students and (b) an observer on the Earth?

56. **Review problem.** An electron is traveling through water at a speed 10.0% faster than the speed of light in water. Determine the electron's (a) total energy, (b) kinetic energy, and (c) momentum. The electron gives off Cerenkov radiation, the electromagnetic equivalent of a bow wave or a sonic boom. (d) Find the angle between the shock wave and the electron's direction of motion. Around the core of a nuclear reactor shielded by a large pool of water, Cerenkov radiation appears as a blue glow.

57. An alien spaceship traveling at 0.600c toward the Earth launches a landing craft with an advance guard of purchasing agents and environmental educators. The landing craft travels in the same direction with a speed of 0.800c relative to the mother ship. As observed on the Earth, the spaceship is 0.200 ly from the Earth when the landing craft is launched. (a) What speed do the Earth-based observers measure for the approaching landing craft? (b) What is the distance to the Earth at the moment of the landing craft's launch as observed by the aliens? (c) What travel time is required for the landing craft to reach the Earth as observed by the aliens on the mother ship? (d) If the landing craft has a mass of 4.00×10^5 kg, what is its kinetic energy as observed in the Earth reference frame?

58. ● *Speed of light in a moving medium.* The motion of a transparent medium influences the speed of light. This effect was first observed by Fizeau in 1851. Consider a light beam in water that moves with speed v in a horizontal pipe. Assume the light travels in the same direction as the water. The speed of light with respect to the water is c/n, where $n = 1.33$ is the index of refraction of water. (a) Use the velocity transformation equation to show that the speed of the light measured in the laboratory frame is

$$u = \frac{c}{n}\left(\frac{1 + nv/c}{1 + v/nc}\right)$$

(b) Show that for $v \ll c$, the expression from part (a) becomes, to a good approximation,

$$u \approx \frac{c}{n} + v - \frac{v}{n^2}$$

Argue for or against the view that we should expect the result to be $u = (c/n) + v$ according to the Galilean transformation and that the presence of the term $-v/n^2$ represents a relativistic effect appearing even at "nonrelativistic" speeds. (c) Evaluate u in the limit as the speed of the water approaches c.

59. A supertrain (proper length 100 m) travels at a speed of 0.950c as it passes through a tunnel (proper length 50.0 m). As seen by a trackside observer, is the train ever completely within the tunnel? If so, how much space is there to spare?

60. Imagine that the entire Sun collapses to a sphere of radius R_g such that the work required to remove a small mass m from the surface would be equal to its rest energy mc^2. This radius is called the *gravitational radius* for the Sun. Find R_g. The ultimate fate of very massive stars is thought to be collapsing beyond their gravitational radii into black holes.

61. ● A particle with electric charge q moves along a straight line in a uniform electric field $\vec{\mathbf{E}}$ with a speed of u. The electric force exerted on the charge is $q\vec{\mathbf{E}}$. The motion and the electric field are both in the x direction. (a) Show that the acceleration of the particle in the x direction is given by

$$a = \frac{du}{dt} = \frac{qE}{m}\left(1 - \frac{u^2}{c^2}\right)^{3/2}$$

(b) Discuss the significance of the dependence of the acceleration on the speed. (c) **What If?** If the particle starts from rest at $x = 0$ at $t = 0$, how would you proceed to find the speed of the particle and its position at time t?

62. An observer in a coasting spacecraft moves toward a mirror at speed v relative to the reference frame labeled by S in Figure P39.62. The mirror is stationary with respect to S. A light pulse emitted by the spacecraft travels toward the mirror and is reflected back to the spacecraft. The front of the spacecraft is a distance d from the mirror (as measured by observers in S) at the moment the light pulse leaves the spacecraft. What is the total travel time of the pulse as measured by observers in (a) the S frame and (b) the front of the spacecraft?

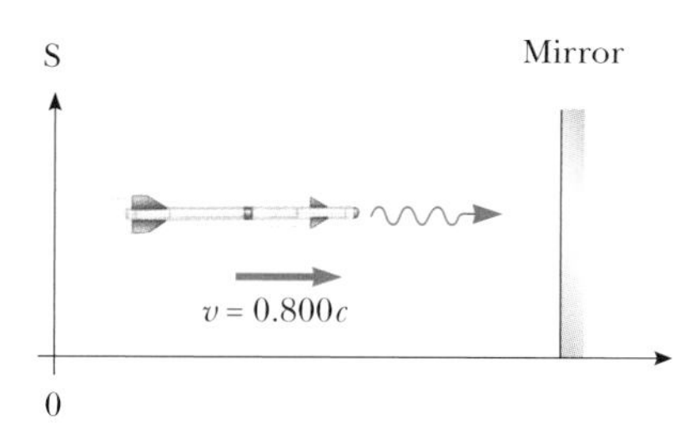

Figure P39.62

63. ● Massive stars ending their lives in supernova explosions produce the nuclei of all the atoms in the bottom half of the periodic table, by fusion of smaller nuclei. This problem roughly models that process. A particle of mass m moving along the x axis with a velocity component $+u$ collides head-on and sticks to a particle of mass $m/3$ moving along the x axis with the velocity component $-u$. (a) What is the mass M of the resulting particle? (b) Evaluate the expression from part (a) in the limit $u \to 0$. Explain whether the result agrees with what you should expect from nonrelativistic physics.

64. The creation and study of new elementary particles is an important part of contemporary physics. Especially interesting is the discovery of a very massive particle. To create a particle of mass M requires an energy Mc^2. With enough energy, an exotic particle can be created by allowing a fast-moving particle of ordinary matter, such as a proton, to collide with a similar target particle. Consider a perfectly inelastic collision between two protons: an incident proton with mass m_p, kinetic energy K, and momentum magnitude p joins with an originally stationary target proton to form a single product particle of mass M. You might think that the creation of a new product particle, nine times more massive than in a previous experiment, would require only nine times more energy for the incident proton. Unfortunately, not all the kinetic energy of the incoming proton is available to create the product particle because conservation of momentum requires that the system as a whole still must have some kinetic energy after the collision. Only a fraction of the energy of the incident particle is therefore available to create a new particle. In this problem, you must determine how the energy available for particle creation depends on the energy of the moving proton. Show that the energy available to create a product particle is given by

$$Mc^2 = 2m_pc^2\sqrt{1 + \frac{K}{2m_pc^2}}$$

This result shows that when the kinetic energy K of the incident proton is large compared with its rest energy m_pc^2, then M approaches $(2m_pK)^{1/2}/c$. Therefore, if the energy of the incoming proton is increased by a factor of 9, the mass you can create increases only by a factor of 3. This disappointing result is the main reason that most modern accelerators such as those at CERN (in Europe), at Fermilab (near Chicago), at SLAC (at Stanford), and at DESY (in Germany) use *colliding beams*. Here the total momentum of a pair of interacting particles can be zero. The center of mass can be at rest after the collision, so, in principle, all the initial kinetic energy can be used for particle creation, according to

$$Mc^2 = 2mc^2 + K = 2mc^2\left(1 + \frac{K}{2mc^2}\right)$$

where K is the total kinetic energy of two identical colliding particles. Here if $K \gg mc^2$, we have M directly proportional to K as we would desire. These machines are difficult to build and to operate, but they open new vistas in physics.

65. ● Suppose our Sun is about to explode. In an effort to escape, we depart in a spacecraft at $v = 0.800c$ and head toward the star Tau Ceti, 12.0 ly away. When we reach the midpoint of our journey from the Earth, we see our Sun explode, and, unfortunately, at the same instant we see Tau Ceti explode as well. (a) In the spacecraft's frame of reference, should we conclude that the two explosions occurred simultaneously? If not, which occurred first? (b) **What If?** In a frame of reference in which the Sun and Tau Ceti are at rest, did they explode simultaneously? If not, which exploded first?

66. Prepare a graph of the relativistic kinetic energy and the classical kinetic energy, both as a function of speed, for an object with a mass of your choice. At what speed does the classical kinetic energy underestimate the experimental value by 1%? By 5%? By 50%?

67. A ^{57}Fe nucleus at rest emits a 14.0-keV photon. Use conservation of energy and momentum to deduce the kinetic energy of the recoiling nucleus in electron volts. Use $Mc^2 = 8.60 \times 10^{-9}$ J for the final state of the ^{57}Fe nucleus.

Answers to Quick Quizzes

39.1 (c). Although the observers' measurements differ, both are correct.

39.2 (d). The Galilean velocity transformation gives us $u_x = u'_x + v = 90$ mi/h $+ 110$ mi/h $= 200$ mi/h.

39.3 (d). The two events (the pulse leaving the flashlight and the pulse hitting the far wall) take place at different locations for both observers, so neither measures the proper time interval.

39.4 (a). The two events are the beginning and the end of the movie, both of which take place at rest with respect to the spacecraft crew. Therefore, the crew measures the proper time interval of 2 h. Any observer in motion with respect to the spacecraft, which includes the observer on Earth, will measure a longer time interval due to time dilation.

39.5 (a). If their on-duty time is based on clocks that remain on the Earth, the astronauts will have larger paychecks. A shorter time interval will have passed for the astronauts in their frame of reference than for their employer back on the Earth.

39.6 (c). Both your body and your sleeping cabin are at rest in your reference frame; therefore, they will have their proper length according to you. There will be no change in measured lengths of objects, including yourself, within your spacecraft.

39.7 (d). Time dilation and length contraction depend only on the relative speed of one observer relative to another, not on whether the observers are receding or approaching each other.

39.8 **(i),** (c). Because of your motion toward the source of the light, the light beam has a horizontal component of velocity as measured by you. The magnitude of the vector sum of the horizontal and vertical component vectors must be equal to c, so the magnitude of the vertical component must be smaller than c. **(ii),** (a). In this case, there is only a horizontal component of the velocity of the light and you must measure a speed of c.

39.9 (a) $m_3 > m_2 = m_1$; the rest energy of particle 3 is $2E$, whereas it is E for particles 1 and 2. (b) $K_3 = K_2 > K_1$; the kinetic energy is the difference between the total energy and the rest energy. The kinetic energy is $4E - 2E = 2E$ for particle 3, $3E - E = 2E$ for particle 2, and $2E - E = E$ for particle 1. (c) $u_2 > u_3 = u_1$; from Equation 39.26, $E = \gamma E_R$. Solving for the square of the particle speed u, we find that $u^2 = c^2(1 - (E_R/E)^2)$. Therefore, the particle with the smallest ratio of rest energy to total energy will have the largest speed. Particles 1 and 3 have the same ratio as each other, and the ratio of particle 2 is smaller.

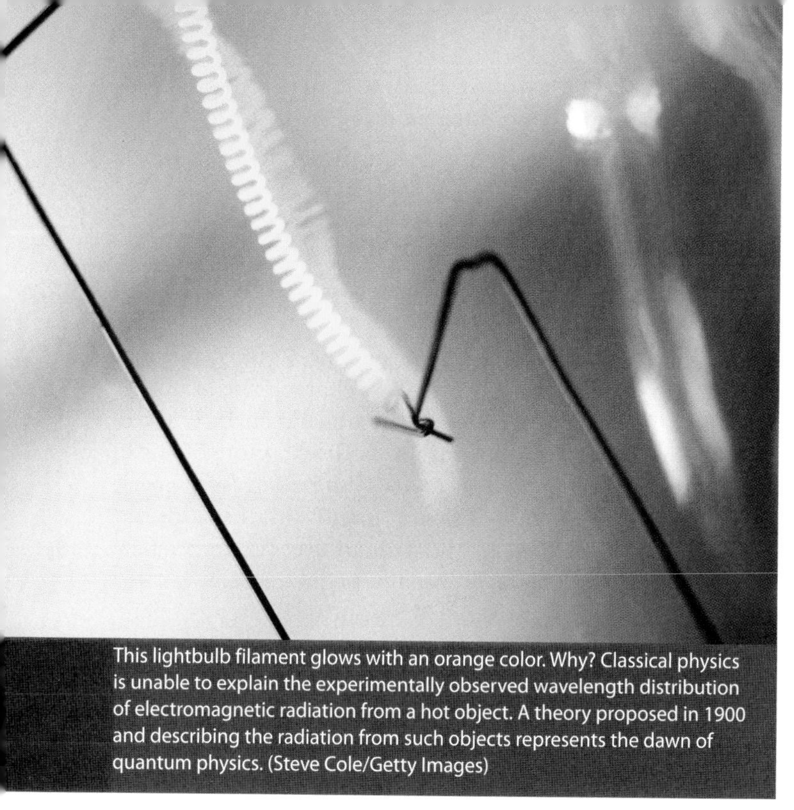

This lightbulb filament glows with an orange color. Why? Classical physics is unable to explain the experimentally observed wavelength distribution of electromagnetic radiation from a hot object. A theory proposed in 1900 and describing the radiation from such objects represents the dawn of quantum physics. (Steve Cole/Getty Images)

40 Introduction to Quantum Physics

In Chapter 39, we discussed that Newtonian mechanics must be replaced by Einstein's special theory of relativity when dealing with particle speeds comparable to the speed of light. As the 20th century progressed, many experimental and theoretical problems were resolved by the special theory of relativity. For many other problems, however, neither relativity nor classical physics could provide a theoretical answer. Attempts to apply the laws of classical physics to explain the behavior of matter on the atomic scale were consistently unsuccessful. For example, the emission of discrete wavelengths of light from atoms in a high-temperature gas could not be explained within the framework of classical physics.

As physicists sought new ways to solve these puzzles, another revolution took place in physics between 1900 and 1930. A new theory called *quantum mechanics* was highly successful in explaining the behavior of particles of microscopic size. Like the special theory of relativity, the quantum theory requires a modification of our ideas concerning the physical world.

The first explanation of a phenomenon using quantum theory was introduced by Max Planck. Many subsequent mathematical developments and interpretations were made by a number of distinguished physicists, including Einstein, Bohr, de Broglie, Schrödinger, and Heisenberg. Despite the great success of the quantum theory, Einstein frequently played the role of its critic, especially with regard to the manner in which the theory was interpreted.

PITFALL PREVENTION 40.1
Expect to Be Challenged

If the discussions of quantum physics in this and subsequent chapters seem strange and confusing to you, it's because your whole life experience has taken place in the macroscopic world, where quantum effects are not evident.

Because an extensive study of quantum theory is beyond the scope of this book, this chapter is simply an introduction to its underlying principles.

40.1 Blackbody Radiation and Planck's Hypothesis

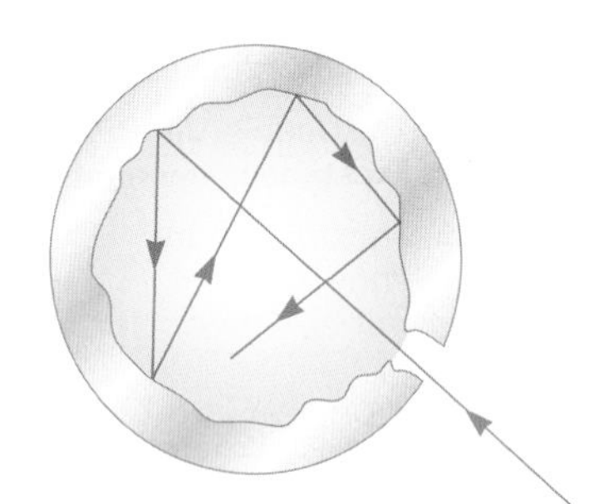

Figure 40.1 The opening to the cavity inside a hollow object is a good approximation of a black body. Light entering the small opening strikes the interior walls, where some is absorbed and some is reflected at a random angle. The cavity walls reradiate at wavelengths corresponding to their temperature, producing standing waves in the cavity. Some of the energy from these standing waves can leave through the opening.

An object at any temperature emits electromagnetic waves in the form of **thermal radiation** from its surface as discussed in Section 20.7. The characteristics of this radiation depend on the temperature and properties of the object's surface. Careful study shows that the radiation consists of a continuous distribution of wavelengths from all portions of the electromagnetic spectrum. If the object is at room temperature, the wavelengths of thermal radiation are mainly in the infrared region and hence the radiation is not detected by the human eye. As the surface temperature of the object increases, the object eventually begins to glow visibly red. At sufficiently high temperatures, the glowing object appears white, as in the hot tungsten filament of a lightbulb.

From a classical viewpoint, thermal radiation originates from accelerated charged particles in the atoms near the surface of the object; those charged particles emit radiation much as small antennas do. The thermally agitated particles can have a distribution of energies, which accounts for the continuous spectrum of radiation emitted by the object. By the end of the 19th century, however, it became apparent that the classical theory of thermal radiation was inadequate. The basic problem was in understanding the observed distribution of wavelengths in the radiation emitted by a black body. As defined in Section 20.7, a **black body** is an ideal system that absorbs all radiation incident on it. The electromagnetic radiation emitted by the black body is called **blackbody radiation.**

A good approximation of a black body is a small hole leading to the inside of a hollow object as shown in Figure 40.1. Any radiation incident on the hole from outside the cavity enters the hole and is reflected a number of times on the interior walls of the cavity; hence, the hole acts as a perfect absorber. The nature of the radiation leaving the cavity through the hole depends only on the temperature of the cavity walls and not on the material of which the walls are made. The spaces between lumps of hot charcoal (Fig. 40.2) emit light that is very much like blackbody radiation.

Corbis

Figure 40.2 The glow emanating from the spaces between these hot charcoal briquettes is, to a close approximation, blackbody radiation. The color of the light depends only on the temperature of the briquettes.

The radiation emitted by oscillators in the cavity walls experiences boundary conditions. As the radiation reflects from the cavity's walls, standing electromagnetic waves are established within the three-dimensional interior of the cavity. Many standing-wave modes are possible, and the distribution of the energy in the cavity among these modes determines the wavelength distribution of the radiation leaving the cavity through the hole.

The wavelength distribution of radiation from cavities was studied experimentally in the late 19th century. Active Figure 40.3 shows how the intensity of blackbody radiation varies with temperature and wavelength. The following two consistent experimental findings were seen as especially significant:

1. **The total power of the emitted radiation increases with temperature.** We discussed this behavior briefly in Chapter 20, where we introduced **Stefan's law:**

Stefan's law ▶

$$\mathcal{P} = \sigma A e T^4 \tag{40.1}$$

where $\mathcal{P}$ is the power in watts radiated at all wavelengths from the surface of an object, σ is the Stefan–Boltzmann constant, equal to $5.670 \times 10^{-8}\ \mathrm{W/m^2 \cdot K^4}$, A is the surface area of the object in square meters, e is the emissivity of the surface, and T is the surface temperature in kelvins. For a black body, the emissivity is $e = 1$ exactly.

2. **The peak of the wavelength distribution shifts to shorter wavelengths as the temperature increases.** This behavior is described by the following relationship, called **Wien's displacement law:**

$$\lambda_{max} T = 2.898 \times 10^{-3} \text{ m} \cdot \text{K} \quad (40.2)$$

◀ Wien's displacement law

where λ_{max} is the wavelength at which the curve peaks and T is the absolute temperature of the surface of the object emitting the radiation. The wavelength at the curve's peak is inversely proportional to the absolute temperature; that is, as the temperature increases, the peak is "displaced" to shorter wavelengths (Active Fig. 40.3).

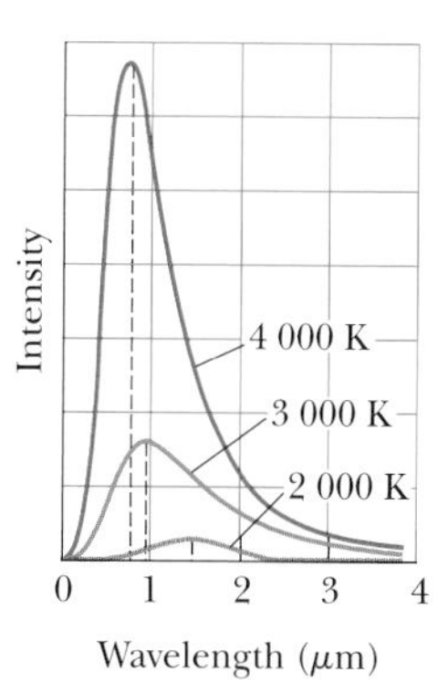

ACTIVE FIGURE 40.3

Intensity of blackbody radiation versus wavelength at three temperatures. The amount of radiation emitted (the area under a curve) increases with increasing temperature. The visible range of wavelengths is between 0.4 μm and 0.7 μm. Therefore, the 4 000-K curve has a peak that is near the visible range and that represents an object that would glow with a yellowish-white appearance. At approximately 6 000 K, the peak is in the center of the visible wavelengths and the object appears white.

Sign in at www.thomsonedu.com and go to ThomsonNOW to adjust the temperature of the black body and study the radiation emitted from it.

Wien's displacement law is consistent with the behavior of the object mentioned at the beginning of this section. At room temperature, the object does not appear to glow because the peak is in the infrared region of the electromagnetic spectrum. At higher temperatures, it glows red because the peak is in the near infrared with some radiation at the red end of the visible spectrum, and at still higher temperatures, it glows white because the peak is in the visible so that all colors are emitted.

Quick Quiz 40.1 Figure 40.4 shows two stars in the constellation Orion. Betelgeuse appears to glow red, whereas Rigel looks blue in color. Which star has a higher surface temperature? (a) Betelgeuse (b) Rigel (c) both the same (d) impossible to determine

Figure 40.4 (Quick Quiz 40.1) Which star is hotter, Betelgeuse or Rigel?

A successful theory for blackbody radiation must predict the shape of the curves in Active Figure 40.3, the temperature dependence expressed in Stefan's law, and the shift of the peak with temperature described by Wien's displacement law. Early attempts to use classical ideas to explain the shapes of the curves in Active Figure 40.3 failed.

Let's consider one of these early attempts. To describe the distribution of energy from a black body, we define $I(\lambda, T)\, d\lambda$ to be the intensity, or power per unit area, emitted in the wavelength interval $d\lambda$. The result of a calculation based on a classical theory of blackbody radiation known as the **Rayleigh–Jeans law** is

$$I(\lambda, T) = \frac{2\pi c k_B T}{\lambda^4} \quad (40.3)$$

◀ Rayleigh–Jeans law

where k_B is Boltzmann's constant. The black body is modeled as the hole leading into a cavity supporting many modes of oscillation of the electromagnetic field caused by accelerated charges in the cavity walls, resulting in the emission of electromagnetic waves at all wavelengths. In the classical theory used to derive Equation 40.3, the average energy for each wavelength of the standing-wave modes is assumed to be proportional to $k_B T$, based on the theorem of equipartition of energy discussed in Section 21.1.

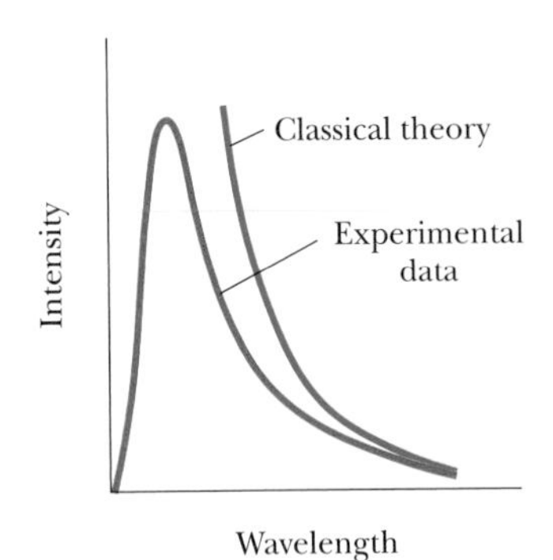

Figure 40.5 Comparison of experimental results and the curve predicted by the Rayleigh–Jeans law for the distribution of blackbody radiation.

An experimental plot of the blackbody radiation spectrum, together with the theoretical prediction of the Rayleigh–Jeans law, is shown in Figure 40.5. At long wavelengths, the Rayleigh–Jeans law is in reasonable agreement with experimental data, but at short wavelengths, major disagreement is apparent.

As λ approaches zero, the function $I(\lambda, T)$ given by Equation 40.3 approaches infinity. Hence, according to classical theory, not only should short wavelengths predominate in a blackbody spectrum, but also the energy emitted by any black body should become infinite in the limit of zero wavelength. In contrast to this prediction, the experimental data plotted in Figure 40.5 show that as λ approaches zero, $I(\lambda, T)$ also approaches zero. This mismatch of theory and experiment was so disconcerting that scientists called it the *ultraviolet catastrophe.* (This "catastrophe"—infinite energy—occurs as the wavelength approaches zero; the word *ultraviolet* was applied because ultraviolet wavelengths are short.)

In 1900, Max Planck developed a theory of blackbody radiation that leads to an equation for $I(\lambda, T)$ that is in complete agreement with experimental results at all wavelengths. Planck assumed the cavity radiation came from atomic oscillators in the cavity walls in Figure 40.1. Planck made two bold and controversial assumptions concerning the nature of the oscillators in the cavity walls:

- The energy of an oscillator can have only certain *discrete* values E_n:

$$E_n = nhf \tag{40.4}$$

  where n is a positive integer called a **quantum number,**[1] f is the oscillator's frequency, and h is a parameter Planck introduced that is now called **Planck's constant**. Because the energy of each oscillator can have only discrete values given by Equation 40.4, we say the energy is **quantized.** Each discrete energy value corresponds to a different **quantum state,** represented by the quantum number n. When the oscillator is in the $n = 1$ quantum state, its energy is hf; when it is in the $n = 2$ quantum state, its energy is $2hf$; and so on.
- The oscillators emit or absorb energy when making a transition from one quantum state to another. The entire energy difference between the initial and final states in the transition is emitted or absorbed as a single quantum of radiation. If the transition is from one state to a lower adjacent state—say, from the $n = 3$ state to the $n = 2$ state—Equation 40.4 shows that the amount of energy emitted by the oscillator and carried by the quantum of radiation is

$$E = hf \tag{40.5}$$

MAX PLANCK
German Physicist (1858–1947)
Planck introduced the concept of "quantum of action" (Planck's constant, h) in an attempt to explain the spectral distribution of blackbody radiation, which laid the foundations for quantum theory. In 1918, he was awarded the Nobel Prize in Physics for this discovery of the quantized nature of energy.

An oscillator emits or absorbs energy only when it changes quantum states. If it remains in one quantum state, no energy is absorbed or emitted. Figure 40.6 is an **energy-level diagram** showing the quantized energy levels and allowed transitions proposed by Planck. This important semigraphical representation is used often in quantum physics.[2] The vertical axis is linear in energy, and the allowed energy levels are represented as horizontal lines. The quantized system can have only the energies represented by the horizontal lines.

The key point in Planck's theory is the radical assumption of quantized energy states. This development—a clear deviation from classical physics—marked the birth of the quantum theory.

In the Rayleigh–Jeans model, the average energy associated with a particular wavelength of standing waves in the cavity is the same for all wavelengths and is equal to k_BT. Planck used the same classical ideas as in the Rayleigh–Jeans model to arrive at the energy density as a product of constants and the average energy for a given wavelength, but the average energy is not given by the equipartition theo-

PITFALL PREVENTION 40.2
***n* Is Again an Integer**

In the preceding chapters on optics, we used the symbol n for the index of refraction, which was not an integer. Here we are again using n as we did in Chapter 18 to indicate the standing-wave mode on a string or in an air column. In quantum physics, n is often used as an integer quantum number to identify a particular quantum state of a system.

[1] A quantum number is generally an integer (although half-integer quantum numbers can occur) that describes an allowed state of a system, such as the values of n describing the normal modes of oscillation of a string fixed at both ends, as discussed in Section 18.3.

[2] We first saw an energy-level diagram in Section 21.4.

rem. A wave's average energy is the average energy difference between levels of the oscillator, *weighted according to the probability of the wave being emitted.* This weighting is based on the occupation of higher-energy states as described by the Boltzmann distribution law, which was discussed in Section 21.5. According to this law, the probability of a state being occupied is proportional to the factor e^{-E/k_BT}, where E is the energy of the state.

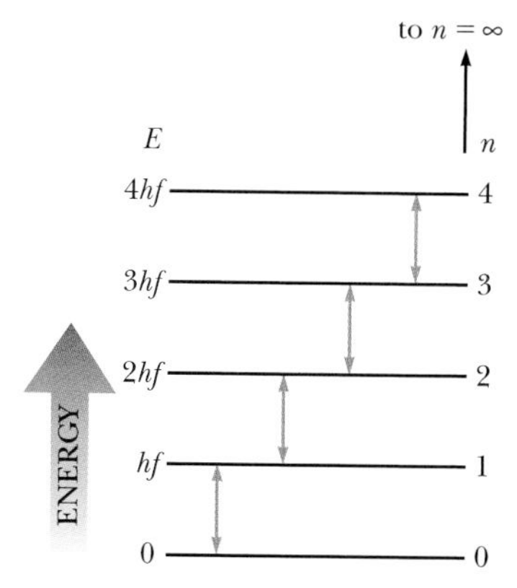

Figure 40.6 Allowed energy levels for an oscillator with frequency f. Allowed transitions are indicated by the double-headed arrows.

At low frequencies, the energy levels are close together as on the right in Active Figure 40.7, and many of the energy states are excited because the Boltzmann factor e^{-E/k_BT} is relatively large for these states. Therefore, there are many contributions to the outgoing radiation, although each contribution has very low energy. Now, consider high-frequency radiation, that is, radiation with short wavelength. To obtain this radiation, the allowed energies are very far apart as on the left in Active Figure 40.7. The probability of thermal agitation exciting these high energy levels is small because of the small value of the Boltzmann factor for large values of E. At high frequencies, the low probability of excitation results in very little contribution to the total energy, even though each quantum is of large energy. This low probability "turns the curve over" and brings it down to zero again at short wavelengths.

Using this approach, Planck generated a theoretical expression for the wavelength distribution that agreed remarkably well with the experimental curves in Active Figure 40.3:

$$I(\lambda, T) = \frac{2\pi hc^2}{\lambda^5\left(e^{hc/\lambda k_BT} - 1\right)} \quad \textbf{(40.6)}$$

◀ Planck's wavelength distribution function

This function includes the parameter h, which Planck adjusted so that his curve matched the experimental data at all wavelengths. The value of this parameter is found to be independent of the material of which the black body is made and independent of the temperature; it is a fundamental constant of nature. The value of h, Planck's constant, which was first introduced in Chapter 35, is

$$h = 6.626 \times 10^{-34}\ \text{J}\cdot\text{s} \quad \textbf{(40.7)}$$

◀ Planck's constant

At long wavelengths, Equation 40.6 reduces to the Rayleigh–Jeans expression, Equation 40.3 (see Problem 13), and at short wavelengths, it predicts an exponential decrease in $I(\lambda, T)$ with decreasing wavelength, in agreement with experimental results.

When Planck presented his theory, most scientists (including Planck!) did not consider the quantum concept to be realistic. They believed it was a mathematical

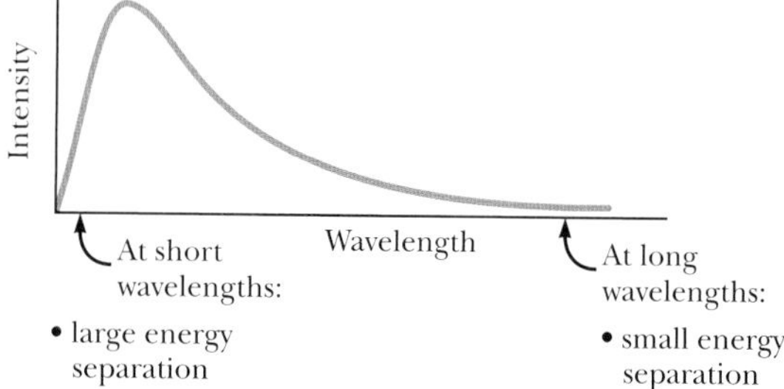

ACTIVE FIGURE 40.7

In Planck's model, the average energy associated with a given wavelength is the product of the energy of a transition and a factor related to the probability of the transition occurring. As the energy levels move farther apart at shorter wavelengths (higher energy), the probability of excitation decreases, as does the probability of a transition from the excited state.

Sign in at www.thomsonedu.com and go to ThomsonNOW to investigate the energy levels and observe the emission of radiation of different wavelengths.

Figure 40.8 An ear thermometer measures a patient's temperature by detecting the intensity of infrared radiation leaving the eardrum.

trick that happened to predict the correct results. Hence, Planck and others continued to search for a more "rational" explanation of blackbody radiation. Subsequent developments, however, showed that a theory based on the quantum concept (rather than on classical concepts) had to be used to explain not only blackbody radiation but also a number of other phenomena at the atomic level.

In 1905, Einstein rederived Planck's results by assuming the cavity oscillations of the electromagnetic field were themselves quantized. In other words, he proposed that quantization is a fundamental property of light and other electromagnetic radiation, which led to the concept of photons as shall be discussed in Section 40.2. Critical to the success of the quantum or photon theory was the relation between energy and frequency, which classical theory completely failed to predict.

You may have had your body temperature measured at the doctor's office by an *ear thermometer,* which can read your temperature in a matter of seconds (Fig. 40.8). In a fraction of a second, this type of thermometer measures the amount of infrared radiation emitted by the eardrum. It then converts the amount of radiation into a temperature reading. This thermometer is very sensitive because temperature is raised to the fourth power in Stefan's law. Suppose you have a fever 1°C above normal. Because absolute temperatures are found by adding 273 to Celsius temperatures, the ratio of your fever temperature to normal body temperature of 37°C is

$$\frac{T_{\text{fever}}}{T_{\text{normal}}} = \frac{38°\text{C} + 273°\text{C}}{37°\text{C} + 273°\text{C}} = 1.003\ 2$$

which is only a 0.32% increase in temperature. The increase in radiated power, however, is proportional to the fourth power of temperature, so

$$\frac{\mathcal{P}_{\text{fever}}}{\mathcal{P}_{\text{normal}}} = \left(\frac{38°\text{C} + 273°\text{C}}{37°\text{C} + 273°\text{C}}\right)^4 = 1.013$$

The result is a 1.3% increase in radiated power, which is easily measured by modern infrared radiation sensors.

EXAMPLE 40.1 Thermal Radiation from Different Objects

(A) Find the peak wavelength of the blackbody radiation emitted by the human body when the skin temperature is 35°C.

SOLUTION

Conceptualize Thermal radiation is emitted from the surface of any object. The peak wavelength is related to the surface temperature through Wien's displacement law (Eq. 40.2).

Categorize We evaluate results using an equation developed in this section, so we categorize this example as a substitution problem.

Solve Equation 40.2 for λ_{max}:

$$(1)\quad \lambda_{\text{max}} = \frac{2.898 \times 10^{-3}\ \text{m}\cdot\text{K}}{T}$$

Substitute the surface temperature:

$$\lambda_{\text{max}} = \frac{2.898 \times 10^{-3}\ \text{m}\cdot\text{K}}{308\ \text{K}} = 9.4\ \mu\text{m}$$

This radiation is in the infrared region of the spectrum and is invisible to the human eye. Some animals (pit vipers, for instance) are able to detect radiation of this wavelength and therefore can locate warm-blooded prey even in the dark.

(B) Find the peak wavelength of the blackbody radiation emitted by the tungsten filament of a lightbulb, which operates at 2 000 K.

SOLUTION

Substitute the filament temperature into Equation (1):

$$\lambda_{max} = \frac{2.898 \times 10^{-3}\ \text{m}\cdot\text{K}}{2\ 000\ \text{K}} = 1.4\ \mu\text{m}$$

This radiation is also in the infrared, meaning that most of the energy emitted by a lightbulb is not visible to us.

(C) Find the peak wavelength of the blackbody radiation emitted by the Sun, which has a surface temperature of approximately 5 800 K.

SOLUTION

Substitute the surface temperature into Equation (1):

$$\lambda_{max} = \frac{2.898 \times 10^{-3}\ \text{m}\cdot\text{K}}{5\ 800\ \text{K}} = 0.50\ \mu\text{m}$$

This radiation is near the center of the visible spectrum, near the color of a yellow-green tennis ball. Because it is the most prevalent color in sunlight, our eyes have evolved to be most sensitive to light of approximately this wavelength.

EXAMPLE 40.2 The Quantized Oscillator

A 2.0-kg block is attached to a massless spring that has a force constant of $k = 25$ N/m. The spring is stretched 0.40 m from its equilibrium position and released from rest.

(A) Find the total energy of the system and the frequency of oscillation according to classical calculations.

SOLUTION

Conceptualize We understand the details of the block's motion from our study of simple harmonic motion in Chapter 15.

Categorize The phrase "according to classical calculations" tells us to categorize this part of the problem as a classical analysis of the oscillator. We model the block as a particle in simple harmonic motion.

Analyze Based on the way that the block is set into motion, its amplitude is 0.40 m.

Evaluate the total energy of the block–spring system using Equation 15.21:

$$E = \tfrac{1}{2}kA^2 = \tfrac{1}{2}(25\ \text{N/m})(0.40\ \text{m})^2 = 2.0\ \text{J}$$

Evaluate the frequency of oscillation from Equation 15.14:

$$f = \frac{1}{2\pi}\sqrt{\frac{k}{m}} = \frac{1}{2\pi}\sqrt{\frac{25\ \text{N/m}}{2.0\ \text{kg}}} = 0.56\ \text{Hz}$$

(B) Assuming the energy of the oscillator is quantized, find the quantum number n for the system oscillating with this amplitude.

SOLUTION

Categorize This part of the problem is categorized as a quantum analysis of the oscillator. We model the block–spring system as a Planck oscillator.

Analyze Solve Equation 40.4 for the quantum number n:

$$n = \frac{E_n}{hf}$$

Substitute numerical values:

$$n = \frac{2.0\ \text{J}}{(6.626 \times 10^{-34}\ \text{J}\cdot\text{s})(0.56\ \text{Hz})} = 5.4 \times 10^{33}$$

Finalize Notice that 5.4×10^{33} is a very large quantum number, which is typical for macroscopic systems. Changes between quantum states for the oscillator are explored next.

What If? Suppose the oscillator makes a transition from the $n = 5.4 \times 10^{33}$ state to the state corresponding to $n = 5.4 \times 10^{33} - 1$. By how much does the energy of the oscillator change in this one-quantum change?

Answer From Equation 40.5, the energy carried away due to the transition between states differing in n by 1 is

$$E = hf = (6.626 \times 10^{-34}\ \text{J}\cdot\text{s})(0.56\ \text{Hz}) = 3.7 \times 10^{-34}\ \text{J}$$

This energy change due to a one-quantum change is fractionally equal to $3.7 \times 10^{-34}\ \text{J}/2.0\ \text{J}$, or on the order of one part in 10^{34}! It is such a small fraction of the total energy of the oscillator that it cannot be detected. Therefore, even though the energy of a macroscopic block–spring system is quantized and does indeed decrease by small quantum jumps, our senses perceive the decrease as continuous. Quantum effects become important and detectable only on the submicroscopic level of atoms and molecules.

40.2 The Photoelectric Effect

Blackbody radiation was the first phenomenon to be explained with a quantum model. In the latter part of the 19th century, at the same time that data were taken on thermal radiation, experiments showed that light incident on certain metallic surfaces causes electrons to be emitted from those surfaces. This phenomenon, which was first discussed in Section 35.1, is known as the **photoelectric effect,** and the emitted electrons are called **photoelectrons.**[3]

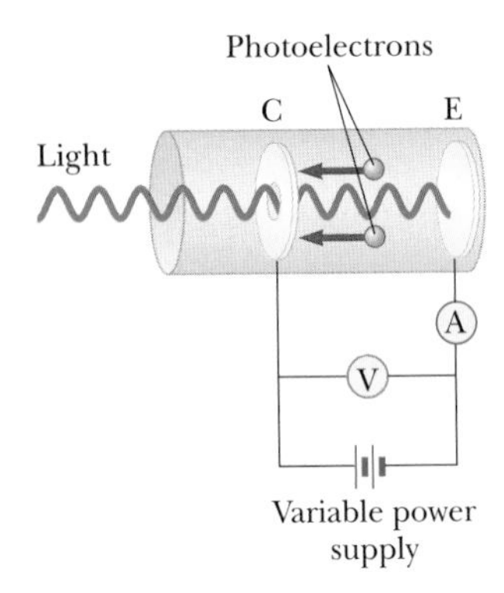

ACTIVE FIGURE 40.9

A circuit diagram for studying the photoelectric effect. When light strikes plate E (the emitter), photoelectrons are ejected from the plate. Electrons moving from plate E to plate C (the collector) constitute a current in the circuit.

Sign in at www.thomsonedu.com and go to ThomsonNOW to observe the motion of electrons for various frequencies of light and plate voltages.

Active Figure 40.9 is a diagram of an apparatus for studying the photoelectric effect. An evacuated glass or quartz tube contains a metallic plate E (the emitter) connected to the negative terminal of a battery and another metallic plate C (the collector) that is connected to the positive terminal of the battery. When the tube is kept in the dark, the ammeter reads zero, indicating no current in the circuit. However, when plate E is illuminated by light having an appropriate wavelength, a current is detected by the ammeter, indicating a flow of charges across the gap between plates E and C. This current arises from photoelectrons emitted from plate E and collected at plate C.

Active Figure 40.10 is a plot of photoelectric current versus potential difference ΔV applied between plates E and C for two light intensities. At large values of ΔV, the current reaches a maximum value; all the electrons emitted from E are collected at C, and the current cannot increase further. In addition, the maximum current increases as the intensity of the incident light increases, as you might expect, and more electrons are ejected by the higher-intensity light. Finally, when ΔV is negative—that is, when the battery in the circuit is reversed to make plate E positive and plate C negative—the current drops because many of the photoelectrons emitted from E are repelled by the now negative plate C. In this situation, only those photoelectrons having a kinetic energy greater than $e|\Delta V|$ reach plate C, where e is the magnitude of the charge on the electron. When ΔV is equal to or more negative than $-\Delta V_s$, where ΔV_s is the **stopping potential,** no photoelectrons reach C and the current is zero.

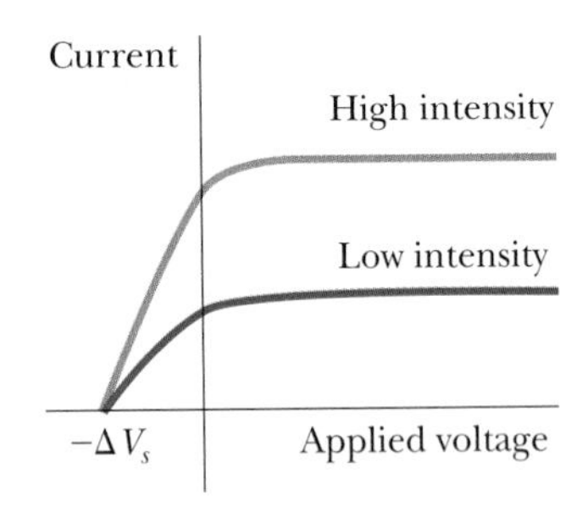

ACTIVE FIGURE 40.10

Photoelectric current versus applied potential difference for two light intensities. The current increases with intensity but reaches a saturation level for large values of ΔV. At voltages equal to or more negative than $-\Delta V_s$, where ΔV_s is the stopping potential, the current is zero.

Sign in at www.thomsonedu.com and go to ThomsonNOW to sweep through the voltage range and observe the current curve for different intensities of radiation.

Let's model the combination of the electric field between the plates and an electron ejected from plate E as an isolated system. Suppose this electron stops just as it reaches plate C. Because the system is isolated, the total mechanical energy of the system must be conserved, so we have

$$K_1 + U_1 = K_2 + U_2$$

where configuration 1 refers to the instant the electron leaves the metal with kinetic energy K_1 and configuration 2 is when the electron stops just before touching plate C. If we define the electric potential energy of the system in configuration 1 to be zero, we have

$$K_1 + 0 = 0 + (-e)(-\Delta V)$$

[3] Photoelectrons are not different from other electrons. They are given this name solely because of their ejection from a metal by light in the photoelectric effect.

Now suppose the potential difference ΔV is increased in the negative direction just until the current is zero. In this case, the electron that stops immediately before reaching plate C has the maximum possible kinetic energy upon leaving the metal surface and ΔV equals the stopping potential ΔV_s. The previous equation can then be written as

$$K_{\max} = e\Delta V_s \qquad \textbf{(40.8)}$$

This equation allows us to measure $K_{\max}$ experimentally by determining the voltage ΔV_s at which the current drops to zero.

Several features of the photoelectric effect are listed below. For each feature, we compare the predictions made by a classical approach, using the wave model for light, with the experimental results.

1. Dependence of photoelectron kinetic energy on light intensity

Classical prediction: Electrons should absorb energy continuously from the electromagnetic waves. As the light intensity incident on a metal is increased, energy should be transferred into the metal at a higher rate and the electrons should be ejected with more kinetic energy.
Experimental result: The maximum kinetic energy of photoelectrons is *independent* of light intensity as shown in Active Figure 40.10 with both curves falling to zero at the *same* negative voltage. (According to Equation 40.8, the maximum kinetic energy is proportional to the stopping potential.)

2. Time interval between incidence of light and ejection of photoelectrons

Classical prediction: At low light intensities, a measurable time interval should pass between the instant the light is turned on and the time an electron is ejected from the metal. This time interval is required for the electron to absorb the incident radiation before it acquires enough energy to escape from the metal.
Experimental result: Electrons are emitted from the surface of the metal almost *instantaneously* (less than 10^{-9} s after the surface is illuminated), even at very low light intensities.

3. Dependence of ejection of electrons on light frequency

Classical prediction: Electrons should be ejected from the metal at any incident light frequency, as long as the light intensity is high enough, because energy is transferred to the metal regardless of the incident light frequency.
Experimental result: No electrons are emitted if the incident light frequency falls below some **cutoff frequency** f_c, whose value is characteristic of the material being illuminated. No electrons are ejected below this cutoff frequency *regardless* of the light intensity.

4. Dependence of photoelectron kinetic energy on light frequency

Classical prediction: There should be *no* relationship between the frequency of the light and the electron kinetic energy. The kinetic energy should be related to the intensity of the light.
Experimental result: The maximum kinetic energy of the photoelectrons increases with increasing light frequency.

For these features, experimental results contradict *all four* classical predictions. A successful explanation of the photoelectric effect was given by Einstein in 1905, the same year he published his special theory of relativity. As part of a general paper on electromagnetic radiation, for which he received a Nobel Prize in Physics in 1921, Einstein extended Planck's concept of quantization to electromagnetic waves as mentioned in Section 40.1. Einstein assumed light (or any other electromagnetic wave) of frequency f can be considered a stream of quanta, regardless of the source of the radiation. Today we call these quanta **photons.** Each photon has an energy E given by Equation 40.5, $E = hf$, and each moves in a vacuum at the speed of light c, where $c = 3.00 \times 10^8$ m/s.

Quick Quiz 40.2 While standing outdoors one evening, you are exposed to the following four types of electromagnetic radiation: yellow light from a sodium street lamp, radio waves from an AM radio station, radio waves from an FM radio station, and microwaves from an antenna of a communications system. Rank these types of waves in terms of increasing photon energy, lowest first.

In Einstein's model of the photoelectric effect, a photon of the incident light gives *all* its energy hf to a *single* electron in the metal. Therefore, the absorption of energy by the electrons is not a continuous absorption process as envisioned in the wave model; rather, it is a discontinuous process in which energy is delivered to the electrons in discrete bundles. The energy transfer is accomplished via a one photon–one electron event.[4]

Electrons ejected from the surface of the metal and not making collisions with other metal atoms before escaping possess the maximum kinetic energy K_{max}. According to Einstein, the maximum kinetic energy for these liberated electrons is

Photoelectric effect equation ▶

$$K_{max} = hf - \phi \tag{40.9}$$

where ϕ is called the **work function** of the metal. **The work function represents the minimum energy with which an electron is bound in the metal** and is on the order of a few electron volts. Table 40.1 lists work functions for various metals.

We can understand Equation 40.9 by rearranging it as follows:

$$K_{max} + \phi = hf$$

In this form, Einstein's equation is equivalent to Equation 8.2 applied to the nonisolated system of the electron and the metal. Here, K_{max} is the change ΔK in kinetic energy of the electron, assuming it begins at rest; ϕ is the change ΔU in potential energy of the system, assuming the potential energy is defined to be zero when the electron is within the metal; and hf is the transfer of energy into the system by electromagnetic radiation (T_{ER}).

Using the photon model of light, one can explain the observed features of the photoelectric effect that could not be understood using classical concepts:

1. Dependence of photoelectron kinetic energy on light intensity

Equation 40.9 shows that K_{max} is independent of the light intensity. The maximum kinetic energy of any one electron, which equals $hf - \phi$, depends only on the light frequency and the work function. If the light intensity is doubled, the number of photons arriving per unit time is doubled, which doubles the rate at which photoelectrons are emitted. The maximum kinetic energy of any one photoelectron, however, is unchanged.

2. Time interval between incidence of light and ejection of photoelectrons

Near-instantaneous emission of electrons is consistent with the photon model of light. The incident energy appears in small packets, and there is a one-to-one interaction between photons and electrons. If the incident light has very low intensity, there are very few photons arriving per unit time interval; each photon, however, can have sufficient energy to eject an electron immediately.

3. Dependence of ejection of electrons on light frequency

Because the photon must have energy greater than the work function ϕ to eject an electron, the photoelectric effect cannot be observed below a certain cutoff frequency. If the energy of an incoming photon does not satisfy this requirement, an electron cannot be ejected from the surface, regardless of light intensity.

TABLE 40.1

Work Functions of Selected Metals

Metal	ϕ (eV)
Na	2.46
Al	4.08
Cu	4.70
Zn	4.31
Ag	4.73
Pt	6.35
Pb	4.14
Fe	4.50

Note: Values are typical for metals listed. Actual values may vary depending on whether the metal is a single crystal or polycrystalline. Values may also depend on the face from which electrons are ejected from crystalline metals. Furthermore, different experimental procedures may produce differing values.

[4] In principle, two photons could combine to provide an electron with their combined energy. That is highly improbable, however, without the high intensity of radiation available from very strong lasers.

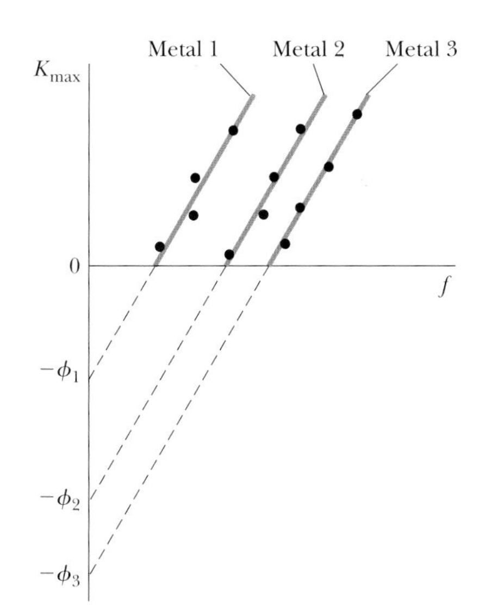

ACTIVE FIGURE 40.11

A plot of K_{max} for photoelectrons versus frequency of incident light in a typical photoelectric effect experiment. Photons with frequency less than the cutoff frequency for a given metal do not have sufficient energy to eject an electron from the metal.

Sign in at www.thomsonedu.com and go to ThomsonNOW to sweep through the frequency range and observe the curve for different target metals.

4. Dependence of photoelectron kinetic energy on light frequency

A photon of higher frequency carries more energy and therefore ejects a higher-energy photoelectron than does a photon of lower frequency.

Einstein's model predicts a linear relationship (Eq. 40.9) between the maximum electron kinetic energy K_{max} and the light frequency f. Experimental observation of a linear relationship between K_{max} and f would be a final confirmation of Einstein's theory. Indeed, such a linear relationship is observed as sketched in Active Figure 40.11, and the slope of the lines in such a plot is Planck's constant h. The intercept on the horizontal axis gives the cutoff frequency below which no photoelectrons are emitted. The cutoff frequency is related to the work function through the relationship $f_c = \phi/h$. The cutoff frequency corresponds to a **cutoff wavelength** λ_c, where

$$\lambda_c = \frac{c}{f_c} = \frac{c}{\phi/h} = \frac{hc}{\phi} \qquad \textbf{(40.10)}$$

◀ Cutoff wavelength

and c is the speed of light. Wavelengths greater than λ_c incident on a material having a work function ϕ do not result in the emission of photoelectrons.

The combination hc in Equation 40.10 often occurs when relating a photon's energy to its wavelength. A common shortcut when solving problems is to express this combination in useful units according to the following approximation:

$$hc = 1\,240 \text{ eV} \cdot \text{nm}$$

One of the first practical uses of the photoelectric effect was as the detector in a camera's light meter. Light reflected from the object to be photographed strikes a photoelectric surface in the meter, causing it to emit photoelectrons that then pass through a sensitive ammeter. The magnitude of the current in the ammeter depends on the light intensity.

The phototube, another early application of the photoelectric effect, acts much like a switch in an electric circuit. It produces a current in the circuit when light of sufficiently high frequency falls on a metal plate in the phototube, but produces no current in the dark. Phototubes were used in burglar alarms and in the detection of the soundtrack on motion picture film. Modern semiconductor devices have now replaced older devices based on the photoelectric effect.

Today, the photoelectric effect is used in the operation of photomultiplier tubes. Figure 40.12 shows the structure of such a device. A photon striking the photocathode ejects an electron by means of the photoelectric effect. This electron accelerates across the potential difference between the photocathode and the first *dynode*, shown as being at +200 V relative to the photocathode in Figure

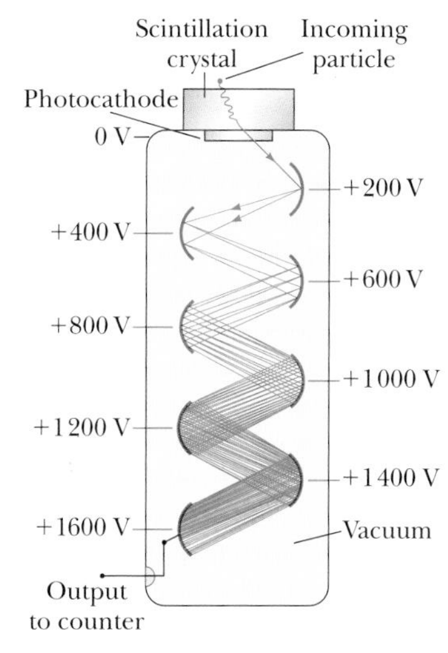

Figure 40.12 The multiplication of electrons in a photomultiplier tube.

40.12. This high-energy electron strikes the dynode and ejects several more electrons. The same process is repeated through a series of dynodes at ever higher potentials until an electrical pulse is produced as millions of electrons strike the last dynode. The tube is therefore called a *multiplier*: one photon at the input has resulted in millions of electrons at the output.

The photomultiplier tube is used in nuclear detectors to detect photons produced by the interaction of energetic charged particles or gamma rays with certain materials. It is also used in astronomy in a technique called *photoelectric photometry*. In that technique, the light collected by a telescope from a single star is allowed to fall on a photomultiplier tube for a time interval. The tube measures the total energy transferred by light during the time interval, which can then be converted to a luminosity of the star.

The photomultiplier tube is being replaced in many astronomical observations with a *charge-coupled device* (CCD), which is the same device used in a digital camera (Section 36.6). In a CCD, an array of pixels is formed on the silicon surface of an integrated circuit (Section 43.7). When the surface is exposed to light from an astronomical scene through a telescope or a terrestrial scene through a digital camera, electrons generated by the photoelectric effect are caught in "traps" beneath the surface. The number of electrons is related to the intensity of the light striking the surface. A signal processor measures the number of electrons associated with each pixel and converts this information into a digital code that a computer can use to reconstruct and display the scene.

The *electron bombardment CCD camera* allows higher sensitivity than a conventional CCD. In this device, electrons ejected from a photocathode by the photoelectric effect are accelerated through a high voltage before striking a CCD array. The higher energy of the electrons results in a very sensitive detector of low-intensity radiation.

Quick Quiz 40.3 Consider one of the curves in Active Figure 40.10. Suppose the intensity of the incident light is held fixed but its frequency is increased. Does the stopping potential in Active Figure 40.10 (a) remain fixed, (b) move to the right, or (c) move to the left?

Quick Quiz 40.4 Suppose classical physicists had the idea of plotting K_{max} versus f as in Active Figure 40.11. Draw a graph of what the expected plot would look like, based on the wave model for light.

EXAMPLE 40.3 The Photoelectric Effect for Sodium

A sodium surface is illuminated with light having a wavelength of 300 nm. The work function for sodium metal is 2.46 eV.

(A) Find the maximum kinetic energy of the ejected photoelectrons.

SOLUTION

Conceptualize Imagine a photon striking the metal surface and ejecting an electron. The electron with the maximum energy is one near the surface that experiences no interactions with other particles in the metal that would reduce its energy on its way out of the metal.

Categorize We evaluate the results using equations developed in this section, so we categorize this example as a substitution problem.

Find the energy of each photon in the illuminating light beam from Equation 40.5:

$$E = hf = \frac{hc}{\lambda} = \frac{1\,240 \text{ eV}\cdot\text{nm}}{300 \text{ nm}} = 4.13 \text{ eV}$$

From Equation 40.9, find the maximum kinetic energy of an electron:

$$K_{max} = hf - \phi = 4.13 \text{ eV} - 2.46 \text{ V} = 1.67 \text{ eV}$$

(B) Find the cutoff wavelength λ_c for sodium.

SOLUTION

Calculate λ_c using Equation 40.10:

$$\lambda_c = \frac{hc}{\phi} = \frac{1\,240\text{ eV}\cdot\text{nm}}{2.46\text{ eV}} = 504\text{ nm}$$

40.3 The Compton Effect

In 1919, Einstein concluded that a photon of energy E travels in a single direction and carries a momentum equal to $E/c = hf/c$. In 1923, Arthur Holly Compton (1892–1962) and Peter Debye (1884–1966) independently carried Einstein's idea of photon momentum further.

Prior to 1922, Compton and his coworkers had accumulated evidence showing that the classical wave theory of light failed to explain the scattering of x-rays from electrons. According to classical theory, electromagnetic waves of frequency f_0 incident on electrons should have two effects: (1) radiation pressure (see Section 34.5) should cause the electrons to accelerate in the direction of propagation of the waves, and (2) the oscillating electric field of the incident radiation should set the electrons into oscillation at the apparent frequency f', where f' is the frequency in the frame of the moving electrons. This apparent frequency is different from the frequency f_0 of the incident radiation because of the Doppler effect (see Section 17.4). Each electron first absorbs radiation as a moving particle and then reradiates as a moving particle, thereby exhibiting two Doppler shifts in the frequency of radiation.

Because different electrons move at different speeds after the interaction, depending on the amount of energy absorbed from the electromagnetic waves, the scattered wave frequency at a given angle to the incoming radiation should show a distribution of Doppler-shifted values. Contrary to this prediction, Compton's experiments showed that at a given angle only *one* frequency of radiation is observed. Compton and his coworkers explained these experiments by treating photons not as waves but rather as point-like particles having energy hf and momentum hf/c and by assuming the energy and momentum of the isolated system of the colliding photon–electron pair are conserved. Compton adopted a particle model for something that was well known as a wave, and today this scattering phenomenon is known as the **Compton effect.** Figure 40.13 shows the quantum picture of the collision between an individual x-ray photon and an electron. In the quantum model, the electron is scattered through an angle ϕ with respect to this direction as in a billiard-ball type of collision. (The symbol ϕ used here is an angle and is not to be confused with the work function, which was discussed in the preceding section.)

Figure 40.14 is a schematic diagram of the apparatus used by Compton. The x-rays, scattered from a graphite target, were analyzed with a rotating crystal spectrometer, and the intensity was measured with an ionization chamber that generated

ARTHUR HOLLY COMPTON
American Physicist (1892–1962)
Compton was born in Wooster, Ohio, and attended Wooster College and Princeton University. He became the director of the laboratory at the University of Chicago, where experimental work concerned with sustained nuclear chain reactions was conducted. This work was of central importance to the construction of the first nuclear weapon. His discovery of the Compton effect led to his sharing of the 1927 Nobel Prize in Physics with Charles Wilson.

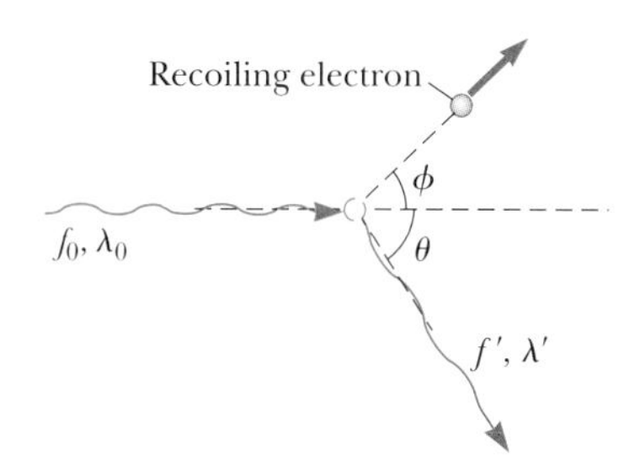

Figure 40.13 The quantum model for x-ray scattering from an electron. The collision of the photon with the electron displays the particle-like nature of the photon.

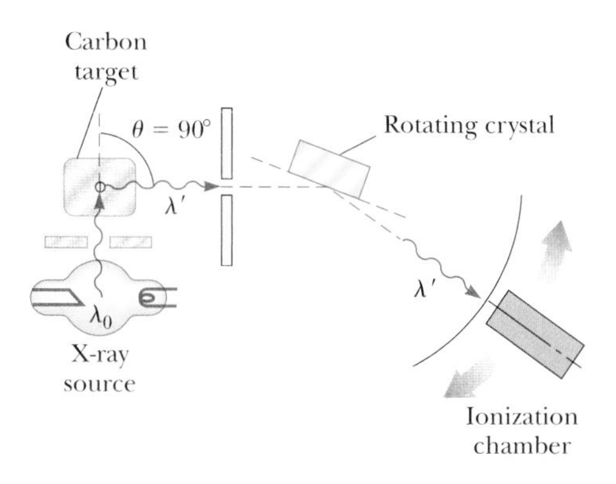

Figure 40.14 Schematic diagram of Compton's apparatus. The wavelength was measured with a rotating crystal spectrometer using graphite (carbon) as the target.

a current proportional to the intensity. The incident beam consisted of monochromatic x-rays of wavelength $\lambda_0 = 0.071$ nm. The experimental intensity-versus-wavelength plots observed by Compton for four scattering angles (corresponding to θ in Fig. 40.13) are shown in Figure 40.15. The graphs for the three nonzero angles show two peaks, one at λ_0 and one at $\lambda' > \lambda_0$. The shifted peak at λ' is caused by the scattering of x-rays from free electrons, which was predicted by Compton to depend on scattering angle as

Compton shift equation ▶

$$\lambda' - \lambda_0 = \frac{h}{m_e c}(1 - \cos\theta) \tag{40.11}$$

where m_e is the mass of the electron. This expression is known as the **Compton shift equation.** The factor $h/m_e c$, called the **Compton wavelength** of the electron, has a currently accepted value of

Compton wavelength ▶

$$\lambda_C = \frac{h}{m_e c} = 0.002\,43 \text{ nm}$$

The unshifted peak at λ_0 in Figure 40.15 is caused by x-rays scattered from electrons tightly bound to the target atoms. This unshifted peak also is predicted by Equation 40.11 if the electron mass is replaced with the mass of a carbon atom, which is approximately 23 000 times the mass of the electron. Therefore, there is a wavelength shift for scattering from an electron bound to an atom, but it is so small that it was undetectable in Compton's experiment.

Compton's measurements were in excellent agreement with the predictions of Equation 40.11. These results were the first to convince many physicists of the fundamental validity of quantum theory.

Quick Quiz 40.5 For any given scattering angle θ, Equation 40.11 gives the same value for the Compton shift for any wavelength. Keeping that in mind, for which of the following types of radiation is the fractional shift in wavelength at a given scattering angle the largest? (a) radio waves (b) microwaves (c) visible light (d) x-rays

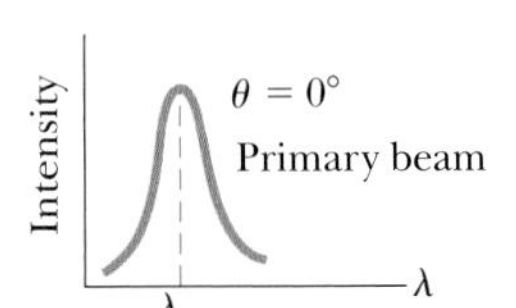

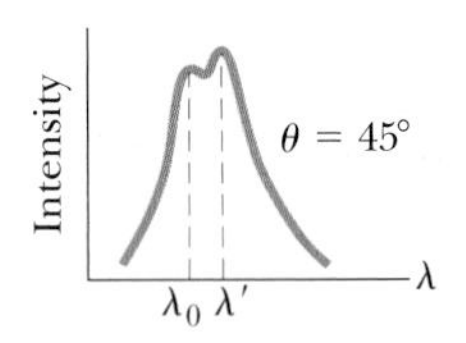

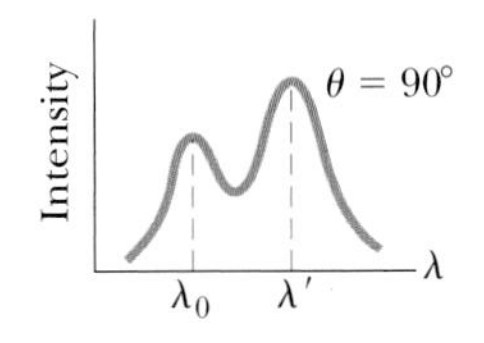

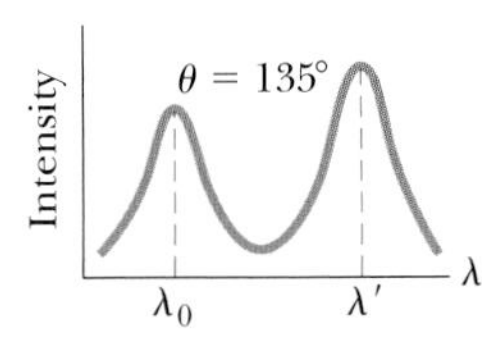

Figure 40.15 Scattered x-ray intensity versus wavelength for Compton scattering at $\theta = 0°, 45°, 90°$, and $135°$.

Derivation of the Compton Shift Equation

We can derive the Compton shift equation by assuming the photon behaves like a particle and collides elastically with a free electron initially at rest as shown in Figure 40.13. The photon is treated as a particle having energy $E = hf = hc/\lambda$ and zero rest energy. We apply the isolated system model to the photon and the electron. In the scattering process, the total energy and total linear momentum of the system are conserved. Applying the principle of conservation of energy to this process gives

$$\frac{hc}{\lambda_0} = \frac{hc}{\lambda'} + K_e$$

where hc/λ_0 is the energy of the incident photon, hc/λ' is the energy of the scattered photon, and K_e is the kinetic energy of the recoiling electron. Because the electron may recoil at a speed comparable to that of light, we must use the relativistic expression $K_e = (\gamma - 1)m_e c^2$ (Eq. 39.23). Therefore,

$$\frac{hc}{\lambda_0} = \frac{hc}{\lambda'} + (\gamma - 1)m_e c^2 \tag{40.12}$$

where $\gamma = 1/\sqrt{1 - (u^2/c^2)}$ and u is the speed of the electron.

Next, let's apply the law of conservation of momentum to this collision, noting that the x and y components of momentum are each conserved independently. Equation 39.28 shows that the momentum of a photon has a magnitude $p = E/c$, and we know from Equation 40.5 that $E = hf$. Therefore, $p = hf/c$. Substituting λf for c (Eq. 16.12) in this expression gives $p = h/\lambda$. Because the relativistic expression for the momentum of the recoiling electron is $p_e = \gamma m_e u$ (Eq. 39.19), we obtain the following expressions for the x and y components of linear momentum, where the angles are as described in Figure 40.13:

$$x \text{ component:} \quad \frac{h}{\lambda_0} = \frac{h}{\lambda'} \cos\theta + \gamma m_e u \cos\phi \qquad \textbf{(40.13)}$$

$$y \text{ component:} \quad 0 = \frac{h}{\lambda'} \sin\theta - \gamma m_e u \sin\phi \qquad \textbf{(40.14)}$$

Eliminating u and ϕ from Equations 40.12 through 40.14 gives a single expression that relates the remaining three variables (λ', λ_0, and θ). After some algebra (see Problem 59), we obtain Equation 40.11.

EXAMPLE 40.4 **Compton Scattering at 45°**

X-rays of wavelength $\lambda_0 = 0.200\ 000$ nm are scattered from a block of material. The scattered x-rays are observed at an angle of 45.0° to the incident beam. Calculate their wavelength.

SOLUTION

Conceptualize Imagine the process in Figure 40.13, with the photon scattered at 45° to its original direction.

Categorize We evaluate the result using an equation developed in this section, so we categorize this example as a substitution problem.

Solve Equation 40.11 for the wavelength of the scattered x-ray:

$$(1) \quad \lambda' = \lambda_0 + \frac{h(1 - \cos\theta)}{m_e c}$$

Substitute numerical values:

$$\lambda' = 0.200\ 000 \times 10^{-9}\ \text{m} + \frac{(6.626 \times 10^{-34}\ \text{J}\cdot\text{s})(1 - \cos 45.0^\circ)}{(9.11 \times 10^{-31}\ \text{kg})(3.00 \times 10^{8}\ \text{m/s})}$$

$$= 0.200\ 000 \times 10^{-9}\ \text{m} + 7.10 \times 10^{-13}\ \text{m} = 0.200\ 710\ \text{nm}$$

What If? What if the detector is moved so that scattered x-rays are detected at an angle larger than 45°? Does the wavelength of the scattered x-rays increase or decrease as the angle θ increases?

Answer In Equation (1), if the angle θ increases, $\cos\theta$ decreases. Consequently, the factor $(1 - \cos\theta)$ increases. Therefore, the scattered wavelength increases.

We could also apply an energy argument to achieve this same result. As the scattering angle increases, more energy is transferred from the incident photon to the electron. As a result, the energy of the scattered photon decreases with increasing scattering angle. Because $E = hf$, the frequency of the scattered photon decreases, and because $\lambda = c/f$, the wavelength increases.

40.4 Photons and Electromagnetic Waves

Phenomena such as the photoelectric effect and the Compton effect offer ironclad evidence that when light (or other forms of electromagnetic radiation) and matter interact, the light behaves as if it were composed of particles having energy hf and momentum h/λ. How can light be considered a photon (in other words, a particle) when we know it is a wave? On the one hand, we describe light in terms of photons having energy and momentum. On the other hand, light and other electromagnetic waves exhibit interference and diffraction effects, which are consistent only with a wave interpretation.

Which model is correct? Is light a wave or a particle? The answer depends on the phenomenon being observed. Some experiments can be explained either better or solely with the photon model, whereas others are explained either better or solely with the wave model. The end result is that **we must accept both models and admit that the true nature of light is not describable in terms of any single classical picture.** The same light beam that can eject photoelectrons from a metal (meaning that the beam consists of photons) can also be diffracted by a grating (meaning that the beam is a wave). In other words, **the particle model and the wave model of light complement each other.**

The success of the particle model of light in explaining the photoelectric effect and the Compton effect raises many other questions. If light is a particle, what is the meaning of the "frequency" and "wavelength" of the particle, and which of these two properties determines its energy and momentum? Is light *simultaneously* a wave and a particle? Although photons have no rest energy (a nonobservable quantity because a photon cannot be at rest), is there a simple expression for the *effective mass* of a moving photon? If photons have effective mass, do they experience gravitational attraction? What is the spatial extent of a photon, and how does an electron absorb or scatter one photon? Some of these questions can be answered, but others demand a view of atomic processes that is too pictorial and literal. Many of them stem from classical analogies such as colliding billiard balls and ocean waves breaking on a seashore. Quantum mechanics gives light a more flexible nature by treating the particle model and the wave model of light as both necessary and complementary. Neither model can be used exclusively to describe all properties of light. A complete understanding of the observed behavior of light can be attained only if the two models are combined in a complementary manner.

40.5 The Wave Properties of Particles

AIP Niels Bohr Library

LOUIS DE BROGLIE
French Physicist (1892–1987)
De Broglie was born in Dieppe, France. At the Sorbonne in Paris, he studied history in preparation for what he hoped would be a career in the diplomatic service. The world of science is lucky he changed his career path to become a theoretical physicist. De Broglie was awarded the Nobel Prize in Physics in 1929 for his prediction of the wave nature of electrons.

Students introduced to the dual nature of light often find the concept difficult to accept. In the world around us, we are accustomed to regarding such things as baseballs solely as particles and other things such as sound waves solely as forms of wave motion. Every large-scale observation can be interpreted by considering either a wave explanation or a particle explanation, but in the world of photons and electrons, such distinctions are not as sharply drawn. Even more disconcerting is that, under certain conditions, the things we unambiguously call "particles" exhibit wave characteristics.

In his 1923 doctoral dissertation, Louis de Broglie postulated that **because photons have both wave and particle characteristics, perhaps all forms of matter have both properties.** This highly revolutionary idea had no experimental confirmation at the time. According to de Broglie, electrons, just like light, have a dual particle–wave nature.

In Section 40.3, we found that the momentum of a photon can be expressed as

$$p = \frac{h}{\lambda}$$

This equation shows that the photon wavelength can be specified by its momentum: $\lambda = h/p$. De Broglie suggested that material particles of momentum p have a characteristic wavelength that is given by the same expression, $\lambda = h/p$. Because the magnitude of the momentum of a particle of mass m and speed u is $p = mu$, the **de Broglie wavelength** of that particle is[5]

[5] The de Broglie wavelength for a particle moving at *any* speed u is $\lambda = h/\gamma mu$, where $\gamma = [1 - (u^2/c^2)]^{-1/2}$.

$$\lambda = \frac{h}{p} = \frac{h}{mu} \tag{40.15}$$

◀ de Broglie wavelength

Furthermore, in analogy with photons, de Broglie postulated that particles obey the Einstein relation $E = hf$, where E is the total energy of the particle. The frequency of a particle is then

$$f = \frac{E}{h} \tag{40.16}$$

The dual nature of matter is apparent in Equations 40.15 and 40.16 because each contains both particle quantities (p and E) and wave quantities (λ and f).

The Davisson–Germer Experiment

De Broglie's 1923 proposal that matter exhibits both wave and particle properties was regarded as pure speculation. If particles such as electrons had wave properties, under the correct conditions they should exhibit diffraction effects. Only three years later, C. J. Davisson (1881–1958) and L. H. Germer (1896–1971) succeeded in measuring the wavelength of electrons. Their important discovery provided the first experimental confirmation of the matter waves proposed by de Broglie.

Interestingly, the intent of the initial Davisson–Germer experiment was not to confirm the de Broglie hypothesis. In fact, their discovery was made by accident (as is often the case). The experiment involved the scattering of low-energy electrons (approximately 54 eV) from a nickel target in a vacuum. During one experiment, the nickel surface was badly oxidized because of an accidental break in the vacuum system. After the target was heated in a flowing stream of hydrogen to remove the oxide coating, electrons scattered by it exhibited intensity maxima and minima at specific angles. The experimenters finally realized that the nickel had formed large crystalline regions upon heating and that the regularly spaced planes of atoms in these regions served as a diffraction grating for electrons. (See the discussion of diffraction of x-rays by crystals in Section 38.5.)

Shortly thereafter, Davisson and Germer performed more extensive diffraction measurements on electrons scattered from single-crystal targets. Their results showed conclusively the wave nature of electrons and confirmed the de Broglie relationship $p = h/\lambda$. In the same year, G. P. Thomson (1892–1975) of Scotland also observed electron diffraction patterns by passing electrons through very thin gold foils. Diffraction patterns have since been observed in the scattering of helium atoms, hydrogen atoms, and neutrons. Hence, the wave nature of particles has been established in various ways.

The problem of understanding the dual nature of matter and radiation is conceptually difficult because the two models seem to contradict each other. This problem as it applies to light was discussed earlier. The **principle of complementarity** states that the **wave and particle models of either matter or radiation complement each other.** Neither model can be used exclusively to describe matter or radiation adequately. Because humans tend to generate mental images based on their experiences from the everyday world (baseballs, water waves, and so forth), we use both descriptions in a complementary manner to explain any given set of data from the quantum world.

PITFALL PREVENTION 40.3
What's Waving?

If particles have wave properties, what's waving? You are familiar with waves on strings, which are very concrete. Sound waves are more abstract, but you are likely comfortable with them. Electromagnetic waves are even more abstract, but at least they can be described in terms of physical variables and electric and magnetic fields. In contrast, waves associated with particles are completely abstract and cannot be associated with a physical variable. In Chapter 41, we describe the wave associated with a particle in terms of probability.

Quick Quiz 40.6 An electron and a proton both moving at nonrelativistic speeds have the same de Broglie wavelength. Which of the following quantities are also the same for the two particles? (a) speed (b) kinetic energy (c) momentum (d) frequency

---

EXAMPLE 40.5 Wavelengths for Microscopic and Macroscopic Objects

(A) Calculate the de Broglie wavelength for an electron ($m_e = 9.11 \times 10^{-31}$ kg) moving at 1.00×10^7 m/s.

SOLUTION

Conceptualize Imagine the electron moving through space. From a classical viewpoint, it is a particle under constant velocity. From the quantum viewpoint, the electron has a wavelength associated with it.

Categorize We evaluate the result using an equation developed in this section, so we categorize this example as a substitution problem.

Evaluate the wavelength using Equation 40.15:

$$\lambda = \frac{h}{m_e u} = \frac{6.63 \times 10^{-34}\ \text{J}\cdot\text{s}}{(9.11 \times 10^{-31}\ \text{kg})(1.00 \times 10^7\ \text{m/s})} = 7.28 \times 10^{-11}\ \text{m}$$

(B) A rock of mass 50 g is thrown with a speed of 40 m/s. What is its de Broglie wavelength?

SOLUTION

Evaluate the de Broglie wavelength using Equation 40.15:

$$\lambda = \frac{h}{mu} = \frac{6.63 \times 10^{-34}\ \text{J}\cdot\text{s}}{(50 \times 10^{-3}\ \text{kg})(40\ \text{m/s})} = 3.32 \times 10^{-34}\ \text{m}$$

This wavelength is much smaller than any aperture through which the rock could possibly pass. Hence, we could not observe diffraction effects, and as a result, the wave properties of large-scale objects cannot be observed.

The Electron Microscope

A practical device that relies on the wave characteristics of electrons is the **electron microscope.** A *transmission* electron microscope, used for viewing flat, thin samples, is shown in Figure 40.16. In many respects, it is similar to an optical micro-

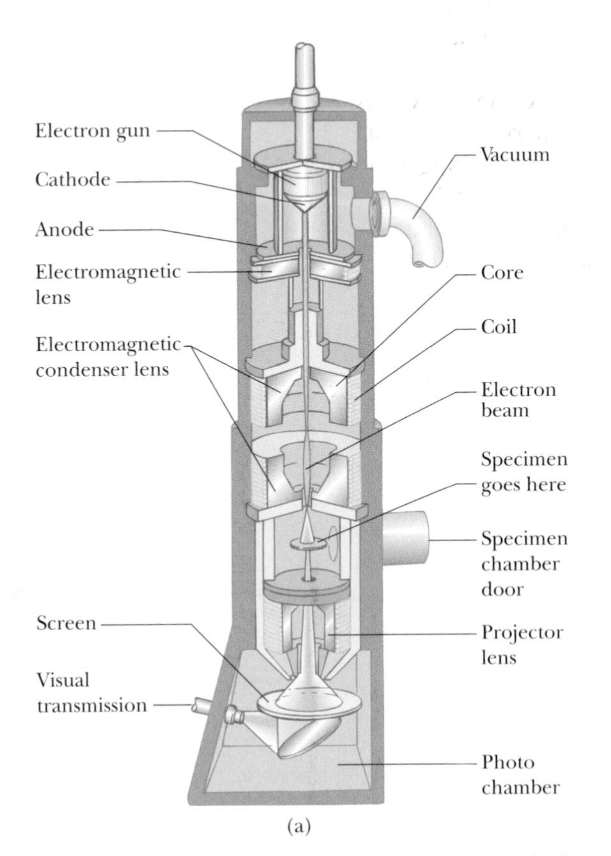

(a)

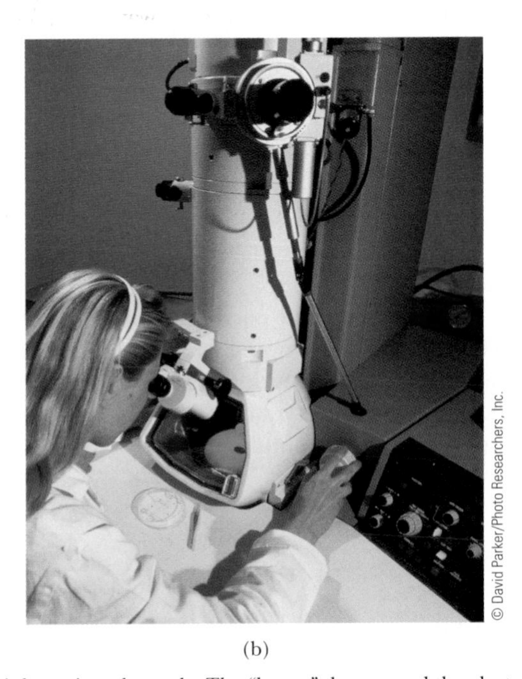

(b)

Figure 40.16 (a) Diagram of a transmission electron microscope for viewing a thinly sectioned sample. The "lenses" that control the electron beam are magnetic deflection coils. (b) An electron microscope.

scope; the electron microscope, however, has a much greater resolving power because it can accelerate electrons to very high kinetic energies, giving them very short wavelengths. No microscope can resolve details that are significantly smaller than the wavelength of the waves used to illuminate the object. Typically, the wavelengths of electrons are approximately 100 times shorter than those of the visible light used in optical microscopes. As a result, an electron microscope with ideal lenses would be able to distinguish details approximately 100 times smaller than those distinguished by an optical microscope. (Electromagnetic radiation of the same wavelength as the electrons in an electron microscope is in the x-ray region of the spectrum.)

The electron beam in an electron microscope is controlled by electrostatic or magnetic deflection, which acts on the electrons to focus the beam and form an image. Rather than examining the image through an eyepiece as in an optical microscope, the viewer looks at an image formed on a monitor or other type of display screen. Figure 40.17 shows the amazing detail available with an electron microscope.

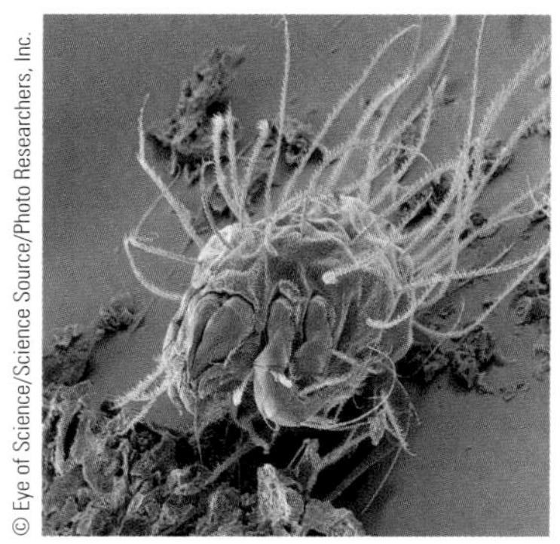

Figure 40.17 A color-enhanced electron microscope photograph shows significant detail of a storage mite, *Lepidoglyphus destructor*. The mite is so small, with a maximum length of 0.75 mm, that ordinary microscopes do not reveal minute anatomical details.

40.6 The Quantum Particle

Because in the past we considered the particle and wave models to be distinct, the discussions presented in previous sections may be quite disturbing. The notion that both light and material particles have both particle and wave properties does not fit with this distinction. Experimental evidence shows, however, that this conclusion is exactly what we must accept. The recognition of this dual nature leads to a new model, the **quantum particle,** which is a combination of the particle model introduced in Chapter 2 and the wave model discussed in Chapter 16. In this new model, entities have both particle and wave characteristics, and we must choose one appropriate behavior—particle or wave—to understand a particular phenomenon.

In this section, we shall explore this model in a way that might make you more comfortable with this idea. We shall do so by demonstrating that an entity that exhibits properties of a particle can be constructed from waves.

Let's first recall some characteristics of ideal particles and ideal waves. An ideal particle has zero size. Therefore, an essential feature of a particle is that it is *localized* in space. An ideal wave has a single frequency and is infinitely long as suggested by Figure 40.18a. Therefore, an ideal wave is *unlocalized* in space. A localized entity can be built from infinitely long waves as follows. Imagine drawing one wave along the x axis, with one of its crests located at $x = 0$, as at the top of Figure 40.18b. Now draw a second wave, of the same amplitude but a different frequency, with one of its crests also at $x = 0$. As a result of the superposition of these two waves, *beats* exist as the waves are alternately in phase and out of phase. (Beats were discussed in Section 18.7.) The bottom curve in Figure 40.18b shows the results of superposing these two waves.

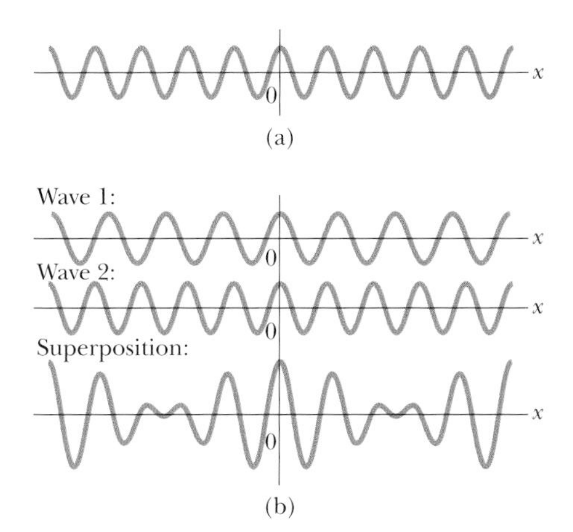

Figure 40.18 (a) An idealized wave of an exact single frequency is the same throughout space and time. (b) If two ideal waves with slightly different frequencies are combined, beats result (Section 18.7). The regions of space at which there is constructive interference are different from those at which there is destructive interference.

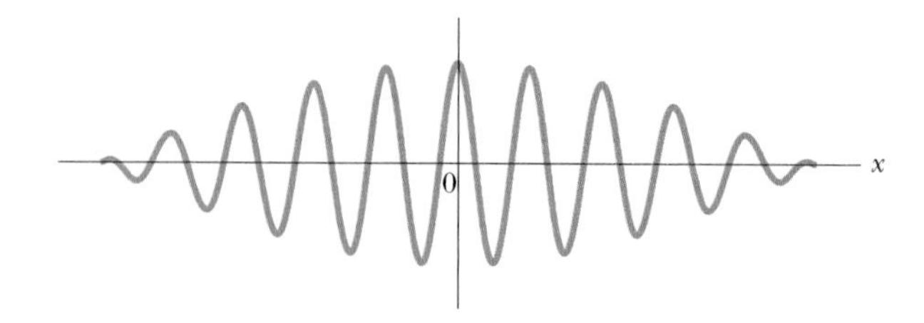

ACTIVE FIGURE 40.19

If a large number of waves are combined, the result is a wave packet, which represents a particle.

Sign in at www.thomsonedu.com and go to ThomsonNOW to choose the number of waves to add together and observe the resulting wave packet.

Notice that we have already introduced some localization by superposing the two waves. A single wave has the same amplitude everywhere in space; no point in space is any different from any other point. By adding a second wave, however, there is something different about the in-phase points compared with the out-of-phase points.

Now imagine that more and more waves are added to our original two, each new wave having a new frequency. Each new wave is added so that one of its crests is at $x = 0$ with the result that all the waves add constructively at $x = 0$. When we add a large number of waves, the probability of a positive value of a wave function at any point $x \neq 0$ is equal to the probability of a negative value, and there is destructive interference *everywhere* except near $x = 0$, where all of the crests are superposed. The result is shown in Active Figure 40.19. The small region of constructive interference is called a **wave packet.** This localized region of space is different from all other regions. We can identify the wave packet as a particle because it has the localized nature of a particle! The location of the wave packet corresponds to the particle's position.

The localized nature of this entity is the *only* characteristic of a particle that was generated with this process. We have not addressed how the wave packet might achieve such particle characteristics as mass, electric charge, and spin. Therefore, you may not be completely convinced that we have built a particle. As further evidence that the wave packet can represent the particle, let's show that the wave packet has another characteristic of a particle.

To simplify the mathematical representation, we return to our combination of two waves. Consider two waves with equal amplitudes but different frequencies f_1 and f_2. We can represent the waves mathematically as

$$y_1 = A\cos(k_1x - \omega_1t) \quad \text{and} \quad y_2 = A\cos(k_2x - \omega_2t)$$

where, as in Chapter 16, $k = 2\pi/\lambda$ and $\omega = 2\pi f$. Using the superposition principle, let's add the waves:

$$y = y_1 + y_2 = A\cos(k_1x - \omega_1t) + A\cos(k_2x - \omega_2t)$$

It is convenient to write this expression in a form that uses the trigonometric identity

$$\cos a + \cos b = 2\cos\left(\frac{a-b}{2}\right)\cos\left(\frac{a+b}{2}\right)$$

Letting $a = k_1x - \omega_1t$ and $b = k_2x - \omega_2t$ gives

$$y = 2A\cos\left[\frac{(k_1x - \omega_1t) - (k_2x - \omega_2t)}{2}\right]\cos\left[\frac{(k_1x - \omega_1t) + (k_2x - \omega_2t)}{2}\right]$$

$$y = \left[2A\cos\left(\frac{\Delta k}{2}x - \frac{\Delta\omega}{2}t\right)\right]\cos\left(\frac{k_1 + k_2}{2}x - \frac{\omega_1 + \omega_2}{2}t\right) \quad \textbf{(40.17)}$$

where $\Delta k = k_1 - k_2$ and $\Delta\omega = \omega_1 - \omega_2$. The second cosine factor represents a wave with a wave number and frequency that are equal to the averages of the values for the individual waves.

In Equation 40.17, the factor in square brackets represents the envelope of the wave as shown by the blue curve in Active Figure 40.20. This factor also has the mathematical form of a wave. **This envelope of the combination can travel through space with a different speed than the individual waves.** As an extreme example of

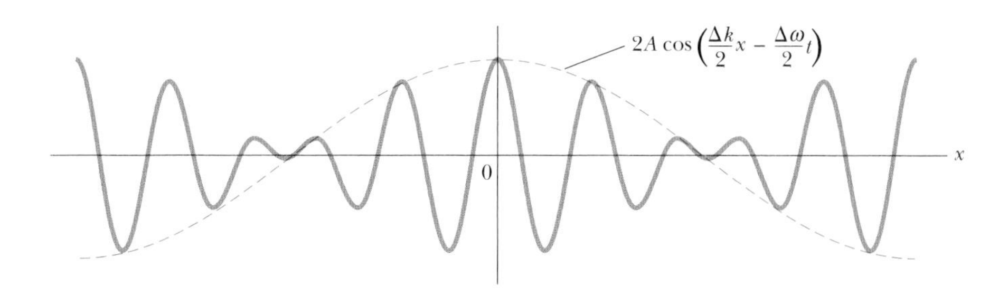

ACTIVE FIGURE 40.20

The beat pattern of Figure 40.18b, with an envelope function (blue curve) superimposed.

Sign in at www.thomsonedu.com and go to ThomsonNOW to observe the movement of the waves and the envelope.

this possibility, imagine combining two identical waves moving in opposite directions. The two waves move with the same speed, but the envelope has a speed of *zero* because we have built a standing wave, which we studied in Section 18.2.

For an individual wave, the speed is given by Equation 16.11,

$$v_{\text{phase}} = \frac{\omega}{k} \qquad \textbf{(40.18)}$$

◀ Phase speed of a wave in a wave packet

This speed is called the **phase speed** because it is the rate of advance of a crest on a single wave, which is a point of fixed phase. Equation 40.18 can be interpreted as follows: the phase speed of a wave is the ratio of the coefficient of the time variable t to the coefficient of the space variable x in the equation for the wave, $y = A\cos(kx - \omega t)$.

The factor in brackets in Equation 40.17 is of the form of a wave, so it moves with a speed given by this same ratio:

$$v_g = \frac{\text{coefficient of time variable } t}{\text{coefficient of space variable } x} = \frac{(\Delta\omega/2)}{(\Delta k/2)} = \frac{\Delta\omega}{\Delta k}$$

The subscript g on the speed indicates that it is commonly called the **group speed,** or the speed of the wave packet (the *group* of waves) we have built. We have generated this expression for a simple addition of two waves. When a large number of waves are superposed to form a wave packet, this ratio becomes a derivative:

$$v_g = \frac{d\omega}{dk} \qquad \textbf{(40.19)}$$

◀ Group speed of a wave packet

Multiplying the numerator and the denominator by $\hbar$, where $\hbar = h/2\pi$, gives

$$v_g = \frac{\hbar\, d\omega}{\hbar\, dk} = \frac{d(\hbar\omega)}{d(\hbar k)} \qquad \textbf{(40.20)}$$

Let's look at the terms in the parentheses of Equation 40.20 separately. For the numerator,

$$\hbar\omega = \frac{h}{2\pi}(2\pi f) = hf = E$$

For the denominator,

$$\hbar k = \frac{h}{2\pi}\left(\frac{2\pi}{\lambda}\right) = \frac{h}{\lambda} = p$$

Therefore, Equation 40.20 can be written as

$$v_g = \frac{d(\hbar\omega)}{d(\hbar k)} = \frac{dE}{dp} \qquad \textbf{(40.21)}$$

Because we are exploring the possibility that the envelope of the combined waves represents the particle, consider a free particle moving with a speed u that is small compared with the speed of light. The energy of the particle is its kinetic energy:

$$E = \tfrac{1}{2}mu^2 = \frac{p^2}{2m}$$

Differentiating this equation with respect to p gives

$$v_g = \frac{dE}{dp} = \frac{d}{dp}\left(\frac{p^2}{2m}\right) = \frac{1}{2m}(2p) = u \tag{40.22}$$

Therefore, the group speed of the wave packet is identical to the speed of the particle that it is modeled to represent, giving us further confidence that the wave packet is a reasonable way to build a particle.

Quick Quiz 40.7 As an analogy to wave packets, consider an "automobile packet" that occurs near the scene of an accident on a freeway. The phase speed is analogous to the speed of individual automobiles as they move through the backup caused by the accident. The group speed can be identified as the speed of the leading edge of the packet of cars. For the automobile packet, is the group speed (a) the same as the phase speed, (b) less than the phase speed, or (c) greater than the phase speed?

40.7 The Double-Slit Experiment Revisited

Wave–particle duality is now a firmly accepted concept reinforced by experimental results, including those of the Davisson–Germer experiment. As with the postulates of special relativity, however, this concept often leads to clashes with familiar thought patterns we hold from everyday experience.

One way to crystallize our ideas about the electron's wave–particle duality is through an experiment in which electrons are fired at a double slit. Consider a parallel beam of mono-energetic electrons incident on a double slit as in Figure 40.21. Let's assume the slit widths are small compared with the electron wavelength so that we need not worry about diffraction maxima and minima as discussed for light in Section 38.2. An electron detector screen is positioned far from the slits at a distance much greater than d, the separation distance of the slits. If the detector screen collects electrons for a long enough time, we find a typical wave interference pattern for the counts per minute, or probability of arrival of electrons. Such an interference pattern would not be expected if the electrons behaved as classical particles, giving clear evidence that electrons are interfering, a distinct wave-like behavior.

If we measure the angles θ at which the maximum intensity of electrons arrives at the detector screen in Figure 40.21, we find they are described by exactly the same equation (Eq. 37.2) as that for light, $d \sin \theta = m\lambda$, where m is the order number and λ is the electron wavelength. Therefore, the dual nature of the electron is

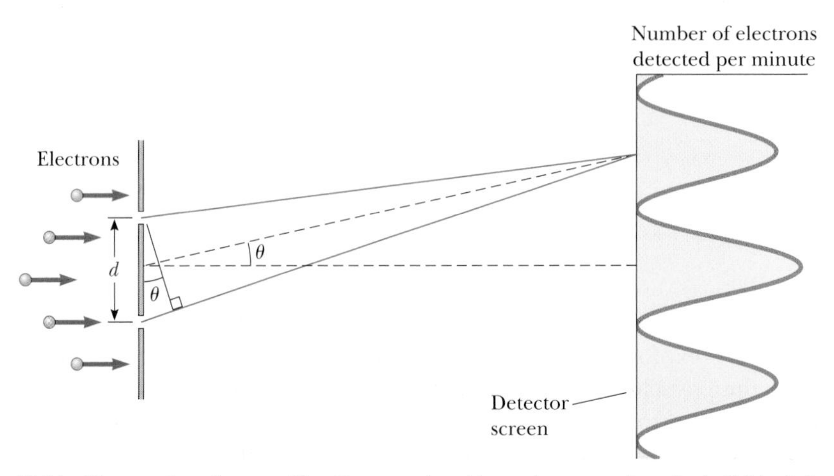

Figure 40.21 Electron interference. The slit separation d is much greater than the individual slit widths and much less than the distance between the slit and the detector screen.

clearly shown in this experiment: **the electrons are detected as particles at a localized spot on the detector screen at some instant of time, but the probability of arrival at that spot is determined by finding the intensity of two interfering waves.**

At extremely low electron beam intensities, one electron at a time arrives at the double slit. It is tempting to assume the electron goes through either slit 1 or slit 2. You might argue that there are no interference effects because there is not a second electron going through the other slit to interfere with the first. This assumption places too much emphasis on the particle model of the electron, however. The interference pattern is still observed if the time interval for the measurement is sufficiently long for many electrons to arrive at the detector screen! This situation is illustrated by the computer-simulated patterns in Active Figure 40.22 where the interference pattern becomes clearer as the number of electrons reaching the detector screen increases. Hence, our assumption that the electron is localized and goes through only one slit when both slits are open must be wrong (a painful conclusion!).

To interpret these results, we are forced to conclude that **an electron interacts with both slits simultaneously.** If you try to determine experimentally which slit the electron goes through, the act of measuring destroys the interference pattern. It is impossible to determine which slit the electron goes through. In effect, we can say only that **the electron passes through both slits!** The same arguments apply to photons.

If we restrict ourselves to a pure particle model, it is an uncomfortable notion that the electron can be present at both slits at once. From the quantum particle model, however, the particle can be considered to be built from waves that exist throughout space. Therefore, the wave components of the electron are present at both slits at the same time, and this model leads to a more comfortable interpretation of this experiment.

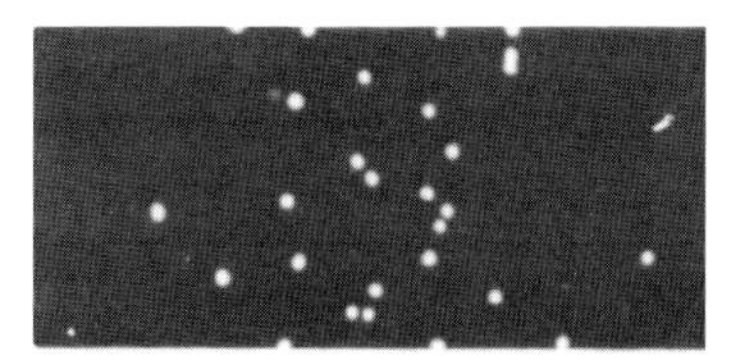

(a) After 28 electrons

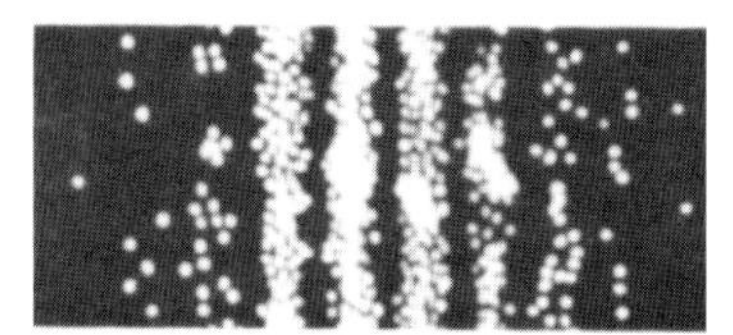

(b) After 1 000 electrons

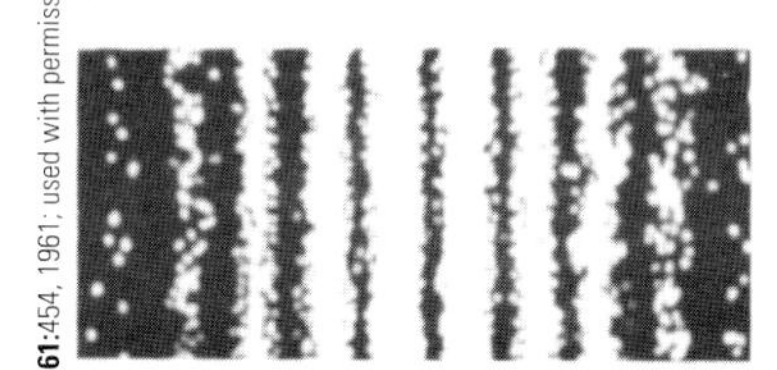

(c) After 10 000 electrons

(d) Two-slit electron pattern

ACTIVE FIGURE 40.22

(a)–(c) Computer-simulated interference patterns for a beam of electrons incident on a double slit. (d) Photograph of a double-slit interference pattern produced by electrons.

Sign in at www.thomsonedu.com and go to ThomsonNOW to watch the interference pattern develop over time and see how it is destroyed by the action of keeping track of which slit an electron goes through.

40.8 The Uncertainty Principle

Whenever one measures the position or velocity of a particle at any instant, experimental uncertainties are built into the measurements. According to classical mechanics, there is no fundamental barrier to an ultimate refinement of the apparatus or experimental procedures. In other words, it is possible, in principle, to make such measurements with arbitrarily small uncertainty. Quantum theory predicts, however, that **it is fundamentally impossible to make simultaneous measurements of a particle's position and momentum with infinite accuracy.**

In 1927, Werner Heisenberg (1901–1976) introduced this notion, which is now known as the **Heisenberg uncertainty principle:**

If a measurement of the position of a particle is made with uncertainty Δx and a simultaneous measurement of its x component of momentum is made with uncertainty Δp_x, the product of the two uncertainties can never be smaller than $\hbar/2$:

$$\Delta x \, \Delta p_x \geq \frac{\hbar}{2} \qquad \textbf{(40.23)}$$

◀ Heisenberg uncertainty principle

That is, **it is physically impossible to measure simultaneously the exact position and exact momentum of a particle.** Heisenberg was careful to point out that the inescapable uncertainties Δx and Δp_x do not arise from imperfections in practical measuring instruments. Rather, **the uncertainties arise from the quantum structure of matter.**

To understand the uncertainty principle, imagine that a particle has a single wavelength that is known *exactly*. According to the de Broglie relation, $\lambda = h/p$, we would therefore know the momentum to be precisely $p = h/\lambda$. In reality, a

WERNER HEISENBERG
German Theoretical Physicist (1901–1976)
Heisenberg obtained his Ph.D. in 1923 at the University of Munich. While other physicists tried to develop physical models of quantum phenomena, Heisenberg developed an abstract mathematical model called *matrix mechanics*. The more widely accepted physical models were shown to be equivalent to matrix mechanics. Heisenberg made many other significant contributions to physics, including his famous uncertainty principle for which he received a Nobel Prize in Physics in 1932, the prediction of two forms of molecular hydrogen, and theoretical models of the nucleus.

PITFALL PREVENTION 40.4
The Uncertainty Principle

Some students incorrectly interpret the uncertainty principle as meaning that a measurement interferes with the system. For example, if an electron is observed in a hypothetical experiment using an optical microscope, the photon used to see the electron collides with it and makes it move, giving it an uncertainty in momentum. This scenario does *not* represent the basis of the uncertainty principle. The uncertainty principle is independent of the measurement process and is based on the wave nature of matter.

single-wavelength wave would exist throughout space. Any region along this wave is the same as any other region (Fig. 40.18a). If we were to ask, Where is the particle this wave represents?, no special location in space along the wave could be identified with the particle; all points along the wave are the same. Therefore, we have *infinite* uncertainty in the position of the particle, and we know nothing about its location. Perfect knowledge of the particle's momentum has cost us all information about its location.

In comparison, now consider a particle whose momentum is uncertain so that it has a range of possible values of momentum. According to the de Broglie relation, the result is a range of wavelengths. Therefore, the particle is not represented by a single wavelength, but rather by a combination of wavelengths within this range. This combination forms a wave packet as we discussed in Section 40.6 and illustrated in Active Figure 40.19. If you were asked to determine the location of the particle, you could only say that it is somewhere in the region defined by the wave packet because there is a distinct difference between this region and the rest of space. Therefore, by losing some information about the momentum of the particle, we have gained information about its position.

If you were to lose *all* information about the momentum, you would be adding together waves of all possible wavelengths, resulting in a wave packet of zero length. Therefore, if you know nothing about the momentum, you know exactly where the particle is.

The mathematical form of the uncertainty principle states that the product of the uncertainties in position and momentum is always larger than some minimum value. This value can be calculated from the types of arguments discussed above, which result in the value of $\hbar/2$ in Equation 40.23.

Another form of the uncertainty principle can be generated by reconsidering Active Figure 40.19. Imagine that the horizontal axis is time rather than spatial position x. We can then make the same arguments that were made about knowledge of wavelength and position in the time domain. The corresponding variables would be frequency and time. Because frequency is related to the energy of the particle by $E = hf$, the uncertainty principle in this form is

$$\Delta E \, \Delta t \geq \frac{\hbar}{2} \qquad \textbf{(40.24)}$$

The form of the uncertainty principle given in Equation 40.24 suggests that energy conservation can appear to be violated by an amount ΔE as long as it is only for a short time interval Δt consistent with that equation. We shall use this notion to estimate the rest energies of particles in Chapter 46.

Quick Quiz 40.8 A particle's location is measured and specified as being exactly at $x = 0$, with *zero* uncertainty in the x direction. How does that location affect the uncertainty of its velocity component in the y direction? (a) It does not affect it. (b) It makes it infinite. (c) It makes it zero.

EXAMPLE 40.6 Locating an Electron

The speed of an electron is measured to be 5.00×10^3 m/s to an accuracy of 0.003 00%. Find the minimum uncertainty in determining the position of this electron.

SOLUTION

Conceptualize The fractional value given for the accuracy of the electron's speed can be interpreted as the fractional uncertainty in its momentum. This uncertainty corresponds to a minimum uncertainty in the electron's position through the uncertainty principle.

Categorize We evaluate the result using concepts developed in this section, so we categorize this example as a substitution problem.

Assume the electron is moving along the x axis and find the x component of its momentum:

$$p_x = mu_x = (9.11 \times 10^{-31}\ \text{kg})(5.00 \times 10^3\ \text{m/s}) = 4.56 \times 10^{-27}\ \text{kg} \cdot \text{m/s}$$

Find the uncertainty in p_x as 0.003 00% of this value:

$$\Delta p_x = (0.000\ 030\ 0)(4.56 \times 10^{-27}\ \text{kg} \cdot \text{m/s}) = 1.37 \times 10^{-31}\ \text{kg} \cdot \text{m/s}$$

Solve Equation 40.23 for the uncertainty in the electron's position:

$$\Delta x \geq \frac{\hbar}{2\Delta p_x} = \frac{1.055 \times 10^{-34}\ \text{J} \cdot \text{s}}{2(1.37 \times 10^{-31}\ \text{kg} \cdot \text{m/s})} = 0.386\ \text{mm}$$

EXAMPLE 40.7 The Line Width of Atomic Emissions

Atoms have quantized energy levels similar to those of Planck's oscillators, although the energy levels of an atom are usually not evenly spaced. When an atom makes a transition between states, energy is emitted in the form of a photon. Although an excited atom can radiate at any time from $t = 0$ to $t = \infty$, the average time interval after excitation during which an atom radiates is called the **lifetime** τ. If $\tau = 1.0 \times 10^{-8}$ s, use the uncertainty principle to compute the line width Δf produced by this finite lifetime.

SOLUTION

Conceptualize The lifetime τ given for the excited state can be interpreted as the uncertainty Δt in the time at which the transition occurs. This uncertainty corresponds to a minimum uncertainty in the frequency of the radiated photon through the uncertainty principle.

Categorize We evaluate the result using concepts developed in this section, so we categorize this example as a substitution problem.

Use Equation 40.5 to relate the uncertainty in the photon's frequency to the uncertainty in its energy:

$$E = hf \rightarrow \Delta E = h\Delta f \rightarrow \Delta f = \frac{\Delta E}{h}$$

Use Equation 40.24 to substitute for the uncertainty in the photon's energy, giving the minimum value of Δf:

$$\Delta f \geq \frac{1}{h}\frac{\hbar}{2\Delta t} = \frac{1}{h}\frac{(h/2\pi)}{2\Delta t} = \frac{1}{4\pi\Delta t} = \frac{1}{4\pi\tau}$$

Substitute for the lifetime of the excited state:

$$\Delta f \geq \frac{1}{4\pi(1.0 \times 10^{-8}\ \text{s})} = 8.0 \times 10^6\ \text{Hz}$$

What If? What if this same lifetime were associated with a transition that emits a radio wave rather than a visible light wave from an atom? Is the fractional line width $\Delta f/f$ larger or smaller than for the visible light?

Answer Because we are assuming the same lifetime for both transitions, Δf is independent of the frequency of radiation. Radio waves have lower frequencies than light waves, so the ratio $\Delta f/f$ will be larger for the radio waves. Assuming a light-wave frequency f of 6.00×10^{14} Hz, the fractional line width is

$$\frac{\Delta f}{f} = \frac{8.0 \times 10^6\ \text{Hz}}{6.00 \times 10^{14}\ \text{Hz}} = 1.3 \times 10^{-8}$$

This narrow fractional line width can be measured with a sensitive interferometer. Usually, however, temperature and pressure effects overshadow the natural line width and broaden the line through mechanisms associated with the Doppler effect and collisions.

Assuming a radio-wave frequency f of 94.7×10^6 Hz, the fractional line width is

$$\frac{\Delta f}{f} = \frac{8.0 \times 10^6\ \text{Hz}}{94.7 \times 10^6\ \text{Hz}} = 8.4 \times 10^{-2}$$

Therefore, for the radio wave, this same absolute line width corresponds to a fractional line width of more than 8%.

Summary

ThomsonNOW™ Sign in at **www.thomsonedu.com** and go to ThomsonNOW to take a practice test for this chapter.

CONCEPTS AND PRINCIPLES

The characteristics of **blackbody radiation** cannot be explained using classical concepts. Planck introduced the quantum concept and Planck's constant h when he assumed atomic oscillators existing only in discrete energy states were responsible for this radiation. In Planck's model, radiation is emitted in single quantized packets whenever an oscillator makes a transition between discrete energy states. The energy of a packet is

$$E = hf \quad (40.5)$$

where f is the frequency of the oscillator. Einstein successfully extended Planck's quantum hypothesis to the standing waves of electromagnetic radiation in a cavity used in the blackbody radiation model.

The **photoelectric effect** is a process whereby electrons are ejected from a metal surface when light is incident on that surface. In Einstein's model, light is viewed as a stream of particles, or **photons,** each having energy $E = hf$, where h is Planck's constant and f is the frequency. The maximum kinetic energy of the ejected photoelectron is

$$K_{\max} = hf - \phi \quad (40.9)$$

where ϕ is the **work function** of the metal.

X-rays are scattered at various angles by electrons in a target. In such a scattering event, a shift in wavelength is observed for the scattered x-rays, a phenomenon known as the **Compton effect.** Classical physics does not predict the correct behavior in this effect. If the x-ray is treated as a photon, conservation of energy and linear momentum applied to the photon–electron collisions yields, for the Compton shift,

$$\lambda' - \lambda_0 = \frac{h}{m_e c}(1 - \cos\theta) \quad (40.11)$$

where m_e is the mass of the electron, c is the speed of light, and θ is the scattering angle.

Light has a dual nature in that it has both wave and particle characteristics. Some experiments can be explained either better or solely by the particle model, whereas others can be explained either better or solely by the wave model.

Every object of mass m and momentum $p = mu$ has wave properties, with a **de Broglie wavelength** given by

$$\lambda = \frac{h}{p} = \frac{h}{mu} \quad (40.15)$$

By combining a large number of waves, a single region of constructive interference, called a **wave packet,** can be created. The wave packet carries the characteristic of localization like a particle does, but it has wave properties because it is built from waves. For an individual wave in the wave packet, the **phase speed** is

$$v_{\text{phase}} = \frac{\omega}{k} \quad (40.18)$$

For the wave packet as a whole, the **group speed** is

$$v_g = \frac{d\omega}{dk} \quad (40.19)$$

For a wave packet representing a particle, the group speed can be shown to be the same as the speed of the particle.

The **Heisenberg uncertainty principle** states that if a measurement of the position of a particle is made with uncertainty Δx and a simultaneous measurement of its linear momentum is made with uncertainty Δp_x, the product of the two uncertainties is restricted to

$$\Delta x\,\Delta p_x \geq \frac{\hbar}{2} \quad (40.23)$$

Another form of the uncertainty principle relates measurements of energy and time:

$$\Delta E\,\Delta t \geq \frac{\hbar}{2} \quad (40.24)$$

Questions

☐ denotes answer available in *Student Solutions Manual/Study Guide;* **O** denotes objective question

1. The classical model of blackbody radiation given by the Rayleigh–Jeans law has two major flaws. Identify them and explain how Planck's law deals with them.

2. All objects radiate energy. Why, then, are we not able to see all objects in a dark room?

3. O In a certain experiment, a filament in an evacuated lightbulb carries a current I_1 and you measure the spectrum of light emitted by the filament, which behaves as a black body at temperature T_1. The wavelength emitted with highest intensity (symbolized by $\lambda_{\max}$) has the value λ_1. You then increase the potential difference across the filament by a factor of 8, and the current increases by a factor of 2. **(i)** After this change, what is the new value of the temperature of the filament? (a) $16T_1$ (b) $8T_1$ (c) $4T_1$ (d) $2T_1$ (e) still T_1 **(ii)** What is the new value of the wavelength emitted with highest intensity? (a) $4\lambda_1$ (b) $2\lambda_1$ (c) $\sqrt{2}\lambda_1$ (d) λ_1 (e) $\lambda_1/\sqrt{2}$ (f) $\lambda_1/2$ (g) $\lambda_1/4$

4. ☐ If the photoelectric effect is observed for one metal, can you conclude that the effect will also be observed for another metal under the same conditions? Explain.

5. What does the slope of the lines in Active Figure 40.11 represent? What does the y intercept represent? How would such graphs for different metals compare with one another?

6. ☐ Why does the existence of a cutoff frequency in the photoelectric effect favor a particle theory for light over a wave theory?

7. In the photoelectric effect, explain why the stopping potential depends on the frequency of light but not on the intensity.

8. Which has more energy, a photon of ultraviolet radiation or a photon of yellow light?

9. O Which of the following is most likely to cause sunburn by delivering more energy to individual molecules in skin cells? (a) infrared light (b) visible light (c) ultraviolet light (d) microwaves (e) Choices (a) through (d) are equally likely.

10. How does the Compton effect differ from the photoelectric effect?

11. ☐ **O** An x-ray photon is scattered by an originally stationary electron. What happens to the frequency of the scattered photon relative to the frequency of the incident photon? Is the frequency of the scattered photon (a) lower, (b) higher, or (c) unchanged?

12. Suppose a photograph were made of a person's face using only a few photons. Would the result be simply a very faint image of the face? Explain your answer.

13. O Consider (a) an electron, (b) a photon, and (c) a proton, all moving in vacuum. Choose all correct answers for each question. **(i)** Which of the three possess rest energy? **(ii)** Which have charge? **(iii)** Which carry energy? **(iv)** Which carry momentum? **(v)** Which move at the speed of light? **(vi)** Which have a wavelength characterizing their motion?

14. Is light a wave or a particle? Support your answer by citing specific experimental evidence.

15. Is an electron a wave or a particle? Support your answer by citing some experimental results.

16. Why was the demonstration of electron diffraction by Davisson and Germer an important experiment?

17. O An electron and a proton, moving in opposite directions, are accelerated from rest through the same potential difference. Which particle has the longer wavelength? (a) The electron does. (b) The proton does. (c) Both are the same. (d) Neither has a wavelength

18. ☐ If matter has a wave nature, why is this wave-like characteristic not observable in our daily experiences?

19. O Rank the wavelengths of the following quantum particles from the largest to the smallest. If any have equal wavelengths, display the equality in your ranking. (a) a photon with energy 3 eV (b) an electron with kinetic energy 3 eV (c) a proton with kinetic energy 3 eV (d) a photon with energy 0.3 eV (e) a photon with momentum $3\text{ eV}/c = 1.6 \times 10^{-27}\text{ kg} \cdot \text{m/s}$ (f) an electron with momentum $3\text{ eV}/c$ (g) a proton with momentum $3\text{ eV}/c$

20. In describing the passage of electrons through a slit and arriving at a screen, physicist Richard Feynman said that "electrons arrive in lumps, like particles, but the probability of arrival of these lumps is determined as the intensity of the waves would be. It is in this sense that the electron behaves sometimes like a particle and sometimes like a wave." Elaborate on this point in your own words. For further discussion, see R. Feynman, *The Character of Physical Law* (Cambridge, MA: MIT Press, 1980), chap. 6.

21. Why is an electron microscope more suitable than an optical microscope for "seeing" objects less than 1 μm in size?

22. O Both an electron and a proton are accelerated to the same speed, and the experimental uncertainty in the speed is the same for the two particles. The positions of the two particles are also measured. Is the minimum possible uncertainty in the electron's position (a) less than the minimum possible uncertainty in the proton's position, (b) the same as that for the proton, (c) more than that for the proton, or (d) impossible to tell from the given information?

23. Shown in the photograph opening Chapter 37 and in Figure P38.52a in Chapter 38, *iridescence* is the phenomenon that gives shining colors to the feathers of peacocks, hummingbirds, resplendent quetzals, and even ducks and grackles. Without pigments, it colors Morpho butterflies, Urania moths, some beetles and flies, rainbow trout, and mother-of-pearl in abalone shells. Iridescent colors change as you turn an object. They can look different to your two eyes so that the object appears to have metallic luster. Iridescent colors were first described in print not by an artist or biologist, but by a physicist, Isaac Newton. They are produced by a wide variety of intricate structures in different species; Problem 52 in Chapter 38 describes those in a peacock feather. These structures were all unknown until the invention of the electron microscope. Explain why light microscopes cannot reveal them.

24. *Blacker than black, brighter than white.* (a) Take a large, closed, empty cardboard box. Cut a slot a few millimeters wide in one side. Use black pens, markers, and black material to make some stripes next to the slot as shown in Figure Q40.24a. Inspect the slot and the stripes with care and choose which is blackest; the figure may not show enough contrast to reveal which it is. Explain why it is blackest. (b) Locate an intricately shaped compact fluorescent light fixture. Look at it through dark glasses and describe where it appears brightest. Explain why it is brightest there. Figure Q40.24b shows two such light fixtures held near each other. *Suggestion:* Gustav Kirchhoff, professor at Heidelberg and master of the obvious, gave the same answer to part (a) as you likely will. His answer to part (b) would begin as follows. When electromagnetic radiation falls on its surface, an object reflects some fraction r of the energy and absorbs the rest. Whether the fraction reflected is 0.8 or 0.001, the fraction absorbed is $a = 1 - r$. Suppose the object and its surroundings are at the same temperature. The energy the object absorbs joins its fund of internal energy, but the second law of thermodynamics implies that the absorbed energy cannot raise the object's temperature. It does not produce a temperature increase because the object's energy budget has one more term: energy radiated. . . .

You still have to make the observations and answer questions (a) and (b), but you can incorporate some of Kirchhoff's ideas into your answer if you wish.

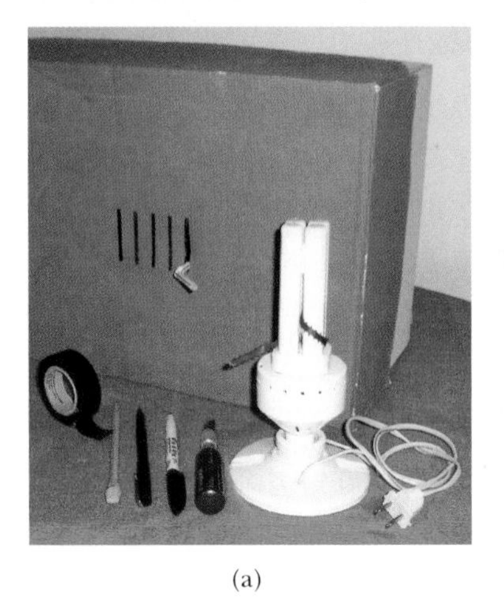

(a)

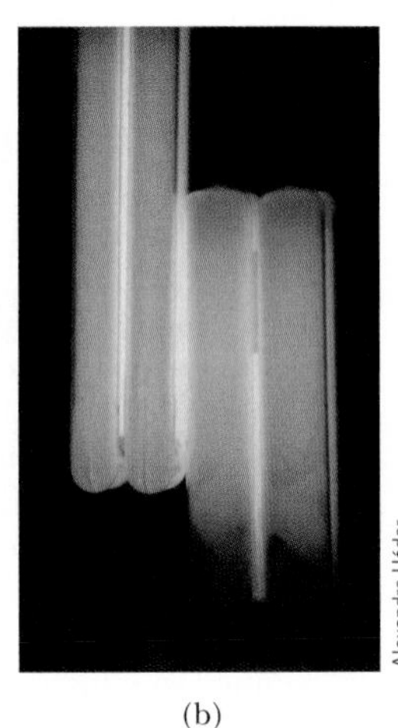

(b)

Alexandra Héder

Figure Q40.24

Problems

WebAssign The Problems from this chapter may be assigned online in WebAssign.

ThomsonNOW Sign in at **www.thomsonedu.com** and go to ThomsonNOW to assess your understanding of this chapter's topics with additional quizzing and conceptual questions.

1, 2, 3 denotes straightforward, intermediate, challenging; □ denotes full solution available in *Student Solutions Manual/Study Guide;* ▲ denotes coached solution with hints available at **www.thomsonedu.com;** denotes developing symbolic reasoning; ● denotes asking for qualitative reasoning; denotes computer useful in solving problem

Section 40.1 Blackbody Radiation and Planck's Hypothesis

1. The human eye is most sensitive to 560-nm light. What is the temperature of a black body that would radiate most intensely at this wavelength?

2. ● (a) Model the tungsten filament of a lightbulb as a blackbody at temperature 2 900 K. Determine the wavelength of light it emits most strongly. (b) Explain why the answer to part (a) suggests that more energy from the lightbulb goes into infrared radiation than into visible light.

3. ● Figure P40.3 shows the spectrum of light emitted by a firefly. Determine the temperature of a black body that would emit radiation peaked at the same wavelength. Based on your result, explain whether firefly radiation is blackbody radiation.

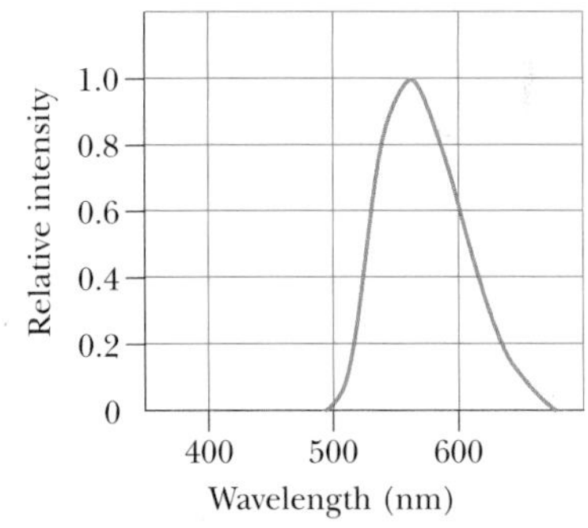

Figure P40.3

4. (a) Lightning produces a maximum air temperature on the order of 10^4 K, whereas (b) a nuclear explosion produces a temperature on the order of 10^7 K. Use Wien's displacement law to find the order of magnitude of the wavelength of the thermally produced photons radiated with greatest intensity by each of these sources. Name the part of the electromagnetic spectrum where you would expect each to radiate most strongly.

5. A black body at 7 500 K consists of an opening of diameter 0.050 0 mm, looking into an oven. Find the number of photons per second escaping the hole and having wavelengths between 500 nm and 501 nm.

6. Consider a black body of surface area 20.0 cm^2 and temperature 5 000 K. (a) How much power does it radiate? (b) At what wavelength does it radiate most intensely? Find the spectral power per wavelength at (c) this wavelength and at wavelengths of (d) 1.00 nm (an x- or gamma ray), (e) 5.00 nm (ultraviolet light or an x-ray), (f) 400 nm (at the boundary between UV and visible light), (g) 700 nm (at the boundary between visible and infrared light), (h) 1.00 mm (infrared light or a microwave), and (i) 10.0 cm (a microwave or radio wave).

(j) Approximately how much power does the object radiate as visible light?

7. The radius of our Sun is 6.96×10^8 m, and its total power output is 3.85×10^{26} W. (a) Assuming the Sun's surface emits as a black body, calculate its surface temperature. (b) Using the result of part (a), find λ_{max} for the Sun.

8. The average threshold of dark-adapted (scotopic) vision is 4.00×10^{-11} W/m^2 at a central wavelength of 500 nm. If light having this intensity and wavelength enters the eye and the pupil is open to its maximum diameter of 8.50 mm, how many photons per second enter the eye?

9. ▲ Calculate the energy, in electron volts, of a photon whose frequency is (a) 620 THz, (b) 3.10 GHz, and (c) 46.0 MHz. (d) Determine the corresponding wavelengths for these photons and state the classification of each on the electromagnetic spectrum.

10. A simple pendulum has a length of 1.00 m and a mass of 1.00 kg. The amplitude of oscillations of the pendulum is 3.00 cm. Calculate the quantum number for the pendulum.

11. An FM radio transmitter has a power output of 150 kW and operates at a frequency of 99.7 MHz. How many photons per second does the transmitter emit?

12. **Review problem.** This problem is about how strongly matter is coupled to radiation, the subject with which quantum mechanics began. For a simple model, consider a solid iron sphere 2.00 cm in radius. Assume its temperature is always uniform throughout its volume. (a) Find the mass of the sphere. (b) Assume the sphere is at 20°C and has emissivity 0.860. Find the power with which it radiates electromagnetic waves. (c) If it were alone in the Universe, at what rate would the sphere's temperature be changing? (d) Assume Wien's law describes the sphere. Find the wavelength λ_{max} of electromagnetic radiation it emits most strongly. Although it emits a spectrum of waves having all different wavelengths, assume its power output is carried by photons of wavelength λ_{max}. Find (e) the energy of one photon and (f) the number of photons it emits each second. *Note:* The answer to part (f) gives an indication of how fast the object is emitting and also absorbing photons when it is in thermal equilibrium with its surroundings at 20°C.

13. Show that at long wavelengths, Planck's radiation law (Eq. 40.6) reduces to the Rayleigh–Jeans law (Eq. 40.3).

Section 40.2 The Photoelectric Effect

14. Molybdenum has a work function of 4.20 eV. (a) Find the cutoff wavelength and cutoff frequency for the photoelectric effect. (b) What is the stopping potential if the incident light has a wavelength of 180 nm?

15. Two light sources are used in a photoelectric experiment to determine the work function for a particular metal surface. When green light from a mercury lamp (λ = 546.1 nm) is used, a stopping potential of 0.376 V reduces the photocurrent to zero. (a) Based on this measurement, what is the work function for this metal? (b) What stopping potential would be observed when using the yellow light from a helium discharge tube (λ = 587.5 nm)?

16. Electrons are ejected from a metallic surface with speeds ranging up to 4.60×10^5 m/s when light with a wavelength of 625 nm is used. (a) What is the work function of the surface? (b) What is the cutoff frequency for this surface?

17. ● Lithium, beryllium, and mercury have work functions of 2.30 eV, 3.90 eV, and 4.50 eV, respectively. Light with a wavelength of 400 nm is incident on each of these metals. (a) Determine which of these metals exhibit the photoelectric effect. Explain your reasoning. (b) Find the maximum kinetic energy for the photoelectrons in each case.

18. ● From the scattering of sunlight, J. J. Thomson calculated the classical radius of the electron as having the value 2.82×10^{-15} m. Sunlight with an intensity of 500 W/m^2 falls on a disk with this radius. Calculate the time interval required to accumulate 1.00 eV of energy. Assume light is a classical wave and the light striking the disk is completely absorbed. Explain how your result compares with the observation that photoelectrons are emitted promptly (within 10^{-9} s).

19. **Review problem.** An isolated copper sphere of radius 5.00 cm, initially uncharged, is illuminated by ultraviolet light of wavelength 200 nm. What charge does the photoelectric effect induce on the sphere? The work function for copper is 4.70 eV.

20. ● **Review problem.** A light source emitting radiation at 7.00×10^{14} Hz is incapable of ejecting photoelectrons from a certain metal. In an attempt to use this source to eject photoelectrons from the metal, the source is given a velocity toward the metal. (a) Explain how this procedure can produce photoelectrons. (b) When the speed of the light source is equal to $0.280c$, photoelectrons just begin to be ejected from the metal. What is the work function of the metal? (c) When the speed of the light source is increased to $0.900c$, determine the maximum kinetic energy of the photoelectrons.

Section 40.3 The Compton Effect

21. Calculate the energy and momentum of a photon of wavelength 700 nm.

22. ● X-rays with a wavelength of 120.0 pm undergo Compton scattering. (a) Find the wavelengths of the photons scattered at angles of 30.0°, 60.0°, 90.0°, 120°, 150°, and 180°. (b) Find the energy of the scattered electron in each case. (c) Which of the scattering angles provides the electron with the greatest energy? Explain whether you could answer this question without doing any calculations.

23. ▲ A 0.001 60-nm photon scatters from a free electron. For what (photon) scattering angle does the recoiling electron have kinetic energy equal to the energy of the scattered photon?

24. X-rays having an energy of 300 keV undergo Compton scattering from a target. The scattered rays are detected at 37.0° relative to the incident rays. Find (a) the Compton shift at this angle, (b) the energy of the scattered x-ray, and (c) the energy of the recoiling electron.

25. A 0.880-MeV photon is scattered by a free electron initially at rest such that the scattering angle of the scattered electron is equal to that of the scattered photon ($\theta = \phi$ in Fig. 40.13). (a) Determine the angles θ and ϕ. (b) Determine the energy and momentum of the scattered photon. (c) Determine the kinetic energy and momentum of the scattered electron.

26. A photon having energy E_0 is scattered by a free electron initially at rest such that the scattering angle of the scattered electron is equal to that of the scattered photon ($\theta = \phi$ in Fig. 40.13). (a) Determine the angles θ and ϕ. (b) Determine the energy and momentum of the scattered photon. (c) Determine the kinetic energy and momentum of the scattered electron.

27. After a 0.800-nm x-ray photon scatters from a free electron, the electron recoils at 1.40×10^6 m/s. (a) What is the Compton shift in the photon's wavelength? (b) Through what angle is the photon scattered?

28. In a Compton scattering experiment, an x-ray photon scatters through an angle of 17.4° from a free electron that is initially at rest. The electron recoils with a speed of 2 180 km/s. Calculate (a) the wavelength of the incident photon and (b) the angle through which the electron scatters.

29. ● In a Compton scattering experiment, a photon is scattered through an angle of 90.0° and the electron is set into motion in a direction at an angle of 20.0° to the original direction of the photon. Explain whether this information is sufficient to determine uniquely the wavelength of the scattered photon. If it is, find this wavelength.

30. A photon having wavelength λ scatters off a free electron at A (Fig. P40.30), producing a second photon having wavelength λ'. This photon then scatters off another free electron at B, producing a third photon having wavelength λ'' and moving in a direction directly opposite the original photon as shown in the figure. Determine the numerical value of $\Delta\lambda = \lambda'' - \lambda$.

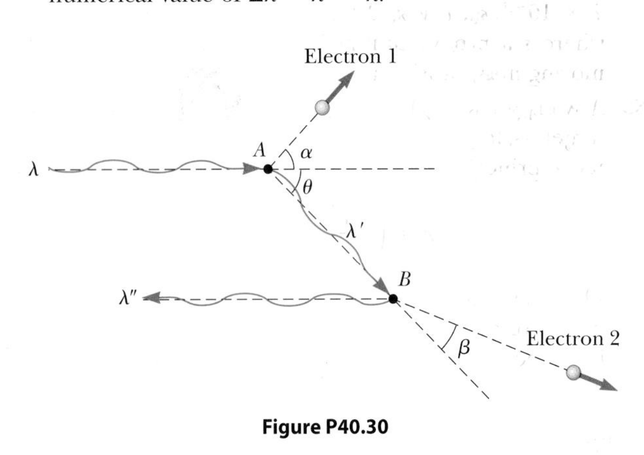

Figure P40.30

31. Find the maximum fractional energy loss for a 0.511-MeV gamma ray that is Compton scattered from (a) a free electron and (b) a free proton.

Section 40.4 Photons and Electromagnetic Waves

32. ● An electromagnetic wave is called *ionizing radiation* if its photon energy is larger than, say, 10.0 eV so that a single photon has enough energy to break apart an atom. With reference to Figure 34.11, explain what region or regions of the electromagnetic spectrum fit this definition of ionizing radiation and what do not.

33. **Review problem.** A helium–neon laser produces a beam of diameter 1.75 mm, delivering 2.00×10^{18} photons/s. Each photon has a wavelength of 633 nm. (a) Calculate the amplitudes of the electric and magnetic fields inside the beam. (b) If the beam shines perpendicularly onto a perfectly reflecting surface, what force does it exert on the surface? (c) If the beam is absorbed by a block of ice at 0°C for 1.50 h, what mass of ice is melted?

Section 40.5 The Wave Properties of Particles

34. Calculate the de Broglie wavelength for a proton moving with a speed of 1.00×10^6 m/s.

35. Calculate the de Broglie wavelength for an electron that has kinetic energy (a) 50.0 eV and (b) 50.0 keV.

36. (a) An electron has a kinetic energy of 3.00 eV. Find its wavelength. (b) **What If?** A photon has energy 3.00 eV. Find its wavelength.

37. ● ▲ The nucleus of an atom is on the order of 10^{-14} m in diameter. For an electron to be confined to a nucleus, its de Broglie wavelength would have to be on this order of magnitude or smaller. (a) What would be the kinetic energy of an electron confined to this region? (b) Make an order-of-magnitude estimate of the electric potential energy of a system of an electron inside an atomic nucleus. Would you expect to find an electron in a nucleus? Explain.

38. In the Davisson–Germer experiment, 54.0-eV electrons were diffracted from a nickel lattice. If the first maximum in the diffraction pattern was observed at $\phi = 50.0°$ (Fig. P40.38), what was the lattice spacing a between the vertical rows of atoms in the figure? (It is not the same as the spacing between the horizontal rows of atoms.)

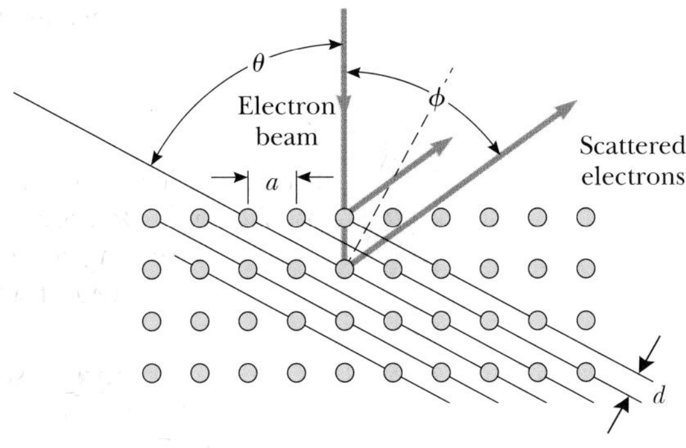

Figure P40.38

39. ● (a) Show that the frequency f and wavelength λ of a freely moving quantum particle with mass are related by the expression

$$\left(\frac{f}{c}\right)^2 = \frac{1}{\lambda^2} + \frac{1}{\lambda_C^{\,2}}$$

where $\lambda_C = h/mc$ is the Compton wavelength of the particle. (b) Is it ever possible for a particle having nonzero mass to have the same wavelength *and* frequency as a photon? Explain.

40. A photon has an energy equal to the kinetic energy of an electron with speed u, which may be close to the speed of light. (a) Calculate the ratio of the wavelength of the photon to the wavelength of the electron. (b) Evaluate the ratio for the particle speed $u = 0.900c$. (c) **What If?** What would happen to the answer if the material particle were a proton instead of an electron? (d) Evaluate the ratio for the particle speed $u = 0.001\,00c$. (e) What value does the ratio of the wavelengths approach at high particle speeds? (f) At low particle speeds?

41. The resolving power of a microscope depends on the wavelength used. If you wanted to "see" an atom, a resolution of approximately 1.00×10^{-11} m would be required. (a) If electrons are used (in an electron microscope), what minimum kinetic energy is required for the electrons? (b) **What If?** If photons are used, what minimum photon energy is needed to obtain the required resolution?

42. ● After learning about de Broglie's hypothesis that material particles of momentum p move as waves with wavelength $\lambda = h/p$, an 80.0-kg student has grown concerned about being diffracted when passing through a 75.0-cm-wide doorway. Assume significant diffraction occurs when the width of the diffraction aperture is less that 10.0 times the wavelength of the wave being diffracted. (a) Determine the maximum speed at which the student can pass through the doorway if he is to be significantly diffracted. (b) With that speed, over what time interval does the student pass through the doorway if it is in a wall 15.0 cm thick? State how your answer compares with the age of the Universe, which is about 4×10^{17} s. (c) Explain whether this student should worry about being diffracted.

43. ● Robert Hofstadter won the 1961 Nobel Prize in Physics for his pioneering work in studying the scattering of 20-GeV electrons from nuclei. (a) What is the γ factor for an electron with total energy 20.0 GeV, defined by $\gamma = 1/\sqrt{1 - u^2/c^2}$? (b) Find the momentum of the electron. (c) Find the wavelength of the electron. State how it compares with the diameter of an atomic nucleus, typically on the order of 10^{-14} m.

Section 40.6 The Quantum Particle

44. Consider a freely moving quantum particle with mass m and speed u. Its energy is $E = K = \frac{1}{2}mu^2$. Determine the phase speed of the quantum wave representing the particle and show that it is different from the speed at which the particle transports mass and energy.

45. For a free relativistic quantum particle moving with speed u, the total energy is $E = hf = \hbar\omega = \sqrt{p^2c^2 + m^2c^4}$ and the momentum is $p = h/\lambda = \hbar k = \gamma mu$. For the quantum wave representing the particle, the group speed is $v_g = d\omega/dk$. Prove that the group speed of the wave is the same as the speed of the particle.

Section 40.7 The Double-Slit Experiment Revisited

46. A modified oscilloscope is used to perform an electron interference experiment. Electrons are incident on a pair of narrow slits 0.060 0 μm apart. The bright bands in the interference pattern are separated by 0.400 mm on a screen 20.0 cm from the slits. Determine the potential difference through which the electrons were accelerated to give this pattern.

47. ▲ Neutrons traveling at 0.400 m/s are directed through a pair of slits having a 1.00-mm separation. An array of detectors is placed 10.0 m from the slits. (a) What is the de Broglie wavelength of the neutrons? (b) How far off axis is the first zero-intensity point on the detector array? (c) When a neutron reaches a detector, can we say which slit the neutron passed through? Explain.

48. In a certain vacuum tube, electrons evaporate from a hot cathode at a slow, steady rate and accelerate from rest through a potential difference of 45.0 V. Then they travel 28.0 cm as they pass through an array of slits and fall on a screen to produce an interference pattern. If the beam current is below a certain value, only one electron at a time will be in flight in the tube. What is this value? In this situation, the interference pattern still appears, showing that each individual electron can interfere with itself.

Section 40.8 The Uncertainty Principle

49. ▲ An electron ($m_e = 9.11 \times 10^{-31}$ kg) and a bullet ($m = 0.020\,0$ kg) each have a velocity with a magnitude of 500 m/s, accurate to within 0.010 0%. Within what limits could we determine the position of the objects along the direction of the velocity?

50. Suppose Fuzzy, a quantum–mechanical duck, lives in a world in which $h = 2\pi$ J · s. Fuzzy has a mass of 2.00 kg and is initially known to be within a pond 1.00 m wide. (a) What is the minimum uncertainty in the component of the duck's velocity parallel to the width of the pond? (b) Assuming that this uncertainty in speed prevails for 5.00 s, determine the uncertainty in Fuzzy's position after this time interval.

51. ● An air rifle is used to shoot 1.00-g particles at 100 m/s through a hole of diameter 2.00 mm. How far from the rifle must an observer be to see the beam spread by 1.00 cm because of the uncertainty principle? State how this answer compares with the diameter of the visible part of the Universe, 2×10^{26} m.

52. Use the uncertainty principle to show that if an electron were confined inside an atomic nucleus of diameter 2×10^{-15} m, it would have to be moving relativistically, whereas a proton confined to the same nucleus can be moving nonrelativistically.

53. A woman on a ladder drops small pellets toward a point target on the floor. (a) Show that according to the uncertainty principle, the average miss distance must be at least

$$\Delta x_f = \left(\frac{2\hbar}{m}\right)^{1/2}\left(\frac{2H}{g}\right)^{1/4}$$

where m is the mass of each pellet and H is the initial height of each pellet above the floor. Assume the spread in impact points is given by $\Delta x_f = \Delta x_i + (\Delta v_x)t$. (b) If $H = 2.00$ m and $m = 0.500$ g, what is Δx_f?

Additional Problems

54. **Review problem.** Design an incandescent lamp filament. Specify the length and radius a tungsten wire can have to radiate electromagnetic waves with power 75.0 W when its ends are connected across a 120-V power supply. Assume its constant operating temperature is 2 900 K and its emissivity is 0.450. Also assume it takes in energy only by electric transmission and loses energy only by electromagnetic radiation. From Table 27.2, you may take the resistivity of tungsten at 2 900 K as

$$5.6 \times 10^{-8}\,\Omega\cdot\text{m}[1 + (4.5 \times 10^{-3}/^\circ\text{C})(2\,607^\circ\text{C})]$$
$$= 7.13 \times 10^{-7}\,\Omega\cdot\text{m}$$

55. ▲ The accompanying table shows data obtained in a photoelectric experiment. (a) Using these data, make a graph similar to Active Figure 40.11 that plots as a straight line. From the graph, determine (b) an experimental value for

Planck's constant (in joule-seconds) and (c) the work function (in electron volts) for the surface. Two significant figures for each answer are sufficient.

Wavelength (nm)	Maximum Kinetic Energy of Photoelectrons (eV)
588	0.67
505	0.98
445	1.35
399	1.63

56. Figure P40.56 shows the stopping potential versus the incident photon frequency for the photoelectric effect for sodium. Use the graph to find (a) the work function of sodium, (b) the ratio h/e, and (c) the cutoff wavelength. The data are taken from R. A. Millikan, *Physical Review* 7:362 (1916).

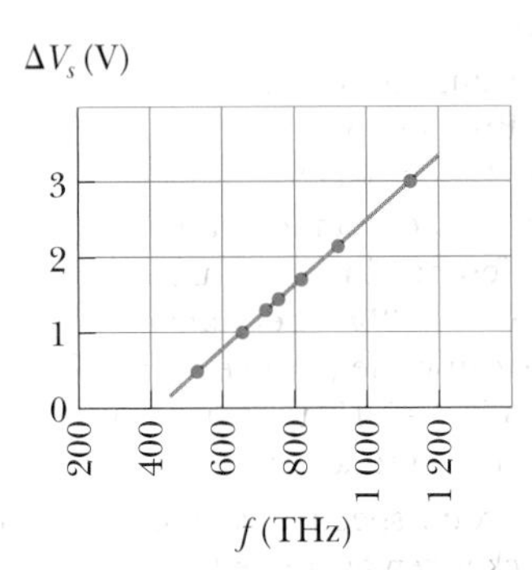

Figure P40.56

57. **Review problem.** Photons of wavelength λ are incident on a metal. The most energetic electrons ejected from the metal are bent into a circular arc of radius R by a magnetic field having a magnitude B. What is the work function of the metal?

58. ● Ultraviolet light with a single wavelength and with intensity 550 W/m² is incident normally on the surface of a metal that has a work function of 3.44 eV. Photoelectrons are emitted with a maximum speed of 420 km/s. (a) Find the maximum possible rate of photoelectron emission from 1 cm² of the surface by imagining that every photon produces one photoelectron. (b) Find the electric current these electrons constitute. (c) How do you suppose the actual current compares with this maximum possible current?

59. Derive the equation for the Compton shift (Eq. 40.11) from Equations 40.12 through 40.14.

60. Show that a photon cannot transfer all of its energy to a free electron. *Suggestion:* Note that system energy and momentum must be conserved.

61. The total power per unit area radiated by a black body at a temperature T is the area under the $I(\lambda, T)$-versus-λ curve as shown in Active Figure 40.3. (a) Show that this power per unit area is

$$\int_0^\infty I(\lambda, T)\,d\lambda = \sigma T^4$$

where $I(\lambda, T)$ is given by Planck's radiation law and σ is a constant independent of T. This result is known as Stefan's law. (See Section 20.7.) To carry out the integration, you should make the change of variable $x = hc/\lambda k_B T$ and use

$$\int_0^\infty \frac{x^3\,dx}{e^x - 1} = \frac{\pi^4}{15}$$

(b) Show that the Stefan–Boltzmann constant σ has the value

$$\sigma = \frac{2\pi^5 k_B{}^4}{15c^2h^3} = 5.67 \times 10^{-8}\ \text{W/m}^2 \cdot \text{K}^4$$

62. Derive Wien's displacement law from Planck's law. Proceed as follows. In Active Figure 40.3, notice that the wavelength at which a black body radiates with greatest intensity is the wavelength for which the graph of $I(\lambda, T)$ versus λ has a horizontal tangent. From Equation 40.6, evaluate the derivative $dI/d\lambda$. Set it equal to zero. Solve the resulting transcendental equation numerically to prove $hc/\lambda_{max} k_B T = 4.965\ldots$ or $\lambda_{max} T = hc/4.965k_B$. Evaluate the constant as precisely as possible and compare it with Wien's experimental value.

63. ● The neutron has a mass of 1.67×10^{-27} kg. Neutrons emitted in nuclear reactions can be slowed down by collisions with matter. They are referred to as thermal neutrons after they come into thermal equilibrium with the environment. The average kinetic energy ($3k_BT/2$) of a thermal neutron is approximately 0.04 eV. Calculate the de Broglie wavelength of a neutron with a kinetic energy of 0.040 0 eV. How does your answer compare with the characteristic atomic spacing in a crystal? Explain whether you expect thermal neutrons to exhibit diffraction effects when scattered by a crystal.

64. Johnny Jumper's favorite trick is to step out of his 16th-story window and fall 50.0 m into a pool. A news reporter takes a picture of 75.0-kg Johnny just before he makes a splash, using an exposure time of 5.00 ms. Find (a) Johnny's de Broglie wavelength at this moment, (b) the uncertainty of his kinetic energy measurement during such an interval of time, and (c) the percent error caused by such an uncertainty.

65. Show that the ratio of the Compton wavelength λ_C to the de Broglie wavelength $\lambda = h/p$ for a relativistic electron is

$$\frac{\lambda_C}{\lambda} = \left[\left(\frac{E}{m_e c^2}\right)^2 - 1\right]^{1/2}$$

where E is the total energy of the electron and m_e is its mass.

66. A photon of initial energy E_0 undergoes Compton scattering at an angle θ from a free electron (mass m_e) initially at rest. Using relativistic equations for energy and momentum conservation, derive the following relationship for the final energy E' of the scattered photon:

$$E' = \frac{E_0}{1 + \left(\dfrac{E_0}{m_e c^2}\right)(1 - \cos\theta)}$$

67. A π^0 meson is an unstable particle produced in high-energy particle collisions. Its rest energy is approximately 135 MeV, and it exists for a lifetime of only 8.70×10^{-17} s before decaying into two gamma rays. Using the uncertainty principle, estimate the fractional uncertainty $\Delta m/m$ in its mass determination.

68. A photon with wavelength λ_0 moves toward a free electron that is moving with speed u in the same direction as the photon (Fig. P40.68a). The photon scatters at an angle θ (Fig. P40.68b). Show that the wavelength of the scattered photon is

$$\lambda' = \lambda_0\left(\frac{1-(u/c)\cos\theta}{1-(u/c)}\right) + \frac{h}{m_e c}\sqrt{\frac{1+(u/c)}{1-(u/c)}}(1-\cos\theta)$$

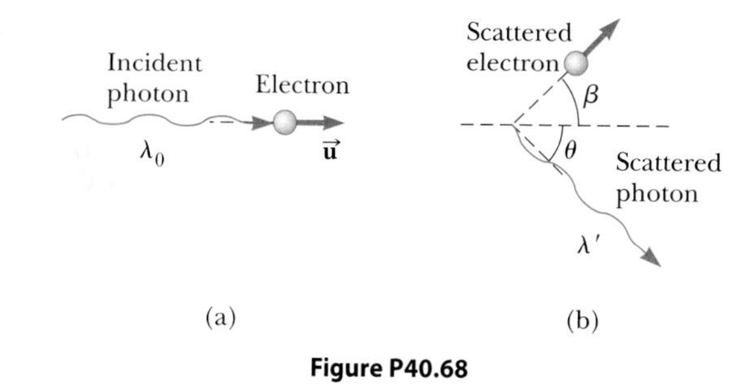

Figure P40.68

Answers to Quick Quizzes

40.1 (b). A very hot star has its peak in the blackbody intensity distribution curve at wavelengths shorter than the visible. As a result, more blue light is emitted than red light.

40.2 AM radio, FM radio, microwaves, sodium light. The order of photon energy will be the same as the order of frequency. See Figure 34.11 for a pictorial representation of electromagnetic radiation in order of frequency.

40.3 (c). When the frequency is increased, the photons each carry more energy, so a stopping potential larger in magnitude is required for the current to fall to zero.

40.4 Classical physics predicts that light of sufficient intensity causes emission of photoelectrons, independent of frequency and without a cutoff frequency. Also, the greater the intensity, the larger the maximum kinetic energy of the electrons, with some time delay in emission at low intensities. Therefore, the classical expectation (which did not match experiment) yields a graph like the following drawing:

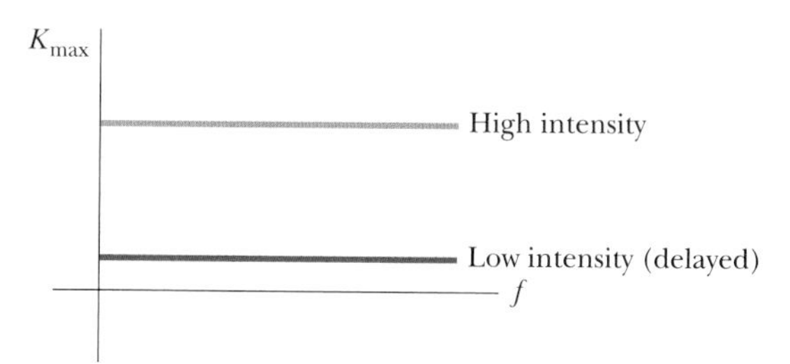

40.5 (d). The shift $\Delta\lambda$ is independent of λ. Therefore, the largest fractional shift will correspond to the smallest wavelength.

40.6 (c). According to Equation 40.15, two particles with the same de Broglie wavelength have the same momentum $p = mu$. If the electron and proton have the same momentum, they cannot have the same speed because of the difference in their masses. For the same reason, because $K = p^2/2m$, they cannot have the same kinetic energy. Because the particles have different kinetic energies, Equation 40.16 tells us that the particles do not have the same frequency.

40.7 (b). The group speed is zero because the leading edge of the packet remains fixed at the location of the accident.

40.8 (a). The uncertainty principle relates uncertainty in position and velocity along the same axis. The zero uncertainty in position along the x axis results in infinite uncertainty in its velocity component in the x direction, but it is unrelated to the y direction.

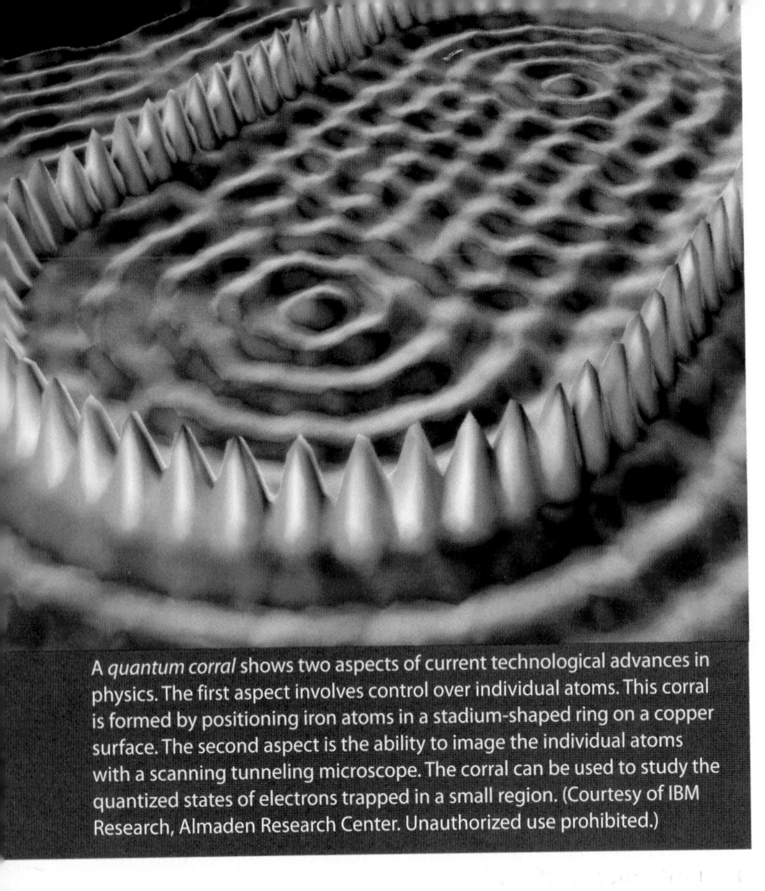

A *quantum corral* shows two aspects of current technological advances in physics. The first aspect involves control over individual atoms. This corral is formed by positioning iron atoms in a stadium-shaped ring on a copper surface. The second aspect is the ability to image the individual atoms with a scanning tunneling microscope. The corral can be used to study the quantized states of electrons trapped in a small region. (Courtesy of IBM Research, Almaden Research Center. Unauthorized use prohibited.)

41 Quantum Mechanics

In this chapter, we introduce *quantum mechanics*, an extremely successful theory for explaining the behavior of microscopic particles. This theory, developed in the 1920s by Erwin Schrödinger, Werner Heisenberg, and others, enables us to understand a host of phenomena involving atoms, molecules, nuclei, and solids. The discussion in this chapter follows from the quantum particle model that was developed in Chapter 40 and incorporates some of the features of the waves under boundary conditions model that was explored in Chapter 18. We also discuss practical applications of quantum mechanics, including the scanning tunneling microscope and nanoscale devices that may be used in future quantum computers. Finally, we shall return to the simple harmonic oscillator that was introduced in Chapter 15 and examine it from a quantum mechanical point of view.

41.1 An Interpretation of Quantum Mechanics

In Chapter 40, we introduced some new and strange ideas. In particular, we concluded on the basis of experimental evidence that both matter and electromagnetic radiation are sometimes best modeled as particles and sometimes as waves, depending on the phenomenon being observed. We can improve our understanding of quantum physics by making another connection between particles and waves using the notion of probability, a concept that was introduced in Chapter 40.

We begin by discussing electromagnetic radiation using the particle model. The probability per unit volume of finding a photon in a given region of space at an instant of time is proportional to the number N of photons per unit volume at that time:

$$\frac{\text{Probability}}{V} \propto \frac{N}{V}$$

The number of photons per unit volume is proportional to the intensity of the radiation:

$$\frac{N}{V} \propto I$$

Now, let's form a connection between the particle model and the wave model by recalling that the intensity of electromagnetic radiation is proportional to the square of the electric field amplitude E for the electromagnetic wave (Eq. 34.24):

$$I \propto E^2$$

Equating the beginning and the end of this series of proportionalities gives

$$\frac{\text{Probability}}{V} \propto E^2 \tag{41.1}$$

Therefore, for electromagnetic radiation, the probability per unit volume of finding a particle associated with this radiation (the photon) is proportional to the square of the amplitude of the associated electromagnetic wave.

Recognizing the wave–particle duality of both electromagnetic radiation and matter, we should suspect a parallel proportionality for a material particle: the probability per unit volume of finding the particle is proportional to the square of the amplitude of a wave representing the particle. In Chapter 40, we learned that there is a de Broglie wave associated with every particle. The amplitude of the de Broglie wave associated with a particle is not a measurable quantity because the wave function representing a particle is generally a complex function as we discuss below. In contrast, the electric field for an electromagnetic wave is a real function. The matter analog to Equation 41.1 relates the square of the amplitude of the wave to the probability per unit volume of finding the particle. Hence, the amplitude of the wave associated with the particle is called the **probability amplitude,** or the **wave function,** and it has the symbol Ψ.

In general, the complete wave function Ψ for a system depends on the positions of all the particles in the system and on time; therefore, it can be written $\Psi(\vec{\mathbf{r}}_1, \vec{\mathbf{r}}_2, \vec{\mathbf{r}}_3, \ldots, \vec{\mathbf{r}}_j, \ldots, t)$, where $\vec{\mathbf{r}}_j$ is the position vector of the jth particle in the system. For many systems of interest, including all those we study in this text, the wave function Ψ is mathematically separable in space and time and can be written as a product of a space function ψ for one particle of the system and a complex time function:[1]

$$\Psi(\vec{\mathbf{r}}_1, \vec{\mathbf{r}}_2, \vec{\mathbf{r}}_3, \ldots, \vec{\mathbf{r}}_j, \ldots, t) = \psi(\vec{\mathbf{r}}_j)e^{-i\omega t} \tag{41.2}$$

◀ Space- and time-dependent wave function Ψ

where ω $(= 2\pi f)$ is the angular frequency of the wave function and $i = \sqrt{-1}$.

For any system in which the potential energy is time-independent and depends only on the positions of particles within the system, the important information about the system is contained within the space part of the wave function. The time

[1] The standard form of a complex number is $a + ib$. The notation $e^{i\theta}$ is equivalent to the standard form as follows:

$$e^{i\theta} = \cos\theta + i\sin\theta$$

Therefore, the notation $e^{-i\omega t}$ in Equation 41.2 is equivalent to $\cos(-\omega t) + i\sin(-\omega t) = \cos\omega t - i\sin\omega t$.

part is simply the factor $e^{-i\omega t}$. Therefore, an understanding of ψ is the critical aspect of a given problem.

The wave function ψ is often complex-valued. The absolute square $|\psi|^2 = \psi^*\psi$, where ψ^* is the complex conjugate[2] of ψ, is always real and positive and is proportional to the probability per unit volume of finding a particle at a given point at some instant. The wave function contains within it all the information that can be known about the particle.

Probability density $|\psi|^2$ ▶

Although ψ cannot be measured, we can measure the real quantity $|\psi|^2$, which can be interpreted as follows. If ψ represents a single particle, then $|\psi|^2$—called the **probability density**—is the relative probability per unit volume that the particle will be found at any given point in the volume. This interpretation can also be stated in the following manner. If dV is a small volume element surrounding some point, the probability of finding the particle in that volume element is

$$P(x, y, z)\,dV = |\psi|^2 dV \quad \textbf{(41.3)}$$

PITFALL PREVENTION 41.1

The Wave Function Belongs to a System

The common language in quantum mechanics is to associate a wave function with a particle. The wave function, however, is determined by the particle *and* its interaction with its environment, so it more rightfully belongs to a system. In many cases, the particle is the only part of the system that experiences a change, which is why the common language has developed. You will see examples in the future in which it is more proper to think of the system wave function rather than the particle wave function.

This probabilistic interpretation of the wave function was first suggested by Max Born (1882–1970) in 1928. In 1926, Erwin Schrödinger proposed a wave equation that describes the manner in which the wave function changes in space and time. The *Schrödinger wave equation,* which we shall examine in Section 41.3, represents a key element in the theory of quantum mechanics.

The concepts of quantum mechanics, strange as they sometimes may seem, developed from classical ideas. In fact, when the techniques of quantum mechanics are applied to macroscopic systems, the results are essentially identical to those of classical physics. This blending of the two approaches occurs when the de Broglie wavelength is small compared with the dimensions of the system. The situation is similar to the agreement between relativistic mechanics and classical mechanics when $v \ll c$.

In Section 40.5, we found that the de Broglie equation relates the momentum of a particle to its wavelength through the relation $p = h/\lambda$. If an ideal free particle has a precisely known momentum p_x, its wave function is an infinitely long sinusoidal wave of wavelength $\lambda = h/p_x$ and the particle has equal probability of being at any point along the x axis (Fig. 40.18a). The wave function ψ for such a free particle moving along the x axis can be written as

Wave function for a free particle ▶

$$\psi(x) = Ae^{ikx} \quad \textbf{(41.4)}$$

where A is a constant amplitude and $k = 2\pi/\lambda$ is the angular wave number (Eq. 16.8) of the wave representing the particle.[3]

One-Dimensional Wave Functions and Expectation Values

This section discusses only one-dimensional systems, where the particle must be located along the x axis, so the probability $|\psi|^2\,dV$ in Equation 41.3 is modified to become $|\psi|^2\,dx$. The probability that the particle will be found in the infinitesimal interval dx around the point x is

$$P(x)\,dx = |\psi|^2\,dx \quad \textbf{(41.5)}$$

Although it is not possible to specify the position of a particle with complete certainty, it is possible through $|\psi|^2$ to specify the probability of observing it in a

[2] For a complex number $z = a + ib$, the complex conjugate is found by changing i to $-i$: $z^* = a - ib$. The product of a complex number and its complex conjugate is always real and positive. That is, $z^*z = (a - ib)(a + ib) = a^2 - (ib)^2 = a^2 - (i)^2b^2 = a^2 + b^2$.

[3] For the free particle, the full wave function, based on Equation 41.2, is

$$\Psi(x, t) = Ae^{ikx}e^{-i\omega t} = Ae^{i(kx-\omega t)} = A[\cos(kx - \omega t) + i\sin(kx - \omega t)]$$

The real part of this wave function has the same form as the waves we added together to form wave packets in Section 40.6.

region surrounding a given point x. **The probability of finding the particle in the arbitrary interval $a \le x \le b$ is**

$$P_{ab} = \int_a^b |\psi|^2 \, dx \qquad \textbf{(41.6)}$$

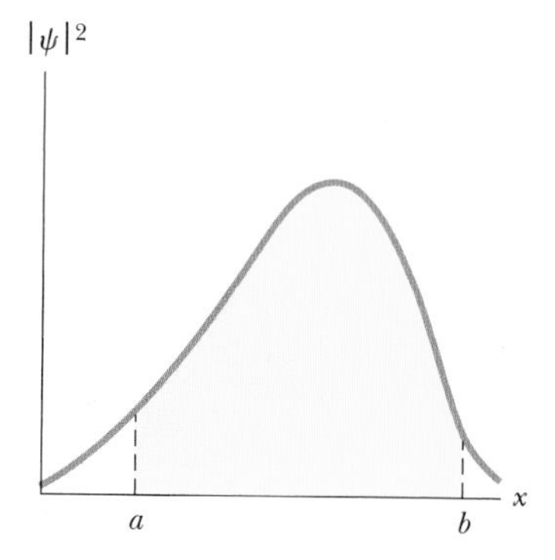

Figure 41.1 The probability of a particle being in the interval $a \le x \le b$ is the area under the probability density curve from a to b.

The probability P_{ab} is the area under the curve of $|\psi|^2$ versus x between the points $x = a$ and $x = b$ as in Figure 41.1.

Experimentally, there is a finite probability of finding a particle in an interval near some point at some instant. The value of that probability must lie between the limits 0 and 1. For example, if the probability is 0.30, there is a 30% chance of finding the particle in the interval.

Because the particle must be somewhere along the x axis, the sum of the probabilities over all values of x must be 1:

$$\int_{-\infty}^{\infty} |\psi|^2 \, dx = 1 \qquad \textbf{(41.7)}$$

◀ Normalization condition on ψ

Any wave function satisfying Equation 41.7 is said to be **normalized.** Normalization is simply a statement that the particle exists at some point in space.

Once the wave function for a particle is known, it is possible to calculate the average position at which you would expect to find the particle after many measurements. This average position is called the **expectation value** of x and is defined by the equation

$$\langle x \rangle \equiv \int_{-\infty}^{\infty} \psi^* x \psi \, dx \qquad \textbf{(41.8)}$$

◀ Expectation value for position x

(Brackets, $\langle \ldots \rangle$, are used to denote expectation values.) Furthermore, one can find the expectation value of any function $f(x)$ associated with the particle by using the following equation:[4]

$$\langle f(x) \rangle \equiv \int_{-\infty}^{\infty} \psi^* f(x) \psi \, dx \qquad \textbf{(41.9)}$$

◀ Expectation value for a function $f(x)$

Quick Quiz 41.1 Consider the wave function for the free particle, Equation 41.4. At what value of x is the particle most likely to be found at a given time? (a) at $x = 0$ (b) at small nonzero values of x (c) at large values of x (d) anywhere along the x axis

EXAMPLE 41.1 A Wave Function for a Particle

Consider a particle whose wave function is graphed in Figure 41.2 and is given by

$$\psi(x) = Ae^{-ax^2}$$

(A) What is the value of A if this wave function is normalized?

SOLUTION

Conceptualize The particle is not a free particle because the wave function is not a sinusoidal function. Figure 41.2 indicates that the particle is constrained to remain close to $x = 0$ at all times. Think of a physical system in which the particle

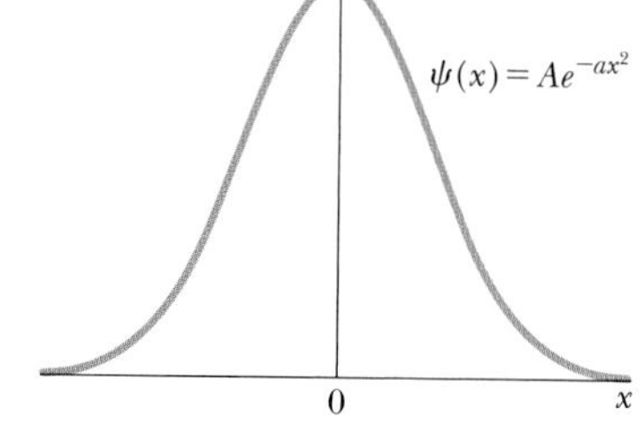

Figure 41.2 (Example 41.1) A symmetric wave function for a particle, given by $\psi(x) = Ae^{-ax^2}$.

[4] Expectation values are analogous to "weighted averages," in which each possible value of a function is multiplied by the probability of the occurrence of that value before summing over all possible values. We write the expectation value as $\int_{-\infty}^{\infty} \psi^* f(x) \psi \, dx$ rather than $\int_{-\infty}^{\infty} f(x) |\psi|^2 \, dx$ because $f(x)$ may be represented by an operator (such as a derivative) rather than a simple multiplicative function in more advanced treatments of quantum mechanics. In these situations, the operator is applied only to ψ and not to ψ^*.

always stays close to a given point. Examples of such systems are a block on a spring, a marble at the bottom of a bowl, and the bob of a simple pendulum.

Categorize Because the statement of the problem describes the wave nature of a particle, this example requires a quantum approach rather than a classical approach.

Analyze Apply the normalization condition, Equation 41.7, to the wave function:

$$\int_{-\infty}^{\infty} |\psi|^2\, dx = \int_{-\infty}^{\infty} (Ae^{-ax^2})^2\, dx = A^2 \int_{-\infty}^{\infty} e^{-2ax^2}\, dx = 1$$

Express the integral as the sum of two integrals:

$$(1)\quad A^2 \int_{-\infty}^{\infty} e^{-2ax^2}\, dx = A^2\left(\int_{0}^{\infty} e^{-2ax^2}\, dx + \int_{-\infty}^{0} e^{-2ax^2}\, dx\right) = 1$$

Change the integration variable from x to $-x$ in the second integral:

$$\int_{-\infty}^{0} e^{-2ax^2}\, dx = \int_{\infty}^{0} e^{-2a(-x)^2}(-dx) = -\int_{\infty}^{0} e^{-2ax^2}\, dx$$

Reverse the order of the limits, which introduces a negative sign:

$$-\int_{\infty}^{0} e^{-2ax^2}\, dx = \int_{0}^{\infty} e^{-2ax^2}\, dx$$

Substitute this expression for the second integral in Equation (1):

$$A^2\left(\int_{0}^{\infty} e^{-2ax^2}\, dx + \int_{0}^{\infty} e^{-2ax^2}\, dx\right) = 1$$

$$(2)\quad 2A^2 \int_{0}^{\infty} e^{-2ax^2}\, dx = 1$$

Evaluate the integral with the help of Table B.6 in Appendix B:

$$\int_{0}^{\infty} e^{-2ax^2}\, dx = \tfrac{1}{2}\sqrt{\frac{\pi}{2a}}$$

Substitute this result into Equation (2) and solve for A:

$$2A^2\left(\tfrac{1}{2}\sqrt{\frac{\pi}{2a}}\right) = 1 \quad\rightarrow\quad A = \left(\frac{2a}{\pi}\right)^{1/4}$$

(B) What is the expectation value of x for this particle?

SOLUTION

Evaluate the expectation value using Equation 41.8:

$$\langle x\rangle \equiv \int_{-\infty}^{\infty} \psi^* x \psi\, dx = \int_{-\infty}^{\infty} (Ae^{-ax^2})x(Ae^{-ax^2})\, dx$$

$$= A^2 \int_{-\infty}^{\infty} xe^{-2ax^2}\, dx$$

As in part (A), express the integral as a sum of two integrals:

$$(3)\quad \langle x\rangle = A^2\left(\int_{0}^{\infty} xe^{-2ax^2}\, dx + \int_{-\infty}^{0} xe^{-2ax^2}\, dx\right)$$

Change the integration variable from x to $-x$ in the second integral:

$$\int_{-\infty}^{0} xe^{-2ax^2}\, dx = \int_{\infty}^{0} -xe^{-2a(-x)^2}(-dx) = \int_{\infty}^{0} xe^{-2ax^2}\, dx$$

Reverse the order of the limits, which introduces a negative sign:

$$\int_{\infty}^{0} xe^{-2ax^2}\, dx = -\int_{0}^{\infty} xe^{-2ax^2}\, dx$$

Substitute this expression for the second integral in Equation (3):

$$\langle x \rangle = A^2\left(\int_0^\infty xe^{-2ax^2}\,dx - \int_0^\infty xe^{-2ax^2}\,dx\right) = 0$$

Finalize Given the symmetry of the wave function around $x = 0$ in Figure 41.2, it is not surprising that the average position of the particle is at $x = 0$. In Section 41.7, we show that the wave function studied in this example represents the lowest-energy state of the quantum harmonic oscillator.

41.2 The Quantum Particle Under Boundary Conditions

The free particle discussed in Section 41.1 has no boundary conditions; it can be anywhere in space. The particle in Example 41.1 is not a free particle. Figure 41.2 shows that the particle is always restricted to positions near $x = 0$. In this section, we shall investigate the effects of restrictions on the motion of a quantum particle.

A Particle in a Box

We begin by applying some of the ideas we have developed to a simple physical problem, a particle confined to a one-dimensional region of space, called the *particle-in-a-box* problem (even though the "box" is one-dimensional!). From a classical viewpoint, if a particle is bouncing elastically back and forth along the x axis between two impenetrable walls separated by a distance L as in Figure 41.3a, it can be modeled as a particle under constant speed. If the speed of the particle is u, the magnitude of its momentum mu remains constant as does its kinetic energy. (Recall that in Chapter 39 we used u for particle speed to distinguish it from v, the speed of a reference frame.) Classical physics places no restrictions on the values of a particle's momentum and energy. The quantum-mechanical approach to this problem is quite different and requires that we find the appropriate wave function consistent with the conditions of the situation.

Because the walls are impenetrable, there is zero probability of finding the particle outside the box, so the wave function $\psi(x)$ must be zero for $x < 0$ and $x > L$. To be a mathematically well-behaved function, $\psi(x)$ must be continuous in space. There must be no discontinuous jumps in the value of the wave function at any point.[5] Therefore, if ψ is zero outside the walls, it must also be zero *at* the walls; that is, $\psi(0) = 0$ and $\psi(L) = 0$. Only those wave functions that satisfy these boundary conditions are allowed.

Figure 41.3b, a graphical representation of the particle-in-a-box problem, shows the potential energy of the particle–environment system as a function of the position of the particle. As long as the particle is inside the box, the potential energy of the system does not depend on the location of the particle and we can choose its constant value to be zero. Outside the box, we must ensure that the wave function is zero. We can do so by defining the system's potential energy as infinitely large if the particle were outside the box. Therefore, the only way a particle could be outside the box is if the system has an infinite amount of energy, which is impossible.

The wave function for a particle in the box can be expressed as a real sinusoidal function:[6]

$$\psi(x) = A\sin\left(\frac{2\pi x}{\lambda}\right) \qquad \textbf{(41.10)}$$

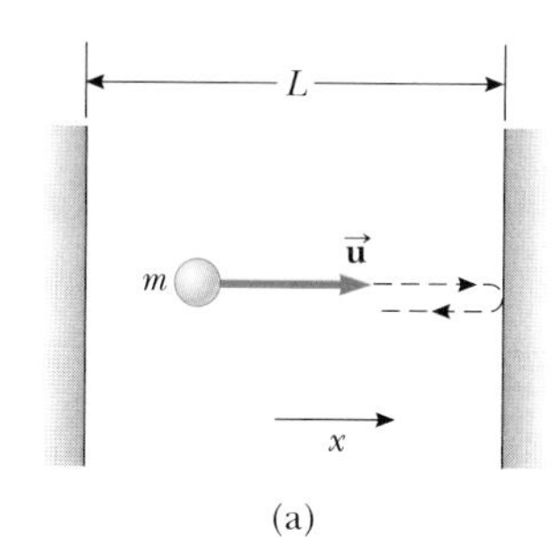

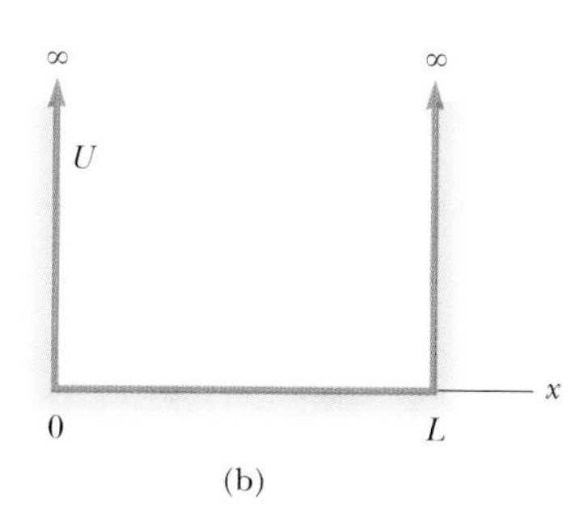

Figure 41.3 (a) A particle of mass m and speed u, confined to bouncing between two impenetrable walls separated by a distance L. (b) The potential energy function for the system.

[5] If the wave function were not continuous at a point, the derivative of the wave function at that point would be infinite. This result leads to difficulties in the Schrödinger equation, for which the wave function is a solution as discussed in Section 41.3.

[6] We shall show this result explicitly in Section 41.3.

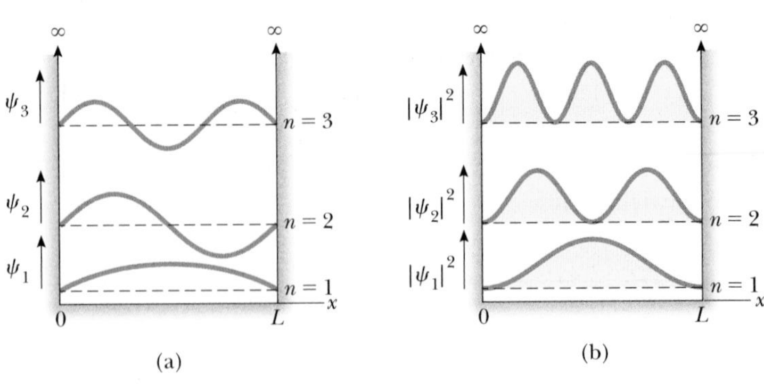

ACTIVE FIGURE 41.4

The first three allowed states for a particle confined to a one-dimensional box. The states are shown superimposed on the potential energy function of Figure 41.3b. (a) The wave functions ψ for $n = 1, 2,$ and 3. (b) The probability densities $|\psi|^2$ for $n = 1, 2,$ and 3. The wave functions and probability densities are plotted vertically from separate axes that are offset vertically for clarity. The positions of these axes on the potential energy function suggest the relative energies of the states.

Sign in at www.thomsonedu.com and go to ThomsonNOW to measure the probability of a particle being between two points for the three quantum states in the figure.

where λ is the de Broglie wavelength associated with the particle. This wave function must satisfy the boundary conditions at the walls. The boundary condition $\psi(0) = 0$ is satisfied already because the sine function is zero when $x = 0$. The boundary condition $\psi(L) = 0$ gives

$$\psi(L) = 0 = A \sin\left(\frac{2\pi L}{\lambda}\right)$$

which can only be true if

$$\frac{2\pi L}{\lambda} = n\pi \quad \rightarrow \quad \lambda = \frac{2L}{n} \qquad \textbf{(41.11)}$$

where $n = 1, 2, 3, \ldots$. Therefore, only certain wavelengths for the particle are allowed! Each of the allowed wavelengths corresponds to a quantum state for the system, and n is the quantum number. Incorporating Equation 41.11 in Equation 41.10 gives

Wave functions for a particle in a box ▶

$$\psi(x) = A \sin\left(\frac{2\pi x}{2L/n}\right) = A \sin\left(\frac{n\pi x}{L}\right) \qquad \textbf{(41.12)}$$

Normalizing this wave function shows that $A = \sqrt{2/L}$. (See Problem 15.) Therefore, the normalized wave function for the particle in a box is

Normalized wave function for a particle in a box ▶

$$\psi_n(x) = \sqrt{\frac{2}{L}} \sin\left(\frac{n\pi x}{L}\right) \qquad \textbf{(41.13)}$$

Active Figures 41.4a and b are graphical representations of ψ versus x and $|\psi|^2$ versus x for $n = 1, 2,$ and 3 for the particle in a box.[7] Although ψ can be positive or negative, $|\psi|^2$ is always positive. Because $|\psi|^2$ represents a probability density, a negative value for $|\psi|^2$ would be meaningless.

Further inspection of Active Figure 41.4b shows that $|\psi|^2$ is zero at the boundaries, satisfying our boundary conditions. In addition, $|\psi|^2$ is zero at other points, depending on the value of n. For $n = 2$, $|\psi|^2 = 0$ at $x = L/2$; for $n = 3$, $|\psi|^2 = 0$ at $x = L/3$ and at $x = 2L/3$. The number of zero points increases by one each time the quantum number increases by one.

[7] Note that $n = 0$ is not allowed because, according to Equation 41.12, the wave function would be $\psi = 0$, which is not a physically reasonable wave function. For example, it cannot be normalized because $\int_{-\infty}^{\infty} |\psi|^2\, dx = \int_{-\infty}^{\infty} (0)\, dx = 0$, but Equation 41.7 tells us that this integral must equal 1.

Because the wavelengths of the particle are restricted by the condition $\lambda = 2L/n$, the magnitude of the momentum of the particle is also restricted to specific values, which can be found from the expression for the de Broglie wavelength, Equation 40.15:

$$p = \frac{h}{\lambda} = \frac{h}{2L/n} = \frac{nh}{2L}$$

We have chosen the potential energy of the system to be zero when the particle is inside the box. Therefore, the energy of the system is simply the kinetic energy of the particle and the allowed values are given by

$$E_n = \tfrac{1}{2}mu^2 = \frac{p^2}{2m} = \frac{(nh/2L)^2}{2m}$$

$$E_n = \left(\frac{h^2}{8mL^2}\right)n^2 \quad n = 1, 2, 3, \ldots \qquad \textbf{(41.14)}$$

◀ Quantized energies for a particle in a box

This expression shows that **the energy of the particle is quantized.** The lowest allowed energy corresponds to the **ground state,** which is the lowest energy state for any system. For the particle in a box, the ground state corresponds to $n = 1$, for which $E_1 = h^2/8mL^2$. Because $E_n = n^2E_1$, the **excited states** corresponding to $n = 2, 3, 4, \ldots$ have energies given by $4E_1, 9E_1, 16E_1, \ldots$.

Active Figure 41.5 is an energy-level diagram describing the energy values of the allowed states. Because the lowest energy of the particle in a box is not zero, then, **according to quantum mechanics, the particle can never be at rest.** The smallest energy it can have, corresponding to $n = 1$, is called the **ground-state energy.** This result contradicts the classical viewpoint, in which $E = 0$ is an acceptable state, as are *all* positive values of E.

PITFALL PREVENTION 41.2
Reminder: Energy Belongs to a System

We often refer to the energy of a particle in commonly used language. As in Pitfall Prevention 41.1, we are actually describing the energy of the *system* of the particle and whatever environment is establishing the impenetrable walls. For the particle in a box, the only type of energy is kinetic energy belonging to the particle, which is the origin of the common description.

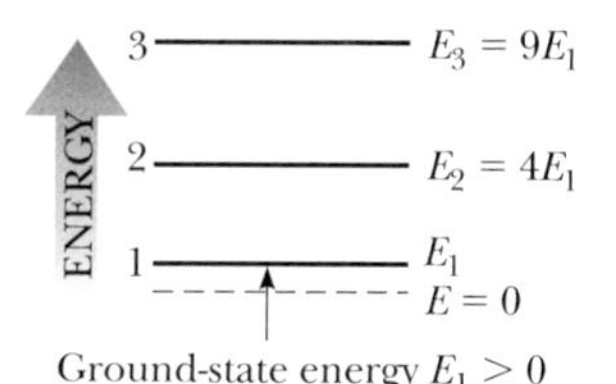

ACTIVE FIGURE 41.5

Energy-level diagram for a particle confined to a one-dimensional box of length L. The lowest allowed energy is $E_1 = h^2/8mL^2$.

Sign in at www.thomsonedu.com and go to ThomsonNOW to adjust the length of the box and the mass of the particle and see the effect on the energy levels.

Quick Quiz 41.2 Consider an electron, a proton, and an alpha particle (a helium nucleus), each trapped separately in identical boxes. **(i)** Which particle corresponds to the highest ground-state energy? (a) the electron (b) the proton (c) the alpha particle (d) The ground-state energy is the same in all three cases. **(ii)** Which particle has the longest wavelength when the system is in the ground state? (a) the electron (b) the proton (c) the alpha particle (d) All three particles have the same wavelength.

Quick Quiz 41.3 A particle is in a box of length L. Suddenly, the length of the box is increased to $2L$. What happens to the energy levels shown in Active Figure 41.5? (a) Nothing; they are unaffected. (b) They move farther apart. (c) They move closer together.

EXAMPLE 41.2 Microscopic and Macroscopic Particles in Boxes

(A) An electron is confined between two impenetrable walls 0.200 nm apart. Determine the energy levels for the states $n = 1$, 2, and 3.

SOLUTION

Conceptualize In Figure 41.3a, imagine that the particle is an electron and the walls are very close together.

Categorize We evaluate the energy levels using an equation developed in this section, so we categorize this example as a substitution problem.

Use Equation 41.14 for the $n = 1$ state:

$$E_1 = \frac{h^2}{8m_eL^2}(1)^2 = \frac{(6.63 \times 10^{-34}\,\text{J}\cdot\text{s})^2}{8(9.11 \times 10^{-31}\,\text{kg})(2.00 \times 10^{-10}\,\text{m})^2}$$

$$= 1.51 \times 10^{-18}\,\text{J} = 9.42\,\text{eV}$$

Using $E_n = n^2E_1$, find the energies of the $n = 2$ and $n = 3$ states:

$$E_2 = (2)^2E_1 = 4(9.42 \text{ eV}) = 37.7 \text{ eV}$$

$$E_3 = (3)^2E_1 = 9(9.42 \text{ eV}) = 84.8 \text{ eV}$$

(B) Find the speed of the electron in the $n = 1$ state.

SOLUTION

Solve the classical expression for kinetic energy for the particle speed:

$$K = \tfrac{1}{2}m_e u^2 \rightarrow u = \sqrt{\frac{2K}{m_e}}$$

Recognize that the kinetic energy of the particle is equal to the system energy and substitute E_n for K:

$$(1) \quad u = \sqrt{\frac{2E_n}{m_e}}$$

Substitute numerical values from part (A):

$$u = \sqrt{\frac{2(1.51 \times 10^{-18}\text{ J})}{9.11 \times 10^{-31}\text{ kg}}} = 1.82 \times 10^6 \text{ m/s}$$

Simply placing the electron in the box results in a *minimum* speed of the electron equal to 0.6% of the speed of light!

(C) A 0.500-kg baseball is confined between two rigid walls of a stadium that can be modeled as a box of length 100 m. Calculate the minimum speed of the baseball.

SOLUTION

Conceptualize In Figure 41.3a, imagine that the particle is a baseball and the walls are those of the stadium.

Categorize This part of the example is a substitution problem in which we apply a quantum approach to a macroscopic object.

Use Equation 41.14 for the $n = 1$ state:

$$E_1 = \frac{h^2}{8mL^2}(1)^2 = \frac{(6.63 \times 10^{-34}\text{ J}\cdot\text{s})^2}{8(0.500\text{ kg})(100\text{ m})^2} = 1.10 \times 10^{-71}\text{ J}$$

Use Equation (1) to find the speed:

$$u = \sqrt{\frac{2(1.10 \times 10^{-71}\text{ J})}{0.500\text{ kg}}} = 6.63 \times 10^{-36}\text{ m/s}$$

This speed is so small that the object can be considered to be at rest, which is what one would expect for the minimum speed of a macroscopic object.

What If? What if a sharp line drive is hit so that the baseball is moving with a speed of 150 m/s? What is the quantum number of the state in which the baseball now resides?

Answer We expect the quantum number to be very large because the baseball is a macroscopic object.

Evaluate the kinetic energy of the baseball:

$$\tfrac{1}{2}mu^2 = \tfrac{1}{2}(0.500\text{ kg})(150\text{ m/s})^2 = 5.62 \times 10^3\text{ J}$$

From Equation 41.14, calculate the quantum number n:

$$n = \sqrt{\frac{8mL^2E_n}{h^2}} = \sqrt{\frac{8(0.500\text{ kg})(100\text{ m})^2(5.62 \times 10^3\text{ J})}{(6.63 \times 10^{-34}\text{ J}\cdot\text{s})^2}} = 2.26 \times 10^{37}$$

This result is a tremendously large quantum number. As the baseball pushes air out of the way, hits the ground, and rolls to a stop, it moves through more than 10^{37} quantum states. These states are so close together in energy that we cannot observe the transitions from one state to the next. Rather, we see what appears to be a smooth variation in the speed of the ball. The quantum nature of the universe is simply not evident in the motion of macroscopic objects.

EXAMPLE 41.3 The Expectation Values for the Particle in a Box

A particle of mass m is confined to a one-dimensional box between $x = 0$ and $x = L$. Find the expectation value of the position x of the particle in the state characterized by quantum number n.

SOLUTION

Conceptualize Active Figure 41.4b shows that the probability for the particle to be at a given location varies with position within the box. Can you predict what the expectation value of x will be from the symmetry of the wave functions?

Categorize The statement of the example categorizes the problem for us: we focus on a quantum particle in a box and on the calculation of its expectation value of x.

Analyze In Equation 41.8, the integration from $-\infty$ to ∞ reduces to the limits 0 to L because $\psi = 0$ everywhere except in the box.

Substitute Equation 41.13 into Equation 41.8 to find the expectation value for x:

$$\langle x \rangle = \int_{-\infty}^{\infty} \psi_n^* x \psi_n \, dx = \int_0^L x\left[\sqrt{\frac{2}{L}} \sin\left(\frac{n\pi x}{L}\right)\right]^2 dx$$

$$= \frac{2}{L}\int_0^L x \sin^2\left(\frac{n\pi x}{L}\right) dx$$

Evaluate the integral by consulting an integral table or by mathematical integration:[8]

$$\langle x \rangle = \frac{2}{L}\left[\frac{x^2}{4} - \frac{x \sin\left(2\dfrac{n\pi x}{L}\right)}{4\dfrac{n\pi}{L}} - \frac{\cos\left(2\dfrac{n\pi x}{L}\right)}{8\left(\dfrac{n\pi}{L}\right)^2}\right]_0^L$$

$$= \frac{2}{L}\left[\frac{L^2}{4}\right] = \frac{L}{2}$$

Finalize This result shows that the expectation value of x is at the center of the box for all values of n, which you would expect from the symmetry of the square of the wave functions (the probability density) about the center (Active Fig. 41.4b).

The $n = 2$ wave function in Active Figure 41.4b has a value of zero at the midpoint of the box. Can the expectation value of the particle be at a position at which the particle has zero probability of existing? Remember that the expectation value is the *average* position. Therefore, the particle is as likely to be found to the right of the midpoint as to the left, so its average position is at the midpoint even though its probability of being there is zero. As an analogy, consider a group of students for whom the average final examination score is 50%. There is no requirement that some student achieve a score of exactly 50% for the average of all students to be 50%.

Boundary Conditions on Particles in General

The discussion of the particle in a box is very similar to the discussion in Chapter 18 of standing waves on strings:

- Because the ends of the string must be nodes, the wave functions for allowed waves must be zero at the boundaries of the string. Because the particle in a box cannot exist outside the box, the allowed wave functions for the particle must be zero at the boundaries.
- The boundary conditions on the string waves lead to quantized wavelengths and frequencies of the waves. The boundary conditions on the wave function for the particle in a box lead to quantized wavelengths and frequencies of the particle.

[8] To integrate this function, first replace $\sin^2(n\pi x/L)$ with $\frac{1}{2}(1 - \cos 2n\pi x/L)$ (refer to Table B.3 in Appendix B), which allows $\langle x \rangle$ to be expressed as two integrals. The second integral can then be evaluated by partial integration (Section B.7 in Appendix B).

In quantum mechanics, it is very common for particles to be subject to boundary conditions. We therefore introduce a new analysis model, the **quantum particle under boundary conditions.** In many ways, this model is similar to the waves under boundary conditions model studied in Section 18.3. In fact, the allowed wavelengths for the wave function of a particle in a box (Eq. 41.11) are identical in form to the allowed wavelengths for mechanical waves on a string fixed at both ends (Eq. 18.4).

The quantum particle under boundary conditions model *differs* in some ways from the waves under boundary conditions model:

- In most cases of quantum particles, the wave function is *not* a simple sinusoidal function like the wave function for waves on strings. Furthermore, the wave function for a quantum particle may be a complex function.
- For a quantum particle, frequency is related to energy through $E = hf$, so the quantized frequencies lead to quantized energies.
- There may be no stationary "nodes" associated with the wave function of a quantum particle under boundary conditions. Systems more complicated than the particle in a box have more complicated wave functions, and some boundary conditions may not lead to zeroes of the wave function at fixed points.

In general, **an interaction of a quantum particle with its environment represents one or more boundary conditions, and, if the interaction restricts the particle to a finite region of space, results in quantization of the energy of the system.**

Boundary conditions on quantum wave functions are related to the coordinates describing the problem. For the particle in a box, the wave function must be zero at two values of x. In the case of a three-dimensional system such as the hydrogen atom we shall discuss in Chapter 42, the problem is best presented in *spherical coordinates.* These coordinates, an extension of the plane polar coordinates introduced in Section 3.1, consist of a radial coordinate r and two angular coordinates. The generation of the wave function and application of the boundary conditions for the hydrogen atom are beyond the scope of this book. We shall, however, examine the behavior of some of the hydrogen-atom wave functions in Chapter 42.

Boundary conditions on wave functions that exist for all values of x require that the wave function approach zero as $x \to \infty$ (so that the wave function can be normalized) and remain finite as $x \to 0$. One boundary condition on any angular parts of wave functions is that adding 2π radians to the angle must return the wave function to the same value because an addition of 2π results in the same angular position.

Quick Quiz 41.4 Which of the following exhibit quantized energy levels? (a) an atom in a crystal (b) an electron and a proton in a hydrogen atom (c) a proton in the nucleus of a heavy atom (d) all of the above (e) none of the above

41.3 The Schrödinger Equation

In Section 34.3, we discussed a wave equation for electromagnetic radiation that follows from Maxwell's equations. The waves associated with particles also satisfy a wave equation. The wave equation for material particles is different from that associated with photons because material particles have a nonzero rest energy. The appropriate wave equation was developed by Schrödinger in 1926. In analyzing the behavior of a quantum system, the approach is to determine a solution to this equation and then apply the appropriate boundary conditions to the solution. This process yields the allowed wave functions and energy levels of the system under consideration. Proper manipulation of the wave function then enables one to calculate all measurable features of the system.

The Schrödinger equation as it applies to a particle of mass m confined to moving along the x axis and interacting with its environment through a potential energy function $U(x)$ is

$$-\frac{\hbar^2}{2m}\frac{d^2\psi}{dx^2} + U\psi = E\psi \qquad (41.15)$$

◀ Time-independent Schrödinger equation

where E is a constant equal to the total energy of the system (the particle and its environment). Because this equation is independent of time, it is commonly referred to as the **time-independent Schrödinger equation.** (We shall not discuss the time-dependent Schrödinger equation in this book.)

The Schrödinger equation is consistent with the principle of conservation of mechanical energy of a system. Problem 25 shows, both for a free particle and a particle in a box, that the first term in the Schrödinger equation reduces to the kinetic energy of the particle multiplied by the wave function. Therefore, Equation 41.15 indicates that the total energy of the system is the sum of the kinetic energy and the potential energy and that the total energy is a constant: $K + U = E =$ constant.

In principle, if the potential energy function U for a system is known, one can solve Equation 41.15 and obtain the wave functions and energies for the allowed states of the system. Because U may be discontinuous in position, it may be necessary to obtain solutions to the equation for different regions of the x axis. Solutions to the Schrödinger equation in different regions must join smoothly at the boundaries, so $\psi(x)$ must be *continuous.* In addition, $d\psi/dx$ must also be continuous for finite values of the potential energy.[9]

The task of solving the Schrödinger equation may be very difficult, depending on the form of the potential energy function. As it turns out, the Schrödinger equation is extremely successful in explaining the behavior of atomic and nuclear systems, whereas classical physics fails to explain this behavior. Furthermore, when quantum mechanics is applied to macroscopic objects, the results agree with classical physics.

ERWIN SCHRÖDINGER
Austrian Theoretical Physicist (1887–1961)
Schrödinger is best known as one of the creators of quantum mechanics. His approach to quantum mechanics was demonstrated to be mathematically equivalent to the more abstract matrix mechanics developed by Heisenberg. Schrödinger also produced important papers in the fields of statistical mechanics, color vision, and general relativity.

The Particle in a Box Revisited

To see how the quantum particle under boundary conditions model is applied to a problem, let's return to our particle in a one-dimensional box of length L (see Fig. 41.3) and analyze it with the Schrödinger equation. Figure 41.3b is the potential-energy diagram that describes this problem. Potential-energy diagrams are a useful representation for understanding and solving problems with the Schrödinger equation.

Because of the shape of the curve in Figure 41.3b, the particle in a box is sometimes said to be in a **square well,**[10] where a **well** is an upward-facing region of the curve in a potential-energy diagram. (A downward-facing region is called a *barrier,* which we investigate in Section 41.5.) Figure 41.3b shows an infinite square well.

In the region $0 < x < L$, where $U = 0$, we can express the Schrödinger equation in the form

$$\frac{d^2\psi}{dx^2} = -\frac{2mE}{\hbar^2}\psi = -k^2\psi \qquad (41.16)$$

where

$$k = \frac{\sqrt{2mE}}{\hbar}$$

PITFALL PREVENTION 41.3
Potential Wells

A potential well such as that in Figure 41.3b is a graphical representation of energy, not a pictorial representation, so you would not see this shape if you were able to observe the situation. A particle moves *only horizontally* at a fixed vertical position in a potential-energy diagram, representing the conserved energy of the system of the particle and its environment.

[9] If $d\psi/dx$ were not continuous, we would not be able to evaluate $d^2\psi/dx^2$ in Equation 41.15 at the point of discontinuity.

[10] It is called a square well even if it has a rectangular shape in a potential energy diagram.

The solution to Equation 41.16 is a function ψ whose second derivative is the negative of the same function multiplied by a constant k^2. Both the sine and cosine functions satisfy this requirement. Therefore, the most general solution to the equation is a linear combination of both solutions:

$$\psi(x) = A \sin kx + B \cos kx$$

where A and B are constants that are determined by the boundary and normalization conditions.

The first boundary condition on the wave function is that $\psi(0) = 0$:

$$\psi(0) = A \sin 0 + B \cos 0 = 0 + B = 0$$

which means that $B = 0$. Therefore, our solution reduces to

$$\psi(x) = A \sin kx$$

The second boundary condition, $\psi(L) = 0$, when applied to the reduced solution gives

$$\psi(L) = A \sin kL = 0$$

This equation could be satisfied by setting $A = 0$, but that would mean that $\psi = 0$ everywhere, which is not a valid wave function. The boundary condition is satisfied if kL is an integral multiple of π, that is, if $kL = n\pi$, where n is an integer. Substituting $k = \sqrt{2mE}/\hbar$ into this expression gives

$$kL = \frac{\sqrt{2mE}}{\hbar} L = n\pi$$

Each value of the integer n corresponds to a quantized energy that we call E_n. Solving for the allowed energies E_n gives

$$E_n = \left(\frac{h^2}{8mL^2}\right) n^2 \qquad \textbf{(41.17)}$$

which are identical to the allowed energies in Equation 41.14.

Substituting the values of k in the wave function, the allowed wave functions $\psi_n(x)$ are given by

$$\psi_n(x) = A \sin\left(\frac{n\pi x}{L}\right) \qquad \textbf{(41.18)}$$

which is the wave function (Eq. 41.12) used in our initial discussion of the particle in a box.

41.4 A Particle in a Well of Finite Height

Now consider a particle in a *finite* potential well, that is, a system having a potential energy that is zero when the particle is in the region $0 < x < L$ and a finite value U when the particle is outside this region as in Figure 41.6. Classically, if the total energy E of the system is less than U, the particle is permanently bound in the potential well. If the particle were outside the well, its kinetic energy would have to be negative, which is an impossibility. According to quantum mechanics, however, **a finite probability exists that the particle can be found outside the well even if $E < U$.** That is, the wave function ψ is generally nonzero outside the well—regions I and III in Figure 41.6—so the probability density $|\psi|^2$ is also nonzero in these regions. Although this notion may be uncomfortable to accept, the uncertainty principle indicates that the energy of the system is uncertain. This uncertainty allows the particle to be outside the well as long as the apparent violation of conservation of energy does not exist in any measurable way.

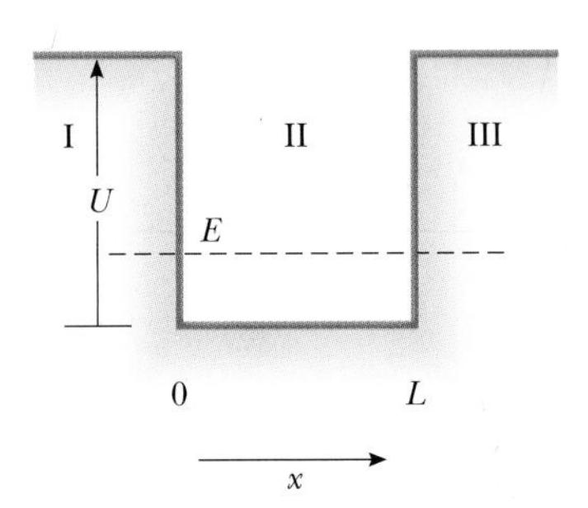

Figure 41.6 Potential energy diagram of a well of finite height U and length L. A particle is trapped in the well. The total energy E of the particle–well system is less than U.

In region II, where $U = 0$, the allowed wave functions are again sinusoidal because they represent solutions of Equation 41.16. The boundary conditions, however, no longer require that ψ be zero at the ends of the well, as was the case with the infinite square well.

The Schrödinger equation for regions I and III may be written

$$\frac{d^2\psi}{dx^2} = \frac{2m(U - E)}{\hbar^2}\psi \tag{41.19}$$

Because $U > E$, the coefficient of ψ on the right-hand side is necessarily positive. Therefore, we can express Equation 41.19 as

$$\frac{d^2\psi}{dx^2} = C^2\psi \tag{41.20}$$

where $C^2 = 2m(U - E)/\hbar^2$ is a positive constant in regions I and III. As you can verify by substitution, the general solution of Equation 41.20 is

$$\psi = Ae^{Cx} + Be^{-Cx} \tag{41.21}$$

where A and B are constants.

We can use this general solution as a starting point for determining the appropriate solution for regions I and III. The solution must remain finite as $x \rightarrow \pm\infty$. Therefore, in region I, where $x < 0$, the function ψ cannot contain the term Be^{-Cx}. This requirement is handled by taking $B = 0$ in this region to avoid an infinite value for ψ for large negative values of x. Likewise, in region III, where $x > L$, the function ψ cannot contain the term Ae^{Cx}. This requirement is handled by taking $A = 0$ in this region to avoid an infinite value for ψ for large positive x values. Hence, the solutions in regions I and III are

$$\psi_{\text{I}} = Ae^{Cx} \qquad \text{for } x < 0$$

$$\psi_{\text{III}} = Be^{-Cx} \qquad \text{for } x > L$$

In region II, the wave function is sinusoidal and has the general form

$$\psi_{\text{II}}(x) = F\sin kx + G\cos kx$$

where F and G are constants.

These results show that the wave functions outside the potential well (where classical physics forbids the presence of the particle) decay exponentially with distance. At large negative x values, ψ_{I} approaches zero; at large positive x values, ψ_{III} approaches zero. These functions, together with the sinusoidal solution in region II, are shown in Active Figure 41.7a for the first three energy states. In evaluating the complete wave function, we impose the following boundary conditions:

$$\psi_{\text{I}} = \psi_{\text{II}} \quad \text{and} \quad \frac{d\psi_{\text{I}}}{dx} = \frac{d\psi_{\text{II}}}{dx} \quad \text{at } x = 0$$

$$\psi_{\text{II}} = \psi_{\text{III}} \quad \text{and} \quad \frac{d\psi_{\text{II}}}{dx} = \frac{d\psi_{\text{III}}}{dx} \quad \text{at } x = L$$

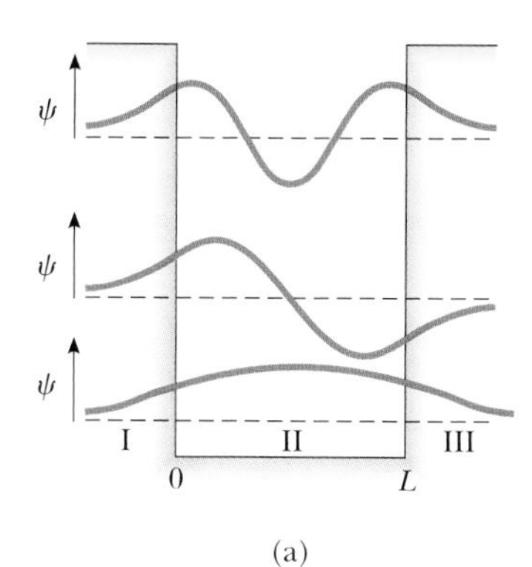

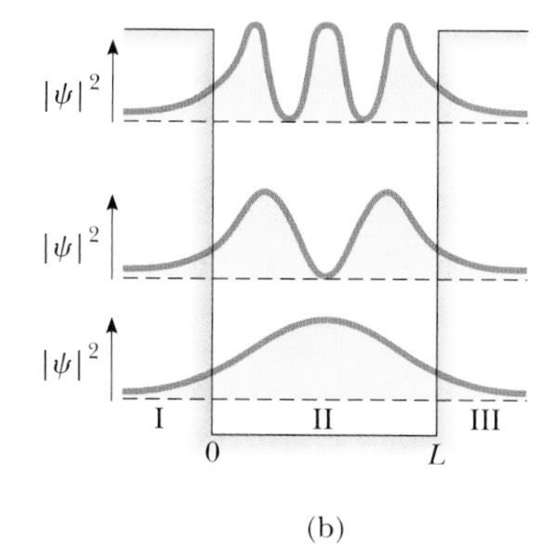

ACTIVE FIGURE 41.7

(a) Wave functions ψ and (b) probability densities $|\psi|^2$ for the lowest three energy states for a particle in a potential well of finite height. The states are shown superimposed on the potential energy function of Figure 41.6. The wave functions and probability densities are plotted vertically from separate axes that are offset vertically for clarity. The positions of these axes on the potential energy function suggest the relative energies of the states.

Sign in at www.thomsonedu.com and go to ThomsonNOW to adjust the length of the box and see the effect on the quantized states.

These four boundary conditions and the normalization condition (Eq. 41.7) are sufficient to determine the four constants A, B, F, and G and the allowed values of the energy E. Active Figure 41.7b plots the probability densities for these states. In each case, the wave functions inside and outside the potential well join smoothly at the boundaries.

The notion of trapping particles in potential wells is used in the burgeoning field of **nanotechnology,** which refers to the design and application of devices having dimensions ranging from 1 to 100 nm. The fabrication of these devices often involves manipulating single atoms or small groups of atoms to form structures such as the quantum corral shown in the opening photograph of this chapter.

One area of nanotechnology of interest to researchers is the **quantum dot,** a small region that is grown in a silicon crystal and acts as a potential well. This region can trap electrons into states with quantized energies. The wave functions for a particle in a quantum dot look similar to those in Active Figure 41.7a if L is on the order of nanometers. The storage of binary information using quantum dots is an active field of research. A simple binary scheme would involve associating a one with a quantum dot containing an electron and a zero with an empty dot. Other schemes involve cells of multiple dots such that arrangements of electrons among the dots correspond to ones and zeroes. Several research laboratories are studying the properties and potential applications of quantum dots. Information should be forthcoming from these laboratories at a steady rate in the next few years.

41.5 Tunneling Through a Potential Energy Barrier

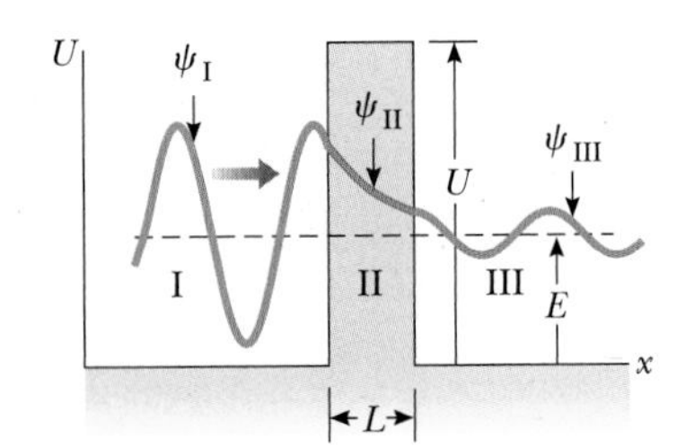

Figure 41.8 Wave function ψ for a particle incident from the left on a barrier of height U and width L. The wave function is sinusoidal in regions I and III, but is exponentially decaying in region II. The wave function is plotted vertically from an axis positioned at the energy of the particle.

Consider the potential energy function shown in Figure 41.8. In this situation, the potential energy has a constant value of U in the region of width L and is zero in all other regions.[11] A potential energy function of this shape is called a **square barrier,** and U is called the **barrier height.** A very interesting and peculiar phenomenon occurs when a moving particle encounters such a barrier of finite height and width. Suppose a particle of energy $E < U$ is incident on the barrier from the left (Fig. 41.8). Classically, the particle is reflected by the barrier. If the particle were located in region II, its kinetic energy would be negative, which is not classically allowed. Consequently, region II and therefore region III are both classically *forbidden* to the particle incident from the left. According to quantum mechanics, however, **all regions are accessible to the particle, regardless of its energy.** (Although all regions are accessible, the probability of the particle being in a classically forbidden region is very low.) According to the uncertainty principle, the particle could be within the barrier as long as the time interval during which it is in the barrier is short and consistent with Equation 40.24. If the barrier is relatively narrow, this short time interval can allow the particle to move through the barrier.

Let's approach this situation using a mathematical representation. The Schrödinger equation has valid solutions in all three regions. The solutions in regions I and III are sinusoidal like Equation 41.12. In region II, the solution is exponential like Equation 41.21. Applying the boundary conditions that the wave functions in the three regions and their derivatives must join smoothly at the boundaries, a full solution, such as the one represented by the curve in Figure 41.8, can be found. Because the probability of locating the particle is proportional to $|\psi|^2$, the probability of finding the particle beyond the barrier in region III is nonzero. This result is in complete disagreement with classical physics. The movement of the particle to the far side of the barrier is called **tunneling** or **barrier penetration.**

The probability of tunneling can be described with a **transmission coefficient** T and a **reflection coefficient** R. The transmission coefficient represents the probability that the particle penetrates to the other side of the barrier, and the reflection

PITFALL PREVENTION 41.4
"Height" on an Energy Diagram

The word *height* (as in *barrier height*) refers to an energy in discussions of barriers in potential energy diagrams. For example, we might say the height of the barrier is 10 eV. On the other hand, the barrier *width* refers to the traditional usage of such a word and is an actual physical length measurement between the two locations of the vertical sides of the barrier.

[11] It is common in physics to refer to L as the *length* of a well but the *width* of a barrier.

coefficient is the probability that the particle is reflected by the barrier. Because the incident particle is either reflected or transmitted, we require that $T + R = 1$. An approximate expression for the transmission coefficient that is obtained in the case of $T \ll 1$ (a very wide barrier or a very high barrier, that is, $U \gg E$) is

$$T \approx e^{-2CL} \quad \textbf{(41.22)}$$

where

$$C = \frac{\sqrt{2m(U - E)}}{\hbar} \quad \textbf{(41.23)}$$

This quantum model of barrier penetration and specifically Equation 41.22 show that T can be nonzero. That the phenomenon of tunneling is observed experimentally provides further confidence in the principles of quantum physics.

Quick Quiz 41.5 Which of the following changes would increase the probability of transmission of a particle through a potential barrier? (You may choose more than one answer.) (a) decreasing the width of the barrier (b) increasing the width of the barrier (c) decreasing the height of the barrier (d) increasing the height of the barrier (e) decreasing the kinetic energy of the incident particle (f) increasing the kinetic energy of the incident particle

---

EXAMPLE 41.4 Transmission Coefficient for an Electron

A 30-eV electron is incident on a square barrier of height 40 eV.

(A) What is the probability that the electron tunnels through the barrier if its width is 1.0 nm?

SOLUTION

Conceptualize Because the particle energy is smaller than the height of the potential barrier, we expect the electron to reflect from the barrier with a probability of 100% according to classical physics. Because of the tunneling phenomenon, however, there is a finite probability that the particle can appear on the other side of the barrier.

Categorize We evaluate the probability using an equation developed in this section, so we categorize this example as a substitution problem.

Evaluate the quantity $U - E$ that appears in Equation 41.23:

$$U - E = 40 \text{ eV} - 30 \text{ eV} = 10 \text{ eV}\left(\frac{1.6 \times 10^{-19}\text{ J}}{1 \text{ eV}}\right) = 1.6 \times 10^{-18}\text{ J}$$

Evaluate the quantity $2CL$ using Equation 41.23:

$$(1)\quad 2CL = 2\frac{\sqrt{2(9.11 \times 10^{-31}\text{ kg})(1.6 \times 10^{-18}\text{ J})}}{1.055 \times 10^{-34}\text{ J}\cdot\text{s}}(1.0 \times 10^{-9}\text{ m}) = 32.4$$

From Equation 41.22, find the probability of tunneling through the barrier:

$$T \approx e^{-2CL} = e^{-32.4} = 8.5 \times 10^{-15}$$

(B) What is the probability that the electron tunnels through the barrier if its width is 0.10 nm?

SOLUTION

In this case, the width L in Equation (1) is one-tenth as large, so evaluate the new value of $2CL$:

$$2CL = (0.1)(32.4) = 3.24$$

From Equation 41.22, find the new probability of tunneling through the barrier:

$$T \approx e^{-2CL} = e^{-3.24} = 0.039$$

In part (A), the electron has approximately 1 chance in 10^{14} of tunneling through the barrier. In part (B), however, the electron has a much higher probability (3.9%) of penetrating the barrier. Therefore, reducing the width of the barrier by only one order of magnitude increases the probability of tunneling by about 12 orders of magnitude!

41.6 Applications of Tunneling

As we have seen, tunneling is a quantum phenomenon, a manifestation of the wave nature of matter. Many examples exist (on the atomic and nuclear scales) for which tunneling is very important.

Alpha Decay

One form of radioactive decay is the emission of alpha particles (the nuclei of helium atoms) by unstable, heavy nuclei (Chapter 44). To escape from the nucleus, an alpha particle must penetrate a barrier whose height is several times larger than the energy of the nucleus–alpha particle system. The barrier results from a combination of the attractive nuclear force (discussed in Chapter 44) and the Coulomb repulsion (discussed in Chapter 23) between the alpha particle and the rest of the nucleus. Occasionally, an alpha particle tunnels through the barrier, which explains the basic mechanism for this type of decay and the large variations in the mean lifetimes of various radioactive nuclei.

Figure 41.8 shows the wave function of a particle tunneling through a barrier in one dimension. A similar wave function having spherical symmetry describes the barrier penetration of an alpha particle leaving a radioactive nucleus. The wave function exists both inside and outside the nucleus, and its amplitude is constant in time. In this way, the wave function correctly describes the small but constant probability that the nucleus will decay. The moment of decay cannot be predicted. In general, quantum mechanics implies that the future is indeterminate. This feature is in contrast to classical mechanics, from which the trajectory of an object can be calculated to arbitrarily high precision from precise knowledge of its initial position and velocity and of the forces exerted on it. Do not think that the future is undetermined simply because we have incomplete information about the present. The wave function contains all the information about the state of a system. Sometimes precise predictions can be made, such as the energy of a bound system, but sometimes only probabilities can be calculated about the future. The fundamental laws of nature are probabilistic. Therefore, it appears that Einstein's famous statement about quantum mechanics, "God does not roll dice," was wrong.

A radiation detector can be used to show that a nucleus decays by emitting a particle at a particular moment and in a particular direction. To point out the contrast between this experimental result and the wave function describing it, Schrödinger imagined a box containing a cat, a radioactive sample, a radiation counter, and a vial of poison. When a nucleus in the sample decays, the counter triggers the administration of lethal poison to the cat. Quantum mechanics correctly predicts the probability of finding the cat dead when the box is opened. Before the box is opened, does the cat have a wave function describing it as fractionally dead, with some chance of being alive?

This question is under continuing investigation, never with actual cats but sometimes with interference experiments building upon the experiment described in Section 40.7. Does the act of measurement change the system from a probabilistic to a definite state? When a particle emitted by a radioactive nucleus is detected at one particular location, does the wave function describing the particle drop instantaneously to zero everywhere else in the Universe? (Einstein called such a state change a "spooky action at a distance.") Is there a fundamental difference between a quantum system and a macroscopic system? The answers to these questions are unknown.

Nuclear Fusion

The basic reaction that powers the Sun and, indirectly, almost everything else in the solar system is fusion, which we shall study in Chapter 45. In one step of the process that occurs at the core of the Sun, protons must approach one another to within such a small distance that they fuse and form a deuterium nucleus. (See Section 45.4.) According to classical physics, these protons cannot overcome and penetrate the barrier caused by their mutual electrical repulsion. Quantum mechanically, however, the protons are able to tunnel through the barrier and fuse together.

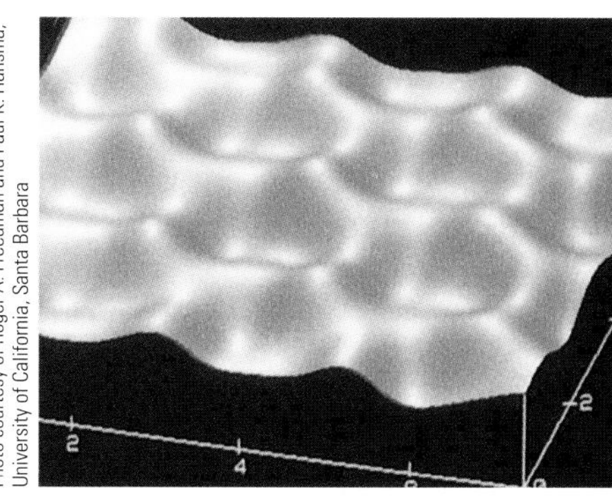

Photo courtesy of Roger A. Freedman and Paul K. Hansma, University of California, Santa Barbara

Figure 41.9 The surface of graphite as "viewed" with a scanning tunneling microscope. This type of microscope enables scientists to see details with a lateral resolution of about 0.2 nm and a vertical resolution of 0.001 nm. The contours seen here represent the ring-like arrangement of individual carbon atoms on the crystal surface.

Scanning Tunneling Microscopes

The scanning tunneling microscope (STM) enables scientists to obtain highly detailed images of surfaces at resolutions comparable to the size of a *single atom.* Figure 41.9, showing the surface of a piece of graphite, demonstrates what STMs can do. What makes this image so remarkable is that its resolution is approximately 0.2 nm. For an optical microscope, the resolution is limited by the wavelength of the light used to make the image. Therefore, an optical microscope has a resolution no better than 200 nm, about half the wavelength of visible light, and so could never show the detail displayed in Figure 41.9.

Scanning tunneling microscopes achieve such high resolution by using the basic idea shown in Figure 41.10. An electrically conducting probe with a very sharp tip is brought near the surface to be studied. The empty space between tip and surface represents the "barrier" we have been discussing, and the tip and surface are the two walls of the "potential well." Because electrons obey quantum rules rather than Newtonian rules, they can "tunnel" across the barrier of empty space. If a voltage is applied between surface and tip, electrons in the atoms of the surface material can tunnel preferentially from surface to tip to produce a tunneling current. In this way, the tip samples the distribution of electrons immediately above the surface.

In the empty space between tip and surface, the electron wave function falls off exponentially (see region II in Fig. 41.8 and Example 41.4). For tip-to-surface distances $z > 1$ nm (that is, beyond a few atomic diameters), essentially no tunneling takes place. This exponential behavior causes the current of electrons tunneling from surface to tip to depend very strongly on z. By monitoring the tunneling current as the tip is scanned over the surface, scientists obtain a sensitive measure of the topography of the electron distribution on the surface. The result of this scan is used to make images like that in Figure 41.9. In this way, the STM can measure the height of surface features to within 0.001 nm, approximately 1/100 of an atomic diameter!

You can appreciate the sensitivity of STMs by examining Figure 41.9. Of the six carbon atoms in each ring, three appear lower than the other three. In fact, all six atoms are at the same height, but all have slightly different electron distributions. The three atoms that appear lower are bonded to other carbon atoms directly beneath them in the underlying atomic layer; as a result, their electron distributions, which are responsible for the bonding, extend downward beneath the surface. The atoms in the surface layer that appear higher do not lie directly over subsurface atoms and hence are not bonded to any underlying atoms. For these higher-appearing atoms, the electron distribution extends upward into the space above the surface. Because STMs map the topography of the electron distribution, this extra electron density makes these atoms appear higher in Figure 41.9.

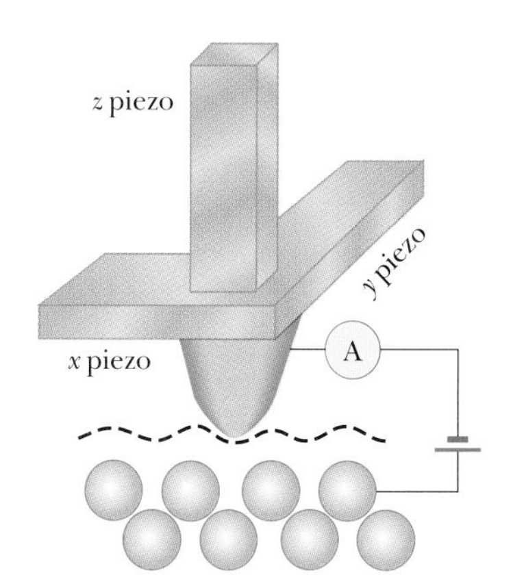

Figure 41.10 Schematic view of a scanning tunneling microscope. A scan of the tip over the sample can reveal surface contours down to the atomic level. An STM image is composed of a series of scans displaced laterally from one another. (Based on a drawing from P. K. Hansma, V. B. Elings, O. Marti, and C. Bracker, *Science* **242:** 209, 1988. © 1988 by the AAAS.)

The STM has one serious limitation: Its operation depends on the electrical conductivity of the sample and the tip. Unfortunately, most materials are not electrically conductive at their surfaces. Even metals, which are usually excellent electrical conductors, are covered with nonconductive oxides. A newer microscope, the atomic force microscope, or AFM, overcomes this limitation.

Resonant Tunneling Devices

Let's expand on the quantum-dot discussion in Section 41.4 by exploring the **resonant tunneling device.** Active Figure 41.11a (page 1204) shows the physical

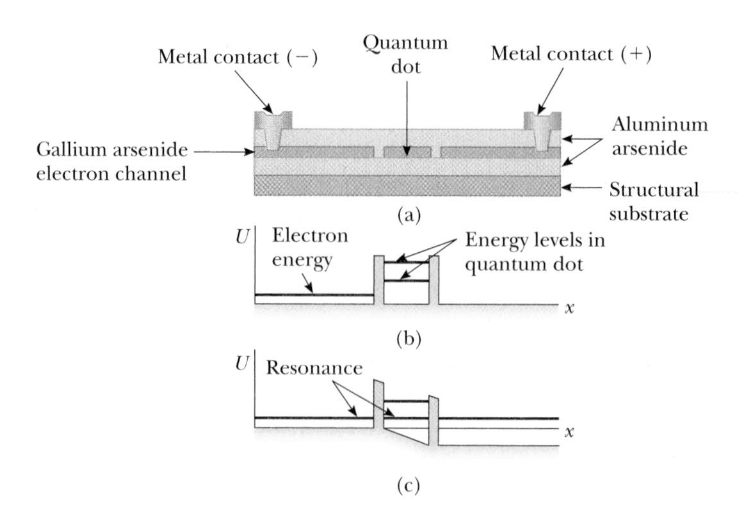

ACTIVE FIGURE 41.11

(a) The physical structure of a resonant tunneling device. (b) A potential energy diagram showing the double barrier representing the walls of the quantum dot. (c) A voltage is applied across the device. The distortion of the potential energy curve causes one of the states in the quantum dot to resonate with the incident electron energy.

Sign in at www.thomsonedu.com and go to ThomsonNOW to vary the voltage across the resonant tunneling device.

construction of such a device. The island of gallium arsenide in the center is a quantum dot located between two barriers formed from the thin extensions of aluminum arsenide. Active Figure 41.11b shows both the potential barriers encountered by electrons incident from the left and the quantized energy levels in the quantum dot. This situation differs from the one shown in Figure 41.8 in that there are quantized energy levels on the right of the first barrier. In Figure 41.8, an electron that tunnels through the barrier is considered a free particle and can have any energy. In contrast, in Active Figure 41.11b, as the electron with the energy shown encounters the first barrier it has no energy levels available on the right side of the barrier, which greatly reduces the probability of tunneling.

Active Figure 41.11c shows the effect of applying a voltage: the potential decreases with position as we move to the right across the device. The deformation of the potential barrier results in an energy level in the quantum dot coinciding with the energy of the incident electrons. This "resonance" of energies gives the device its name. When the voltage is applied, the probability of tunneling increases tremendously and the device carries current. In this manner, the device can be used as a very fast switch on a nanotechnological scale.

Resonant Tunneling Transistors

Figure 41.12a shows the addition of a gate electrode at the top of the resonant tunneling device over the quantum dot. This electrode turns the device into a **resonant tunneling transistor.** The basic function of a transistor is amplification, converting a small varying voltage into a large varying voltage. Figure 41.12b, representing the potential-energy diagram for the tunneling transistor, has a slope at the bottom of the quantum dot due to the differing voltages at the source and

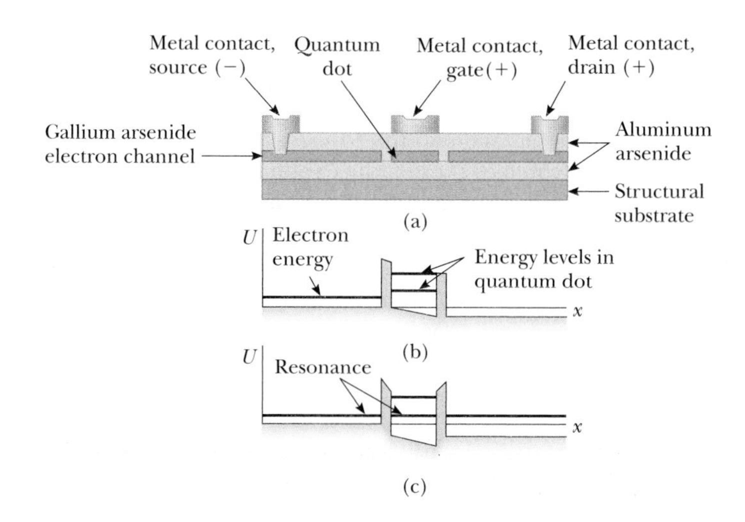

Figure 41.12 (a) The addition of a gate electrode to the structure in Active Figure 41.11 converts it to a resonant tunneling transistor. (b) A potential energy diagram showing the double barrier representing the walls of the quantum dot. (c) A voltage is applied to the gate electrode. The potential in the region of the quantum dot drops, along with the quantized energy levels.

drain electrodes. In this configuration, there is no resonance between the electron energies outside the quantum dot and the quantized energies within the dot. By applying a small voltage to the gate electrode as in Figure 41.12c, the quantized energies can be brought into resonance with the electron energy outside the well and resonant tunneling occurs. The resulting current causes a voltage across an external resistor that is much larger than that of the gate voltage; hence, the device amplifies the input signal to the gate electrode.

41.7 The Simple Harmonic Oscillator

Consider a particle that is subject to a linear restoring force $F = -kx$, where k is a constant and x is the position of the particle relative to equilibrium ($x = 0$). The classical motion of a particle subject to such a force is simple harmonic motion, which was discussed in Chapter 15. The potential energy of the system is, from Equation 15.20,

$$U = \tfrac{1}{2}kx^2 = \tfrac{1}{2}m\omega^2x^2$$

where the angular frequency of vibration is $\omega = \sqrt{k/m}$. Classically, if the particle is displaced from its equilibrium position and released, it oscillates between the points $x = -A$ and $x = A$, where A is the amplitude of the motion. Furthermore, its total energy E is, from Equation 15.21,

$$E = K + U = \tfrac{1}{2}kA^2 = \tfrac{1}{2}m\omega^2A^2$$

In the classical model, any value of E is allowed, including $E = 0$, which is the total energy when the particle is at rest at $x = 0$.

Let's investigate how the simple harmonic oscillator is treated from a quantum point of view. The Schrödinger equation for this problem is obtained by substituting $U = \frac{1}{2}m\omega^2x^2$ into Equation 41.15:

$$-\frac{\hbar^2}{2m}\frac{d^2\psi}{dx^2} + \tfrac{1}{2}m\omega^2x^2\psi = E\psi \qquad \textbf{(41.24)}$$

The mathematical technique for solving this equation is beyond the level of this book; nonetheless, it is instructive to guess at a solution. We take as our guess the following wave function:

$$\psi = Be^{-Cx^2} \qquad \textbf{(41.25)}$$

Substituting this function into Equation 41.24 shows that it is a satisfactory solution to the Schrödinger equation, provided that

$$C = \frac{m\omega}{2\hbar} \quad \text{and} \quad E = \tfrac{1}{2}\hbar\omega$$

It turns out that the solution we have guessed corresponds to the ground state of the system, which has an energy $\frac{1}{2}\hbar\omega$. Because $C = m\omega/2\hbar$, it follows from Equation 41.25 that the wave function for this state is

$$\psi = Be^{-(m\omega/2\hbar)x^2} \qquad \textbf{(41.26)}$$

◀ Wave function for the ground state of a simple harmonic oscillator

where B is a constant to be determined from the normalization condition. This result is but one solution to Equation 41.24. The remaining solutions that describe the excited states are more complicated, but all solutions include the exponential factor e^{-Cx^2}.

The energy levels of a harmonic oscillator are quantized as we would expect because the oscillating particle is bound to stay near $x = 0$. The energy of a state having an arbitrary quantum number n is

$$E_n = (n + \tfrac{1}{2})\hbar\omega \quad n = 0, 1, 2, \ldots \qquad \textbf{(41.27)}$$

◀ Quantized energies for a simple harmonic oscillator

The state $n = 0$ corresponds to the ground state, whose energy is $E_0 = \frac{1}{2}\hbar\omega$; the state $n = 1$ corresponds to the first excited state, whose energy is $E_1 = \frac{3}{2}\hbar\omega$; and so

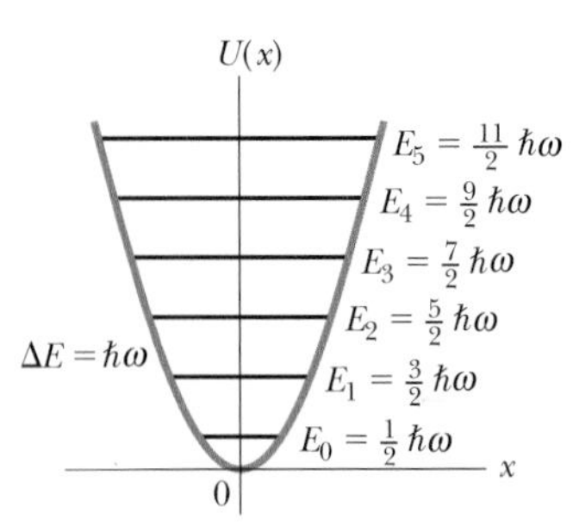

Figure 41.13 Energy-level diagram for a simple harmonic oscillator, superimposed on the potential energy function. The levels are equally spaced, with separation $\hbar\omega$. The ground-state energy is $E_0 = \frac{1}{2}\hbar\omega$.

on. The energy-level diagram for this system is shown in Figure 41.13. The separations between adjacent levels are equal and given by

$$\Delta E = \hbar\omega \quad \textbf{(41.28)}$$

Notice that the energy levels for the harmonic oscillator in Figure 41.13 are equally spaced, just as Planck proposed for the oscillators in the walls of the cavity that was used in the model for blackbody radiation in Section 40.1. Planck's Equation 40.4 for the energy levels of the oscillators differs from Equation 41.27 only in the term $\frac{1}{2}$ added to n. This additional term does not affect the energy emitted in a transition, given by Equation 40.5, which is equivalent to Equation 41.28. That Planck generated these concepts without the benefit of the Schrödinger equation is testimony to his genius.

EXAMPLE 41.5 Molar Specific Heat of Hydrogen Gas

In Figure 21.7 (Section 21.4), which shows the molar specific heat of hydrogen as a function of temperature, vibration does not contribute to the molar specific heat at room temperature. Explain why, modeling the hydrogen molecule as a simple harmonic oscillator. The effective spring constant for the bond in the hydrogen molecule is 573 N/m.

SOLUTION

Conceptualize Imagine the only mode of vibration available to a diatomic molecule. This mode (shown in Fig. 21.6c) consists of the two atoms always moving in opposite directions with equal speeds.

Categorize We categorize this example as a quantum harmonic oscillator problem, with the molecule modeled as a two-particle system.

Analyze The motion of the particles relative to the center of mass can be analyzed by considering the oscillation of a single particle with reduced mass μ. (See Problem 39.)

Use the result of Problem 39 to evaluate the reduced mass of the hydrogen molecule, in which the masses of the two particles are the same:

$$\mu = \frac{m_1 m_2}{m_1 + m_2} = \frac{m^2}{2m} = \frac{1}{2}m$$

Using Equation 41.28, calculate the energy necessary to excite the molecule from its ground vibrational state to its first excited vibrational state:

$$\Delta E = \hbar\omega = \hbar\sqrt{\frac{k}{\mu}} = \hbar\sqrt{\frac{k}{\frac{1}{2}m}} = \hbar\sqrt{\frac{2k}{m}}$$

Substitute numerical values, noting that m is the mass of a hydrogen atom:

$$\Delta E = (1.055 \times 10^{-34}\,\text{J}\cdot\text{s})\sqrt{\frac{2(573\,\text{N/m})}{1.67 \times 10^{-27}\,\text{kg}}} = 8.74 \times 10^{-20}\,\text{J}$$

Set this energy equal to $\frac{3}{2}k_BT$ from Equation 21.4 and find the temperature at which the average molecular translational kinetic energy is equal to that required to excite the first vibrational state of the molecule:

$$\tfrac{3}{2}k_BT = 8.74 \times 10^{-20}\,\text{J}$$

$$T = \tfrac{2}{3}\left(\frac{8.74 \times 10^{-20}\,\text{J}}{k_B}\right) = \tfrac{2}{3}\left(\frac{8.74 \times 10^{-20}\,\text{J}}{1.38 \times 10^{-23}\,\text{J/K}}\right) = 4.22 \times 10^3\,\text{K}$$

Finalize The temperature of the gas must be more than 4 000 K for the translational kinetic energy to be comparable to the energy required to excite the first vibrational state. This excitation energy must come from collisions between molecules, so if the molecules do not have sufficient translational kinetic energy, they cannot be excited to the first vibrational state and vibration does not contribute to the molar specific heat. Hence, the curve in Figure 21.7 does not rise to a value corresponding to the contribution of vibration until the hydrogen gas has been raised to thousands of kelvins.

Figure 21.7 shows that rotational energy levels must be more closely spaced in energy than vibrational levels because they are excited at a lower temperature than the vibrational levels. The translational energy levels are those of a particle in a three-dimensional box, where the box is the container holding the gas. These levels are given by an expression similar to Equation 41.14. Because the box is macroscopic in size, L is very large and the energy levels are very close together. In fact, they are so close together that translational energy levels are excited at a fraction of a kelvin.

Summary

ThomsonNOW™ Sign in at **www.thomsonedu.com** and go to ThomsonNOW to take a practice test for this chapter.

DEFINITIONS

The **wave function** Ψ for a system is a mathematical function that can be written as a product of a space function ψ for one particle of the system and a complex time function:

$$\Psi(\vec{\mathbf{r}}_1, \vec{\mathbf{r}}_2, \vec{\mathbf{r}}_3, \ldots, \vec{\mathbf{r}}_j, \ldots, t) = \psi(\vec{\mathbf{r}}_j)e^{-i\omega t} \qquad \textbf{(41.2)}$$

where ω (= $2\pi f$) is the angular frequency of the wave function and $i = \sqrt{-1}$. The wave function contains within it all the information that can be known about the particle.

The measured position x of a particle, averaged over many trials, is called the **expectation value** of x and is defined by

$$\langle x \rangle \equiv \int_{-\infty}^{\infty} \psi^* x \psi \, dx \qquad \textbf{(41.8)}$$

CONCEPTS AND PRINCIPLES

In quantum mechanics, a particle in a system can be represented by a wave function $\psi(x, y, z)$. The probability per unit volume (or probability density) that a particle will be found at a point is $|\psi|^2 = \psi^*\psi$, where ψ^* is the complex conjugate of ψ. If the particle is confined to moving along the x axis, the probability that it is located in an interval dx is $|\psi|^2\, dx$. Furthermore, the sum of all these probabilities over all values of x must be 1:

$$\int_{-\infty}^{\infty} |\psi|^2 dx = 1 \qquad \textbf{(41.7)}$$

This expression is called the **normalization condition.**

If a particle of mass m is confined to moving in a one-dimensional box of length L whose walls are impenetrable, then ψ must be zero at the walls and outside the box. The wave functions for this system are given by

$$\psi(x) = A \sin\left(\frac{n\pi x}{L}\right) \qquad n = 1, 2, 3, \ldots \qquad \textbf{(41.12)}$$

where A is the maximum value of ψ. The allowed states of a particle in a box have quantized energies given by

$$E_n = \left(\frac{h^2}{8mL^2}\right)n^2 \qquad n = 1, 2, 3, \ldots \qquad \textbf{(41.14)}$$

The wave function for a system must satisfy the **Schrödinger equation.** The time-independent Schrödinger equation for a particle confined to moving along the x axis is

$$-\frac{\hbar^2}{2m}\frac{d^2\psi}{dx^2} + U\psi = E\psi \qquad \textbf{(41.15)}$$

where U is the potential energy of the system and E is the total energy.

(continued)

ANALYSIS MODEL FOR PROBLEM SOLVING

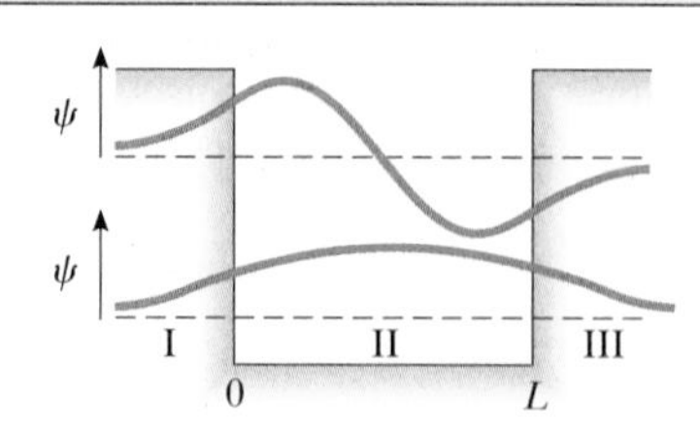

Quantum Particle Under Boundary Conditions. An interaction of a quantum particle with its environment represents one or more boundary conditions. If the interaction restricts the particle to a finite region of space, the energy of the system is quantized. All wave functions must satisfy the following four boundary conditions: (1) $\psi(x)$ must remain finite as x approaches 0, (2) $\psi(x)$ must approach zero as x approaches $\pm\infty$, (3) $\psi(x)$ must be continuous for all values of x, and (4) $d\psi/dx$ must be continuous for all finite values of $U(x)$. If the solution to Equation 41.15 is piecewise, conditions (3) and (4) must be applied at the boundaries between regions of x in which Equation 41.15 has been solved.

Questions

☐ denotes answer available in *Student Solutions Manual/Study Guide;* **O** denotes objective question

1. What is the significance of the wave function ψ?

2. O The probability of finding a certain quantum particle in the section of the x axis between $x = 4$ nm and $x = 7$ nm is 48%. The particle's wave function $\psi(x)$ is constant over this range. **(i)** What numerical value can be attributed to $\psi(x)$? (a) 0.48 (b) 0.16 (c) 0.12 (d) 0.69 (e) 0.40 (f) 0.35 **(ii)** What is its unit? (a) nm (b) $(\text{nm})^{1/2}$ (c) $(\text{nm})^{1/3}$ (d) $(\text{nm})^{-1}$ (e) $(\text{nm})^{-1/2}$ (f) $(\text{nm})^{-1/3}$

3. O (i) Is each one of the following statements (a) through (j) true or false for a photon? **(ii)** Is each one of the statements true or false for an electron? Make a list of your answers.

(a) It is a quantum particle, behaving in some experiments like a classical particle and in some experiments like a classical wave.
(b) Its rest energy is zero.
(c) Its rest energy is nonzero.
(d) It carries energy in its motion.
(e) It carries momentum in its motion.
(f) Its motion is described by a wave function that has a wavelength and satisfies a wave equation.
(g) For one-dimensional motion, the wave equation is

$$\frac{\partial^2 E}{\partial x^2} = \mu_0 \epsilon_0 \frac{\partial^2 E}{\partial t^2}$$

where the wave function E is the electric field magnitude, with a similar equation for $B = E/c$.
(h) For one-dimensional motion, the wave equation is

$$-\frac{\hbar^2}{2m}\frac{d^2\psi}{dx^2} + U\psi = E\psi$$

where E is the constant energy.
(i) The intensity of the wave is proportional to the square of its wave function.
(j) The intensity of the wave is measured from the rate at which the quantum particles bombard a detector.

4. O A quantum particle of mass m_1 is in a square well with infinitely high walls and length 3 nm. Rank the situations (a) through (e) according to the particle's energy from highest to lowest, noting any cases of equality. (a) The particle of mass m_1 is in the ground state of the well. (b) The same particle is in the $n = 2$ excited state of the same well. (c) A particle with mass $2m_1$ is in the ground state of the same well. (d) A particle of mass m_1 is in the ground state of a well of length 6 nm. (e) A particle of mass m_1 is in the ground state of a well of length 3 nm, and the uncertainty principle has become inoperative; that is, Planck's constant has been reduced to zero.

5. For a quantum particle in a box, the probability density at certain points is zero as seen in Active Figure 41.4b. Does this value imply that the particle cannot move across these points? Explain.

6. Discuss the relationship between ground-state energy and the uncertainty principle.

7. O Two square wells have the same length. Well 1 has walls of finite height, and well 2 has walls of infinite height. Both wells contain identical quantum particles, one in each well. **(i)** Is the wavelength of the ground-state wave function (a) greater for well 1, (b) greater for well 2, or (c) equal for both wells? **(ii)** Is the magnitude of the ground-state momentum (a) greater for well 1, (b) greater for well 2, or (c) equal for both wells? **(iii)** Is the ground-state energy of the particle (a) greater for well 1, (b) greater for well 2, or (c) equal for both wells?

8. What is the Schrödinger equation? How is it useful in describing quantum phenomena?

9. **O** A beam of quantum particles with kinetic energy 2 eV is reflected from a potential barrier of small width and original height 3 eV. How does the fraction of the particles that are reflected change as the barrier height is reduced to 2.01 eV? (a) It increases. (b) It decreases. (c) It stays constant at zero. (d) It stays constant at 1. (e) It stays constant with some other value.

10. **O** Suppose a tunneling current in an electronic device goes through a potential-energy barrier. The tunneling current is small because the width of the barrier is large and the barrier is high. To increase the current most effectively, what should you do? (a) Reduce the width of the barrier. (b) Reduce the height of the barrier. (c) Either choice (a) or choice (b) is equally effective. (d) Neither choice (a) nor choice (b) increases the current.

11. A philosopher once said that "it is necessary for the very existence of science that the same conditions always produce the same results." In view of what has been discussed in this chapter, present an argument showing that this statement is false. How might the statement be reworded to make it true?

12. In quantum mechanics, it is possible for the energy E of a particle to be less than the potential energy, but classically this condition is not possible. Explain.

13. **O** Unlike the idealized diagram of Figure 41.10, a typical tip used for a scanning tunneling microscope is rather jagged on the atomic scale, with several irregularly spaced points. For such a tip, does most of the tunneling current occur between the sample and (a) all the points of the tip equally, (b) the most centrally located point, (c) the point closest to the sample, or (d) the point farthest from the sample?

Problems

WebAssign The Problems from this chapter may be assigned online in WebAssign.

ThomsonNOW Sign in at **www.thomsonedu.com** and go to ThomsonNOW to assess your understanding of this chapter's topics with additional quizzing and conceptual questions.

1, 2, 3 denotes straightforward, intermediate, challenging; □ denotes full solution available in *Student Solutions Manual/Study Guide;* ▲ denotes coached solution with hints available at **www.thomsonedu.com;** denotes developing symbolic reasoning; ● denotes asking for qualitative reasoning; denotes computer useful in solving problem

Section 41.1 An Interpretation of Quantum Mechanics

1. ▲ A free electron has a wave function

$$\psi(x) = Ae^{i(5.00 \times 10^{10}x)}$$

where x is in meters. Find (a) its de Broglie wavelength, (b) its momentum, and (c) its kinetic energy in electron volts.

2. The wave function for a quantum particle is

$$\psi(x) = \sqrt{\frac{a}{\pi(x^2 + a^2)}}$$

for $a > 0$ and $-\infty < x < +\infty$. Determine the probability that the particle is located somewhere between $x = -a$ and $x = +a$.

Section 41.2 The Quantum Particle Under Boundary Conditions

3. An electron is confined to a one-dimensional region in which its ground-state ($n = 1$) energy is 2.00 eV. (a) What is the length L of the region? (b) What energy input is required to promote the electron to its first excited state?

4. An electron that has an energy of approximately 6 eV moves between infinitely high walls 1.00 nm apart. Find (a) the quantum number n for the energy state that the electron occupies and (b) the precise energy of the electron.

5. ▲ An electron is contained in a one-dimensional box of length 0.100 nm. (a) Draw an energy-level diagram for the electron for levels up to $n = 4$. (b) Find the wavelengths of all photons that can be emitted by the electron in making downward transitions that could eventually carry it from the $n = 4$ state to the $n = 1$ state.

6. A bead of mass 5.00 g slides on a horizontal wire 20.0 cm long. Its speed is 0.100 nm/yr, so it is apparently at rest. Treat this system as a quantum particle in a one-dimensional well with infinitely high walls. Calculate the quantum number of the state described.

7. A ruby laser emits 694.3-nm light. Assume light of this wavelength is due to a transition of an electron in a box from its $n = 2$ state to its $n = 1$ state. Find the length of the box.

8. A laser emits light of wavelength λ. Assume this light is due to a transition of an electron in a box from its $n = 2$ state to its $n = 1$ state. Find the length of the box.

9. ● The nuclear potential energy that binds protons and neutrons in a nucleus is often approximated by a square well. Imagine a proton confined in an infinitely high square well of length 10.0 fm, a typical nuclear diameter. Calculate the wavelength and energy associated with the photon emitted when the proton moves from the $n = 2$ state to the ground state. Identify the region of the electromagnetic spectrum to which this wavelength belongs.

10. ● A proton is confined to move in a one-dimensional box of length 0.200 nm. (a) Find the lowest possible energy of the proton. (b) **What If?** What is the lowest possible energy of an electron confined to the same box? (c) How do you account for the great difference in your results for parts (a) and (b)?

11. ● Use the quantum-particle-in-a-box model to calculate the first three energy levels of a neutron trapped in an atomic nucleus of diameter 20.0 fm. Explain whether the energy-level differences have a realistic order of magnitude.

12. A photon with wavelength λ is absorbed by an electron confined to a box. As a result, the electron moves from state $n = 1$ to $n = 4$. (a) Find the length of the box. (b) What is the wavelength of the photon emitted in the transition of that electron from the state $n = 4$ to the state $n = 2$?

13. ● For a quantum particle of mass m in the ground state of a square well with length L and infinitely high walls, the uncertainty in position is $\Delta x \approx L$. (a) Use the uncertainty principle to estimate the uncertainty in its momentum. (b) Because the particle stays inside the box, its average momentum must be zero. Its average squared momentum is then $\langle p^2 \rangle \approx (\Delta p)^2$. Estimate the energy of the particle. State how the result compares with the actual ground-state energy.

14. ● A quantum particle in an infinitely deep square well has a wave function given by

$$\psi_2(x) = \sqrt{\frac{2}{L}} \sin\left(\frac{2\pi x}{L}\right)$$

for $0 \leq x \leq L$ and zero otherwise. (a) Determine the expectation value of x. (b) Determine the probability of finding the particle near $L/2$ by calculating the probability that the particle lies in the range $0.490L \leq x \leq 0.510L$. (c) **What If?** Determine the probability of finding the particle near $L/4$ by calculating the probability that the particle lies in the range $0.240L \leq x \leq 0.260L$. (d) Argue that the result of part (a) does not contradict the results of parts (b) and (c).

15. The wave function for a quantum particle confined to moving in a one-dimensional box is

$$\psi(x) = A \sin\left(\frac{n\pi x}{L}\right)$$

Use the normalization condition on ψ to show that

$$A = \sqrt{\frac{2}{L}}$$

Suggestion: Because the length of the box is L, the wave function is zero for $x < 0$ and for $x > L$, so the normalization condition (Eq. 41.7) reduces to

$$\int_0^L |\psi|^2 \, dx = 1$$

16. An electron is trapped in an infinitely deep potential well 0.300 nm in length. (a) If the electron is in its ground state, what is the probability of finding it within 0.100 nm of the left-hand wall? (b) Identify the classical probability of finding the electron in this interval and state how it compares with the answer to part (a). (c) Repeat parts (a) and (b) assuming the particle is in the 99th energy state.

17. ● An electron in an infinitely deep square well has a wave function that is given by

$$\psi_2(x) = \sqrt{\frac{2}{L}} \sin\left(\frac{2\pi x}{L}\right)$$

for $0 \leq x \leq L$ and is zero otherwise. What are the most probable positions of the electron? Explain how you identify them.

18. A quantum particle is in the $n = 1$ state of an infinitely deep square well with walls at $x = 0$ and $x = L$. Let ℓ be an arbitrary value of x between $x = 0$ and $x = L$. (a) Find an expression for the probability, as a function of ℓ, that the particle will be found between $x = 0$ and $x = \ell$. (b) Sketch the probability as a function of the variable ℓ/L. Choose values of ℓ/L ranging from 0 to 1.00 in steps of 0.100. (c) Explain why the probability function must have particular values at $\ell/L = 0$ and at $\ell/L = 1$. (d) Find the value of ℓ for which the probability of finding the particle between $x = 0$ and $x = \ell$ is twice the probability of finding the particle between $x = \ell$ and $x = L$. You can solve the transcendental equation for ℓ/L numerically.

19. ● A quantum particle in an infinitely deep square well has a wave function

$$\psi_1(x) = \sqrt{\frac{2}{L}} \sin\left(\frac{\pi x}{L}\right)$$

for $0 \leq x \leq L$ and is zero otherwise. (a) Determine the probability of finding the particle between $x = 0$ and $x = L/3$. (b) Use the result of this calculation and a symmetry argument to find the probability of finding the particle between $x = L/3$ and $x = 2L/3$. Do not re-evaluate the integral. (c) **What If?** State how the result of part (a) compares with the classical probability.

Section 41.3 The Schrödinger Equation

20. The wave function of a quantum particle is

$$\psi(x) = A\cos(kx) + B\sin(kx)$$

where A, B, and k are constants. Show that ψ is a solution of the Schrödinger equation (Eq. 41.15), assuming the particle is free ($U = 0$), and find the corresponding energy E of the particle.

21. Show that the wave function $\psi = Ae^{i(kx - \omega t)}$ is a solution to the Schrödinger equation (Eq. 41.15), where $k = 2\pi/\lambda$ and $U = 0$.

22. In a region of space, a quantum particle with zero total energy has a wave function

$$\psi(x) = Axe^{-x^2/L^2}$$

(a) Find the potential energy U as a function of x. (b) Make a sketch of $U(x)$ versus x.

23. A quantum particle of mass m moves in a potential well of length $2L$. Its potential energy is infinite for $x < -L$ and for $x > +L$. Inside the region $-L < x < L$, its potential energy is given by

$$U(x) = -\frac{\hbar^2 x^2}{mL^2(L^2 - x^2)}$$

In addition, the particle is in a stationary state that is described by the wave function $\psi(x) = A(1 - x^2/L^2)$ for $-L < x < +L$ and by $\psi(x) = 0$ elsewhere. (a) Determine the energy of the particle in terms of $\hbar$, m, and L. *Suggestion:* Use the Schrödinger equation, Eq. 41.15. (b) Show that

$$A = \left(\frac{15}{16L}\right)^{1/2}$$

(c) Determine the probability that the particle is located between $x = -L/3$ and $x = +L/3$.

24. Consider a quantum particle moving in a one-dimensional box for which the walls are at $x = -L/2$ and $x = L/2$. (a) Write the wave functions and probability densities for $n = 1$, $n = 2$, and $n = 3$. (b) Sketch the wave functions and probability densities. *Suggestion:* Make an analogy to the case of a particle in a box for which the walls are at $x = 0$ and $x = L$.

25. Prove that the first term in the Schrödinger equation, $-(\hbar^2/2m)(d^2\psi/dx^2)$, reduces to the kinetic energy of the quantum particle multiplied by the wave function (a) for a freely moving particle, with the wave function given by Equation 41.4, and (b) for a particle in a box, with the wave function given by Equation 41.13.

Section 41.4 A Particle in a Well of Finite Height

26. Sketch the wave function $\psi(x)$ and the probability density $|\psi(x)|^2$ for the $n = 4$ state of a quantum particle in a finite potential well. (See Active Fig. 41.7.)

27. Suppose a quantum particle is trapped in its ground state in a box that has infinitely high walls (see Active Fig. 41.4a). Now suppose the left-hand wall is suddenly lowered to a finite height and width. (a) Qualitatively sketch the wave function for the particle a short time later. (b) If the box has a length L, what is the wavelength of the wave that penetrates the left-hand wall?

Section 41.5 Tunneling Through a Potential Energy Barrier

28. An electron having total energy $E = 4.50$ eV approaches a rectangular energy barrier with $U = 5.00$ eV and $L = 950$ pm as shown in Figure P41.28. Classically, the electron cannot pass through the barrier because $E < U$. Quantum-mechanically, however, the probability of tunneling is not zero. Calculate this probability, which is the transmission coefficient.

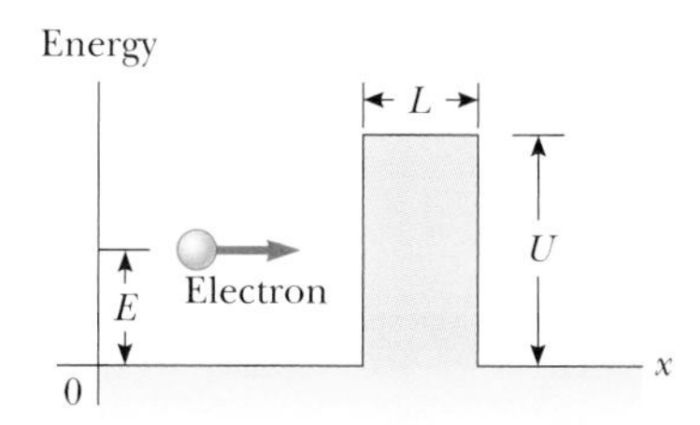

Figure P41.28 Problems 28, 29, and 31.

29. **What If?** In Problem 28, by how much would the width L of the potential barrier have to be increased for the chance of an incident 4.50-eV electron tunneling through the barrier to be one in a million?

30. Calculate the transmission probability for quantum-mechanical tunneling in each of the following cases. (a) An electron with an energy deficit of $U - E = 0.010\,0$ eV is incident on a barrier of width $L = 0.100$ nm. (b) An electron with an energy deficit of 1.00 eV is incident on the same barrier. (c) An alpha particle (mass 6.65×10^{-27} kg) with an energy deficit of 1.00 MeV is incident on a barrier of width 1.00 fm. (d) An 8.00-kg bowling ball with an energy deficit of 1.00 J is incident on a barrier of width 2.00 cm.

31. An electron with kinetic energy $E = 5.00$ eV is incident on a barrier with thickness $L = 0.200$ nm and height $U = 10.0$ eV (Fig. P41.28). What is the probability that the electron (a) tunnels through the barrier? (b) Is reflected?

32. An electron has a kinetic energy of 12.0 eV. The electron is incident upon a rectangular barrier of height 20.0 eV and thickness 1.00 nm. By what factor would the electron's probability of tunneling through the barrier increase if the electron absorbed all the energy of a photon of green light (with wavelength 546 nm) at the instant it reached the barrier?

Section 41.6 Applications of Tunneling

33. A scanning tunneling microscope (STM) can precisely determine the depths of surface features because the current through its tip is very sensitive to differences in the width of the gap between the tip and the sample surface. Assume the electron wave function falls off exponentially in this direction with a decay length of 0.100 nm, that is, with $C = 10.0$/nm. Determine the ratio of the current when the STM tip is 0.500 nm above a surface feature to the current when the tip is 0.515 nm above the surface.

34. The design criterion for a typical scanning tunneling microscope specifies that it must be able to detect, on the sample below its tip, surface features that differ in height by only 0.002 00 nm. To achieve this resolution, what percentage change in electron transmission must the electronics of the STM be able to detect? Assume the electron transmission coefficient is e^{-2CL} with $C = 10.0$/nm.

Section 41.7 The Simple Harmonic Oscillator

> *Note:* Problem 41 in Chapter 16 can be assigned with this section.

35. Show that Equation 41.26 is a solution of Equation 41.24 with energy $E = \frac{1}{2}\hbar\omega$.

36. A one-dimensional harmonic oscillator wave function is

$$\psi = Axe^{-bx^2}$$

(a) Show that ψ satisfies Equation 41.24. (b) Find b and the total energy E. (c) Is this wave function for the ground state or for the first excited state?

37. A quantum simple harmonic oscillator consists of an electron bound by a restoring force proportional to its position relative to a certain equilibrium point. The proportionality constant is 8.99 N/m. What is the longest wavelength of light that can excite the oscillator?

38. (a) Normalize the wave function for the ground state of a simple harmonic oscillator. That is, apply Equation 41.7 to Equation 41.26 and find the required value for the constant B in terms of m, ω, and fundamental constants. (b) Determine the probability of finding the oscillator in a narrow interval $-\delta/2 < x < \delta/2$ around its equilibrium position.

39. Two particles with masses m_1 and m_2 are joined by a light spring of force constant k. They vibrate along a straight line with their center of mass fixed. (a) Show that the total energy

$$\tfrac{1}{2}m_1u_1^2 + \tfrac{1}{2}m_2u_2^2 + \tfrac{1}{2}kx^2$$

can be written as $\frac{1}{2}\mu u^2 + \frac{1}{2}kx^2$, where $u = |u_1| + |u_2|$ is the *relative* speed of the particles and $\mu = m_1m_2/(m_1 + m_2)$ is the reduced mass of the system. This result demonstrates

2 = intermediate; 3 = challenging; ☐ = SSM/SG; ▲ = ThomsonNOW; = symbolic reasoning; ● = qualitative reasoning

that the pair of freely vibrating particles can be precisely modeled as a single particle vibrating on one end of a spring that has its other end fixed. (b) Differentiate the equation

$$\tfrac{1}{2}\mu u^2 + \tfrac{1}{2}kx^2 = \text{constant}$$

with respect to x. Proceed to show that the system executes simple harmonic motion. Find its frequency.

40. The total energy of a particle–spring system in which the particle moves with simple harmonic motion along the x axis is

$$E = \frac{p_x^2}{2m} + \frac{kx^2}{2}$$

where p_x is the momentum of the quantum particle and k is the spring constant. (a) Using the uncertainty principle, show that this expression can also be written as

$$E \geq \frac{p_x^2}{2m} + \frac{k\hbar^2}{8p_x^2}$$

(b) Show that the minimum energy of the harmonic oscillator is

$$E_{\min} = K + U = \tfrac{1}{4}\hbar\sqrt{\frac{k}{m}} + \frac{\hbar\omega}{4} = \frac{\hbar\omega}{2}$$

Additional Problems

41. A marble rolls back and forth across a shoebox at a constant speed of 0.8 m/s. Make an order-of-magnitude estimate of the probability of its escaping through the wall of the box by quantum tunneling. State the quantities you take as data and the values you measure or estimate for them.

42. A particle of mass 2.00×10^{-28} kg is confined to a one-dimensional box of length 1.00×10^{-10} m. For $n = 1$, what are (a) the particle's wavelength, (b) its momentum, and (c) its ground-state energy?

43. ▲ ● An electron is represented by the time-independent wave function

$$\psi(x) = \begin{cases} Ae^{-\alpha x} & \text{for } x > 0 \\ Ae^{+\alpha x} & \text{for } x < 0 \end{cases}$$

(a) Sketch the wave function as a function of x. (b) Sketch the probability density representing the likelihood that the electron is found between x and $x + dx$. (c) Only an infinite value of potential energy could produce the discontinuity in the derivative of the wave function at $x = 0$. Aside from this feature, argue that $\psi(x)$ can be a physically reasonable wave function. (d) Normalize the wave function. (e) Determine the probability of finding the electron somewhere in the range of

$$x_1 = -\frac{1}{2\alpha} \quad \text{to} \quad x_2 = \frac{1}{2\alpha}$$

44. Prove that assuming $n = 0$ for a quantum particle in an infinitely deep potential well leads to a violation of the uncertainty principle $\Delta p_x \Delta x \geq \hbar/2$.

45. ▲ An electron in an infinitely deep potential well has a ground-state energy of 0.300 eV. (a) Show that the photon emitted in a transition from the $n = 3$ state to the $n = 1$ state has a wavelength of 517 nm, which makes it green visible light. (b) Find the wavelength and the spectral region for each of the other five transitions that take place among the four lowest energy levels.

46. Particles incident from the left in Figure P41.46 are confronted with a step in potential energy. The step has a height U at $x = 0$. The particles have energy $E > U$. Classically, all the particles would continue moving forward with reduced speed. According to quantum mechanics, however, a fraction of the particles are reflected at the step. (a) Prove that the reflection coefficient R for this case is

$$R = \frac{(k_1 - k_2)^2}{(k_1 + k_2)^2}$$

where $k_1 = 2\pi/\lambda_1$ and $k_2 = 2\pi/\lambda_2$ are the wave numbers for the incident and transmitted particles. Proceed as follows. Show that the wave function $\psi_1 = Ae^{ik_1x} + Be^{-ik_1x}$ satisfies the Schrödinger equation in region 1, for $x < 0$. Here Ae^{ik_1x} represents the incident beam and Be^{-ik_1x} represents the reflected particles. Show that $\psi_2 = Ce^{ik_2x}$ satisfies the Schrödinger equation in region 2, for $x > 0$. Impose the boundary conditions $\psi_1 = \psi_2$ and $d\psi_1/dx = d\psi_2/dx$, at $x = 0$, to find the relationship between B and A. Then evaluate $R = B^2/A^2$. (b) A particle that has kinetic energy $E = 7.00$ eV is incident from a region where the potential energy is zero onto one in which $U = 5.00$ eV. Find its probability of being reflected and its probability of being transmitted.

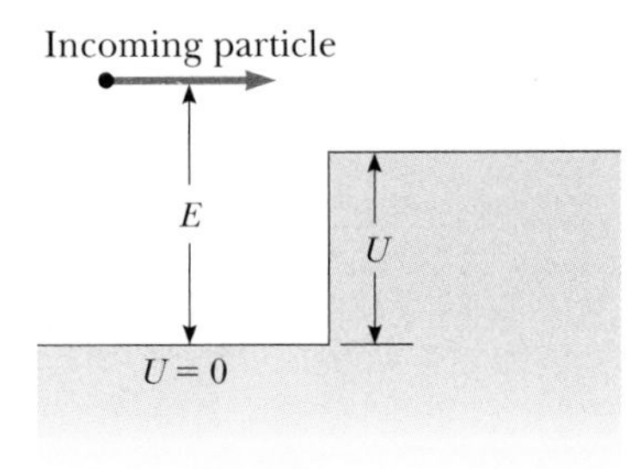

Figure P41.46 Problems 46 and 47.

47. Particles incident from the left in Figure P41.46 are confronted with a step in potential energy. The step has a height U at $x = 0$, and the particles have energy $E = 2U$. Classically, all the particles would pass into the region of higher potential energy at the right. According to quantum mechanics, however, a fraction of the particles are reflected at the barrier. Use the result of Problem 46 to determine the fraction of the incident particles that are reflected. This situation is analogous to the partial reflection and transmission of light striking an interface between two different media.

48. An electron is trapped in a quantum dot. The quantum dot may be modeled as a one-dimensional, rigid-walled box of length 1.00 nm. (a) Sketch the wave functions and probability densities for the $n = 1$ and $n = 2$ states. (b) For the $n = 1$ state, calculate the probability of finding the electron between $x_1 = 0.150$ nm and $x_2 = 0.350$ nm, where $x = 0$ is the left side of the box. (c) Repeat part (b) for the $n = 2$ state. (d) Calculate the energies in electron

volts of the $n = 1$ and $n = 2$ states. *Suggestion:* For parts (b) and (c), use Equation 41.6 and note that

$$\int \sin^2 ax\, dx = \tfrac{1}{2}x - \frac{1}{4a}\sin 2ax$$

49. An atom in an excited state 1.80 eV above the ground state remains in that excited state 2.00 μs before moving to the ground state. Find (a) the frequency and (b) the wavelength of the emitted photon. (c) Find the approximate uncertainty in energy of the photon.

50. An electron is confined to move in the xy plane in a rectangle whose dimensions are L_x and L_y. That is, the electron is trapped in a two-dimensional potential well having lengths of L_x and L_y. In this situation, the allowed energies of the electron depend on two quantum numbers n_x and n_y and are given by

$$E = \frac{h^2}{8m_e}\left(\frac{n_x^{\,2}}{L_x^{\,2}} + \frac{n_y^{\,2}}{L_y^{\,2}}\right)$$

(a) Assuming $L_x = L_y = L$, find the energies of the lowest four energy levels for the electron. (b) Construct an energy-level diagram for the electron and determine the energy difference between the second excited state and the ground state.

51. For a quantum particle described by a wave function $\psi(x)$, the expectation value of a physical quantity $f(x)$ associated with the particle is defined by

$$\langle f(x)\rangle \equiv \int_{-\infty}^{\infty} \psi^* f(x)\psi\, dx$$

For a particle in an infinitely deep one-dimensional box extending from $x = 0$ to $x = L$, show that

$$\langle x^2\rangle = \frac{L^2}{3} - \frac{L^2}{2n^2\pi^2}$$

52. A quantum particle is described by the wave function

$$\psi(x) = \begin{cases} A\cos\left(\dfrac{2\pi x}{L}\right) & \text{for } -\dfrac{L}{4} \le x \le \dfrac{L}{4} \\ 0 & \text{for other values of } x \end{cases}$$

(a) Determine the normalization constant A. (b) What is the probability that the particle will be found between $x = 0$ and $x = L/8$ if its position is measured? *Suggestion:* Use Equation 41.6.

53. A quantum particle has a wave function

$$\psi(x) = \begin{cases} \sqrt{\dfrac{2}{a}}\, e^{-x/a} & \text{for } x > 0 \\ 0 & \text{for } x < 0 \end{cases}$$

(a) Find and sketch the probability density. (b) Find the probability that the particle will be at any point where $x < 0$. (c) Show that ψ is normalized and then find the probability of finding the particle between $x = 0$ and $x = a$.

54. A quantum particle of mass m is placed in a one-dimensional box of length L. **What If?** Assume the box is so small that the particle's motion is *relativistic* and $K = p^2/2m$ is not valid. (a) Derive an expression for the kinetic energy levels of the particle. (b) Assume the particle is an electron in a box of length $L = 1.00 \times 10^{-12}$ m. Find its lowest possible kinetic energy. By what percent is the nonrelativistic equation in error? *Suggestion:* See Equation 39.23.

55. ● Consider a "crystal" consisting of two nuclei and two electrons as shown in Figure P41.55. (a) Taking into account all the pairs of interactions, find the potential energy of the system as a function of d. (b) Assuming the electrons to be restricted to a one-dimensional box of length $3d$, find the minimum kinetic energy of the two electrons. (c) Find the value of d for which the total energy is a minimum. (d) State how this value of d compares with the spacing of atoms in lithium, which has a density of 0.530 g/cm^3 and an atomic mass of 7 u. This type of calculation can be used to estimate the density of crystals and certain stars.

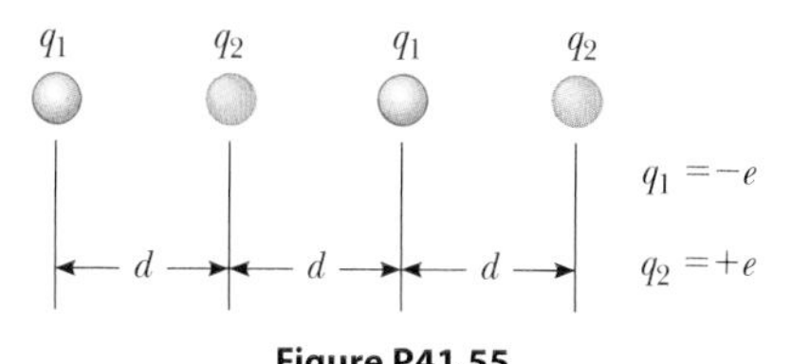

Figure P41.55

56. *The simple harmonic oscillator excited.* The wave function

$$\psi(x) = Bxe^{-(m\omega/2\hbar)x^2}$$

is a solution to the simple harmonic oscillator problem. (a) Find the energy of this state. (b) At what position are you least likely to find the particle? (c) At what positions are you most likely to find the particle? (d) Determine the value of B required to normalize the wave function. (e) **What If?** Determine the classical probability of finding the particle in an interval of small length δ centered at the position $x = 2(\hbar/m\omega)^{1/2}$. (f) What is the actual probability of finding the particle in this interval?

57. *Normalization of wave functions.* (a) Find the normalization constant A for a wave function made up of the two lowest states of a quantum particle in a box:

$$\psi(x) = A\left[\sin\left(\frac{\pi x}{L}\right) + 4\sin\left(\frac{2\pi x}{L}\right)\right]$$

(b) A particle is described in the space $-a \le x \le a$ by the wave function

$$\psi(x) = A\cos\left(\frac{\pi x}{2a}\right) + B\sin\left(\frac{\pi x}{a}\right)$$

Determine the relationship between the values of A and B required for normalization. *Suggestion:* Use the identity $\sin 2\theta = 2\sin\theta\cos\theta$.

58. The normalized wave functions for the ground state, $\psi_0(x)$, and the first excited state, $\psi_1(x)$, of a quantum harmonic oscillator are

$$\psi_0(x) = \left(\frac{a}{\pi}\right)^{1/4} e^{-ax^2/2} \qquad \psi_1(x) = \left(\frac{4a^3}{\pi}\right)^{1/4} xe^{-ax^2/2}$$

where $a = m\omega/\hbar$. A mixed state, $\psi_{01}(x)$, is constructed from these states:

$$\psi_{01}(x) = \frac{1}{\sqrt{2}}[\psi_0(x) + \psi_1(x)]$$

The symbol $\langle q \rangle_s$ denotes the expectation value of the quantity q for the state $\psi_s(x)$. Calculate the following expectation values: (a) $\langle x \rangle_0$ (b) $\langle x \rangle_1$ (c) $\langle x \rangle_{01}$

59. A two-slit electron diffraction experiment is done with slits of *unequal* widths. When only slit 1 is open, the number of electrons reaching the screen per second is 25.0 times the number of electrons reaching the screen per second when only slit 2 is open. When both slits are open, an interference pattern results in which the destructive interference is not complete. Find the ratio of the probability of an electron arriving at an interference maximum to the probability of an electron arriving at an adjacent interference minimum. *Suggestion:* Use the superposition principle.

Answers to Quick Quizzes

41.1 (d). The probability density for this wave function is $|\psi|^2 = \psi^*\psi = (Ae^{-ikx})(Ae^{ikx}) = A^2$, which is independent of x. Consequently, the particle is equally likely to be found at any value of x, which is consistent with the uncertainty principle. If the wavelength is known precisely (based on a specific value of k in Eq. 41.4), we have no knowledge of the position of the particle.

41.2 **(i),** (a). In Equation 41.14, setting $n = 1$ for the ground-state energy shows that the energy is inversely proportional to the particle mass. **(ii),** (d). The wavelength is determined by the length L of the well.

41.3 (c). According to Equation 41.14, if L is increased, all quantized energies become smaller. Therefore, the energy levels move closer together. As L becomes macroscopic, the energy levels are so close together that the quantized behavior cannot be observed.

41.4 (d). The particles in all three parts (a), (b), and (c) are part of a bound system.

41.5 (a), (c), (f). Decreasing the barrier height and increasing the particle energy both reduce the value of C in Equation 41.23, increasing the transmission coefficient in Equation 41.22. Decreasing the width L of the barrier increases the transmission coefficient in Equation 41.22.

This street in the Ginza district in Tokyo displays many signs formed from neon lamps of varying bright colors. The light from these lamps has its origin in transitions between quantized energy states in the atoms contained in the lamps. In this chapter we investigate those transitions. (© Ken Straiton/Corbis)

42 Atomic Physics

In Chapter 41, we introduced some basic concepts and techniques used in quantum mechanics along with their applications to various one-dimensional systems. In this chapter, we apply quantum mechanics to atomic systems. A large portion of the chapter is focused on the application of quantum mechanics to the study of the hydrogen atom. Understanding the hydrogen atom, the simplest atomic system, is important for several reasons:

- The hydrogen atom is the only atomic system that can be solved exactly.
- Much of what was learned in the 20th century about the hydrogen atom, with its single electron, can be extended to such single-electron ions as He^+ and Li^{2+}.
- The hydrogen atom is an ideal system for performing precise tests of theory against experiment and for improving our overall understanding of atomic structure.
- The quantum numbers that are used to characterize the allowed states of hydrogen can also be used to investigate more complex atoms, and such a description enables us to understand the periodic table of the elements. This understanding is one of the greatest triumphs of quantum mechanics.
- The basic ideas about atomic structure must be well understood before we attempt to deal with the complexities of molecular structures and the electronic structure of solids.

The full mathematical solution of the Schrödinger equation applied to the hydrogen atom gives a complete and beautiful description of the atom's properties. Because the mathematical procedures involved are beyond the scope of this text, however, many details are omitted. The solutions for some states of hydrogen are discussed, together with the quantum numbers used to characterize various allowed states. We also discuss the physical significance of the quantum numbers and the effect of a magnetic field on certain quantum states.

A new physical idea, the *exclusion principle,* is presented in this chapter. This principle is extremely important for understanding the properties of multielectron atoms and the arrangement of elements in the periodic table.

Finally, we apply our knowledge of atomic structure to describe the mechanisms involved in the production of x-rays and in the operation of a laser.

42.1 Atomic Spectra of Gases

As pointed out in Section 40.1, all objects emit thermal radiation characterized by a continuous distribution of wavelengths. In sharp contrast to this continuous-distribution spectrum is the discrete **line spectrum** observed when a low-pressure gas is subject to an electric discharge. (Electric discharge occurs when the gas is subjected to a potential difference that creates an electric field greater than the dielectric strength of the gas.) Observation and analysis of these spectral lines is called **emission spectroscopy.**

PITFALL PREVENTION 42.1
Why Lines?

The phrase "spectral lines" is often used when discussing the radiation from atoms. Lines are seen because the light passes through a long and very narrow slit before being separated by wavelength. You will see many references to these "lines" in both physics and chemistry.

When the light from a gas discharge is examined using a spectrometer (see Active Fig. 38.15), it is found to consist of a few bright lines of color on a generally dark background. This discrete line spectrum contrasts sharply with the continuous rainbow of colors seen when a glowing solid is viewed through the same instrument. Figure 42.1a shows that the wavelengths contained in a given line spectrum are characteristic of the element emitting the light. The simplest line spectrum is that for atomic hydrogen, and we describe this spectrum in detail. Because no two elements have the same line spectrum, this phenomenon represents a practical and sensitive technique for identifying the elements present in unknown samples.

Another form of spectroscopy very useful in analyzing substances is **absorption spectroscopy.** An absorption spectrum is obtained by passing white light from a continuous source through a gas or a dilute solution of the element being analyzed. The absorption spectrum consists of a series of dark lines superimposed on

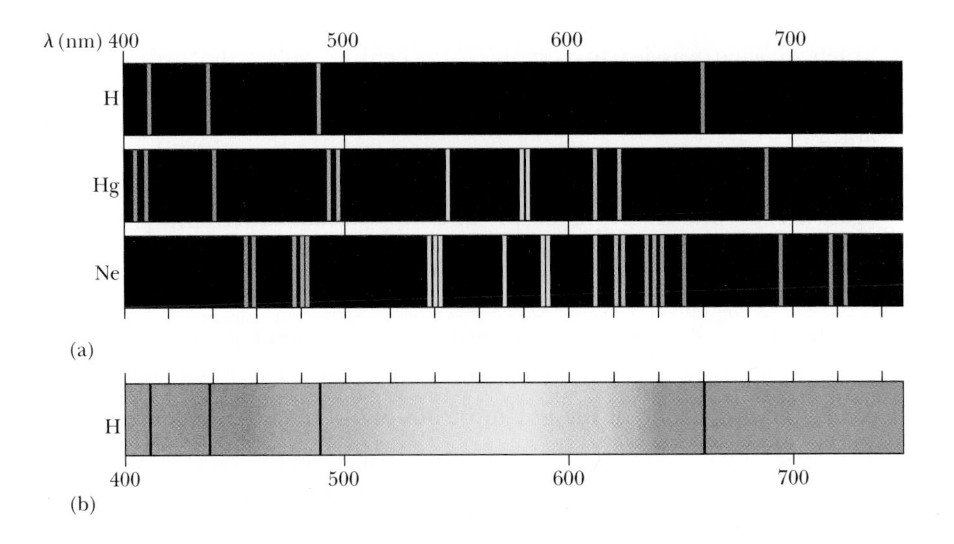

Figure 42.1 (a) Emission line spectra for hydrogen, mercury, and neon. (b) The absorption spectrum for hydrogen. Notice that the dark absorption lines occur at the same wavelengths as the hydrogen emission lines in (a). (K. W. Whitten, R. E. Davis, M. L. Peck, and G. G. Stanley, *General Chemistry,* 7th ed., Belmont, CA, Brooks/Cole, 2004.)

the continuous spectrum of the light source as shown in Figure 42.1b for atomic hydrogen.

The absorption spectrum of an element has many practical applications. For example, the continuous spectrum of radiation emitted by the Sun must pass through the cooler gases of the solar atmosphere. The various absorption lines observed in the solar spectrum have been used to identify elements in the solar atmosphere. In early studies of the solar spectrum, experimenters found some lines that did not correspond to any known element. A new element had been discovered! The new element was named helium, after the Greek word for Sun, *helios.* Helium was subsequently isolated from subterranean gas on the Earth.

Using this technique, scientists have examined the light from stars other than our Sun and have never detected elements other than those present on the Earth. Absorption spectroscopy has also been useful in analyzing heavy-metal contamination of the food chain. For example, the first determination of high levels of mercury in tuna was made with the use of atomic absorption spectroscopy.

The discrete emissions of light from gas discharges are used in "neon" signs such as those in the opening photograph of this chapter. Neon, the first gas used in these types of signs and the gas after which these signs are named, emits strongly in the red region. As a result, a glass tube filled with neon gas emits bright red light when an applied voltage causes a continuous discharge. Early signs used different gases to provide different colors, although the brightness of these signs was generally very low. Many present-day "neon" signs contain mercury vapor, which emits strongly in the ultraviolet range of the electromagnetic spectrum. The inside of a present-day sign's glass tube is coated with a material that emits a particular color when it absorbs ultraviolet radiation from the mercury. The color of the light from the tube results from the particular material chosen. A household fluorescent light operates in the same manner, with a white-emitting material coating the inside of the glass tube.

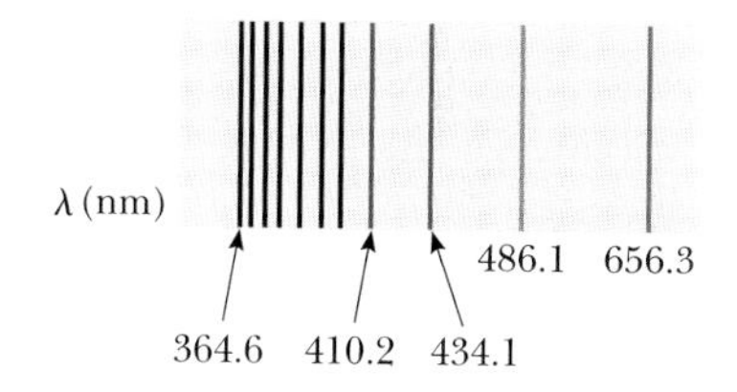

Figure 42.2 The Balmer series of spectral lines for atomic hydrogen, with several lines marked with the wavelength in nanometers. (The horizontal wavelength axis is not to scale.) The line labeled 364.6 is the shortest-wavelength line and is in the ultraviolet region of the electromagnetic spectrum. The other labeled lines are in the visible region.

From 1860 to 1885, scientists accumulated a great deal of data on atomic emissions using spectroscopic measurements. In 1885, a Swiss schoolteacher, Johann Jacob Balmer (1825–1898), found an empirical equation that correctly predicted the wavelengths of four visible emission lines of hydrogen: H_α(red), H_β (blue-green), H_γ (blue-violet), and H_δ (violet). Figure 42.2 shows these and other lines (in the ultraviolet) in the emission spectrum of hydrogen. The four visible lines occur at the wavelengths 656.3 nm, 486.1 nm, 434.1 nm, and 410.2 nm. The complete set of lines is called the **Balmer series.** The wavelengths of these lines can be described by the following equation, which is a modification made by Johannes Rydberg (1854–1919) of Balmer's original equation:

$$\frac{1}{\lambda} = R_{\mathrm{H}}\left(\frac{1}{2^2} - \frac{1}{n^2}\right) \qquad n = 3, 4, 5, \ldots \qquad \textbf{(42.1)}$$

◀ Balmer series

where R_{H} is a constant now called the **Rydberg constant** with a value of $1.097\,373\,2 \times 10^7\ \mathrm{m}^{-1}$. The integer values of n from 3 to 6 give the four visible lines from 656.3 nm (red) down to 410.2 nm (violet). Equation 42.1 also describes the ultraviolet spectral lines in the Balmer series if n is carried out beyond $n = 6$. The **series limit** is the shortest wavelength in the series and corresponds to $n \to \infty$, with a wavelength of 364.6 nm as in Figure 42.2. The measured spectral lines agree with the empirical equation, Equation 42.1, to within 0.1%.

Other lines in the spectrum of hydrogen were found following Balmer's discovery. These spectra are called the Lyman, Paschen, and Brackett series after their discoverers. The wavelengths of the lines in these series can be calculated through the use of the following empirical equations:

$$\frac{1}{\lambda} = R_{\mathrm{H}}\left(1 - \frac{1}{n^2}\right) \qquad n = 2, 3, 4, \ldots \qquad \textbf{(42.2)}$$

◀ Lyman series

$$\frac{1}{\lambda} = R_{\mathrm{H}}\left(\frac{1}{3^2} - \frac{1}{n^2}\right) \qquad n = 4, 5, 6, \ldots \qquad \textbf{(42.3)}$$

◀ Paschen series

Brackett series ▶

$$\frac{1}{\lambda} = R_{\mathrm{H}}\left(\frac{1}{4^2} - \frac{1}{n^2}\right) \quad n = 5, 6, 7, \ldots \tag{42.4}$$

No theoretical basis existed for these equations; they simply worked. The same constant R_{H} appears in each equation, and all equations involve small integers. In Section 42.3, we shall discuss the remarkable achievement of a theory for the hydrogen atom that provided an explanation for these equations.

42.2 Early Models of the Atom

The model of the atom in the days of Newton was a tiny, hard, indestructible sphere. Although this model provided a good basis for the kinetic theory of gases, new models had to be devised when experiments revealed the electrical nature of atoms. In 1897, J. J. Thomson established the charge-to-mass ratio for electrons. (See Fig. 29.14 in Section 29.3.) The following year, he suggested a model that describes the atom as a region in which positive charge is spread out in space with electrons embedded throughout the region, much like the seeds in a watermelon or raisins in thick pudding (Fig. 42.3). The atom as a whole would then be electrically neutral.

JOSEPH JOHN THOMSON
English physicist (1856–1940)
The recipient of a Nobel Prize in Physics in 1906, Thomson is usually considered the discoverer of the electron. He opened up the field of subatomic particle physics with his extensive work on the deflection of cathode rays (electrons) in an electric field.

In 1911, Ernest Rutherford (1871–1937) and his students Hans Geiger and Ernest Marsden performed a critical experiment that showed that Thomson's model could not be correct. In this experiment, a beam of positively charged alpha particles (helium nuclei) was projected into a thin metallic foil such as the target in Figure 42.4a. Most of the particles passed through the foil as if it were empty space, but some of the results of the experiment were astounding. Many of the particles deflected from their original direction of travel were scattered through *large* angles. Some particles were even deflected backward, completely reversing their direction of travel! When Geiger informed Rutherford that some alpha particles were scattered backward, Rutherford wrote, "It was quite the most incredible event that has ever happened to me in my life. It was almost as incredible as if you fired a 15-inch [artillery] shell at a piece of tissue paper and it came back and hit you."

Such large deflections were not expected on the basis of Thomson's model. According to that model, the positive charge of an atom in the foil is spread out over such a great volume (the entire atom) that there is no concentration of positive charge strong enough to cause any large-angle deflections of the positively charged alpha particles. Furthermore, the electrons are so much less massive than the alpha particles that they would not cause large-angle scattering either. Rutherford explained his astonishing results by developing a new atomic model, one that

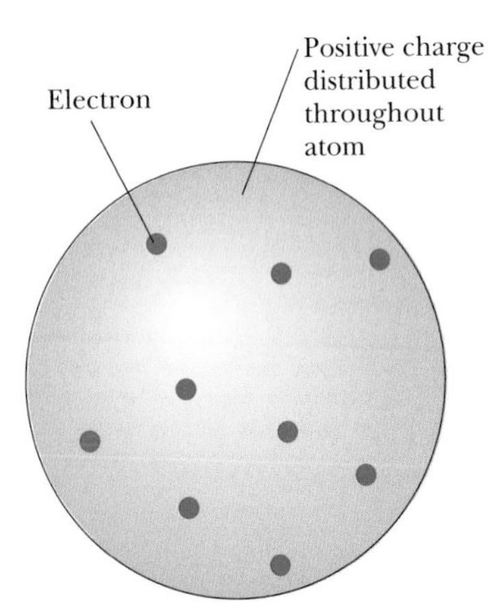

Figure 42.3 Thomson's model of the atom: negatively charged electrons in a volume of continuous positive charge.

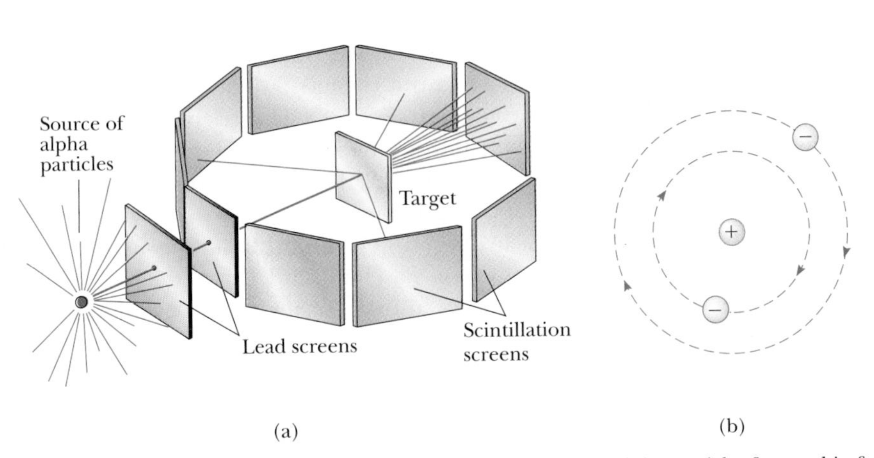

Figure 42.4 (a) Rutherford's technique for observing the scattering of alpha particles from a thin foil target. The source is a naturally occurring radioactive substance, such as radium. (b) Rutherford's planetary model of the atom.

assumed the positive charge in the atom was concentrated in a region that was small relative to the size of the atom. He called this concentration of positive charge the **nucleus** of the atom. Any electrons belonging to the atom were assumed to be in the relatively large volume outside the nucleus. To explain why these electrons were not pulled into the nucleus by the attractive electric force, Rutherford modeled them as moving in orbits around the nucleus in the same manner as the planets orbit the Sun (Fig. 42.4b). For this reason, this model is often referred to as the planetary model of the atom.

Two basic difficulties exist with Rutherford's planetary model. As we saw in Section 42.1, an atom emits (and absorbs) certain characteristic frequencies of electromagnetic radiation and no others, but the Rutherford model cannot explain this phenomenon. A second difficulty is that Rutherford's electrons are undergoing a centripetal acceleration. According to Maxwell's theory of electromagnetism, centripetally accelerated charges revolving with frequency f should radiate electromagnetic waves of frequency f. Unfortunately, this classical model leads to a prediction of self-destruction when applied to the atom. As the electron radiates, energy is carried away from the atom, the radius of the electron's orbit steadily decreases, and its frequency of revolution increases. This process would lead to an ever-increasing frequency of emitted radiation and an ultimate collapse of the atom as the electron plunges into the nucleus (Fig. 42.5).

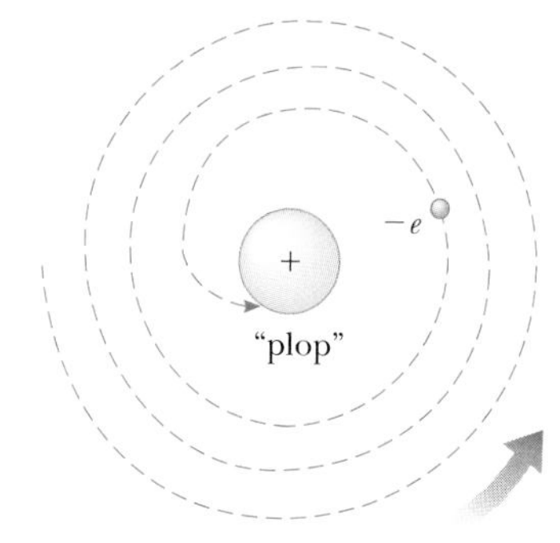

Figure 42.5 The classical model of the nuclear atom. Because the accelerating electron radiates energy, the orbit decays until the electron falls into the nucleus.

42.3 Bohr's Model of the Hydrogen Atom

Given the situation described at the end of Section 42.2, the stage was set for Niels Bohr in 1913 when he presented a new model of the hydrogen atom that circumvented the difficulties of Rutherford's planetary model. Bohr applied Planck's ideas of quantized energy levels (Section 40.1) to orbiting atomic electrons. Bohr's theory was historically important to the development of quantum physics, and it appeared to explain the spectral line series described by Equations 42.1 through 42.4. Although Bohr's model is now considered obsolete and has been completely replaced by a probabilistic quantum-mechanical theory, we can use this model to develop the notions of energy quantization and angular momentum quantization as applied to atomic-sized systems.

Bohr combined ideas from Planck's original quantum theory, Einstein's concept of the photon, Rutherford's planetary model of the atom, and Newtonian mechanics to arrive at a semiclassical model based on some revolutionary postulates. The basic ideas of the Bohr theory as it applies to the hydrogen atom are as follows:

1. The electron moves in circular orbits around the proton under the influence of the electric force of attraction as shown in Figure 42.6.
2. Only certain electron orbits are stable. When in one of these **stationary states,** as Bohr called them, the electron does not emit energy in the form of radiation. Hence, the total energy of the atom remains constant and classical mechanics can be used to describe the electron's motion. Bohr's model claims that the centripetally accelerated electron does not continuously emit radiation, losing energy and eventually spiraling into the nucleus, as predicted by classical physics in the form of Rutherford's planetary model.
3. The atom emits radiation when the electron makes a transition from a more energetic initial orbit to a lower-energy orbit. This transition cannot be visualized or treated classically. In particular, the frequency f of the photon emitted in the transition is related to the change in the atom's energy and is not equal to the frequency of the electron's orbital motion. The frequency of the emitted radiation is found from the energy-conservation expression

$$E_i - E_f = hf \quad (42.5)$$

where E_i is the energy of the initial state, E_f is the energy of the final state, and $E_i > E_f$. In addition, energy of an incident photon can be absorbed by

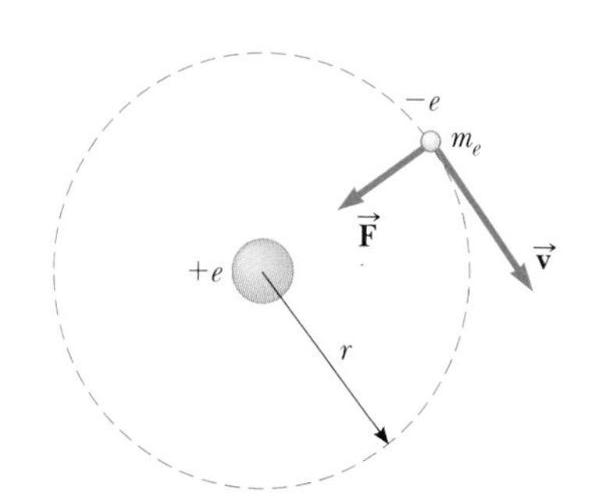

Figure 42.6 Diagram representing Bohr's model of the hydrogen atom. The orbiting electron is allowed to be only in specific orbits of discrete radii.

NIELS BOHR
Danish Physicist (1885–1962)
Bohr was an active participant in the early development of quantum mechanics and provided much of its philosophical framework. During the 1920s and 1930s, he headed the Institute for Advanced Studies in Copenhagen. The institute was a magnet for many of the world's best physicists and provided a forum for the exchange of ideas. When Bohr visited the United States in 1939 to attend a scientific conference, he brought news that the fission of uranium had been observed by Hahn and Strassman in Berlin. The results were the foundations of the nuclear weapon developed in the United States during World War II. Bohr was awarded the 1922 Nobel Prize in Physics for his investigation of the structure of atoms and the radiation emanating from them.

the atom, but only if the photon has an energy that exactly matches the difference in energy between an allowed state of the atom and a higher-energy state. Upon absorption, the photon disappears and the atom makes a transition to the higher-energy state.

4. The size of an allowed electron orbit is determined by a condition imposed on the electron's orbital angular momentum: the allowed orbits are those for which the electron's orbital angular momentum about the nucleus is quantized and equal to an integral multiple of $\hbar = h/2\pi$,

$$m_e vr = n\hbar \qquad n = 1, 2, 3, \ldots \tag{42.6}$$

where m_e is the electron mass, v is the electron's speed in its orbit, and r is the orbital radius.

Assumption 3 implies qualitatively the existence of a characteristic discrete emission line spectrum *and also* a corresponding absorption line spectrum of the kind shown in Figure 42.1 for hydrogen. Using these four assumptions, we can calculate the allowed energy levels and find quantitative values of the emission wavelengths of the hydrogen atom.

The electric potential energy of the system shown in Figure 42.6 is given by Equation 25.13, $U = k_e q_1 q_2/r = -k_e e^2/r$, where k_e is the Coulomb constant and the negative sign arises from the charge $-e$ on the electron. Therefore, the *total* energy of the atom, which consists of the electron's kinetic energy and the system's potential energy, is

$$E = K + U = \tfrac{1}{2}m_e v^2 - k_e \frac{e^2}{r} \tag{42.7}$$

The electron is a particle in uniform circular motion, so the electric force $k_e e^2/r^2$ exerted on the electron must equal the product of its mass and its centripetal acceleration ($a_c = v^2/r$):

$$\frac{k_e e^2}{r^2} = \frac{m_e v^2}{r}$$

$$v^2 = \frac{k_e e^2}{m_e r} \tag{42.8}$$

From Equation 42.8, we find that the kinetic energy of the electron is

$$K = \tfrac{1}{2}m_e v^2 = \frac{k_e e^2}{2r}$$

Substituting this value of K into Equation 42.7 gives the following expression for the total energy of the atom:[1]

$$E = -\frac{k_e e^2}{2r} \tag{42.9}$$

Because the total energy is *negative*, which indicates a bound electron–proton system, energy in the amount of $k_e e^2/2r$ must be added to the atom to remove the electron and make the total energy of the system zero.

We can obtain an expression for r, the radius of the allowed orbits, by solving Equation 42.6 for v^2 and equating it to Equation 42.8:

$$v^2 = \frac{n^2\hbar^2}{m_e^{\,2} r^2} = \frac{k_e e^2}{m_e r}$$

$$r_n = \frac{n^2\hbar^2}{m_e k_e e^2} \qquad n = 1, 2, 3, \ldots \tag{42.10}$$

[1] Compare Equation 42.9 with its gravitational counterpart, Equation 13.18.

Equation 42.10 shows that the radii of the allowed orbits have discrete values: they are quantized. The result is based on the *assumption* that the electron can exist only in certain allowed orbits determined by the integer n.

The orbit with the smallest radius, called the **Bohr radius** a_0, corresponds to $n = 1$ and has the value

$$a_0 = \frac{\hbar^2}{m_e k_e e^2} = 0.052\,9 \text{ nm} \qquad \textbf{(42.11)}$$

◀ Bohr radius

Substituting Equation 42.11 into Equation 42.10 gives a general expression for the radius of any orbit in the hydrogen atom:

$$r_n = n^2 a_0 = n^2(0.052\,9 \text{ nm}) \qquad \textbf{(42.12)}$$

◀ Radii of Bohr orbits in hydrogen

Bohr's theory predicts a value for the radius of a hydrogen atom on the right order of magnitude, based on experimental measurements. This result was a striking triumph for Bohr's theory. The first three Bohr orbits are shown to scale in Active Figure 42.7.

The quantization of orbit radii leads to energy quantization. Substituting $r_n = n^2 a_0$ into Equation 42.9 gives

$$E_n = -\frac{k_e e^2}{2a_0}\left(\frac{1}{n^2}\right) \qquad n = 1, 2, 3, \ldots \qquad \textbf{(42.13)}$$

◀ Allowed energies of the Bohr hydrogen atom

Inserting numerical values into this expression, we find that

$$E_n = -\frac{13.606 \text{ eV}}{n^2} \qquad n = 1, 2, 3, \ldots \qquad \textbf{(42.14)}$$

Only energies satisfying this equation are permitted. The lowest allowed energy level, the ground state, has $n = 1$ and energy $E_1 = -13.606$ eV. The next energy level, the first excited state, has $n = 2$ and energy $E_2 = E_1/2^2 = -3.401$ eV. Active Figure 42.8 is an energy-level diagram showing the energies of these discrete energy states and the corresponding quantum numbers n. The uppermost level corresponds to $n = \infty$ (or $r = \infty$) and $E = 0$.

Notice how the allowed energies of the hydrogen atom differ from those of the particle in a box. The particle-in-a-box energies (Eq. 41.14) increase as n^2, so they become farther apart in energy as n increases. On the other hand, the energies of the hydrogen atom (Eq. 42.14) vary inversely with n^2, so their separation in energy becomes smaller as n increases. The separation between energy levels approaches zero as n approaches infinity and the energy approaches zero.

Zero energy represents the boundary between a bound system of an electron and a proton and an unbound system. If the energy of the atom is raised from that of the ground state to any energy larger than zero, the atom is **ionized.** The minimum energy required to ionize the atom in its ground state is called the **ionization energy.** As can be seen from Active Figure 42.8, the ionization energy for hydrogen

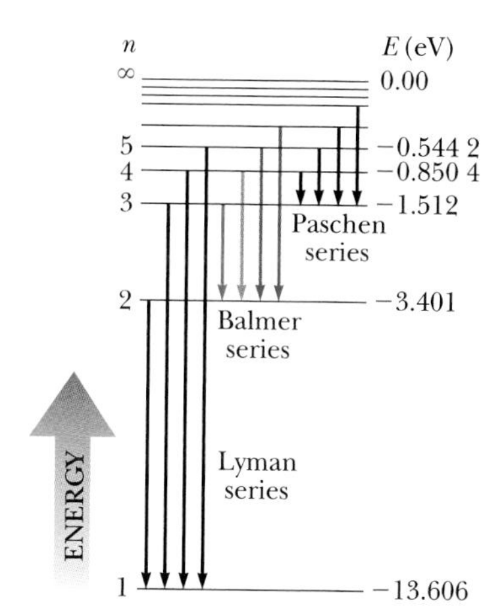

ACTIVE FIGURE 42.8

An energy-level diagram for the hydrogen atom. Quantum numbers are given on the left, and energies (in electron volts) are given on the right. Vertical arrows represent the four lowest-energy transitions for each of the spectral series shown. The colored arrows for the Balmer series indicate that this series results in visible light.

Sign in at www.thomsonedu.com and go to ThomsonNOW to choose the initial and final states of the hydrogen atom and observe the transitions in this figure and in Active Figure 42.7.

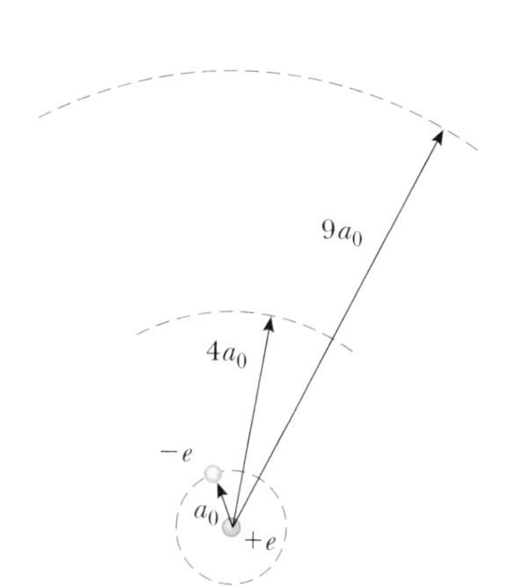

ACTIVE FIGURE 42.7

The first three circular orbits predicted by the Bohr model of the hydrogen atom.

Sign in at www.thomsonedu.com and go to ThomsonNOW to choose the initial and final states of the hydrogen atom and observe the transitions in this figure and in Active Figure 42.8.

in the ground state, based on Bohr's calculation, is 13.6 eV. This finding constituted another major achievement for the Bohr theory because the ionization energy for hydrogen had already been measured to be 13.6 eV.

Equations 42.5 and 42.13 can be used to calculate the frequency of the photon emitted when the electron makes a transition from an outer orbit to an inner orbit:

Frequency of a photon emitted from hydrogen ▶

$$f = \frac{E_i - E_f}{h} = \frac{k_e e^2}{2a_0 h}\left(\frac{1}{n_f^2} - \frac{1}{n_i^2}\right) \tag{42.15}$$

Because the quantity measured experimentally is wavelength, it is convenient to use $c = f\lambda$ to express Equation 42.15 in terms of wavelength:

$$\frac{1}{\lambda} = \frac{f}{c} = \frac{k_e e^2}{2a_0 hc}\left(\frac{1}{n_f^2} - \frac{1}{n_i^2}\right) \tag{42.16}$$

Remarkably, this expression, which is purely theoretical, is *identical* to the general form of the empirical relationships discovered by Balmer and Rydberg and given by Equations 42.1 to 42.4:

$$\frac{1}{\lambda} = R_{\mathrm{H}}\left(\frac{1}{n_f^2} - \frac{1}{n_i^2}\right) \tag{42.17}$$

provided the constant $k_e e^2/2a_0 hc$ is equal to the experimentally determined Rydberg constant. Soon after Bohr demonstrated that these two quantities agree to within approximately 1%, this work was recognized as the crowning achievement of his new quantum theory of the hydrogen atom. Furthermore, Bohr showed that all the spectral series for hydrogen have a natural interpretation in his theory. The different series correspond to transitions to different final states characterized by the quantum number n_f. Active Figure 42.8 shows the origin of these spectral series as transitions between energy levels.

Bohr extended his model for hydrogen to other elements in which all but one electron had been removed. These systems have the same structure as the hydrogen atom except that the nuclear charge is larger. Ionized elements such as He^+, Li^{2+}, and Be^{3+} were suspected to exist in hot stellar atmospheres, where atomic collisions frequently have enough energy to completely remove one or more atomic electrons. Bohr showed that many mysterious lines observed in the spectra of the Sun and several other stars could not be due to hydrogen but were correctly predicted by his theory if attributed to singly ionized helium. In general, the number of protons in the nucleus of an atom is called the **atomic number** of the element and is given the symbol Z. To describe a single electron orbiting a fixed nucleus of charge $+Ze$, Bohr's theory gives

$$r_n = (n^2)\frac{a_0}{Z} \tag{42.18}$$

$$E_n = -\frac{k_e e^2}{2a_0}\left(\frac{Z^2}{n^2}\right) \quad n = 1, 2, 3, \ldots \tag{42.19}$$

PITFALL PREVENTION 42.2

The Bohr Model Is Great, But . . .

The Bohr model correctly predicts the ionization energy and general features of the spectrum for hydrogen, but it cannot account for the spectra of more complex atoms and is unable to predict many subtle spectral details of hydrogen and other simple atoms. Scattering experiments show that the electron in a hydrogen atom does not move in a flat circle around the nucleus. Instead, the atom is spherical. The ground-state angular momentum of the atom is zero and not $\hbar$.

Although the Bohr theory was triumphant in its agreement with some experimental results on the hydrogen atom, it suffered from some difficulties. One of the first indications that the Bohr theory needed to be modified arose when improved spectroscopic techniques were used to examine the spectral lines of hydrogen. It was found that many of the lines in the Balmer and other series were not single lines at all. Instead, each was a group of lines spaced very close together. An additional difficulty arose when it was observed that in some situations certain single spectral lines were split into three closely spaced lines when the atoms were placed in a strong magnetic field. Efforts to explain these and other deviations from the Bohr model led to modifications in the theory and ultimately to a replacement theory that will be discussed in Section 42.4.

Bohr's Correspondence Principle

In our study of relativity, we found that Newtonian mechanics is a special case of relativistic mechanics and is usable only for speeds much less than c. Similarly, **quantum physics agrees with classical physics when the difference between quantized levels becomes vanishingly small.** This principle, first set forth by Bohr, is called the **correspondence principle.**[2]

For example, consider an electron orbiting the hydrogen atom with $n > 10\ 000$. For such large values of n, the energy differences between adjacent levels approach zero; therefore, the levels are nearly continuous. Consequently, the classical model is reasonably accurate in describing the system for large values of n. According to the classical picture, the frequency of the light emitted by the atom is equal to the frequency of revolution of the electron in its orbit about the nucleus. Calculations show that for $n > 10\ 000$, this frequency is different from that predicted by quantum mechanics by less than 0.015%.

Quick Quiz 42.1 A hydrogen atom is in its ground state. Incident on the atom are many photons each having an energy of 10.5 eV. What is the result? (a) The atom is excited to a higher allowed state. (b) The atom is ionized. (c) The photons pass by the atom without interaction.

Quick Quiz 42.2 A hydrogen atom makes a transition from the $n = 3$ level to the $n = 2$ level. It then makes a transition from the $n = 2$ level to the $n = 1$ level. Which transition results in emission of the longest-wavelength photon? (a) the first transition (b) the second transition (c) neither transition because the wavelengths are the same for both

---

EXAMPLE 42.1 Electronic Transitions in Hydrogen

(A) The electron in a hydrogen atom makes a transition from the $n = 2$ energy level to the ground level ($n = 1$). Find the wavelength and frequency of the emitted photon.

SOLUTION

Conceptualize Imagine the electron in a circular orbit about the nucleus as in the Bohr model in Figure 42.6. When the electron makes a transition to a lower stationary state, it emits a photon with a given frequency.

Categorize We evaluate the results using equations developed in this section, so we categorize this example as a substitution problem.

Use Equation 42.17 to obtain λ, with $n_i = 2$ and $n_f = 1$:

$$\frac{1}{\lambda} = R_{\mathrm{H}}\left(\frac{1}{1^2} - \frac{1}{2^2}\right) = \frac{3R_{\mathrm{H}}}{4}$$

$$\lambda = \frac{4}{3R_{\mathrm{H}}} = \frac{4}{3(1.097 \times 10^7\ \mathrm{m}^{-1})} = 1.22 \times 10^{-7}\ \mathrm{m} = 122\ \mathrm{nm}$$

Use Equation 34.20 to find the frequency of the photon:

$$f = \frac{c}{\lambda} = \frac{3.00 \times 10^8\ \mathrm{m/s}}{1.22 \times 10^{-7}\ \mathrm{m}} = 2.47 \times 10^{15}\ \mathrm{Hz}$$

(B) In interstellar space, highly excited hydrogen atoms called Rydberg atoms have been observed. Find the wavelength to which radio astronomers must tune to detect signals from electrons dropping from the $n = 273$ level to the $n = 272$ level.

[2] In reality, the correspondence principle is the starting point for Bohr's postulate 4 on angular momentum quantization. To see how postulate 4 arises from the correspondence principle, see J. W. Jewett, *Physics Begins with Another M . . . Mysteries, Magic, Myth, and Modern Physics* (Boston: Allyn & Bacon, 1996), pp. 353–356.

SOLUTION

Use Equation 42.17, this time with $n_i = 273$ and $n_f = 272$:

$$\frac{1}{\lambda} = R_H\left(\frac{1}{n_f^2} - \frac{1}{n_i^2}\right) = R_H\left(\frac{1}{(272)^2} - \frac{1}{(273)^2}\right) = 9.88 \times 10^{-8} R_H$$

Solve for λ:

$$\lambda = \frac{1}{9.88 \times 10^{-8} R_H} = \frac{1}{(9.88 \times 10^{-8})(1.097 \times 10^7 \text{ m}^{-1})} = 0.922 \text{ m}$$

(C) What is the radius of the electron orbit for a Rydberg atom for which $n = 273$?

SOLUTION

Use Equation 42.12 to find the radius of the orbit:

$$r_{273} = (273)^2(0.052\,9 \text{ nm}) = 3.94\ \mu\text{m}$$

This radius is large enough that the atom is on the verge of becoming macroscopic!

(D) How fast is the electron moving in a Rydberg atom for which $n = 273$?

SOLUTION

Solve Equation 42.8 for the electron's speed:

$$v = \sqrt{\frac{k_e e^2}{m_e r}} = \sqrt{\frac{(8.99 \times 10^9 \text{ N}\cdot\text{m}^2/\text{C}^2)(1.60 \times 10^{-19}\text{ C})^2}{(9.11 \times 10^{-31}\text{ kg})(3.94 \times 10^{-6}\text{ m})}}$$

$$= 8.02 \times 10^3 \text{ m/s}$$

What If? What if radiation from the Rydberg atom in part (B) is treated classically? What is the wavelength of radiation emitted by the atom in the $n = 273$ level?

Answer Classically, the frequency of the emitted radiation is that of the rotation of the electron around the nucleus.

Calculate this frequency using the period defined in Equation 4.15:

$$f = \frac{1}{T} = \frac{v}{2\pi r}$$

Substitute the radius and speed from parts (C) and (D):

$$f = \frac{v}{2\pi r} = \frac{8.02 \times 10^3 \text{ m/s}}{2\pi(3.94 \times 10^{-6}\text{ m})} = 3.24 \times 10^8 \text{ Hz}$$

Find the wavelength of the radiation from Equation 34.20:

$$\lambda = \frac{c}{f} = \frac{3.00 \times 10^8 \text{ m/s}}{3.24 \times 10^8 \text{ Hz}} = 0.926 \text{ m}$$

This value is less than 0.5% different from the wavelength calculated in part (B). As indicated in the discussion of Bohr's correspondence principle, this difference becomes even smaller for higher values of n.

42.4 The Quantum Model of the Hydrogen Atom

In the preceding section, we described how the Bohr model views the electron as a particle orbiting the nucleus in nonradiating, quantized energy levels. This approach leads to an analysis that combines both classical and quantum concepts. Although the model demonstrates excellent agreement with some experimental results, it cannot explain others. These difficulties are removed when a full quantum model involving the Schrödinger equation is used to describe the hydrogen atom.

The formal procedure for solving the problem of the hydrogen atom is to substitute the appropriate potential energy function into the Schrödinger equation, find solutions to the equation, and apply boundary conditions as we did for the particle in a box in Chapter 41. The potential energy function for the hydrogen atom is that due to the electrical interaction between the electron and the proton:

$$U(r) = -k_e \frac{e^2}{r} \quad \textbf{(42.20)}$$

where r is the radial distance from the proton (situated at $r = 0$) to the electron and $k_e = 8.99 \times 10^9 \text{ N} \cdot \text{m}^2/\text{C}^2$ is the Coulomb constant.

The mathematics for the hydrogen atom problem is more complicated than that for the particle in a box because the atom is three-dimensional and U depends on the radial coordinate r. If the time-independent Schrödinger equation (Eq. 41.15) is extended to three-dimensional rectangular coordinates, the result is

$$-\frac{\hbar^2}{2m}\left(\frac{\partial^2\psi}{\partial x^2} + \frac{\partial^2\psi}{\partial y^2} + \frac{\partial^2\psi}{\partial z^2}\right) + U\psi = E\psi$$

It is easier to solve this equation for the hydrogen atom if rectangular coordinates are converted to spherical polar coordinates, an extension of the plane polar coordinates introduced in Section 3.1. In spherical polar coordinates, a point in space is represented by the three variables r, θ, and ϕ, where r is the radial distance from the origin, $r = \sqrt{x^2 + y^2 + z^2}$. With the point represented at the end of a position vector $\vec{\mathbf{r}}$ as shown in Figure 42.9, the angular coordinate θ specifies its angular position relative to the z axis. Once that position vector is projected onto the xy plane, the angular coordinate ϕ specifies the projection's (and therefore the point's) angular position relative to the x axis.

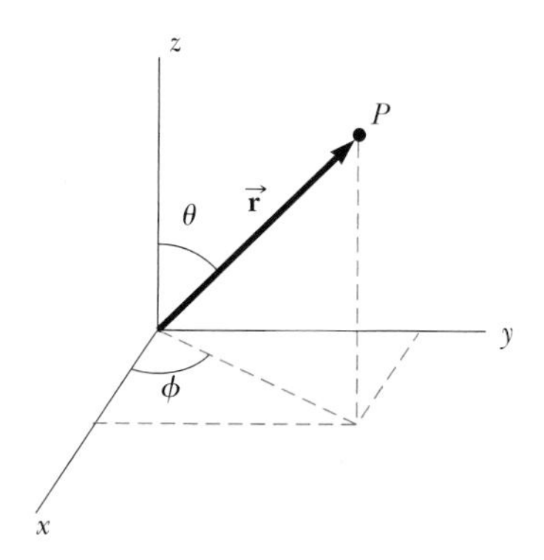

Figure 42.9 A point P in space is located by means of a position vector $\vec{\mathbf{r}}$. In Cartesian coordinates, the components of this vector are x, y, and z. In spherical polar coordinates, the point is described by r, the distance from the origin; θ, the angle between $\vec{\mathbf{r}}$ and the z axis; and ϕ, the angle between the x axis and a projection of $\vec{\mathbf{r}}$ onto the xy plane.

The conversion of the three-dimensional time-independent Schrödinger equation for $\psi(x, y, z)$ to the equivalent form for $\psi(r, \theta, \phi)$ is straightforward but very tedious, so we omit the details.[3] In Chapter 41, we separated the time dependence from the space dependence in the general wave function Ψ. In this case of the hydrogen atom, the three space variables in $\psi(x, y, z)$ can be similarly separated by writing the wave function as a product of functions of each single variable:

$$\psi(r, \theta, \phi) = R(r)f(\theta)g(\phi)$$

In this way, Schrödinger's equation, which is a three-dimensional partial differential equation, can be transformed into three separate ordinary differential equations: one for $R(r)$, one for $f(\theta)$, and one for $g(\phi)$. Each of these functions is subject to boundary conditions. For example, $R(r)$ must remain finite as $r \to 0$ and $r \to \infty$; furthermore, $g(\phi)$ must have the same value as $g(\phi + 2\pi)$.

The potential energy function given in Equation 42.20 depends *only* on the radial coordinate r and not on either of the angular coordinates; therefore, it appears only in the equation for $R(r)$. As a result, the equations for θ and ϕ are independent of the particular system and their solutions are valid for *any* system exhibiting rotation.

When the full set of boundary conditions is applied to all three functions, three different quantum numbers are found for each allowed state of the hydrogen atom, one for each of the separate differential equations. These quantum numbers are restricted to integer values and correspond to the three independent degrees of freedom (three space dimensions).

The first quantum number, associated with the radial function $R(r)$ of the full wave function, is called the **principal quantum number** and is assigned the symbol n. The differential equation for $R(r)$ leads to functions giving the probability of finding the electron at a certain radial distance from the nucleus. In Section 42.5,

[3] Descriptions of the solutions to the Schrödinger equation for the hydrogen atom are available in modern physics textbooks such as R. A. Serway, C. Moses, and C. A. Moyer, *Modern Physics*, 3rd ed. (Belmont, CA: Brooks/Cole, 2005).

we will describe two of these radial wave functions. The energies of the allowed states for the hydrogen atom are found to be related to n as follows:

Allowed energies of the quantum hydrogen atom ▶

$$E_n = -\left(\frac{k_e e^2}{2a_0}\right)\frac{1}{n^2} = -\frac{13.606 \text{ eV}}{n^2} \qquad n = 1, 2, 3, \ldots \qquad (42.21)$$

This result is in exact agreement with that obtained in the Bohr theory (Eqs. 42.13 and 42.14)! This agreement is remarkable because the Bohr theory and the full quantum theory arrive at the result from completely different starting points.

The **orbital quantum number,** symbolized ℓ, comes from the differential equation for $f(\theta)$ and is associated with the orbital angular momentum of the electron. The **orbital magnetic quantum number** m_ℓ arises from the differential equation for $g(\phi)$. Both ℓ and m_ℓ are integers. We will expand our discussion of these two quantum numbers in Section 42.6, where we also introduce a fourth (nonintegral) quantum number, resulting from a relativistic treatment of the hydrogen atom.

The application of boundary conditions on the three parts of the full wave function leads to important relationships among the three quantum numbers as well as certain restrictions on their values:

PITFALL PREVENTION 42.3
Energy Depends on n Only for Hydrogen

The implication in Equation 42.21 that the energy depends only on the quantum number n is true only for the hydrogen atom. For more complicated atoms, we will use the same quantum numbers developed here for hydrogen. The energy levels for these atoms depend primarily on n, but they also depend to a lesser degree on other quantum numbers.

Restrictions on the values of hydrogen-atom quantum numbers ▶

The values of n are integers that can range from 1 to ∞.

The values of ℓ are integers that can range from 0 to $n - 1$.

The values of m_ℓ are integers that can range from $-\ell$ to ℓ.

For example, if $n = 1$, only $\ell = 0$ and $m_\ell = 0$ are permitted. If $n = 2$, then ℓ may be 0 or 1; if $\ell = 0$, then $m_\ell = 0$; but if $\ell = 1$, then m_ℓ may be 1, 0, or -1. Table 42.1 summarizes the rules for determining the allowed values of ℓ and m_ℓ for a given n.

For historical reasons, **all states having the same principal quantum number are said to form a shell.** Shells are identified by the letters K, L, M, . . ., which designate the states for which $n = 1, 2, 3, \ldots$. Likewise, **all states having the same values of n and ℓ are said to form a subshell.** The letters[4] *s, p, d, f, g, h,* . . . are used to designate the subshells for which $\ell = 0, 1, 2, 3, \ldots$. The state designated by $3p$, for example, has the quantum numbers $n = 3$ and $\ell = 1$; the $2s$ state has the quantum numbers $n = 2$ and $\ell = 0$. These notations are summarized in Table 42.2.

States that violate the rules given in Table 42.1 do not exist. (They do not satisfy the boundary conditions on the wave function.) For instance, the $2d$ state, which would have $n = 2$ and $\ell = 2$, cannot exist because the highest allowed value of ℓ is $n - 1$, which in this case is 1. Therefore, for $n = 2$, the $2s$ and $2p$ states are allowed but $2d$, $2f$, . . . are not. For $n = 3$, the allowed subshells are $3s$, $3p$, and $3d$.

PITFALL PREVENTION 42.4
Quantum Numbers Describe a System

It is common to assign the quantum numbers to an electron. Remember, however, that these quantum numbers arise from the Schrödinger equation, which involves a potential energy function for the *system* of the electron and the nucleus. Therefore, it is more *proper* to assign the quantum numbers to the atom, but it is more *popular* to assign them to an electron. We follow this latter usage because it is so common.

TABLE 42.1
Three Quantum Numbers for the Hydrogen Atom

Quantum Number	Name	Allowed Values	Number of Allowed States
n	Principal quantum number	$1, 2, 3, \ldots$	Any number
ℓ	Orbital quantum number	$0, 1, 2, \ldots, n - 1$	n
m_ℓ	Orbital magnetic quantum number	$-\ell, -\ell + 1, \ldots, 0, \ldots, \ell - 1, \ell$	$2\ell + 1$

[4] The first four of these letters come from early classifications of spectral lines: sharp, principal, diffuse, and fundamental. The remaining letters are in alphabetical order.

TABLE 42.2

Atomic Shell and Subshell Notations

n	Shell Symbol	ℓ	Subshell Symbol
1	K	0	s
2	L	1	p
3	M	2	d
4	N	3	f
5	O	4	g
6	P	5	h

Quick Quiz 42.3 How many possible subshells are there for the $n = 4$ level of hydrogen? (a) 5 (b) 4 (c) 3 (d) 2 (e) 1

Quick Quiz 42.4 When the principal quantum number is $n = 5$, how many different values of (a) ℓ and (b) m_ℓ are possible?

EXAMPLE 42.2 The *n* = 2 Level of Hydrogen

For a hydrogen atom, determine the allowed states corresponding to the principal quantum number $n = 2$ and calculate the energies of these states.

SOLUTION

Conceptualize Think about the atom in the $n = 2$ quantum state. There is only one such state in the Bohr theory, but our discussion of the quantum theory allows for more states because of the possible values of ℓ and m_ℓ.

Categorize We evaluate the results using rules discussed in this section, so we categorize this example as a substitution problem.

From Table 42.1, we find that when $n = 2$, ℓ can be 0 or 1. Find the possible values of m_ℓ from Table 42.1:

$$\ell = 0 \quad \rightarrow \quad m_\ell = 0$$

$$\ell = 1 \quad \rightarrow \quad m_\ell = -1, 0, \text{ or } 1$$

Hence, we have one state, designated as the $2s$ state, that is associated with the quantum numbers $n = 2$, $\ell = 0$, and $m_\ell = 0$, and we have three states, designated as $2p$ states, for which the quantum numbers are $n = 2$, $\ell = 1$, $m_\ell = -1$; $n = 2$, $\ell = 1$, $m_\ell = 0$; and $n = 2$, $\ell = 1$, $m_\ell = 1$.

Find the energy for all four of these states with $n = 2$ from Equation 42.21:

$$E_2 = -\frac{13.606 \text{ eV}}{2^2} = -3.401 \text{ eV}$$

42.5 The Wave Functions for Hydrogen

Because the potential energy of the hydrogen atom depends only on the radial distance r between nucleus and electron, some of the allowed states for this atom can be represented by wave functions that depend only on r. For these states, $f(\theta)$ and $g(\phi)$ are constants. The simplest wave function for hydrogen is the one that describes the $1s$ state and is designated $\psi_{1s}(r)$:

$$\psi_{1s}(r) = \frac{1}{\sqrt{\pi a_0^{\,3}}} e^{-r/a_0} \qquad (42.22)$$

◀ Wave function for hydrogen in its ground state

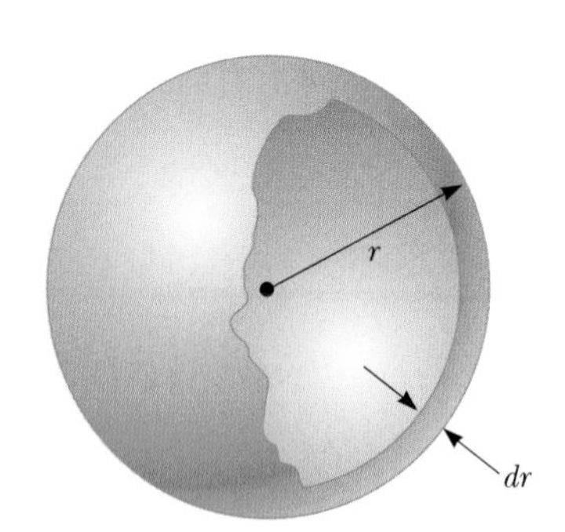

Figure 42.10 A spherical shell of radius r and thickness dr has a volume equal to $4\pi r^2\,dr$.

where a_0 is the Bohr radius. (In Problem 19, you can verify that this function satisfies the Schrödinger equation.) Note that ψ_{1s} approaches zero as r approaches ∞ and is normalized as presented (see Eq. 41.7). Furthermore, because ψ_{1s} depends only on r, it is *spherically symmetric.* This symmetry exists for all s states.

Recall that the probability of finding a particle in any region is equal to an integral of the probability density $|\psi|^2$ for the particle over the region. The probability density for the $1s$ state is

$$|\psi_{1s}|^2 = \left(\frac{1}{\pi a_0{}^3}\right) e^{-2r/a_0} \qquad \textbf{(42.23)}$$

Because we imagine the nucleus to be fixed in space at $r = 0$, we can assign this probability density to the question of locating the electron. According to Equation 41.3, the probability of finding the electron in a volume element dV is $|\psi|^2\,dV$. It is convenient to define the *radial probability density function* $P(r)$ as the probability per unit radial length of finding the electron in a spherical shell of radius r and thickness dr. Therefore, $P(r)\,dr$ is the probability of finding the electron in this shell. The volume dV of such an infinitesimally thin shell equals its surface area $4\pi r^2$ multiplied by the shell thickness dr (Fig. 42.10), so we can write this probability as

$$P(r)\,dr = |\psi|^2 dV = |\psi|^2 4\pi r^2 dr$$

Therefore, the radial probability density function is

$$P(r) = 4\pi r^2 |\psi|^2 \qquad \textbf{(42.24)}$$

Substituting Equation 42.23 into Equation 42.24 gives the radial probability density function for the hydrogen atom in its ground state:

Radial probability density for the $1s$ state of hydrogen ▶

$$P_{1s}(r) = \left(\frac{4r^2}{a_0{}^3}\right) e^{-2r/a_0} \qquad \textbf{(42.25)}$$

A plot of the function $P_{1s}(r)$ versus r is presented in Figure 42.11a. The peak of the curve corresponds to the most probable value of r for this particular state. We show in Example 42.3 that this peak occurs at the Bohr radius, the radial position of the electron when the hydrogen atom is in its ground state in the Bohr theory, another remarkable agreement between the Bohr theory and the quantum theory.

According to quantum mechanics, the atom has no sharply defined boundary as suggested by the Bohr theory. The probability distribution in Figure 42.11a suggests that the charge of the electron can be modeled as being extended throughout a region of space, commonly referred to as an *electron cloud.* Figure 42.11b shows the probability density of the electron in a hydrogen atom in the $1s$ state as a function of position in the xy plane. The darkness of the blue color corresponds to the value of the probability density. The darkest portion of the distribution appears at $r = a_0$, corresponding to the most probable value of r for the electron.

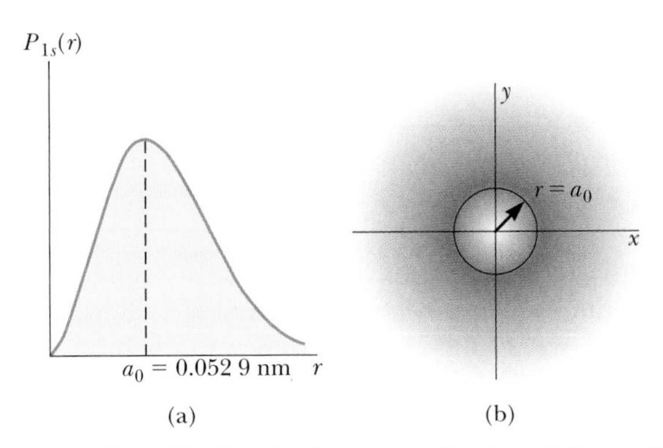

Figure 42.11 (a) The probability of finding the electron as a function of distance from the nucleus for the hydrogen atom in the $1s$ (ground) state. The probability has its maximum value when r equals the Bohr radius a_0. (b) The cross section in the xy plane of the spherical electronic charge distribution for the hydrogen atom in its $1s$ state.

EXAMPLE 42.3 The Ground State of Hydrogen

(A) Calculate the most probable value of r for an electron in the ground state of the hydrogen atom.

SOLUTION

Conceptualize Do not imagine the electron in orbit around the proton as in the Bohr theory of the hydrogen atom. Instead, imagine the charge of the electron spread out in space around the proton.

Categorize Because the statement of the problem asks for the "most probable value of r," we categorize this example as a problem in which the quantum approach is used. (In the Bohr atom, the electron moves in an orbit with an *exact* value of r.)

Analyze The most probable value of r corresponds to the maximum in the plot of $P_{1s}(r)$ versus r. We can evaluate the most probable value of r by setting $dP_{1s}/dr = 0$ and solving for r.

Differentiate Equation 42.25 and set the result equal to zero:

$$\frac{dP_{1s}}{dr} = \frac{d}{dr}\left[\left(\frac{4r^2}{a_0^{\,3}}\right)e^{-2r/a_0}\right] = 0$$

$$e^{-2r/a_0}\frac{d}{dr}(r^2) + r^2\frac{d}{dr}(e^{-2r/a_0}) = 0$$

$$2re^{-2r/a_0} + r^2(-2/a_0)e^{-2r/a_0} = 0$$

$$(1) \quad 2r[1 - (r/a_0)]e^{-2r/a_0} = 0$$

Set the bracketed expression equal to zero and solve for r:

$$1 - \frac{r}{a_0} = 0 \quad \rightarrow \quad r = a_0$$

Finalize The most probable value of r is the Bohr radius! Equation (1) is also satisfied at $r = 0$ and as $r \rightarrow \infty$. These points are locations of the *minimum* probability, which is equal to zero as seen in Figure 42.11a.

(B) Calculate the probability that the electron in the ground state of hydrogen will be found outside the first Bohr radius.

SOLUTION

Analyze The probability is found by integrating the radial probability density function $P_{1s}(r)$ for this state from the Bohr radius a_0 to ∞.

Set up this integral using Equation 42.25:

$$P = \int_{a_0}^{\infty} P_{1s}(r)\,dr = \frac{4}{a_0^{\,3}}\int_{a_0}^{\infty} r^2 e^{-2r/a_0}\,dr$$

Put the integral in dimensionless form by changing variables from r to $z = 2r/a_0$, noting that $z = 2$ when $r = a_0$ and that $dr = (a_0/2)\,dz$:

$$P = \frac{4}{a_0^{\,3}}\int_{a_0}^{\infty}\left(\frac{za_0}{2}\right)^2 e^{-z}\left(\frac{a_0}{2}\right)dz = \frac{1}{2}\int_{2}^{\infty} z^2 e^{-z}\,dz$$

Evaluate the integral using partial integration (see Appendix B.7):

$$P = -\tfrac{1}{2}(z^2 + 2z + 2)e^{-z}\Big|_{2}^{\infty}$$

Evaluate between the limits:

$$P = -\tfrac{1}{2}(0) - \left[-\tfrac{1}{2}(4 + 4 + 2)e^{-2}\right] = 5e^{-2} = 0.677 \text{ or } 67.7\%$$

Finalize This probability is larger than 50%. The reason for this value is the asymmetry in the radial probability density function (Fig. 42.11a), which has more area to the right of the peak than to the left.

What If? What if you were asked for the *average* value of r for the electron in the ground state rather than the most probable value?

Answer The average value of r is the same as the expectation value for r.

Use Equation 42.25 to evaluate the average value of r:

$$r_{avg} = \langle r \rangle = \int_0^\infty rP(r)\, dr = \int_0^\infty r\left(\frac{4r^2}{a_0^{\,3}}\right)e^{-2r/a_0}\, dr$$

$$= \left(\frac{4}{a_0^{\,3}}\right)\int_0^\infty r^3 e^{-2r/a_0}\, dr$$

Evaluate the integral with the help of the first integral listed in Table B.6 in Appendix B:

$$r_{avg} = \left(\frac{4}{a_0^{\,3}}\right)\left(\frac{3!}{(2/a_0)^4}\right) = \tfrac{3}{2}a_0$$

Again, the average value is larger than the most probable value because of the asymmetry in the wave function as seen in Figure 42.11a.

The next-simplest wave function for the hydrogen atom is the one corresponding to the 2s state ($n = 2$, $\ell = 0$). The normalized wave function for this state is

Wave function for hydrogen in the 2s state ▶

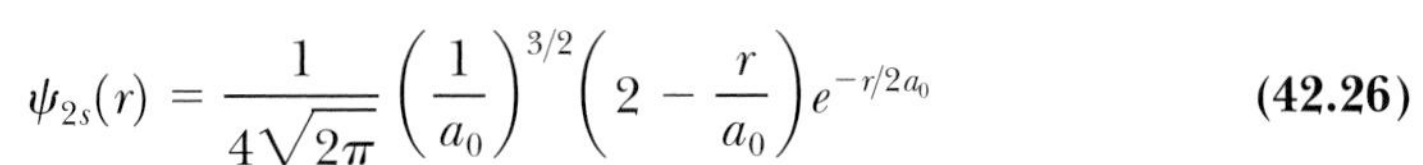

$$\psi_{2s}(r) = \frac{1}{4\sqrt{2\pi}}\left(\frac{1}{a_0}\right)^{3/2}\left(2 - \frac{r}{a_0}\right)e^{-r/2a_0} \qquad \textbf{(42.26)}$$

Again notice that ψ_{2s} depends only on r and is spherically symmetric. The energy corresponding to this state is $E_2 = -(13.606/4)$ eV $= -3.401$ eV. This energy level represents the first excited state of hydrogen. A plot of the radial probability density function for this state in comparison to the 1s state is shown in Active Figure 42.12. The plot for the 2s state has two peaks. In this case, the most probable value corresponds to that value of r that has the highest value of P ($\approx 5a_0$). An electron in the 2s state would be much farther from the nucleus (on the average) than an electron in the 1s state.

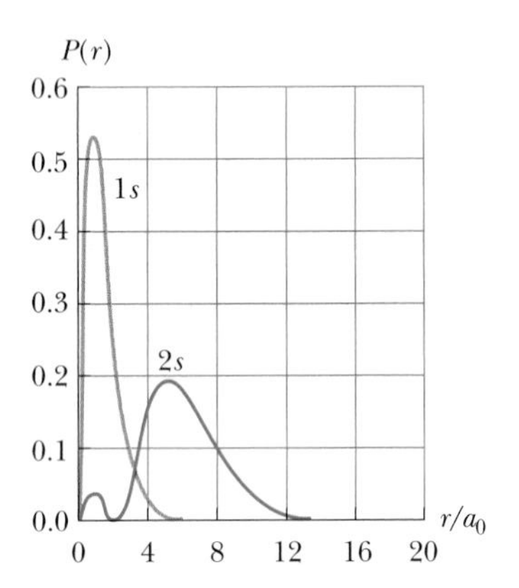

ACTIVE FIGURE 42.12

The radial probability density function versus r/a_0 for the 1s and 2s states of the hydrogen atom.

Sign in at www.thomsonedu.com and go to ThomsonNOW to choose values of r/a_0 and find the probability that the electron is located between two values.

42.6 Physical Interpretation of the Quantum Numbers

The principal quantum number n of a particular state in the hydrogen atom determines the energy of the atom according to Equation 42.21. Now let's see what the other quantum numbers in our atomic model correspond to physically.

The Orbital Quantum Number ℓ

We begin this discussion by returning briefly to the Bohr model of the atom. If the electron moves in a circle of radius r, the magnitude of its angular momentum relative to the center of the circle is $L = m_e vr$. The direction of $\vec{\mathbf{L}}$ is perpendicular to the plane of the circle and is given by a right-hand rule. According to classical physics, the magnitude L of the orbital angular momentum can have any value. The Bohr model of hydrogen, however, postulates that the magnitude of the angular momentum of the electron is restricted to multiples of $\hbar$; that is, $L = n\hbar$. This model must be modified because it predicts (incorrectly) that the ground state of hydrogen has one unit of angular momentum. Furthermore, if L is taken to be zero in the Bohr model, the electron must be pictured as a particle oscillating along a straight line through the nucleus, which is a physically unacceptable situation.

These difficulties are resolved with the quantum-mechanical model of the atom, although we must give up the convenient mental representation of an electron orbiting in a well-defined circular path. Despite the absence of this representation, the atom does indeed possess an angular momentum and it is still called orbital angular momentum. According to quantum mechanics, an atom in a state whose principal quantum number is n can take on the following *discrete* values of the magnitude of the orbital angular momentum:[5]

$$L = \sqrt{\ell(\ell + 1)}\hbar \qquad \ell = 0, 1, 2, \ldots, n - 1 \tag{42.27}$$

◀ Allowed values of L

Given these allowed values of ℓ, we see that $L = 0$ (corresponding to $\ell = 0$) is an acceptable value of the magnitude of the angular momentum. That L can be zero in this model serves to point out the inherent difficulties in any attempt to describe results based on quantum mechanics in terms of a purely particle-like (classical) model. In the quantum-mechanical interpretation, the electron cloud for the $L = 0$ state is spherically symmetric and has no fundamental rotation axis.

The Orbital Magnetic Quantum Number m_ℓ

Because angular momentum is a vector, its direction must be specified. Recall from Chapter 29 that a current loop has a corresponding magnetic moment $\vec{\boldsymbol{\mu}} = I\vec{\mathbf{A}}$ (Eq. 29.15), where I is the current in the loop and $\vec{\mathbf{A}}$ is a vector perpendicular to the loop whose magnitude is the area of the loop. Such a moment placed in a magnetic field $\vec{\mathbf{B}}$ interacts with the field. Suppose a weak magnetic field applied along the z axis defines a direction in space. According to classical physics, the energy of the loop–field system depends on the direction of the magnetic moment of the loop with respect to the magnetic field as described by Equation 29.18, $U = -\vec{\boldsymbol{\mu}} \cdot \vec{\mathbf{B}}$. Any energy between $-\mu B$ and $+\mu B$ is allowed by classical physics.

In the Bohr theory, the circulating electron represents a current loop. In the quantum-mechanical approach to the hydrogen atom, we abandon the circular orbit viewpoint of the Bohr theory, but the atom still possesses an orbital angular momentum. Therefore, there is some sense of rotation of the electron around the nucleus and a magnetic moment is present due to this angular momentum.

As mentioned in Section 42.3, spectral lines from some atoms are observed to split into groups of three closely spaced lines when the atoms are placed in a magnetic field. Suppose the hydrogen atom is located in a magnetic field. According to quantum mechanics, **there are discrete directions allowed for the magnetic moment vector $\vec{\boldsymbol{\mu}}$ with respect to the magnetic field vector $\vec{\mathbf{B}}$.** This situation is very different from that in classical physics, in which all directions are allowed.

Because the magnetic moment $\vec{\boldsymbol{\mu}}$ of the atom can be related[6] to the angular momentum vector $\vec{\mathbf{L}}$, the discrete directions of $\vec{\boldsymbol{\mu}}$ translate to the direction of $\vec{\mathbf{L}}$ being quantized. This quantization means that L_z (the projection of $\vec{\mathbf{L}}$ along the z axis) can have only discrete values. The orbital magnetic quantum number m_ℓ specifies the allowed values of the z component of the orbital angular momentum according to the expression[7]

$$L_z = m_\ell \hbar \tag{42.28}$$

◀ Allowed values of L_z

The quantization of the possible orientations of $\vec{\mathbf{L}}$ with respect to an external magnetic field is often referred to as **space quantization.**

Let's look at the possible orientations of $\vec{\mathbf{L}}$ for a given value of ℓ. Recall that m_ℓ can have values ranging from $-\ell$ to ℓ. If $\ell = 0$, then $m_\ell = 0$ and $L_z = 0$. If $\ell = 1$, the possible values of m_ℓ are -1, 0, and 1, so L_z may be $-\hbar$, 0, or $\hbar$. If $\ell = 2$, then m_ℓ can be -2, -1, 0, 1, or 2, corresponding to L_z values of $-2\hbar$, $-\hbar$, 0, $\hbar$, or $2\hbar$, and so on.

[5] Equation 42.27 is a direct result of the mathematical solution of the Schrödinger equation and the application of angular boundary conditions. This development, however, is beyond the scope of this book.

[6] See Equation 30.22 for this relationship as derived from a classical viewpoint. Quantum mechanics arrives at the same result.

[7] As with Equation 42.27, the relationship expressed in Equation 42.28 arises from the solution to the Schrödinger equation and application of boundary conditions.

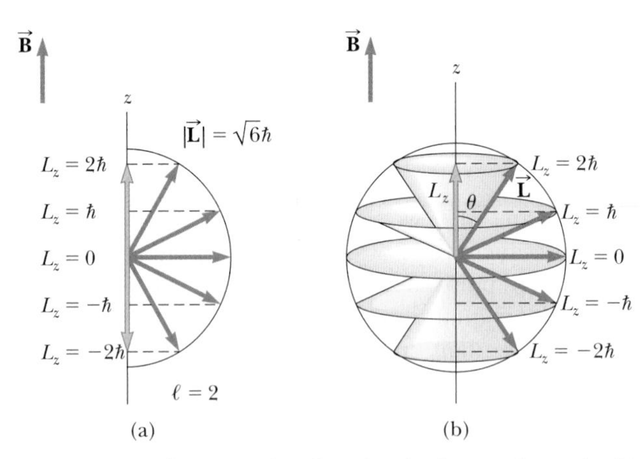

Figure 42.13 A vector model for $\ell = 2$. (a) The allowed projections on the z axis of the orbital angular momentum $\vec{\mathbf{L}}$. (b) The orbital angular momentum vector $\vec{\mathbf{L}}$ lies on the surface of a cone.

Figure 42.13a shows a **vector model** that describes space quantization for the case $\ell = 2$. Notice that $\vec{\mathbf{L}}$ can never be aligned parallel or antiparallel to $\vec{\mathbf{B}}$ because the maximum value of L_z is $\ell\hbar$, which is less than the magnitude of the angular momentum $L = \sqrt{\ell(\ell + 1)}\hbar$. The angular momentum vector $\vec{\mathbf{L}}$ is allowed to be perpendicular to $\vec{\mathbf{B}}$, which corresponds to the case of $L_z = 0$ and $\ell = 0$.

The vector $\vec{\mathbf{L}}$ does not point in one specific direction even though its z component is fixed. If $\vec{\mathbf{L}}$ were known exactly, all three components L_x, L_y, and L_z would be specified, which is inconsistent with the uncertainty principle. How can the magnitude and z component of a vector be specified, but the vector not be completely specified? To answer, imagine that L_x and L_y are completely unspecified so that $\vec{\mathbf{L}}$ lies anywhere on the surface of a cone that makes an angle θ with the z axis as shown in Figure 42.13b. From the figure, we see that θ is also quantized and that its values are specified through the relationship

$$\cos\theta = \frac{L_z}{L} = \frac{m_\ell}{\sqrt{\ell(\ell + 1)}} \quad \textbf{(42.29)}$$

If the atom is placed in a magnetic field, the energy $U = -\vec{\boldsymbol{\mu}} \cdot \vec{\mathbf{B}}$ is additional energy for the atom–field system beyond that described in Equation 42.21. Because the directions of $\vec{\boldsymbol{\mu}}$ are quantized, there are discrete total energies for the atom corresponding to different values of m_ℓ. Figure 42.14a shows a transition between two atomic levels in the absence of a magnetic field. In Figure 42.14b, a magnetic field is applied and the upper level, with $\ell = 1$, splits into three levels

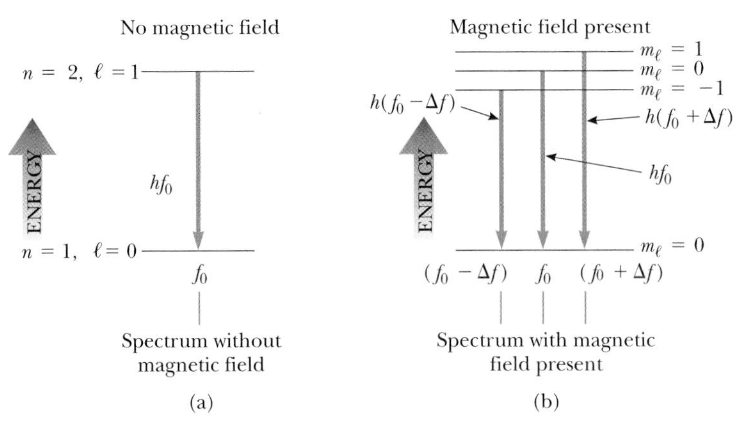

Figure 42.14 The Zeeman effect. (a) Energy levels for the ground and first excited states of a hydrogen atom. When $\vec{\mathbf{B}} = 0$, only a single spectral line at f_0 is observed. (b) When the atom is immersed in a magnetic field $\vec{\mathbf{B}}$, the state with $\ell = 1$ splits into three states. Atoms in the excited states decay to the ground state with the emission of photons with three different energies, giving rise to emission lines at f_0, $f_0 + \Delta f$, and $f_0 - \Delta f$, where Δf is the frequency shift of the emission caused by the magnetic field.

corresponding to the different directions of $\vec{\mu}$. There are now three possible transitions from the $\ell = 1$ subshell to the $\ell = 0$ subshell. Therefore, in a collection of atoms, there are atoms in all three states and the single spectral line in Figure 42.14a splits into three spectral lines. This phenomenon is called the *Zeeman effect.*

The Zeeman effect can be used to measure extraterrestrial magnetic fields. For example, the splitting of spectral lines in light from hydrogen atoms in the surface of the Sun can be used to calculate the magnitude of the magnetic field at that location. The Zeeman effect is one of many phenomena that cannot be explained with the Bohr model but are successfully explained by the quantum model of the atom.

EXAMPLE 42.4 Space Quantization for Hydrogen

Consider the hydrogen atom in the $\ell = 3$ state. Calculate the magnitude of $\vec{\mathbf{L}}$, the allowed values of L_z, and the corresponding angles θ that $\vec{\mathbf{L}}$ makes with the z axis.

SOLUTION

Conceptualize Consider Figure 42.13, which is a vector model for $\ell = 2$. Draw such a vector model for $\ell = 3$ to help with this problem.

Categorize We evaluate results using equations developed in this section, so we categorize this example as a substitution problem.

Calculate the magnitude of the orbital angular momentum using Equation 42.27:

$$L = \sqrt{\ell(\ell + 1)}\hbar = \sqrt{3(3 + 1)}\hbar = 2\sqrt{3}\hbar$$

Calculate the allowed values of L_z using Equation 42.28 with $m_\ell = -3, -2, -1, 0, 1, 2,$ and 3:

$$L_z = -3\hbar, -2\hbar, -\hbar, 0, \hbar, 2\hbar, 3\hbar$$

Calculate the allowed values of $\cos\theta$ using Equation 42.29:

$$\cos\theta = \frac{\pm 3}{2\sqrt{3}} = \pm 0.866 \qquad \cos\theta = \frac{\pm 2}{2\sqrt{3}} = \pm 0.577$$

$$\cos\theta = \frac{\pm 1}{2\sqrt{3}} = \pm 0.289 \qquad \cos\theta = \frac{0}{2\sqrt{3}} = 0$$

Find the angles corresponding to these values of $\cos\theta$:

$$\theta = 30.0°, 54.7°, 73.2°, 90.0°, 107°, 125°, 150°$$

What If? What if the value of ℓ is an arbitrary integer? For an arbitrary value of ℓ, how many values of m_ℓ are allowed?

Answer For a given value of ℓ, the values of m_ℓ range from $-\ell$ to $+\ell$ in steps of 1. Therefore, there are 2ℓ nonzero values of m_ℓ (specifically, $\pm 1, \pm 2, \ldots, \pm\ell$). In addition, one more value of $m_\ell = 0$ is possible, for a total of $2\ell + 1$ values of m_ℓ. This result is critical in understanding the results of the Stern–Gerlach experiment described below with regard to spin.

The Spin Magnetic Quantum Number m_s

The three quantum numbers n, ℓ, and m_ℓ discussed so far are generated by applying boundary conditions to solutions of the Schrödinger equation, and we can assign a physical interpretation to each quantum number. Let's now consider **electron spin,** which does *not* come from the Schrödinger equation.

Wolfgang Pauli and Niels Bohr watch a spinning top. The spin of the electron is analogous to the spin of the top but is different in many ways.

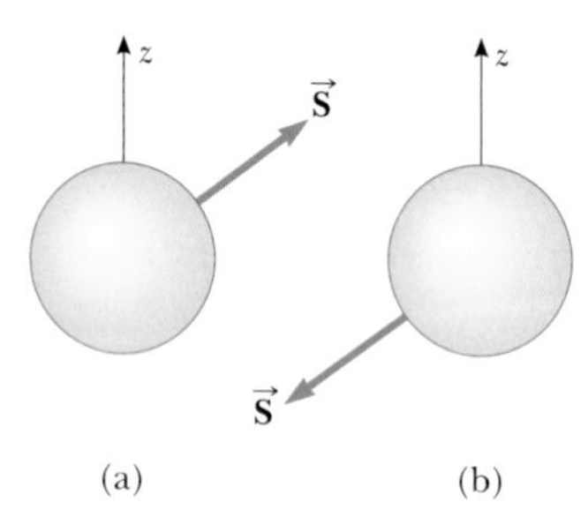

Figure 42.15 The spin of an electron can be either (a) up or (b) down relative to a specified z axis. The spin can never be aligned with the axis.

PITFALL PREVENTION 42.5

The Electron Is Not Spinning

Although the concept of a spinning electron is conceptually useful, it should not be taken literally. The spin of the Earth is a mechanical rotation. On the other hand, electron spin is a purely quantum effect that gives the electron an angular momentum as if it were physically spinning.

In Example 42.2, we found four quantum states corresponding to $n = 2$. In reality, however, eight such states occur. The additional four states can be explained by requiring a fourth quantum number for each state, the **spin magnetic quantum number m_s.**

The need for this new quantum number arises because of an unusual feature observed in the spectra of certain gases, such as sodium vapor. Close examination of one prominent line in the emission spectrum of sodium reveals that the line is, in fact, two closely spaced lines called a *doublet*.[8] The wavelengths of these lines occur in the yellow region of the electromagnetic spectrum at 589.0 nm and 589.6 nm. In 1925, when this doublet was first observed, it could not be explained with the existing atomic theory. To resolve this dilemma, Samuel Goudsmit (1902–1978) and George Uhlenbeck (1900–1988), following a suggestion made by Austrian physicist Wolfgang Pauli, proposed the spin quantum number.

To describe this new quantum number, it is convenient (but technically incorrect) to imagine the electron spinning about its axis as it orbits the nucleus as described in Section 30.6. As illustrated in Figure 42.15, only two directions exist for the electron spin. If the direction of spin is as shown in Figure 42.15a, the electron is said to have *spin up*. If the direction of spin is as shown in Figure 42.15b, the electron is said to have *spin down*. In the presence of a magnetic field, the energy of the electron is slightly different for the two spin directions. This energy difference accounts for the sodium doublet.

The classical description of electron spin—as resulting from a spinning electron—is incorrect. More recent theory indicates that the electron is a point particle, without spatial extent. Therefore, the electron cannot be considered to be spinning. Despite this conceptual difficulty, all experimental evidence supports the idea that an electron does have some intrinsic angular momentum that can be described by the spin magnetic quantum number. Paul Dirac (1902–1984) showed that this fourth quantum number originates in the relativistic properties of the electron.

In 1921, Otto Stern (1888–1969) and Walter Gerlach (1889–1979) performed an experiment that demonstrated space quantization. Their results, however, were not in quantitative agreement with the atomic theory that existed at that time. In their experiment, a beam of silver atoms sent through a nonuniform magnetic field was split into two discrete components (Fig. 42.16). Stern and Gerlach repeated the experiment using other atoms, and in each case the beam split into two or more components. The classical argument is as follows. If the z direction is chosen to be the direction of the maximum nonuniformity of $\vec{\mathbf{B}}$, the net magnetic force on the atoms is along the z axis and is proportional to the component of the magnetic moment $\vec{\boldsymbol{\mu}}$ of the atom in the z direction. Classically, $\vec{\boldsymbol{\mu}}$ can have any orientation, so the deflected beam should be spread out continuously. According to quantum mechanics, however, the deflected beam has an integral number of discrete components and the number of components determines the number of possible values of μ_z. Therefore, because the Stern–Gerlach experiment showed split beams, space quantization was at least qualitatively verified.

For the moment, let's assume the magnetic moment $\vec{\boldsymbol{\mu}}$ of the atom is due to the orbital angular momentum. Because μ_z is proportional to m_ℓ, the number of possible values of μ_z is $2\ell + 1$ as found in the **What If?** section of Example 42.4. Furthermore, because ℓ is an integer, the number of values of μ_z is always odd. This prediction is not consistent with Stern and Gerlach's observation of two components (an *even* number) in the deflected beam of silver atoms. Hence, either quantum mechanics is incorrect or the model is in need of refinement.

In 1927, T. E. Phipps and J. B. Taylor repeated the Stern–Gerlach experiment using a beam of hydrogen atoms. Their experiment was important because it

[8] This phenomenon is a Zeeman effect for spin and is identical in nature to the Zeeman effect for orbital angular momentum discussed before Example 42.4 except that no external magnetic field is required. The magnetic field for this Zeeman effect is internal to the atom and arises from the relative motion of the electron and the nucleus.

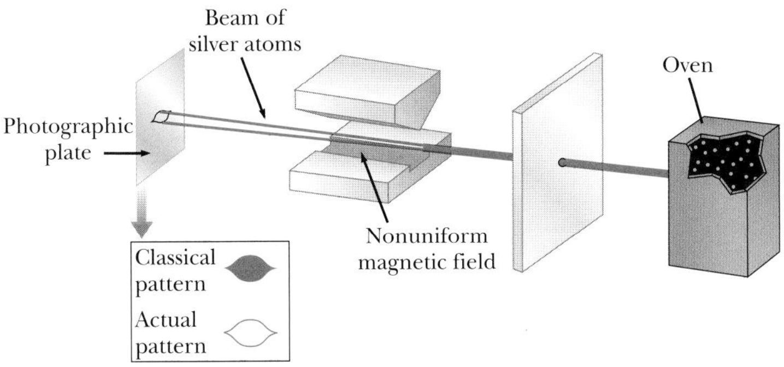

Figure 42.16 The technique used by Stern and Gerlach to verify space quantization. A beam of silver atoms is split in two by a nonuniform magnetic field.

involved an atom containing a single electron in its ground state, for which the quantum theory makes reliable predictions. Recall that $\ell = 0$ for hydrogen in its ground state, so $m_\ell = 0$. Therefore, we would not expect the beam to be deflected by the magnetic field because the magnetic moment $\vec{\boldsymbol{\mu}}$ of the atom is zero. The beam in the Phipps–Taylor experiment, however, was again split into two components. On the basis of that result, we must conclude that something other than the electron's orbital motion is contributing to the atomic magnetic moment.

As we learned earlier, Goudsmit and Uhlenbeck had proposed that the electron has an intrinsic angular momentum, spin, apart from its orbital angular momentum. In other words, the total angular momentum of the electron in a particular electronic state contains both an orbital contribution $\vec{\mathbf{L}}$ and a spin contribution $\vec{\mathbf{S}}$. The Phipps–Taylor result confirmed the hypothesis of Goudsmit and Uhlenbeck.

In 1929, Dirac used the relativistic form of the total energy of a system to solve the relativistic wave equation for the electron in a potential well. His analysis confirmed the fundamental nature of electron spin. (Spin, like mass and charge, is an *intrinsic* property of a particle, independent of its surroundings.) Furthermore, the analysis showed that electron spin[9] can be described by a single quantum number s, whose value can be only $s = \frac{1}{2}$. The spin angular momentum of the electron *never changes.* This notion contradicts classical laws, which dictate that a rotating charge slows down in the presence of an applied magnetic field because of the Faraday emf that accompanies the changing field. Furthermore, if the electron is viewed as a spinning ball of charge subject to classical laws, parts of the electron near its surface would be rotating with speeds exceeding the speed of light. Therefore, the classical picture must not be pressed too far; ultimately, spin of an electron is a quantum entity defying any simple classical description.

Because spin is a form of angular momentum, it must follow the same quantum rules as orbital angular momentum. In accordance with Equation 42.27, the magnitude of the **spin angular momentum** $\vec{\mathbf{S}}$ for the electron is

$$S = \sqrt{s(s+1)}\,\hbar = \frac{\sqrt{3}}{2}\,\hbar \qquad \textbf{(42.30)}$$

◀ Magnitude of the spin angular momentum of an electron

Like orbital angular momentum $\vec{\mathbf{L}}$, spin angular momentum $\vec{\mathbf{S}}$ exhibits space quantization as described in Figure 42.17. It can have two orientations relative to a z axis, specified by the **spin magnetic quantum number** $m_s = \pm\frac{1}{2}$. Similar to Equation 42.28 for orbital angular momentum, the z component of spin angular momentum is

$$S_z = m_s\hbar = \pm\tfrac{1}{2}\hbar \qquad \textbf{(42.31)}$$

◀ Allowed values of S_z

[9] Scientists often use the word *spin* when referring to the spin angular momentum quantum number. For example, it is common to say, "The electron has a spin of one half."

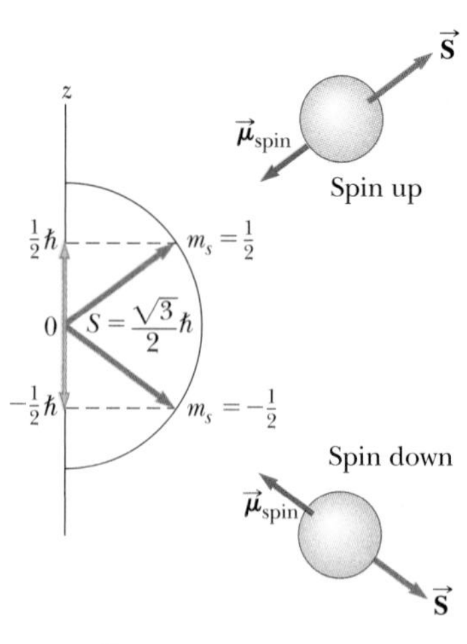

Figure 42.17 Spin angular momentum $\vec{\mathbf{S}}$ exhibits space quantization. This figure shows the two allowed orientations of the spin angular momentum vector $\vec{\mathbf{S}}$ and the spin magnetic moment $\vec{\mu}_{spin}$ for a spin-$\frac{1}{2}$ particle, such as the electron.

The two values $\pm\hbar/2$ for S_z correspond to the two possible orientations for $\vec{\mathbf{S}}$ shown in Figure 42.17. The value $m_s = +\frac{1}{2}$ refers to the spin-up case, and $m_s = -\frac{1}{2}$ refers to the spin-down case. Notice that Equations 42.30 and 42.31 do not allow the spin vector to lie along the z axis. The actual direction of $\vec{\mathbf{S}}$ is at a relatively large angle with respect to the z axis as shown in Figures 42.15 and 42.17.

The spin magnetic moment $\vec{\mu}_{spin}$ of the electron is related to its spin angular momentum $\vec{\mathbf{S}}$ by the expression

$$\vec{\mu}_{spin} = -\frac{e}{m_e}\vec{\mathbf{S}} \tag{42.32}$$

where e is the electronic charge and m_e is the mass of the electron. Because $S_z = \pm\frac{1}{2}\hbar$, the z component of the spin magnetic moment can have the values

$$\mu_{spin,z} = \pm\frac{e\hbar}{2m_e} \tag{42.33}$$

As we learned in Section 30.6, the quantity $e\hbar/2m_e$ is the Bohr magneton $\mu_B = 9.27 \times 10^{-24}$ J/T. The ratio of magnetic moment to angular momentum is twice as great for spin angular momentum (Eq. 42.32) as it is for orbital angular momentum (Eq. 30.22). The factor of 2 is explained in a relativistic treatment first carried out by Dirac.

Today, physicists explain the Stern–Gerlach and Phipps–Taylor experiments as follows. The observed magnetic moments for both silver and hydrogen are due to spin angular momentum only, with no contribution from orbital angular momentum. A single-electron atom such as hydrogen has its electron spin quantized in the magnetic field in such a way that the z component of spin angular momentum is either $\frac{1}{2}\hbar$ or $-\frac{1}{2}\hbar$, corresponding to $m_s = \pm\frac{1}{2}$. Electrons with spin $+\frac{1}{2}$ are deflected downward, and those with spin $-\frac{1}{2}$ are deflected upward.

The Stern–Gerlach experiment provided two important results. First, it verified the concept of space quantization. Second, it showed that spin angular momentum exists, even though this property was not recognized until four years after the experiments were performed.

As mentioned earlier, there are eight quantum states corresponding to $n = 2$ in the hydrogen atom, not four as found in Example 42.2. Each of the four states in Example 42.2 is actually two states because of the two possible values of m_s. Table 42.3 shows the quantum states corresponding to these eight states.

TABLE 42.3

Quantum Numbers for the $n = 2$ State of Hydrogen

n	ℓ	m_ℓ	m_s	Subshell	Shell	Number of States in Subshell
2	0	0	$\frac{1}{2}$	$2s$	L	2
2	0	0	$-\frac{1}{2}$			
2	1	1	$\frac{1}{2}$	$2p$	L	6
2	1	1	$-\frac{1}{2}$			
2	1	0	$\frac{1}{2}$			
2	1	0	$-\frac{1}{2}$			
2	1	-1	$\frac{1}{2}$			
2	1	-1	$-\frac{1}{2}$			

42.7 The Exclusion Principle and the Periodic Table

We have found that the state of a hydrogen atom is specified by four quantum numbers: n, ℓ, m_ℓ, and m_s. As it turns out, the number of states available to other atoms may also be predicted by this same set of quantum numbers. In fact, these four quantum numbers can be used to describe all the electronic states of an atom, regardless of the number of electrons in its structure.

For our discussion of atoms with many electrons, it is often easiest to assign the quantum numbers to the electrons in the atom as opposed to the entire atom. An obvious question that arises here is, "How many electrons can be in a particular quantum state?" Pauli answered this important question in 1925, in a statement known as the **exclusion principle:**

> No two electrons can ever be in the same quantum state; therefore, no two electrons in the same atom can have the same set of quantum numbers.

If this principle were not valid, an atom could radiate energy until every electron in the atom is in the lowest possible energy state and therefore the chemical behavior of the elements would be grossly modified. Nature as we know it would not exist.

In reality, we can view the electronic structure of complex atoms as a succession of filled levels increasing in energy. As a general rule, the order of filling of an atom's subshells is as follows. Once a subshell is filled, the next electron goes into the lowest-energy vacant subshell. We can understand this behavior by recognizing that if the atom were not in the lowest energy state available to it, it would radiate energy until it reached this state.

Before we discuss the electronic configuration of various elements, it is convenient to define an *orbital* as the atomic state characterized by the quantum numbers n, ℓ, and m_ℓ. The exclusion principle tells us that **only two electrons can be present in any orbital.** One of these electrons has a spin magnetic quantum number $m_s = +\frac{1}{2}$, and the other has $m_s = -\frac{1}{2}$. Because each orbital is limited to two electrons, the number of electrons that can occupy the various shells is also limited.

Table 42.4 shows the allowed quantum states for an atom up to $n = 3$. The arrows pointing upward indicate an electron described by $m_s = +\frac{1}{2}$, and those pointing downward indicate that $m_s = -\frac{1}{2}$. The $n = 1$ shell can accommodate only two electrons because $m_\ell = 0$ means that only one orbital is allowed. (The three quantum numbers describing this orbital are $n = 1$, $\ell = 0$, and $m_\ell = 0$.) The $n = 2$ shell has two subshells, one for $\ell = 0$ and one for $\ell = 1$. The $\ell = 0$ subshell is limited to two electrons because $m_\ell = 0$. The $\ell = 1$ subshell has three allowed

PITFALL PREVENTION 42.6

The Exclusion Principle Is More General

A more general form of the exclusion principle, discussed in Chapter 46, states that no two *fermions* can be in the same quantum state. Fermions are particles with half-integral spin ($\frac{1}{2}$, $\frac{3}{2}$, $\frac{5}{2}$, etc.).

WOLFGANG PAULI

Austrian Theoretical Physicist (1900–1958)

An extremely talented theoretician who made important contributions in many areas of modern physics, Pauli gained public recognition at the age of 21 with a masterful review article on relativity that is still considered one of the finest and most comprehensive introductions to the subject. His other major contributions were the discovery of the exclusion principle, the explanation of the connection between particle spin and statistics, theories of relativistic quantum electrodynamics, the neutrino hypothesis, and the hypothesis of nuclear spin.

TABLE 42.4

Allowed Quantum States for an Atom Up to $n = 3$

Shell	n	1	2				3								
Subshell	ℓ	0	0	1			0	1			2				
Orbital	m_ℓ	0	0	1	0	−1	0	1	0	−1	2	1	0	−1	−2
	m_s	↑↓	↑↓	↑↓	↑↓	↑↓	↑↓	↑↓	↑↓	↑↓	↑↓	↑↓	↑↓	↑↓	↑↓

orbitals, corresponding to $m_\ell = 1, 0,$ and -1. Because each orbital can accommodate two electrons, the $\ell = 1$ subshell can hold six electrons. Therefore, the $n = 2$ shell can contain eight electrons as shown in Table 42.3. The $n = 3$ shell has three subshells ($\ell = 0, 1, 2$) and nine orbitals, accommodating up to 18 electrons. In general, each shell can accommodate up to $2n^2$ electrons.

The exclusion principle can be illustrated by examining the electronic arrangement in a few of the lighter atoms. The atomic number Z of any element is the number of protons in the nucleus of an atom of that element. A neutral atom of that element has Z electrons. Hydrogen ($Z = 1$) has only one electron, which, in the ground state of the atom, can be described by either of two sets of quantum numbers n, ℓ, m_ℓ, m_s: $1, 0, 0, \frac{1}{2}$ or $1, 0, 0, -\frac{1}{2}$. This electronic configuration is often written $1s^1$. The notation $1s$ refers to a state for which $n = 1$ and $\ell = 0$, and the superscript indicates that one electron is present in the s subshell.

Helium ($Z = 2$) has two electrons. In the ground state, their quantum numbers are $1, 0, 0, \frac{1}{2}$ and $1, 0, 0, -\frac{1}{2}$. No other possible combinations of quantum numbers exist for this level, and we say that the K shell is filled. This electronic configuration is written $1s^2$.

Lithium ($Z = 3$) has three electrons. In the ground state, two of them are in the $1s$ subshell. The third is in the $2s$ subshell because this subshell is slightly lower in energy than the $2p$ subshell.[10] Hence, the electronic configuration for lithium is $1s^2 2s^1$.

The electronic configurations of lithium and the next several elements are provided in Figure 42.18. The electronic configuration of beryllium ($Z = 4$), with its four electrons, is $1s^2 2s^2$, and boron ($Z = 5$) has a configuration of $1s^2 2s^2 2p^1$. The $2p$ electron in boron may be described by any of the six equally probable sets of quantum numbers listed in Table 42.3. In Figure 42.18, we show this electron in the leftmost $2p$ box with spin up, but it is equally likely to be in any $2p$ box with spin either up or down.

Carbon ($Z = 6$) has six electrons, giving rise to a question concerning how to assign the two $2p$ electrons. Do they go into the same orbital with paired spins (↑↓), or do they occupy different orbitals with unpaired spins (↑↑)? Experimental data show that the most stable configuration (that is, the one with the lowest energy) is the latter, in which the spins are unpaired. Hence, the two $2p$ electrons in carbon and the three $2p$ electrons in nitrogen ($Z = 7$) have unpaired spins as Figure 42.18 shows. The general rule that governs such situations, called **Hund's rule,** states that

> when an atom has orbitals of equal energy, the order in which they are filled by electrons is such that a maximum number of electrons have unpaired spins.

Some exceptions to this rule occur in elements having subshells that are close to being filled or half-filled.

[10] To a first approximation, energy depends only on the quantum number n, as we have discussed. Because of the effect of the electronic charge shielding the nuclear charge, however, energy depends on ℓ also in multielectron atoms. We shall discuss these shielding effects in Section 42.8.

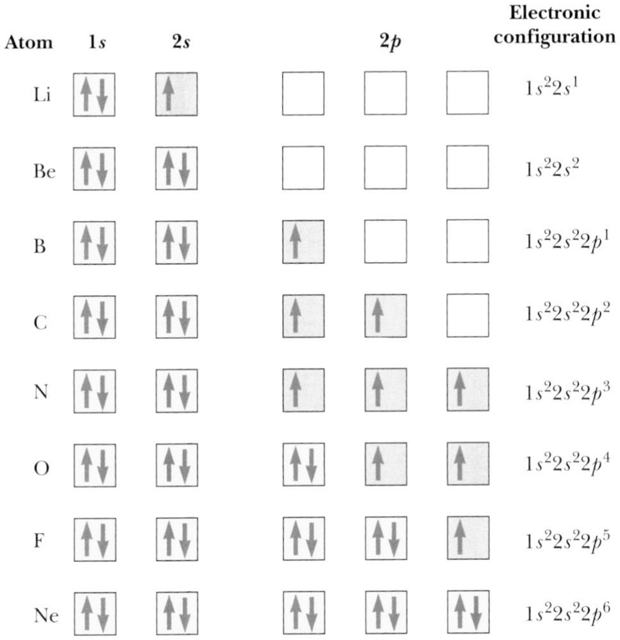

Atom	1s	2s	2p			Electronic configuration
Li	↑↓	↑				$1s^2 2s^1$
Be	↑↓	↑↓				$1s^2 2s^2$
B	↑↓	↑↓	↑			$1s^2 2s^2 2p^1$
C	↑↓	↑↓	↑	↑		$1s^2 2s^2 2p^2$
N	↑↓	↑↓	↑	↑	↑	$1s^2 2s^2 2p^3$
O	↑↓	↑↓	↑↓	↑	↑	$1s^2 2s^2 2p^4$
F	↑↓	↑↓	↑↓	↑↓	↑	$1s^2 2s^2 2p^5$
Ne	↑↓	↑↓	↑↓	↑↓	↑↓	$1s^2 2s^2 2p^6$

Figure 42.18 The filling of electronic states must obey both the exclusion principle and Hund's rule.

In 1871, long before quantum mechanics was developed, the Russian chemist Dmitri Mendeleev (1834–1907) made an early attempt at finding some order among the chemical elements. He was trying to organize the elements for the table of contents of a book he was writing. He arranged the atoms in a table similar to that shown in Figure 42.19 (page 1240), according to their atomic masses and chemical similarities. The first table Mendeleev proposed contained many blank spaces, and he boldly stated that the gaps were there only because the elements had not yet been discovered. By noting the columns in which some missing elements should be located, he was able to make rough predictions about their chemical properties. Within 20 years of this announcement, most of these elements were indeed discovered.

The elements in the **periodic table** (Fig. 42.19) are arranged so that all those in a column have similar chemical properties. For example, consider the elements in the last column, which are all gases at room temperature: He (helium), Ne (neon), Ar (argon), Kr (krypton), Xe (xenon), and Rn (radon). The outstanding characteristic of all these elements is that they do not normally take part in chemical reactions; that is, they do not readily join with other atoms to form molecules. They are therefore called *noble gases.*

We can partially understand this behavior by looking at the electronic configurations in Figure 42.19. The chemical behavior of an element depends on the outermost shell that contains electrons. The electronic configuration for helium is $1s^2$, and the $n = 1$ shell (which is the outermost shell because it is the only shell) is filled. Also, the energy of the atom in this configuration is considerably lower than the energy for the configuration in which an electron is in the next available level, the $2s$ subshell. Next, look at the electronic configuration for neon, $1s^2 2s^2 2p^6$. Again, the outermost shell ($n = 2$ in this case) is filled and a wide gap in energy occurs between the filled $2p$ subshell and the next available one, the $3s$ subshell. Argon has the configuration $1s^2 2s^2 2p^6 3s^2 3p^6$. Here, it is only the $3p$ subshell that is filled, but again a wide gap in energy occurs between the filled $3p$ subshell and the next available one, the $3d$ subshell. This pattern continues through all the noble gases. Krypton has a filled $4p$ subshell, xenon a filled $5p$ subshell, and radon a filled $6p$ subshell.

The column to the left of the noble gases in the periodic table consists of a group of elements called the *halogens*: fluorine, chlorine, bromine, iodine, and

Group I	Group II			Transition elements				Transition elements				Group III	Group VI	Group V	Group VI	Group VII	Group 0
H 1 $1s^1$																H 1 $1s^1$	He 2 $1s^2$
Li 3 $2s^1$	Be 4 $2s^2$											B 5 $2p^1$	C 6 $2p^2$	N 7 $2p^3$	O 8 $2p^4$	F 9 $2p^5$	Ne 10 $2p^6$
Na 11 $3s^1$	Mg 12 $3s^2$											Al 13 $3p^1$	Si 14 $3p^2$	P 15 $3p^3$	S 16 $3p^4$	Cl 17 $3p^5$	Ar 18 $3p^6$
K 19 $4s^1$	Ca 20 $4s^2$	Sc 21 $3d^1 4s^2$	Ti 22 $3d^2 4s^2$	V 23 $3d^3 4s^2$	Cr 24 $3d^5 4s^1$	Mn 25 $3d^5 4s^2$	Fe 26 $3d^6 4s^2$	Co 27 $3d^7 4s^2$	Ni 28 $3d^8 4s^2$	Cu 29 $3d^{10} 4s^1$	Zn 30 $3d^{10} 4s^2$	Ga 31 $4p^1$	Ge 32 $4p^2$	As 33 $4p^3$	Se 34 $4p^4$	Br 35 $4p^5$	Kr 36 $4p^6$
Rb 37 $5s^1$	Sr 38 $5s^2$	Y 39 $4d^1 5s^2$	Zr 40 $4d^2 5s^2$	Nb 41 $4d^4 5s^1$	Mo 42 $4d^5 5s^1$	Tc 43 $4d^5 5s^2$	Ru 44 $4d^7 5s^1$	Rh 45 $4d^8 5s^1$	Pd 46 $4d^{10}$	Ag 47 $4d^{10} 5s^1$	Cd 48 $4d^{10} 5s^2$	In 49 $5p^1$	Sn 50 $5p^2$	Sb 51 $5p^3$	Te 52 $5p^4$	I 53 $5p^5$	Xe 54 $5p^6$
Cs 55 $6s^1$	Ba 56 $6s^2$	57-71*	Hf 72 $5d^2 6s^2$	Ta 73 $5d^3 6s^2$	W 74 $5d^4 6s^2$	Re 75 $5d^5 6s^2$	Os 76 $5d^6 6s^2$	Ir 77 $5d^7 6s^2$	Pt 78 $5d^9 6s^1$	Au 79 $5d^{10} 6s^1$	Hg 80 $5d^{10} 6s^2$	Tl 81 $6p^1$	Pb 82 $6p^2$	Bi 83 $6p^3$	Po 84 $6p^4$	At 85 $6p^5$	Rn 86 $6p^6$
Fr 87 $7s^1$	Ra 88 $7s^2$	89- 103**	Rf 104 $6d^2 7s^2$	Db 105 $6d^3 7s^2$	Sg 106 $6d^4 7s^2$	Bh 107 $6d^5 7s^2$	Hs 108 $6d^6 7s^2$	Mt 109 $6d^7 7s^2$	Ds 110 $6d^9 7s^1$	Rg 111	112		114				

*Lanthanide series	La 57 $5d^1 6s^2$	Ce 58 $5d^1 4f^1 6s^2$	Pr 59 $4f^3 6s^2$	Nd 60 $4f^4 6s^2$	Pm 61 $4f^5 6s^2$	Sm 62 $4f^6 6s^2$	Eu 63 $4f^7 6s^2$	Gd 64 $5d^1 4f^7 6s^2$	Tb 65 $5d^1 4f^8 6s^2$	Dy 66 $4f^{10} 6s^2$	Ho 67 $4f^{11} 6s^2$	Er 68 $4f^{12} 6s^2$	Tm 69 $4f^{13} 6s^2$	Yb 70 $4f^{14} 6s^2$	Lu 71 $5d^1 4f^{14} 6s^2$
**Actinide series	Ac 89 $6d^1 7s^2$	Th 90 $6d^2 7s^2$	Pa 91 $5f^2 6d^1 7s^2$	U 92 $5f^3 6d^1 7s^2$	Np 93 $5f^4 6d^1 7s^2$	Pu 94 $5f^6 6d^0 7s^2$	Am 95 $5f^7 6d^0 7s^2$	Cm 96 $5f^7 6d^1 7s^2$	Bk 97 $5f^8 6d^1 7s^2$	Cf 98 $5f^{10} 6d^0 7s^2$	Es 99 $5f^{11} 6d^0 7s^2$	Fm 100 $5f^{12} 6d^0 7s^2$	Md 101 $5f^{13} 6d^0 7s^2$	No 102 $5f^{14} 6d^0 7s^2$	Lr 103 $5f^{14} 7s^2 7p^1$

Figure 42.19 The periodic table of the elements is an organized tabular representation of the elements that shows their periodic chemical behavior. Elements in a given column have similar chemical behavior. This table shows the element name, the atomic number, and the electron configuration. A more complete periodic table is available in Appendix C.

astatine. At room temperature, fluorine and chlorine are gases, bromine is a liquid, and iodine and astatine are solids. In each of these atoms, the outer subshell is one electron short of being filled. As a result, the halogens are chemically very active, readily accepting an electron from another atom to form a closed shell. The halogens tend to form strong ionic bonds with atoms at the other side of the periodic table. (We shall discuss ionic bonds in Chapter 43.) In a halogen lightbulb, bromine or iodine atoms combine with tungsten atoms evaporated from the filament and return them to the filament, resulting in a longer-lasting lightbulb. In addition, the filament can be operated at a higher temperature than in ordinary lightbulbs, giving a brighter and whiter light.

At the left side of the periodic table, the Group I elements consist of hydrogen and the *alkali metals*: lithium, sodium, potassium, rubidium, cesium, and francium. Each of these atoms contains one electron in a subshell outside of a closed subshell. Therefore, these elements easily form positive ions because the lone electron is bound with a relatively low energy and is easily removed. Therefore, the alkali metal atoms are chemically active and form very strong bonds with halogen atoms. For example, table salt, NaCl, is a combination of an alkali metal and a halogen. Because the outer electron is weakly bound, pure alkali metals tend to be good electrical conductors. Because of their high chemical activity, however, they are not generally found in nature in pure form.

It is interesting to plot ionization energy versus atomic number Z as in Figure 42.20. Notice the pattern of $\Delta Z = 2, 8, 8, 18, 18, 32$ for the various peaks. This pattern follows from the exclusion principle and helps explain why the elements repeat their chemical properties in groups. For example, the peaks at $Z = 2, 10, 18$, and 36 correspond to the noble gases helium, neon, argon, and krypton,

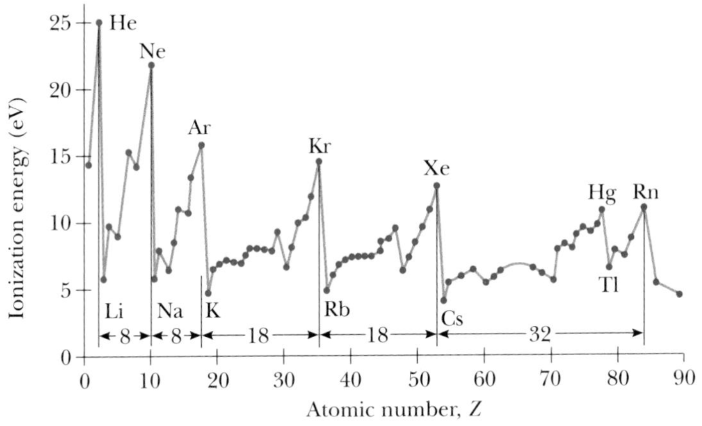

Figure 42.20 Ionization energy of the elements versus atomic number.

respectively, which, as we have mentioned, all have filled outermost shells. These elements have relatively high ionization energies and similar chemical behavior.

42.8 More on Atomic Spectra: Visible and X-Ray

In Section 42.1, we discussed the observation and early interpretation of visible spectral lines from gases. These spectral lines have their origin in transitions between quantized atomic states. We shall investigate these transitions more deeply in these final three sections of this chapter.

A modified energy-level diagram for hydrogen is shown in Figure 42.21. In this diagram, the allowed values of ℓ for each shell are separated horizontally. Figure 42.21 shows only those states up to $\ell = 2$; the shells from $n = 4$ upward would have more sets of states to the right, which are not shown. Transitions for which ℓ does not change are very unlikely to occur and are called *forbidden transitions.* (Such transitions actually can occur, but their probability is very low relative to the probability of "allowed" transitions.) The various diagonal lines represent allowed transitions between stationary states. Whenever an atom makes a transition from a higher energy state to a lower one, a photon of light is emitted. The frequency of this photon is $f = \Delta E/h$, where ΔE is the energy difference between the two states and h is Planck's constant. The **selection rules** for the *allowed transitions* are

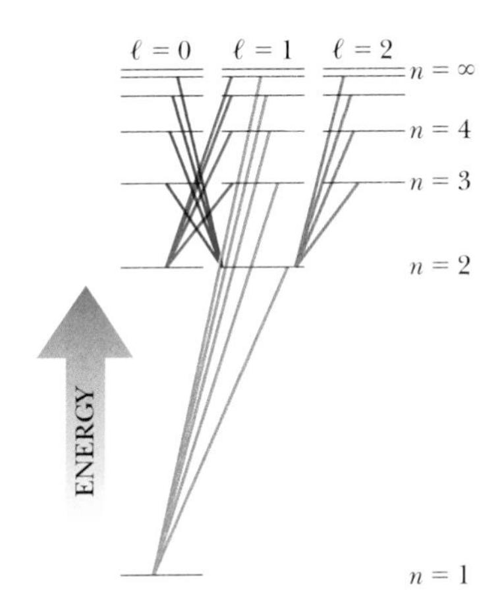

Figure 42.21 Some allowed electronic transitions for hydrogen, represented by the colored lines. These transitions must obey the selection rule $\Delta\ell = \pm 1$.

$$\Delta\ell = \pm 1 \quad \text{and} \quad \Delta m_\ell = 0, \pm 1 \tag{42.34}$$

◀ Selection rules for allowed atomic transitions

Because the orbital angular momentum of an atom changes when a photon is emitted or absorbed (that is, as a result of a transition between states) and because angular momentum of the isolated atom–photon system must be conserved, we conclude that **the photon involved in the process must carry angular momentum**. In fact, the photon has an angular momentum equivalent to that of a particle having a spin of 1. Therefore, a photon has energy, linear momentum, and angular momentum.

Recall from Equation 42.19 that the allowed energies for one-electron atoms and ions, such as hydrogen and He^+, are

$$E_n = -\frac{k_e e^2}{2a_0}\left(\frac{Z^2}{n^2}\right) = -\frac{(13.6\text{ eV})Z^2}{n^2} \tag{42.35}$$

This equation was developed from the Bohr theory, but it serves as a good first approximation in quantum theory as well. For multielectron atoms, the positive nuclear charge Ze is largely shielded by the negative charge of the inner-shell electrons. Therefore, the outer electrons interact with a net charge that is smaller than

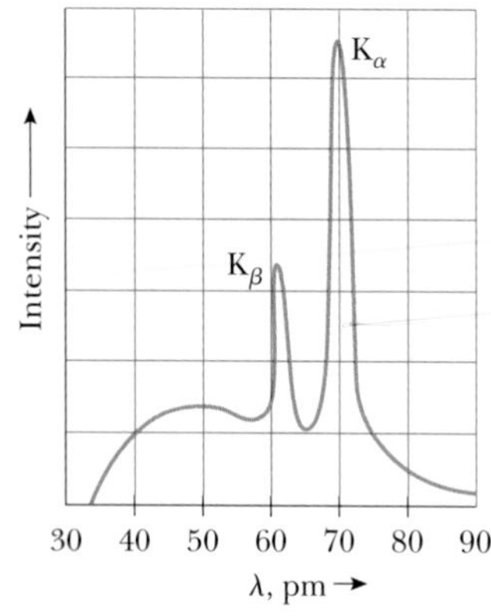

Figure 42.22 The x-ray spectrum of a metal target consists of a broad continuous spectrum (*bremsstrahlung*) plus a number of sharp lines, which are due to *characteristic x-rays*. The data shown were obtained when 37-keV electrons bombarded a molybdenum target.

the nuclear charge. The expression for the allowed energies for multielectron atoms has the same form as Equation 42.35 with Z replaced by an effective atomic number Z_{eff}:

$$E_n = -\frac{(13.6 \text{ eV})Z_{\text{eff}}^2}{n^2} \tag{42.36}$$

where Z_{eff} depends on n and ℓ.

X-Ray Spectra

X-rays are emitted when high-energy electrons or any other charged particles bombard a metal target. The x-ray spectrum typically consists of a broad continuous band containing a series of sharp lines as shown in Figure 42.22. In Section 34.6, we mentioned that an accelerated electric charge emits electromagnetic radiation. The x-rays in Figure 42.22 are the result of the slowing down of high-energy electrons as they strike the target. It may take several interactions with the atoms of the target before the electron loses all its kinetic energy. The amount of kinetic energy lost in any given interaction can vary from zero up to the entire kinetic energy of the electron. Therefore, the wavelength of radiation from these interactions lies in a continuous range from some minimum value up to infinity. It is this general slowing down of the electrons that provides the continuous curve in Figure 42.22, which shows the cutoff of x-rays below a minimum wavelength value that depends on the kinetic energy of the incoming electrons. X-ray radiation with its origin in the slowing down of electrons is called **bremsstrahlung,** the German word for "braking radiation."

The discrete lines in Figure 42.22, called **characteristic x-rays** and discovered in 1908, have a different origin. Their origin remained unexplained until the details of atomic structure were understood. The first step in the production of characteristic x-rays occurs when a bombarding electron collides with a target atom. The electron must have sufficient energy to remove an inner-shell electron from the atom. The vacancy created in the shell is filled when an electron in a higher level drops down into the level containing the vacancy. The time interval for that to happen is very short, less than 10^{-9} s. This transition is accompanied by the emission of a photon whose energy equals the difference in energy between the two levels. Typically, the energy of such transitions is greater than 1 000 eV and the emitted x-ray photons have wavelengths in the range of 0.01 nm to 1 nm.

Let's assume the incoming electron has dislodged an atomic electron from the innermost shell, the K shell. If the vacancy is filled by an electron dropping from the next higher shell—the L shell—the photon emitted has an energy corresponding to the K_α characteristic x-ray line on the curve of Figure 42.22. In this notation, K refers to the final level of the electron and the subscript α, as the *first* letter of the Greek alphabet, refers to the initial level as the *first* one above the final level. Figure 42.23 shows this transition as well as others discussed below. If the vacancy in the K shell is filled by an electron dropping from the M shell, the K_β line in Figure 42.22 is produced.

Other characteristic x-ray lines are formed when electrons drop from upper levels to vacancies other than those in the K shell. For example, L lines are produced when vacancies in the L shell are filled by electrons dropping from higher shells. An L_α line is produced as an electron drops from the M shell to the L shell, and an L_β line is produced by a transition from the N shell to the L shell.

Although multielectron atoms cannot be analyzed exactly with either the Bohr model or the Schrödinger equation, we can apply Gauss's law from Chapter 24 to make some surprisingly accurate estimates of expected x-ray energies and wavelengths. Consider an atom of atomic number Z in which one of the two electrons in the K shell has been ejected. Imagine drawing a gaussian sphere immediately inside the most probable radius of the L electrons. The electric field at the position of the L electrons is a combination of the fields created by the nucleus, the single K electron, the other L electrons, and the outer electrons. The wave func-

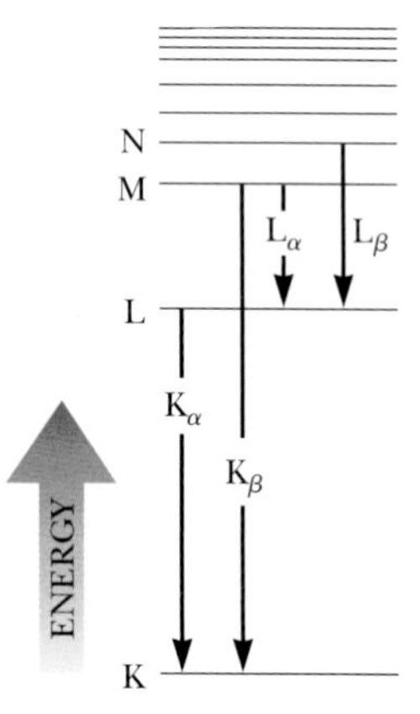

Figure 42.23 Transitions between higher and lower atomic energy levels that give rise to x-ray photons from heavy atoms when they are bombarded with high-energy electrons.

tions of the outer electrons are such that the electrons have a very high probability of being farther from the nucleus than the L electrons are. Therefore, the outer electrons are much more likely to be outside the gaussian surface than inside and, on average, do not contribute significantly to the electric field at the position of the L electrons. The effective charge inside the gaussian surface is the positive nuclear charge and one negative charge due to the single K electron. Ignoring the interactions between L electrons, a single L electron behaves as if it experiences an electric field due to a charge $(Z - 1)e$ enclosed by the gaussian surface. The nuclear charge is shielded by the electron in the K shell such that Z_{eff} in Equation 42.36 is $Z - 1$. For higher-level shells, the nuclear charge is shielded by electrons in all of the inner shells.

We can now use Equation 42.36 to estimate the energy associated with an electron in the L shell:

$$E_L = -(Z-1)^2 \frac{13.6 \text{ eV}}{2^2}$$

After the atom makes the transition, there are two electrons in the K shell. We can approximate the energy associated with one of these electrons as that of a one-electron atom. (In reality, the nuclear charge is reduced somewhat by the negative charge of the other electron, but let's ignore this effect.) Therefore,

$$E_K \approx -Z^2(13.6 \text{ eV}) \qquad \textbf{(42.37)}$$

As Example 42.5 shows, the energy of the atom with an electron in an M shell can be estimated in a similar fashion. Taking the energy difference between the initial and final levels, we can then calculate the energy and wavelength of the emitted photon.

In 1914, Henry G. J. Moseley (1887–1915) plotted $\sqrt{1/\lambda}$ versus the Z values for a number of elements where λ is the wavelength of the K_α line of each element. He found that the plot is a straight line as in Figure 42.24, which is consistent with rough calculations of the energy levels given by Equation 42.37. From this plot, Moseley determined the Z values of elements that had not yet been discovered and produced a periodic table in excellent agreement with the known chemical properties of the elements. Until that experiment, atomic numbers had been merely placeholders for the elements that appeared in the periodic table, the elements being ordered according to mass.

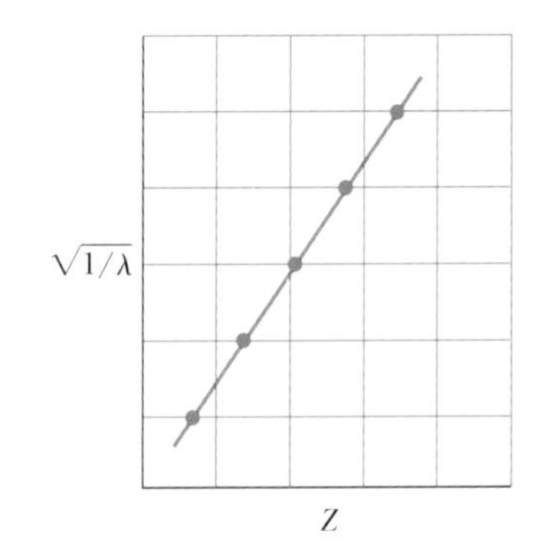

Figure 42.24 A Moseley plot of $\sqrt{1/\lambda}$ versus Z, where λ is the wavelength of the K_α x-ray line of the element of atomic number Z.

Quick Quiz 42.5 In an x-ray tube, as you increase the energy of the electrons striking the metal target, do the wavelengths of the characteristic x-rays (a) increase, (b) decrease, or (c) do not change?

Quick Quiz 42.6 True or False: It is possible for an x-ray spectrum to show the continuous spectrum of x-rays without the presence of the characteristic x-rays.

EXAMPLE 42.5 Estimating the Energy of an X-Ray

Estimate the energy of the characteristic x-ray emitted from a tungsten target when an electron drops from an M shell ($n = 3$ state) to a vacancy in the K shell ($n = 1$ state). The atomic number for tungsten is $Z = 74$.

SOLUTION

Conceptualize Imagine an accelerated electron striking a tungsten atom and ejecting an electron from the K shell. Subsequently, an electron in the M shell drops down to fill the vacancy and the energy difference between the states is emitted as an x-ray photon.

Categorize We estimate the results using equations developed in this section, so we categorize this example as a substitution problem.

Use Equation 42.37 and Z = 74 for tungsten to estimate the energy associated with the electron in the K shell:

$$E_K \approx -(74)^2(13.6 \text{ eV}) = -7.4 \times 10^4 \text{ eV}$$

Use Equation 42.36 and that nine electrons shield the nuclear charge (eight electrons in the $n = 2$ state and one electron in the $n = 1$ state) to estimate the energy of the M shell:

$$E_M \approx -\frac{(13.6 \text{ eV})(74 - 9)^2}{(3)^2} \approx -6.4 \times 10^3 \text{ eV}$$

Find the energy of the emitted x-ray photon:

$$hf = E_M - E_K \approx -6.4 \times 10^3 \text{ eV} - (-7.4 \times 10^4 \text{ eV})$$

$$\approx 6.8 \times 10^4 \text{ eV} = 68 \text{ keV}$$

Consultation of x-ray tables shows that the M–K transition energies in tungsten vary from 66.9 keV to 67.7 keV, where the range of energies is due to slightly different energy values for states of different ℓ. Therefore, our estimate differs from the midpoint of this experimentally measured range by approximately 1%.

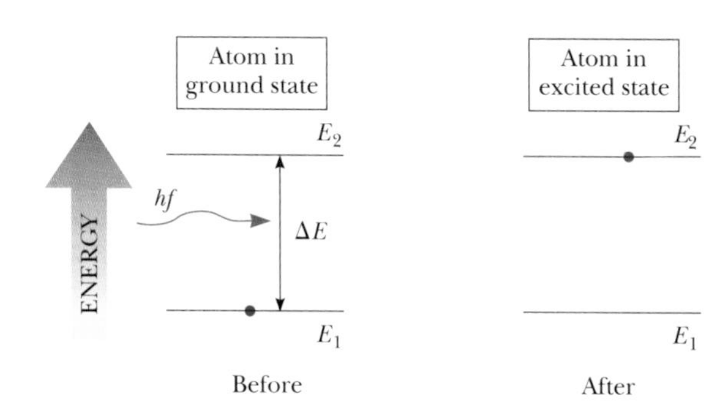

ACTIVE FIGURE 42.25

Stimulated absorption of a photon. The blue dot represents an electron. The electron is transferred from the ground state to the excited state when the atom absorbs a photon of energy $hf = E_2 - E_1$.

Sign in at www.thomsonedu.com and go to ThomsonNOW to adjust the energy difference between states and observe stimulated absorption.

42.9 Spontaneous and Stimulated Transitions

We have seen that an atom absorbs and emits electromagnetic radiation only at frequencies that correspond to the energy differences between allowed states. Let's now examine more details of these processes. Consider an atom having the allowed energy levels labeled E_1, E_2, E_3, ... When radiation is incident on the atom, only those photons whose energy hf matches the energy separation ΔE between two energy levels can be absorbed by the atom as represented in Active Figure 42.25. This process is called **stimulated absorption** because the photon stimulates the atom to make the upward transition. At ordinary temperatures, most of the atoms in a sample are in the ground state. If a vessel containing many atoms of a gaseous element is illuminated with radiation of all possible photon frequencies (that is, a continuous spectrum), only those photons having energy $E_2 - E_1$, $E_3 - E_1$, $E_4 - E_1$, and so on are absorbed by the atoms. As a result of this absorption, some of the atoms are raised to excited states.

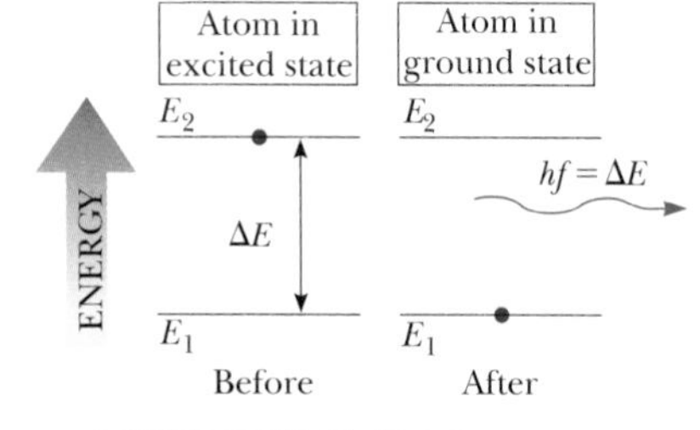

ACTIVE FIGURE 42.26

Spontaneous emission of a photon by an atom that is initially in the excited state E_2. When the atom falls to the ground state, it emits a photon of energy $hf = E_2 - E_1$.

Sign in at www.thomsonedu.com and go to ThomsonNOW to adjust the energy difference between states and observe spontaneous emission.

Once an atom is in an excited state, the excited atom can make a transition back to a lower energy level, emitting a photon in the process as in Active Figure 42.26. This process is known as **spontaneous emission** because it happens naturally, without requiring an event to trigger the transition. Typically, an atom remains in an excited state for only about 10^{-8} s.

In addition to spontaneous emission, **stimulated emission** occurs. Suppose an atom is in an excited state E_2 as in Active Figure 42.27. If the excited state is a *metastable state*—that is, if its lifetime is much longer than the typical 10^{-8} s lifetime of excited states—the time interval until spontaneous emission occurs is relatively long. Let's imagine that during that interval a photon of energy $hf = E_2 - E_1$ is incident on the atom. One possibility is that the photon energy is sufficient for the

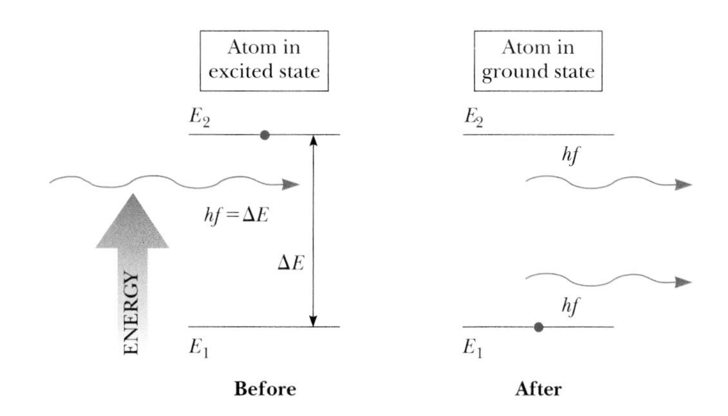

ACTIVE FIGURE 42.27

Stimulated emission of a photon by an incoming photon of energy $hf = E_2 - E_1$. Initially, the atom is in the excited state. The incoming photon stimulates the atom to emit a second photon of energy given by $hf = E_2 - E_1$.

Sign in at www.thomsonedu.com and go to ThomsonNOW to adjust the energy difference between states and observe stimulated emission.

photon to ionize the atom. Another possibility is that the interaction between the incoming photon and the atom causes the atom to return to the ground state[11] and thereby emit a second photon with energy $hf = E_2 - E_1$. In this process, the incident photon is not absorbed; therefore, after the stimulated emission, two photons with identical energy exist: the incident photon and the emitted photon. The two are in phase and travel in the same direction, which is an important consideration in lasers, discussed next.

42.10 Lasers

In this section, we explore the nature of laser light and a variety of applications of lasers in our technological society. The primary properties of laser light that make it useful in these technological applications are the following:

- Laser light is coherent. The individual rays of light in a laser beam maintain a fixed phase relationship with each other.
- Laser light is monochromatic. Light in a laser beam has a very narrow range of wavelengths.
- Laser light has a small angle of divergence. The beam spreads out very little, even over large distances.

To understand the origin of these properties, let's combine our knowledge of atomic energy levels from this chapter with some special requirements for the atoms that emit laser light.

We have described how an incident photon can cause atomic energy transitions either upward (stimulated absorption) or downward (stimulated emission). The two processes are equally probable. When light is incident on a collection of atoms, a net absorption of energy usually occurs because when the system is in thermal equilibrium, many more atoms are in the ground state than in excited states. If the situation can be inverted so that more atoms are in an excited state than in the ground state, however, a net emission of photons can result. Such a condition is called **population inversion.**

Population inversion is, in fact, the fundamental principle involved in the operation of a **laser** (an acronym for *l*ight *a*mplification by *s*timulated *e*mission of *r*adiation). The full name indicates one of the requirements for laser light: to achieve laser action, the process of stimulated emission must occur.

Suppose an atom is in the excited state E_2 as in Active Figure 42.27 and a photon with energy $hf = E_2 - E_1$ is incident on it. As described in Section 42.9, the

[11] This phenomenon is fundamentally due to *resonance.* The incoming photon has a frequency and drives the system of the atom at that frequency. Because the driving frequency matches that associated with a transition between states—one of the natural frequencies of the atom—there is a large response: the atom makes the transition.

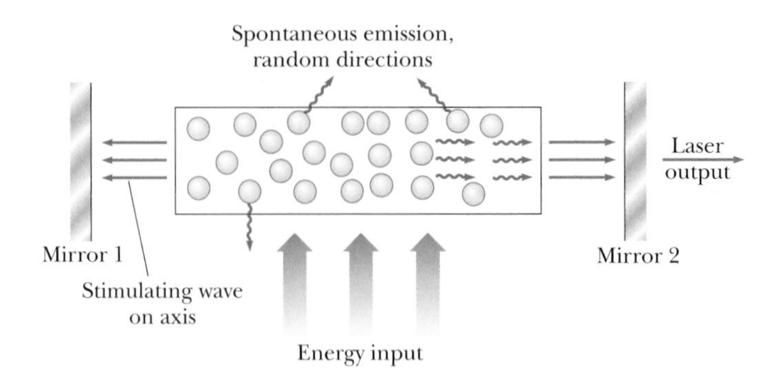

Figure 42.28 Schematic diagram of a laser design. The tube contains the atoms that are the active medium. An external source of energy (for example, an optical or electrical device) "pumps" the atoms to excited states. The parallel end mirrors confine the photons to the tube, but mirror 2 is only partially reflective.

incoming photon can stimulate the excited atom to return to the ground state and thereby emit a second photon having the same energy hf and traveling in the same direction. The incident photon is not absorbed, so after the stimulated emission, there are two identical photons: the incident photon and the emitted photon. The emitted photon is in phase with the incident photon. These photons can stimulate other atoms to emit photons in a chain of similar processes. The many photons produced in this fashion are the source of the intense, coherent light in a laser.

For the stimulated emission to result in laser light, there must be a buildup of photons in the system. The following three conditions must be satisfied to achieve this buildup:

- The system must be in a state of population inversion: there must be more atoms in an excited state than in the ground state. That must be true because the number of photons emitted must be greater than the number absorbed.
- The excited state of the system must be a *metastable state,* meaning that its lifetime must be long compared with the usually short lifetimes of excited states, which are typically 10^{-8} s. In this case, the population inversion can be established and stimulated emission is likely to occur before spontaneous emission.
- The emitted photons must be confined in the system long enough to enable them to stimulate further emission from other excited atoms. That is achieved by using reflecting mirrors at the ends of the system. One end is made totally reflecting, and the other is partially reflecting. A fraction of the light intensity passes through the partially reflecting end, forming the beam of laser light (Fig. 42.28).

One device that exhibits stimulated emission of radiation is the helium–neon gas laser. Figure 42.29 is an energy-level diagram for the neon atom in this system. The mixture of helium and neon is confined to a glass tube that is sealed at the ends by mirrors. A voltage applied across the tube causes electrons to sweep through the tube, colliding with the atoms of the gases and raising them into excited states. Neon atoms are excited to state $E_3{}^*$ through this process (the asterisk indicates a metastable state) and also as a result of collisions with excited helium atoms. Stimulated emission occurs, causing neon atoms to make transitions to state E_2. Neighboring excited atoms are also stimulated. The result is the production of coherent light at a wavelength of 632.8 nm.

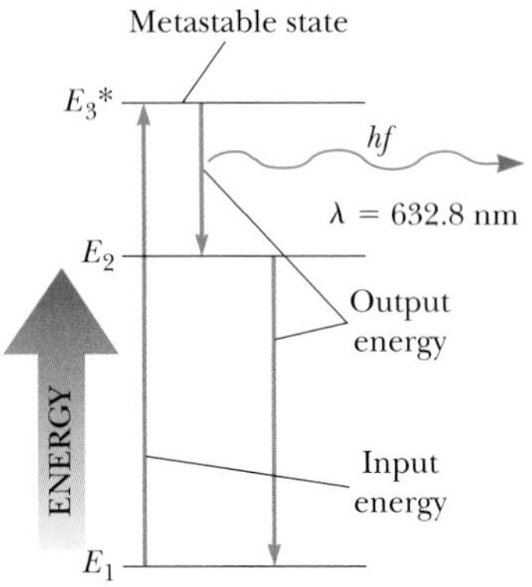

Metastable state
$E_3{}^*$
hf
$\lambda = 632.8$ nm
E_2
Output energy
ENERGY
Input energy
E_1

Figure 42.29 Energy-level diagram for a neon atom in a helium–neon laser. The atom emits 632.8-nm photons through stimulated emission in the transition $E_3{}^* - E_2$. That is the source of coherent light in the laser.

Applications

Since the development of the first laser in 1960, tremendous growth has occurred in laser technology. Lasers that cover wavelengths in the infrared, visible, and ultraviolet regions are now available. Applications include surgical "welding" of detached retinas, precision surveying and length measurement, precision cutting of metals and other materials (such as the fabric in Fig. 42.30), and telephone communication along optical fibers. These and other applications are possible because of the unique characteristics of laser light. In addition to being highly monochromatic, laser light is also highly directional and can be sharply focused to produce regions of extremely intense light energy (with energy densities 10^{12} times the density in the flame of a typical cutting torch).

Lasers are used in precision long-range distance measurement (range finding). In recent years, it has become important in astronomy and geophysics to measure as precisely as possible the distances from various points on the surface of the Earth to a point on the Moon's surface. To facilitate these measurements, the *Apollo* astronauts set up a 0.5-m square of reflector prisms on the Moon, which enables laser pulses directed from an Earth-based station to be retroreflected to the same station (see Fig. 35.8a). Using the known speed of light and the measured round-trip travel time of a laser pulse, the Earth–Moon distance can be determined to a precision of better than 10 cm.

Figure 42.30 This robot carrying laser scissors, which can cut up to 50 layers of fabric at a time, is one of the many applications of laser technology.

Because various laser wavelengths can be absorbed in specific biological tissues, lasers have a number of medical applications. For example, certain laser procedures have greatly reduced blindness in patients with glaucoma and diabetes. Glaucoma is a widespread eye condition characterized by a high fluid pressure in the eye, a condition that can lead to destruction of the optic nerve. A simple laser operation (iridectomy) can "burn" open a tiny hole in a clogged membrane, relieving the destructive pressure. A serious side effect of diabetes is neovascularization, the proliferation of weak blood vessels, which often leak blood. When neovascularization occurs in the retina, vision deteriorates (diabetic retinopathy) and finally is destroyed. Today, it is possible to direct the green light from an argon ion laser through the clear eye lens and eye fluid, focus on the retina edges, and photocoagulate the leaky vessels. Even people who have only minor vision defects such as nearsightedness are benefiting from the use of lasers to reshape the cornea, changing its focal length and reducing the need for eyeglasses.

Laser surgery is now an everyday occurrence at hospitals and medical clinics around the world. Infrared light at 10 μm from a carbon dioxide laser can cut through muscle tissue, primarily by vaporizing the water contained in cellular material. Laser power of approximately 100 W is required in this technique. The advantage of the "laser knife" over conventional methods is that laser radiation cuts tissue and coagulates blood at the same time, leading to a substantial reduction in blood loss. In addition, the technique virtually eliminates cell migration, an important consideration when tumors are being removed.

A laser beam can be trapped in fine optical fiber light guides (endoscopes) by means of total internal reflection. An endoscope can be introduced through natural orifices, conducted around internal organs, and directed to specific interior body locations, eliminating the need for invasive surgery. For example, bleeding in the gastrointestinal tract can be optically cauterized by endoscopes inserted through the patient's mouth.

In biological and medical research, it is often important to isolate and collect unusual cells for study and growth. A laser cell separator exploits the tagging of specific cells with fluorescent dyes. All cells are then dropped from a tiny charged nozzle and laser-scanned for the dye tag. If triggered by the correct light-emitting tag, a small voltage applied to parallel plates deflects the falling electrically charged cell into a collection beaker.

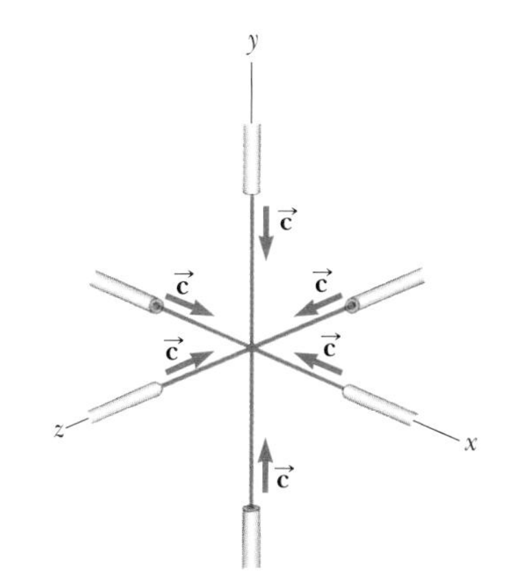

Figure 42.31 An optical trap for atoms is formed at the intersection point of six counterpropagating laser beams along mutually perpendicular axes. The frequency of the laser light is tuned to be immediately below that for absorption by the trapped atoms. If an atom moves away from the trap, it absorbs the Doppler-shifted laser light and the momentum of the light pushes the atom back into the trap.

An exciting area of research and technological applications arose in the 1990s with the development of *laser trapping* of atoms. One scheme, called *optical molasses* and developed by Steven Chu of Stanford University and his colleagues, involves focusing six laser beams onto a small region in which atoms are to be trapped. Each pair of lasers is along one of the x, y, and z axes and emits light in opposite directions (Fig. 42.31). The frequency of the laser light is tuned to be slightly below the absorption frequency of the subject atom. Imagine that an atom has been placed into the trap region and moves along the positive x axis toward the laser that is emitting light toward it (the rightmost laser in Fig. 42.31). Because the atom is moving, the light from the laser appears Doppler-shifted upward in frequency in the reference frame of the atom. Therefore, a match between the Doppler-shifted laser frequency and the absorption frequency of the atom exists

Courtesy of Mark Helfer/National Institute of Standards and Technology

Figure 42.32 A staff member of the National Institute of Standards and Technology views a sample of trapped sodium atoms (the small yellow dot in the center of the vacuum chamber) cooled to a temperature of less than 1 mK.

and the atom absorbs photons.[12] The momentum carried by these photons results in the atom being pushed back to the center of the trap. By incorporating six lasers, the atoms are pushed back into the trap regardless of which way they move along any axis.

In 1986, Chu developed *optical tweezers,* a device that uses a single tightly focused laser beam to trap and manipulate small particles. In combination with microscopes, optical tweezers have opened up many new possibilities for biologists. Optical tweezers have been used to manipulate live bacteria without damage, move chromosomes within a cell nucleus, and measure the elastic properties of a single DNA molecule. Chu shared the 1997 Nobel Prize in Physics with two of his colleagues for the development of the techniques of optical trapping.

An extension of laser trapping, *laser cooling,* is possible because the normal high speeds of the atoms are reduced when they are restricted to the region of the trap. As a result, the temperature of the collection of atoms can be reduced to a few microkelvins. The technique of laser cooling allows scientists to study the behavior of atoms at extremely low temperatures (Fig. 42.32).

[12] The laser light traveling in the same direction as the atom is Doppler-shifted further downward in frequency, so there is no absorption. Therefore, the atom is not pushed out of the trap by the diametrically opposed laser.

Summary

ThomsonNOW™ Sign in at **www.thomsonedu.com** and go to ThomsonNOW to take a practice test for this chapter.

CONCEPTS AND PRINCIPLES

The wavelengths of spectral lines from hydrogen, called the **Balmer series,** can be described by the equation

$$\frac{1}{\lambda} = R_H\left(\frac{1}{2^2} - \frac{1}{n^2}\right) \quad n = 3, 4, 5, \ldots \qquad \textbf{(42.1)}$$

where R_H is the **Rydberg constant.** The spectral lines corresponding to values of n from 3 to 6 are in the visible range of the electromagnetic spectrum. Values of n higher than 6 correspond to spectral lines in the ultraviolet region of the spectrum.

The Bohr model of the atom is successful in describing the spectra of atomic hydrogen and hydrogen-like ions. One basic assumption of the model is that the electron can exist only in discrete orbits such that the angular momentum of the electron is an integral multiple of $h/2\pi = \hbar$. When we assume circular orbits and a simple Coulomb attraction between electron and proton, the energies of the quantum states for hydrogen are calculated to be

$$E_n = -\frac{k_e e^2}{2a_0}\left(\frac{1}{n^2}\right) \quad n = 1, 2, 3, \ldots \qquad \textbf{(42.13)}$$

where n is an integer called the **quantum number,** k_e is the Coulomb constant, e is the electronic charge, and $a_0 = 0.052\,9$ nm is the **Bohr radius.**

If the electron in a hydrogen atom makes a transition from an orbit whose quantum number is n_i to one whose quantum number is n_f, where $n_f < n_i$, a photon is emitted by the atom. The frequency of this photon is

$$f = \frac{k_e e^2}{2a_0 h}\left(\frac{1}{n_f^2} - \frac{1}{n_i^2}\right) \qquad \textbf{(42.15)}$$

(*continued*)

Quantum mechanics can be applied to the hydrogen atom by the use of the potential energy function $U(r) = -k_e e^2/r$ in the Schrödinger equation. The solution to this equation yields wave functions for allowed states and allowed energies:

$$E_n = -\left(\frac{k_e e^2}{2a_0}\right)\frac{1}{n^2} = -\frac{13.606 \text{ eV}}{n^2} \qquad n = 1, 2, 3, \ldots \qquad \textbf{(42.21)}$$

where n is the **principal quantum number.** The allowed wave functions depend on three quantum numbers: n, ℓ, and m_ℓ, where ℓ is the **orbital quantum number** and m_ℓ is the **orbital magnetic quantum number.** The restrictions on the quantum numbers are

$$n = 1, 2, 3, \ldots$$

$$\ell = 0, 1, 2, \ldots, n - 1$$

$$m_\ell = -\ell, -\ell+1, \ldots, \ell - 1, \ell$$

All states having the same principal quantum number n form a **shell,** identified by the letters K, L, M, . . . (corresponding to $n = 1, 2, 3, \ldots$). All states having the same values of n and ℓ form a **subshell,** designated by the letters *s, p, d, f,* . . . (corresponding to $\ell = 0, 1, 2, 3, \ldots$).

An atom in a state characterized by a specific value of n can have the following values of L, the magnitude of the atom's orbital angular momentum $\vec{\mathbf{L}}$:

$$L = \sqrt{\ell(\ell + 1)}\hbar$$

$$\ell = 0, 1, 2, \ldots, n - 1 \qquad \textbf{(42.27)}$$

The allowed values of the projection of $\vec{\mathbf{L}}$ along the z axis are

$$L_z = m_\ell \hbar \qquad \textbf{(42.28)}$$

Only discrete values of L_z are allowed as determined by the restrictions on m_ℓ. This quantization of L_z is referred to as **space quantization.**

The electron has an intrinsic angular momentum called the **spin angular momentum.** Electron spin can be described by a single quantum number $s = \frac{1}{2}$. To describe a quantum state completely, it is necessary to include a fourth quantum number m_s, called the **spin magnetic quantum number.** This quantum number can have only two values, $\pm\frac{1}{2}$. The magnitude of the spin angular momentum is

$$S = \frac{\sqrt{3}}{2}\hbar \qquad \textbf{(42.30)}$$

and the z component of $\vec{\mathbf{S}}$ is

$$S_z = m_s\hbar = \pm\tfrac{1}{2}\hbar \qquad \textbf{(42.31)}$$

That is, the spin angular momentum is also quantized in space, as specified by the spin magnetic quantum number $m_s = \pm\frac{1}{2}$.

The **exclusion principle** states that **no two electrons in an atom can be in the same quantum state.** In other words, no two electrons can have the same set of quantum numbers n, ℓ, m_ℓ, and m_s. Using this principle, the electronic configurations of the elements can be determined. This principle serves as a basis for understanding atomic structure and the chemical properties of the elements.

The magnetic moment $\vec{\boldsymbol{\mu}}_{\text{spin}}$ associated with the spin angular momentum of an electron is

$$\vec{\boldsymbol{\mu}}_{\text{spin}} = -\frac{e}{m_e}\vec{\mathbf{S}} \qquad \textbf{(42.32)}$$

The z component of $\vec{\boldsymbol{\mu}}_{\text{spin}}$ can have the values

$$\mu_{\text{spin},z} = \pm\frac{e\hbar}{2m_e} \qquad \textbf{(42.33)}$$

The x-ray spectrum of a metal target consists of a set of sharp characteristic lines superimposed on a broad continuous spectrum. **Bremsstrahlung** is x-radiation with its origin in the slowing down of high-energy electrons as they encounter the target. **Characteristic x-rays** are emitted by atoms when an electron undergoes a transition from an outer shell to a vacancy in an inner shell.

Atomic transitions can be described with three processes: **stimulated absorption,** in which an incoming photon raises the atom to a higher energy state; **spontaneous emission,** in which the atom makes a transition to a lower energy state, emitting a photon; and **stimulated emission,** in which an incident photon causes an excited atom to make a downward transition, emitting a photon identical to the incident one.

Questions

☐ denotes answer available in *Student Solutions Manual/Study Guide;* **O** denotes objective question

1. Suppose the electron in the hydrogen atom obeyed classical mechanics rather than quantum mechanics. Why should a gas of such hypothetical atoms emit a continuous spectrum rather than the observed line spectrum?

2. O (a) Can a hydrogen atom in the ground state absorb a photon of energy less than 13.6 eV? (b) Can this atom absorb a photon of energy greater than 13.6 eV?

3. O Compare this question with Quick Quiz 42.1. When an electron collides with an atom, it can transfer all or some of its energy to the atom. A hydrogen atom is in its ground state. Incident on the atom are several electrons each having a kinetic energy of 10.5 eV. What is the result? (a) The atom is excited to a higher allowed state. (b) The atom is ionized. (c) The electrons pass by the atom without interaction.

4. O Let $-E$ represent the energy of a hydrogen atom. **(i)** What is the kinetic energy of the electron? (a) $2E$ (b) E (c) $E/2$ (d) 0 (e) $-E/2$ (f) $-E$ (g) $-2E$ **(ii)** What is the potential energy of the atom? Choose from the same possibilities (a) through (g).

5. O (i) Rank the following transitions for a hydrogen atom from the transition with the greatest gain in energy to that with the greatest loss, showing any cases of equality: (a) $n_i = 2$; $n_f = 5$ (b) $n_i = 5$; $n_f = 3$ (c) $n_i = 7$; $n_f = 4$ (d) $n_i = 4$; $n_f = 7$ **(ii)** Rank the same transitions according to the wavelength of the photon absorbed or emitted by an otherwise isolated atom from greatest wavelength to smallest.

6. O (a) In the hydrogen atom, can the quantum number n increase without limit? (b) Can the frequency of possible lines in the spectrum of hydrogen increase without limit? (c) Can the wavelength of possible lines in the spectrum of hydrogen increase without limit?

7. According to Bohr's model of the hydrogen atom, what is the uncertainty in the radial coordinate of the electron? What is the uncertainty in the radial component of the velocity of the electron? In what way does the model violate the uncertainty principle?

8. O Consider the quantum numbers (a) n, (b) ℓ, (c) m_ℓ, and (d) m_s. **(i)** Which of these quantum numbers are fractional as opposed to being integers? **(ii)** Which can be negative as opposed to being always positive? **(iii)** Which can be zero?

9. Why are three quantum numbers needed to describe the state of a one-electron atom (ignoring spin)?

10. Compare the Bohr theory and the Schrödinger treatment of the hydrogen atom. Comment on the total energy and orbital angular momentum.

[11.] Could the Stern–Gerlach experiment be performed with ions rather than neutral atoms? Explain.

12. Why is a *nonuniform* magnetic field used in the Stern–Gerlach experiment?

[13.] Discuss some consequences of the exclusion principle.

14. An energy of about 21 eV is required to excite an electron in a helium atom from the 1*s* state to the 2*s* state. The same transition for the He^+ ion requires approximately twice as much energy. Explain.

[15.] Why do lithium, potassium, and sodium exhibit similar chemical properties?

16. It is easy to understand how two electrons (one spin up, one spin down) fill the $n = 1$ or K shell for a helium atom. How is it possible that eight more electrons can fit into the $n = 2$ shell, filling the K and L shells for a neon atom?

17. O (i) What is the principal quantum number of the initial state of an atom as it emits an M_β line in an x-ray spectrum? (a) 1 (b) 2 (c) 3 (d) 4 (e) 5 (f) none of these answers **(ii)** What is the principal quantum number of the final state for this transition? Choose from the same possibilities.

18. Does the intensity of light from a laser fall off as $1/r^2$? Explain.

19. Why is stimulated emission so important in the operation of a laser?

20. (a) "As soon as I define a particular direction as the z axis, precisely one half of the electrons in this part of the Universe have their magnetic moment vectors oriented at 54.735 61° to that axis, and all the rest have their magnetic moments at 125.264 39°." Argue for or against this statement. (b) "The Universe is not simply stranger than we suppose; it is stranger than we *can* suppose." Argue for or against this statement.

21. A message reads, "*All your base are belong to us!*" Argue for or against the view that a scientific discovery is like a communication from an utterly alien source, in need of interpretation and susceptible to misunderstanding. Argue for or against the view that the human mind is not necessarily well adapted to understand the Universe. Argue for or against the view that education in science is the best preparation for life in a rapidly changing world.

Problems

WebAssign The Problems from this chapter may be assigned online in WebAssign.

ThomsonNOW Sign in at **www.thomsonedu.com** and go to ThomsonNOW to assess your understanding of this chapter's topics with additional quizzing and conceptual questions.

1, 2, 3 denotes straightforward, intermediate, challenging; □ denotes full solution available in *Student Solutions Manual/Study Guide;* ▲ denotes coached solution with hints available at **www.thomsonedu.com;** denotes developing symbolic reasoning; ● denotes asking for qualitative reasoning; denotes computer useful in solving problem

Section 42.1 Atomic Spectra of Gases

1. (a) What value of n_i is associated with the 94.96-nm spectral line in the Lyman series of hydrogen? (b) **What If?** Could this wavelength be associated with the Paschen series? The Balmer series?

2. (a) An isolated atom of a certain element emits light of wavelength 520 nm when the atom falls from its fifth excited state into its second excited state. The atom emits a photon of wavelength 410 nm when it drops from its sixth excited state into its second excited state. Find the wavelength of the light radiated when the atom makes a transition from its sixth to its fifth excited state. (b) Solve the same problem again in symbolic terms. Letting λ_{BA} represent the wavelength emitted in the transition B to A and λ_{CA} represent the shorter wavelength emitted in the transition C to A, find λ_{CB}. This problem exemplifies the *Ritz combination principle,* an empirical rule formulated in 1908.

Section 42.2 Early Models of the Atom

3. ▲ According to classical physics, a charge e moving with an acceleration a radiates at a rate

$$\frac{dE}{dt} = -\frac{1}{6\pi\epsilon_0}\frac{e^2a^2}{c^3}$$

(a) Show that an electron in a classical hydrogen atom (see Fig. 42.5) spirals into the nucleus at a rate

$$\frac{dr}{dt} = -\frac{e^4}{12\pi^2\epsilon_0^{\,2}r^2m_e^{\,2}c^3}$$

(b) Find the time interval over which the electron reaches $r = 0$, starting from $r_0 = 2.00 \times 10^{-10}$ m.

4. **Review problem.** In the Rutherford scattering experiment, 4.00-MeV alpha particles (^{4}He nuclei containing 2 protons and 2 neutrons) scatter off gold nuclei (containing 79 protons and 118 neutrons). Assume a particular alpha particle makes a direct head-on collision with the gold nucleus and scatters backward at 180°. Determine (a) the distance of closest approach of the alpha particle to the gold nucleus and (b) the maximum force exerted on the alpha particle. Assume the gold nucleus remains fixed throughout the entire process.

Section 42.3 Bohr's Model of the Hydrogen Atom

5. The Balmer series for the hydrogen atom corresponds to electronic transitions that terminate in the state with quantum number $n = 2$ as shown in Figure 42.2 and Active Figure 42.8. (a) Consider the photon of longest wavelength. Determine its energy and wavelength. (b) Consider the spectral line of shortest wavelength. Find its photon energy and wavelength.

6. For a hydrogen atom in its ground state, use the Bohr model to compute (a) the orbital speed of the electron, (b) the kinetic energy of the electron, and (c) the electric potential energy of the atom.

7. ▲ A hydrogen atom is in its first excited state ($n = 2$). Using the Bohr theory of the atom, calculate (a) the radius of the orbit, (b) the linear momentum of the electron, (c) the angular momentum of the electron, (d) the kinetic energy of the electron, (e) the potential energy of the system, and (f) the total energy of the system.

8. ● A monochromatic beam of light is absorbed by a collection of ground-state hydrogen atoms in such a way that six different wavelengths are observed when the hydrogen relaxes back to the ground state. (a) What is the wavelength of the incident beam? Explain the steps in your solution. (b) What is the longest wavelength in the emission spectrum of these atoms? To what portion of the electromagnetic spectrum and to what series does it belong? (c) What is the shortest wavelength? To what series does it belong?

9. (a) Construct an energy-level diagram for the He^+ ion, for which $Z = 2$. (b) What is the ionization energy for He^+?

10. A photon with energy 2.28 eV is barely capable of causing a photoelectric effect when it strikes a sodium plate. Suppose the photon is instead absorbed by hydrogen. Find (a) the minimum n for a hydrogen atom that can be ionized by such a photon and (b) the speed of the released electron far from the nucleus.

11. The positron is the antiparticle to the electron. It has the same mass and a positive electric charge of the same magnitude as that of the electron. Positronium is a hydrogen-like atom consisting of a positron and an electron revolving around each other. Using the Bohr model, find the allowed distances between the two particles and the allowed energies of the system.

12. ● An electron is in the nth Bohr orbit of the hydrogen atom. (a) Show that the period of the electron is $T = t_0n^3$ and determine the numerical value of t_0. (b) On average, an electron remains in the $n = 2$ orbit for approximately 10 μs before it jumps down to the $n = 1$ (ground-state) orbit. How many revolutions does the electron make in the excited state? (c) Define the period of one revolution as an electron year, analogous to an Earth year being the period of the Earth's motion around the Sun. Explain whether we should think of the electron in the $n = 2$ orbit as "living for a long time."

Section 42.4 The Quantum Model of the Hydrogen Atom

13. A general expression for the energy levels of one-electron atoms and ions is

$$E_n = -\frac{\mu k_e^2 q_1^2 q_2^2}{2\hbar^2 n^2}$$

Here μ is the reduced mass of the atom, given by $\mu = m_1 m_2/(m_1 + m_2)$, where m_1 is the mass of the electron and m_2 is the mass of the nucleus; k_e is the Coulomb constant; and q_1 and q_2 are the charges of the electron and the nucleus, respectively. The wavelength for the $n = 3$ to $n = 2$ transition of the hydrogen atom is 656.3 nm (visible red light). **What If?** What are the wavelengths for this same transition in (a) positronium, which consists of an electron and a positron, and (b) singly ionized helium? *Note:* A positron is a positively charged electron.

14. Ordinary hydrogen gas is a mixture of two kinds of atoms (isotopes) containing either one- or two-particle nuclei. These isotopes are hydrogen-1 with a proton nucleus, and hydrogen-2, called deuterium, with a deuteron nucleus. A deuteron is one proton and one neutron bound together. Hydrogen-1 and deuterium have identical chemical properties, but they can be separated via an ultracentrifuge or by other methods. Their emission spectra show lines of the same colors at very slightly different wavelengths. (a) Use the equation given in Problem 13 to show that the difference in wavelength between the hydrogen-1 and deuterium spectral lines associated with a particular electron transition is given by

$$\lambda_H - \lambda_D = \left(1 - \frac{\mu_H}{\mu_D}\right)\lambda_H$$

(b) Evaluate the wavelength difference for the Balmer alpha line of hydrogen, with wavelength 656.3 nm, emitted by an atom making a transition from an $n = 3$ state to an $n = 2$ state. Harold Urey observed this wavelength difference in 1931 and so confirmed his discovery of deuterium.

15. ● An electron of momentum p is at a distance r from a stationary proton. The electron has kinetic energy $K = p^2/2m_e$. The atom has potential energy $U = -k_e e^2/r$ and total energy $E = K + U$. If the electron is bound to the proton to form a hydrogen atom, its average position is at the proton but the uncertainty in its position is approximately equal to the radius r of its orbit. The electron's average vector momentum is zero, but its average squared momentum is approximately equal to the squared uncertainty in its momentum as given by the uncertainty principle. Treating the atom as a one-dimensional system, (a) estimate the uncertainty in the electron's momentum in terms of r. (b) Estimate the electron's kinetic, potential, and total energies in terms of r. (c) The actual value of r is the one that *minimizes the total energy*, resulting in a stable atom. Find that value of r and the resulting total energy. State how your answers compare with the predictions of the Bohr theory.

Section 42.5 The Wave Functions for Hydrogen

16. Plot the wave function $\psi_{1s}(r)$ versus r (see Eq. 42.22) and the radial probability density function $P_{1s}(r)$ versus r (see Eq. 42.25) for hydrogen. Let r range from 0 to $1.5a_0$, where a_0 is the Bohr radius.

17. The ground-state wave function for the electron in a hydrogen atom is

$$\psi_{1s}(r) = \frac{1}{\sqrt{\pi a_0^3}} e^{-r/a_0}$$

where r is the radial coordinate of the electron and a_0 is the Bohr radius. (a) Show that the wave function as given is normalized. (b) Find the probability of locating the electron between $r_1 = a_0/2$ and $r_2 = 3a_0/2$.

18. The wave function for an electron in the $2p$ state of hydrogen is

$$\psi_{2p} = \frac{1}{\sqrt{3}(2a_0)^{3/2}} \frac{r}{a_0} e^{-r/2a_0}$$

What is the most likely distance from the nucleus to find an electron in the $2p$ state?

19. ▲ For a spherically symmetric state of a hydrogen atom, the Schrödinger equation in spherical coordinates is

$$-\frac{\hbar^2}{2m_e}\left(\frac{d^2\psi}{dr^2} + \frac{2}{r}\frac{d\psi}{dr}\right) - \frac{k_e e^2}{r}\psi = E\psi$$

Show that the 1s wave function for an electron in hydrogen,

$$\psi_{1s}(r) = \frac{1}{\sqrt{\pi a_0^3}} e^{-r/a_0}$$

satisfies the Schrödinger equation.

20. In an experiment, electrons are fired at a sample of neutral hydrogen atoms and observations are made of how the incident particles scatter. A large set of trials can be thought of as containing 1 000 observations of the electron in the ground state of a hydrogen atom being momentarily at a distance $a_0/2$ from the nucleus. In this set of trials, how many times is the atomic electron observed at a distance $2a_0$ from the nucleus?

Section 42.6 Physical Interpretation of the Quantum Numbers

21. List the possible sets of quantum numbers for the hydrogen atom associated with (a) the 3d subshell and (b) the 3p subshell.

22. Calculate the orbital angular momentum for a hydrogen atom in (a) the 4d state and (b) the 6f state.

23. (a) State a problem for which the following equation appears in the solution:

$$4.714 \times 10^{-34}\ \text{J}\cdot\text{s} = \frac{\sqrt{\ell(\ell+1)}(6.626 \times 10^{-34}\ \text{J}\cdot\text{s})}{2\pi}$$

(b) Solve the equation for the unknown appearing in it and state the name of this quantity.

24. A hydrogen atom is in its fifth excited state, with principal quantum number 6. The atom emits a photon with a wavelength of 1 090 nm. Determine the maximum possible orbital angular momentum of the atom after emission.

25. ▲ How many sets of quantum numbers are possible for a hydrogen atom for which (a) $n = 1$, (b) $n = 2$, (c) $n = 3$, (d) $n = 4$, and (e) $n = 5$? Check your results to show that they agree with the general rule that the number of sets of quantum numbers for a shell is equal to $2n^2$.

26. Find all possible values of L, L_z, and θ for a hydrogen atom in a $3d$ state.

27. ● (a) Find the mass density of a proton, modeling it as a solid sphere of radius 1.00×10^{-15} m. (b) **What If?** Consider a classical model of an electron as a uniform solid sphere with the same density as the proton. Find its radius. (c) Imagine that this electron possesses spin angular momentum $I\omega = \hbar/2$ because of classical rotation about the z axis. Determine the speed of a point on the equator of the electron. (d) State how this speed compares to the speed of light.

28. An electron is in the N shell. Determine the maximum value the z component of its angular momentum could have.

29. The ρ^- meson has a charge of $-e$, a spin quantum number of 1, and a mass 1 507 times that of the electron. The possible values for its spin magnetic quantum number are -1, 0, and 1. **What If?** Imagine that the electrons in atoms were replaced by ρ^- mesons. List the possible sets of quantum numbers for ρ^- mesons in the $3d$ subshell.

Section 42.7 The Exclusion Principle and the Periodic Table

30. (a) What is the electronic configuration for the ground state of oxygen ($Z = 8$)? (b) Write out a set of possible values for the quantum numbers n, ℓ, m_ℓ, and m_s for each electron in oxygen.

31. ● As we go down the periodic table, which subshell is filled first, the $3d$ or the $4s$ subshell? Which electronic configuration has a lower energy, $[\text{Ar}]3d^44s^2$ or $[\text{Ar}]3d^54s^1$? Which has the greater number of unpaired spins? Identify this element and discuss Hund's rule in this case. *Note:* The notation [Ar] represents the filled configuration for argon.

32. Devise a table similar to that shown in Figure 42.18 for atoms containing 11 through 19 electrons. Use Hund's rule and educated guesswork.

33. A certain element has its outermost electron in a $3p$ subshell. It has valence +3 because it has three more electrons than a certain noble gas. What element is it?

34. Two electrons in the same atom both have $n = 3$ and $\ell = 1$. (a) List the quantum numbers for the possible states of the atom. (b) **What If?** How many states would be possible if the exclusion principle were inoperative?

35. ● ▲ (a) Scanning through Figure 42.19 in order of increasing atomic number, notice that the electrons usually fill the subshells in such a way that those subshells with the lowest values of $n + \ell$ are filled first. If two subshells have the same value of $n + \ell$, the one with the lower value of n is generally filled first. Using these two rules, write the order in which the subshells are filled through $n + \ell = 7$. (b) Predict the chemical valence for the elements that have atomic numbers 15, 47, and 86, and compare your predictions with the actual valences (which may be found in a chemistry textbook).

36. For a neutral atom of element 110, what would be the probable ground-state electronic configuration?

37. **Review problem.** For an electron with magnetic moment $\vec{\mu}_s$ in a magnetic field $\vec{\mathbf{B}}$, Section 29.5 showed the following. The electron-field system can be in a higher energy state with the z component of the electron's magnetic moment opposite to the field or a lower energy state with the z component of the magnetic moment in the direction of the field. The difference in energy between the two states is $2\mu_B B$.

Under high resolution, many spectral lines are observed to be doublets. The most famous of these are the two yellow lines in the spectrum of sodium (the D lines), with wavelengths of 588.995 nm and 589.592 nm. Their existence was explained in 1925 by Goudsmit and Uhlenbeck, who postulated that an electron has intrinsic spin angular momentum. When the sodium atom is excited with its outermost electron in a $3p$ state, the orbital motion of the outermost electron creates a magnetic field. The atom's energy is somewhat different depending on whether the electron is itself spin-up or spin-down in this field. Then the photon energy the atom radiates as it falls back into its ground state depends on the energy of the excited state. Calculate the magnitude of the internal magnetic field mediating this so-called spin-orbit coupling.

Section 42.8 More on Atomic Spectra: Visible and X-Ray

38. (a) Determine the possible values of the quantum numbers ℓ and m_ℓ for the He^+ ion in the state corresponding to $n = 3$. (b) What is the energy of this state?

39. ● The $3p$ level of sodium has an energy of -3.0 eV, and the $3d$ level has an energy of -1.5 eV. Determine Z_{eff} for each of these states. Explain the difference.

40. ● In x-ray production, electrons are accelerated through a high voltage and then decelerated by striking a target. (a) To make possible the production of x-rays of wavelength λ, what is the minimum potential difference ΔV through which the electrons must be accelerated? (b) State how the required potential difference depends on the wavelength. (c) Explain whether your result agrees with the information in Figure 42.22. (d) Does the relationship from part (a) apply to other kinds of electromagnetic radiation besides x-rays? What does the potential difference approach as λ goes to zero? What does the potential difference approach as λ increases without limit?

41. Use the method illustrated in Example 42.5 to calculate the wavelength of the x-ray emitted from a molybdenum target ($Z = 42$) when an electron moves from the L shell ($n = 2$) to the K shell ($n = 1$).

42. The K series of the discrete x-ray spectrum of tungsten contains wavelengths of 0.018 5 nm, 0.020 9 nm, and 0.021 5 nm. The K-shell ionization energy is 69.5 keV. Determine the ionization energies of the L, M, and N shells. Draw a diagram of the transitions.

43. The wavelength of characteristic x-rays in the K_β line from a particular source is 0.152 nm. Determine the material in the target.

Section 42.9 Spontaneous and Stimulated Transitions

Section 42.10 Lasers

44. Figure P42.44 shows portions of the energy-level diagrams of the helium and neon atoms. An electrical discharge excites the He atom from its ground state to its excited state of 20.61 eV. The excited He atom collides with a Ne atom in its ground state and excites this atom to the state at 20.66 eV. Lasing action takes place for electron transitions from $E_3{}^*$ to E_2 in the Ne atoms. From the data in the figure, show that the wavelength of the red He–Ne laser light is approximately 633 nm.

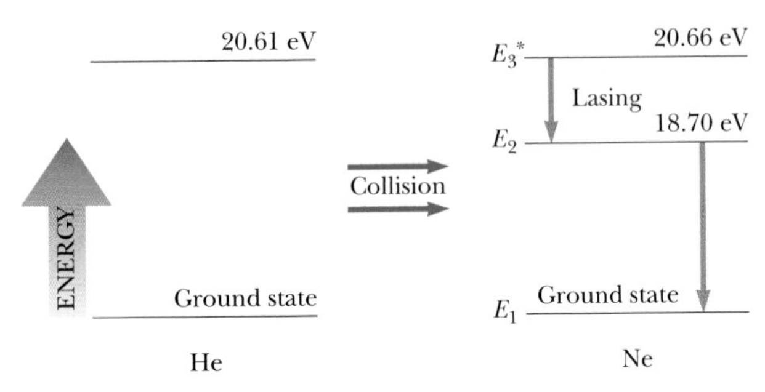

Figure P42.44

45. The carbon dioxide laser is one of the most powerful developed. The energy difference between the two laser levels is 0.117 eV. Determine the frequency and wavelength of the radiation emitted by this laser. In what portion of the electromagnetic spectrum is this radiation?

46. A neodymium-yttrium-aluminum garnet laser used in eye surgery emits a 3.00-mJ pulse in 1.00 ns, focused to a spot 30.0 μm in diameter on the retina. (a) Find (in SI units) the power per unit area at the retina. (In the optics industry, this quantity is called the *irradiance.*) (b) What energy is delivered to an area of molecular size, taken as a circular area 0.600 nm in diameter?

47. ▲ A ruby laser delivers a 10.0-ns pulse of 1.00 MW average power. If the photons have a wavelength of 694.3 nm, how many are contained in the pulse?

48. ● The number N of atoms in a particular state is called the population of that state. This number depends on the energy of that state and the temperature. In thermal equilibrium, the population of atoms in a state of energy E_n is given by a Boltzmann distribution expression

$$N = N_g e^{-(E_n - E_g)/k_B T}$$

where N_g is the population of the ground state of energy E_g and T is the absolute temperature. For simplicity, assume each energy level has only one quantum state associated with it. (a) Before the power is switched on, the neon atoms in a laser are in thermal equilibrium at 27.0°C. Find the equilibrium ratio of the populations of the states $E_3{}^*$ and E_2 shown in Figure 42.29. Lasers operate by a clever artificial production of a "population inversion" between the upper and lower atomic energy states involved in the lasing transition. This term means that more atoms are in the upper excited state than in the lower one. Consider the helium–neon laser transition at 632.8 nm. Assume 2% more atoms occur in the upper state than in the lower. (b) To demonstrate how unnatural such a situation is, find the temperature for which the Boltzmann distribution describes a 2.00% population inversion. (c) Why does such a situation not occur naturally?

49. Review problem. A helium–neon laser can produce a green laser beam instead of a red one. Refer to Figure 42.29, which omits some energy levels between E_2 and E_1. After a population inversion is established, neon atoms make a variety of downward transitions in falling from the state labeled $E_3{}^*$ down eventually to level E_1. The atoms emit both red light with a wavelength of 632.8 nm and green light with a wavelength of 543 nm in a competing transition. Assume the atoms are in a cavity between mirrors designed to reflect the green light with high efficiency but to allow the red light to leave the cavity immediately. Then stimulated emission can lead to the buildup of a collimated beam of green light between the mirrors having a greater intensity than that of the red light. To constitute the radiated laser beam, a small fraction of the green light is permitted to escape by transmission through one mirror. The mirrors forming the resonant cavity can be made of layers of silicon dioxide and titanium dioxide. (a) How thick a layer of silicon dioxide, between layers of titanium dioxide, would minimize reflection of the red light? (b) What should be the thickness of a similar but separate layer of silicon dioxide to maximize reflection of the green light?

Additional Problems

50. ● As the Earth moves around the Sun, its orbits are quantized. (a) Follow the steps of Bohr's analysis of the hydrogen atom to show that the allowed radii of the Earth's orbit are given by

$$r = \frac{n^2 \hbar^2}{G M_S M_E{}^2}$$

where n is an integer quantum number, M_S is the mass of the Sun, and M_E is the mass of the Earth. (b) Calculate the numerical value of n. (c) Find the distance between the orbit for quantum number n and the next orbit out from the Sun corresponding to the quantum number $n + 1$. Discuss the significance of your results.

51. LENINGRAD, 1930—Four years after publication of the Schrödinger equation, Lev Davidovich Landau, age 23, solved the equation for a charged particle moving in a uniform magnetic field. A single electron moving perpendicular to a field $\vec{\mathbf{B}}$ can be considered as a model atom without a nucleus or as the irreducible quantum limit of the cyclotron. Landau proved its energy is quantized in uniform steps of $e\hbar B/m_e$.

CAMBRIDGE, MA, 1999—Gerald Gabrielse trapped a single electron in an evacuated centimeter-size metal can cooled to a temperature of 80 mK. In a magnetic field of magnitude 5.26 T, the electron circulated for hours in its lowest energy level, generating a measurable signal as it moved. (a) Evaluate the size of a quantum jump in the electron's energy. (b) For comparison, evaluate $k_B T$ as a measure of the energy available to the electron in blackbody radiation from the walls of its container. (c) Microwave radiation was introduced to excite the electron. Calculate the frequency and wavelength of the photon that the electron absorbed as it jumped to its second energy level. Measurement of the resonant absorption frequency verified the theory and permitted precise determination of properties of the electron.

52. Example 42.3 calculates the most probable value and the average value for the radial coordinate r of the electron in the ground state of a hydrogen atom. **What If?** For comparison with these modal and mean values, find the median value of r. Proceed as follows. (a) Derive an expression for the probability, as a function of r, that the electron in the ground state of hydrogen will be found outside a sphere of radius r centered on the nucleus. (b) Make a graph of the probability as a function of r/a_0. Choose values of r/a_0 ranging from 0 to 4.00 in steps of 0.250. (c) Find the value of r for which the probability of finding the electron outside a sphere of radius r is equal to the probability of finding the electron inside this sphere. You must solve a transcendental equation numerically, and your graph is a good starting point.

53. *An example of the correspondence principle.* Use Bohr's model of the hydrogen atom to show that when the electron moves from the state n to the state $n - 1$, the frequency of the emitted light is

$$f = \left(\frac{2\pi^2 m_e k_e^{\,2} e^4}{h^3 n^2}\right) \frac{2n - 1}{(n - 1)^2}$$

Show that as $n \to \infty$, this expression varies as $1/n^3$ and reduces to the classical frequency one expects the atom to emit. *Suggestion:* To calculate the classical frequency, note that the frequency of revolution is $v/2\pi r$, where v is the speed of the electron and r is given by Eq. 42.10.

54. **Review problem.** (a) How much energy is required to cause an electron in hydrogen to move from the $n = 1$ state to the $n = 2$ state? (b) Suppose the atom gains this energy through collisions among hydrogen atoms at a high temperature. At what temperature would the average atomic kinetic energy $3k_B T/2$ be great enough to excite the electron? Here k_B is Boltzmann's constant.

55. ● Astronomers observe a series of spectral lines in the light from a distant galaxy. On the hypothesis that the lines form the Lyman series for a (new?!) one-electron atom, they start to construct the energy-level diagram shown in Figure P42.55, which gives the wavelengths of the first four lines and the short-wavelength limit of this series. Based on this information, calculate (a) the energies of the ground state and first four excited states for this one-electron atom and (b) the wavelengths of the first three lines and the short-wavelength limit in the Balmer series for this atom. (c) Show that the wavelengths of the first four lines and the short-wavelength limit of the Lyman series for the hydrogen atom are all 60.0% of the wavelengths for the Lyman series in the one-electron atom described in part (b). (d) Based on this observation, explain why this atom could be hydrogen.

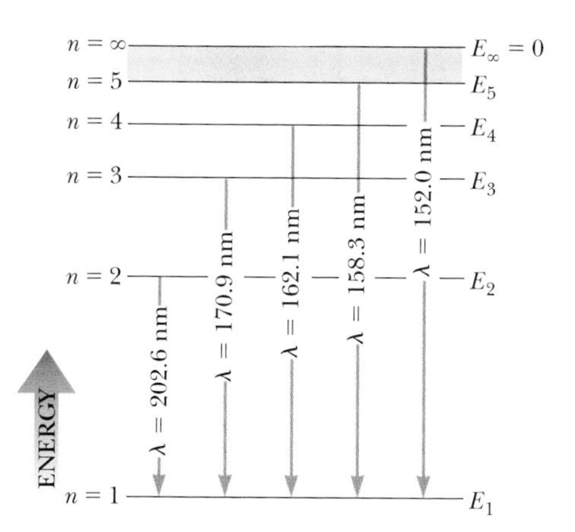

Figure P42.55

56. (a) For a hydrogen atom making a transition from the $n = 4$ state to the $n = 2$ state, determine the wavelength of the photon created in the process. (b) Assuming that the atom was initially at rest, determine the recoil speed of the hydrogen atom when it emits this photon.

57. Suppose a hydrogen atom is in the $2s$ state, with its wave function given by Equation 42.26. Taking $r = a_0$, calculate values for (a) $\psi_{2s}(a_0)$, (b) $|\psi_{2s}(a_0)|^2$, and (c) $P_{2s}(a_0)$.

58. An elementary theorem in statistics states that the root-mean-square uncertainty in a quantity r is given by $\Delta r = \sqrt{\langle r^2\rangle - \langle r\rangle^2}$. Evaluate the uncertainty in the radial position of the electron in the ground state of the hydrogen atom. Use the average value of r found in Example 42.3: $\langle r\rangle = 3a_0/2$. The average value of the squared distance between the electron and the proton is given by

$$\langle r^2\rangle = \int_{\text{all space}} |\psi|^2\, r^2\, dV = \int_0^\infty P(r)\, r^2\, dr$$

59. A pulsed ruby laser emits light at 694.3 nm. For a 14.0-ps pulse containing 3.00 J of energy, find (a) the physical length of the pulse as it travels through space and (b) the number of photons in it. (c) The beam has a circular cross section of diameter 0.600 cm. Find the number of photons per cubic millimeter.

60. A pulsed laser emits light of wavelength λ. For a pulse of duration Δt having energy E, find (a) the physical length of the pulse as it travels through space and (b) the number of photons in it. (c) The beam has a circular cross section having diameter d. Find the number of photons per unit volume.

61. Assume three identical uncharged particles of mass m and spin $\frac{1}{2}$ are contained in a one-dimensional box of length L. What is the ground-state energy of this system?

62. The force on a magnetic moment μ_z in a nonuniform magnetic field B_z is given by $F_z = \mu_z(dB_z/dz)$. If a beam of silver atoms travels a horizontal distance of 1.00 m through such a field and each atom has a speed of 100 m/s, how strong must be the field gradient dB_z/dz to deflect the beam 1.00 mm?

63. (a) Show that the most probable radial position for an electron in the $2s$ state of hydrogen is $r = 5.236a_0$. (b) Show that the wave function given by Equation 42.26 is normalized.

64. **Review problem.** Steven Chu, Claude Cohen-Tannoudji, and William Phillips received the 1997 Nobel Prize in Physics for "the development of methods to cool and trap atoms with laser light." One part of their work was with a beam of atoms (mass ~ 10^{-25} kg) that move at a speed on the order of 1 km/s, similar to the speed of molecules in air at room temperature. An intense laser light beam tuned to a visible atomic transition (assume 500 nm) is directed straight into the atomic beam; that is, the atomic beam and light beam are traveling in opposite directions. An atom in the ground state immediately absorbs a photon. Total system momentum is conserved in the absorption process. After a lifetime on the order of 10^{-8} s, the excited atom radiates by spontaneous emission. It has an equal probability of emitting a photon in any direction. Therefore, the average "recoil" of the atom is zero over many absorption and emission cycles. (a) Estimate the

average deceleration of the atomic beam. (b) What is the order of magnitude of the distance over which the atoms in the beam are brought to a halt?

65. An electron in chromium moves from the $n = 2$ state to the $n = 1$ state without emitting a photon. Instead, the excess energy is transferred to an outer electron (one in the $n = 4$ state), which is then ejected by the atom. In this Auger (pronounced "ohjay") process, the ejected electron is referred to as an Auger electron. Use the Bohr theory to find the kinetic energy of the Auger electron.

66. Suppose the ionization energy of an atom is 4.10 eV. In the spectrum of this same atom, we observe emission lines with wavelengths 310 nm, 400 nm, and 1 377.8 nm. Use this information to construct the energy-level diagram with the fewest levels. Assume the higher levels are closer together.

67. For hydrogen in the $1s$ state, what is the probability of finding the electron farther than $2.50a_0$ from the nucleus?

68. All atoms have the same size, to an order of magnitude. (a) To demonstrate this fact, estimate the diameters for aluminum (with molar mass 27.0 g/mol and density 2.70 g/cm^3) and uranium (molar mass 238 g/mol and density 18.9 g/cm^3). (b) What do the results of part (a) imply about the wave functions for inner-shell electrons as we progress to higher and higher atomic mass atoms? *Suggestion:* The molar volume is approximately D^3N_A, where D is the atomic diameter and N_A is Avogadro's number.

69. In the technique known as electron spin resonance (ESR), a sample containing unpaired electrons is placed in a magnetic field. Consider the simplest situation, in which only one electron is present and therefore only two energy states are possible, corresponding to $m_s = \pm\frac{1}{2}$. In ESR, the absorption of a photon causes the electron's spin magnetic moment to flip from the lower energy state to the higher energy state. According to Section 29.5, the change in energy is $2\mu_B B$. (The lower energy state corresponds to the case in which the z component of the magnetic moment $\vec{\mu}_{spin}$ is aligned with the magnetic field, and the higher energy state corresponds to the case in which the z component of $\vec{\mu}_{spin}$ is aligned opposite to the field.) What is the photon frequency required to excite an ESR transition in a 0.350-T magnetic field?

Answers to Quick Quizzes

42.1 (c). Because the energy of 10.5 eV does not correspond to raising the atom from the ground state to an allowed excited state, there is no interaction between the photon and the atom.

42.2 (a). The longest-wavelength photon is associated with the lowest energy transition, which is $n = 3$ to $n = 2$.

42.3 (b). The number of subshells is the same as the number of allowed values of ℓ. The allowed values of ℓ for $n = 4$ are $\ell = 0, 1, 2,$ and 3, so there are four subshells.

42.4 (a). Five values (0, 1, 2, 3, 4) of ℓ and (b) nine different values ($-4, -3, -2, -1, 0, 1, 2, 3, 4$) of m_ℓ as follows:

ℓ	m_ℓ
0	0
1	$-1, 0, 1$
2	$-2, -1, 0, 1, 2$
3	$-3, -2, -1, 0, 1, 2, 3$
4	$-4, -3, -2, -1, 0, 1, 2, 3, 4$

42.5 (c). The wavelengths of the characteristic x-rays are determined by the separation between energy levels in the atoms of the target, which is unrelated to the energy with which electrons are fired at the target. The only dependence is that the incoming electrons must have enough energy to eject an atomic electron from an inner shell.

42.6 True. If the electrons arrive at the target with very low energy, atomic electrons cannot be ejected and characteristic x-rays do not appear. Because the incoming electrons experience accelerations, the continuous spectrum appears.

An understanding of the physics of solids has led to the technology of integrated circuits, found in countless electronic devices used by consumers in today's society. In this photograph, the microchip sitting on a fingertip contains millions of electrical components. (Bruce Dale/Getty Images)

43 Molecules and Solids

The most random atomic arrangement, that of a gas, was well understood in the 1800s as discussed in Chapter 21. In a crystalline solid, the atoms are not randomly arranged; rather, they form a regular array. The symmetry of the arrangement of atoms both stimulated and allowed rapid progress in the field of solid-state physics in the 20th century. Recently, our understanding of liquids and amorphous solids has advanced. (In an amorphous solid such as glass or paraffin, the atoms do not form a regular array.) The recent interest in the physics of low-cost amorphous materials has been driven by their use in such devices as solar cells, memory elements, and fiber-optic waveguides.

We begin this chapter by studying the aggregates of atoms known as molecules. We describe the bonding mechanisms in molecules, the various modes of molecular excitation, and the radiation emitted or absorbed by molecules. Next, we show how molecules combine to form solids. Then, by examining their energy-level structure, we explain the differences between insulating, conducting, semiconducting, and superconducting materials. The chapter also includes discussions of semiconducting junctions and several semiconductor devices.

43.1 Molecular Bonds

The bonding mechanisms in a molecule are fundamentally due to electric forces between atoms (or ions). The forces between atoms in the system of a molecule are related to a potential energy function. A stable molecule is expected at a configuration for which the potential energy function has its minimum value. (See Section 7.9.)

A potential energy function that can be used to model a molecule should account for two known features of molecular bonding:

1. The force between atoms is repulsive at very small separation distances. When two atoms are brought close to each other, some of their electron shells overlap, resulting in repulsion between the shells. This repulsion is partly electrostatic in origin and partly the result of the exclusion principle. Because all electrons must obey the exclusion principle, some electrons in the overlapping shells are forced into higher energy states and the system energy increases as if a repulsive force existed between the atoms.
2. At somewhat larger separations, the force between atoms is attractive. If that were not true, the atoms in a molecule would not be bound together.

Taking into account these two features, the potential energy for a system of two atoms can be represented by an expression of the form

$$U(r) = -\frac{A}{r^n} + \frac{B}{r^m} \quad (43.1)$$

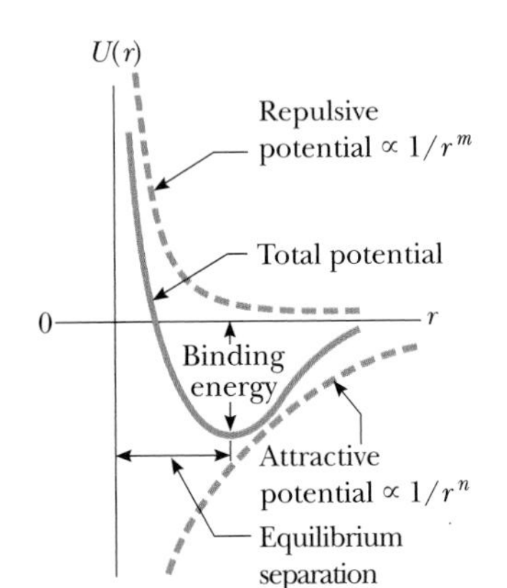

Figure 43.1 Total potential energy as a function of internuclear separation distance for a system of two atoms.

where r is the internuclear separation distance between the two atoms and n and m are small integers. The parameter A is associated with the attractive force and B with the repulsive force. Example 7.9 gives one common model for such a potential energy function, the Lennard–Jones potential.

Potential energy versus internuclear separation distance for a two-atom system is graphed in Figure 43.1. At large separation distances between the two atoms, the slope of the curve is positive, corresponding to a net attractive force. At the equilibrium separation distance, the attractive and repulsive forces just balance. At this point, the potential energy has its minimum value and the slope of the curve is zero.

A complete description of the bonding mechanisms in molecules is highly complex because bonding involves the mutual interactions of many particles. In this section, we discuss only some simplified models.

Ionic Bonding

When two atoms combine in such a way that one or more outer electrons are transferred from one atom to the other, the bond formed is called an **ionic bond.** Ionic bonds are fundamentally caused by the Coulomb attraction between oppositely charged ions.

A familiar example of an ionically bonded solid is sodium chloride, NaCl, which is common table salt. Sodium, which has the electronic configuration $1s^22s^22p^63s^1$, is ionized relatively easily, giving up its $3s$ electron to form a Na^+ ion. The energy required to ionize the atom to form Na^+ is 5.1 eV. Chlorine, which has the electronic configuration $1s^22s^22p^5$, is one electron short of the filled-shell structure of argon. If we compare the energy of a system of a free electron and a Cl atom with one in which the electron joins the atom to make the Cl^- ion, we find that the energy of the ion is lower. When the electron makes a transition from the $E = 0$ state to the negative energy state associated with the available shell in the atom, energy is released. This amount of energy is called the **electron affinity** of the atom. For chlorine, the electron affinity is 3.6 eV. Therefore, the energy required to form Na^+ and Cl^- from isolated atoms is $5.1 - 3.6 = 1.5$ eV. It costs 5.1 eV to remove the electron from the Na atom, but 3.6 eV of it is gained back when that electron is allowed to join with the Cl atom.

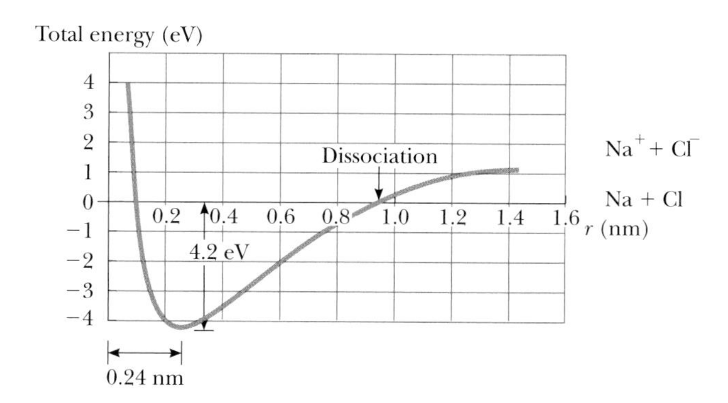

Figure 43.2 Total energy versus internuclear separation distance for Na^+ and Cl^- ions. The horizontal axis is labeled Na + Cl because we define zero energy as that for the system of neutral sodium and chlorine *atoms.* The asymptote of the curve for large values of r is marked $Na^+ + Cl^-$ because that is the energy of the system of sodium and chloride *ions.*

Now imagine that these two charged ions interact with one another to form a NaCl "molecule."[1] The total energy of the NaCl molecule versus internuclear separation distance is graphed in Figure 43.2. At very large separation distances, the energy of the system of ions is 1.5 eV as calculated above. The total energy has a minimum value of −4.2 eV at the equilibrium separation distance, which is approximately 0.24 nm. Hence, the energy required to break the $Na^+{-}Cl^-$ bond and form neutral sodium and chlorine atoms, called the **dissociation energy,** is 4.2 eV. The energy of the molecule is lower than that of the system of two neutral atoms. Consequently, it is **energetically favorable** for the molecule to form: if a lower energy state of a system exists, the system tends to emit energy to achieve this lower energy state. The system of neutral sodium and chlorine atoms can reduce its total energy by transferring energy out of the system (by electromagnetic radiation, for example) and forming the NaCl molecule.

PITFALL PREVENTION 43.1
Ionic and Covalent Bonds

In practice, these descriptions of ionic and covalent bonds represent extreme ends of a spectrum of bonds involving electron transfer. In a real bond, the electron may not be *completely* transferred as in an ionic bond or *equally* shared as in a covalent bond. Therefore, real bonds lie somewhere between these extremes.

Covalent Bonding

A **covalent bond** between two atoms is one in which electrons supplied by either one or both atoms are shared by the two atoms. Many diatomic molecules—such as H_2, F_2, and CO—owe their stability to covalent bonds. The bond between two hydrogen atoms can be described by using atomic wave functions. The ground-state wave function for a hydrogen atom (Chapter 42) is

$$\psi_{1s}(r) = \frac{1}{\sqrt{\pi a_0^{\,3}}}\, e^{-r/a_0}$$

This wave function is graphed in Active Figure 43.3a for two hydrogen atoms that are far apart. There is very little overlap of the wave functions $\psi_1(r)$ for atom 1, located at $r = 0$, and $\psi_2(r)$ for atom 2, located some distance away. Suppose now the two atoms are brought close together. As that happens, their wave functions overlap and form the compound wave function $\psi_1(r) + \psi_2(r)$ shown in Active Figure 43.3b. Notice that the probability amplitude is larger between the atoms than it is on either side of the combination of atoms. As a result, the probability is higher that the electrons associated with the atoms will be located between the atoms than on the outer regions of the system. Consequently, the average position of negative charge in the system is halfway between the atoms. This scenario can be modeled as if there were a fixed negative charge between the atoms, exerting attractive Coulomb forces on both nuclei. Therefore, there is an overall attractive force between the atoms, resulting in a covalent bond.

Because of the exclusion principle, the two electrons in the ground state of H_2 must have antiparallel spins. Also because of the exclusion principle, if a third H atom is brought near the H_2 molecule, the third electron would have to occupy a

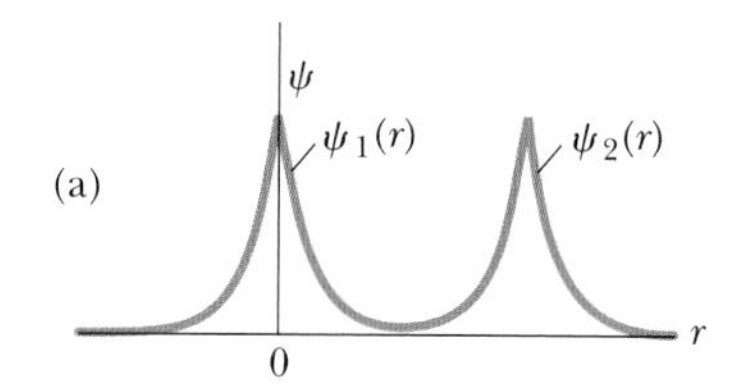

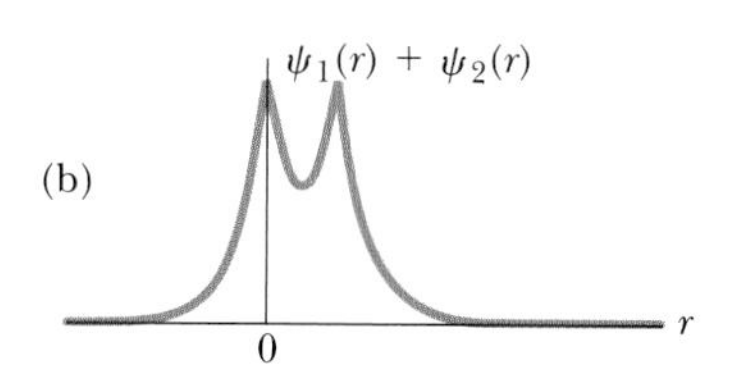

ACTIVE FIGURE 43.3

Ground-state wave functions $\psi_1(r)$ and $\psi_2(r)$ for two atoms making a covalent bond. (a) The atoms are far apart, and their wave functions overlap minimally. (b) The atoms are close together, forming a composite wave function $\psi_1(r) + \psi_2(r)$ for the system. The probability amplitude for an electron to be between the atoms is high.

Sign in at www.thomsonedu.com and go to ThomsonNOW to move the individual wave functions and observe the composite wave function.

[1] NaCl does not exist as an isolated molecule. In the solid state, NaCl forms a crystalline array of ions as described in Section 43.3. In the liquid state or in solution with water, the Na^+ and Cl^- ions dissociate and are free to move relative to each other.

higher energy level, which is not an energetically favorable situation. For this reason, the H_3 molecule is not stable and does not form.

Van der Waals Bonding

Ionic and covalent bonds occur between atoms to form molecules or ionic solids, so they can be described as bonds *within* molecules. Two additional types of bonds, van der Waals bonds and hydrogen bonds, can occur *between* molecules.

You might think that two neutral molecules would not interact by means of the electric force because they each have zero net charge. They are attracted to each other, however, by weak electrostatic forces called **van der Waals forces.** Likewise, atoms that do not form ionic or covalent bonds are attracted to each other by van der Waals forces. Noble gas atoms, for example, because of their filled shell structure, do not generally form molecules or bond to each other to form a liquid. Because of van der Waals forces, however, at sufficiently low temperatures at which thermal excitations are negligible, noble gases first condense to liquids and then solidify. (The exception is helium, which does not solidify at atmospheric pressure.)

The van der Waals force results from the following situation. While being electrically neutral, a molecule has a charge distribution with positive and negative centers at different positions in the molecule. As a result, the molecule may act as an electric dipole. (See Section 23.4.) Because of the dipole electric fields, two molecules can interact such that there is an attractive force between them.

There are three types of van der Waals forces. The first type, called the *dipole–dipole force,* is an interaction between two molecules each having a permanent electric dipole moment. For example, polar molecules such as HCl have permanent electric dipole moments and attract other polar molecules.

The second type, the *dipole–induced dipole force,* results when a polar molecule having a permanent electric dipole moment induces a dipole moment in a nonpolar molecule. In this case, the electric field of the polar molecule creates the dipole moment in the nonpolar molecule, which then results in an attractive force between the molecules.

The third type is called the *dispersion force,* an attractive force that occurs between two nonpolar molecules. In this case, although the average dipole moment of a nonpolar molecule is zero, the average of the square of the dipole moment is nonzero because of charge fluctuations. Two nonpolar molecules near each other tend to have dipole moments that are correlated in time so as to produce an attractive van der Waals force.

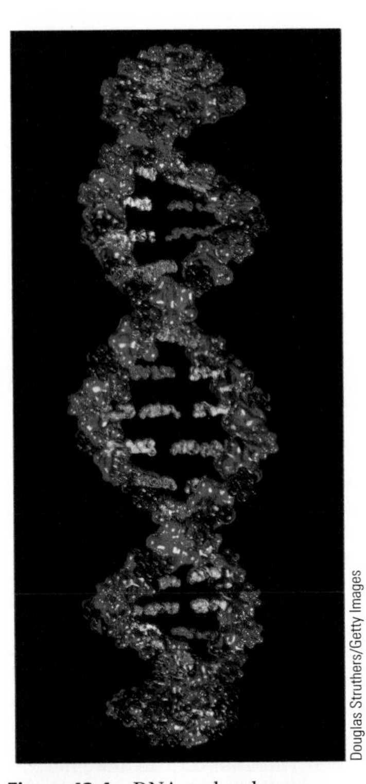

Douglas Struthers/Getty Images

Figure 43.4 DNA molecules are held together by hydrogen bonds.

Hydrogen Bonding

Because hydrogen has only one electron, it is expected to form a covalent bond with only one other atom within a molecule. A hydrogen atom in a given molecule can also form a second type of bond between molecules called a **hydrogen bond.** Let's use the water molecule H_2O as an example. In the two covalent bonds in this molecule, the electrons from the hydrogen atoms are more likely to be found near the oxygen atom than near the hydrogen atoms, leaving essentially bare protons at the positions of the hydrogen atoms. This unshielded positive charge can be attracted to the negative end of another polar molecule. Because the proton is unshielded by electrons, the negative end of the other molecule can come very close to the proton to form a bond strong enough to form a solid crystalline structure, such as that of ordinary ice. The bonds within a water molecule are covalent, but the bonds between water molecules in ice are hydrogen bonds.

The hydrogen bond is relatively weak compared with other chemical bonds and can be broken with an input energy of approximately 0.1 eV. Because of this weakness, ice melts at the low temperature of 0°C. Even though this bond is weak, however, hydrogen bonding is a critical mechanism responsible for the linking of biological molecules and polymers. For example, in the case of the DNA (deoxyribonucleic acid) molecule, which has a double-helix structure (Fig. 43.4), hydrogen bonds formed by the sharing of a proton between two atoms create linkages between the turns of the helix.

Quick Quiz 43.1 For each of the following atoms or molecules, identify the most likely type of bonding that occurs between the atoms or between the molecules. Choose from the following list: ionic, covalent, van der Waals, hydrogen. (a) atoms of krypton (b) potassium and chlorine atoms (c) hydrogen fluoride (HF) molecules (d) chlorine and oxygen atoms in a hypochlorite ion (ClO^-)

43.2 Energy States and Spectra of Molecules

Consider an individual molecule in the gaseous phase of a substance. The energy E of the molecule can be divided into four categories: (1) electronic energy, due to the interactions between the molecule's electrons and nuclei; (2) translational energy, due to the motion of the molecule's center of mass through space; (3) rotational energy, due to the rotation of the molecule about its center of mass; and (4) vibrational energy, due to the vibration of the molecule's constituent atoms:

$$E = E_{\text{el}} + E_{\text{trans}} + E_{\text{rot}} + E_{\text{vib}}$$

◀ Total energy of a molecule

We explored the roles of translational, rotational, and vibrational energy of molecules in determining the molar specific heats of gases in Sections 21.2 and 21.4. Because the translational energy is unrelated to internal structure, this molecular energy is unimportant in interpreting molecular spectra. The electronic energy of a molecule is very complex because it involves the interaction of many charged particles, but various techniques have been developed to approximate its values. Although the electronic energies can be studied, significant information about a molecule can be determined by analyzing its quantized rotational and vibrational energy states. Transitions between these states give spectral lines in the microwave and infrared regions of the electromagnetic spectrum, respectively.

Rotational Motion of Molecules

Let's consider the rotation of a molecule around its center of mass, confining our discussion to the diatomic molecule (Active Fig. 43.5a, page 1262) but noting that the same ideas can be extended to polyatomic molecules. A diatomic molecule aligned along an x axis has only two rotational degrees of freedom, corresponding to rotations about the y and z axes passing through the molecule's center of mass. If ω is the angular frequency of rotation about one of these axes, the rotational kinetic energy of the molecule about that axis can be expressed as

$$E_{\text{rot}} = \tfrac{1}{2}I\omega^2 \qquad \textbf{(43.2)}$$

In this equation, I is the moment of inertia of the molecule about its center of mass, given by

$$I = \left(\frac{m_1 m_2}{m_1 + m_2}\right) r^2 = \mu r^2 \qquad \textbf{(43.3)}$$

◀ Moment of inertia for a diatomic molecule

where m_1 and m_2 are the masses of the atoms that form the molecule, r is the atomic separation, and μ is the **reduced mass** of the molecule (see Example 41.5 and Problem 39 in Chapter 41):

$$\mu = \frac{m_1 m_2}{m_1 + m_2} \qquad \textbf{(43.4)}$$

◀ Reduced mass of a diatomic molecule

The magnitude of the molecule's angular momentum about its center of mass is $L = I\omega$, which classically can have any value. Quantum mechanics, however,

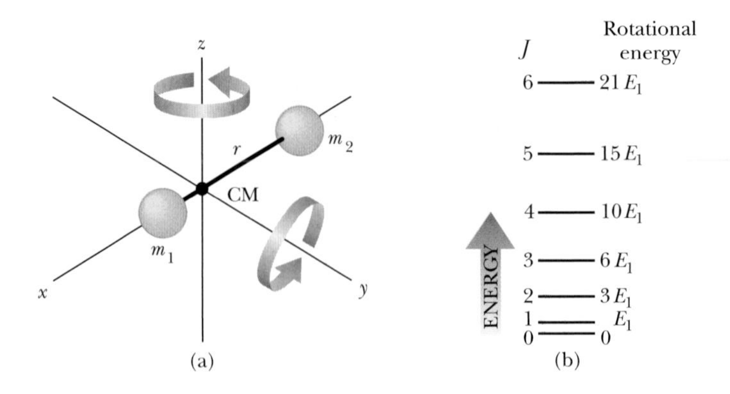

ACTIVE FIGURE 43.5

Rotation of a diatomic molecule around its center of mass. (a) A diatomic molecule oriented along the x axis has two rotational degrees of freedom, corresponding to rotation about the y and z axes. (b) Allowed rotational energies of a diatomic molecule calculated with Equation 43.6, where $E_1 = \hbar^2/I$.

Sign in at www.thomsonedu.com and go to ThomsonNOW to adjust the distance between the atoms and choose the initial rotational energy state of the molecule. Then observe transitions of the molecule to lower energy states.

restricts the molecule to certain quantized rotational frequencies such that the angular momentum of the molecule has the values[2]

Allowed values of rotational angular momentum ▶

$$L = \sqrt{J(J+1)}\hbar \quad J = 0, 1, 2, \ldots \tag{43.5}$$

where J is an integer called the **rotational quantum number.** Combining Equations 43.5 and 43.2, we obtain an expression for the allowed values of the rotational kinetic energy of the molecule:

$$E_{\text{rot}} = \tfrac{1}{2}I\omega^2 = \frac{1}{2I}(I\omega)^2 = \frac{L^2}{2I} = \frac{(\sqrt{J(J+1)}\hbar)^2}{2I}$$

Allowed values of rotational energy ▶

$$E_{\text{rot}} = E_J = \frac{\hbar^2}{2I}J(J+1) \qquad J = 0, 1, 2, \ldots \tag{43.6}$$

The allowed rotational energies of a diatomic molecule are plotted in Active Figure 43.5b. As the quantum number J goes up, the states become farther apart as displayed earlier for rotational energy levels in Figure 21.8.

For most molecules, transitions between adjacent rotational energy levels result in radiation that lies in the microwave range of frequencies ($f \sim 10^{11}$ Hz). When a molecule absorbs a microwave photon, the molecule jumps from a lower rotational energy level to a higher one. The allowed rotational transitions of linear molecules are regulated by the selection rule $\Delta J = \pm 1$. Given this selection rule, all absorption lines in the spectrum of a linear molecule correspond to energy separations equal to $E_J - E_{J-1}$, where $J = 1, 2, 3, \ldots$. From Equation 43.6, we see that the energies of the absorbed photons are given by

$$E_{\text{photon}} = \Delta E_{\text{rot}} = E_J - E_{J-1} = \frac{\hbar^2}{2I}[J(J+1) - (J-1)J]$$

Energy of a photon absorbed in a transition between adjacent rotational levels ▶

$$E_{\text{photon}} = \frac{\hbar^2}{I}J = \frac{h^2}{4\pi^2 I}J \qquad J = 1, 2, 3, \ldots \tag{43.7}$$

where J is the rotational quantum number of the higher energy state. Because $E_{\text{photon}} = hf$, where f is the frequency of the absorbed photon, we see that the allowed frequency for the transition $J = 0$ to $J = 1$ is $f_1 = h/4\pi^2 I$. The frequency corresponding to the $J = 1$ to $J = 2$ transition is $2f_1$, and so on. These predictions are in excellent agreement with the observed frequencies.

Quick Quiz 43.2 A gas of identical diatomic molecules absorbs electromagnetic radiation over a wide range of frequencies. Molecule 1 is in the $J = 0$ rotation state and makes a transition to the $J = 1$ state. Molecule 2 is in the $J = 2$ state and

[2] Equation 43.5 is similar to Equation 42.27 for orbital angular momentum in an atom. The relationship between the magnitude of the angular momentum of a system and the associated quantum number is the same as it is in these equations for *any* system that exhibits rotation as long as the potential energy function for the system is spherically symmetric.

makes a transition to the $J = 3$ state. Is the ratio of the frequency of the photon that excited molecule 2 to that of the photon that excited molecule 1 equal to (a) 1, (b) 2, (c) 3, (d) 4, or (e) impossible to determine?

EXAMPLE 43.1 Rotation of the CO Molecule

The $J = 0$ to $J = 1$ rotational transition of the CO molecule occurs at a frequency of 1.15×10^{11} Hz.

(A) Use this information to calculate the moment of inertia of the molecule.

SOLUTION

Conceptualize Imagine that the two atoms in Active Figure 43.5a are carbon and oxygen. The center of mass of the molecule is not midway between the atoms because of the difference in masses of the C and O atoms.

Categorize The statement of the problem tells us to categorize this example as one involving a quantum-mechanical treatment and to restrict our investigation to the rotational motion of a diatomic molecule.

Analyze Use Equation 43.7 to find the energy of a photon that excites the molecule from the $J = 0$ to the $J = 1$ rotational level:

$$E_{\text{photon}} = \frac{\hbar^2}{4\pi^2 I}(1) = \frac{\hbar^2}{4\pi^2 I}$$

Equate this energy to $E = hf$ for the absorbed photon and solve for I:

$$\frac{\hbar^2}{4\pi^2 I} = hf \quad \rightarrow \quad I = \frac{h}{4\pi^2 f}$$

Substitute the frequency given in the problem statement:

$$I = \frac{6.626 \times 10^{-34}\ \text{J}\cdot\text{s}}{4\pi^2(1.15 \times 10^{11}\ \text{s}^{-1})} = 1.46 \times 10^{-46}\ \text{kg}\cdot\text{m}^2$$

(B) Calculate the bond length of the molecule.

SOLUTION

Find the reduced mass μ of the CO molecule:

$$\mu = \frac{m_1 m_2}{m_1 + m_2} = \frac{(12\ \text{u})(16\ \text{u})}{12\ \text{u} + 16\ \text{u}} = 6.86\ \text{u}$$

$$= (6.86\ \text{u})\left(\frac{1.66 \times 10^{-27}\ \text{kg}}{1\ \text{u}}\right) = 1.14 \times 10^{-26}\ \text{kg}$$

Solve Equation 43.3 for r and substitute for the reduced mass and the moment of inertia from part (A):

$$r = \sqrt{\frac{I}{\mu}} = \sqrt{\frac{1.46 \times 10^{-46}\ \text{kg}\cdot\text{m}^2}{1.14 \times 10^{-26}\ \text{kg}}}$$

$$= 1.13 \times 10^{-10}\ \text{m} = 0.113\ \text{nm}$$

Finalize The moment of inertia of the molecule and the separation distance between the atoms are both very small, as expected for a microscopic system.

What If? What if another photon of frequency 1.15×10^{11} Hz is incident on the CO molecule while that molecule is in the $J = 1$ state? What happens?

Answer Because the rotational quantum states are not equally spaced in energy, the $J = 1$ to $J = 2$ transition does not have the same energy as the $J = 0$ to $J = 1$ transition. Therefore, the molecule will *not* be excited to the $J = 2$ state. Two possibilities exist. The photon could pass by the molecule with no interaction, or the photon could induce a stimulated emission, similar to that for atoms and discussed in Section 42.9. In this case, the molecule makes a transition back to the $J = 0$ state and the original photon and a second identical photon leave the scene of the interaction.

Vibrational Motion of Molecules

If we consider a molecule to be a flexible structure in which the atoms are bonded together by "effective springs" as shown in Active Figure 43.6a, we can model the molecule as a simple harmonic oscillator as long as the atoms in the molecule are not too far from their equilibrium positions. Recall from Section 15.3 that the potential energy function for a simple harmonic oscillator is parabolic, varying as the square of the displacement from equilibrium. (See Eq. 15.20.) Active Figure 43.6b shows a plot of potential energy versus atomic separation for a diatomic molecule, where r_0 is the equilibrium atomic separation. For separations close to r_0, the shape of the potential energy curve closely resembles a parabola.

According to classical mechanics, the frequency of vibration for the system shown in Active Figure 43.6a is given by Equation 15.14:

$$f = \frac{1}{2\pi}\sqrt{\frac{k}{\mu}} \tag{43.8}$$

where k is the effective spring constant and μ is the reduced mass given by Equation 43.4.

Quantum mechanics predicts that a molecule vibrates in quantized states as described in Section 41.7. The vibrational motion and quantized vibrational energy can be altered if the molecule acquires energy of the proper value to cause a transition between quantized vibrational states. As discussed in Section 41.7, the allowed vibrational energies are

$$E_{\text{vib}} = \left(v + \tfrac{1}{2}\right)hf \qquad v = 0, 1, 2, \ldots \tag{43.9}$$

where v is an integer called the **vibrational quantum number**. (We used n in Section 41.7 for a general harmonic oscillator, but v is often used for the quantum number when discussing molecular vibrations.) If the system is in the lowest vibrational state, for which $v = 0$, its ground-state energy is $\frac{1}{2}hf$. In the first excited vibrational state, $v = 1$ and the energy is $\frac{3}{2}hf$, and so on.

Substituting Equation 43.8 into Equation 43.9 gives the following expression for the allowed vibrational energies:

Allowed values of vibrational energy ▶

$$E_{\text{vib}} = \left(v + \tfrac{1}{2}\right)\frac{h}{2\pi}\sqrt{\frac{k}{\mu}} \qquad v = 0, 1, 2, \ldots \tag{43.10}$$

The selection rule for the allowed vibrational transitions is $\Delta v = \pm 1$. Transitions between vibrational levels are caused by absorption of photons in the infrared region of the spectrum. The energy of an absorbed photon is equal to the energy difference between any two successive vibrational levels. Therefore, the photon energy is given by

$$E_{\text{photon}} = \Delta E_{\text{vib}} = \frac{h}{2\pi}\sqrt{\frac{k}{\mu}} \tag{43.11}$$

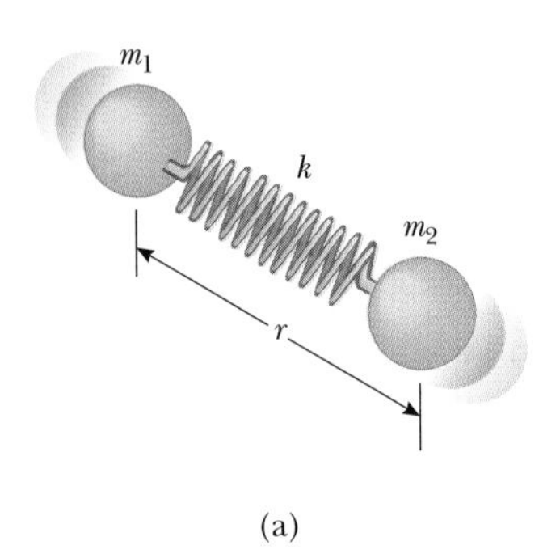

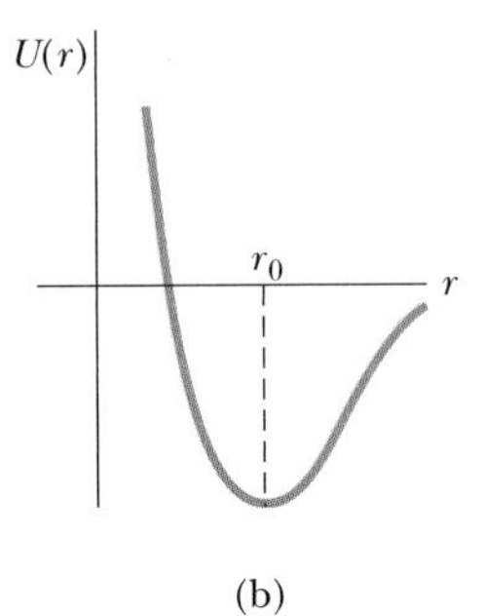

ACTIVE FIGURE 43.6

(a) Effective-spring model of a diatomic molecule. The vibration is along the molecular axis. (b) Plot of the potential energy of a diatomic molecule versus atomic separation distance, where r_0 is the equilibrium separation distance of the atoms. Compare with Figure 15.11.

Sign in at www.thomsonedu.com and go to ThomsonNOW to adjust the spring constant and choose the initial vibrational energy state of the molecule. Then observe transitions of the molecule to lower energy states in Active Figure 43.7.

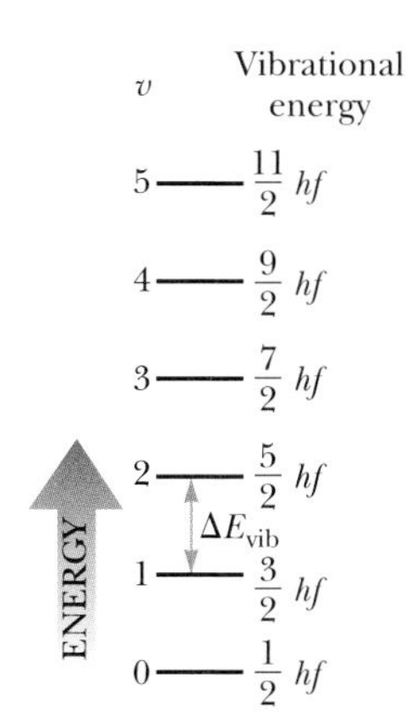

ACTIVE FIGURE 43.7

Allowed vibrational energies of a diatomic molecule, where f is the frequency of vibration of the molecule, given by Equation 43.8. The spacings between adjacent vibrational levels are equal if the molecule behaves as a harmonic oscillator.

Sign in at www.thomsonedu.com and go to ThomsonNOW to adjust the spring constant in Active Figure 43.6 and choose the initial vibrational energy state of the molecule. Then observe transitions of the molecule to other energy states.

The vibrational energies of a diatomic molecule are plotted in Active Figure 43.7. At ordinary temperatures, most molecules have vibrational energies corresponding to the $v = 0$ state because the spacing between vibrational states is much greater than $k_B T$, where k_B is Boltzmann's constant and T is the temperature.

Quick Quiz 43.3 A gas of identical diatomic molecules absorbs electromagnetic radiation over a wide range of frequencies. Molecule 1, initially in the $v = 0$ vibrational state, makes a transition to the $v = 1$ state. Molecule 2, initially in the $v = 2$ state, makes a transition to the $v = 3$ state. What is the ratio of the frequency of the photon that excited molecule 2 to that of the photon that excited molecule 1? (a) 1 (b) 2 (c) 3 (d) 4 (e) impossible to determine

---

EXAMPLE 43.2 Vibration of the CO Molecule

The frequency of the photon that causes the $v = 0$ to $v = 1$ transition in the CO molecule is 6.42×10^{13} Hz. We ignore any changes in the rotational energy for this example.

(A) Calculate the force constant k for this molecule.

SOLUTION

Conceptualize Imagine that the two atoms in Active Figure 43.6a are carbon and oxygen. As the molecule vibrates, a given point on the imaginary spring is at rest. This point is not midway between the atoms because of the difference in masses of the C and O atoms.

Categorize The statement of the problem tells us to categorize this example as one involving a quantum-mechanical treatment and to restrict our investigation to the vibrational motion of a diatomic molecule.

Analyze Set Equation 43.11 equal to the photon energy hf and solve for the force constant:

$$\frac{h}{2\pi}\sqrt{\frac{k}{\mu}} = hf \rightarrow k = 4\pi^2\mu f^2$$

Substitute the frequency given in the problem statement and the reduced mass from Example 43.1:

$$k = 4\pi^2(1.14 \times 10^{-26}\text{ kg})(6.42 \times 10^{13}\text{ s}^{-1})^2 = 1.85 \times 10^3\text{ N/m}$$

(B) What is the classical amplitude A of vibration for this molecule in the $v = 0$ vibrational state?

SOLUTION

Equate the maximum elastic potential energy $\frac{1}{2}kA^2$ in the molecule (Eq. 15.21) to the vibrational energy given by Equation 43.10 with $v = 0$ and solve for A:

$$\tfrac{1}{2}kA^2 = \frac{h}{4\pi}\sqrt{\frac{k}{\mu}} \rightarrow A = \sqrt{\frac{h}{2\pi}}\left(\frac{1}{\mu k}\right)^{1/4}$$

Substitute the value for k from part (A) and the value for μ:

$$A = \sqrt{\frac{6.626 \times 10^{-34}\ \text{J}\cdot\text{s}}{2\pi}}\left[\frac{1}{(1.14 \times 10^{-26}\ \text{kg})(1.85 \times 10^{3}\ \text{N/m})}\right]^{1/4}$$

$$= 4.79 \times 10^{-12}\ \text{m} = 0.004\ 79\ \text{nm}$$

Finalize Comparing this result with the bond length of 0.113 nm we calculated in Example 43.1 shows that the classical amplitude of vibration is approximately 4% of the bond length.

Molecular Spectra

In general, a molecule vibrates and rotates simultaneously. To a first approximation, these motions are independent of each other, so the total energy of the molecule for these motions is the sum of Equations 43.6 and 43.9:

$$E = (v + \tfrac{1}{2})hf + \frac{\hbar^2}{2I}J(J+1) \qquad \textbf{(43.12)}$$

The energy levels of any molecule can be calculated from this expression, and each level is indexed by the two quantum numbers v and J. From these calculations, an energy-level diagram like the one shown in Active Figure 43.8a can be constructed. For each allowed value of the vibrational quantum number v, there is a complete set of rotational levels corresponding to $J = 0, 1, 2, \ldots$. The energy separation between successive rotational levels is much smaller than the separation between successive vibrational levels. As noted earlier, most molecules at ordinary temperatures are in the $v = 0$ vibrational state; these molecules can be in various rotational states as Active Figure 43.8a shows.

When a molecule absorbs a photon with the appropriate energy, the vibrational quantum number v increases by one unit while the rotational quantum number J either increases or decreases by one unit as can be seen in Active Figure 43.8. Therefore, the molecular absorption spectrum consists of two groups of lines: one group to the right of center and satisfying the selection rules $\Delta J = +1$ and $\Delta v = +1$; the other group to the left of center and satisfying the selection rules $\Delta J = -1$ and $\Delta v = +1$.

The energies of the absorbed photons can be calculated from Equation 43.12:

$$E_{\text{photon}} = \Delta E = hf + \frac{\hbar^2}{I}(J+1) \quad J = 0, 1, 2, \ldots \quad (\Delta J = +1) \qquad \textbf{(43.13)}$$

$$E_{\text{photon}} = \Delta E = hf - \frac{\hbar^2}{I}J \quad J = 1, 2, 3, \ldots \quad (\Delta J = -1) \qquad \textbf{(43.14)}$$

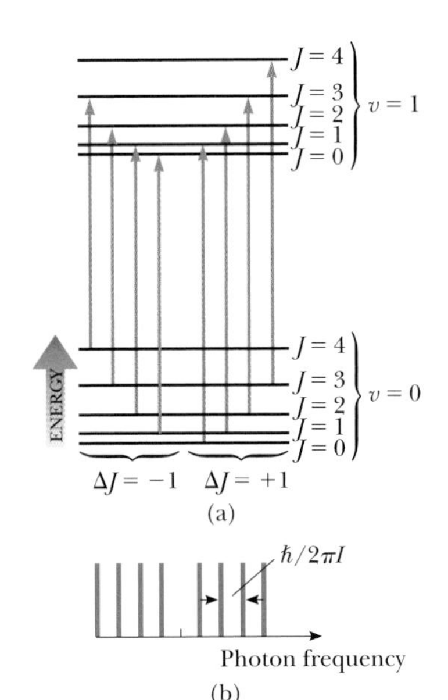

ACTIVE FIGURE 43.8

(a) Absorptive transitions between the $v = 0$ and $v = 1$ vibrational states of a diatomic molecule. The transitions obey the selection rule $\Delta J = \pm 1$ and fall into two sequences, those for $\Delta J = +1$ and those for $\Delta J = -1$. The transition energies are given by Equations 43.13 and 43.14. Compare the energy levels in this figure with those in Figure 21.8. (b) Expected lines in the absorption spectrum of a molecule. The lines to the right of the center mark correspond to transitions in which J changes by +1; the lines to the left of the center mark correspond to transitions for which J changes by −1. These same lines appear in the emission spectrum.

Sign in at www.thomsonedu.com and go to ThomsonNOW to adjust the spring constant and the moment of inertia of the molecule and observe the effect on the energy levels and the spectral lines.

where J is the rotational quantum number of the *initial* state. Equation 43.13 generates the series of equally spaced lines *higher* than the frequency f, whereas Equation 43.14 generates the series *lower* than this frequency. Adjacent lines are separated in frequency by the fundamental unit $\hbar/2\pi I$. Active Figure 43.8b shows the expected frequencies in the absorption spectrum of the molecule; these same frequencies appear in the emission spectrum.

The experimental absorption spectrum of the HCl molecule shown in Figure 43.9 follows this pattern very well and reinforces our model. One peculiarity is apparent, however: each line is split into a doublet. This doubling occurs because two chlorine isotopes (see Section 44.1) were present in the sample used to obtain this spectrum. Because the isotopes have different masses, the two HCl molecules have different values of I.

The intensity of the spectral lines in Figure 43.9 follows an interesting pattern, rising first as one moves away from the central gap (at about 8.65×10^{13} Hz, corresponding to the forbidden $J = 0$ to $J = 0$ transition) and then falling. This intensity is determined by a product of two functions of J. The first function corresponds to the number of available states for a given value of J. This function is $2J + 1$, corresponding to the number of values of m_J, the molecular rotation ana-

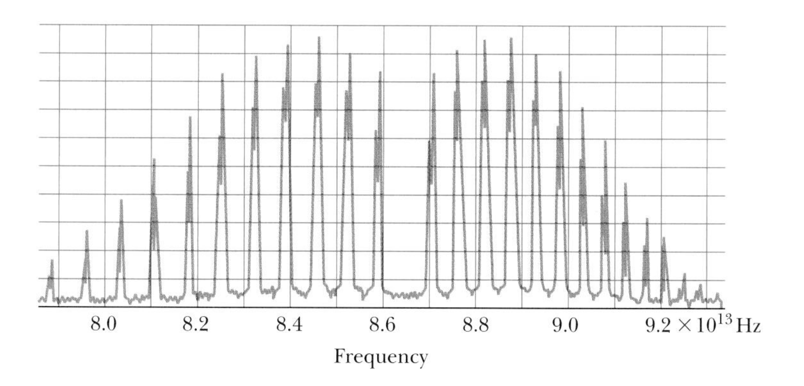

Figure 43.9 Absorption spectrum of the HCl molecule. Each line is split into a doublet because the sample contains two chlorine isotopes that have different masses and therefore different moments of inertia.

log to m_ℓ for atomic states. For example, the $J = 2$ state has five substates with five values of m_J ($m_J = -2, -1, 0, 1, 2$), whereas the $J = 1$ state has only three substates ($m_J = -1, 0, 1$). Therefore, on average and without regard for the second function described below, five-thirds as many molecules make the transition from the $J = 2$ state as from the $J = 1$ state.

The second function determining the envelope of the intensity of the spectral lines is the Boltzmann factor, introduced in Section 21.5. The number of molecules in an excited rotational state is given by

$$n = n_0 e^{-\hbar^2 J(J+1)/(2Ik_BT)}$$

where n_0 is the number of molecules in the $J = 0$ state.

Multiplying these factors together indicates that the intensity of spectral lines should be described by a function of J as follows:

$$I \propto (2J + 1)e^{-\hbar^2 J(J+1)/(2Ik_BT)} \quad \textbf{(43.15)}$$

◀ Intensity variation in the vibration–rotation spectrum of a molecule

The factor $(2J + 1)$ increases with J while the exponential second factor decreases. The product of the two factors gives a behavior that closely describes the envelope of the spectral lines in Figure 43.9.

The excitation of rotational and vibrational energy levels is an important consideration in current models of global warming. Most of the absorption lines for CO_2 are in the infrared portion of the spectrum. Therefore, visible light from the Sun is not absorbed by atmospheric CO_2 but instead strikes the Earth's surface, warming it. In turn, the surface of the Earth, being at a much lower temperature than the Sun, emits thermal radiation that peaks in the infrared portion of the electromagnetic spectrum (Section 40.1). This infrared radiation is absorbed by the CO_2 molecules in the air instead of radiating out into space. Atmospheric CO_2 acts like a one-way valve for energy from the Sun and is responsible, along with some other atmospheric molecules, for raising the temperature of the Earth's surface above its value in the absence of an atmosphere. This phenomenon is commonly called the "greenhouse effect." The burning of fossil fuels in today's industrialized society adds more CO_2 to the atmosphere. This addition of CO_2 increases the absorption of infrared radiation, raising the Earth's temperature further. In turn, this increase in temperature causes substantial climatic changes. The increased temperature also results in melting of ice from Arctic ice sheets, raising sea levels worldwide and endangering coastal population centers. In February, 2007, the Intergovernmental Panel on Climate Change of the United Nations issued "Climate Change 2007: The Physical Science Basis." This survey, based on the work of over 2 500 scientists from more than 130 countries, contains chilling new statements that global warming is clearly linked to human activity and that the global warming issue is no longer a matter of debate.

CONCEPTUAL EXAMPLE 43.3 Comparing Figures 43.8 and 43.9

In Active Figure 43.8a, the transitions indicated correspond to spectral lines that are equally spaced as shown in Active Figure 43.8b. The actual spectrum in Figure 43.9, however, shows lines that move closer together as the frequency increases. Why does the spacing of the actual spectral lines differ from the diagram in Active Figure 43.8?

SOLUTION

In Active Figure 43.8, we modeled the rotating diatomic molecule as a rigid object (Chapter 10). In reality, however, as the molecule rotates faster and faster, the effective spring in Active Figure 43.6a stretches and provides the increased force associated with the larger centripetal acceleration of each atom. As the molecule stretches along its length, its moment of inertia I increases. Therefore, the rotational part of the energy expression in Equation 43.12 has an extra dependence on J in the moment of inertia I. Because the increasing moment of inertia is in the denominator, as J increases, the energies do not increase as rapidly with J as indicated in Equation 43.12. With each higher energy level being lower than indicated by Equation 43.12, the energy associated with a transition to that level is smaller, as is the frequency of the absorbed photon, destroying the even spacing of the spectral lines and giving the uneven spacing seen in Figure 43.9.

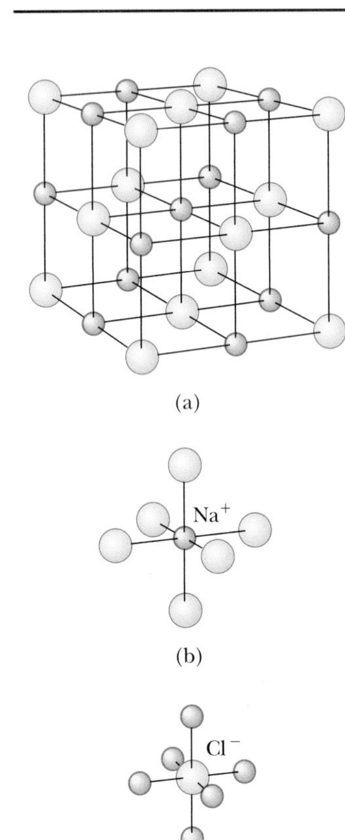

Figure 43.10 (a) Crystalline structure of NaCl. (b) Each positive sodium ion (orange spheres) is surrounded by six negative chloride ions (blue spheres). (c) Each chloride ion is surrounded by six sodium ions.

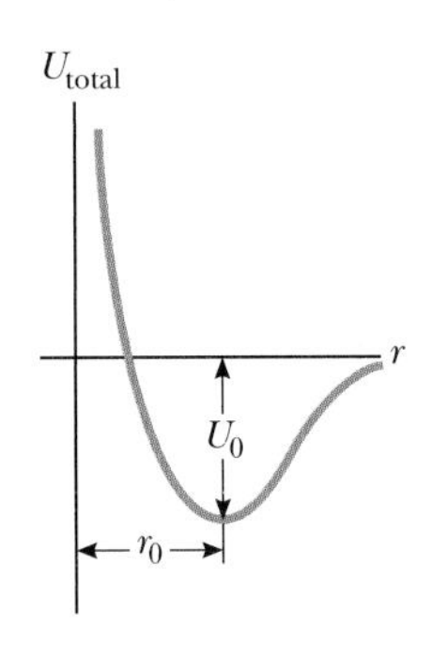

Figure 43.11 Total potential energy versus ion separation distance for an ionic solid, where U_0 is the ionic cohesive energy and r_0 is the equilibrium separation distance between ions.

43.3 Bonding in Solids

A crystalline solid consists of a large number of atoms arranged in a regular array, forming a periodic structure. The ions in the NaCl crystal are ionically bonded, as already noted, and the carbon atoms in diamond form covalent bonds with one another. The metallic bond described at the end of this section is responsible for the cohesion of copper, silver, sodium, and other solid metals.

Ionic Solids

Many crystals are formed by ionic bonding, in which the dominant interaction between ions is the Coulomb force. Consider the NaCl crystal in Figure 43.10. Each Na^+ ion has six nearest-neighbor Cl^- ions, and each Cl^- ion has six nearest-neighbor Na^+ ions. Each Na^+ ion is attracted to its six Cl^- neighbors. The corresponding potential energy is $-6k_e e^2/r$, where k_e is the Coulomb constant and r is the separation distance between each Na^+ and Cl^-. In addition, there are 12 next-nearest-neighbor Na^+ ions at a distance of $\sqrt{2}r$ from the Na^+ ion, and these 12 positive ions exert weaker repulsive forces on the central Na^+. Furthermore, beyond these 12 Na^+ ions are more Cl^- ions that exert an attractive force, and so on. The net effect of all these interactions is a resultant negative electric potential energy

$$U_{\text{attractive}} = -\alpha k_e \frac{e^2}{r} \tag{43.16}$$

where α is a dimensionless number known as the **Madelung constant.** The value of α depends only on the particular crystalline structure of the solid. For example, $\alpha = 1.747\ 6$ for the NaCl structure. When the constituent ions of a crystal are brought close together, a repulsive force exists because of electrostatic forces and the exclusion principle as discussed in Section 43.1. The potential energy term B/r^m in Equation 43.1 accounts for this repulsive force. We do not include neighbors other than nearest neighbors here because the repulsive forces occur only for ions that are very close together. (Electron shells must overlap for exclusion-principle effects to become important.) Therefore, we can express the total potential energy of the crystal as

$$U_{\text{total}} = -\alpha k_e \frac{e^2}{r} + \frac{B}{r^m} \tag{43.17}$$

where m in this expression is some small integer.

A plot of total potential energy versus ion separation distance is shown in Figure 43.11. The potential energy has its minimum value U_0 at the equilibrium separation, when $r = r_0$. It is left as a problem (Problem 57) to show that

$$U_0 = -\alpha k_e \frac{e^2}{r_0}\left(1 - \frac{1}{m}\right) \tag{43.18}$$

This minimum energy U_0 is called the **ionic cohesive energy** of the solid, and its absolute value represents the energy required to separate the solid into a collection of isolated positive and negative ions. Its value for NaCl is -7.84 eV per ion pair.

To calculate the **atomic cohesive energy,** which is the binding energy relative to the energy of the neutral atoms, 5.14 eV must be added to the ionic cohesive energy value to account for the transition from Na^+ to Na and 3.62 eV must be subtracted to account for the conversion of Cl^- to Cl. Therefore, the atomic cohesive energy of NaCl is

$$-7.84 \text{ eV} + 5.14 \text{ eV} - 3.62 \text{ eV} = -6.32 \text{ eV}$$

In other words, 6.32 eV of energy per ion pair is needed to separate the solid into isolated neutral atoms of Na and Cl.

◀ Properties of ionic crystals

Ionic crystals form relatively stable, hard crystals. They are poor electrical conductors because they contain no free electrons; each electron in the solid is bound tightly to one of the ions, so it is not sufficiently mobile to carry current. Ionic crystals have high melting points; for example, the melting point of NaCl is 801°C. Ionic crystals are transparent to visible radiation because the shells formed by the electrons in ionic solids are so tightly bound that visible radiation does not possess sufficient energy to promote electrons to the next allowed shell. Infrared radiation is absorbed strongly because the vibrations of the ions have natural resonant frequencies in the low-energy infrared region.

Covalent Solids

Solid carbon, in the form of diamond, is a crystal whose atoms are covalently bonded. Because atomic carbon has the electronic configuration $1s^22s^22p^2$, it is four electrons short of filling its $n = 2$ shell, which can accommodate eight electrons. Hence, two carbon atoms have a strong attraction for each other, with a cohesive energy of 7.37 eV. In the diamond structure, each carbon atom is covalently bonded to four other carbon atoms located at four corners of a cube as shown in Figure 43.12a.

The crystalline structure of diamond is shown in Figure 43.12b. Notice that each carbon atom forms covalent bonds with four nearest-neighbor atoms. The basic structure of diamond is called tetrahedral (each carbon atom is at the center of a regular tetrahedron), and the angle between the bonds is 109.5°. Other crystals such as silicon and germanium have the same structure.

Carbon is interesting in that it can form several different types of structures. In addition to the diamond structure, it forms graphite, with completely different properties. In this form, the carbon atoms form flat layers with hexagonal arrays of atoms. A very weak interaction between the layers allows the layers to be removed easily under friction, as occurs in the graphite used in pencil lead.

Carbon atoms can also form a large hollow structure; in this case, the compound is called **buckminsterfullerene** after the famous architect R. Buckminster Fuller, who invented the geodesic dome. The unique shape of this molecule (Fig.

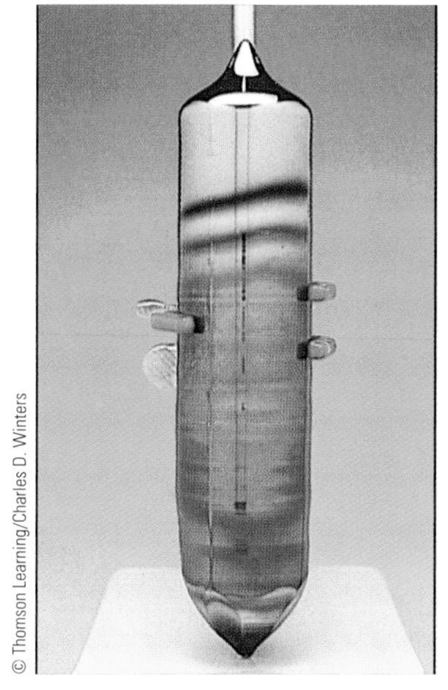

A cylinder of nearly pure crystalline silicon (Si), approximately 25 cm long. Such crystals are cut into wafers and processed to make various semiconductor devices.

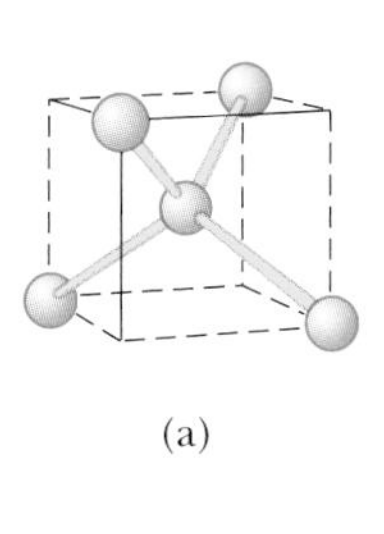
(a)

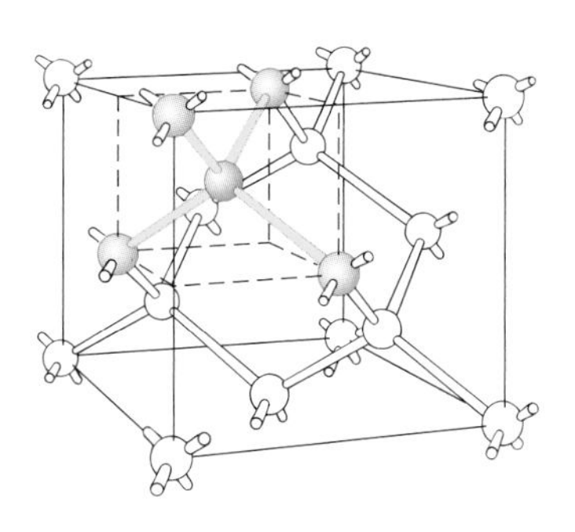
(b)

Figure 43.12 (a) Each carbon atom in a diamond crystal is covalently bonded to four other carbon atoms so that a tetrahedral structure is formed. (b) The crystal structure of diamond, showing the tetrahedral bond arrangement.

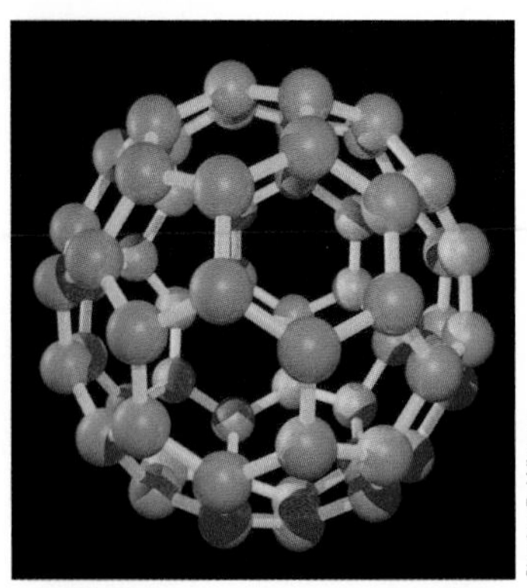

Figure 43.13 Computer rendering of a "buckyball," short for the molecule buckminsterfullerene. These nearly spherical molecular structures that look like soccer balls were named for the inventor of the geodesic dome. This form of carbon, C_{60}, was discovered by astrophysicists investigating the carbon gas that exists between stars. Scientists are actively studying the properties and potential uses of buckminsterfullerene and related molecules.

TABLE 43.1

Atomic Cohesive Energies of Some Covalent Solids

Solid	Cohesive Energy (eV per ion pair)
C (diamond)	7.37
Si	4.63
Ge	3.85
InAs	5.70
SiC	6.15
ZnS	6.32
CuCl	9.24

43.13) provides a "cage" to hold other atoms or molecules. Related structures, called "buckytubes" because of their long, narrow cylindrical arrangements of carbon atoms, may provide the basis for extremely strong, yet lightweight materials.

The atomic cohesive energies of some covalent solids are given in Table 43.1. The large energies account for the hardness of covalent solids. Diamond is particularly hard and has an extremely high melting point (about 4 000 K). Covalently bonded solids are usually very hard, have high bond energies and high melting points, and are good electrical insulators.

Metallic Solids

Metallic bonds are generally weaker than ionic or covalent bonds. The outer electrons in the atoms of a metal are relatively free to move throughout the material, and the number of such mobile electrons in a metal is large. The metallic structure can be viewed as a "sea" or a "gas" of nearly free electrons surrounding a lattice of positive ions (Fig. 43.14). The bonding mechanism in a metal is the attractive force between the entire collection of positive ions and the electron gas. Metals have a cohesive energy in the range of 1 to 3 eV per atom, which is less than the cohesive energies of ionic or covalent solids.

Light interacts strongly with the free electrons in metals. Hence, visible light is absorbed and re-emitted quite close to the surface of a metal, which accounts for the shiny nature of metal surfaces. In addition to the high electrical conductivity of metals produced by the free electrons, the nondirectional nature of the metallic bond allows many different types of metal atoms to be dissolved in a host metal in varying amounts. The resulting *solid solutions,* or *alloys,* may be designed to have particular properties, such as tensile strength, ductility, electrical and thermal conductivity, and resistance to corrosion.

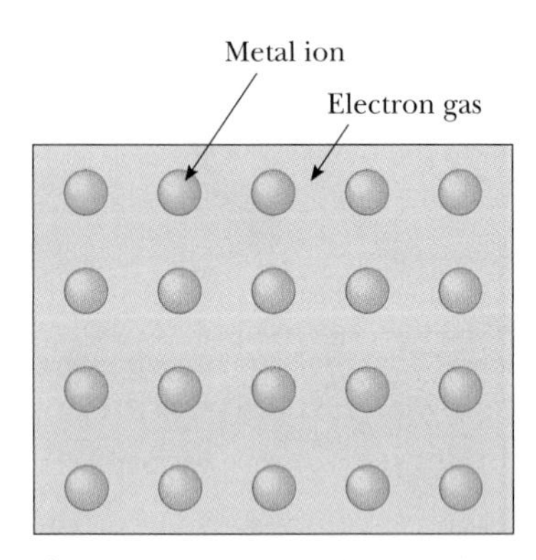

Figure 43.14 Highly schematic diagram of a metal. The blue area represents the electron gas, and the orange circles represent the positive metal ions.

Because the bonding in metals is between all the electrons and all the positive ions, metals tend to bend when stressed. This bending is in contrast to nonmetallic solids, which tend to fracture when stressed. Fracturing results because bonding in nonmetallic solids is primarily with nearest-neighbor ions or atoms. When the distortion causes sufficient stress between some set of nearest neighbors, fracture occurs.

43.4 Free-Electron Theory of Metals

In Section 27.3, we described a classical free-electron theory of electrical conduction in metals that led to Ohm's law. According to this theory, a metal is modeled as a classical gas of conduction electrons moving through a fixed lattice of ions.

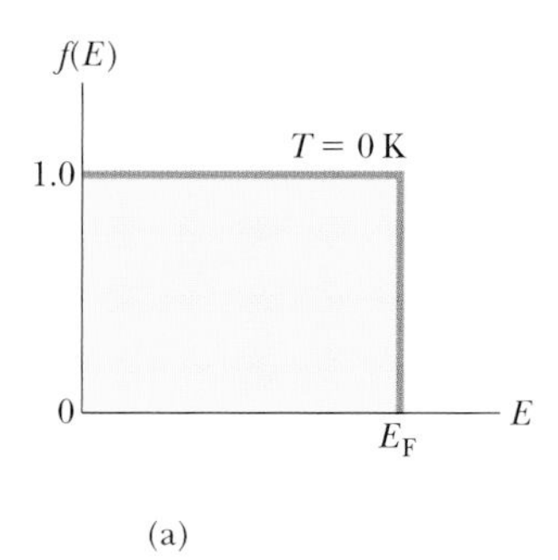

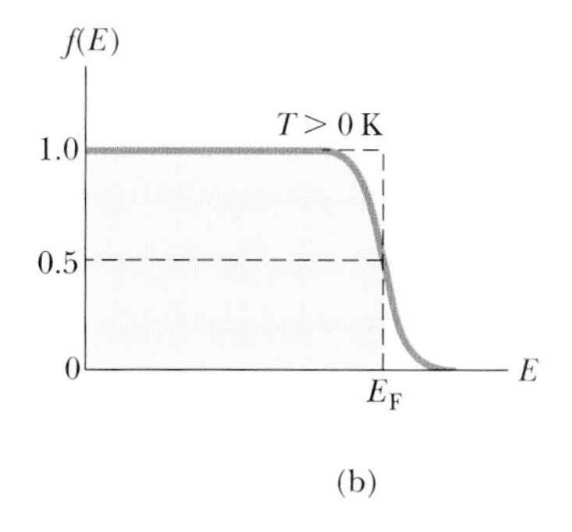

ACTIVE FIGURE 43.15

Plot of the Fermi–Dirac distribution function $f(E)$ versus energy at (a) $T = 0$ K and (b) $T > 0$ K. The energy E_F is the Fermi energy.

Sign in at www.thomsonedu.com and go to ThomsonNOW to adjust the temperature and observe the effect on the Fermi–Dirac distribution function.

Although this theory predicts the correct functional form of Ohm's law, it does not predict the correct values of electrical and thermal conductivities.

A quantum-based free-electron theory of metals remedies the shortcomings of the classical model by taking into account the wave nature of the electrons. In this model, the outer-shell electrons are free to move through the metal but are trapped within a three-dimensional box formed by the metal surfaces. Therefore, each electron is represented as a particle in a box. As discussed in Section 41.2, particles in a box are restricted to quantized energy levels.

Statistical physics can be applied to a collection of particles in an effort to relate microscopic properties to macroscopic properties as we saw with kinetic theory of gases in Chapter 21. In the case of electrons, it is necessary to use *quantum statistics*, with the requirement that each state of the system can be occupied by only two electrons (one with spin up and the other with spin down) as a consequence of the exclusion principle. The probability that a particular state having energy E is occupied by one of the electrons in a solid is

$$f(E) = \frac{1}{e^{(E-E_F)/k_B T} + 1} \qquad \textbf{(43.19)}$$

◀ Fermi–Dirac distribution function

where $f(E)$ is called the **Fermi–Dirac distribution function** and E_F is called the **Fermi energy.** A plot of $f(E)$ versus E at $T = 0$ K is shown in Active Figure 43.15a. Notice that $f(E) = 1$ for $E < E_F$ and $f(E) = 0$ for $E > E_F$. That is, at 0 K, all states having energies less than the Fermi energy are occupied and all states having energies greater than the Fermi energy are vacant. A plot of $f(E)$ versus E at some temperature $T > 0$ K is shown in Active Figure 43.15b. This curve shows that as T increases, the distribution rounds off slightly. Because of thermal excitation, states near and below E_F lose population and states near and above E_F gain population. The Fermi energy E_F also depends on temperature, but the dependence is weak in metals.

Let's now follow up on our discussion of the particle in a box in Chapter 41 to generalize the results to a three-dimensional box. Recall that if a particle of mass m is confined to move in a one-dimensional box of length L, the allowed states have quantized energy levels given by Equation 41.14:

$$E_n = \frac{h^2}{8mL^2}n^2 = \frac{\hbar^2\pi^2}{2mL^2}n^2 \qquad n = 1, 2, 3, \ldots$$

Now imagine a piece of metal in the shape of a solid cube of sides L and volume L^3 and focus on one electron that is free to move anywhere in this volume. Therefore, the electron is modeled as a particle in a three-dimensional box. In this model, we require that $\psi(x, y, z) = 0$ at the boundaries of the metal. It can be shown (see Problem 38) that the energy for such an electron is

$$E = \frac{\hbar^2\pi^2}{2m_e L^2}(n_x^2 + n_y^2 + n_z^2) \qquad \textbf{(43.20)}$$

where m_e is the mass of the electron and n_x, n_y, and n_z are quantum numbers. As we expect, the energies are quantized, and each allowed value of the energy is

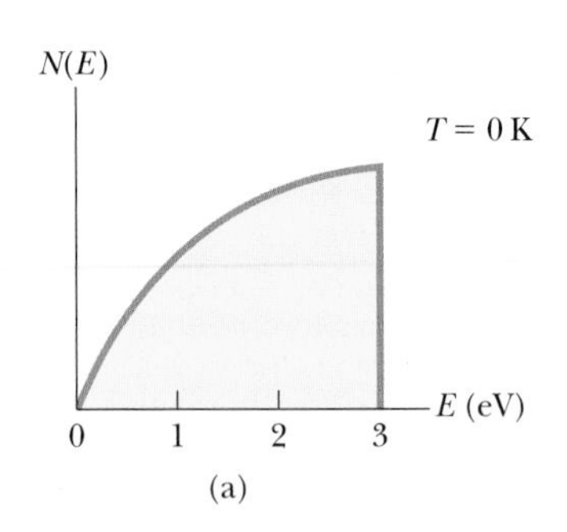

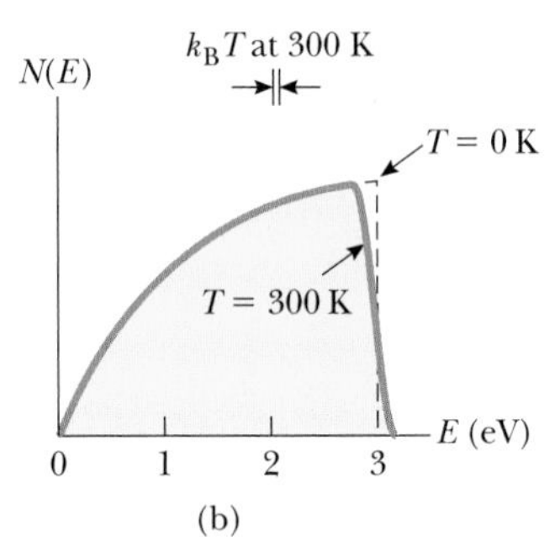

Figure 43.16 Plot of the electron distribution function versus energy in a metal at (a) $T = 0$ K and (b) $T = 300$ K. The Fermi energy E_F is 3 eV.

characterized by this set of three quantum numbers (one for each degree of freedom) and the spin quantum number m_s. For example, the ground state, corresponding to $n_x = n_y = n_z = 1$, has an energy equal to $3\hbar^2\pi^2/2m_eL^2$ and can be occupied by two electrons, corresponding to spin-up and spin-down.

Because of the macroscopic size L of the box, the energy levels for the electrons are very close together. As a result, we can treat the quantum numbers as continuous variables. Under this assumption, the number of allowed states per unit volume that have energies between E and $E + dE$ is

$$g(E)\,dE = \frac{8\sqrt{2}\pi m_e^{3/2}}{h^3} E^{1/2}\,dE \quad \textbf{(43.21)}$$

(See Example 43.5.) The function $g(E)$ is called the **density-of-states function.**

If a metal is in thermal equilibrium, the number of electrons per unit volume $N(E)\,dE$ that have energy between E and $E + dE$ is equal to the product of the number of allowed states and the probability that a state is occupied; that is, $N(E)\,dE = g(E)f(E)\,dE$:

$$N(E)\,dE = \left(\frac{8\sqrt{2}\pi m_e^{3/2}}{h^3} E^{1/2}\right)\left(\frac{1}{e^{(E-E_F)/k_BT} + 1}\right)dE \quad \textbf{(43.22)}$$

Plots of $N(E)$ versus E for two temperatures are given in Figure 43.16.

If n_e is the total number of electrons per unit volume, we require that

$$n_e = \int_0^\infty N(E)\,dE = \frac{8\sqrt{2}\pi m_e^{3/2}}{h^3}\int_0^\infty \frac{E^{1/2}\,dE}{e^{(E-E_F)/k_BT} + 1} \quad \textbf{(43.23)}$$

We can use this condition to calculate the Fermi energy. At $T = 0$ K, the Fermi–Dirac distribution function $f(E) = 1$ for $E < E_F$ and $f(E) = 0$ for $E > E_F$. Therefore, at $T = 0$ K, Equation 43.23 becomes

$$n_e = \frac{8\sqrt{2}\pi m_e^{3/2}}{h^3}\int_0^{E_F} E^{1/2}\,dE = \tfrac{2}{3}\frac{8\sqrt{2}\pi m_e^{3/2}}{h^3} E_F^{3/2} \quad \textbf{(43.24)}$$

Solving for the Fermi energy at 0 K gives

Fermi energy at $T = 0$ K ▶

$$E_F(0) = \frac{h^2}{2m_e}\left(\frac{3n_e}{8\pi}\right)^{2/3} \quad \textbf{(43.25)}$$

The order of magnitude of the Fermi energy for metals is approximately 5 eV. Representative values for various metals are given in Table 43.2. It is left as a problem (Problem 35) to show that the average energy of a free electron in a metal at 0 K is

$$E_{avg} = \tfrac{3}{5}E_F \quad \textbf{(43.26)}$$

TABLE 43.2

Calculated Values of the Fermi Energy for Metals at 300 K Based on the Free-Electron Theory

Metal	Electron Concentration (m^{-3})	Fermi Energy (eV)
Li	4.70×10^{28}	4.72
Na	2.65×10^{28}	3.23
K	1.40×10^{28}	2.12
Cu	8.46×10^{28}	7.05
Ag	5.85×10^{28}	5.48
Au	5.90×10^{28}	5.53

In summary, we can consider a metal to be a system comprising a very large number of energy levels available to the free electrons. These electrons fill the levels in accordance with the Pauli exclusion principle, beginning with $E = 0$ and ending with E_F. At $T = 0$ K, all levels below the Fermi energy are filled and all levels above the Fermi energy are empty. At 300 K, a small fraction of the free electrons are excited above the Fermi energy.

EXAMPLE 43.4 The Fermi Energy of Gold

Each atom of gold (Au) contributes one free electron to the metal. Compute the Fermi energy for gold.

SOLUTION

Conceptualize Imagine electrons filling available levels at $T = 0$ K in gold until the solid is neutral. The highest energy filled is the Fermi energy.

Categorize We evaluate the result using a result from this section, so we categorize this example as a substitution problem.

Substitute the concentration of free electrons in gold from Table 43.2 into Equation 43.25 to calculate the Fermi energy at 0 K:

$$E_F(0) = \frac{(6.626 \times 10^{-34}\ \text{J}\cdot\text{s})^2}{2(9.11 \times 10^{-31}\ \text{kg})}\left[\frac{3(5.90 \times 10^{28}\ \text{m}^{-3})}{8\pi}\right]^{2/3}$$

$$= 8.85 \times 10^{-19}\ \text{J} = 5.53\ \text{eV}$$

EXAMPLE 43.5 Deriving Equation 43.21

Based on the allowed states of a particle in a three-dimensional box, derive Equation 43.21.

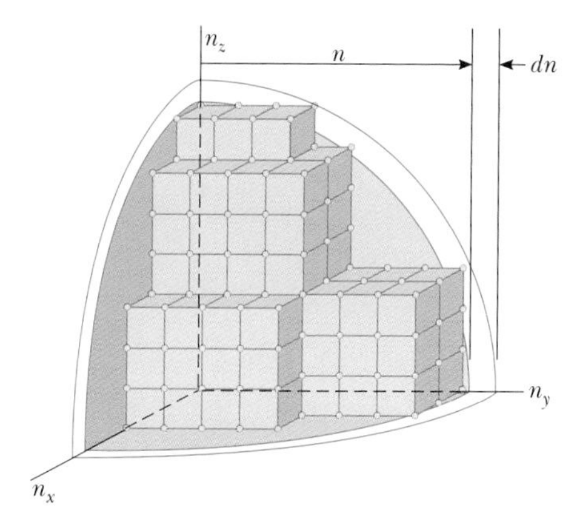

Figure 43.17 (Example 43.5) The allowed states of particles in a three-dimensional box can be represented by dots (blue circles) in a quantum number space. This space is not traditional space in which a location is specified by coordinates x, y, and z; rather, it is a space in which allowed states can be specified by coordinates representing the quantum numbers. The dots representing the allowed states are located at integer values of n_x, n_y, and n_z and are therefore at the corners of cubes with sides of "length" 1. The number of allowed states having energies between E and $E + dE$ corresponds to the number of dots in the spherical shell of radius n and thickness dn.

SOLUTION

Conceptualize Imagine a particle confined to a three-dimensional box, subject to boundary conditions in three dimensions.

Categorize We categorize this problem as that of a quantum system in which the energies of the particle are quantized. Furthermore, we can base the solution to the problem on our understanding of the particle in a one-dimensional box.

Analyze As noted previously, the allowed states of the particle in a three-dimensional box are described by three quantum numbers n_x, n_y, and n_z. Imagine a three-dimensional *quantum number space* whose axes represent n_x, n_y, and n_z. The allowed states in this space can be represented as dots located at integral values of the three quantum numbers as in Figure 43.17.

Defining $E_0 = \hbar^2\pi^2/2m_eL^2$ and $n = (E/E_0)^{1/2}$, rewrite Equation 43.20:

$$(1)\quad n_x^2 + n_y^2 + n_z^2 = \frac{2m_eL^2}{\hbar^2\pi^2}E = \frac{E}{E_0} = n^2$$

In the quantum number space, Equation (1) is the equation of a sphere of radius n. Therefore, the number of allowed states having energies between E and $E + dE$ is equal to the number of points in a spherical shell of radius n and thickness dn.

Find the "volume" of this shell, which represents the total number of states $G(E)\,dE$:

$$(2)\quad G(E)\,dE = \tfrac{1}{8}(4\pi n^2 dn) = \tfrac{1}{2}\pi n^2 dn$$

We have taken one-eighth of the total volume because we are restricted to the octant of a three-dimensional space in which all three quantum numbers are positive.

Replace n in Equation (2) with its equivalent in terms of E using the relation $n^2 = E/E_0$ from Equation (1):

$$G(E)\,dE = \tfrac{1}{2}\pi\left(\frac{E}{E_0}\right)d\left[\left(\frac{E}{E_0}\right)^{1/2}\right] = \tfrac{1}{2}\pi\,\frac{E}{(E_0)^{3/2}}\,d[(E)^{1/2}]$$

Evaluate the differential:

$$G(E)\,dE = \tfrac{1}{2}\pi\left[\frac{E}{(E_0)^{3/2}}\right](\tfrac{1}{2}E^{-1/2}dE) = \tfrac{1}{4}\pi E_0^{-3/2}E^{1/2}dE$$

Substitute for E_0 from its definition above:

$$G(E)\,dE = \tfrac{1}{4}\pi\left(\frac{\hbar^2\pi^2}{2m_eL^2}\right)^{-3/2}E^{1/2}dE$$

$$= \frac{\sqrt{2}}{2}\,\frac{m_e^{3/2}L^3}{\hbar^3\pi^2}\,E^{1/2}dE$$

Letting $g(E)$ represent the number of states per unit volume, where L^3 is the volume V of the cubical box in normal space, find $g(E) = G(E)/V$:

$$g(E)\,dE = \frac{G(E)}{V}\,dE = \frac{\sqrt{2}}{2}\,\frac{m_e^{3/2}}{\hbar^3\pi^2}\,E^{1/2}dE$$

Substitute $\hbar = h/2\pi$:

$$g(E)\,dE = \frac{4\sqrt{2}\pi m_e^{3/2}}{h^3}\,E^{1/2}\,dE$$

Multiply by 2 for the two possible spin states in each particle-in-a-box state:

$$g(E)\,dE = \frac{8\sqrt{2}\pi m_e^{3/2}}{h^3}\,E^{1/2}\,dE$$

Finalize This result is Equation 43.21, which is what we set out to derive.

43.5 Band Theory of Solids

In Section 43.4, the electrons in a metal were modeled as particles free to move around inside a three-dimensional box and we ignored the influence of the parent atoms. In this section, we make the model more sophisticated by incorporating the contribution of the parent atoms that form the crystal.

Recall from Section 41.1 that the probability density $|\psi|^2$ for a system is physically significant, but the probability amplitude ψ is not. Let's consider as an example an atom that has a single *s* electron outside of a closed shell. Both of the following wave functions are valid for such an atom with atomic number Z:

$$\psi_s^{+}(r) = +Af(r)e^{-Zr/na_0} \qquad \psi_s^{-}(r) = -Af(r)e^{-Zr/na_0}$$

where A is the normalization constant and $f(r)$ is a function[3] of r that varies with the value of n. Choosing either of these wave functions leads to the same value of $|\psi|^2$, so both choices are equivalent. A difference arises, however, when two atoms are combined.

If two identical atoms are very far apart, they do not interact and their electronic energy levels can be considered to be those of isolated atoms. Suppose the two atoms are sodium, each having a lone 3*s* electron that is in a well-defined

[3] The functions $f(r)$ are called *Laguerre polynomials.* They can be found in the quantum treatment of the hydrogen atom in modern physics textbooks.

quantum state. As the two sodium atoms are brought closer together, their wave functions begin to overlap as we discussed for covalent bonding in Section 43.1. The properties of the combined system differ depending on whether the two atoms are combined with wave functions $\psi_s^{+}(r)$ as in Figure 43.18a or whether one is combined with wave function $\psi_s^{+}(r)$ and the other with $\psi_s^{-}(r)$ as in Figure 43.18b. The choice of two atoms with wave function $\psi_s^{-}(r)$ is physically equivalent to that with two positive wave functions, so we do not consider it separately. When two wave functions $\psi_s^{+}(r)$ are combined, the result is a composite wave function in which the probability amplitudes add between the atoms. If $\psi_s^{+}(r)$ combines with $\psi_s^{-}(r)$, however, the wave functions between the nuclei subtract. Therefore, the composite probability amplitudes for the two possibilities are different. **These two possible combinations of wave functions represent two possible states of the two-atom system.** We interpret these curves as representing the probability amplitude of finding an electron. The positive–positive curve shows some probability of finding the electron at the midpoint between the atoms. The positive–negative function shows no such probability. A state with a high probability of an electron *between* two positive nuclei must have a different energy than a state with a high probability of the electron being elsewhere! Therefore, the states are *split* into two energy levels due to the two ways of combining the wave functions. The energy difference is relatively small, so the two states are close together on an energy scale.

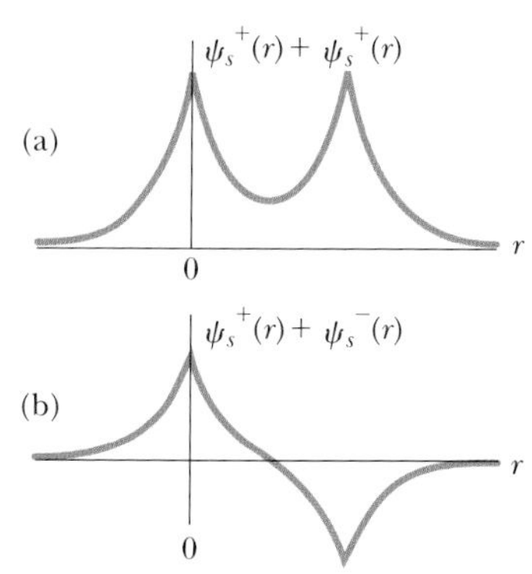

Figure 43.18 The wave functions of two atoms combine to form a composite wave function for the two-atom system when the atoms are close together. (a) Two atoms with wave functions $\psi_s^{+}(r)$ combine. (b) Two atoms with wave functions $\psi_s^{+}(r)$ and $\psi_s^{-}(r)$ combine.

Figure 43.19a shows this splitting effect as a function of separation distance. For large separations r, the electron clouds do not overlap and there is no splitting. As the atoms are brought closer so that r decreases, the electron clouds overlap and we need to consider the system of two atoms.

When a large number of atoms are brought together to form a solid, a similar phenomenon occurs. The individual wave functions can be brought together in various combinations of $\psi_s^{+}(r)$ and $\psi_s^{-}(r)$, each possible combination corresponding to a different energy. As the atoms are brought close together, the various isolated-atom energy levels split into multiple energy levels for the composite system. This splitting in levels for five atoms in close proximity is shown in Figure 43.19b. In this case, there are five energy levels corresponding to five different combinations of isolated-atom wave functions.

If we extend this argument to the large number of atoms found in solids (on the order of 10^{23} atoms per cubic centimeter), we obtain a large number of levels of varying energy so closely spaced that they may be regarded as a continuous **band** of energy levels as shown in Figure 43.19c. In the case of sodium, it is customary to refer to the continuous distributions of allowed energy levels as s bands because the bands originate from the s levels of the individual sodium atoms.

Figure 43.20 (page 1276) shows the allowed energy bands of sodium. Notice that energy gaps, corresponding to *forbidden energies*, occur between the allowed bands. In addition, some bands exhibit sufficient spreading in energy that there is an overlap between bands arising from different quantum states ($3s$ and $3p$).

As indicated by the blue-shaded areas in Figure 43.20, the $1s$, $2s$, and $2p$ bands of sodium are each full of electrons because the $1s$, $2s$, and $2p$ states of each atom

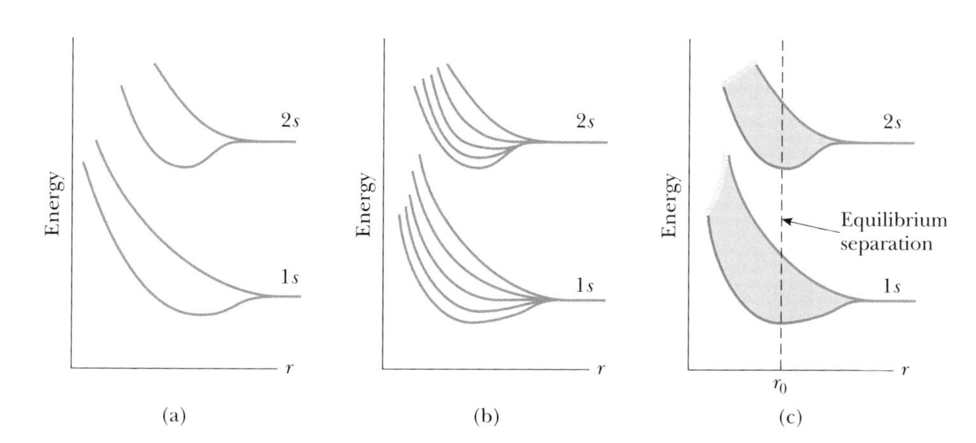

Figure 43.19 (a) Splitting of the $1s$ and $2s$ levels when two sodium atoms are brought together. (b) Splitting of the $1s$ and $2s$ levels when five sodium atoms are brought together. (c) Formation of energy bands when a large number of sodium atoms are assembled to form a solid.

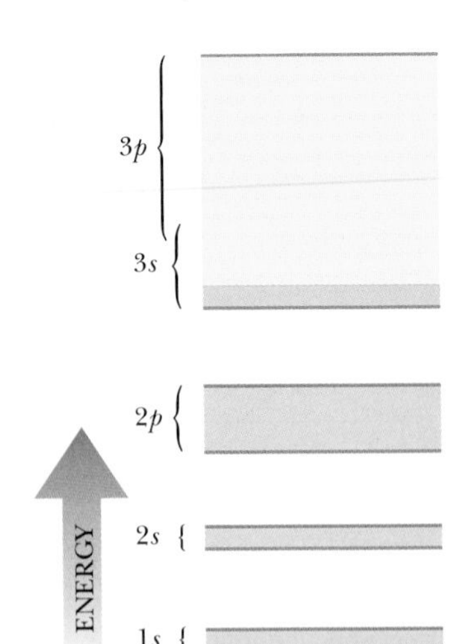

Figure 43.20 Energy bands of a sodium crystal. Notice the energy gaps (white regions) between the allowed bands; electrons cannot occupy states that lie in these gaps. Blue represents energy bands occupied by the sodium electrons when the atom is in its ground state. Gold represents energy bands that are empty.

are full. An energy level in which the orbital angular momentum is ℓ can hold $2(2\ell + 1)$ electrons. The factor 2 arises from the two possible electron spin orientations, and the factor $2\ell + 1$ corresponds to the number of possible orientations of the orbital angular momentum. The capacity of each band for a system of N atoms is $2(2\ell + 1)N$ electrons. Therefore, the 1s and 2s bands each contain $2N$ electrons ($\ell = 0$), and the 2p band contains $6N$ electrons ($\ell = 1$). Because sodium has only one 3s electron and there are a total of N atoms in the solid, the 3s band contains only N electrons and is partially full as indicated by the blue coloring in Figure 43.20. The 3p band, which is the higher region of the overlapping bands, is completely empty (all gold in the figure).

Band theory allows us to build simple models to understand the behavior of conductors, insulators, and semiconductors as well as that of semiconductor devices, as we shall discuss in the following sections.

43.6 Electrical Conduction in Metals, Insulators, and Semiconductors

Good electrical conductors contain a high density of free charge carriers, and the density of free charge carriers in insulators is nearly zero. Semiconductors, first introduced in Section 23.2, are a class of technologically important materials in which charge-carrier densities are intermediate between those of insulators and those of conductors. In this section, we discuss the mechanisms of conduction in these three classes of materials in terms of a model based on energy bands.

Metals

If a material is to be a good electrical conductor, the charge carriers in the material must be free to move in response to an applied electric field. Let's consider the electrons in a metal as the charge carriers. The motion of the electrons in response to an electric field represents an increase in energy of the system (the metal lattice and the free electrons) corresponding to the additional kinetic energy of the moving electrons. Therefore, when an electric field is applied to a conductor, electrons must move upward to an available higher energy state on an energy-level diagram.

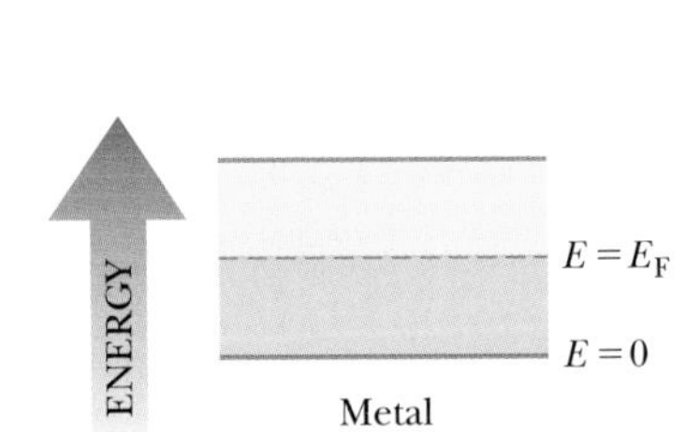

Figure 43.21 Half-filled band of a metal, an electrical conductor. At $T = 0$ K, the Fermi energy lies in the middle of the band.

Figure 43.21 shows a half-filled band in a metal at $T = 0$ K, where the blue region represents levels filled with electrons. Because electrons obey Fermi–Dirac statistics, all levels below the Fermi energy are filled with electrons and all levels above the Fermi energy are empty. The Fermi energy lies in the band at the highest filled state. At temperatures slightly greater than 0 K, some electrons are thermally excited to levels above E_F, but overall there is little change from the 0 K case. **If a potential difference is applied to the metal, however, electrons having energies near the Fermi energy require only a small amount of additional energy from the applied electric field to reach nearby empty energy states above the Fermi energy.** Therefore, electrons in a metal experiencing only a weak applied electric field are free to move because many empty levels are available close to the occupied energy levels. The model of metals based on band theory demonstrates that metals are excellent electrical conductors.

Insulators

Now consider the two outermost energy bands of a material in which the lower band is filled with electrons and the higher band is empty at 0 K (Fig. 43.22). The lower, filled band is called the **valence band,** and the upper, empty band is the **conduction band.** (The conduction band is the one that is partially filled in a metal.) It is common to refer to the energy separation between the valence and conduction bands as the **energy gap** E_g of the material. The Fermi energy lies somewhere in the energy gap[4] as shown in Figure 43.22.

Suppose a material has a relatively large energy gap of, for example, approximately 5 eV. At 300 K (room temperature), $k_B T = 0.025$ eV, which is much smaller than the energy gap. At such temperatures, the Fermi–Dirac distribution predicts that very few electrons are thermally excited into the conduction band. There are no available states that lie close in energy and into which electrons can move upward to account for the extra kinetic energy associated with motion through the material in response to an electric field. Consequently, the electrons do not move; the material is an insulator. Although an insulator has many vacant states in its conduction band that can accept electrons, these states are separated from the filled states by a large energy gap. Only a few electrons occupy these states, so the overall electrical conductivity of insulators is very small.

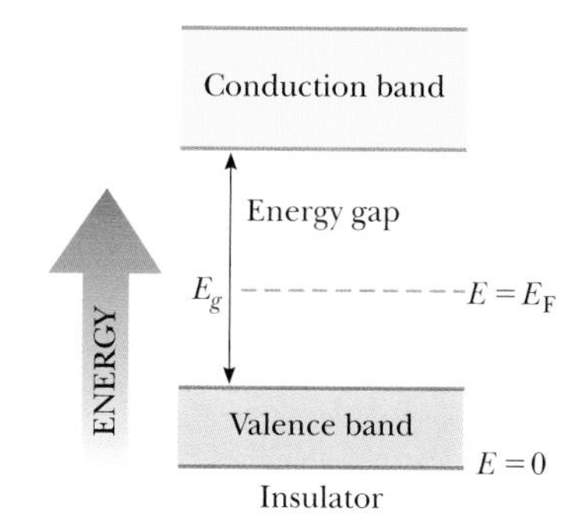

Figure 43.22 An electrical insulator at $T = 0$ K has a filled valence band and an empty conduction band. The Fermi level lies somewhere between these bands in the region known as the energy gap.

Semiconductors

Semiconductors have the same type of band structure as an insulator, but the energy gap is much smaller, on the order of 1 eV. Table 43.3 shows the energy gaps for some representative materials. The band structure of a semiconductor is shown in Figure 43.23. Because the Fermi level is located near the middle of the gap for a semiconductor and E_g is small, appreciable numbers of electrons are thermally excited from the valence band to the conduction band. Because of the many empty levels above the thermally filled levels in the conduction band, a small applied potential difference can easily raise the energy of the electrons in the conduction band, resulting in a moderate current.

At $T = 0$ K, all electrons in these materials are in the valence band and no energy is available to excite them across the energy gap. Therefore, semiconductors are poor conductors at very low temperatures. Because the thermal excitation of electrons across the narrow gap is more probable at higher temperatures, the conductivity of semiconductors increases rapidly with temperature, contrasting sharply with the conductivity of metals, which decreases slowly with increasing temperature.

Charge carriers in a semiconductor can be negative, positive, or both. When an electron moves from the valence band into the conduction band, it leaves behind

TABLE 43.3

Energy-Gap Values for Some Semiconductors

Crystal	E_g (eV) 0 K	300 K
Si	1.17	1.14
Ge	0.74	0.67
InP	1.42	1.34
GaP	2.32	2.26
GaAs	1.52	1.42
CdS	2.58	2.42
CdTe	1.61	1.56
ZnO	3.44	3.2
ZnS	3.91	3.6

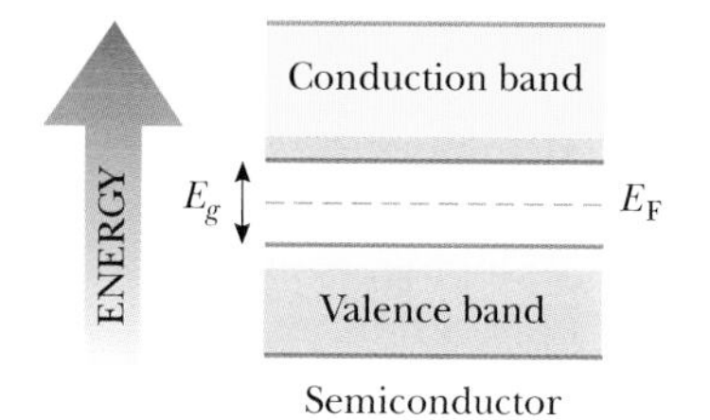

Figure 43.23 Band structure of a semiconductor at ordinary temperatures ($T \approx 300$ K). The energy gap is much smaller than in an insulator, and some electrons from the valence band occupy states in the conduction band.

[4] We defined the Fermi energy as the energy of the highest filled state at $T = 0$, which might suggest that the Fermi energy should be at the top of the valence band in Figure 43.22. A more sophisticated general treatment of the Fermi energy, however, shows that it is located at that energy at which the probability of occupation is one-half (see Active Fig. 43.15b). According to this definition, the Fermi energy lies in the energy gap between the bands.

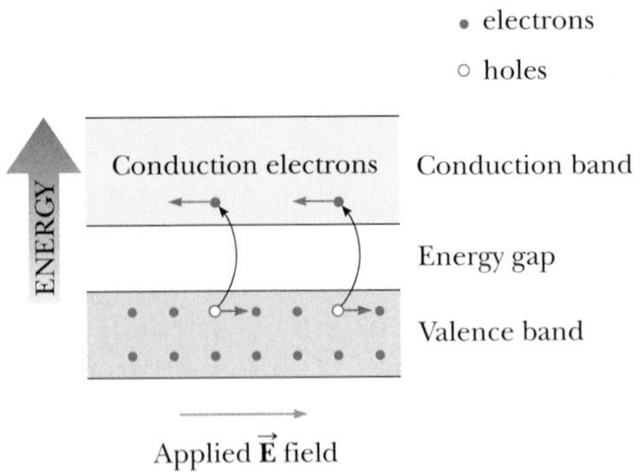

Figure 43.24 Movement of charges (holes and electrons) in an intrinsic semiconductor. The electrons move in the direction opposite the direction of the external electric field, and the holes move in the direction of the field.

a vacant site, called a **hole,** in the otherwise filled valence band. This hole (electron-deficient site) acts as a charge carrier in the sense that a free electron from a nearby site can transfer into the hole. Whenever an electron does so, it creates a new hole at the site it abandoned. Therefore, the net effect can be viewed as the hole migrating through the material in the direction opposite the direction of electron movement. The hole behaves as if it were a particle with a positive charge $+e$.

A pure semiconductor crystal containing only one element or one compound is called an **intrinsic semiconductor.** In these semiconductors, there are equal numbers of conduction electrons and holes. Such combinations of charges are called **electron–hole pairs.** In the presence of an external electric field, the holes move in the direction of the field and the conduction electrons move in the direction opposite the field (Fig. 43.24). Because the electrons and holes have opposite signs, both motions correspond to a current in the same direction.

Quick Quiz 43.4 Consider the data on three materials given in the table.

Material	Conduction Band	E_g
A	Empty	1.2 eV
B	Half full	1.2 eV
C	Empty	8.0 eV

Identify these materials as a conductor, an insulator, or a semiconductor.

Doped Semiconductors

When impurities are added to a semiconductor, both the band structure of the semiconductor and its resistivity are modified. The process of adding impurities, called **doping,** is important in controlling the conductivity of semiconductors. For example, when an atom containing five outer-shell electrons, such as arsenic, is added to a Group IV semiconductor, four of the electrons form covalent bonds with atoms of the semiconductor and one is left over (Fig. 43.25a). This extra electron is nearly free of its parent atom and can be modeled as having an energy level that lies in the energy gap, immediately below the conduction band (Fig. 43.25b). Such a pentavalent atom in effect donates an electron to the structure and hence is referred to as a **donor atom.** Because the spacing between the energy level of the electron of the donor atom and the bottom of the conduction band is very small (typically, approximately 0.05 eV), only a small amount of thermal excitation is needed to cause this electron to move into the conduction band. (Recall that the average energy of an electron at room temperature is approximately $k_B T \approx 0.025$ eV.) Semiconductors doped with donor atoms are called ***n*-type semiconductors** because the majority of charge carriers are electrons, which are *nega*tively charged.

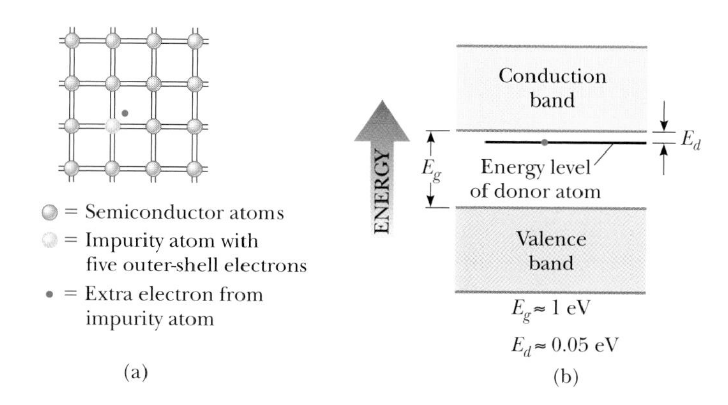

Figure 43.25 (a) Two-dimensional representation of a semiconductor consisting of Group IV atoms (gray) and an impurity atom (yellow) that has five outer-shell electrons. Each double line between atoms represents a covalent bond in which two electrons are shared. (b) Energy-band diagram for a semiconductor in which the nearly free electron of the impurity atom lies in the energy gap, immediately below the bottom of the conduction band. A small amount of energy can excite the electron into the conduction band.

If a Group IV semiconductor is doped with atoms containing three outer-shell electrons, such as indium and aluminum, the three electrons form covalent bonds with neighboring semiconductor atoms, leaving an electron deficiency—a hole—where the fourth bond would be if an impurity-atom electron were available to form it (Fig. 43.26a). This situation can be modeled by placing an energy level in the energy gap, immediately above the valence band, as in Figure 43.26b. An electron from the valence band has enough energy at room temperature to fill this impurity level, leaving behind a hole in the valence band. This hole can carry current in the presence of an electric field. Because a trivalent atom accepts an electron from the valence band, such impurities are referred to as **acceptor atoms.** A semiconductor doped with trivalent (acceptor) impurities is known as a ***p*-type semiconductor** because the majority of charge carriers are *p*ositively charged holes.

When conduction in a semiconductor is the result of acceptor or donor impurities, the material is called an **extrinsic semiconductor.** The typical range of doping densities for extrinsic semiconductors is 10^{13} to 10^{19} cm^{-3}, whereas the electron density in a typical semiconductor is roughly 10^{21} cm^{-3}.

43.7 Semiconductor Devices

The electronics of the first half of the 20th century was based on vacuum tubes, in which electrons pass through empty space between a cathode and an anode. We have seen vacuum tubes in Figure 29.5 (the television picture tube), Figure 29.9 (circular electron beam), Figure 29.14 (Thomson's apparatus for measuring e/m_e for the electron), and Figure 40.9 (photoelectric effect apparatus).

The transistor was invented in 1948, leading to a shift away from vacuum tubes and toward semiconductors as the basis of electronic devices. This phase of electronics has been under way for several decades. As discussed in Chapter 41, there may be a new phase of electronics in the near future using nanotechnological devices employing quantum dots and other nanoscale structures.

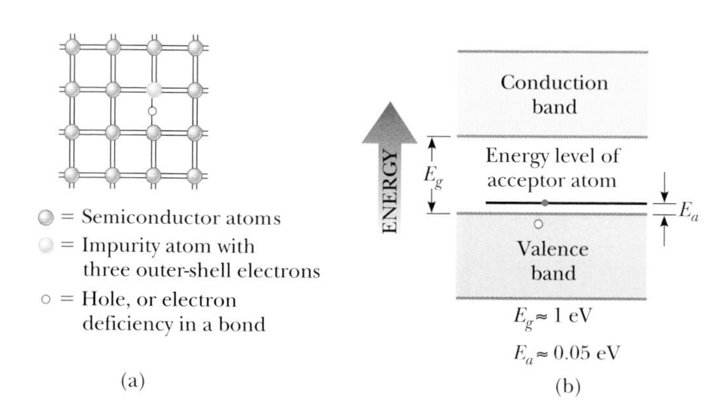

Figure 43.26 (a) Two-dimensional representation of a semiconductor consisting of Group IV atoms (gray) and an impurity atom (yellow) having three outer-shell electrons. The single line between the impurity atom and the semiconductor atom below it represents that there is only one electron shared in this bond.(b) Energy-band diagram for a semiconductor in which the energy level associated with the trivalent impurity atom lies in the energy gap, immediately above the top of the valence band. This diagram shows an electron excited into the energy level of the acceptor atom, leaving a hole in the valence band.

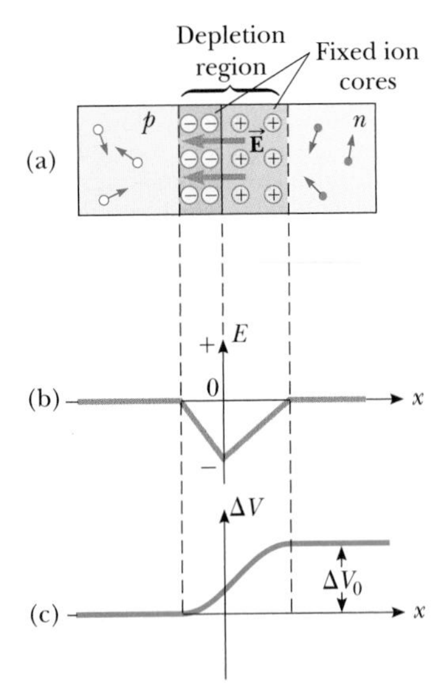

Figure 43.27 (a) Physical arrangement of a p–n junction. (b) Internal electric field magnitude versus x for the p–n junction. (c) Internal electric potential difference ΔV versus x for the p–n junction. The potential difference ΔV_0 represents the potential difference across the junction in the absence of an applied electric field.

In this section, we discuss electronic devices based on semiconductors, which are still in wide use and will be for many years to come.

The Junction Diode

A fundamental unit of a semiconductor device is formed when a p-type semiconductor is joined to an n-type semiconductor to form a ***p–n* junction**. A **junction diode** is a device that is based on a single p–n junction. The role of a diode of any type is to pass current in one direction but not the other. Therefore, it acts as a one-way valve for current.

The p–n junction shown in Figure 43.27a consists of three distinct regions: a p region, an n region, and a small area that extends several micrometers to either side of the interface, called a *depletion region.*

The depletion region may be visualized as arising when the two halves of the junction are brought together. The mobile n-side donor electrons nearest the junction (deep-blue area in Fig. 43.27a) diffuse to the p side and fill holes located there, leaving behind immobile positive ions. While this process occurs, we can model the holes that are being filled as diffusing to the n side, leaving behind a region (brown area in Fig. 43.27a) of fixed negative ions.

Because the two sides of the depletion region each carry a net charge, an internal electric field on the order of 10^4 to 10^6 V/cm exists in the depletion region (see Fig. 43.27b). This field produces an electric force on any remaining mobile charge carriers that sweeps them out of the depletion region. The depletion region is so named because it is depleted of mobile charge carriers. This internal electric field creates an internal potential difference ΔV_0 that prevents further diffusion of holes and electrons across the junction and thereby ensures zero current in the junction when no potential difference is applied.

The operation of the junction as a diode is easiest to understand in terms of the potential difference graph shown in Figure 43.27c. If a voltage ΔV is applied to the junction such that the p side is connected to the positive terminal of a voltage source as shown in Figure 43.28a, the internal potential difference ΔV_0 across the junction decreases; the decrease results in a current that increases exponentially with increasing forward voltage, or *forward bias.* For *reverse bias* (where the n side of the junction is connected to the positive terminal of a voltage source), the internal potential difference ΔV_0 increases with increasing reverse bias; the increase results in a very small reverse current that quickly reaches a saturation value I_0. The current–voltage relationship for an ideal diode is

$$I = I_0(e^{e\Delta V/k_B T} - 1) \tag{43.27}$$

where the first e is the base of the natural logarithm, the second e represents the magnitude of the electron charge, k_B is Boltzmann's constant, and T is the absolute temperature. Figure 43.28b shows a circuit diagram for a diode under forward bias, and Figure 43.28c shows an I–ΔV plot characteristic of a real p–n junction, demonstrating the diode behavior.

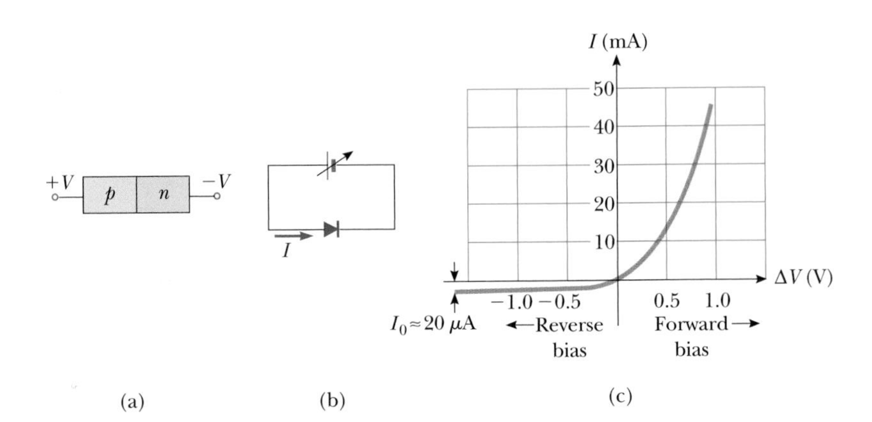

Figure 43.28 (a) A p–n junction under forward bias. (b) The circuit diagram for a diode under forward bias, showing a battery with an adjustable voltage. Both positive and negative voltages can be applied to the diode to study its nonlinear behavior. (c) The characteristic curve for a real p–n junction.

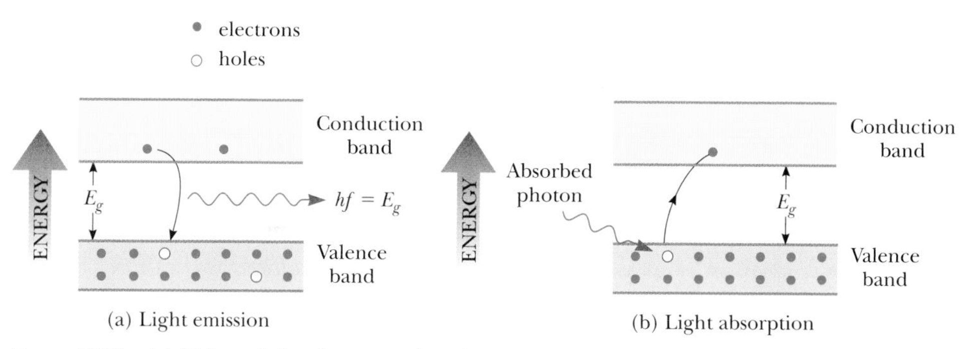

Figure 43.29 (a) Light emission from a semiconductor. (b) Light absorption by a semiconductor.

Light-Emitting and Light-Absorbing Diodes

Light-emitting diodes (LEDs) and semiconductor lasers are common examples of devices that depend on the behavior of semiconductors. LEDs are used in traffic signals, in electronic displays, and as indicator lights for electronic equipment. Semiconductor lasers are often used for pointers in presentations and in compact disc and DVD playback equipment.

Light emission and absorption in semiconductors is similar to light emission and absorption by gaseous atoms except that in the discussion of semiconductors we must incorporate the concept of energy bands rather than the discrete energy levels in single atoms. As shown in Figure 43.29a, an electron excited electrically into the conduction band can easily recombine with a hole (especially if the electron is injected into a p region). As this recombination takes place, a photon of energy E_g is emitted. With proper design of the semiconductor and the associated plastic envelope or mirrors, the light from a large number of these transitions serves as the source of an LED or a semiconductor laser.

Conversely, an electron in the valence band may absorb an incoming photon of light and be promoted to the conduction band, leaving a hole behind (Fig. 43.29b). This absorbed energy can be used to operate an electrical circuit. One device that operates on this principle is the photovoltaic solar cell, which appears in many handheld calculators. Arrays of solar cells are used in generating electric power in space vehicles and in remote areas on the Earth.

EXAMPLE 43.6 Where's the Remote?

Estimate the band gap of the semiconductor in the infrared LED of a typical television remote control.

SOLUTION

Conceptualize Imagine electrons in Figure 43.29a falling from the conduction band to the valence band, emitting infrared photons in the process.

Categorize We use concepts discussed in this section, so we categorize this example as a substitution problem.

In Chapter 34, we learned that the wavelength of infrared light ranges from 700 nm to 1 mm. Let's pick a number that is easy to work with, such as 1 000 nm (which is not a bad estimate because remote controls typically operate in the range of 880 to 950 nm.)

Estimate the energy hf of the photons from the remote control:

$$E = hf = \frac{hc}{\lambda} = \frac{1\ 240\ \text{eV}\cdot\text{nm}}{1\ 000\ \text{nm}} = 1.2\ \text{eV}$$

This value corresponds to an energy gap E_g of approximately 1.2 eV in the LED's semiconductor.

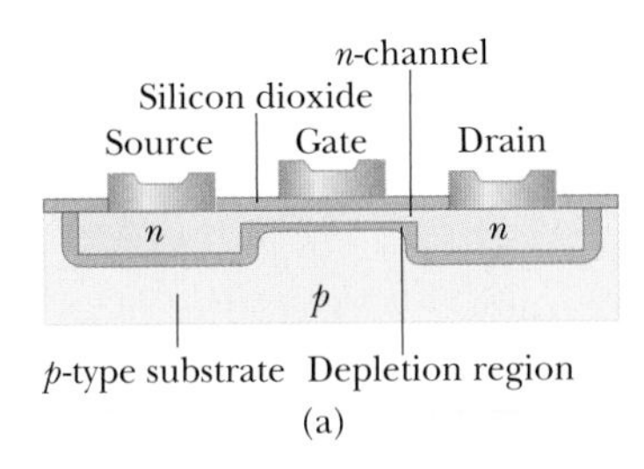

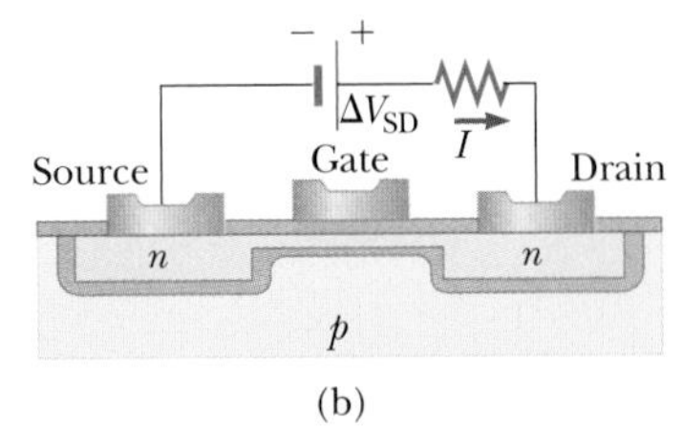

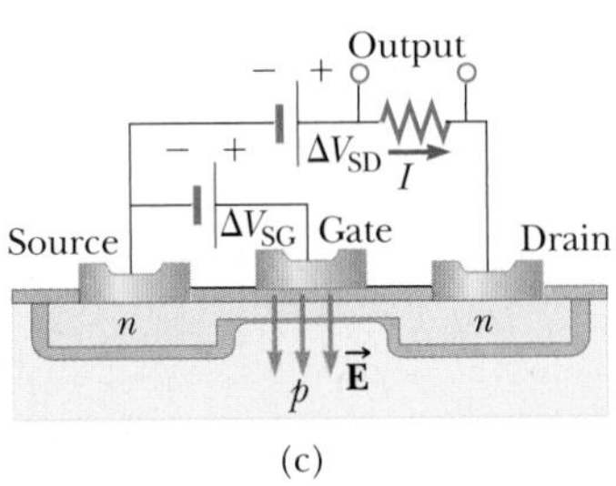

Figure 43.30 (a) The structure of a metal-oxide-semiconductor field-effect transistor (MOSFET). (b) A source–drain voltage is applied, with the result that current exists in the circuit. (c) A gate voltage is applied. The gate voltage can be used to control the source–drain current so that the MOSFET acts as an amplifier.

The Transistor

The invention of the transistor by John Bardeen (1908–1991), Walter Brattain (1902–1987), and William Shockley (1910–1989) in 1948 totally revolutionized the world of electronics. For this work, these three men shared the Nobel Prize in Physics in 1956. By 1960, the transistor had replaced the vacuum tube in many electronic applications. The advent of the transistor created a multitrillion-dollar industry that produces such popular devices as MP3 players, handheld calculators, computers, television receivers, wireless telephones, and electronic games.

A **junction transistor** consists of a semiconducting material in which a very narrow n region is sandwiched between two p regions or a p region is sandwiched between two n regions. In either case, the transistor is formed from two p–n junctions. These types of transistors were used widely in the early days of semiconductor electronics.

During the 1960s, the electronics industry converted many electronic applications from the junction transistor to the **field-effect transistor,** which is much easier to manufacture and just as effective. Figure 43.30a shows the structure of a very common device, the **MOSFET,** or **metal-oxide-semiconductor field-effect transistor.** You are likely using millions of MOSFET devices when you are working on your computer.

There are three metal connections (the M in MOSFET) to the transistor: the *source, drain,* and *gate.* The source and drain are connected to n-type semiconductor regions (the S in MOSFET) at either end of the structure. These regions are connected by a narrow channel of additional n-type material, the n channel. The source and drain regions and the n channel are embedded in a p-type substrate material, which forms a depletion region, as in the junction diode, along the bottom of the n channel. (Depletion regions also exist at the junctions underneath the source and drain regions, but we will ignore them because the operation of the device depends primarily on the behavior in the channel.)

The gate is separated from the n channel by a layer of insulating silicon dioxide (the O in MOSFET, for oxide). Therefore, it does not make electrical contact with the rest of the semiconducting material.

Imagine that a voltage source ΔV_{SD} is applied across the source and drain as shown in Figure 43.30b. In this situation, electrons flow through the upper region of the n channel. Electrons cannot flow through the depletion region in the lower part of the n channel because this region is depleted of charge carriers. Now a second voltage ΔV_{SG} is applied across the source and gate as in Figure 43.30c. The positive potential on the gate electrode results in an electric field below the gate that is directed downward in the n channel (the field in "field-effect"). This electric field exerts upward forces on electrons in the region below the gate, causing them to move into the n channel. Consequently, the depletion region becomes smaller, widening the area through which there is current between the top of the n channel and the depletion region. As the area becomes wider, the current increases.

If a varying voltage, such as that generated from music stored on a compact disc, is applied to the gate, the area through which the source–drain current exists varies in size according to the varying gate voltage. A small variation in gate voltage results in a large variation in current and a correspondingly large voltage across the resistor in Figure 43.30c. Therefore, the MOSFET acts as a voltage amplifier. A circuit consisting of a chain of such transistors can result in a very small initial signal from a microphone being amplified enough to drive powerful speakers at an outdoor concert.

The Integrated Circuit

Invented independently by Jack Kilby (b. 1923, Nobel Prize in Physics, 2000) at Texas Instruments in late 1958 and by Robert Noyce (1927–1990) at Fairchild Camera and Instrument in early 1959, the integrated circuit has been justly called "the most remarkable technology ever to hit mankind." Kilby's first device is shown in Figure 43.31. Integrated circuits have indeed started a "second industrial revolution" and are found at the heart of computers, watches, cameras, automobiles, aircraft, robots, space vehicles, and all sorts of communication and switching networks.

In simplest terms, an **integrated circuit** is a collection of interconnected transistors, diodes, resistors, and capacitors fabricated on a single piece of silicon known as a *chip*. Contemporary electronic devices often contain many integrated circuits (Fig. 43.32). State-of-the-art chips easily contain several million components within a 1-cm^2 area, and the number of components per square inch has increased steadily since the integrated circuit was invented. Figure 43.33 illustrates the dramatic advances made in chip technology since Intel introduced the first microprocessor in 1971.

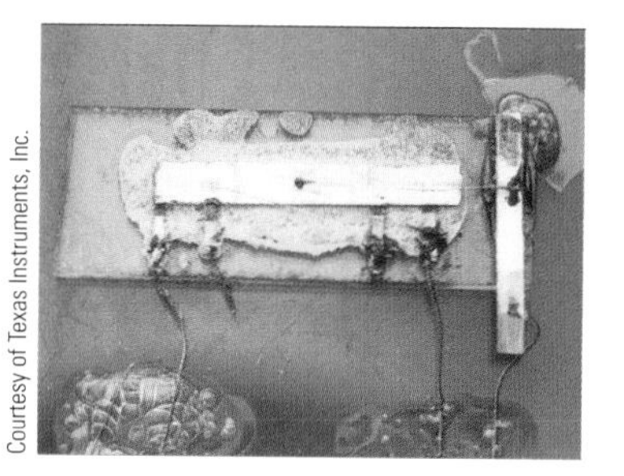
Courtesy of Texas Instruments, Inc.

Figure 43.31 Jack Kilby's first integrated circuit, tested on September 12, 1958.

Integrated circuits were invented partly to solve the interconnection problem spawned by the transistor. In the era of vacuum tubes, power and size considerations of individual components set modest limits on the number of components that could be interconnected in a given circuit. With the advent of the tiny, low-power, highly reliable transistor, design limits on the number of components disappeared and were replaced by the problem of wiring together hundreds of thousands of components. The magnitude of this problem can be appreciated when we consider that second-generation computers (consisting of discrete transistors rather than integrated circuits) contained several hundred thousand components requiring more than a million joints that had to be hand-soldered and tested.

In addition to solving the interconnection problem, integrated circuits possess the advantages of miniaturization and fast response, two attributes critical for high-speed computers. Because the response time of a circuit depends on the time interval required for electrical signals traveling at the speed of light to pass from one component to another, miniaturization and close packing of components result in fast response times.

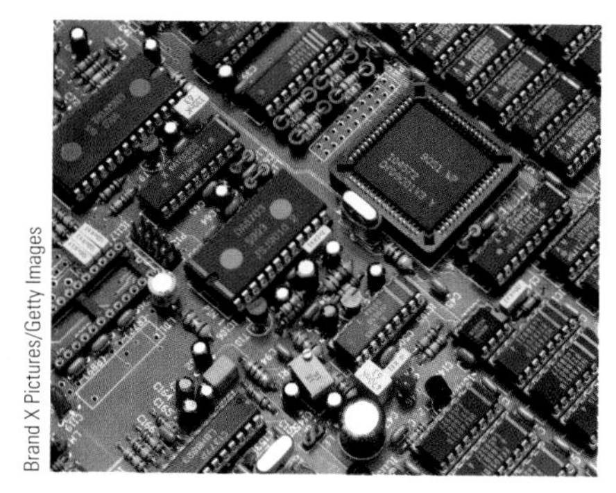
Brand X Pictures/Getty Images

Figure 43.32 Integrated circuits are prevalent in many electronic devices. All the flat circuit elements with black-topped surfaces in this photograph are integrated circuits.

43.8 Superconductivity

We learned in Section 27.5 that there is a class of metals and compounds known as **superconductors** whose electrical resistance decreases to virtually zero below a certain temperature T_c called the *critical temperature* (Table 27.3). Let's now look at these amazing materials in greater detail, using what we know about the properties of solids to help us understand the behavior of superconductors.

Let's start by examining the Meissner effect, introduced in Section 30.6 as the exclusion of magnetic flux from the interior of superconductors. The Meissner

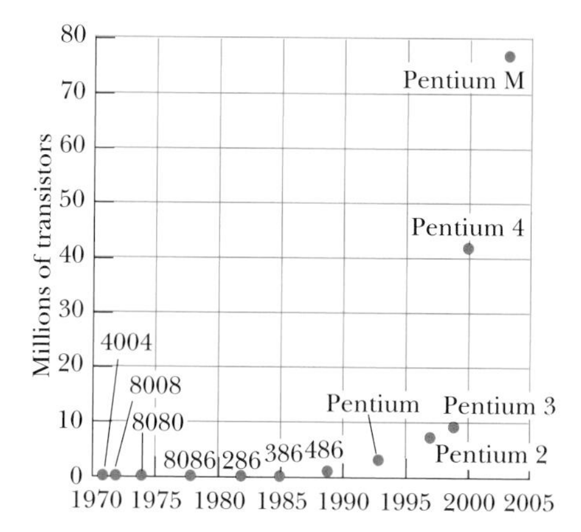

Figure 43.33 Dramatic advances in chip technology related to computer microchips manufactured by Intel, shown by a plot of the number of transistors on a single computer chip versus year of manufacture.

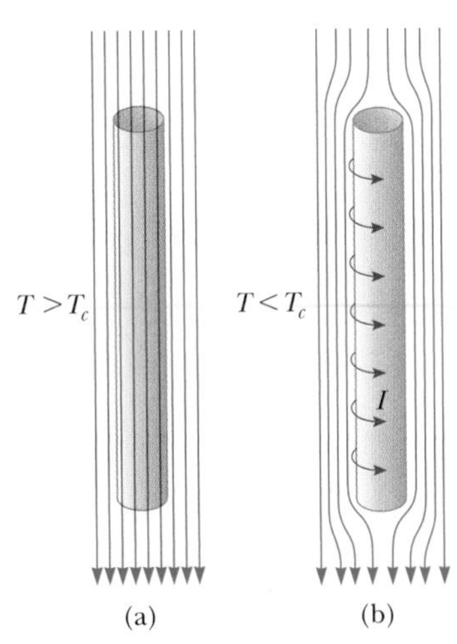

Figure 43.34 A superconductor in the form of a long cylinder in the presence of an external magnetic field. (a) At temperatures above T_c, the field lines penetrate the cylinder because it is in its normal state. (b) When the cylinder is cooled to $T < T_c$ and becomes superconducting, magnetic flux is excluded from its interior by the induction of surface currents.

effect is illustrated in Figure 43.34 for a superconducting material in the shape of a long cylinder. Notice that the magnetic field penetrates the cylinder when its temperature is greater than T_c (Fig. 43.34a). As the temperature is lowered to below T_c, however, the field lines are spontaneously expelled from the interior of the superconductor (Fig. 43.34b). Therefore, a superconductor is more than a perfect conductor (resistivity $\rho = 0$); it is also a perfect diamagnet ($\vec{\mathbf{B}} = 0$). The property that $\vec{\mathbf{B}} = 0$ in the interior of a superconductor is as fundamental as the property of zero resistance. If the magnitude of the applied magnetic field exceeds a critical value B_c, defined as the value of B that destroys a material's superconducting properties, the field again penetrates the sample.

Because a superconductor is a perfect diamagnet, it repels a permanent magnet. In fact, one can perform a demonstration of the Meissner effect by floating a small permanent magnet above a superconductor and achieving magnetic levitation as seen in Figure 30.27 in Chapter 30.

Recall from our study of electricity that a good conductor expels static electric fields by moving charges to its surface. In effect, the surface charges produce an electric field that exactly cancels the externally applied field inside the conductor. In a similar manner, a superconductor expels magnetic fields by forming surface currents. To see why that happens, consider again the superconductor shown in Figure 43.34. Let's assume the sample is initially at a temperature $T > T_c$ as illustrated in Figure 43.34a so that the magnetic field penetrates the cylinder. As the cylinder is cooled to a temperature $T < T_c$, the field is expelled as shown in Figure 43.34b. Surface currents induced on the superconductor's surface produce a magnetic field that exactly cancels the externally applied field inside the superconductor. As you would expect, the surface currents disappear when the external magnetic field is removed.

A successful theory for superconductivity in metals was published in 1957 by J. Bardeen, L. N. Cooper, and J. R. Schrieffer; it is generally called BCS theory, based on the first letters of their last names. This theory led to a Nobel Prize in Physics for the three scientists in 1972. In this theory, two electrons can interact via distortions in the array of lattice ions so that there is a net attractive force between the electrons.[5] As a result, the two electrons are bound into an entity called a *Cooper pair,* which behaves like a particle with integral spin. Particles with integral spin are called *bosons.* (As noted in Pitfall Prevention 42.6, *fermions* make up another class of particles, those with half-integral spin.) An important feature of bosons is that they do not obey the Pauli exclusion principle. Consequently, at very low temperatures, it is possible for all bosons in a collection of such particles to be in the lowest quantum state. The entire collection of Cooper pairs in the metal is described by a single wave function. Above the energy level associated with this wave function is an energy gap equal to the binding energy of a Cooper pair. Under the action of an applied electric field, the Cooper pairs experience an electric force and move through the metal. A random scattering event of a Cooper pair from a lattice ion would represent resistance to the electric current. Such a collision would change the energy of the Cooper pair because some energy would be transferred to the lattice ion. There are no available energy levels below that of the Cooper pair (it is already in the lowest state), however, and none available above because of the energy gap. As a result, collisions do not occur and there is no resistance to the movement of Cooper pairs.

An important development in physics that elicited much excitement in the scientific community was the discovery of high-temperature copper oxide–based superconductors. The excitement began with a 1986 publication by J. Georg Bednorz (b. 1950) and K. Alex Müller (b. 1927), scientists at the IBM Zurich Research

[5] A highly simplified explanation of this attraction between electrons is as follows. The attractive Coulomb force between one electron and the surrounding positively charged lattice ions causes the ions to move inward slightly toward the electron. As a result, there is a higher concentration of positive charge in this region than elsewhere in the lattice. A second electron is attracted to the higher concentration of positive charge.

Laboratory in Switzerland. In their seminal paper,[6] Bednorz and Müller reported strong evidence for superconductivity at 30 K in an oxide of barium, lanthanum, and copper. They were awarded the Nobel Prize in Physics in 1987 for their remarkable discovery. Shortly thereafter, a new family of compounds was open for investigation and research activity in the field of superconductivity proceeded vigorously. In early 1987, groups at the University of Alabama at Huntsville and the University of Houston announced superconductivity at approximately 92 K in an oxide of yttrium, barium, and copper ($YBa_2Cu_3O_7$). Later that year, teams of scientists from Japan and the United States reported superconductivity at 105 K in an oxide of bismuth, strontium, calcium, and copper. More recently, scientists have reported superconductivity at temperatures as high as 150 K in an oxide containing mercury. Today, one cannot rule out the possibility of room-temperature superconductivity, and the mechanisms responsible for the behavior of high-temperature superconductors are still under investigation. The search for novel superconducting materials continues both for scientific reasons and because practical applications become more probable and widespread as the critical temperature is raised.

Although BCS theory was very successful in explaining superconductivity in metals, there is currently no widely accepted theory for high-temperature superconductivity. It remains an area of active research.

Summary

ThomsonNOW™ Sign in at **www.thomsonedu.com** and go to ThomsonNOW to take a practice test for this chapter.

CONCEPTS AND PRINCIPLES

Two or more atoms combine to form molecules because of a net attractive force between the atoms. The mechanisms responsible for molecular bonding can be classified as follows:

- **Ionic bonds** form primarily because of the Coulomb attraction between oppositely charged ions. Sodium chloride (NaCl) is one example.
- **Covalent bonds** form when the constituent atoms of a molecule share electrons. For example, the two electrons of the H_2 molecule are equally shared between the two nuclei.
- **Van der Waals bonds** are weak electrostatic bonds between molecules or between atoms that do not form ionic or covalent bonds. These bonds are responsible for the condensation of noble gas atoms and nonpolar molecules into the liquid phase.
- **Hydrogen bonds** form between the center of positive charge in a polar molecule that includes one or more hydrogen atoms and the center of negative charge in another polar molecule.

The allowed values of the rotational energy of a diatomic molecule are

$$E_{\text{rot}} = E_J = \frac{\hbar^2}{2I} J(J+1) \quad J = 0, 1, 2, \ldots \quad \textbf{(43.6)}$$

where I is the moment of inertia of the molecule and J is an integer called the **rotational quantum number.** The selection rule for transitions between rotational states is $\Delta J = \pm 1$.

The allowed values of the vibrational energy of a diatomic molecule are

$$E_{\text{vib}} = \left(v + \tfrac{1}{2}\right)\frac{h}{2\pi}\sqrt{\frac{k}{\mu}} \quad v = 0, 1, 2, \ldots \quad \textbf{(43.10)}$$

where v is the **vibrational quantum number,** k is the force constant of the "effective spring" bonding the molecule, and μ is the **reduced mass** of the molecule. The selection rule for allowed vibrational transitions is $\Delta v = \pm 1$, and the energy difference between any two adjacent levels is the same, regardless of which two levels are involved.

(*continued*)

[6] J. G. Bednorz and K. A. Müller, *Z. Phys. B* **64:**189, 1986.

Bonding mechanisms in solids can be classified in a manner similar to the schemes for molecules. For example, the Na^+ and Cl^- ions in NaCl form **ionic bonds,** whereas the carbon atoms in diamond form **covalent bonds.** The **metallic bond** is characterized by a net attractive force between positive ion cores and the mobile free electrons of a metal.

In the **free-electron theory of metals,** the free electrons fill the quantized levels in accordance with the Pauli exclusion principle. The number of states per unit volume available to the conduction electrons having energies between E and $E + dE$ is

$$N(E)\,dE = \left(\frac{8\sqrt{2}\pi m_e^{3/2}}{h^3}E^{1/2}\right)\left(\frac{1}{e^{(E-E_F)/k_BT}+1}\right)dE \qquad \textbf{(43.22)}$$

where E_F is the **Fermi energy.** At $T = 0$ K, all levels below E_F are filled, all levels above E_F are empty and

$$E_F(0) = \frac{h^2}{2m_e}\left(\frac{3n_e}{8\pi}\right)^{2/3} \qquad \textbf{(43.25)}$$

where n_e is the total number of conduction electrons per unit volume. Only those electrons having energies near E_F can contribute to the electrical conductivity of the metal.

In a crystalline solid, the energy levels of the system form a set of **bands.** Electrons occupy the lowest energy states, with no more than one electron per state. Energy gaps are present between the bands of allowed states.

A **semiconductor** is a material having an energy gap of approximately 1 eV and a valence band that is filled at $T = 0$ K. Because of the small energy gap, a significant number of electrons can be thermally excited from the valence band into the conduction band. The band structures and electrical properties of a Group IV semiconductor can be modified by the addition of either donor atoms containing five outer-shell electrons or acceptor atoms containing three outer-shell electrons. A semiconductor **doped** with donor impurity atoms is called an ***n*-type semiconductor,** and one doped with acceptor impurity atoms is called a ***p*-type semiconductor.**

Questions

☐ denotes answer available in *Student Solutions Manual/Study Guide;* **O** denotes objective question

Note: Questions 6 and 8 in Chapter 27 can be assigned with this chapter.

1. Discuss models for the different types of bonds that form stable molecules.
2. Discuss the three major forms of excitation of a molecule (other than translational motion) and the relative energies associated with these three forms.
3. **O** Consider a typical material composed of covalently bonded diatomic molecules. Rank the following energies from the largest in magnitude to the smallest in magnitude. (a) the latent heat of fusion per molecule (b) the molecular binding energy (c) the energy of the first excited state of molecular rotation (d) the energy of the first excited state of molecular vibration
4. How can the analysis of the rotational spectrum of a molecule lead to an estimate of the size of that molecule?
5. **O** An infrared absorption spectrum like that in Figure 43.9 is taken when the sample of HCl is at a much higher temperature. Compared with Figure 43.9, in this new spectrum are the highest absorption peaks (a) at the same frequencies, (b) farther from the gap, or (c) closer to the gap?
6. **O** What kind of bonding likely holds the atoms together in the following solids **(i), (ii),** and **(iii)**? Choose your answers from these possibilities: (a) ionic bonding, (b) covalent bonding, and (c) metallic bonding. **(i)** The crystal is transparent, brittle, and soluble in water. It is a poor conductor of electricity. **(ii)** The crystal is opaque, brittle, very hard, and a good electric insulator. **(iii)** The solid is opaque, shiny, flexible, and a good electric conductor.
7. **O** The Fermi energy for silver is 5.48 eV. In a piece of solid silver, free-electron energy levels are measured near 2 eV and near 6 eV. **(i)** Near which of these energies are the energy levels closer together? (a) 2 eV (b) 6 eV (c) The spacing is the same. **(ii)** Near which of these energies are more electrons occupying energy levels? (a) 2 eV (b) 6 eV (c) The number of electrons is the same.
8. Discuss the differences in the band structures of metals, insulators, and semiconductors. How does the band-structure model enable you to understand the electrical properties of these materials better?
9. **O** As discussed in Chapter 27, the conductivity of metals decreases with increasing temperature due to electron collisions with vibrating atoms. In contrast, the conductivity of semiconductors increases with increasing tempera-

ture. What property of a semiconductor is responsible for this behavior? (a) Atomic vibrations decrease as temperature increases. (b) The number of conduction electrons and the number of holes increase steeply with increasing temperature. (c) The energy gap decreases with increasing temperature. (d) Electrons do not collide with atoms in a semiconductor.

10. When a photon is absorbed by a semiconductor, an electron–hole pair is created. Give a physical explanation of this statement using the energy-band model as the basis for your description.

11. Pentavalent atoms such as arsenic are donor atoms in a semiconductor such as silicon, whereas trivalent atoms such as indium are acceptors. Inspect the periodic table in Appendix C and determine what other elements might make good donors or acceptors.

12. What essential assumptions are made in the free-electron theory of metals? How does the energy-band model differ from the free-electron theory in describing the properties of metals?

13. How do the vibrational and rotational levels of heavy hydrogen (D_2) molecules compare with those of H_2 molecules?

14. The energies of photons of visible light range between 1.8 and 3.1 eV. Does that explain why silicon, with an energy gap of 1.1 eV (see Table 43.3), appears opaque, whereas diamond, with an energy gap of 5.5 eV, appears transparent?

15. **O** **(i)** Should you expect an *n*-type doped semiconductor to have (a) higher, (b) lower, or (c) the same conductivity as an intrinsic (pure) semiconductor? **(ii)** Should you expect a *p*-type doped semiconductor to have (a) higher, (b) lower, or (c) the same conductivity as an intrinsic (pure) semiconductor?

16. **O** Is each one of the following statements true or false for a superconductor below its critical temperature? (a) It can carry infinite current. (b) It must carry some nonzero current. (c) Its interior electric field must be zero. (d) Its internal magnetic field must be zero. (e) No internal energy appears when it carries electric current.

Problems

WebAssign The Problems from this chapter may be assigned online in WebAssign.

ThomsonNOW Sign in at **www.thomsonedu.com** and go to ThomsonNOW to assess your understanding of this chapter's topics with additional quizzing and conceptual questions.

1, 2, 3 denotes straightforward, intermediate, challenging; □ denotes full solution available in *Student Solutions Manual/Study Guide;* ▲ denotes coached solution with hints available at **www.thomsonedu.com;** denotes developing symbolic reasoning; ● denotes asking for qualitative reasoning; denotes computer useful in solving problem

Section 43.1 Molecular Bonds

1. ▲ **Review problem.** A K^+ ion and a Cl^- ion are separated by a distance of 5.00×10^{-10} m. Assuming the two ions act like charged particles, determine (a) the force each ion exerts on the other and (b) the potential energy of the two-ion system in electron volts.

2. Potassium chloride is an ionically bonded molecule, sold as a salt substitute for use in a low-sodium diet. The electron affinity of chlorine is 3.6 eV. An energy input of 0.7 eV is required to form separate K^+ and Cl^- ions from separate K and Cl atoms. What is the ionization energy of K?

3. One description of the potential energy of a diatomic molecule is given by the Lennard–Jones potential,

$$U = \frac{A}{r^{12}} - \frac{B}{r^6}$$

where A and B are constants. Find, in terms of A and B, (a) the value r_0 at which the energy is a minimum and (b) the energy E required to break up a diatomic molecule. (c) Evaluate r_0 in meters and E in electron volts for the H_2 molecule, taking $A = 0.124 \times 10^{-120}$ eV · m^{12} and $B = 1.488 \times 10^{-60}$ eV · m^6. *Note:* Although this potential is widely used for modeling, it is known to have serious defects. For example, its behavior at both small and large values of r is significantly in error.

4. In the potassium iodide molecule, assume the K and I atoms bond ionically by the transfer of one electron from K to I. (a) The ionization energy of K is 4.34 eV, and the electron affinity of I is 3.06 eV. What energy is needed to transfer an electron from K to I, to form K^+ and I^- ions from neutral atoms? This quantity is sometimes called the activation energy E_a. (b) A model potential energy function for the KI molecule is the Lennard–Jones potential:

$$U(r) = 4\epsilon\left[\left(\frac{\sigma}{r}\right)^{12} - \left(\frac{\sigma}{r}\right)^{6}\right] + E_a$$

where r is the internuclear separation distance and σ and ϵ are adjustable parameters. The E_a term is added to ensure the correct asymptotic behavior at large r. At the equilibrium separation distance, $r = r_0 = 0.305$ nm, $U(r)$ is a minimum, and $dU/dr = 0$. Now $U(r_0)$ is the negative of the dissociation energy: $U(r_0) = -3.37$ eV. Evaluate σ and ϵ. (c) Calculate the force needed to break up a KI molecule. (d) Calculate the force constant for small oscillations about $r = r_0$. *Suggestion:* Set $r = r_0 + s$, where $s/r_0 \ll 1$, and expand $U(r)$ in powers of s/r_0 up to second-order terms.

5. A van der Waals dispersion force between helium atoms produces a very shallow potential well, with a depth on the order of 1 meV. At approximately what temperature would you expect helium to condense?

Section 43.2 Energy States and Spectra of Molecules

6. Assume the distance between the protons in the H_2 molecule is 0.750×10^{-10} m. (a) Find the energy of the first rotational state, with $J = 1$. (b) Find the wavelength of radiation emitted in the transition from $J = 1$ to $J = 0$.

7. The cesium iodide (CsI) molecule has an atomic separation of 0.127 nm. (a) Determine the energy of the lowest excited rotational state and the frequency of the photon absorbed in the $J = 0$ to $J = 1$ transition. (b) **What If?** What would be the fractional change in this frequency if the estimate of the atomic separation is off by 10%?

8. ● The photon frequency that would be absorbed by the NO molecule in a transition from vibration state $v = 0$ to $v = 1$, with no change in rotation state, is 56.3 THz. The bond between the atoms has an effective spring constant of 1 530 N/m. Use this information to calculate the reduced mass of the NO molecule. Then compute a value for μ using Equation 43.4. State how the two results compare.

9. ▲ An HCl molecule is excited to its first rotational energy level, corresponding to $J = 1$. If the distance between its nuclei is 0.127 5 nm, what is the angular speed of the molecule about its center of mass?

10. The CO molecule makes a transition from the $J = 1$ to the $J = 2$ rotational state when it absorbs a photon of frequency 2.30×10^{11} Hz. Find the moment of inertia of this molecule from these data.

11. A diatomic molecule consists of two atoms having masses m_1 and m_2 and separated by a distance r. Show that the moment of inertia about an axis through the center of mass of the molecule is given by Equation 43.3, $I = \mu r^2$.

12. (a) Calculate the moment of inertia of an NaCl molecule about its center of mass. The atoms are separated by a distance $r = 0.28$ nm. (b) Calculate the wavelength of radiation emitted when an NaCl molecule undergoes a transition from the $J = 2$ state to the $J = 1$ state.

13. The rotational spectrum of the HCl molecule contains lines with wavelengths of 0.060 4, 0.069 0, 0.080 4, 0.096 4, and 0.120 4 mm. What is the moment of inertia of the molecule?

14. ● The effective spring constant describing the potential energy of the HI molecule is 320 N/m. That for the HF molecule is 970 N/m. Calculate the minimum amplitude of vibration for (a) the HI molecule and (b) the HF molecule. (c) Which molecule has the weaker bond? Explain how you can tell.

15. ▲ Taking the effective force constant of a vibrating HCl molecule as $k = 480$ N/m, find the energy difference between the ground state and the first excited vibrational level.

16. The nuclei of the O_2 molecule are separated by 1.20×10^{-10} m. The mass of each oxygen atom in the molecule is 2.66×10^{-26} kg. (a) Determine the rotational energies of an oxygen molecule in electron volts for the levels corresponding to $J = 0$, 1, and 2. (b) The effective force constant k between the atoms in the oxygen molecule is 1 177 N/m. Determine the vibrational energies (in electron volts) corresponding to $v = 0$, 1, and 2.

17. Figure P43.17 is a model of a benzene molecule. All atoms lie in a plane, and the carbon atoms form a regular hexagon, as do the hydrogen atoms. The carbon atoms are 0.110 nm apart center to center. Determine the allowed energies of rotation about an axis perpendicular to the plane of the paper through the center point O. Hydrogen and carbon atoms have masses of 1.67×10^{-27} kg and 1.99×10^{-26} kg, respectively.

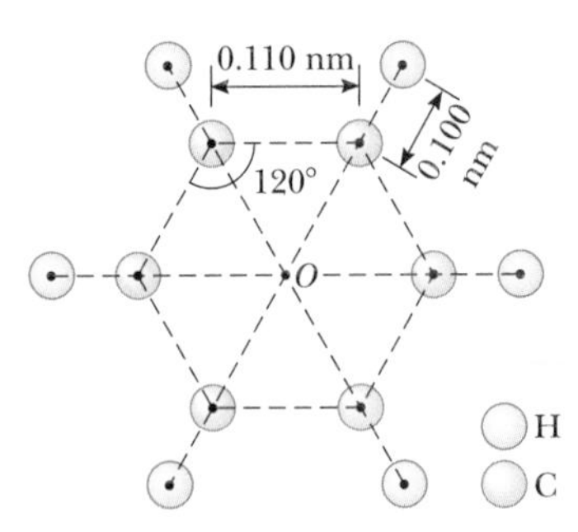

Figure P43.17

18. (a) Calculate the longest wavelength in the rotational spectrum of HCl. Take the Cl atom to be the isotope ^{35}Cl. The equilibrium separation of the H and Cl atoms is 0.127 46 nm. The atomic mass of the H atom is 1.007 825 u, and that of the ^{35}Cl atom is 34.968 853 u. (b) **What If?** Repeat the calculation in part (a), but take the Cl atom to be the isotope ^{37}Cl, which has atomic mass 36.965 903 u. The equilibrium separation distance is the same as in part (a). (c) Naturally occurring chlorine contains approximately three parts of ^{35}Cl to one part of ^{37}Cl. Because of the two different Cl masses, each line in the microwave rotational spectrum of HCl is split into a doublet. Calculate the doublet separation for the longest wavelength. (Figure 43.9 shows the doublets in the infrared vibrational spectrum).

19. Calculate the moment of inertia of an HCl molecule from its infrared absorption spectrum shown in Figure 43.9.

20. An H_2 molecule is in its vibrational and rotational ground states. It absorbs a photon of wavelength 2.211 2 μm and jumps to the $v = 1$, $J = 1$ energy level. It then drops to the $v = 0$, $J = 2$ energy level while emitting a photon of wavelength 2.405 4 μm. Calculate (a) the moment of inertia of the H_2 molecule about an axis through its center of mass and perpendicular to the H–H bond, (b) the vibrational frequency of the H_2 molecule, and (c) the equilibrium separation distance for this molecule.

21. Photons of what frequencies can be spontaneously emitted by CO molecules in the state with $v = 1$ and $J = 0$?

22. Most of the mass of an atom is in its nucleus. Model the mass distribution in a diatomic molecule as two spheres, each of radius 2.00×10^{-15} m and mass 1.00×10^{-26} kg, located at points along the x axis in Active Figure 43.5a, and separated by 2.00×10^{-10} m. Rotation about the axis joining the nuclei in the diatomic molecule is ordinarily ignored because the first excited state would have an energy that is too high to access. To see why, calculate the ratio of the energy of the first excited state for rotation about the x axis to the energy of the first excited state for rotation about the y axis.

Section 43.3 Bonding in Solids

23. Use Equation 43.18 to calculate the ionic cohesive energy for NaCl. Take $\alpha = 1.747\ 6$, $r_0 = 0.281$ nm, and $m = 8$.

24. Use a magnifying glass to look at the table salt that comes out of a salt shaker. Compare what you see with Figure 43.10a. The distance between a sodium ion and a nearest-neighbor chlorine ion is 0.261 nm. (a) Make an order-of-

magnitude estimate of the number N of atoms in a typical grain of salt. (b) **What If?** Suppose you had a number of grains of salt equal to this number N. What would be the volume of this quantity of salt?

25. Consider a one-dimensional chain of alternating positive and negative ions. Show that the potential energy associated with one of the ions and its interactions with the rest of this hypothetical crystal is

$$U(r) = -k_e \alpha \frac{e^2}{r}$$

where the Madelung constant is $\alpha = 2 \ln 2$ and r is the interionic spacing. *Suggestion:* Use the series expansion for $\ln(1 + x)$.

26. The distance between the K^+ and Cl^- ions in a KCl crystal is 0.314 nm. Calculate the distances from one K^+ ion to its nearest-neighbor K^+ ions, to its second-nearest-neighbor K^+ ions, and to its third-nearest-neighbor K^+ ions.

Section 43.4 Free-Electron Theory of Metals

Section 43.5 Band Theory of Solids

27. The Fermi energy for silver is 5.48 eV. Silver has a density of 10.6×10^3 kg/m^3 and an atomic mass of 108 u. Use this information to show that silver has one free electron per atom.

28. ● (a) State what the Fermi energy depends on according to the free-electron theory of metals and how the Fermi energy depends on that quantity. (b) Show that Equation 43.25 can be expressed as $E_F = (3.65 \times 10^{-19}) n_e^{2/3}$ eV, where E_F is in electron volts when n_e is in electrons per cubic meter. (c) According to Table 43.2, by what factor does the free-electron concentration in copper exceed that in potassium? Which of these metals has the larger Fermi energy? By what factor is the Fermi energy larger? Explain whether this behavior is predicted by Equation 43.25.

29. ● (a) Find the typical speed of a conduction electron in copper, taking its kinetic energy as equal to the Fermi energy, 7.05 eV. (b) State how this speed compares with a drift speed of 0.1 mm/s.

30. For copper at 300 K, calculate the probability that a state with an energy equal to 99.0% of the Fermi energy is occupied.

31. The Fermi energy of copper at 300 K is 7.05 eV. (a) What is the average energy of a conduction electron in copper at 300 K? (b) At what temperature would the average energy of a molecule in an ideal gas be equal to the energy calculated in part (a)?

32. Sodium is a monovalent metal having a density of 0.971 g/cm^3 and a molar mass of 23.0 g/mol. Use this information to calculate (a) the density of charge carriers and (b) the Fermi energy of sodium.

33. ▲ Calculate the energy of a conduction electron in silver at 800 K, assuming the probability of finding an electron in that state is 0.950. The Fermi energy of silver is 5.48 eV at this temperature.

34. When solid silver starts to melt, what is the approximate fraction of the conduction electrons that are thermally excited above the Fermi level?

35. Show that the average kinetic energy of a conduction electron in a metal at 0 K is $E_{avg} = \frac{3}{5} E_F$. *Suggestion:* In general, the average kinetic energy is

$$E_{avg} = \frac{1}{n_e} \int EN(E)\, dE$$

where n_e is the density of particles, $N(E)\, dE$ is given by Equation 43.22, and the integral is over all possible values of the energy.

36. Consider a cube of gold 1.00 mm on an edge. Calculate the approximate number of conduction electrons in this cube whose energies lie in the range 4.000 to 4.025 eV.

37. ● (a) Consider a system of electrons confined to a three-dimensional box. Calculate the ratio of the number of allowed energy levels at 8.50 eV to the number at 7.00 eV. (b) **What If?** Copper has a Fermi energy of 7.0 eV at 300 K. Calculate the ratio of the number of occupied levels in copper at an energy of 8.50 eV to the number at the Fermi energy. State how your answer compares with that obtained in part (a).

38. **Review problem.** An electron moves in a three-dimensional box of edge length L and volume L^3. The wave function of the particle is $\psi = A \sin(k_x x) \sin(k_y y) \sin(k_z z)$. Show that its energy is given by Equation 43.20,

$$E = \frac{\hbar^2 \pi^2}{2m_e L^2} (n_x^2 + n_y^2 + n_z^2)$$

where the quantum numbers (n_x, n_y, n_z) are integers ≥ 1. *Suggestions:* The Schrödinger equation in three dimensions may be written

$$\frac{\hbar^2}{2m}\left(\frac{\partial^2 \psi}{\partial x^2} + \frac{\partial^2 \psi}{\partial y^2} + \frac{\partial^2 \psi}{\partial z^2}\right) = (U - E)\psi$$

To confine the electron inside the box, take $U = 0$ inside and $U = \infty$ outside.

Section 43.6 Electrical Conduction in Metals, Insulators, and Semiconductors

39. The energy gap for silicon at 300 K is 1.14 eV. (a) Find the lowest-frequency photon that can promote an electron from the valence band to the conduction band. (b) What is the wavelength of this photon?

40. Light from a hydrogen discharge tube is incident on a CdS crystal. Which spectral lines from the Balmer series are absorbed and which are transmitted?

41. ▲ ● Most solar radiation has a wavelength of 1 μm or less. What energy gap should the material in a solar cell have if it is to absorb this radiation? Is silicon appropriate (see Table 43.3)? Explain your answer.

42. A light-emitting diode (LED) made of the semiconductor GaAsP emits red light ($\lambda = 650$ nm). Determine the energy-band gap E_g in the semiconductor.

43. You are asked to build a scientific instrument that is thermally isolated from its surroundings. You wish to use an external laser to raise the temperature of a target inside the instrument. (It might be a calorimeter, but these design criteria could apply to other devices as well.) Because you know that diamond is transparent and a

good thermal insulator, you decide to use a diamond window in the apparatus. Diamond has an energy gap of 5.5 eV between its valence and conduction bands. What is the shortest laser wavelength you can use to warm the sample inside?

44. **Review problem.** Silicon is a semiconductor widely used in computer chips and other electronic devices. Its most important properties result from doping it with impurities so as to control its electrical conductivity. Phosphorus, which is adjacent to silicon in the periodic table, has five outer valence electrons, compared with four for silicon. When a phosphorus atom is substituted for a silicon atom in a crystal, four of the phosphorus valence electrons form bonds with neighboring atoms and the remaining electron is much more loosely bound. You can model the electron as free to move through the crystal lattice. The phosphorus nucleus has one more positive charge than does the silicon nucleus, however, so the extra electron provided by the phosphorus atom is attracted to this single nuclear charge $+e$. The energy levels of the extra electron are similar to those of the electron in the Bohr hydrogen atom with two important exceptions. First, the Coulomb attraction between the electron and the positive charge on the phosphorus nucleus is reduced by a factor of $1/\kappa$ from what it would be in free space (see Eq. 26.21), so the orbit radii are greatly increased. Here κ is the dielectric constant of the crystal, with a value of 11.7 in silicon. Second, the influence of the periodic electric potential of the lattice causes the electron to move as if it had an effective mass m^*, which is quite different from the mass m_e of a free electron. You can use the Bohr model of hydrogen to obtain fairly accurate values for the allowed energy levels of the extra electron. These energy levels, called donor states, play an important role in semiconductor devices. Assume $m^* = 0.220m_e$. Calculate the energy and the radius for an extra electron in the first Bohr orbit around a donor atom in silicon.

Section 43.7 Semiconductor Devices

45. (a) For what value of the bias voltage ΔV in Equation 43.27 does $I = 9.00I_0$? (b) **What If?** What if $I = -0.900I_0$? Assume $T = 300$ K.

46. The diode shown in Figure 43.28 is connected in series with a battery and a 150-Ω resistor. What battery emf is required for a current of 25.0 mA?

47. You put a diode in a microelectronic circuit to protect the system in case an untrained person installs the battery backward. In the correct forward-bias situation, the current is 200 mA with a potential difference of 100 mV across the diode at room temperature (300 K). If the battery were reversed, what would the magnitude of the current through the diode be?

48. ● A diode, a resistor, and a battery are connected in a series circuit. The diode is at a temperature for which $k_BT = 25.0$ meV, and the saturation value I_0 of the current is 1.00 μA. The resistance of the resistor is 745 Ω, and the battery maintains a constant potential difference between its terminals of 2.42 V. (a) Find graphically the current in the loop. Proceed as follows. On the same axes, draw graphs of the diode current I_D and the current in the wire I_W versus the voltage across the diode ΔV. Choose values of ΔV ranging from 0 to 0.250 V in steps of 0.025 V. Determine the value of ΔV at the intersection of the two graph lines and calculate the corresponding currents I_D and I_W. Explain whether they agree. (b) Find the ohmic resistance of the diode, defined as the ratio $\Delta V/I_D$. (c) Find the dynamic resistance of the diode, which is defined as the derivative $d(\Delta V)/dI_D$.

Section 43.8 Superconductivity

Note: Problem 30 in Chapter 30 and Problems 64 through 67 in Chapter 32 can also be assigned with this section.

49. Determine the current generated in a superconducting ring of niobium metal 2.00 cm in diameter when a 0.020 0-T magnetic field directed perpendicular to the ring is suddenly decreased to zero. The inductance of the ring is 3.10×10^{-8} H.

50. ● *A convincing demonstration of zero resistance.* A direct and relatively simple demonstration of zero DC resistance can be carried out using the four-point probe method. The probe shown in Figure P43.50 consists of a disk of $YBa_2Cu_3O_7$ (a high-T_c superconductor) to which four wires are attached by indium solder or some other suitable contact material. Current is maintained through the sample by applying a DC voltage between points a and b, and it is measured with a DC ammeter. The current can be varied with the variable resistance R. The potential difference ΔV_{cd} between c and d is measured with a digital voltmeter. When the probe is immersed in liquid nitrogen, the sample quickly cools to 77 K, below the critical temperature of the material, 92 K. The current remains approximately constant, but *ΔV_{cd} drops abruptly to zero.* (a) Explain this observation on the basis of what you know about superconductors. (b) The data in the accompanying table represent actual values of ΔV_{cd} for different values of I taken on the sample at room temperature in the senior author's laboratory. A 6-V battery in series with a variable resistor R supplied the current. The values of R ranged from 10 Ω to 100 Ω. Make an I–ΔV plot of the data and determine whether the sample behaves in a linear manner. From the data, obtain a value for the DC resistance of the sample at room temperature. (c) At room temperature, it was found that $\Delta V_{cd} = 2.234$ mV for $I = 100.3$ mA, but after the sample was cooled to 77 K, $\Delta V_{cd} = 0$ and $I = 98.1$ mA. What do you think might have caused the slight decrease in current?

Current Versus Potential Difference ΔV_{cd} Measured in a Bulk Ceramic Sample of $YBa_2Cu_3O_{7-\delta}$ at Room Temperature

I (mA)	ΔV_{cd} (mV)
57.8	1.356
61.5	1.441
68.3	1.602
76.8	1.802
87.5	2.053
102.2	2.398
123.7	2.904
155	3.61

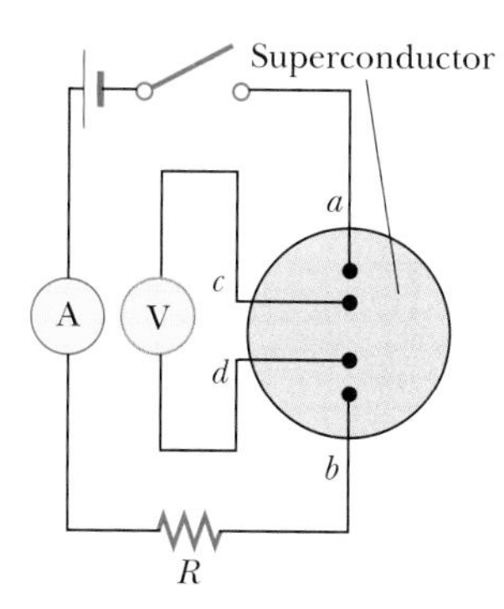

Figure P43.50

51. A thin rod of superconducting material 2.50 cm long is placed into a 0.540-T magnetic field with its cylindrical axis along the magnetic field lines. (a) Sketch the directions of the applied field and the induced surface current. (b) Find the magnitude of the surface current on the curved surface of the rod.

Additional Problems

52. The effective spring constant associated with bonding in the N_2 molecule is 2 297 N/m. The nitrogen atoms each have a mass of 2.32×10^{-26} kg, and their nuclei are 0.120 nm apart. Assume the molecule is rigid and in the ground vibrational state. Calculate the J value of the rotational state that has the same energy as the first excited vibrational state.

53. As you will learn in Chapter 44, carbon-14 (^{14}C) is an isotope of carbon. It has the same chemical properties and electronic structure as the much more abundant isotope carbon-12 (^{12}C), but it has different nuclear properties. Its mass is 14 u, greater than that of carbon-12 because of the two extra neutrons in the carbon-14 nucleus. Assume the CO molecular potential energy is the same for both isotopes of carbon and the examples in Section 43.2 contain accurate data and results for carbon monoxide with carbon-12 atoms. (a) What is the vibrational frequency of ^{14}CO? (b) What is the moment of inertia of ^{14}CO? (c) What wavelengths of light can be absorbed by ^{14}CO in the ($v = 0, J = 10$) state that cause it to end up in the $v = 1$ level?

54. Under pressure, liquid helium can solidify as each atom bonds with four others, and each bond has an average energy of 1.74×10^{-23} J. Find the latent heat of fusion for helium in joules per gram. (The molar mass of He is 4.00 g/mol.)

55. The hydrogen molecule comes apart (dissociates) when it is excited internally by 4.5 eV. Assuming this molecule behaves like a harmonic oscillator having classical angular frequency $\omega = 8.28 \times 10^{14}$ rad/s, find the highest vibrational quantum number for a state below the 4.5-eV dissociation energy.

56. The dissociation energy of ground-state molecular hydrogen is 4.48 eV, but it only takes 3.96 eV to dissociate it when it starts in the first excited vibrational state with $J = 0$. Using this information, determine the depth of the H_2 molecular potential-energy function.

57. Show that the ionic cohesive energy of an ionically bonded solid is given by Equation 43.18. *Suggestion:* Start with Equation 43.17 and note that $dU/dr = 0$ at $r = r_0$.

58. (a) Show that the force exerted on an ion in an ionic solid can by written as

$$F = -k_e \alpha \frac{e^2}{r^2}\left[1 - \left(\frac{r_0}{r}\right)^{m-1}\right]$$

where α is the Madelung constant and r_0 is the equilibrium separation. *Suggestion:* Start with Equation 43.17 and note that $F = -dU/dr = 0$ at $r = r_0$. (b) Imagine that an ion in the solid is displaced a small distance x from r_0. Show that the ion experiences a restoring force $F = -Kx$, where

$$K = \frac{k_e \alpha e^2}{r_0^{\,3}}(m - 1)$$

(c) Use the result of part (b) to find the frequency of vibration of a Na^+ ion in NaCl. Take $m = 8$ and use the value $\alpha = 1.747\ 6$.

59. A particle moves in one-dimensional motion through a field for which the potential energy of the particle–field system is

$$U(x) = \frac{A}{x^3} - \frac{B}{x}$$

where $A = 0.150 \text{ eV} \cdot \text{nm}^3$ and $B = 3.68 \text{ eV} \cdot \text{nm}$. The general shape of this function is shown in Figure 43.11, where x replaces r. (a) Find the static equilibrium position x_0 of the particle. (b) Determine the depth U_0 of this potential well. (c) In moving along the x axis, what maximum force toward the negative x direction does the particle experience?

60. A particle of mass m moves in one-dimensional motion through a field for which the potential energy of the particle–field system is

$$U(x) = \frac{A}{x^3} - \frac{B}{x}$$

where A and B are constants with appropriate units. The general shape of this function is shown in Figure 43.11, where x replaces r. (a) Find the static equilibrium position x_0 of the particle in terms of m, A, and B. (b) Determine the depth U_0 of this potential well. (c) In moving along the x axis, what maximum force toward the negative x direction does the particle experience?

61. As an alternative to Equation 43.1, another useful model for the potential energy of a diatomic molecule is the Morse potential

$$U(r) = B[e^{-a(r-r_0)} - 1]^2$$

where B, a, and r_0 are parameters used to adjust the shape of the potential and its depth. (a) What is the equilibrium separation of the nuclei? (b) What is the depth of the potential well, defined as the difference in energy between the potential's minimum value and its asymptote as r approaches infinity? (c) If μ is the reduced mass of the system of two nuclei, what is the vibrational frequency of the diatomic molecule in its ground state? (Assume the potential is nearly parabolic about the well minimum). (d) What amount of energy needs to be supplied to the ground-state molecule to separate the two nuclei to infinity?

62. The Fermi–Dirac distribution function can be written as

$$f(E) = \frac{1}{e^{(E-E_F)/k_BT} + 1} = \frac{1}{e^{(E/E_F-1)T_F/T} + 1}$$

where T_F is the *Fermi temperature*, defined according to

$$k_B T_F \equiv E_F$$

Write a spreadsheet to calculate and plot $f(E)$ versus E/E_F at a fixed temperature T. Examine the curves obtained for $T = 0.1T_F$, $0.2T_F$, and $0.5T_F$.

Answers to Quick Quizzes

43.1 (a) van der Waals (b) ionic (c) hydrogen (d) covalent

43.2 (c). Equation 43.7 shows that the energy spacing between adjacent states is proportional to J, the quantum number of the higher state in the transition. Because the frequency of the absorbed photon is proportional to the energy separation of the states, the frequencies are in the same ratio as the energy separations.

43.3 (a). This situation is similar to Quick Quiz 43.2 except that the vibrational states are all separated by the same energy difference.

43.4 A: semiconductor; B: conductor; C: insulator

The Ice Man, discovered in 1991 when an Italian glacier melted enough to expose his remains. His possessions, particularly his tools, have shed light on the way people lived in the Bronze Age. Radioactivity was used to determine how long ago this person lived. (Paul Hanny/Gamma Liaison)

44 Nuclear Structure

The year 1896 marks the birth of nuclear physics when French physicist Antoine-Henri Becquerel (1852–1908) discovered radioactivity in uranium compounds. This discovery prompted scientists to investigate the details of radioactivity and, ultimately, the structure of the nucleus. Pioneering work by Ernest Rutherford showed that the radiation emitted from radioactive substances is of three types: alpha, beta, and gamma rays, classified according to the nature of their electric charge and their ability to penetrate matter and ionize air. Later experiments showed that alpha rays are helium nuclei, beta rays are electrons, and gamma rays are high-energy photons.

In 1911, Rutherford, Hans Geiger, and Ernest Marsden performed the alpha-particle scattering experiments described in Section 42.2. These experiments established that the nucleus of an atom can be modeled as a point mass and point charge and that most of the atomic mass is contained in the nucleus. Subsequent studies revealed the presence of a new type of force, the short-range nuclear force, which is predominant at particle separation distances less than approximately 10^{-14} m and is zero for large distances.

In this chapter, we discuss the properties and structure of the atomic nucleus. We start by describing the basic properties of nuclei, followed by a discussion of nuclear forces and binding energy, nuclear models, and the phenomenon of radioactivity. Finally, we explore nuclear reactions and the various processes by which nuclei decay.

44.1 Some Properties of Nuclei

All nuclei are composed of two types of particles: protons and neutrons. The only exception is the ordinary hydrogen nucleus, which is a single proton. We describe the atomic nucleus by the number of protons and neutrons it contains, using the following quantities:

- the **atomic number** Z, which equals the number of protons in the nucleus (sometimes called the *charge number*)
- the **neutron number** N, which equals the number of neutrons in the nucleus
- the **mass number** $A = Z + N$, which equals the number of **nucleons** (neutrons plus protons) in the nucleus

PITFALL PREVENTION 44.1
Mass Number Is Not Atomic Mass

The mass number A should not be confused with the atomic mass. Mass number is an integer specific to an isotope and has no units; it is simply a count of the number of nucleons. Atomic mass has units and is generally not an integer because it is an average of the masses of a given element's naturally occurring isotopes.

In representing nuclei, it is convenient to use the symbol $^A_Z\mathrm{X}$ to show how many protons and neutrons are present, where X represents the chemical symbol of the element. For example, $^{56}_{26}\mathrm{Fe}$ (iron) has mass number 56 and atomic number 26; therefore, it contains 26 protons and 30 neutrons. When no confusion is likely to arise, we omit the subscript Z because the chemical symbol can always be used to determine Z. Therefore, $^{56}_{26}\mathrm{Fe}$ is the same as $^{56}\mathrm{Fe}$ and can also be expressed as "iron-56."

The nuclei of all atoms of a particular element contain the same number of protons but often contain different numbers of neutrons. Nuclei related in this way are called **isotopes.** The isotopes of an element have the same Z value but different N and A values.

The natural abundance of isotopes can differ substantially. For example $^{11}_{6}\mathrm{C}$, $^{12}_{6}\mathrm{C}$, $^{13}_{6}\mathrm{C}$, and $^{14}_{6}\mathrm{C}$ are four isotopes of carbon. The natural abundance of the $^{12}_{6}\mathrm{C}$ isotope is approximately 98.9%, whereas that of the $^{13}_{6}\mathrm{C}$ isotope is only about 1.1%. Some isotopes, such as $^{11}_{6}\mathrm{C}$ and $^{14}_{6}\mathrm{C}$, do not occur naturally but can be produced by nuclear reactions in the laboratory or by cosmic rays.

Even the simplest element, hydrogen, has isotopes: $^1_1\mathrm{H}$, the ordinary hydrogen nucleus; $^2_1\mathrm{H}$, deuterium; and $^3_1\mathrm{H}$, tritium.

Quick Quiz 44.1 For each part of this Quick Quiz, choose from the following answers: (a) protons (b) neutrons (c) nucleons. **(i)** The three nuclei $^{12}\mathrm{C}$, $^{13}\mathrm{N}$, and $^{14}\mathrm{O}$ have the same number of what type of particle? **(ii)** The three nuclei $^{12}\mathrm{N}$, $^{13}\mathrm{N}$, and $^{14}\mathrm{N}$ have the same number of what type of particle? **(iii)** The three nuclei $^{14}\mathrm{C}$, $^{14}\mathrm{N}$, and $^{14}\mathrm{O}$ have the same number of what type of particle?

Charge and Mass

The proton carries a single positive charge e, equal in magnitude to the charge $-e$ on the electron ($e = 1.6 \times 10^{-19}$ C). The neutron is electrically neutral as its name implies. Because the neutron has no charge, it was difficult to detect with early experimental apparatus and techniques. Today, neutrons are easily detected with devices such as plastic scintillators.

Nuclear masses can be measured with great precision using a mass spectrometer (see Section 29.3) and by the analysis of nuclear reactions. The proton is approximately 1 836 times as massive as the electron, and the masses of the proton and the neutron are almost equal. The **atomic mass unit** u is defined in such a way that

TABLE 44.1

Masses of Selected Particles in Various Units

Particle	Mass kg	u	MeV/c^2
Proton	$1.672\,62 \times 10^{-27}$	1.007 276	938.28
Neutron	$1.674\,93 \times 10^{-27}$	1.008 665	939.57
Electron	$9.109\,39 \times 10^{-31}$	$5.485\,79 \times 10^{-4}$	0.510 999
${}^{1}_{1}$H atom	$1.673\,53 \times 10^{-27}$	1.007 825	938.783
${}^{4}_{2}$He nucleus	$6.644\,66 \times 10^{-27}$	4.001 506	3 727.38
${}^{12}_{6}$C atom	$1.992\,65 \times 10^{-27}$	12.000 000	11 177.9

the mass of one atom of the isotope ^{12}C is exactly 12 u, where 1 u is equal to $1.660\,539 \times 10^{-27}$ kg. According to this definition, the proton and neutron each have a mass of approximately 1 u and the electron has a mass that is only a small fraction of this value. The masses of these particles and others important to the phenomena discussed in this chapter are given in Table 44.1.

You might wonder how six protons and six neutrons, each having a mass larger than 1 u, can be combined with six electrons to form a carbon-12 atom having a mass of exactly 12 u. The bound system of ^{12}C has a lower rest energy (Section 39.8) than that of six separate protons and six separate neutrons. According to Equation 39.24, $E_R = mc^2$, this lower rest energy corresponds to a smaller mass for the bound system. The difference in mass accounts for the binding energy when the particles are combined to form the nucleus. We shall discuss this point in more detail in Section 44.2.

It is often convenient to express the atomic mass unit in terms of its *rest-energy equivalent*. For one atomic mass unit,

$$E_R = mc^2 = (1.660\,539 \times 10^{-27}\ \text{kg})(2.997\,92 \times 10^8\ \text{m/s})^2 = 931.494\ \text{MeV}$$

where we have used the conversion $1\ \text{eV} = 1.602\,177 \times 10^{-19}$ J.

Based on the rest-energy expression in Equation 39.24, nuclear physicists often express mass in terms of the unit MeV/c^2.

EXAMPLE 44.1 The Atomic Mass Unit

Use Avogadro's number to show that $1\ \text{u} = 1.66 \times 10^{-27}$ kg.

SOLUTION

Conceptualize From the definition of the mole given in Section 19.5, we know that exactly 12 g (= 1 mol) of ^{12}C contains Avogadro's number of atoms.

Categorize We evaluate the atomic mass unit that was introduced in this section, so we categorize this example as a substitution problem.

Find the mass m of one ^{12}C atom:

$$m = \frac{0.012\ \text{kg}}{6.02 \times 10^{23}\ \text{atoms}} = 1.99 \times 10^{-26}\ \text{kg}$$

Because one atom of ^{12}C is defined to have a mass of 12.0 u, divide by 12.0 to find the mass equivalent to 1 u:

$$1\ \text{u} = \frac{1.99 \times 10^{-26}\ \text{kg}}{12.0} = 1.66 \times 10^{-27}\ \text{kg}$$

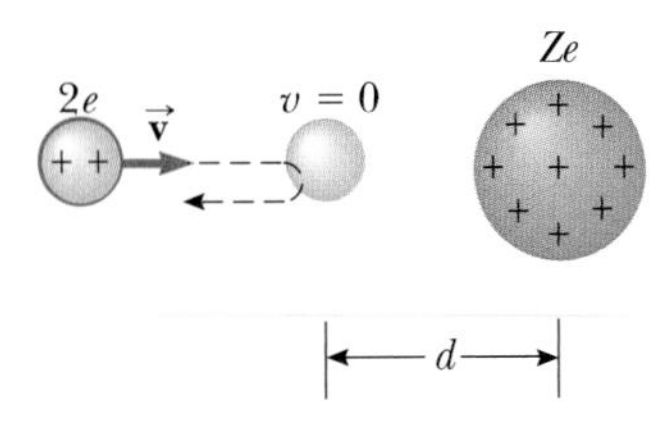

ACTIVE FIGURE 44.1

An alpha particle on a head-on collision course with a nucleus of charge *Ze*. Because of the Coulomb repulsion between the charges of the same sign, the alpha particle approaches to a distance *d* from the nucleus, called the distance of closest approach.

Sign in at www.thomsonedu.com and go to ThomsonNOW to adjust the atomic number of the target nucleus and the kinetic energy of the alpha particle. Then observe the approach of the alpha particle toward the nucleus.

The Size and Structure of Nuclei

In Rutherford's scattering experiments, positively charged nuclei of helium atoms (alpha particles) were directed at a thin piece of metallic foil. As the alpha particles moved through the foil, they often passed near a metal nucleus. Because of the positive charge on both the incident particles and the nuclei, the particles were deflected from their straight-line paths by the Coulomb repulsive force.

Rutherford used conservation of energy for an isolated system to find an expression for the separation distance d at which an alpha particle approaching a nucleus head-on is turned around by Coulomb repulsion. In such a head-on collision, the kinetic energy of the incoming particle must be converted completely to electric potential energy of the alpha particle–nucleus system when the particle stops momentarily at the point of closest approach (the final configuration of the system) before moving back along the same path (Active Fig. 44.1). Applying the conservation of energy principle to the system gives

$$K_i + U_i = K_f + U_f$$

$$\tfrac{1}{2}mv^2 + 0 = 0 + k_e \frac{q_1 q_2}{d}$$

where m is the mass of the alpha particle and v is its initial speed. Solving for d gives

$$d = 2k_e \frac{q_1 q_2}{mv^2} = 2k_e \frac{(2e)(Ze)}{mv^2} = 4k_e \frac{Ze^2}{mv^2}$$

where Z is the atomic number of the target nucleus. From this expression, Rutherford found that the alpha particles approached nuclei to within 3.2×10^{-14} m when the foil was made of gold. Therefore, the radius of the gold nucleus must be less than this value. From the results of his scattering experiments, Rutherford concluded that the positive charge in an atom is concentrated in a small sphere, which he called the nucleus, whose radius is no greater than approximately 10^{-14} m.

Because such small lengths are common in nuclear physics, an often-used convenient length unit is the femtometer (fm), which is sometimes called the **fermi** and is defined as

$$1 \text{ fm} \equiv 10^{-15} \text{ m}$$

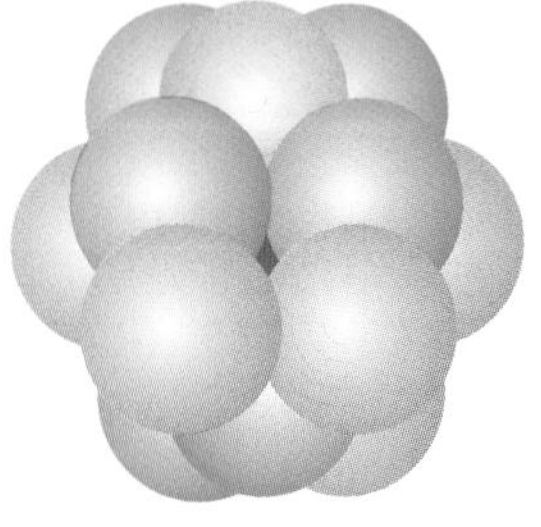

Figure 44.2 A nucleus can be modeled as a cluster of tightly packed spheres, where each sphere is a nucleon.

In the early 1920s, it was known that the nucleus of an atom contains Z protons and has a mass nearly equivalent to that of A protons, where on average $A \approx 2Z$ for lighter nuclei ($Z \leq 20$) and $A > 2Z$ for heavier nuclei. To account for the nuclear mass, Rutherford proposed that each nucleus must also contain $A - Z$ neutral particles that he called neutrons. In 1932, British physicist James Chadwick (1891–1974) discovered the neutron, and he was awarded the Nobel Prize in Physics in 1935 for this important work.

Since the time of Rutherford's scattering experiments, a multitude of other experiments have shown that most nuclei are approximately spherical and have an average radius given by

Nuclear radius ▶

$$r = r_0 A^{1/3} \tag{44.1}$$

where r_0 is a constant equal to 1.2×10^{-15} m and A is the mass number. Because the volume of a sphere is proportional to the cube of its radius, it follows from Equation 44.1 that the volume of a nucleus (assumed to be spherical) is directly proportional to A, the total number of nucleons. This proportionality suggests that *all nuclei have nearly the same density.* When nucleons combine to form a nucleus, they combine as though they were tightly packed spheres (Fig. 44.2). This fact has led to an analogy between the nucleus and a drop of liquid, in which the density of the drop is independent of its size. We shall discuss the liquid-drop model of the nucleus in Section 44.3.

EXAMPLE 44.2 The Volume and Density of a Nucleus

Consider a nucleus of mass number A.

(A) Find an approximate expression for the mass of the nucleus.

SOLUTION

Categorize Let's assume A is large enough that we can imagine the nucleus to be spherical.

Analyze The mass of the proton is approximately equal to that of the neutron. Therefore, if the mass of one of these particles is m, the mass of the nucleus is approximately Am.

(B) Find an expression for the volume of this nucleus in terms of A.

SOLUTION

Assume the nucleus is spherical and use Equation 44.1:

$$(1) \quad V_{\text{nucleus}} = \tfrac{4}{3}\pi r^3 = \tfrac{4}{3}\pi {r_0}^3 A$$

(C) Find a numerical value for the density of this nucleus.

SOLUTION

Use Equation 1.1 and substitute Equation (1):

$$\rho = \frac{m_{\text{nucleus}}}{V_{\text{nucleus}}} = \frac{Am}{\tfrac{4}{3}\pi {r_0}^3 A} = \frac{3m}{4\pi {r_0}^3}$$

Substitute numerical values:

$$\rho = \frac{3(1.67 \times 10^{-27}\text{ kg})}{4\pi(1.2 \times 10^{-15}\text{ m})^3} = 2.3 \times 10^{17}\text{ kg/m}^3$$

Finalize The nuclear density is approximately 2.3×10^{14} times the density of water ($\rho_{\text{water}} = 1.0 \times 10^3\text{ kg/m}^3$).

What If? What if the Earth could be compressed until it had this density? How large would it be?

Answer Because this density is so large, we predict that an Earth of this density would be very small.

Use Equation 1.1 and the mass of the Earth to find the volume of the compressed Earth:

$$V = \frac{M_E}{\rho} = \frac{5.98 \times 10^{24}\text{ kg}}{2.3 \times 10^{17}\text{ kg/m}^3} = 2.6 \times 10^7\text{ m}^3$$

From this volume, find the radius:

$$V = \tfrac{4}{3}\pi r^3 \rightarrow r = \left(\frac{3V}{4\pi}\right)^{1/3} = \left[\frac{3(2.6 \times 10^7\text{ m}^3)}{4\pi}\right]^{1/3}$$

$$= 1.8 \times 10^2\text{ m}$$

An Earth of this radius is indeed a small Earth!

Nuclear Stability

You might expect that the very large repulsive Coulomb forces between the close-packed protons in a nucleus should cause the nucleus to fly apart. Because that does not happen, there must be a counteracting attractive force. The **nuclear force** is a very short range (about 2 fm) attractive force that acts between all nuclear particles. The protons attract each other by means of the nuclear force, and, at the same time, they repel each other through the Coulomb force. The nuclear force also acts between pairs of neutrons and between neutrons and protons. The nuclear force dominates the Coulomb repulsive force within the nucleus (at short ranges), so stable nuclei can exist.

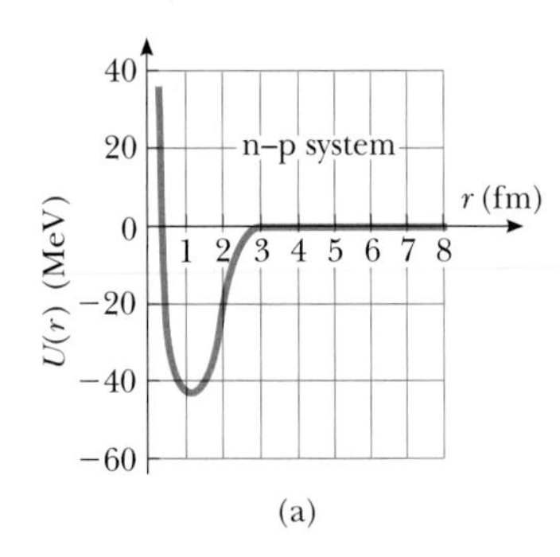

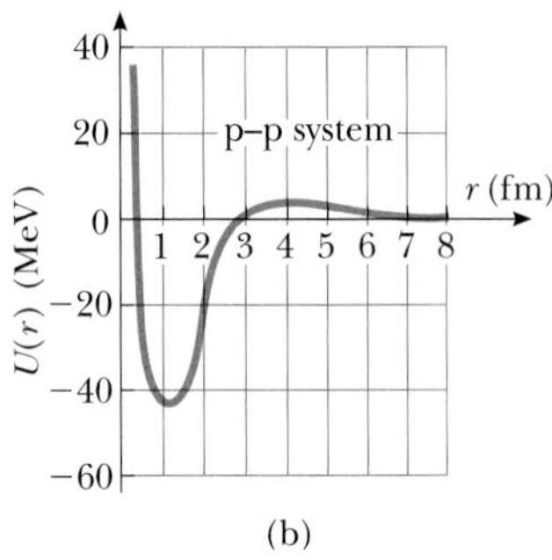

Figure 44.3 (a) Potential energy versus separation distance for a neutron–proton system. (b) Potential energy versus separation distance for a proton–proton system. The difference in the two curves is due to the large Coulomb repulsion in the case of the proton–proton interaction. To display the difference in the curves on this scale, the height of the peak for the proton–proton curve has been exaggerated by a factor of 10.

The nuclear force is independent of charge. In other words, the forces associated with the proton–proton, proton–neutron, and neutron–neutron interactions are the same, apart from the additional repulsive Coulomb force for the proton–proton interaction.

Evidence for the limited range of nuclear forces comes from scattering experiments and from studies of nuclear binding energies. The short range of the nuclear force is shown in the neutron–proton (n–p) potential energy plot of Figure 44.3a obtained by scattering neutrons from a target containing hydrogen. The depth of the n–p potential energy well is 40 to 50 MeV, and there is a strong repulsive component that prevents the nucleons from approaching much closer than 0.4 fm.

The nuclear force does not affect electrons, enabling energetic electrons to serve as point-like probes of nuclei. The charge independence of the nuclear force also means that the main difference between the n–p and p–p interactions is that the p–p potential energy consists of a *superposition* of nuclear and Coulomb interactions as shown in Figure 44.3b. At distances less than 2 fm, both p–p and n–p potential energies are nearly identical, but for distances of 2 fm or greater, the p–p potential has a positive energy barrier with a maximum at 4 fm.

The existence of the nuclear force results in approximately 270 stable nuclei; hundreds of other nuclei have been observed, but they are unstable. A plot of neutron number N versus atomic number Z for a number of stable nuclei is given in Figure 44.4. The stable nuclei are represented by the blue dots, which lie in a narrow range called the *line of stability.* Notice that the light stable nuclei contain an equal number of protons and neutrons; that is, $N = Z$. Also notice that in heavy stable nuclei, the number of neutrons exceeds the number of protons: above $Z = 20$, the line of stability deviates upward from the line representing $N = Z$. This deviation can be understood by recognizing that as the number of protons increases, the strength of the Coulomb force increases, which tends to break the nucleus apart. As a result, more neutrons are needed to keep the nucleus stable because neutrons experience only the attractive nuclear force. Eventually, the repulsive Coulomb forces between protons cannot be compensated by the addi-

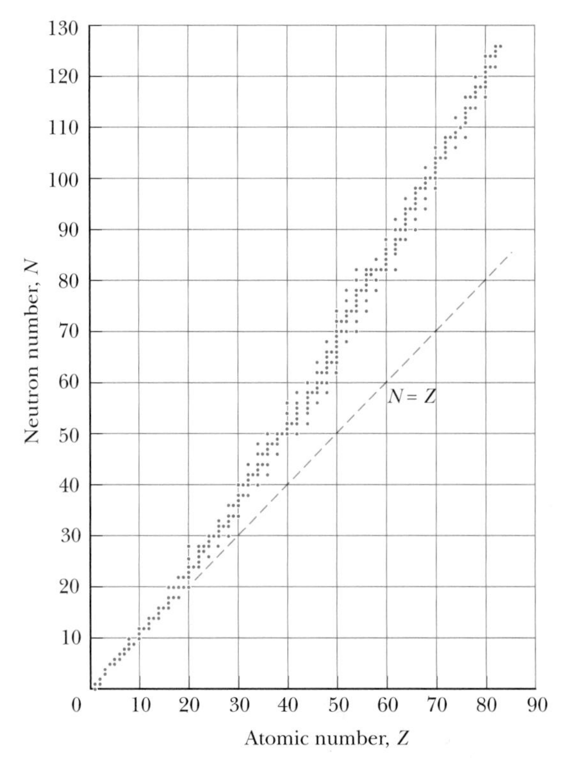

Figure 44.4 Neutron number N versus atomic number Z for stable nuclei (blue dots). These nuclei lie in a narrow band called the line of stability. The dashed line corresponds to the condition $N = Z$.

tion of more neutrons. This point occurs at $Z = 83$, meaning that elements that contain more than 83 protons do not have stable nuclei.

44.2 Nuclear Binding Energy

As mentioned in the discussion of ^{12}C in Section 44.1, the total mass of a nucleus is less than the sum of the masses of its individual nucleons. Therefore, **the rest energy of the bound system (the nucleus) is less than the combined rest energy of the separated nucleons.** This difference in energy is called the **binding energy** of the nucleus and can be interpreted as the energy that must be added to a nucleus to break it apart into its components. Therefore, to separate a nucleus into protons and neutrons, energy must be delivered to the system.

Conservation of energy and the Einstein mass–energy equivalence relationship show that the binding energy E_b of any nucleus of mass M_A is

$$E_b\ (\text{MeV}) = [ZM(\text{H}) + Nm_n - M(^A_Z\text{X})] \times 931.494\ \text{MeV/u} \qquad (44.2)$$

◀ Binding energy of a nucleus

where $M(\text{H})$ is the atomic mass of the neutral hydrogen atom, $M(^A_Z\text{X})$ represents the atomic mass of an atom of the isotope ^A_ZX, m_n is the mass of the neutron, and the masses are all in atomic mass units. The mass of the Z electrons included in $M(\text{H})$ cancels with the mass of the Z electrons included in the term $M(^A_Z\text{X})$ within a small difference associated with the atomic binding energy of the electrons. Because atomic binding energies are typically several electron volts and nuclear binding energies are several million electron volts, this difference is negligible.

A plot of binding energy per nucleon E_b/A as a function of mass number A for various stable nuclei is shown in Figure 44.5. Notice that the curve in Figure 44.5 peaks in the vicinity of $A = 60$. That is, nuclei having mass numbers either greater or less than 60 are not as strongly bound as those near the middle of the periodic table. The decrease in binding energy per nucleon for $A > 60$ implies that energy is released when a heavy nucleus splits, or *fissions*, into two lighter nuclei. Energy is released in fission because the nucleons in each product nucleus are more tightly

PITFALL PREVENTION 44.2
Binding Energy

When separate nucleons are combined to form a nucleus, the energy of the system is reduced. Therefore, the change in energy is negative. The absolute value of this change is called the binding energy. This difference in sign may be confusing. For example, an *increase* in binding energy corresponds to a *decrease* in the energy of the system.

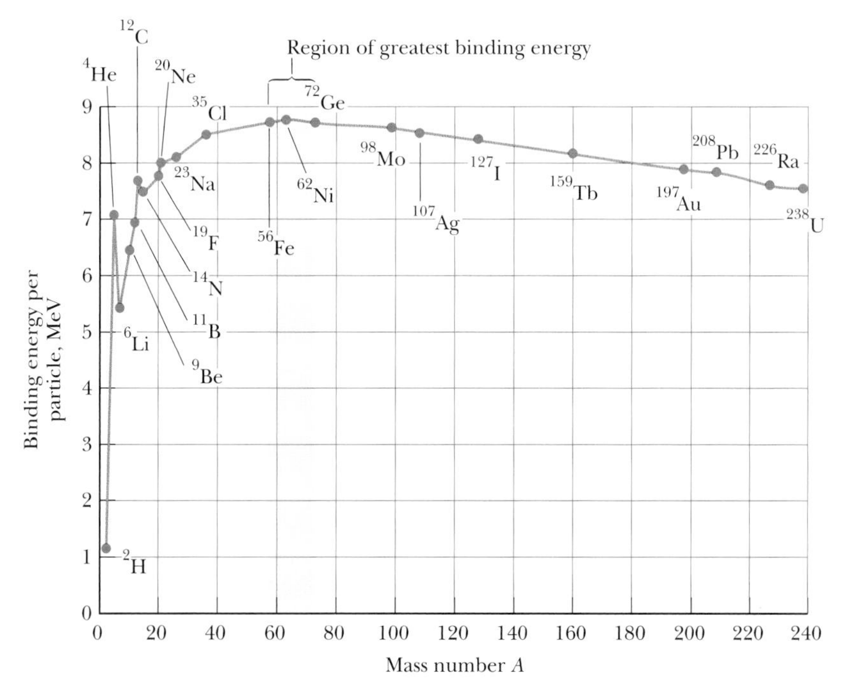

Figure 44.5 Binding energy per nucleon versus mass number for nuclei that lie along the line of stability in Figure 44.4. Some representative nuclei appear as blue dots with labels. (Nuclei to the right of ^{208}Pb are unstable. The curve represents the binding energy for the most stable isotopes.)

bound to one another than are the nucleons in the original nucleus. The important process of fission and a second important process of *fusion,* in which energy is released as light nuclei combine, shall be considered in detail in Chapter 45.

Another important feature of Figure 44.5 is that the binding energy per nucleon is approximately constant at around 8 MeV per nucleon for all nuclei with $A > 50$. For these nuclei, the nuclear forces are said to be *saturated,* meaning that in the closely packed structure shown in Figure 44.2, a particular nucleon can form attractive bonds with only a limited number of other nucleons.

Figure 44.5 provides insight into fundamental questions about the origin of the chemical elements. In the early life of the Universe, the only elements that existed were hydrogen and helium. Clouds of cosmic gas coalesced under gravitational forces to form stars. As a star ages, it produces heavier elements from the lighter elements contained within it, beginning by fusing hydrogen atoms to form helium. This process continues as the star becomes older, generating atoms having larger and larger atomic numbers, up to the peak of the curve shown in Figure 44.5.

The nuclide $^{62}_{28}Ni$ has the largest binding energy per nucleon of 8.794 5 MeV. It takes additional energy to create elements with mass numbers larger than 62 because of their lower binding energies per nucleon. This energy comes from the supernova explosion that occurs at the end of some large stars' lives. Therefore, all the heavy atoms in your body were produced from the explosions of ancient stars. You are literally made of stardust!

44.3 Nuclear Models

The details of the nuclear force are still an area of active research. Several nuclear models have been proposed that are useful in understanding general features of nuclear experimental data and the mechanisms responsible for binding energy. Two such models, the liquid-drop model and the shell model, are discussed below.

Liquid-Drop Model

In 1936, Bohr proposed treating nucleons like molecules in a drop of liquid. In this **liquid-drop model,** the nucleons interact strongly with one another and undergo frequent collisions as they jiggle around within the nucleus. This jiggling motion is analogous to the thermally agitated motion of molecules in a drop of liquid.

Four major effects influence the binding energy of the nucleus in the liquid-drop model:

- **The volume effect.** Figure 44.5 shows that for $A > 50$, the binding energy per nucleon is approximately constant, which indicates that the nuclear force on a given nucleon is due only to a few nearest neighbors and not to all the other nucleons in the nucleus. On average, then, the binding energy associated with the nuclear force for each nucleon is the same in all nuclei: that associated with an interaction with a few neighbors. This property indicates that the total binding energy of the nucleus is proportional to A and therefore proportional to the nuclear volume. The contribution to the binding energy of the entire nucleus is C_1A, where C_1 is an adjustable constant that can be determined by fitting the prediction of the model to experimental results.
- **The surface effect.** Because nucleons on the surface of the drop have fewer neighbors than those in the interior, surface nucleons reduce the binding energy by an amount proportional to their number. Because the number of surface nucleons is proportional to the surface area $4\pi r^2$ of the nucleus (modeled as a sphere) and because $r^2 \propto A^{2/3}$ (Eq. 44.1), the surface term can be expressed as $-C_2A^{2/3}$, where C_2 is a second adjustable constant.

- **The Coulomb repulsion effect.** Each proton repels every other proton in the nucleus. The corresponding potential energy per pair of interacting protons is $k_e e^2/r$, where k_e is the Coulomb constant. The total electric potential energy is equivalent to the work required to assemble Z protons, initially infinitely far apart, into a sphere of volume V. This energy is proportional to the number of proton pairs $Z(Z - 1)/2$ and inversely proportional to the nuclear radius. Consequently, the reduction in binding energy that results from the Coulomb effect is $-C_3 Z(Z - 1)/A^{1/3}$, where C_3 is yet another adjustable constant.
- **The symmetry effect.** Another effect that lowers the binding energy is related to the symmetry of the nucleus in terms of values of N and Z. For small values of A, stable nuclei tend to have $N \approx Z$. Any large asymmetry between N and Z for light nuclei reduces the binding energy and makes the nucleus less stable. For larger A, the value of N for stable nuclei is naturally larger than Z. This effect can be described by a binding-energy term of the form $-C_4(N - Z)^2/A$, where C_4 is another adjustable constant.[1] For small A, any large asymmetry between values of N and Z makes this term relatively large and reduces the binding energy. For large A, this term is small and has little effect on the overall binding energy.

Adding these contributions gives the following expression for the total binding energy:

$$E_b = C_1 A - C_2 A^{2/3} - C_3 \frac{Z(Z-1)}{A^{1/3}} - C_4 \frac{(N-Z)^2}{A} \qquad \textbf{(44.3)}$$

◀ Semiempirical binding-energy formula

This equation, often referred to as the **semiempirical binding-energy formula,** contains four constants that are adjusted to fit the theoretical expression to experimental data. For nuclei having $A \geq 15$, the constants have the values

$$C_1 = 15.7 \text{ MeV} \qquad C_2 = 17.8 \text{ MeV}$$

$$C_3 = 0.71 \text{ MeV} \qquad C_4 = 23.6 \text{ MeV}$$

Equation 44.3, together with these constants, fits the known nuclear mass values very well. The liquid-drop model does not, however, account for some finer details of nuclear structure, such as stability rules and angular momentum. Equation 44.3 is a *theoretical* equation for the binding energy, based on the liquid-drop model, whereas binding energies calculated from Equation 44.2 are *experimental* values based on mass measurements.

EXAMPLE 44.3 Applying the Semiempirical Binding-Energy Formula

The nucleus ^{64}Zn has a tabulated binding energy of 559.09 MeV. Use the semiempirical binding-energy formula to generate a theoretical estimate of the binding energy for this nucleus.

SOLUTION

Conceptualize Imagine bringing the separate protons and neutrons together to form a ^{64}Zn nucleus. The rest energy of the nucleus is smaller than the rest energy of the individual particles. The difference in rest energy is the binding energy.

Categorize From the text of the problem, we know to apply the liquid-drop model. This example is a substitution problem.

[1] The liquid-drop model *describes* that heavy nuclei have $N > Z$. The shell model, as we shall see shortly, *explains* why that is true with a physical argument.

For the ^{64}Zn nucleus, $Z = 30$, $N = 34$, and $A = 64$. Evaluate the four terms of the semiempirical binding-energy formula:

$$C_1A = (15.7\text{ MeV})(64) = 1\,005\text{ MeV}$$

$$C_2A^{2/3} = (17.8\text{ MeV})(64)^{2/3} = 285\text{ MeV}$$

$$C_3\frac{Z(Z-1)}{A^{1/3}} = (0.71\text{ MeV})\frac{(30)(29)}{(64)^{1/3}} = 154\text{ MeV}$$

$$C_4\frac{(N-Z)^2}{A} = (23.6\text{ MeV})\frac{(34-30)^2}{64} = 5.90\text{ MeV}$$

Substitute these values into Equation 44.3:

$$E_b = 1\,005\text{ MeV} - 285\text{ MeV} - 154\text{ MeV} - 5.90\text{ MeV} = 560\text{ MeV}$$

This value differs from the tabulated value by less than 0.2%. Notice how the sizes of the terms decrease from the first to the fourth term. The fourth term is particularly small for this nucleus, which does not have an excessive number of neutrons.

The Shell Model

The liquid-drop model describes the general behavior of nuclear binding energies relatively well. When the binding energies are studied more closely, however, we find the following features:

- Most stable nuclei have an even value of A. Furthermore, only eight stable nuclei have odd values for both Z and N.
- Figure 44.6 shows a graph of the difference between the binding energy per nucleon calculated by Equation 44.3 and the measured binding energy. There is evidence for regularly spaced peaks in the data that are not described by the semiempirical binding-energy formula. The peaks occur at values of N or Z that have become known as **magic numbers:**

Magic numbers ▶

$$Z \text{ or } N = 2, 8, 20, 28, 50, 82 \qquad \textbf{(44.4)}$$

- High-precision studies of nuclear radii show deviations from the simple expression in Equation 44.1. Graphs of experimental data show peaks in the curve of radius versus N at values of N equal to the magic numbers.
- A group of *isotones* is a collection of nuclei having the same value of N and varying values of Z. When the number of stable isotones is graphed as function of N, there are peaks in the graph, again at the magic numbers in Equation 44.4.

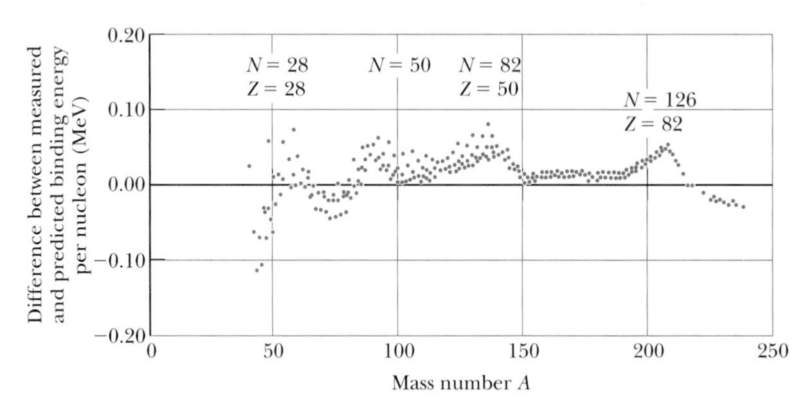

Figure 44.6 The difference between measured binding energies and those calculated from the liquid-drop model as a function of A. The appearance of regular peaks in the experimental data suggests behavior that is not predicted in the liquid-drop model. (Adapted from R. A. Dunlap, *The Physics of Nuclei and Particles,* Brooks/Cole, Belmont CA, 2004.)

- Several other nuclear measurements show anomalous behavior at the magic numbers.[2]

These peaks in graphs of experimental data are reminiscent of the peaks in Figure 42.20 for the ionization energy of atoms, which arose because of the shell structure of the atom. The **shell model** of the nucleus, also called the **independent-particle model,** was developed independently by two German scientists: Maria Goeppert-Mayer in 1949 and Hans Jensen in 1950. Goeppert-Mayer and Jensen shared the 1963 Nobel Prize in Physics for their work. In this model, each nucleon is assumed to exist in a shell, similar to an atomic shell for an electron. The nucleons exist in quantized energy states, and there are few collisions between nucleons. Obviously, the assumptions of this model differ greatly from those made in the liquid-drop model.

The quantized states occupied by the nucleons can be described by a set of quantum numbers. Because both the proton and the neutron have spin $\frac{1}{2}$, the exclusion principle can be applied to describe the allowed states (as it was for electrons in Chapter 42). That is, each state can contain only two protons (or two neutrons) having *opposite* spins (Fig. 44.7). The protons have a set of allowed states, and these states differ from those of the neutrons because the two species move in different potential wells. The proton energy levels are farther apart than the neutron levels because the protons experience a superposition of the Coulomb force and the nuclear force, whereas the neutrons experience only the nuclear force.

One factor influencing the observed characteristics of nuclear ground states is *nuclear spin–orbit* effects. The spin–orbit interaction between the spin of an electron and its orbital motion in an atom gives rise to the sodium doublet discussed in Section 42.6 and is magnetic in origin. In contrast, the spin–orbit effect for nucleons in a nucleus is due to the nuclear force. It is much stronger than in the atomic case, and it has opposite sign. When these effects are taken into account, the shell model is able to account for the observed magic numbers.

The shell model helps us understand why nuclei containing an even number of protons and neutrons are more stable than other nuclei. (There are 160 stable even–even isotopes.) Any particular state is filled when it contains two protons (or two neutrons) having opposite spins. An extra proton or neutron can be added to the nucleus only at the expense of increasing the energy of the nucleus. This increase in energy leads to a nucleus that is less stable than the original nucleus. A careful inspection of the stable nuclei shows that the majority have a special stability when their nucleons combine in pairs, which results in a total angular momentum of zero.

The shell model also helps us understand why nuclei tend to have more neutrons than protons. As in Figure 44.7, the proton energy levels are higher than those for neutrons due to the extra energy associated with Coulomb repulsion. This effect becomes more pronounced as Z increases. Consequently, as Z increases and higher states are filled, a proton level for a given quantum number will be much higher in energy than the neutron level for the same quantum number. In fact, it will be even higher in energy than neutron levels for higher quantum numbers. Hence, it is more energetically favorable for the nucleus to form with neutrons in the lower energy levels rather than protons in the higher energy levels, so the number of neutrons is greater than the number of protons.

More sophisticated models of the nucleus have been and continue to be developed. For example, the *collective model* combines features of the liquid-drop and shell models. The development of theoretical models of the nucleus continues to be an active area of research.

MARIA GOEPPERT-MAYER
German Scientist (1906–1972)
Goeppert-Mayer was born and educated in Germany. She is best known for her development of the shell model (independent-particle model) of the nucleus, published in 1950. A similar model was simultaneously developed by Hans Jensen, another German scientist. Goeppert-Mayer and Jensen were awarded the Nobel Prize in Physics in 1963 for their extraordinary work in understanding the structure of the nucleus.

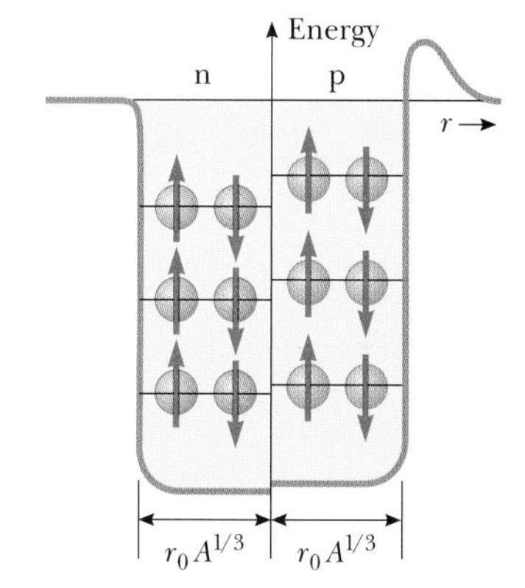

Figure 44.7 A square potential well containing 12 nucleons. The orange circles represent protons, and the green circles represent neutrons. The energy levels for the protons are slightly higher than those for the neutrons because of the electric potential energy associated with the system of protons. The difference in the levels increases as Z increases. Notice that only two nucleons having opposite spins can occupy a given level, as required by the exclusion principle.

[2] For further details, see chapter 5 of R. A. Dunlap, *The Physics of Nuclei and Particles* (Belmont, CA: Brooks/Cole, 2004).

MARIE CURIE
Polish Scientist (1867–1934)
In 1903, Marie Curie shared the Nobel Prize in Physics with her husband, Pierre, and with Becquerel for their studies of radioactive substances. In 1911, she was awarded a Nobel Prize in Chemistry for the discovery of radium and polonium. "I persist in believing that the ideas that then guided us are the only ones which can lead to true social progress. We cannot hope to build a better world without improving the individual. Toward this end, each of us must work toward his own highest development, accepting at the same time his share of responsibility in the general life of humanity."

44.4 Radioactivity

In 1896, Becquerel accidentally discovered that uranyl potassium sulfate crystals emit an invisible radiation that can darken a photographic plate when the plate is covered to exclude light. After a series of experiments, he concluded that the radiation emitted by the crystals was of a new type, one that requires no external stimulation and was so penetrating that it could darken protected photographic plates and ionize gases. This process of spontaneous emission of radiation by uranium was soon to be called **radioactivity.**

Subsequent experiments by other scientists showed that other substances were more powerfully radioactive. The most significant early investigations of this type were conducted by Marie and Pierre Curie. After several years of careful and laborious chemical separation processes on tons of pitchblende, a radioactive ore, the Curies reported the discovery of two previously unknown elements, both radioactive, named polonium and radium. Additional experiments, including Rutherford's famous work on alpha-particle scattering, suggested that radioactivity is the result of the *decay,* or disintegration, of unstable nuclei.

Three types of radioactive decay occur in radioactive substances: alpha (α) decay, in which the emitted particles are ^{4}He nuclei; beta (β) decay, in which the emitted particles are either electrons or positrons; and gamma (γ) decay, in which the emitted particles are high-energy photons. A **positron** is a particle like the electron in all respects except that the positron has a charge of $+e$. (The positron is the *antiparticle* of the electron; see Section 46.2.) The symbol e^- is used to designate an electron, and e^+ designates a positron.

We can distinguish among these three forms of radiation by using the scheme described in Figure 44.8. The radiation from radioactive samples that emit all three types of particles is directed into a region in which there is a magnetic field. The radiation beam splits into three components, two bending in opposite directions and the third experiencing no change in direction. This simple observation shows that the radiation of the undeflected beam carries no charge (the gamma ray), the component deflected upward corresponds to positively charged particles (alpha particles), and the component deflected downward corresponds to negatively charged particles (e^-). If the beam includes a positron (e^+), it is deflected upward like the alpha particle, but it follows a different trajectory due to its smaller mass.

The three types of radiation have quite different penetrating powers. Alpha particles barely penetrate a sheet of paper, beta particles can penetrate a few millimeters of aluminum, and gamma rays can penetrate several centimeters of lead.

The decay process is probabilistic in nature and can be described with statistical calculations for a radioactive substance of macroscopic size containing a large number of radioactive nuclei. For such large numbers, the rate at which a particular decay process occurs in a sample is proportional to the number of radioactive nuclei present (that is, the number of nuclei that have not yet decayed). If N is the

PITFALL PREVENTION 44.3
Rays or Particles?

Early in the history of nuclear physics, the term *radiation* was used to describe the emanations from radioactive nuclei. We now know that alpha radiation and beta radiation involve the emission of particles with nonzero rest energy. Even though they are not examples of electromagnetic radiation, the use of the term *radiation* for all three types of emission is deeply entrenched in our language and in the physics community.

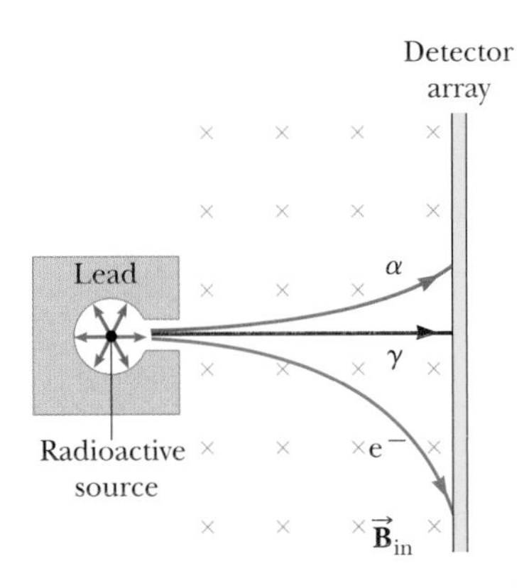

Figure 44.8 The radiation from radioactive sources can be separated into three components by using a magnetic field to deflect the charged particles. The detector array at the right records the events. The gamma ray is not deflected by the magnetic field.

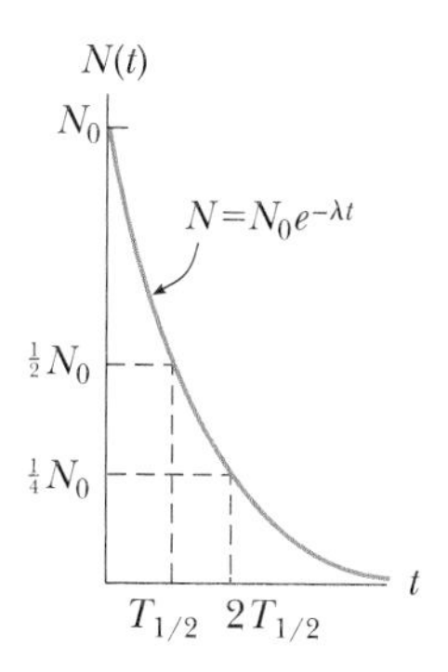

ACTIVE FIGURE 44.9

Plot of the exponential decay of radioactive nuclei. The vertical axis represents the number of undecayed radioactive nuclei present at any time t, and the horizontal axis is time. The time $T_{1/2}$ is the half-life of the sample.

Sign in at www.thomsonedu.com and go to ThomsonNOW to observe the decay curves for nuclei with varying half-lives.

number of undecayed radioactive nuclei present at some instant, the rate of change of N with time is

$$\frac{dN}{dt} = -\lambda N \tag{44.5}$$

where λ, called the **decay constant,** is the probability of decay per nucleus per second. The negative sign indicates that dN/dt is negative; that is, N decreases in time.

Equation 44.5 can be written in the form

$$\frac{dN}{N} = -\lambda dt$$

which, upon integration, gives

$$N = N_0 e^{-\lambda t} \tag{44.6}$$

◀ Exponential behavior of the number of undecayed nuclei

where the constant N_0 represents the number of undecayed radioactive nuclei at $t = 0$. Equation 44.6 shows that the number of undecayed radioactive nuclei in a sample decreases exponentially with time. The plot of N versus t shown in Active Figure 44.9 illustrates the exponential nature of the decay.

The **decay rate** R, which is the number of decays per second, can be obtained by combining Equations 44.5 and 44.6:

$$R = \left|\frac{dN}{dt}\right| = \lambda N = \lambda N_0 e^{-\lambda t} = R_0 e^{-\lambda t} \tag{44.7}$$

◀ Exponential behavior of the decay rate

where $R_0 = \lambda N_0$ is the decay rate at $t = 0$. The decay rate R of a sample is often referred to as its **activity.** Note that both N and R decrease exponentially with time.

Another parameter useful in characterizing nuclear decay is the **half-life** $T_{1/2}$:

> The **half-life** of a radioactive substance is the time interval during which half of a given number of radioactive nuclei decay.

To find an expression for the half-life, we first set $N = N_0/2$ and $t = T_{1/2}$ in Equation 44.6 to give

$$\frac{N_0}{2} = N_0 e^{-\lambda T_{1/2}}$$

Canceling the N_0 factors and then taking the reciprocal of both sides, we obtain $e^{\lambda T_{1/2}} = 2$. Taking the natural logarithm of both sides gives

$$T_{1/2} = \frac{\ln 2}{\lambda} = \frac{0.693}{\lambda} \tag{44.8}$$

◀ Half-life

After a time interval equal to one half-life, there are $N_0/2$ radioactive nuclei remaining (by definition); after two half-lives, half of these remaining nuclei have decayed and $N_0/4$ radioactive nuclei are left; after three half-lives, $N_0/8$ are left;

PITFALL PREVENTION 44.4

Notation Warning

In Section 44.1, we introduced the symbol N as an integer representing the number of neutrons in a nucleus. In this discussion, the symbol N represents the number of undecayed nuclei in a radioactive sample remaining after some time interval. As you read further, be sure to consider the context to determine the appropriate meaning for the symbol N.

PITFALL PREVENTION 44.5

Half-life

It is *not* true that all the original nuclei have decayed after two half-lives! In one half-life, half of the original nuclei will decay. In the second half-life, half of those remaining will decay, leaving $\frac{1}{4}$ of the original number.

and so on. In general, after n half-lives, the number of undecayed radioactive nuclei remaining is $N_0(\frac{1}{2})^n$.

A frequently used unit of activity is the **curie** (Ci), defined as

The curie ▶

$$1 \text{ Ci} \equiv 3.7 \times 10^{10} \text{ decays/s}$$

This value was originally selected because it is the approximate activity of 1 g of radium. The SI unit of activity is the **becquerel** (Bq):

The becquerel ▶

$$1 \text{ Bq} \equiv 1 \text{ decay/s}$$

Therefore, $1 \text{ Ci} = 3.7 \times 10^{10}$ Bq. The curie is a rather large unit, and the more frequently used activity units are the millicurie and the microcurie.

Quick Quiz 44.2 On your birthday, you measure the activity of a sample of ^{210}Bi, which has a half-life of 5.01 days. The activity you measure is 1.000 μCi. What is the activity of this sample on your next birthday? (a) 1.000 μCi (b) 0 (c) $\sim$ 0.2 μCi (d) $\sim$ 0.01 μCi (e) $\sim 10^{-22}$ μCi

EXAMPLE 44.4 How Many Nuclei Are Left?

The isotope carbon-14, $^{14}_{6}$C, is radioactive and has a half-life of 5 730 years. If you start with a sample of 1 000 carbon-14 nuclei, how many nuclei will still be undecayed in 25 000 years?

SOLUTION

Conceptualize The time interval of 25 000 years is much longer than the half-life, so only a small fraction of the originally undecayed nuclei will remain.

Categorize The text of the problem allows us to categorize this example as a radioactive decay problem.

Analyze Divide the time interval by the half-life to determine the number of half-lives:

$$n = \frac{25\ 000 \text{ yr}}{5\ 730 \text{ yr}} = 4.363$$

Determine how many undecayed nuclei are left after 4.363 half-lives:

$$N = N_0(\tfrac{1}{2})^n = 1\ 000(\tfrac{1}{2})^{4.363} = 49$$

Finalize As we have mentioned, radioactive decay is a probabilistic process and accurate statistical predictions are possible only with a very large number of atoms. The original sample in this example contains only 1 000 nuclei, which is certainly not a very large number. Therefore, if you counted the number of undecayed nuclei remaining after 25 000 years, it might not be exactly 49.

EXAMPLE 44.5 The Activity of Carbon

At time $t = 0$, a radioactive sample contains 3.50 μg of pure $^{11}_{6}$C, which has a half-life of 20.4 min.

(A) Determine the number N_0 of nuclei in the sample at $t = 0$.

SOLUTION

Conceptualize The half-life is relatively short, so the number of undecayed nuclei drops rapidly. The molar mass of $^{11}_{6}$C is approximately 11.0 g/mol.

Categorize We evaluate results using equations developed in this section, so we categorize this example as a substitution problem.

Find the number of moles in 3.50 μg of pure $^{11}_{6}C$:

$$n = \frac{3.50 \times 10^{-6}\text{ g}}{11.0\text{ g/mol}} = 3.18 \times 10^{-7}\text{ mol}$$

Find the number of undecayed nuclei in this amount of pure $^{11}_{6}C$:

$$N_0 = 3.18 \times 10^{-7}\text{ mol}(6.02 \times 10^{23}\text{ nuclei/mol}) = \boxed{1.92 \times 10^{17}\text{ nuclei}}$$

(B) What is the activity of the sample initially and after 8.00 h?

SOLUTION

Find the decay constant:

$$\lambda = \frac{0.693}{T_{1/2}} = \frac{0.693}{20.4\text{ min}}\left(\frac{1\text{ min}}{60\text{ s}}\right) = 5.66 \times 10^{-4}\text{ s}^{-1}$$

Find the initial activity of the sample:

$$R_0 = \lambda N_0 = (5.66 \times 10^{-4}\text{ s}^{-1})(1.92 \times 10^{17}) = \boxed{1.08 \times 10^{14}\text{ Bq}}$$

Use Equation 44.7 to find the activity at $t = 8.00\text{ h} = 2.88 \times 10^4$ s:

$$R = R_0 e^{-\lambda t} = (1.08 \times 10^{14}\text{ Bq})e^{-(5.66\times10^{-4}\text{ s}^{-1})(2.88\times10^4\text{ s})} = \boxed{8.96 \times 10^6\text{ Bq}}$$

EXAMPLE 44.6 A Radioactive Isotope of Iodine

A sample of the isotope ^{131}I, which has a half-life of 8.04 days, has an activity of 5.0 mCi at the time of shipment. Upon receipt of the sample at a medical laboratory, the activity is 2.1 mCi. How much time has elapsed between the two measurements?

SOLUTION

Conceptualize The sample is continuously decaying as it is in transit. The decrease in the activity is 58% during the time interval between shipment and receipt, so we expect the elapsed time to be greater than the half-life of 8.04 d.

Categorize The stated activity corresponds to many decays per second, so N is large and we can categorize this problem as one in which we can use our statistical analysis of radioactivity.

Analyze Solve Equation 44.7 for the ratio of the final activity to the initial activity:

$$\frac{R}{R_0} = e^{-\lambda t}$$

Take the natural logarithm of both sides:

$$\ln\left(\frac{R}{R_0}\right) = -\lambda t$$

Solve for the time t:

$$(1)\quad t = -\frac{1}{\lambda}\ln\left(\frac{R}{R_0}\right)$$

Use Equation 44.8 to find λ:

$$(2)\quad \lambda = \frac{0.693}{T_{1/2}} = \frac{0.693}{8.04\text{ d}} = 8.62 \times 10^{-2}\text{ d}^{-1}$$

Substitute Equation (2) into Equation (1):

$$t = -\left(\frac{1}{8.62 \times 10^{-2}\text{ d}^{-1}}\right)\ln\left(\frac{2.1\text{ mCi}}{5.0\text{ mCi}}\right) = \boxed{10\text{ d}}$$

Finalize This result is indeed greater than the half-life, as expected. This example demonstrates the difficulty in shipping radioactive samples with short half-lives. If the shipment is delayed by several days, only a small fraction of the sample might remain upon receipt. This difficulty can be addressed by shipping a combination of isotopes in which the desired isotope is the product of a decay occurring within the sample. It is possible for the desired isotope to be in *equilibrium*, in which case it is created at the same rate as it decays. Therefore, the amount of the desired isotope remains constant during the shipping process and subsequent storage. When needed, the desired isotope can be separated from the rest of the sample; its decay from the initial activity begins at this point rather than upon shipment.

44.5 The Decay Processes

As we stated in Section 44.4, a radioactive nucleus spontaneously decays by one of three processes: alpha decay, beta decay, or gamma decay. Active Figure 44.10 shows a close-up view of a portion of Figure 44.4 from $Z = 65$ to $Z = 80$. The blue circles are the stable nuclei seen in Figure 44.4. In addition, unstable nuclei above and below the line of stability for each value of Z are shown. Above the line of stability, the red circles show unstable nuclei that are neutron-rich and undergo a beta decay process in which an electron is emitted. Below the blue circles are green circles corresponding to proton-rich unstable nuclei that primarily undergo a beta-decay process in which a positron is emitted or a competing process called electron capture. Beta decay and electron capture are described in more detail below. Further below the line of stability (with a few exceptions) are yellow circles that represent very proton-rich nuclei for which the primary decay mechanism is alpha decay, which we discuss first.

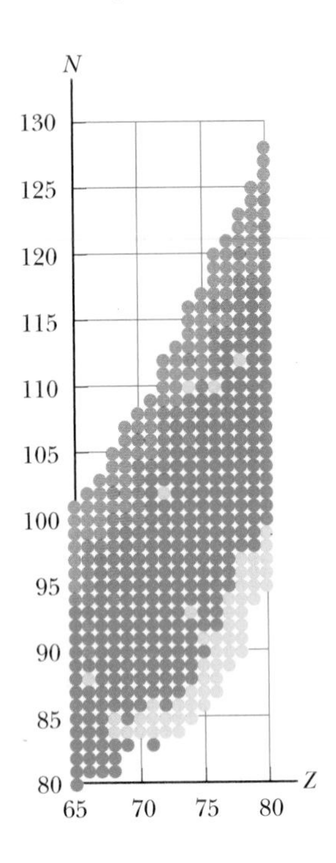

ACTIVE FIGURE 44.10

A close-up view of the line of stability in Figure 44.4 from $Z = 65$ to $Z = 80$. The blue dots represent stable nuclei as in Figure 44.4. The other colored dots represent unstable isotopes above and below the line of stability, with the color of the dot indicating the primary means of decay.

Sign in at www.thomsonedu.com and go to ThomsonNOW to study the decay modes and decay energies by clicking on any of the colored dots.

Alpha Decay

A nucleus emitting an alpha particle ($^4_2$He) loses two protons and two neutrons. Therefore, the atomic number Z decreases by 2, the mass number A decreases by 4, and the neutron number decreases by 2. The decay can be written

$$^A_Z\text{X} \rightarrow \, ^{A-4}_{Z-2}\text{Y} + \, ^4_2\text{He} \quad \textbf{(44.9)}$$

where X is called the **parent nucleus** and Y the **daughter nucleus.** As a general rule in any decay expression such as this one, (1) the sum of the mass numbers A must be the same on both sides of the decay and (2) the sum of the atomic numbers Z must be the same on both sides of the decay. As examples, ^{238}U and ^{226}Ra are both alpha emitters and decay according to the schemes

$$^{238}_{92}\text{U} \rightarrow \, ^{234}_{90}\text{Th} + \, ^4_2\text{He} \quad \textbf{(44.10)}$$

$$^{226}_{88}\text{Ra} \rightarrow \, ^{222}_{86}\text{Rn} + \, ^4_2\text{He} \quad \textbf{(44.11)}$$

The decay of ^{226}Ra is shown in Active Figure 44.11.

When the nucleus of one element changes into the nucleus of another as happens in alpha decay, the process is called **spontaneous decay.** In any spontaneous decay, relativistic energy and momentum of the isolated parent nucleus must be conserved. If we call M_X the mass of the parent nucleus, M_Y the mass of the daughter nucleus, and M_α the mass of the alpha particle, we can define the **disintegration energy** Q of the system as

$$Q = (M_X - M_Y - M_\alpha)c^2 \quad \textbf{(44.12)}$$

The energy Q is in joules when the masses are in kilograms and c is the speed of light, 3.00×10^8 m/s. When the masses are expressed in atomic mass units u, however, Q can be calculated in MeV using the expression

$$Q = (M_X - M_Y - M_\alpha) \times 931.494 \text{ MeV/u} \quad \textbf{(44.13)}$$

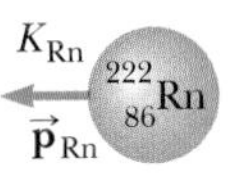

ACTIVE FIGURE 44.11

The alpha decay of radium-226. The radium nucleus is initially at rest. After the decay, the radon nucleus has kinetic energy K_{Rn} and momentum $\vec{p}_{Rn}$ and the alpha particle has kinetic energy K_α and momentum $\vec{p}_\alpha$.

Sign in at www.thomsonedu.com and go to ThomsonNOW to observe the decay of radium-226. For a large number of decays, observe the development of the graph in Active Figure 44.14b (page 1313).

Table 44.2 (page 1310) contains information on selected isotopes, including masses of neutral atoms that can be used in Equation 44.13 and similar equations.

The disintegration energy Q appears in the form of kinetic energy in the daughter nucleus and the alpha particle and is sometimes referred to as the Q value of the nuclear decay. Consider the case of the ^{226}Ra decay described in Active Figure 44.11. If the parent nucleus is at rest before the decay, the total kinetic energy of the products is 4.87 MeV. (See Example 44.7.) Most of this kinetic energy is associated with the alpha particle because this particle is much less massive than the daughter nucleus ^{222}Rn. That is, because momentum must be conserved, the lighter alpha particle recoils with a much higher speed than does the daughter nucleus. Generally, less massive particles carry off most of the energy in nuclear decays.

Experimental observations of alpha-particle energies show a number of discrete energies rather than a single energy because the daughter nucleus may be left in an excited quantum state after the decay. As a result, not all the disintegration energy is available as kinetic energy of the alpha particle and daughter nucleus. The emission of an alpha particle is followed by one or more gamma-ray photons (see below) as the excited nucleus decays to the ground state. The observed discrete alpha-particle energies represent evidence of the quantized nature of the nucleus and allow a determination of the energies of the quantum states.

If one assumes ^{238}U (or any other alpha emitter) decays by emitting either a proton or a neutron, the mass of the decay products would exceed that of the parent nucleus, corresponding to a negative Q value. A negative Q value indicates that such a proposed decay does not occur spontaneously.

PITFALL PREVENTION 44.6
Another Q

We have seen the symbol Q before, but this use is a brand new meaning for this symbol: the disintegration energy. In this context, it is not heat, charge, or quality factor for a resonance, for which we have used Q before.

Quick Quiz 44.3 Which of the following is the correct daughter nucleus associated with the alpha decay of $^{157}_{72}Hf$? (a) $^{153}_{72}Hf$ (b) $^{153}_{70}Yb$ (c) $^{157}_{70}Yb$

EXAMPLE 44.7 The Energy Liberated When Radium Decays

The ^{226}Ra nucleus undergoes alpha decay according to Equation 44.11. Calculate the Q value for this process. From Table 44.2, the masses are 226.025 403 u for ^{226}Ra, 222.017 570 u for ^{222}Rn, and 4.002 603 u for $^{4}_{2}He$.

SOLUTION

Conceptualize Study Active Figure 44.11 to understand the process of alpha decay in this nucleus.

Categorize We use an equation developed in this section, so we categorize this example as a substitution problem.

Evaluate Q using Equation 44.13:

$$Q = (M_X - M_Y - M_\alpha) \times 931.494 \text{ MeV/u}$$
$$= (226.025\,403 \text{ u} - 222.017\,570 \text{ u} - 4.002\,603 \text{ u}) \times 931.494 \text{ MeV/u}$$
$$= (0.005\,230 \text{ u}) \times 931.494 \text{ MeV/u} = 4.87 \text{ MeV}$$

What If? Suppose you measured the kinetic energy of the alpha particle from this decay. Would you measure 4.87 MeV?

Answer The value of 4.87 MeV is the disintegration energy for the decay. It includes the kinetic energy of both the alpha particle and the daughter nucleus after the decay. Therefore, the kinetic energy of the alpha particle would be *less* than 4.87 MeV.

Let's determine this kinetic energy mathematically. The parent nucleus is an isolated system that decays into an alpha particle and a daughter nucleus. Therefore, momentum must be conserved for the system.

Set up a conservation of momentum equation, noting that the initial momentum of the system is zero:

$$(1)\quad 0 = M_Y v_Y - M_\alpha v_\alpha$$

Set the disintegration energy equal to the sum of the kinetic energies of the alpha particle and the daughter nucleus (assuming the daughter nucleus is left in the ground state):

$$(2)\quad Q = \tfrac{1}{2}M_\alpha {v_\alpha}^2 + \tfrac{1}{2}M_Y {v_Y}^2$$

Solve Equation (1) for v_Y and substitute into Equation (2):

$$Q = \tfrac{1}{2}M_\alpha {v_\alpha}^2 + \tfrac{1}{2}M_Y\left(\frac{M_\alpha v_\alpha}{M_Y}\right)^2 = \tfrac{1}{2}M_\alpha {v_\alpha}^2\left(1 + \frac{M_\alpha}{M_Y}\right)$$
$$= K_\alpha\left(\frac{M_Y + M_\alpha}{M_Y}\right)$$

Solve for the kinetic energy of the alpha particle:

$$K_\alpha = Q\left(\frac{M_Y}{M_Y + M_\alpha}\right)$$

Evaluate this kinetic energy for the specific decay of ^{226}Ra that we are exploring in this example:

$$K_\alpha = (4.87 \text{ MeV})\left(\frac{222}{222 + 4}\right) = 4.79 \text{ MeV}$$

TABLE 44.2

Chemical and Nuclear Information for Selected Isotopes

Atomic Number Z	Element	Chemical Symbol	Mass Number A (* means) radioactive)	Mass of Neutral Atom (u)	Percent Abundance	Half-life, if Radioactive $T_{1/2}$
−1	electron	e^-	0	0.000 549		
0	neutron	n	1*	1.008 665		614 s
1	hydrogen	$^1H = p$	1	1.007 825	99.988 5	
	[deuterium	$^2H = D$]	2	2.014 102	0.011 5	
	[tritium	$^3H = T$]	3*	3.016 049		12.33 yr
2	helium	He	3	3.016 029	0.000 137	
	[alpha particle	$\alpha = {}^4He$]	4	4.002 603	99.999 863	
			6*	6.018 888		0.81 s
3	lithium	Li	6	6.015 122	7.5	
			7	7.016 004	92.5	
4	beryllium	Be	7*	7.016 929		53.3 d
			9	9.012 182	100	
5	boron	B	10	10.012 937	19.9	
			11	11.009 306	80.1	
6	carbon	C	11*	11.011 434		20.4 min
			12	12.000 000	98.93	
			13	13.003 355	1.07	
			14*	14.003 242		5 730 yr
7	nitrogen	N	13*	13.005 739		9.96 min
			14	14.003 074	99.632	
			15	15.000 109	0.368	
8	oxygen	O	14*	14.008 595		70.6 s
			15*	15.003 065		122 s
			16	15.994 915	99.757	
			17	16.999 132	0.038	
			18	17.999 160	0.205	
9	fluorine	F	18*	18.000 938		109.8 min
			19	18.998 403	100	
10	neon	Ne	20	19.992 440	90.48	
11	sodium	Na	23	22.989 770	100	
12	magnesium	Mg	23*	22.994 125		11.3 s
			24	23.985 042	78.99	
13	aluminum	Al	27	26.981 539	100	
14	silicon	Si	27*	26.986 705		4.2 s
15	phosphorus	P	30*	29.978 314		2.50 min
			31	30.973 762	100	
			32*	31.973 907		14.26 d
16	sulfur	S	32	31.972 071	94.93	
19	potassium	K	39	38.963 707	93.258 1	
			40*	39.963 999	0.011 7	1.28×10^9 yr
20	calcium	Ca	40	39.962 591	96.941	
			42	41.958 618	0.647	
			43	42.958 767	0.135	
25	manganese	Mn	55	54.938 050	100	
26	iron	Fe	56	55.934 942	91.754	
			57	56.935 399	2.119	

(continued)

TABLE 44.2

Chemical and Nuclear Information for Selected Isotopes (*Continued*)

Atomic Number Z	Element	Chemical Symbol	Mass Number A (* means) radioactive)	Mass of Neutral Atom (u)	Percent Abundance	Half-life, if Radioactive $T_{1/2}$
27	cobalt	Co	57*	56.936 296		270 d
			59	58.933 200	100	
			60*	59.933 822		5.27 yr
28	nickel	Ni	58	57.935 348	68.076 9	
			60	59.930 790	26.223 1	
29	copper	Cu	63	62.929 601	69.17	
			64*	63.929 599		12.7 h
			65	64.927 794	30.83	
30	zinc	Zn	64	63.929 147	48.63	
37	rubidium	Rb	87*	86.909 184	27.83	
38	strontium	Sr	87	86.908 880	7.00	
			88	87.905 614	82.58	
			90*	89.907 738		29.1 yr
41	niobium	Nb	93	92.906 378	100	
42	molybdenum	Mo	94	93.905 088	9.25	
44	ruthenium	Ru	98	97.905 287	1.87	
55	cesium	Cs	137*	136.907 074		30 yr
56	barium	Ba	137	136.905 821	11.232	
58	cerium	Ce	140	139.905 434	88.450	
59	praseodymium	Pr	141	140.907 648	100	
60	neodymium	Nd	144*	143.910 083	23.8	2.3×10^{15} yr
61	promethium	Pm	145*	144.912 744		17.7 yr
79	gold	Au	197	196.966 552	100	
80	mercury	Hg	198	197.966 752	9.97	
			202	201.970 626	29.86	
82	lead	Pb	206	205.974 449	24.1	
			207	206.975 881	22.1	
			208	207.976 636	52.4	
			214*	213.999 798		26.8 min
83	bismuth	Bi	209	208.980 383	100	
84	polonium	Po	210*	209.982 857		138.38 d
			216*	216.001 905		0.145 s
			218*	218.008 966		3.10 min
86	radon	Rn	220*	220.011 384		55.6 s
			222*	222.017 570		3.823 d
88	radium	Ra	226*	226.025 403		1 600 yr
90	thorium	Th	232*	232.038 050	100	1.40×10^{10} yr
			234*	234.043 596		24.1 d
92	uranium	U	235*	235.043 923	0.720 0	7.04×10^{8} yr
			236*	236.045 562		2.34×10^{7} yr
			238*	238.050 783	99.274 5	4.47×10^{9} yr
93	neptunium	Np	236*	236.046 560		1.15×10^{5} yr
			237*	237.048 167		2.14×10^{6} yr
94	plutonium	Pu	239*	239.052 156		24 120 yr

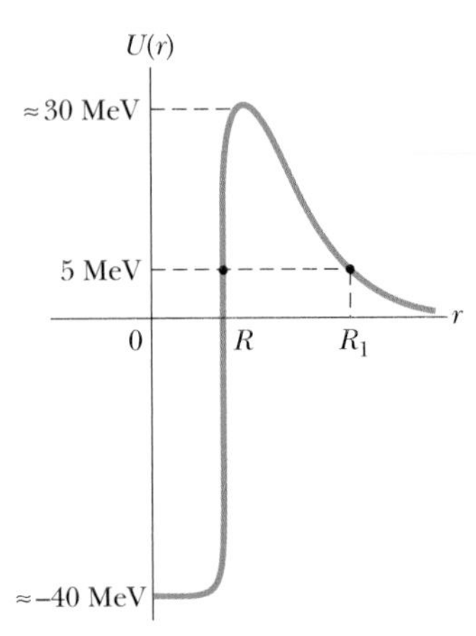

Figure 44.12 Potential energy versus separation distance for a system consisting of an alpha particle and a daughter nucleus. Classically, the energy of the alpha particle is not sufficiently large to overcome the energy barrier, so the particle should not be able to escape from the nucleus. In reality, the alpha particle does escape by tunneling through the barrier.

To understand the mechanism of alpha decay, let's model the parent nucleus as a system consisting of (1) the alpha particle, already formed as an entity within the nucleus, and (2) the daughter nucleus that will result when the alpha particle is emitted. Figure 44.12 shows a plot of potential energy versus separation distance r between the alpha particle and the daughter nucleus, where the distance marked R is the range of the nuclear force. The curve represents the combined effects of (1) the repulsive Coulomb force, which gives the positive part of the curve for $r > R$, and (2) the attractive nuclear force, which causes the curve to be negative for $r < R$. As shown in Example 44.7, a typical disintegration energy Q is approximately 5 MeV, which is the approximate kinetic energy of the alpha particle, represented by the lower dashed line in Figure 44.12.

According to classical physics, the alpha particle is trapped in a potential well. How, then, does it ever escape from the nucleus? The answer to this question was first provided by George Gamow (1904–1968) in 1928 and independently by R. W. Gurney and E. U. Condon in 1929, using quantum mechanics. In the view of quantum mechanics, there is always some probability that a particle can tunnel through a barrier (Section 41.5). That is exactly how we can describe alpha decay: the alpha particle tunnels through the barrier in Figure 44.12, escaping the nucleus. Furthermore, this model agrees with the observation that higher-energy alpha particles come from nuclei with shorter half-lives. For higher-energy alpha particles in Figure 44.12, the barrier is narrower and the probability is higher that tunneling occurs. The higher probability translates to a shorter half-life.

As an example, consider the decays of ^{238}U and ^{226}Ra in Equations 44.10 and 44.11, along with the corresponding half-lives and alpha-particle energies:

$$^{238}\text{U:} \quad T_{1/2} = 4.47 \times 10^9 \text{ yr} \qquad K_\alpha = 4.20 \text{ MeV}$$

$$^{226}\text{Ra:} \quad T_{1/2} = 1.60 \times 10^3 \text{ yr} \qquad K_\alpha = 4.79 \text{ MeV}$$

Notice that a relatively small difference in alpha-particle energy is associated with a tremendous difference of six orders of magnitude in the half-life. The origin of this effect can be understood as follows. Figure 44.12 shows that the curve below an alpha-particle energy of 5 MeV has a slope with a relatively small magnitude. Therefore, a small difference in energy on the vertical axis has a relatively large effect on the width of the potential barrier. Second, recall Equation 41.22, which describes the exponential dependence of the probability of transmission on the barrier width. These two factors combine to give the very sensitive relationship between half-life and alpha-particle energy that the data above suggest.

A life-saving application of alpha decay is the household smoke detector, shown in Figure 44.13. The detector consists of an ionization chamber, a sensitive current detector, and an alarm. A weak radioactive source (usually $^{241}_{95}$Am) ionizes the air in the chamber of the detector, creating charged particles. A voltage is maintained between the plates inside the chamber, setting up a small but detectable current in the external circuit due to the ions acting as charge carriers between the plates. As

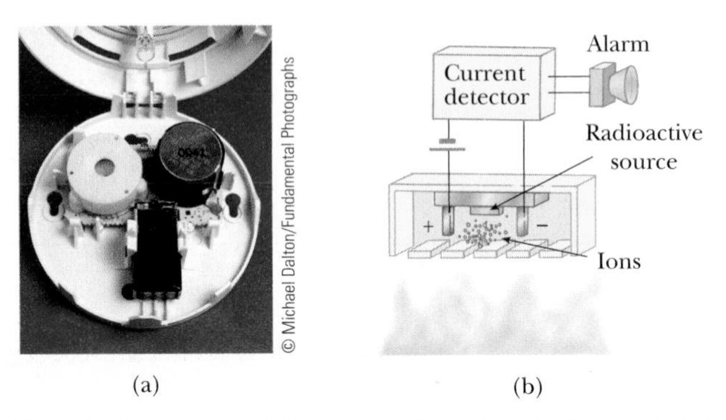

Figure 44.13 (a) A smoke detector uses alpha decay to determine whether smoke is in the air. The alpha source is in the black cylinder at the right. (b) Smoke entering the chamber reduces the detected current, causing the alarm to sound.

long as the current is maintained, the alarm is deactivated. If smoke drifts into the chamber, however, the ions become attached to the smoke particles. These heavier particles do not drift as readily as do the lighter ions, which causes a decrease in the detector current. The external circuit senses this decrease in current and sets off the alarm.

Beta Decay

When a radioactive nucleus undergoes beta decay, the daughter nucleus contains the same number of nucleons as the parent nucleus but the atomic number is changed by 1, which means that the number of protons changes:

$$^{A}_{Z}\text{X} \rightarrow {}_{Z+1}^{\ \ A}\text{Y} + \text{e}^- \quad \text{(incomplete expression)} \qquad \textbf{(44.14)}$$

$$^{A}_{Z}\text{X} \rightarrow {}_{Z-1}^{\ \ A}\text{Y} + \text{e}^+ \quad \text{(incomplete expression)} \qquad \textbf{(44.15)}$$

where, as mentioned in Section 44.4, e^- designates an electron and e^+ designates a positron, with *beta particle* being the general term referring to either. *Beta decay is not described completely by these expressions.* We shall give reasons for this statement shortly.

As with alpha decay, the nucleon number and total charge are both conserved in beta decays. Because A does not change but Z does, we conclude that in beta decay, either a neutron changes to a proton (Eq. 44.14) or a proton changes to a neutron (Eq. 44.15). Note that the electron or positron emitted in these decays is not present beforehand in the nucleus; it is created in the process of the decay from the rest energy of the decaying nucleus. Two typical beta-decay processes are

$$^{14}_{6}\text{C} \rightarrow {}^{14}_{7}\text{N} + \text{e}^- \quad \text{(incomplete expression)} \qquad \textbf{(44.16)}$$

$$^{12}_{7}\text{N} \rightarrow {}^{12}_{6}\text{C} + \text{e}^+ \quad \text{(incomplete expression)} \qquad \textbf{(44.17)}$$

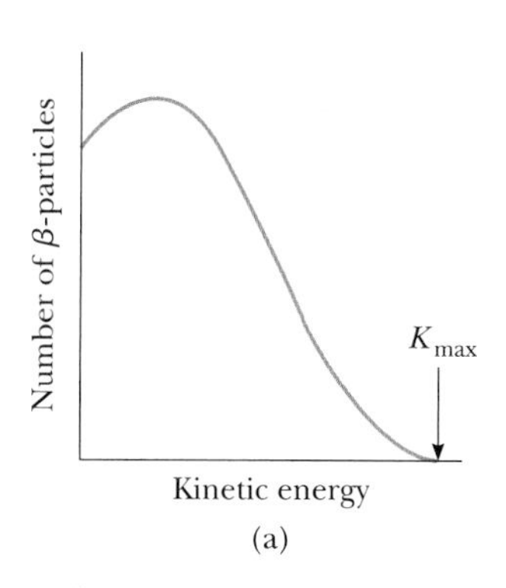

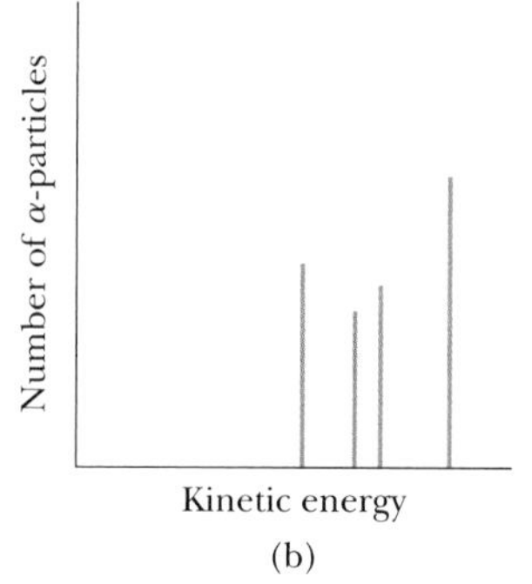

ACTIVE FIGURE 44.14

(a) Distribution of beta particle energies in a typical beta decay. All energies are observed up to a maximum value. (b) In contrast, the energies of alpha particles from an alpha decay are discrete.

Sign in at www.thomsonedu.com and go to ThomsonNOW to observe the development of these graphs for the decays in Active Figures 44.11 and 44.15.

Let's consider the energy of the system undergoing beta decay before and after the decay. As with alpha decay, energy of the isolated system must be conserved. Experimentally, it is found that beta particles from a single type of nucleus are emitted over a continuous range of energies (Active Fig. 44.14a), as opposed to alpha decay, in which the alpha particles are emitted with discrete energies (Active Fig. 44.14b). The kinetic energy of the system after the decay is equal to the decrease in rest energy of the system, that is, the Q value. Because all decaying nuclei in the sample have the same initial mass, however, *the Q value must be the same for each decay.* So, why do the emitted particles have the range of kinetic energies shown in Active Figure 44.14a? The law of conservation of energy seems to be violated! It becomes worse: further analysis of the decay processes described by Equations 44.14 and 44.15 shows that the laws of conservation of angular momentum (spin) and linear momentum are also violated!

After a great deal of experimental and theoretical study, Pauli in 1930 proposed that a third particle must be present in the decay products to carry away the "missing" energy and momentum. Fermi later named this particle the **neutrino** (little neutral one) because it had to be electrically neutral and have little or no mass. Although it eluded detection for many years, the neutrino (symbol ν, Greek nu) was finally detected experimentally in 1956 by Frederick Reines, who received the Nobel Prize in Physics for this work in 1995. The neutrino has the following properties:

◀ Properties of the neutrino

- It has zero electric charge.
- Its mass is either zero (in which case it travels at the speed of light) or very small; much recent persuasive experimental evidence suggests that the neutrino mass is not zero. Current experiments place the upper bound of the mass of the neutrino at approximately 7 eV/c^2.
- It has a spin of $\frac{1}{2}$, which allows the law of conservation of angular momentum to be satisfied in beta decay.
- It interacts very weakly with matter and is therefore very difficult to detect.

We can now write the beta-decay processes (Eqs. 44.14 and 44.15) in their correct and complete form:

Beta decay processes ▶

$$^{A}_{Z}\text{X} \rightarrow {}^{A}_{Z+1}\text{Y} + \text{e}^- + \bar{\nu} \quad \text{(complete expression)} \qquad \textbf{(44.18)}$$

$$^{A}_{Z}\text{X} \rightarrow {}^{A}_{Z-1}\text{Y} + \text{e}^+ + \nu \quad \text{(complete expression)} \qquad \textbf{(44.19)}$$

as well as those for carbon-14 and nitrogen-12 (Eqs. 44.16 and 44.17):

$$^{14}_{6}\text{C} \rightarrow {}^{14}_{7}\text{N} + \text{e}^- + \bar{\nu} \quad \text{(complete expression)} \qquad \textbf{(44.20)}$$

$$^{12}_{7}\text{N} \rightarrow {}^{12}_{6}\text{C} + \text{e}^+ + \nu \quad \text{(complete expression)} \qquad \textbf{(44.21)}$$

where the symbol $\bar{\nu}$ represents the **antineutrino,** the antiparticle to the neutrino. We shall discuss antiparticles further in Chapter 46. For now, it suffices to say that **a neutrino is emitted in positron decay and an antineutrino is emitted in electron decay.** As with alpha decay, the decays listed above are analyzed by applying conservation laws, but relativistic expressions must be used for beta particles because their kinetic energy is large (typically 1 MeV) compared with their rest energy of 0.511 MeV. Active Figure 44.15 shows a pictorial representation of the decays described by Equations 44.20 and 44.21.

PITFALL PREVENTION 44.7
Mass Number of the Electron

An alternative notation for an electron in Equation 44.23 is the symbol ${}^{0}_{-1}\text{e}$, which does not imply that the electron has zero rest energy. The mass of the electron is so much smaller than that of the lightest nucleon, however, that we approximate it as zero in the context of nuclear decays and reactions.

In Equation 44.18, the number of protons has increased by one and the number of neutrons has decreased by one. We can write the fundamental process of e$^-$ decay in terms of a neutron changing into a proton as follows:

$$\text{n} \rightarrow \text{p} + \text{e}^- + \bar{\nu} \qquad \textbf{(44.22)}$$

The electron and the antineutrino are ejected from the nucleus, with the net result that there is one more proton and one fewer neutron, consistent with the changes in Z and $A - Z$. A similar process occurs in e$^+$ decay, with a proton changing into a neutron, a positron, and a neutrino. This latter process can only occur within the nucleus, with the result that the nuclear mass decreases. It cannot occur for an isolated proton because its mass is less than that of the neutron.

A process that competes with e$^+$ decay is **electron capture,** which occurs when a parent nucleus captures one of its own orbital electrons and emits a neutrino. The final product after decay is a nucleus whose charge is $Z - 1$:

Electron capture ▶

$$^{A}_{Z}\text{X} + {}^{0}_{-1}\text{e} \rightarrow {}^{A}_{Z-1}\text{Y} + \nu \qquad \textbf{(44.23)}$$

In most cases, it is a K-shell electron that is captured and the process is therefore referred to as **K capture.** One example is the capture of an electron by ${}^{7}_{4}\text{Be}$:

$$^{7}_{4}\text{Be} + {}^{0}_{-1}\text{e} \rightarrow {}^{7}_{3}\text{Li} + \nu$$

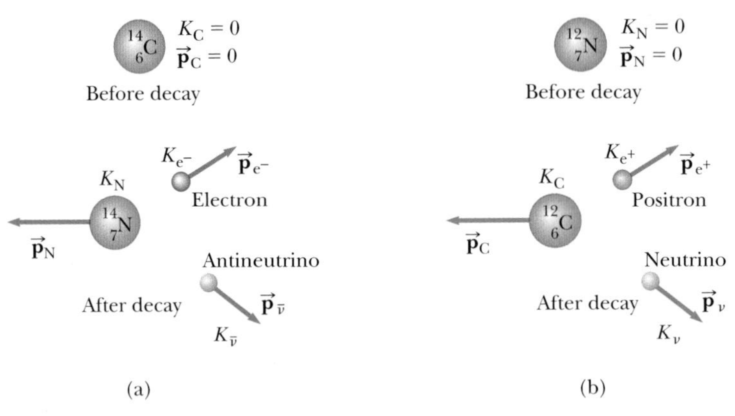

ACTIVE FIGURE 44.15

(a) The beta decay of carbon-14. The final products of the decay are the nitrogen-14 nucleus, an electron, and an antineutrino. (b) The beta decay of nitrogen-12. The final products of the decay are the carbon-12 nucleus, a positron, and a neutrino.

Sign in at www.thomsonedu.com and go to ThomsonNOW to observe the decay of carbon-14. For a large number of decays, observe the development of the graph in Active Figure 44.14a.

Because the neutrino is very difficult to detect, electron capture is usually observed by the x-rays given off as higher-shell electrons cascade downward to fill the vacancy created in the K shell.

Finally, we specify Q values for the beta-decay processes. The Q values for e^- decay and electron capture are given by $Q = (M_X - M_Y)c^2$, where M_X and M_Y are the masses of neutral atoms. In e^- decay, the parent nucleus experiences an increase in atomic number and, for the atom to become neutral, an electron must be absorbed by the atom. If the neutral parent atom and an electron (which will eventually combine with the daughter to form a neutral atom) is the initial system and the final system is the neutral daughter atom and the beta-ejected electron, the system contains a free electron both before and after the decay. Therefore, in subtracting the initial and final masses of the system, this electron mass cancels. The Q values for e^+ decay are given by $Q = (M_X - M_Y - 2m_e)c^2$. The extra term $-2m_ec^2$ in this expression is necessary because the atomic number of the parent decreases by one when the daughter is formed. After it is formed by the decay, the daughter atom sheds one electron to form a neutral atom. Therefore, the final products are the daughter atom, the shed electron, and the ejected positron.

These relationships are useful in determining whether or not a process is energetically possible. For example, the expression for proposed e^+ decay for a particular parent nucleus may turn out to be negative. In that case, this decay does not occur. The expression for electron capture for this parent nucleus, however, may give a positive number, so electron capture can occur even though e^+ decay is not possible. Such is the case for the decay of ${}^{7}_{4}\text{Be}$ shown above.

Quick Quiz 44.4 Which of the following is the correct daughter nucleus associated with the beta decay of ${}^{184}_{72}\text{Hf}$? (a) ${}^{183}_{72}\text{Hf}$ (b) ${}^{183}_{73}\text{Ta}$ (c) ${}^{184}_{73}\text{Ta}$

Carbon Dating

The beta decay of ^{14}C (Eq. 44.20) is commonly used to date organic samples. Cosmic rays in the upper atmosphere cause nuclear reactions (Section 44.7) that create ^{14}C. The ratio of ^{14}C to ^{12}C in the carbon dioxide molecules of our atmosphere has a constant value of approximately 1.3×10^{-12}. The carbon atoms in all living organisms have this same $^{14}\text{C}/^{12}\text{C}$ ratio because the organisms continuously exchange carbon dioxide with their surroundings. When an organism dies, however, it no longer absorbs ^{14}C from the atmosphere, and so the $^{14}\text{C}/^{12}\text{C}$ ratio decreases as the ^{14}C decays with a half-life of 5 730 yr. It is therefore possible to measure the age of a material by measuring its ^{14}C activity. Using this technique, scientists have been able to identify samples of wood, charcoal, bone, and shell as having lived from 1 000 to 25 000 years ago. This knowledge has helped us reconstruct the history of living organisms—including humans—during this time span.

A particularly interesting example is the dating of the Dead Sea Scrolls. This group of manuscripts was discovered by a shepherd in 1947. Translation showed them to be religious documents, including most of the books of the Old Testament. Because of their historical and religious significance, scholars wanted to know their age. Carbon dating applied to the material in which they were wrapped established their age at approximately 1 950 yr.

CONCEPTUAL EXAMPLE 44.8 **The Age of Ice Man**

In 1991, a German tourist discovered the well-preserved remains of a man, now called the "Ice Man," trapped in a glacier in the Italian Alps. (See the photograph at the opening of this chapter.) Radioactive dating with ^{14}C revealed that this person was alive approximately 5 300 years ago. Why did scientists date a sample of the Ice Man using ^{14}C rather than ^{11}C, which is a beta emitter having a half-life of 20.4 min?

SOLUTION

Because ^{14}C has a half-life of 5 730 yr, the fraction of ^{14}C nuclei remaining after one half-life is high enough to allow accurate measurements of changes in the sample's activity. Because ^{11}C has a very short half-life, it is not useful; its activity decreases to a vanishingly small value over the age of the sample, making it impossible to detect.

An isotope used to date a sample must be present in a known amount in the sample when it is formed. As a general rule, the isotope chosen to date a sample should also have a half-life that is on the same order of magnitude as the age of the sample. If the half-life is much less than the age of the sample, there won't be enough activity left to measure because almost all the original radioactive nuclei will have decayed. If the half-life is much greater than the age of the sample, the amount of decay that has taken place since the sample died will be too small to measure. For example, if you have a specimen estimated to have died 50 years ago, neither ^{14}C (5 730 yr) nor ^{11}C (20 min) is suitable. If you know your sample contains hydrogen, however, you can measure the activity of ^{3}H (tritium), a beta emitter that has a half-life of 12.3 yr.

EXAMPLE 44.9 **Radioactive Dating**

A piece of charcoal containing 25.0 g of carbon is found in some ruins of an ancient city. The sample shows a ^{14}C activity R of 250 decays/min. How long has the tree from which this charcoal came been dead?

SOLUTION

Conceptualize Because the charcoal was found in ancient ruins, we expect the current activity to be smaller than the initial activity. If we can determine the initial activity, we can find out how long the wood has been dead.

Categorize The text of the question helps us categorize this example as a carbon dating problem.

Analyze Calculate the decay constant for ^{14}C:

$$\lambda = \frac{0.693}{T_{1/2}} = \frac{0.693}{(5\,730\text{ yr})(3.16 \times 10^7\text{ s/yr})}$$
$$= 3.83 \times 10^{-12}\text{ s}^{-1}$$

Calculate the number of moles in 25.0 g of carbon:

$$n = \frac{25.0\text{ g}}{12.0\text{ g/mol}} = 2.083\text{ mol}$$

Find the number of ^{12}C nuclei in 25.0 g of carbon:

$$N(^{12}\text{C}) = 2.083\text{ mol}(6.02 \times 10^{23}\text{ nuclei/mol}) = 1.25 \times 10^{24}\text{ nuclei}$$

Knowing that the ratio of ^{14}C to ^{12}C in the live sample was 1.3×10^{-12}, find the number of ^{14}C nuclei in 25.0 g *before* decay:

$$N_0(^{14}\text{C}) = (1.3 \times 10^{-12})(1.25 \times 10^{24}) = 1.63 \times 10^{12}\text{ nuclei}$$

Find the initial activity of the sample:

$$R_0 = \lambda N_0 = (3.83 \times 10^{-12}\text{ s}^{-1})(1.63 \times 10^{12}\text{ nuclei})$$
$$= 6.24\text{ decays/s} = 374\text{ decays/min}$$

Solve Equation 44.7 for t:

$$e^{-\lambda t} = \frac{R}{R_0} \rightarrow -\lambda t = \ln\left(\frac{R}{R_0}\right) \rightarrow t = -\frac{1}{\lambda}\ln\left(\frac{R}{R_0}\right)$$

Evaluate t using $R = 250$ decays/min and $R_0 = 368$ decays/min:

$$t = -\frac{1}{3.83 \times 10^{-12}\text{ s}^{-1}}\ln\left(\frac{250\text{ min}^{-1}}{374\text{ min}^{-1}}\right)$$
$$= 1.06 \times 10^{11}\text{ s} = 3.3 \times 10^3\text{ yr}$$

Finalize Note that the time interval found here is on the same order of magnitude as the half-life, so ^{14}C is a valid isotope to use for this sample, as discussed in Conceptual Example 44.8.

Gamma Decay

Very often, a nucleus that undergoes radioactive decay is left in an excited energy state. The nucleus can then undergo a second decay to a lower energy state, perhaps to the ground state, by emitting a high-energy photon:

$$ {}^{A}_{Z}X^* \rightarrow {}^{A}_{Z}X + \gamma \qquad (44.24)$$

◀ Gamma decay

where X* indicates a nucleus in an excited state. The typical half-life of an excited nuclear state is 10^{-10} s. Photons emitted in such a de-excitation process are called gamma rays. Such photons have very high energy (1 MeV to 1 GeV) relative to the energy of visible light (approximately 1 eV). Recall from Section 42.3 that the energy of a photon emitted or absorbed by an atom equals the difference in energy between the two electronic states involved in the transition. Similarly, a gamma-ray photon has an energy hf that equals the energy difference ΔE between two nuclear energy levels. When a nucleus decays by emitting a gamma ray, the only change in the nucleus is that it ends up in a lower energy state. There are no changes in Z, N, or A.

A nucleus may reach an excited state as the result of a violent collision with another particle. More common, however, is for a nucleus to be in an excited state after it has undergone alpha or beta decay. The following sequence of events represents a typical situation in which gamma decay occurs:

$$ {}^{12}_{5}B \rightarrow {}^{12}_{6}C^* + e^- + \bar{\nu} \qquad (44.25)$$

$$ {}^{12}_{6}C^* \rightarrow {}^{12}_{6}C + \gamma \qquad (44.26)$$

Figure 44.16 shows the decay scheme for ^{12}B, which undergoes beta decay to either of two levels of ^{12}C. It can either (1) decay directly to the ground state of ^{12}C by emitting a 13.4-MeV electron or (2) undergo beta decay to an excited state of $^{12}C^*$ followed by gamma decay to the ground state. The latter process results in the emission of a 9.0-MeV electron and a 4.4-MeV photon.

The various pathways by which a radioactive nucleus can undergo decay are summarized in Table 44.3.

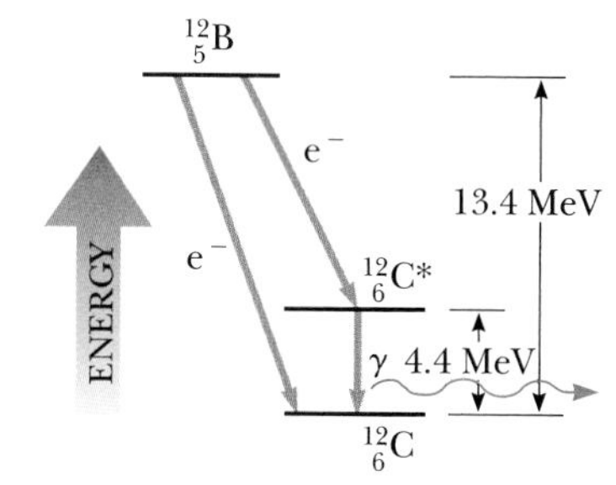

Figure 44.16 An energy-level diagram showing the initial nuclear state of a ^{12}B nucleus and two possible lower-energy states of the ^{12}C nucleus. The beta decay of the ^{12}B nucleus can result in either of two situations: the ^{12}C nucleus is in the ground state or in the excited state, in which case the nucleus is denoted as $^{12}C^*$. In the latter case, the beta decay to $^{12}C^*$ is followed by a gamma decay to ^{12}C as the excited nucleus makes a transition to the ground state.

44.6 Natural Radioactivity

Radioactive nuclei are generally classified into two groups: (1) unstable nuclei found in nature, which give rise to **natural radioactivity,** and (2) unstable nuclei produced in the laboratory through nuclear reactions, which exhibit **artificial radioactivity.**

As Table 44.4 shows, there are three series of naturally occurring radioactive nuclei. Each series starts with a specific long-lived radioactive isotope whose half-life exceeds that of any of its unstable descendants. The three natural series begin with the isotopes ^{238}U, ^{235}U, and ^{232}Th, and the corresponding stable end products are three isotopes of lead: ^{206}Pb, ^{207}Pb, and ^{208}Pb. The fourth series in Table 44.4 begins with ^{237}Np and has as its stable end product ^{209}Bi. The element ^{237}Np is a

TABLE 44.3

Various Decay Pathways

Alpha decay	${}^{A}_{Z}X \rightarrow {}^{A-4}_{Z-2}Y + {}^{4}_{2}He$
Beta decay (e^-)	${}^{A}_{Z}X \rightarrow {}_{Z+1}^{A}Y + e^- + \bar{\nu}$
Beta decay (e^+)	${}^{A}_{Z}X \rightarrow {}_{Z-1}^{A}Y + e^+ + \nu$
Electron capture	${}^{A}_{Z}X + {}^{0}_{-1}e \rightarrow {}_{Z-1}^{A}Y + \nu$
Gamma decay	${}^{A}_{Z}X^* \rightarrow {}^{A}_{Z}X + \gamma$

TABLE 44.4

The Four Radioactive Series

Series		Starting Isotope	Half-life (years)	Stable End Product
Uranium	Natural	${}^{238}_{92}U$	4.47×10^9	${}^{206}_{82}Pb$
Actinium	Natural	${}^{235}_{92}U$	7.04×10^8	${}^{207}_{82}Pb$
Thorium	Natural	${}^{232}_{90}Th$	1.41×10^{10}	${}^{208}_{82}Pb$
Neptunium		${}^{237}_{93}Np$	2.14×10^6	${}^{209}_{83}Bi$

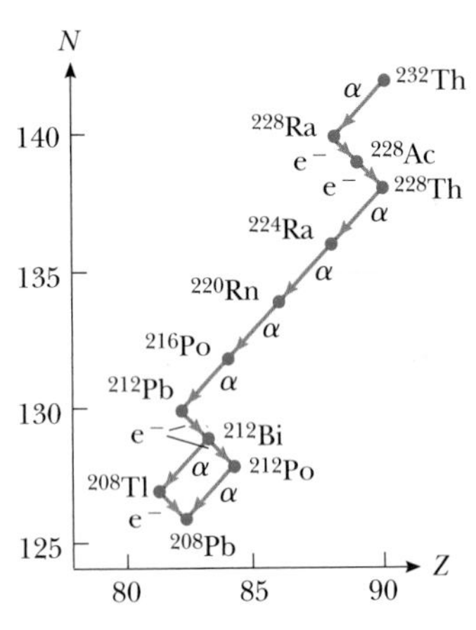

Figure 44.17 Successive decays for the ^{232}Th series.

transuranic element (one having an atomic number greater than that of uranium) not found in nature. This element has a half-life of "only" 2.14×10^6 years.

Figure 44.17 shows the successive decays for the ^{232}Th series. First, ^{232}Th undergoes alpha decay to ^{228}Ra. Next, ^{228}Ra undergoes two successive beta decays to ^{228}Th. The series continues and finally branches when it reaches ^{212}Bi. At this point, there are two decay possibilities. The sequence shown in Figure 44.17 is characterized by a mass-number decrease of either 4 (for alpha decays) or 0 (for beta or gamma decays). The two uranium series are more complex than the ^{232}Th series. In addition, several naturally occurring radioactive isotopes, such as ^{14}C and ^{40}K, are not part of any decay series.

Because of these radioactive series, our environment is constantly replenished with radioactive elements that would otherwise have disappeared long ago. For example, because the Solar System is approximately 5×10^9 years old, the supply of ^{226}Ra (whose half-life is only 1 600 years) would have been depleted by radioactive decay long ago if it were not for the radioactive series starting with ^{238}U.

44.7 Nuclear Reactions

We have studied radioactivity, which is a spontaneous process in which the structure of a nucleus changes. It is also possible to stimulate changes in the structure of nuclei by bombarding them with energetic particles. Such collisions, which change the identity of the target nuclei, are called **nuclear reactions.** Rutherford was the first to observe them, in 1919, using naturally occurring radioactive sources for the bombarding particles. Since then, thousands of nuclear reactions have been observed following the development of charged-particle accelerators in the 1930s. With today's advanced technology in particle accelerators and particle detectors, it is possible to achieve particle energies of at least 1 000 GeV = 1 TeV. These high-energy particles are used to create new particles whose properties are helping to solve the mysteries of the nucleus.

Consider a reaction in which a target nucleus X is bombarded by a particle a, resulting in a daughter nucleus Y and an outgoing particle b:

Nuclear reaction ▶

$$\text{a} + \text{X} \rightarrow \text{Y} + \text{b} \tag{44.27}$$

Sometimes this reaction is written in the more compact form

$$\text{X(a, b)Y}$$

In Section 44.5, the Q value, or disintegration energy, of a radioactive decay was defined as the rest energy transformed to kinetic energy as a result of the decay process. Likewise, we define the **reaction energy** Q associated with a nuclear reaction as *the total change in rest energy resulting from the reaction:*

Reaction energy Q ▶

$$Q = (M_\text{a} + M_\text{X} - M_\text{Y} - M_\text{b})c^2 \tag{44.28}$$

As an example, consider the reaction $^7\text{Li}(\text{p}, \alpha)^4\text{He}$. The notation p indicates a proton, which is a hydrogen nucleus. Therefore, we can write this reaction in the expanded form

$$^1_1\text{H} + ^7_3\text{Li} \rightarrow ^4_2\text{He} + ^4_2\text{He}$$

The Q value for this reaction is 17.3 MeV. A reaction such as this one, for which Q is positive, is called **exothermic.** A reaction for which Q is negative is called **endothermic.** To satisfy conservation of momentum, an endothermic reaction does not occur unless the bombarding particle has a kinetic energy greater than Q. (See Problem 49.) The minimum energy necessary for such a reaction to occur is called the **threshold energy.**

If particles a and b in a nuclear reaction are identical so that X and Y are also necessarily identical, the reaction is called a **scattering event.** If the kinetic energy of the system (a and X) before the event is the same as that of the system (b and

Y) after the event, it is classified as *elastic scattering.* If the kinetic energy of the system after the event is less than that before the event, the reaction is described as *inelastic scattering.* In this case, the target nucleus has been raised to an excited state by the event, which accounts for the difference in energy. The final system now consists of b and an excited nucleus Y*, and eventually it will become b, Y, and γ, where γ is the gamma-ray photon that is emitted when the system returns to the ground state. This elastic and inelastic terminology is identical to that used in describing collisions between macroscopic objects as discussed in Section 9.3.

In addition to energy and momentum, the total charge and total number of nucleons must be conserved in any nuclear reaction. For example, consider the reaction $^{19}\text{F}(\text{p}, \alpha)^{16}\text{O}$, which has a Q value of 8.11 MeV. We can show this reaction more completely as

$$^{1}_{1}\text{H} + ^{19}_{9}\text{F} \rightarrow ^{16}_{8}\text{O} + ^{4}_{2}\text{He} \qquad \textbf{(44.29)}$$

The total number of nucleons before the reaction $(1 + 19 = 20)$ is equal to the total number after the reaction $(16 + 4 = 20)$. Furthermore, the total charge is the same before $(1 + 9)$ and after $(8 + 2)$ the reaction.

44.8 Nuclear Magnetic Resonance and Magnetic Resonance Imaging

In this section, we describe an important application of nuclear physics in medicine called magnetic resonance imaging. To understand this application, we first discuss the spin angular momentum of the nucleus. This discussion has parallels with the discussion of spin for atomic electrons.

In Chapter 42, we discussed that the electron has an intrinsic angular momentum, called spin. Nuclei also have spin because their component particles—neutrons and protons—each have spin $\frac{1}{2}$ as well as orbital angular momentum within the nucleus. All types of angular momentum obey the quantum rules that were outlined for orbital and spin angular momentum in Chapter 42. In particular, two quantum numbers associated with the angular momentum determine the allowed values of the magnitude of the angular momentum vector and its direction in space. The magnitude of the nuclear angular momentum is $\sqrt{I(I+1)}\hbar$, where I is called the **nuclear spin quantum number** and may be an integer or a half-integer, depending on how the individual proton and neutron spins combine. The quantum number I is the analog to ℓ for the electron in an atom as discussed in Section 42.6. Furthermore, there is a quantum number m_I that is the analog to m_ℓ, in that the allowed projections of the nuclear spin angular momentum vector on the z axis are $m_I\hbar$. The values of m_I range from $-I$ to $+I$ in steps of 1. (In fact, for *any* type of spin with a quantum number S, there is a quantum number m_S that ranges in value from $-S$ to $+S$ in steps of 1.) Therefore, the maximum value of the z component of the spin angular momentum vector is $I\hbar$. Figure 44.18 is a vector model (see Section 42.6) illustrating the possible orientations of the nuclear spin vector and its projections along the z axis for the case in which $I = \frac{3}{2}$.

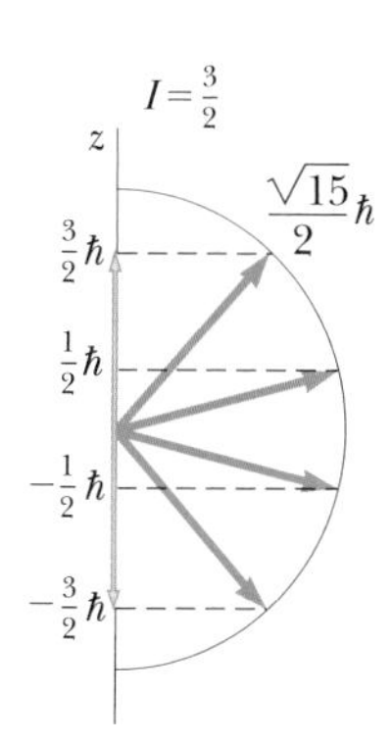

Figure 44.18 A vector model showing possible orientations of the nuclear spin angular momentum vector and its projections along the z axis for the case $I = \frac{3}{2}$.

Nuclear spin has an associated nuclear magnetic moment, similar to that of the electron. The spin magnetic moment of a nucleus is measured in terms of the **nuclear magneton** μ_n, a unit of moment defined as

$$\mu_n \equiv \frac{e\hbar}{2m_p} = 5.05 \times 10^{-27}\ \text{J/T} \qquad \textbf{(44.30)}$$

◀ Nuclear magneton

where m_p is the mass of the proton. This definition is analogous to that of the Bohr magneton μ_B, which corresponds to the spin magnetic moment of a free electron (see Section 42.6). Note that μ_n is smaller than μ_B $(= 9.274 \times 10^{-24}\ \text{J/T})$ by a factor of 1 836 because of the large difference between the proton mass and the electron mass.

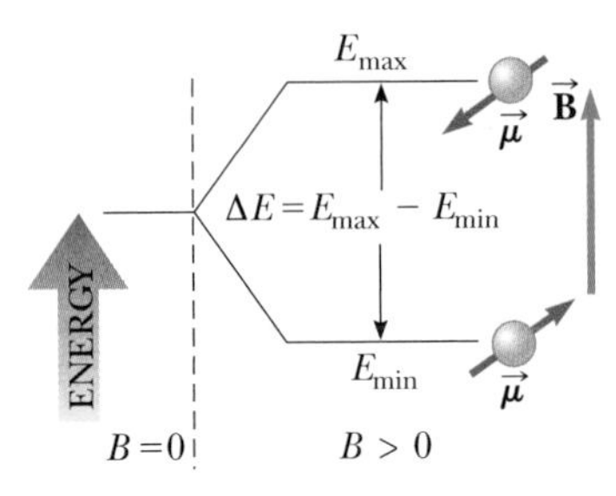

Figure 44.19 A nucleus with spin $\frac{1}{2}$ can occupy one of two energy states when placed in an external magnetic field. The lower-energy state E_{min} corresponds to the case where the spin is aligned with the field as much as possible according to quantum mechanics, and the higher-energy state E_{max} corresponds to the case where the spin is opposite the field as much as possible.

The magnetic moment of a free proton is $2.792\ 8\mu_n$. Unfortunately, there is no general theory of nuclear magnetism that explains this value. The neutron also has a magnetic moment, which has a value of $-1.913\ 5\mu_n$. The negative sign indicates that this moment is opposite the spin angular momentum of the neutron. The existence of a magnetic moment for the neutron is surprising in view of the neutron being uncharged. That suggests that the neutron is not a fundamental particle but rather has an underlying structure consisting of charged constituents. We shall explore this structure in Chapter 46.

The potential energy associated with a magnetic dipole moment $\vec{\mu}$ in an external magnetic field $\vec{B}$ is given by $-\vec{\mu} \cdot \vec{B}$ (Eq. 29.18). When the magnetic moment $\vec{\mu}$ is lined up with the field as closely as quantum physics allows, the potential energy of the dipole–field system has its minimum value E_{min}. When $\vec{\mu}$ is as antiparallel to the field as possible, the potential energy has its maximum value E_{max}. In general, there are other energy states between these values corresponding to the quantized directions of the magnetic moment with respect to the field. For a nucleus with spin $\frac{1}{2}$, there are only two allowed states, with energies E_{min} and E_{max}. These two energy states are shown in Figure 44.19.

It is possible to observe transitions between these two spin states using a technique called **NMR,** for **nuclear magnetic resonance.** A constant magnetic field ($\vec{B}$ in Fig. 44.19) is introduced to define a z axis and split the energies of the spin states. A second, weaker, oscillating magnetic field is then applied perpendicular to $\vec{B}$, creating a cloud of radio-frequency photons around the sample. When the frequency of the oscillating field is adjusted so that the photon energy matches the energy difference between the spin states, there is a net absorption of photons by the nuclei that can be detected electronically.

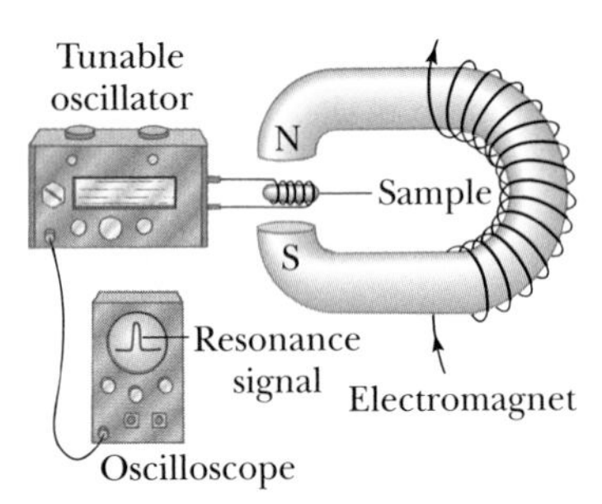

Figure 44.20 Experimental arrangement for nuclear magnetic resonance. The radio-frequency magnetic field created by the coil surrounding the sample and provided by the variable-frequency oscillator is perpendicular to the constant magnetic field created by the electromagnet. When the nuclei in the sample meet the resonance condition, the nuclei absorb energy from the radio-frequency field of the coil; this absorption changes the characteristics of the circuit in which the coil is included. Most modern NMR spectrometers use superconducting magnets at fixed field strengths and operate at frequencies of approximately 200 MHz.

Figure 44.20 is a simplified diagram of the apparatus used in nuclear magnetic resonance. The energy absorbed by the nuclei is supplied by the generator producing the oscillating magnetic field. Nuclear magnetic resonance and a related technique called *electron spin resonance* are extremely important methods for studying nuclear and atomic systems and the ways in which these systems interact with their surroundings.

A widely used medical diagnostic technique called **MRI,** for **magnetic resonance imaging,** is based on nuclear magnetic resonance. Because nearly two-thirds of the atoms in the human body are hydrogen (which gives a strong NMR signal), MRI works exceptionally well for viewing internal tissues. The patient is placed inside a large solenoid that supplies a magnetic field that is constant in time but whose magnitude varies spatially across the body. Because of the variation in the field, hydrogen atoms in different parts of the body have different energy splittings between spin states, so the resonance signal can be used to provide information about the positions of the protons. A computer is used to analyze the position information to provide data for constructing a final image. Contrast in the final image among different types of tissues is created by computer analysis of the time intervals for the nuclei to return to the lower-energy spin state between pulses of radio-frequency photons. Contrast can be enhanced with the use of contrast agents such as gadolinium compounds or iron oxide nanoparticles taken orally or injected intravenously. An MRI scan showing incredible detail in internal body structure is shown in Figure 44.21.

The main advantage of MRI over other imaging techniques is that it causes minimal cellular damage. The photons associated with the radio-frequency signals used in MRI have energies of only about 10^{-7} eV. Because molecular bond strengths are much larger (approximately 1 eV), the radio-frequency radiation causes little cellular damage. In comparison, x-rays have energies ranging from 10^4 to 10^6 eV and can cause considerable cellular damage. Therefore, despite some individuals' fears of the word *nuclear* associated with MRI, the radio-frequency radiation involved is overwhelmingly safer than the x-rays that these individuals might accept more readily. A disadvantage of MRI is that the equipment required to conduct the procedure is very expensive, so MRI images are costly.

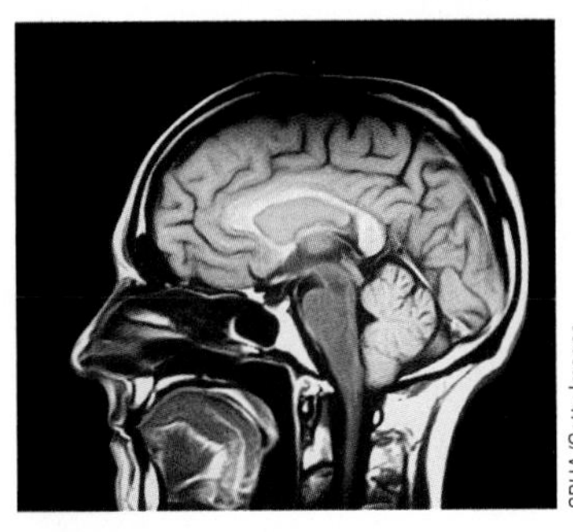

Figure 44.21 A color-enhanced MRI scan of a human brain.

The magnetic field produced by the solenoid is sufficient to lift a car, and the radio signal is about the same magnitude as that from a small commercial broadcasting station. Although MRI is inherently safe in normal use, the strong magnetic field of the solenoid requires diligent care to ensure that no ferromagnetic materials are located in the room near the MRI apparatus. Several accidents have occurred, such as a 2000 incident in which a gun pulled from a police officer's hand discharged upon striking the machine.

Summary

ThomsonNOW™ Sign in at **www.thomsonedu.com** and go to ThomsonNOW to take a practice test for this chapter.

DEFINITIONS

A nucleus is represented by the symbol ${}^{A}_{Z}\text{X}$, where A is the **mass number** (the total number of nucleons) and Z is the **atomic number** (the total number of protons). The total number of neutrons in a nucleus is the **neutron number** N, where $A = N + Z$. Nuclei having the same Z value but different A and N values are **isotopes** of each other.

The magnetic moment of a nucleus is measured in terms of the **nuclear magneton** μ_n, where

$$\mu_n \equiv \frac{e\hbar}{2m_p} = 5.05 \times 10^{-27}\ \text{J/T} \qquad \textbf{(44.30)}$$

CONCEPTS AND PRINCIPLES

Assuming that nuclei are spherical, their radius is given by

$$r = r_0 A^{1/3} \qquad \textbf{(44.1)}$$

where $r_0 = 1.2$ fm.

Nuclei are stable because of the **nuclear force** between nucleons. This short-range force dominates the Coulomb repulsive force at distances of less than about 2 fm and is independent of charge. Light stable nuclei have equal numbers of protons and neutrons. Heavy stable nuclei have more neutrons than protons. The most stable nuclei have Z and N values that are both even.

The difference between the sum of the masses of a group of separate nucleons and the mass of the compound nucleus containing these nucleons, when multiplied by c^2, gives the **binding energy** E_b of the nucleus. The binding energy of the nucleus of mass M_A can be calculated using the expression

$$E_b\ (\text{MeV}) = [ZM(\text{H}) + Nm_n - M({}^{A}_{Z}\text{X})] \times 931.494\ \text{MeV/u} \qquad \textbf{(44.2)}$$

where $M(\text{H})$ is the atomic mass of the neutral hydrogen atom, $M({}^{A}_{Z}\text{X})$ represents the atomic mass of an atom of the isotope ${}^{A}_{Z}\text{X}$, and m_n is the mass of the neutron.

The **liquid-drop model** of nuclear structure treats the nucleons as molecules in a drop of liquid. The four main contributions influencing binding energy are the volume effect, the surface effect, the Coulomb repulsion effect, and the symmetry effect. Summing such contributions results in the **semiempirical binding-energy formula:**

$$E_b = C_1 A - C_2 A^{2/3} - C_3 \frac{Z(Z-1)}{A^{1/3}} - C_4 \frac{(N-Z)^2}{A} \qquad \textbf{(44.3)}$$

The **shell model,** or **independent-particle model,** assumes each nucleon exists in a shell and can only have discrete energy values. The stability of certain nuclei can be explained with this model.

(*continued*)

A radioactive substance decays by **alpha decay, beta decay,** or **gamma decay.** An alpha particle is the ^{4}He nucleus, a beta particle is either an electron (e^-) or a positron (e^+), and a gamma particle is a high-energy photon.

If a radioactive material contains N_0 radioactive nuclei at $t = 0$, the number N of nuclei remaining after a time t has elapsed is

$$N = N_0 e^{-\lambda t} \tag{44.6}$$

where λ is the **decay constant,** a number equal to the probability per second that a nucleus will decay. The **decay rate,** or **activity,** of a radioactive substance is

$$R = \left|\frac{dN}{dt}\right| = R_0 e^{-\lambda t} \tag{44.7}$$

where $R_0 = \lambda N_0$ is the activity at $t = 0$. The **half-life** $T_{1/2}$ is the time interval required for half of a given number of radioactive nuclei to decay, where

$$T_{1/2} = \frac{0.693}{\lambda} \tag{44.8}$$

In alpha decay, a helium nucleus is ejected from the parent nucleus with a discrete set of kinetic energies. A nucleus undergoing beta decay emits either an electron (e^-) and an antineutrino ($\bar{\nu}$) or a positron (e^+) and a neutrino (ν). The electron or positron is ejected with a range of energies. In **electron capture,** the nucleus of an atom absorbs one of its own electrons and emits a neutrino. In gamma decay, a nucleus in an excited state decays to its ground state and emits a gamma ray.

Nuclear reactions can occur when a target nucleus X is bombarded by a particle a, resulting in a daughter nucleus Y and an outgoing particle b:

$$\mathrm{a} + \mathrm{X} \rightarrow \mathrm{Y} + \mathrm{b} \tag{44.27}$$

The mass-energy conversion in such a reaction, called the **reaction energy** Q, is

$$Q = (M_a + M_X - M_Y - M_b)c^2 \tag{44.28}$$

Questions

☐ denotes answer available in *Student Solutions Manual/Study Guide;* **O** denotes objective question

1. In Rutherford's experiment, assume an alpha particle is headed directly toward the nucleus of an atom. Why doesn't the alpha particle make physical contact with the nucleus?
2. **O** Figure 44.17 shows that the nucleus ^{212}Bi can decay both by emitting an alpha particle and by emitting an electron (with a different probability). As formed naturally, this isotope also emits gamma rays. Let W represent such a nucleus. Let X represent the daughter nucleus produced by its alpha decay, Y the daughter nucleus after beta decay, and Z the nucleus after gamma emission. (a) Rank the nuclei W, X, Y, and Z according to their mass numbers from the largest to the smallest. Note in your ranking any cases of equality. (b) Rank the nuclei according to their atomic numbers. (c) Rank the nuclei according to the numbers of neutrons they contain.
3. Explain why nuclei that are well off the line of stability in Figure 44.4 tend to be unstable.
4. Why are very heavy nuclei unstable?
5. **O** Consider two heavy nuclei X and Y having similar mass numbers. If X has the higher binding energy, which nucleus tends to be more unstable?
6. ☐ Why do nearly all the naturally occurring isotopes lie above the $N = Z$ line in Figure 44.4?
7. **O** **(i)** To predict the behavior of a nucleus in a fission reaction, which model would be more appropriate, (a) the liquid-drop model or (b) the shell model? **(ii)** Which model would be more successful in predicting the magnetic moment of a given nucleus? **(iii)** Which could better explain the gamma-ray spectrum of an excited nucleus?
8. "If no more people were to be born, the law of population growth would strongly resemble the radioactive decay law." Discuss this statement.
9. ☐ **O** Two samples of the same radioactive nuclide are prepared. Sample G has twice the initial activity of sample H. **(i)** How does the half-life of G compare with the half-life of H? (a) It is two times larger. (b) It is the same. (c) It is half as large. **(ii)** After each has passed through five half-lives, how do their activities compare? (a) G has more than twice the activity of H. (b) G has twice the activity of H. (c) G and H have the same activity. (d) G has lower activity than H.
10. *Do two halves make a whole?* What fraction of a radioactive sample has decayed after two half-lives have elapsed?

11. The radioactive nucleus $^{226}_{88}Ra$ has a half-life of approximately 1.6×10^3 years. Being that the Solar System is approximately 5 billion years old, why do we still find this nucleus in nature (Fig. Q44.11)?

© 1990 Richard Megna, Fundamental Photographs

Figure Q44.11 Paint on the hands and numbers of this antique watch contains a small amount of natural radium mixed with a phosphorescent material. The decay of the radium causes the phosphor to glow continuously.

12. **O** A free neutron undergoes beta decay by emitting an electron with a half-life of 614 s. Can a free proton undergo a similar decay? (a) yes, the same decay (b) yes, but by emitting a positron (c) yes, but with a very different half-life (d) no

13. If a nucleus such as ^{226}Ra initially at rest undergoes alpha decay, which has more kinetic energy after the decay, the alpha particle or the daughter nucleus? Explain your answer.

14. Can a nucleus emit alpha particles that have different energies? Explain.

15. **O** Which of the following quantities represents the reaction energy of a nuclear reaction?
(a) (final mass − initial mass)/c^2
(b) (initial mass − final mass)/c^2
(c) (final mass − initial mass)c^2
(d) (initial mass − final mass)c^2
(e) none of these choices

16. Suppose it could be shown that the cosmic-ray intensity at the Earth's surface was much greater 10 000 years ago. How would this difference affect what we accept as valid carbon-dated values of the age of ancient samples of once-living matter? Explain your answer.

17. How many values of I_z are possible for $I = \frac{5}{2}$? For $I = 3$?

18. **O** In nuclear magnetic resonance, how does increasing the value of the constant magnetic field change the frequency of the radio-frequency field that excites a particular transition? (a) The frequency is proportional to the square of the constant field. (b) The frequency is directly proportional to the constant field. (c) The frequency is independent of the constant field. (d) The frequency is inversely proportional to the constant field. (e) The frequency is proportional to the reciprocal of the square of the constant field.

19. Do all natural events have causes? Is the Universe intelligible? Give reasons for your answers. *Note:* You may wish to consider again Question 16 in Chapter 6 on whether the future is determinate.

Problems

WebAssign The Problems from this chapter may be assigned online in WebAssign.

ThomsonNOW Sign in at **www.thomsonedu.com** and go to ThomsonNOW to assess your understanding of this chapter's topics with additional quizzing and conceptual questions.

1, 2, 3 denotes straightforward, intermediate, challenging; □ denotes full solution available in *Student Solutions Manual/Study Guide;* ▲ denotes coached solution with hints available at **www.thomsonedu.com;** denotes developing symbolic reasoning; ● denotes asking for qualitative reasoning; denotes computer useful in solving problem

Section 44.1 Some Properties of Nuclei

1. What is the order of magnitude of the number of protons in your body? Of the number of neutrons? Of the number of electrons?

2. **Review problem.** Singly ionized carbon is accelerated through 1 000 V and passed into a mass spectrometer to determine the isotopes present (see Chapter 29). The magnitude of the magnetic field in the spectrometer is 0.200 T. (a) Determine the orbit radii for the ^{12}C and the ^{13}C isotopes as they pass through the field. (b) Show that the ratio of radii may be written in the form

$$\frac{r_1}{r_2} = \sqrt{\frac{m_1}{m_2}}$$

and verify that your radii in part (a) agree with this equation.

3. (a) What fraction of the space in a tank of hydrogen gas at 0°C and 1 atm is occupied by the hydrogen molecules themselves? Assume each hydrogen atom is a sphere with diameter 0.100 nm and a hydrogen molecule consists of two such spheres in contact. (b) What fraction of the space within one hydrogen atom is occupied by its nucleus, of radius 1.20 fm?

4. ● In a Rutherford scattering experiment, alpha particles having kinetic energy of 7.70 MeV are fired toward a gold nucleus. (a) Use energy conservation to determine the distance of closest approach between the alpha particle and gold nucleus. Assume the nucleus remains at rest. (b) **What If?** Calculate the de Broglie wavelength for the 7.70-MeV alpha particle and compare it with the distance obtained in part (a). (c) Based on this comparison, why is it proper to treat the alpha particle as a particle and not as a wave in the Rutherford scattering experiment?

5. (a) Use energy methods to calculate the distance of closest approach for a head-on collision between an alpha particle having an initial energy of 0.500 MeV and a gold nucleus (^{197}Au) at rest. Assume the gold nucleus remains at rest during the collision. (b) What minimum initial speed must the alpha particle have to get as close as 300 fm?

6. Find the radius of (a) a nucleus of $^{4}_{2}\mathrm{He}$ and (b) a nucleus of $^{238}_{92}\mathrm{U}$.

7. A star ending its life with a mass of two times the Sun's mass is expected to collapse, combining its protons and electrons to form a neutron star. Such a star could be thought of as a gigantic atomic nucleus. If a star of mass $2 \times 1.99 \times 10^{30}$ kg collapsed into neutrons ($m_n = 1.67 \times 10^{-27}$ kg), what would its radius be? Assume $r = r_0 A^{1/3}$.

8. **Review problem.** What would be the gravitational force exerted by each of two golf balls on the other if they were made of nuclear matter? Assume each ball has a 4.30-cm diameter and they are 1.00 m apart.

Section 44.2 Nuclear Binding Energy

9. Calculate the binding energy per nucleon for (a) $^{2}\mathrm{H}$, (b) $^{4}\mathrm{He}$, (c) $^{56}\mathrm{Fe}$, and (d) $^{238}\mathrm{U}$.

10. ● The iron isotope $^{56}\mathrm{Fe}$ is near the peak of the stability curve. That is the fundamental reason that iron is more common in the Universe than heavier elements, as the spectra of the Sun and of many other stars reveal. Show that $^{56}\mathrm{Fe}$ has a higher binding energy per nucleon than its neighbors $^{55}\mathrm{Mn}$ and $^{59}\mathrm{Co}$. State how your results compare with Figure 44.5.

11. ▲ Nuclei having the same mass numbers are called *isobars*. The isotope $^{139}_{57}\mathrm{La}$ is stable. A radioactive isobar, $^{139}_{59}\mathrm{Pr}$, is located below the line of stable nuclei in Figure 44.4 and decays by e^+ emission. Another radioactive isobar of $^{139}_{57}\mathrm{La}$, $^{139}_{55}\mathrm{Cs}$, decays by e^- emission and is located above the line of stable nuclei in Figure 44.4. (a) Which of these three isobars has the highest neutron-to-proton ratio? (b) Which has the greatest binding energy per nucleon? (c) Which do you expect to be heavier, $^{139}_{59}\mathrm{Pr}$ or $^{139}_{55}\mathrm{Cs}$?

12. ● Two nuclei having the same mass number are called *isobars*. Calculate the difference in binding energy per nucleon for the isobars $^{23}_{11}\mathrm{Na}$ and $^{23}_{12}\mathrm{Mg}$. How do you account for the difference?

13. ▲ A pair of nuclei for which $Z_1 = N_2$ and $Z_2 = N_1$ are called *mirror isobars* (the atomic and neutron numbers are interchanged). Binding-energy measurements on these nuclei can be used to obtain evidence of the charge independence of nuclear forces (that is, proton–proton, proton–neutron, and neutron–neutron nuclear forces are equal). Calculate the difference in binding energy for the two mirror isobars $^{15}_{8}\mathrm{O}$ and $^{15}_{7}\mathrm{N}$. The electric repulsion among eight protons rather than seven accounts for the difference.

14. ● The energy required to construct a uniformly charged sphere of total charge Q and radius R is $U = 3k_e Q^2/5R$, where k_e is the Coulomb constant (see Problem 64). Assume a $^{40}\mathrm{Ca}$ nucleus contains 20 protons uniformly distributed in a spherical volume. (a) How much energy is required to counter their electrical repulsion according to the above equation? *Suggestion:* First calculate the radius of a $^{40}\mathrm{Ca}$ nucleus. (b) Calculate the binding energy of $^{40}\mathrm{Ca}$. (c) Explain what you can conclude from comparing the result of part (b) with that of part (a).

15. Calculate the minimum energy required to remove a neutron from the $^{43}_{20}\mathrm{Ca}$ nucleus.

Section 44.3 Nuclear Models

16. ● (a) In the liquid-drop model of nuclear structure, why does the surface-effect term $-C_2 A^{2/3}$ have a negative sign? (b) **What If?** The binding energy of the nucleus increases as the volume-to-surface ratio increases. Calculate this ratio for both spherical and cubical shapes and explain which is more plausible for nuclei.

17. Using the graph in Figure 44.5, estimate how much energy is released when a nucleus of mass number 200 fissions into two nuclei each of mass number 100.

18. (a) Use the semiempirical binding-energy formula to compute the binding energy for $^{56}_{26}\mathrm{Fe}$. (b) What percentage is contributed to the binding energy by each of the four terms?

Section 44.4 Radioactivity

19. A sample of radioactive material contains 1.00×10^{15} atoms and has an activity of 6.00×10^{11} Bq. What is its half-life?

20. The half-life of $^{131}\mathrm{I}$ is 8.04 days. On a certain day, the activity of an iodine-131 sample is 6.40 mCi. What is its activity 40.2 days later?

21. A freshly prepared sample of a certain radioactive isotope has an activity of 10.0 mCi. After 4.00 h, its activity is 8.00 mCi. (a) Find the decay constant and half-life. (b) How many atoms of the isotope were contained in the freshly prepared sample? (c) What is the sample's activity 30.0 h after it is prepared?

22. From the equation expressing the law of radioactive decay, derive the following useful formulas for the decay constant and the half-life, in terms of the time interval Δt during which the decay rate decreases from R_0 to R:

$$\lambda = \frac{1}{\Delta t}\ln\left(\frac{R_0}{R}\right) \qquad T_{1/2} = \frac{(\ln 2)\Delta t}{\ln(R_0/R)}$$

23. The radioactive isotope $^{198}\mathrm{Au}$ has a half-life of 64.8 h. A sample containing this isotope has an initial activity ($t = 0$) of 40.0 μCi. Calculate the number of nuclei that decay in the time interval between $t_1 = 10.0$ h and $t_2 = 12.0$ h.

24. A radioactive nucleus has half-life $T_{1/2}$. A sample containing these nuclei has initial activity R_0. Calculate the number of nuclei that decay during the interval between the times t_1 and t_2.

25. Consider a radioactive sample. Determine the ratio of the number of nuclei decaying during the first half of its half-life to the number of nuclei decaying during the second half of its half-life.

26. ● (a) The daughter nucleus formed in radioactive decay is often radioactive. Let N_{10} represent the number of parent nuclei at time $t = 0$, $N_1(t)$ the number of parent nuclei at time t, and λ_1 the decay constant of the parent. Suppose the number of daughter nuclei at time $t = 0$ is zero. Let $N_2(t)$ be the number of daughter nuclei at time t and let λ_2 be the decay constant of the daughter. Show that $N_2(t)$ satisfies the differential equation

$$\frac{dN_2}{dt} = \lambda_1 N_1 - \lambda_2 N_2$$

(b) Verify by substitution that this differential equation has the solution

$$N_2(t) = \frac{N_{10}\lambda_1}{\lambda_1 - \lambda_2}\left(e^{-\lambda_2 t} - e^{-\lambda_1 t}\right)$$

This equation is the law of successive radioactive decays. (c) ^{218}Po decays into ^{214}Pb with a half-life of 3.10 min, and ^{214}Pb decays into ^{214}Bi with a half-life of 26.8 min. On the same axes, plot graphs of $N_1(t)$ for ^{218}Po and $N_2(t)$ for ^{214}Pb. Let N_{10} = 1 000 nuclei and choose values of t from 0 to 36 min in 2-min intervals. The curve for ^{214}Pb at first rises to a maximum and then starts to decay. At what instant t_m is the number of ^{214}Pb nuclei a maximum? (d) By applying the condition for a maximum $dN_2/dt = 0$, derive a symbolic equation for t_m in terms of λ_1 and λ_2. Explain whether the value obtained in part (c) agrees with this equation.

27. In an experiment on the transport of nutrients in the root structure of a plant, two radioactive nuclides X and Y are used. Initially, 2.50 times more nuclei of type X are present than of type Y. Precisely three days later, there are 4.20 times more nuclei of type X than of type Y. Isotope Y has a half-life of 1.60 d. What is the half-life of isotope X?

Section 44.5 The Decay Processes

28. Identify the missing nuclide or particle (X):
(a) $X \rightarrow {}^{65}_{28}\text{Ni} + \gamma$
(b) ${}^{215}_{84}\text{Po} \rightarrow X + \alpha$
(c) $X \rightarrow {}^{55}_{26}\text{Fe} + e^+ + \nu$
(d) ${}^{109}_{48}\text{Cd} + X \rightarrow {}^{109}_{47}\text{Ag} + \nu$
(e) ${}^{14}_{7}\text{N} + {}^{4}_{2}\text{He} \rightarrow X + {}^{17}_{8}\text{O}$

29. Find the energy released in the alpha decay

$${}^{238}_{92}\text{U} \rightarrow {}^{234}_{90}\text{Th} + {}^{4}_{2}\text{He}$$

You will find Table 44.2 useful.

30. A living specimen in equilibrium with the atmosphere contains one atom of ^{14}C (half-life = 5 730 yr) for every 7.7×10^{11} stable carbon atoms. An archeological sample of wood (cellulose, $C_{12}H_{22}O_{11}$) contains 21.0 mg of carbon. When the sample is placed inside a shielded beta counter with 88.0% counting efficiency, 837 counts are accumulated in one week. Assuming the cosmic-ray flux and the Earth's atmosphere have not changed appreciably since the sample was formed, find the age of the sample.

31. ● A sample consists of 1.00×10^6 radioactive nuclei with a half-life of 10.0 h. No other nuclei are present at time $t = 0$. The stable daughter nuclei accumulate in the sample as time goes on. (a) Derive an equation giving the number of daughter nuclei N_d as a function of time. (b) Sketch or describe a graph of the number of daughter nuclei as a function of time. What are the maximum and minimum numbers of daughter nuclei, and when do they occur? What are the maximum and minimum rates of change in the number of daughter nuclei, and when do they occur?

32. A ^{3}H nucleus beta decays into ^{3}He by creating an electron and an antineutrino according to the reaction

$${}^{3}_{1}\text{H} \rightarrow {}^{3}_{2}\text{He} + e^- + \bar{\nu}$$

The symbols in this reaction refer to nuclei. Write the reaction referring to neutral atoms by adding one electron to both sides. Then use Table 44.2 to determine the total energy released in this reaction.

33. The nucleus ${}^{15}_{8}\text{O}$ decays by electron capture. The nuclear reaction is written

$${}^{15}_{8}\text{O} + e^- \rightarrow {}^{15}_{7}\text{N} + \nu$$

(a) Write the process going on for a single particle within the nucleus. (b) Write the decay process referring to neutral atoms. (c) Determine the energy of the neutrino. Disregard the daughter's recoil.

34. Determine which decays can occur spontaneously:
(a) ${}^{40}_{20}\text{Ca} \rightarrow e^+ + {}^{40}_{19}\text{K}$
(b) ${}^{98}_{44}\text{Ru} \rightarrow {}^{4}_{2}\text{He} + {}^{94}_{42}\text{Mo}$
(c) ${}^{144}_{60}\text{Nd} \rightarrow {}^{4}_{2}\text{He} + {}^{140}_{58}\text{Ce}$

Section 44.6 Natural Radioactivity

35. Enter the correct isotope symbol in each open square in Figure P44.35, which shows the sequences of decays in the natural radioactive series starting with the long-lived isotope uranium-235 and ending with the stable nucleus lead-207.

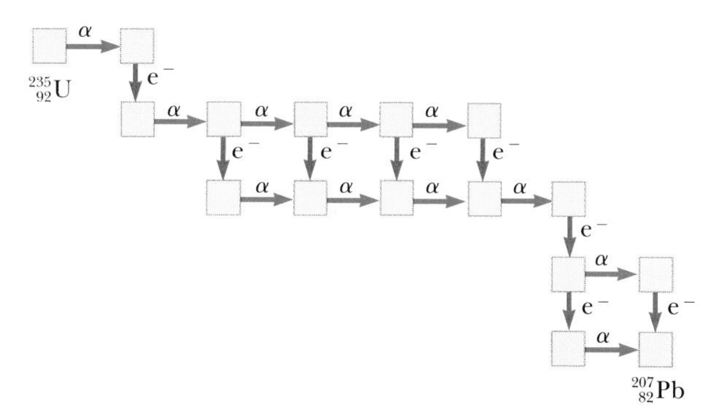

Figure P44.35

36. A rock sample contains traces of ^{238}U, ^{235}U, ^{232}Th, ^{208}Pb, ^{207}Pb, and ^{206}Pb. Analysis shows that the ratio of the amount of ^{238}U to ^{206}Pb is 1.164. (a) Assuming the rock originally contained no lead, determine the age of the rock. (b) What should be the ratios of ^{235}U to ^{207}Pb and of ^{232}Th to ^{208}Pb so that they would yield the same age for the rock? Ignore the minute amounts of the intermediate decay products in the decay chains. *Note:* This form of multiple dating gives reliable geological dates.

37. *Indoor air pollution.* Uranium is naturally present in rock and soil. At one step in its series of radioactive decays, ^{238}U produces the chemically inert gas radon-222, with a half-life of 3.82 days. The radon seeps out of the ground to mix into the atmosphere, typically making open air radioactive with activity 0.3 pCi/L. In homes, ^{222}Rn can be a serious pollutant, accumulating to reach much higher activities in enclosed spaces. If the radon radioactivity exceeds 4 pCi/L, the Environmental Protection Agency suggests taking action to reduce it, such as by reducing infiltration of air from the ground. (a) Convert the activity 4 pCi/L to units of becquerels per cubic meter. (b) How many ^{222}Rn atoms are in one cubic meter of air displaying this activity? (c) What fraction of the mass of the air does the radon constitute?

38. ● The most common isotope of radon is ^{222}Rn, which has half-life 3.82 days. (a) What fraction of the nuclei that were on the Earth one week ago are now undecayed? (b) Of those that existed one year ago? (c) In view of these results, explain why radon remains a problem, contributing significantly to our background radiation exposure.

Section 44.7 Nuclear Reactions

39. Identify the unknown nuclei and particles X and X′ in the following nuclear reactions:
(a) $X + {}^{4}_{2}\text{He} \rightarrow {}^{24}_{12}\text{Mg} + {}^{1}_{0}\text{n}$
(b) ${}^{235}_{92}\text{U} + {}^{1}_{0}\text{n} \rightarrow {}^{90}_{38}\text{Sr} + X + 2{}^{1}_{0}\text{n}$
(c) $2{}^{1}_{1}\text{H} \rightarrow {}^{2}_{1}\text{H} + X + X'$

40. After determining that the Sun has existed for hundreds of millions of years, but before the discovery of nuclear physics, scientists could not explain why the Sun has continued to burn for such a long time interval. For example, if it were a coal fire, it would have burned up in about 3 000 yr. Assume the Sun, whose mass is 1.99×10^{30} kg, originally consisted entirely of hydrogen and its total power output is 3.85×10^{26} W. (a) Assuming the energy-generating mechanism of the Sun is the fusion of hydrogen into helium via the net reaction

$$4({}^{1}_{1}\text{H}) + 2(\text{e}^{-}) \rightarrow {}^{4}_{2}\text{He} + 2\nu + \gamma$$

calculate the energy (in joules) given off by this reaction. (b) Determine how many hydrogen atoms constitute the Sun. Take the mass of one hydrogen atom to be 1.67×10^{-27} kg. (c) If the total power output remains constant, after what time interval will all the hydrogen be converted into helium, making the Sun die? The actual projected lifetime of the Sun is about 10 billion years because only the hydrogen in a relatively small core is available as a fuel. Only in the core are temperatures and densities high enough for the fusion reaction to be self-sustaining.

41. ▲ Natural gold has only one isotope, ${}^{197}_{79}\text{Au}$. If natural gold is irradiated by a flux of slow neutrons, electrons are emitted. (a) Write the reaction equation. (b) Calculate the maximum energy of the emitted electrons.

42. A beam of 6.61-MeV protons is incident on a target of ${}^{27}_{13}\text{Al}$. Those that collide produce the reaction

$$\text{p} + {}^{27}_{13}\text{Al} \rightarrow {}^{27}_{14}\text{Si} + \text{n}$$

Ignoring any recoil of the product nucleus, determine the kinetic energy of the emerging neutrons. You may use Table 44.2.

43. The following reactions are observed:

$${}^{9}_{4}\text{Be} + \text{n} \rightarrow {}^{10}_{4}\text{Be} + \gamma \qquad Q = 6.812 \text{ MeV}$$

$${}^{9}_{4}\text{Be} + \gamma \rightarrow {}^{8}_{4}\text{Be} + \text{n} \qquad Q = -1.665 \text{ MeV}$$

Using the mass of ${}^{9}\text{Be}$ from Table 44.2, calculate the masses of ${}^{8}\text{Be}$ and ${}^{10}\text{Be}$ in unified mass units to four decimal places.

44. (a) Suppose ${}^{10}_{5}\text{B}$ is struck by an alpha particle, releasing a proton and a product nucleus in the reaction. What is the product nucleus? (b) An alpha particle and a product nucleus are produced when ${}^{13}_{6}\text{C}$ is struck by a proton. What is the product nucleus?

Section 44.8 Nuclear Magnetic Resonance and Magnetic Resonance Imaging

45. The radio frequency at which a nucleus displays resonance absorption between spin states is called the Larmor frequency and is given by

$$f = \frac{\Delta E}{h} = \frac{2\mu B}{h}$$

Calculate the Larmor frequency for (a) free neutrons in a magnetic field of 1.00 T, (b) free protons in a magnetic field of 1.00 T, and (c) free protons in the Earth's magnetic field at a location where the magnitude of the field is 50.0 μT.

46. Construct a diagram like that of Figure 44.18 for the cases when I equals (a) $\frac{5}{2}$ and (b) 4.

Additional Problems

47. (a) One method of producing neutrons for experimental use is bombardment of light nuclei with alpha particles. In the method used by James Chadwick in 1932, alpha particles emitted by polonium are incident on beryllium nuclei:

$${}^{4}_{2}\text{He} + {}^{9}_{4}\text{Be} \rightarrow {}^{12}_{6}\text{C} + {}^{1}_{0}\text{n}$$

What is the Q value? (b) Neutrons are also often produced by small-particle accelerators. In one design, deuterons accelerated in a Van de Graaff generator bombard other deuterium nuclei:

$${}^{2}_{1}\text{H} + {}^{2}_{1}\text{H} \rightarrow {}^{3}_{2}\text{He} + {}^{1}_{0}\text{n}$$

Is this reaction exothermic or endothermic? Calculate its Q value.

48. As part of his discovery of the neutron in 1932, James Chadwick determined the mass of the newly identified particle by firing a beam of fast neutrons, all having the same speed, at two different targets and measuring the maximum recoil speeds of the target nuclei. The maximum speeds arise when an elastic head-on collision occurs between a neutron and a stationary target nucleus. (a) Represent the masses and final speeds of the two target nuclei as m_1, v_1, m_2, and v_2 and assume Newtonian mechanics applies. Show that the neutron mass can be calculated from the equation

$$m_n = \frac{m_1 v_1 - m_2 v_2}{v_2 - v_1}$$

(b) Chadwick directed a beam of neutrons (produced from a nuclear reaction) on paraffin, which contains hydrogen. The maximum speed of the protons ejected was found to be 3.3×10^{7} m/s. Because the velocity of the neutrons could not be determined directly, a second experiment was performed using neutrons from the same source and nitrogen nuclei as the target. The maximum recoil speed of the nitrogen nuclei was found to be 4.7×10^{6} m/s. The masses of a proton and a nitrogen nucleus were taken as 1 u and 14 u, respectively. What was Chadwick's value for the neutron mass?

49. When the nuclear reaction represented by Equation 44.27 is endothermic, the reaction energy Q is negative. For the reaction to proceed, the incoming particle must have a minimum energy called the threshold energy, E_{th}. Some fraction of the energy of the incident particle is transferred to the compound nucleus to conserve momentum. Therefore, E_{th} must be greater than the magnitude of Q. (a) Show that

$$E_{\text{th}} = -Q\left(1 + \frac{M_{\text{a}}}{M_{\text{X}}}\right)$$

(b) Calculate the threshold energy of the incident alpha particle in the reaction

$$^{4}_{2}\text{He} + ^{14}_{7}\text{N} \rightarrow ^{17}_{8}\text{O} + ^{1}_{1}\text{H}$$

50. **Review problem.** (a) Is the mass of a hydrogen atom in its ground state larger or smaller than the sum of the masses of a proton and an electron? (b) What is the mass difference? (c) How large is the difference as a percentage of the total mass? (d) Is it large enough to affect the value of the atomic mass listed to six decimal places in Table 44.2?

51. Write the statement of a problem for which the following equation appears in the solution. Determine the value of the unknown in the equation and identify its meaning.

$$10.012\,937\text{ u} + 4.002\,603\text{ u} = 13.003\,355\text{ u} + 1.007\,825\text{ u} + Q/c^2$$

52. A by-product of some fission reactors is the isotope $^{239}_{94}\text{Pu}$, an alpha emitter having a half-life of 24 120 yr:

$$^{239}_{94}\text{Pu} \rightarrow ^{235}_{92}\text{U} + \alpha$$

Consider a sample of 1.00 kg of pure $^{239}_{94}\text{Pu}$ at $t = 0$. Calculate (a) the number of $^{239}_{94}\text{Pu}$ nuclei present at $t = 0$ and (b) the initial activity in the sample. (c) **What If?** For what times interval does the sample have to be stored if a "safe" activity level is 0.100 Bq?

53. (a) Can ^{57}Co decay by e^+ emission? Explain. You may use Table 44.2. (b) **What If?** Can ^{14}C decay by e^- emission? Explain. (c) If either answer is yes, what is the range of kinetic energies available for the beta particle?

54. (a) Find the radius of the $^{12}_{6}\text{C}$ nucleus. (b) Find the force of repulsion between a proton at the surface of a $^{12}_{6}\text{C}$ nucleus and the remaining five protons. (c) How much work (in MeV) has to be done to overcome this electric repulsion in transporting the last proton from a large distance up to the surface of the nucleus? (d) Repeat parts (a), (b), and (c) for $^{238}_{92}\text{U}$.

55. ● (a) Why is the beta decay $p \rightarrow n + e^+ + \nu$ forbidden for a free proton? (b) **What If?** Why is the same reaction possible if the proton is bound in a nucleus? For example, the following reaction occurs:

$$^{13}_{7}\text{N} \rightarrow ^{13}_{6}\text{C} + e^+ + \nu$$

(c) How much energy is released in the reaction given in part (b)? *Suggestion:* Add seven electrons to both sides of the reaction to write it for neutral atoms. You may use Table 44.2.

56. The activity of a radioactive sample was measured over 12 h, with the net count rates shown in the table.

Time (h)	Counting Rate (counts/min)
1.00	3 100
2.00	2 450
4.00	1 480
6.00	910
8.00	545
10.0	330
12.0	200

(a) Plot the logarithm of counting rate as a function of time. (b) Determine the decay constant and half-life of the radioactive nuclei in the sample. (c) What counting rate would you expect for the sample at $t = 0$? (d) Assuming the efficiency of the counting instrument is 10.0%, calculate the number of radioactive atoms in the sample at $t = 0$.

57. After the sudden release of radioactivity from the Chernobyl nuclear reactor accident in 1986, the radioactivity of milk in Poland rose to 2 000 Bq/L due to iodine-131 present in the grass eaten by dairy cattle. Radioactive iodine, with half-life 8.04 days, is particularly hazardous because the thyroid gland concentrates iodine. The Chernobyl accident caused a measurable increase in thyroid cancers among children in Belarus. (a) For comparison, find the activity of milk due to potassium. Assume 1 liter of milk contains 2.00 g of potassium, of which 0.011 7% is the isotope ^{40}K with a half-life 1.28×10^9 yr. (b) After what elapsed time would the activity due to iodine fall below that due to potassium?

58. When, after a reaction or disturbance of any kind, a nucleus is left in an excited state, it can return to its normal (ground) state by emission of a gamma-ray photon (or several photons). This process is illustrated by Equation 44.24. The emitting nucleus must recoil to conserve both energy and momentum. (a) Show that the recoil energy of the nucleus is

$$E_r = \frac{(\Delta E)^2}{2Mc^2}$$

where ΔE is the difference in energy between the excited and ground states of a nucleus of mass M. (b) Calculate the recoil energy of the ^{57}Fe nucleus when it decays by gamma emission from the 14.4-keV excited state. For this calculation, take the mass to be 57 u. *Suggestions:* When writing the equation for conservation of energy, use $(Mv)^2/2M$ for the kinetic energy of the recoiling nucleus. Also, assume $hf \ll Mc^2$ and use the binomial expansion.

59. A theory of nuclear astrophysics proposes that all the elements heavier than iron are formed in supernova explosions ending the lives of massive stars. Assume the amounts of ^{235}U and ^{238}U were equal at the time of the explosion. How long ago did the star(s) explode that released the elements that formed our Earth? The present $^{235}\text{U}/^{238}\text{U}$ ratio is 0.007 25. The half-lives of ^{235}U and ^{238}U are 0.704×10^9 yr and 4.47×10^9 yr.

60. Europeans named a certain direction in the sky as between the horns of Taurus the Bull. On the day they named as July 4, 1054, a brilliant light appeared there. Europeans left no surviving record of the supernova, which could be seen in daylight for some days. As it faded, it remained visible for years, dimming for a time with the 77.1-day half-life of the radioactive cobalt-56 that had been created in the explosion. (a) The remains of the star now form the Crab nebula (see the photographs opening Chapter 34). In it, the cobalt-56 has now decreased to what fraction of its original activity? (b) Suppose an American, of the people called the Anasazi, made a charcoal drawing of the supernova. The carbon-14 in the charcoal has now decayed to what fraction of its original activity?

61. **Review problem.** Consider the Bohr model of the hydrogen atom, with the electron in the ground state. The magnetic field at the nucleus produced by the orbiting electron has a value of 12.5 T. (See Problem 1 in Chapter 30.) The proton can have its magnetic moment aligned in either of two directions perpendicular to the plane of the electron's orbit. The interaction of the proton's magnetic moment with the electron's magnetic field causes a difference in energy between the states with the two different orientations of the proton's magnetic moment. Find that energy difference in eV.

62. *Student determination of the half-life of* ^{137}*Ba.* The radioactive barium isotope ^{137}Ba has a relatively short half-life and can be easily extracted from a solution containing its parent cesium (^{137}Cs). This barium isotope is commonly used in an undergraduate laboratory exercise for demonstrating the radioactive decay law. Undergraduate students using modest experimental equipment took the data presented in Figure P44.62. Determine the half-life for the decay of ^{137}Ba using their data.

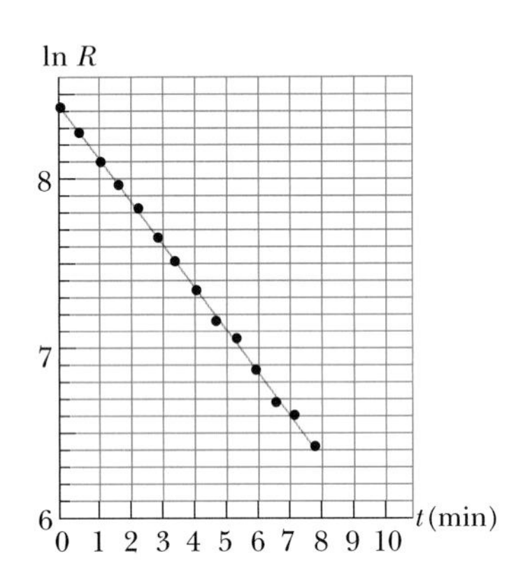

Figure P44.62

63. Free neutrons have a characteristic half-life of 10.4 min. What fraction of a group of free neutrons with kinetic energy 0.040 0 eV decays before traveling a distance of 10.0 km?

64. **Review problem.** Consider a model of the nucleus in which the positive charge (Ze) is uniformly distributed throughout a sphere of radius R. By integrating the energy density $\frac{1}{2}\epsilon_0 E^2$ over all space, show that the electric potential energy may be written

$$U = \frac{3Z^2e^2}{20\pi\epsilon_0 R} = \frac{3k_eZ^2e^2}{5R}$$

Problem 62 in Chapter 25 derived the same result by a different method.

65. ● In a piece of rock from the Moon, the ^{87}Rb content is assayed to be 1.82×10^{10} atoms per gram of material and the ^{87}Sr content is found to be 1.07×10^{9} atoms per gram. (a) Calculate the age of the rock. (b) **What If?** Could the material in the rock actually be much older? What assumption is implicit in using the radioactive dating method? The relevant decay is $^{87}\text{Rb} \rightarrow {}^{87}\text{Sr} + e^- + \bar{\nu}$. The half-life of the decay is 4.75×10^{10} yr.

66. The ground state of $^{93}_{43}$Tc (molar mass 92.910 2 g/mol) decays by electron capture and e^+ emission to energy levels of the daughter (molar mass 92.906 8 g/mol in ground state) at 2.44 MeV, 2.03 MeV, 1.48 MeV, and 1.35 MeV. (a) For which of these levels are electron capture and e^+ decay allowed? (b) Identify the daughter and sketch the decay scheme, assuming all excited states de-excite by direct γ decay to the ground state.

Answers to Quick Quizzes

44.1 **(i),** (b). The value of $N = A - Z$ is the same for all three nuclei. **(ii),** (a). The value of Z is the same for all three nuclei because they are all nuclei of nitrogen. **(iii),** (c). The value of A is the same for all three nuclei as seen by the unchanging superscript.

44.2 (e). A year of 365 days is equivalent to 365 d/5.01 d $\approx$ 73 half-lives. Therefore, the activity will be reduced after one year to approximately $(\frac{1}{2})^{73}$ (1.000 μCi) $\sim 10^{-22}$ μCi.

44.3 (b). In alpha decay, the atomic number decreases by two and the atomic mass number decreases by four.

44.4 (c). In e^- decay, the atomic number increases by one and the atomic mass number stays fixed. None of the choices is consistent with e^+ decay, so we assume the decay must be by e^-.

The San Onofre Nuclear Generating Station, next to the I-5 freeway south of San Clemente, California, is one of dozens of nuclear power plants around the world that provide energy from uranium. These plants operate via a nuclear process called fission. Plants based on a second process, called *fusion*, are years in the future. (Tony Freeman/Index Stock Imagery)

45 Applications of Nuclear Physics

In this chapter, we study two means for deriving energy from nuclear reactions: fis-sion, in which a large nucleus splits into two smaller nuclei, and fusion, in which two small nuclei fuse to form a larger one. In both cases, the released energy can be used either constructively (as in electric power plants) or destructively (as in nuclear weapons). We also examine the ways in which radiation interacts with matter and look at several devices used to detect radiation. The chapter concludes with a discussion of some industrial and biological applications of radiation.

45.1 Interactions Involving Neutrons

Nuclear fission is the process that occurs in present-day nuclear reactors and ultimately results in energy supplied to a community by electrical transmission. Nuclear fusion is an area of active research, but it has not yet been commercially developed for the supply of energy. We will discuss fission first and then explore fusion in Section 45.4.

To understand nuclear fission and the physics of nuclear reactors, we must first understand how neutrons interact with nuclei. Because of their charge neutrality, neutrons are not subject to Coulomb forces and as a result do not interact electrically with electrons or the nucleus. Therefore, neutrons can easily penetrate deep into an atom and collide with the nucleus.

A fast neutron (energy greater than approximately 1 MeV) traveling through matter undergoes many collisions with nuclei. In each collision, the neutron gives up some of its kinetic energy to a nucleus. For fast neutrons in some materials, elastic collisions dominate. Materials for which that occurs are called **moderators** because they slow down (or moderate) the originally energetic neutrons very effectively. Moderator nuclei should be of low mass so that a large amount of kinetic energy is transferred to them in elastic collisions. For this reason, materials that are abundant in hydrogen, such as paraffin and water, are good moderators for neutrons.

Eventually, most neutrons bombarding a moderator become **thermal neutrons,** which means they are in thermal equilibrium with the moderator material. Their average kinetic energy at room temperature is, from Equation 21.4,

$$K_{\text{avg}} = \tfrac{3}{2}k_{\text{B}}T \approx \tfrac{3}{2}(1.38 \times 10^{-23}\ \text{J/K})(300\ \text{K}) = 6.21 \times 10^{-21}\ \text{J} \approx 0.04\ \text{eV}$$

which corresponds to a neutron root-mean-square speed of approximately 2 800 m/s. Thermal neutrons have a distribution of speeds, just as the molecules in a container of gas do (see Chapter 21). High-energy neutrons, those with energy of several MeV, *thermalize* (that is, their average energy reaches K_{avg}) in less than 1 ms when they are incident on a moderator.

Once the neutrons have thermalized and the energy of a particular neutron is sufficiently low, there is a high probability the neutron will be captured by a nucleus, an event that is accompanied by the emission of a gamma ray. This **neutron capture** reaction can be written

Neutron capture reaction ▶

$$^{1}_{0}\text{n} + {}^{A}_{Z}\text{X} \rightarrow {}^{A+1}_{Z}\text{X}^* \rightarrow {}^{A+1}_{Z}\text{X} + \gamma \qquad \textbf{(45.1)}$$

Once the neutron is captured, the nucleus ${}^{A+1}_{Z}\text{X}^*$ is in an excited state for a very short time before it undergoes gamma decay. The product nucleus ${}^{A+1}_{Z}\text{X}$ is usually radioactive and decays by beta emission.

The neutron-capture rate for neutrons passing through any sample depends on the type of atoms in the sample and on the energy of the incident neutrons. The interaction of neutrons with matter increases with decreasing neutron energy because a slow neutron spends a larger time interval in the vicinity of target nuclei.

45.2 Nuclear Fission

PITFALL PREVENTION 45.1
Binding Energy Reminder

Remember from Chapter 44 that binding energy is the absolute value of the system energy and is related to the system mass. Therefore, when considering Figure 44.5, imagine flipping it upside down for a curve representing system mass. In a fission reaction, the system mass decreases. This decrease in mass appears in the system as kinetic energy of the fission products.

As mentioned in Section 44.2, nuclear **fission** occurs when a heavy nucleus, such as ^{235}U, splits into two smaller nuclei. Fission is initiated when a heavy nucleus captures a thermal neutron as described by the first step in Equation 45.1. The absorption of the neutron creates a nucleus that is unstable and can change to a lower-energy configuration by splitting into two smaller nuclei. In such a reaction, the combined mass of the daughter nuclei is less than the mass of the parent nucleus, and the difference in mass is called the **mass defect.** Multiplying the mass defect by c^2 gives the numerical value of the released energy. Energy is released because the binding energy per nucleon of the daughter nuclei is approximately 1 MeV greater than that of the parent nucleus (see Fig. 44.5).

Nuclear fission was first observed in 1938 by Otto Hahn (1879–1968) and Fritz Strassman (1902–1980) following some basic studies by Fermi. After bombarding uranium with neutrons, Hahn and Strassman discovered among the reaction products two medium-mass elements, barium and lanthanum. Shortly thereafter, Lise Meitner (1878–1968) and her nephew Otto Frisch (1904–1979) explained what had happened. After absorbing a neutron, the uranium nucleus had split into two nearly equal fragments plus several neutrons. Such an occurrence was of considerable interest to physicists attempting to understand the nucleus, but it was to have even more far-reaching consequences. Measurements showed that approximately 200 MeV of energy was released in each fission event, and this fact was to affect the course of history.

The fission of ^{235}U by thermal neutrons can be represented by the reaction

$$^{1}_{0}\text{n} + {}^{235}_{92}\text{U} \rightarrow {}^{236}_{92}\text{U}^* \rightarrow \text{X} + \text{Y} + \text{neutrons} \qquad \textbf{(45.2)}$$

◀ General fission reaction for ^{235}U

where ^{236}U* is an intermediate excited state that lasts for approximately 10^{-12} s before splitting into medium-mass nuclei X and Y, which are called **fission fragments.** In any fission reaction, there are many combinations of X and Y that satisfy the requirements of conservation of energy and charge. In the case of uranium, for example, approximately 90 daughter nuclei can be formed.

Fission also results in the production of several neutrons, typically two or three. On average, approximately 2.5 neutrons are released per event. A typical fission reaction for uranium is

$$^{1}_{0}\text{n} + {}^{235}_{92}\text{U} \rightarrow {}^{141}_{56}\text{Ba} + {}^{92}_{36}\text{Kr} + 3({}^{1}_{0}\text{n}) \qquad \textbf{(45.3)}$$

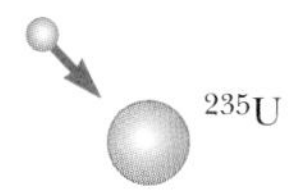

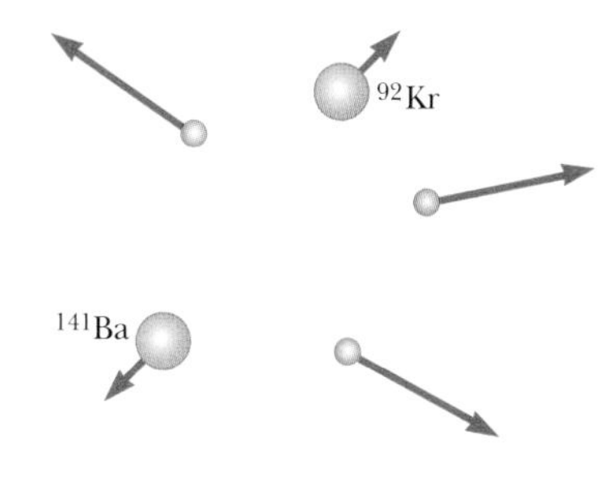

Figure 45.1 A nuclear fission event. Before the event, a slow neutron approaches a ^{235}U nucleus. After the event, there are two lighter nuclei and three neutrons.

Figure 45.1 shows a pictorial representation of the fission event in Equation 45.3.

Figure 45.2 is a graph of the distribution of fission products versus mass number A. The most probable products have mass numbers $A \approx 140$ and $A \approx 95$. Suppose these products are $^{140}_{53}$I (with 87 neutrons) and $^{95}_{39}$Y (with 56 neutrons). If these nuclei are located on the graph of Figure 44.4, it is seen that both are well above the line of stability. Because these fragments are very unstable owing to their unusually high number of neutrons, they almost instantaneously release two or three neutrons.

Let's estimate the disintegration energy Q released in a typical fission process. From Figure 44.5, we see that the binding energy per nucleon is approximately 7.2 MeV for heavy nuclei ($A \approx 240$) and approximately 8.2 MeV for nuclei of intermediate mass. The amount of energy released is 8.2 MeV − 7.2 MeV = 1 MeV per nucleon. Because there are a total of 235 nucleons in $^{235}_{92}$U, the energy released per fission event is approximately 235 MeV, a large amount of energy relative to the amount released in chemical processes. For example, the energy released in the combustion of one molecule of octane used in gasoline engines is about one millionth of the energy released in a single fission event!

Quick Quiz 45.1 When a nucleus undergoes fission, the two daughter nuclei are generally radioactive. By which process are they most likely to decay? (a) alpha decay (b) beta decay (e^-) (c) beta decay (e^+)

Quick Quiz 45.2 Which of the following are possible fission reactions?

(a) $^{1}_{0}\text{n} + {}^{235}_{92}\text{U} \rightarrow {}^{140}_{54}\text{Xe} + {}^{94}_{38}\text{Sr} + 2({}^{1}_{0}\text{n})$

(b) $^{1}_{0}\text{n} + {}^{235}_{92}\text{U} \rightarrow {}^{132}_{50}\text{Sn} + {}^{101}_{42}\text{Mo} + 3({}^{1}_{0}\text{n})$

(c) $^{1}_{0}\text{n} + {}^{239}_{94}\text{Pu} \rightarrow {}^{137}_{53}\text{I} + {}^{97}_{41}\text{Nb} + 3({}^{1}_{0}\text{n})$

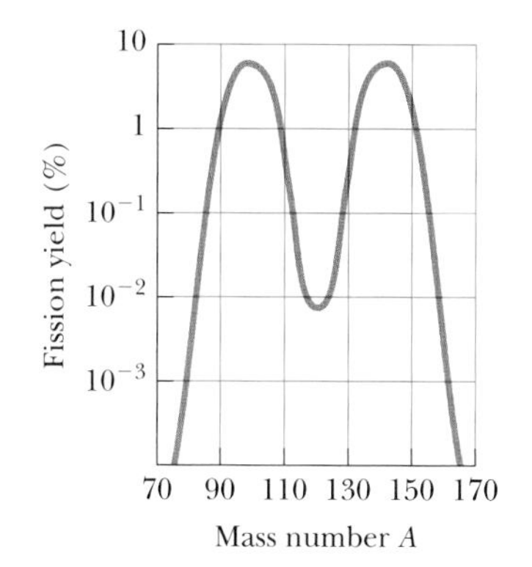

Figure 45.2 Distribution of fission products versus mass number for the fission of ^{235}U bombarded with thermal neutrons. Notice that the vertical axis is logarithmic.

EXAMPLE 45.1 The Energy Released in the Fission of ^{235}U

Calculate the energy released when 1.00 kg of ^{235}U fissions, taking the disintegration energy per event to be Q = 208 MeV.

SOLUTION

Conceptualize Imagine a nucleus of ^{235}U absorbing a neutron and then splitting into two smaller nuclei and several neutrons as in Figure 45.1.

Categorize The problem statement tells us to categorize this example as one involving an energy analysis of nuclear fission.

Analyze Because $A = 235$ for uranium, one mole of this isotope has a mass of 235 g.

From the given mass of ^{235}U, find the number of nuclei in our sample:

$$N = \left(\frac{1.00 \times 10^3 \text{ g}}{235 \text{ g}}\right)(6.02 \times 10^{23} \text{ nuclei})$$
$$= 2.56 \times 10^{24} \text{ nuclei}$$

Find the total energy released when all nuclei undergo fission:

$$E = NQ = (2.56 \times 10^{24} \text{ nuclei})(208 \text{ MeV/nucleus})$$
$$= 5.32 \times 10^{26} \text{ MeV}$$

Finalize Convert this energy to kWh:

$$E = (5.32 \times 10^{26} \text{ MeV})\left(\frac{1.60 \times 10^{-13} \text{ J}}{\text{MeV}}\right)\left(\frac{1 \text{ kWh}}{3.60 \times 10^6 \text{ J}}\right) = 2.37 \times 10^7 \text{ kWh}$$

which is enough energy to keep a 100-W lightbulb operating for 30 000 years! If the available fission energy in 1 kg of ^{235}U were suddenly released, it would be equivalent to detonating about 20 000 tons of TNT.

45.3 Nuclear Reactors

In Section 45.2, we learned that when ^{235}U fissions, one incoming neutron results in an average of 2.5 neutrons emitted per event. These neutrons can trigger other nuclei to fission. Because more neutrons are produced by the event than are absorbed, there is the possibility of an ever-building chain reaction (Active Fig. 45.3). Calculations show that if the chain reaction is not controlled (that is, if it does not proceed slowly), it can result in a violent explosion, with the sudden release of an enormous amount of energy. When the reaction is controlled, however, the energy released can be put to constructive use. In the United States, for example, nearly 20% of the electricity generated each year comes from nuclear power plants, and nuclear power is used extensively in many other countries, including France, Japan, and Germany.

A nuclear reactor is a system designed to maintain what is called a **self-sustained chain reaction.** This important process was first achieved in 1942 by Enrico Fermi and his team at the University of Chicago, using naturally occurring uranium as

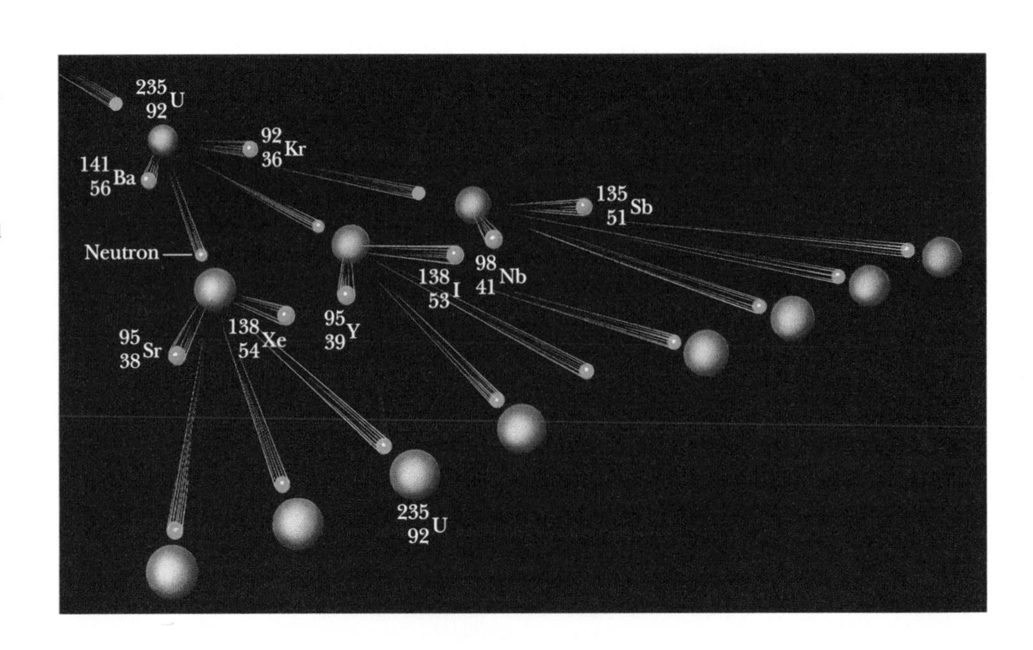

ACTIVE FIGURE 45.3

A nuclear chain reaction initiated by the capture of a neutron. Uranium nuclei are shown in magenta, neutrons in green, and daughter nuclei in orange.

Sign in at www.thomsonedu.com and go to ThomsonNOW to observe the chain reaction.

Courtesy of Chicago Historical Society

Figure 45.4 Artist's rendition of the world's first nuclear reactor. Because of wartime secrecy, there are few photographs of the completed reactor, which was composed of layers of moderating graphite interspersed with uranium. A self-sustained chain reaction was first achieved on December 2, 1942. Word of the success was telephoned immediately to Washington, D.C., with this message: "The Italian navigator has landed in the New World and found the natives very friendly." The historic event took place in an improvised laboratory in the racquet court under the stands of the University of Chicago's Stagg Field, and the Italian navigator was Enrico Fermi.

the fuel.[1] In the first nuclear reactor (Fig. 45.4), Fermi placed bricks of graphite (carbon) between the fuel elements. Carbon nuclei are about 12 times more massive than neutrons, but after several collisions with carbon nuclei, a neutron is slowed sufficiently to increase its likelihood of fission with ^{235}U. In this design, carbon is the moderator; most modern reactors use water as the moderator.

Most reactors in operation today also use uranium as fuel. Naturally occurring uranium contains only 0.7% of the ^{235}U isotope, however, with the remaining 99.3% being ^{238}U. This fact is important to the operation of a reactor because ^{238}U almost never fissions. Instead, it tends to absorb neutrons without a subsequent fission event, producing neptunium and plutonium. For this reason, reactor fuels must be artificially *enriched* to contain at least a few percent ^{235}U.

To achieve a self-sustained chain reaction, an average of one neutron emitted in each ^{235}U fission must be captured by another ^{235}U nucleus and cause that nucleus to undergo fission. A useful parameter for describing the level of reactor operation is the **reproduction constant** K, defined as **the average number of neutrons from each fission event that cause another fission event.** As we have seen, K has an average value of 2.5 in the uncontrolled fission of uranium.

A self-sustained and controlled chain reaction is achieved when $K = 1$. Under this condition, the reactor is said to be **critical.** When $K < 1$, the reactor is subcritical and the reaction dies out. When $K > 1$, the reactor is supercritical and a runaway reaction occurs. In a nuclear reactor used to furnish power to a utility company, it is necessary to maintain a value of K close to 1. If K rises above this value, the internal energy produced in the reaction could melt the reactor.

Several types of reactor systems allow the kinetic energy of fission fragments to be transformed to other types of energy and eventually transferred out of the reactor plant by electrical transmission. The most common reactor in use in the United States is the pressurized-water reactor (Fig. 45.5, page 1334). We shall examine this type because its main parts are common to all reactor designs. Fission events in the uranium **fuel elements** in the reactor core raise the temperature of the water contained in the primary loop, which is maintained at high pressure to keep the water from boiling. (This water also serves as the moderator to slow down the neutrons released in the fission events with energy of approximately 2 MeV.) The hot water is pumped through a heat exchanger, where the internal energy of the water is transferred by conduction to the water contained in the secondary loop. The hot water in the secondary loop is converted to steam, which does work to drive a turbine–generator system to create electric power. The water in the secondary loop is isolated from the water in the primary loop to avoid contamination of the secondary water and the steam by radioactive nuclei from the reactor core.

National Accelerator Laboratory

ENRICO FERMI
Italian Physicist (1901–1954)
Fermi was awarded the Nobel Prize in Physics in 1938 for producing transuranic elements by neutron irradiation and for his discovery of nuclear reactions brought about by thermal neutrons. He made many other outstanding contributions to physics, including his theory of beta decay, the free-electron theory of metals, and the development of the world's first fission reactor in 1942. Fermi was truly a gifted theoretical and experimental physicist. He was also well known for his ability to present physics in a clear and exciting manner. "Whatever Nature has in store for mankind, unpleasant as it may be, men must accept, for ignorance is never better than knowledge."

[1] Although Fermi's reactor was the first manufactured nuclear reactor, there is evidence that a natural fission reaction may have sustained itself for perhaps hundreds of thousands of years in a deposit of uranium in Gabon, West Africa. See G. Cowan, "A Natural Fission Reactor," *Scientific American* **235**(5): 36, 1976.

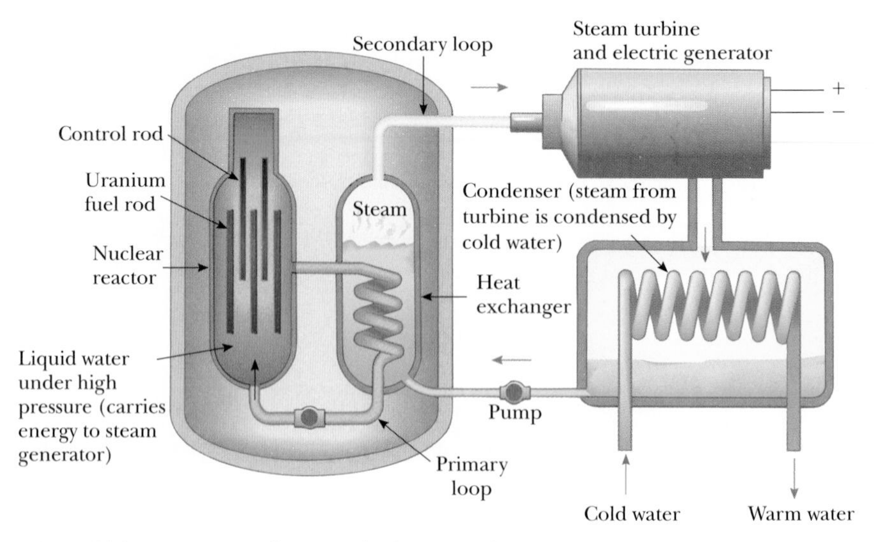

Figure 45.5 Main components of a pressurized-water nuclear reactor.

In any reactor, a fraction of the neutrons produced in fission leak out of the uranium fuel elements before inducing other fission events. If the fraction leaking out is too large, the reactor will not operate. The percentage lost is large if the fuel elements are very small because leakage is a function of the ratio of surface area to volume. Therefore, a critical feature of the reactor design is an optimal surface area–to–volume ratio of the fuel elements.

Control of Power Level

Safety is of critical importance in the operation of a nuclear reactor. The reproduction constant K must not be allowed to rise above 1, lest a runaway reaction occur. Consequently, reactor design must include a means of controlling the value of K.

The basic design of a nuclear reactor core is shown in Figure 45.6. The fuel elements consist of uranium that has been enriched in the ^{235}U isotope. To control the power level, **control rods** are inserted into the reactor core. These rods are made of materials such as cadmium that are very efficient in absorbing neutrons. By adjusting the number and position of the control rods in the reactor core, the K value can be varied and any power level within the design range of the reactor can be achieved.

Quick Quiz 45.3 To reduce the value of the reproduction constant K, do you (a) push the control rods deeper into the core or (b) pull the control rods farther out of the core?

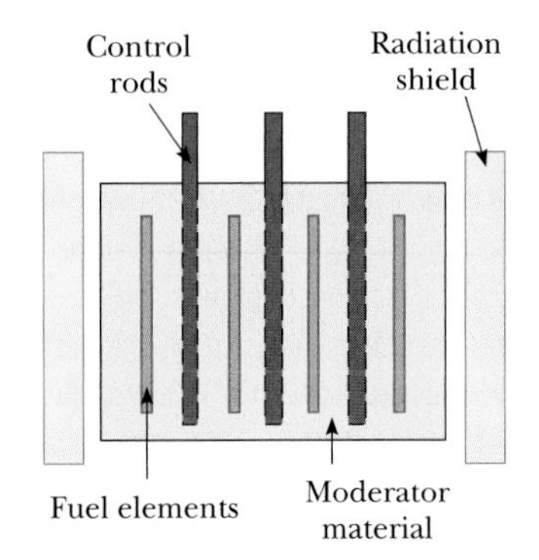

Figure 45.6 Cross section of a reactor core showing the control rods, fuel elements containing enriched fuel, and moderating material, all surrounded by a radiation shield.

Safety and Waste Disposal

The 1979 near-disaster at a nuclear power plant at Three Mile Island in Pennsylvania and the 1986 accident at the Chernobyl reactor in Ukraine rightfully focused attention on reactor safety. The Three Mile Island accident was the result of inadequate control-room instrumentation and poor emergency-response training. There were no injuries or detectable health effects from the event, even though more than one-third of the fuel melted.

Unfortunately, at Chernobyl the activity of the materials released immediately after the accident totaled approximately 1.2×10^{19} Bq and resulted in the evacuation of 135 000 people. Thirty individuals died during the accident or shortly thereafter, and data from the Ukraine Radiological Institute suggest that more than 2 500 deaths could be attributed to the Chernobyl accident. In the period 1986–1997, there was a tenfold increase in the number of children contracting thyroid cancer from the ingestion of radioactive iodine in milk from cows that ate contaminated grass. One conclusion of an international conference studying the Ukraine accident was that the main causes of the Chernobyl accident were the coincidence of severe deficiencies in the reactor physical design and a violation of safety procedures. Most of these deficiencies have since been addressed at plants of similar design in Russia and neighboring countries of the former Soviet Union.

Commercial reactors achieve safety through careful design and rigid operating protocol, and only when these variables are compromised do reactors pose a danger. Radiation exposure and the potential health risks associated with such exposure are controlled by three layers of containment. The fuel and radioactive fission products are contained inside the reactor vessel. Should this vessel rupture, the reactor building acts as a second containment structure to prevent radioactive material from contaminating the environment. Finally, the reactor facilities must be in a remote location to protect the general public from exposure should radiation escape the reactor building.

A continuing concern about nuclear fission reactors is the safe disposal of radioactive material when the reactor core is replaced. This waste material contains long-lived, highly radioactive isotopes and must be stored over long time intervals in such a way that there is no chance of environmental contamination. At present, sealing radioactive wastes in waterproof containers and burying them in deep geologic repositories seems to be the most promising solution.

Transport of reactor fuel and reactor wastes poses additional safety risks. Accidents during transport of nuclear fuel could expose the public to harmful levels of radiation. The U.S. Department of Energy requires stringent crash tests of all containers used to transport nuclear materials. Container manufacturers must demonstrate that their containers will not rupture even in high-speed collisions.

Despite these risks, there are advantages to the use of nuclear power to be weighed against the risks. For example, nuclear power plants do not produce air pollution and greenhouse gases as do fossil fuel plants, and the supply of uranium on the Earth is predicted to last longer than the supply of fossil fuels. For each source of energy—whether nuclear, hydroelectric, fossil fuel, wind, solar or other—the risks must be weighed against the benefits and the availability of the energy source.

45.4 Nuclear Fusion

In Chapter 44, we found that the binding energy for light nuclei ($A < 20$) is much smaller than the binding energy for heavier nuclei, which suggests a process that is the reverse of fission. As mentioned in Section 39.9, when two light nuclei combine to form a heavier nucleus, the process is called nuclear **fusion.** Because the mass of the final nucleus is less than the combined masses of the original nuclei, there is a loss of mass accompanied by a release of energy.

Two examples of such energy-liberating fusion reactions are as follows:

$$ {}^{1}_{1}\mathrm{H} + {}^{1}_{1}\mathrm{H} \rightarrow {}^{2}_{1}\mathrm{H} + \mathrm{e}^{+} + \nu $$

$$ {}^{1}_{1}\mathrm{H} + {}^{2}_{1}\mathrm{H} \rightarrow {}^{3}_{2}\mathrm{He} + \gamma $$

These reactions occur in the core of a star and are responsible for the outpouring of energy from the star. The second reaction is followed by either hydrogen–helium fusion or helium–helium fusion:

$$ {}^{1}_{1}\mathrm{H} + {}^{3}_{2}\mathrm{He} \rightarrow {}^{4}_{2}\mathrm{He} + \mathrm{e}^{+} + \nu $$

$$ {}^{3}_{2}\mathrm{He} + {}^{3}_{2}\mathrm{He} \rightarrow {}^{4}_{2}\mathrm{He} + {}^{1}_{1}\mathrm{H} + {}^{1}_{1}\mathrm{H} $$

These fusion reactions are the basic reactions in the **proton–proton cycle,** believed to be one of the basic cycles by which energy is generated in the Sun and other stars that contain an abundance of hydrogen. Most of the energy production takes place in the Sun's interior, where the temperature is approximately 1.5×10^{7} K.

PITFALL PREVENTION 45.2
Fission and Fusion

The words *fission* and *fusion* sound similar, but they correspond to different processes. Consider the binding-energy curve in Figure 44.5. There are two directions from which you can approach the peak of the curve so that energy is released: combining two light nuclei, or fusion, and separating a heavy nucleus into two lighter nuclei, or fission.

Because such high temperatures are required to drive these reactions, they are called **thermonuclear fusion reactions.** All the reactions in the proton–proton cycle are exothermic. An overview of the cycle is that four protons combine to generate an alpha particle, positrons, gamma rays, and neutrinos.

Quick Quiz 45.4 In the core of a star, hydrogen nuclei combine in fusion reactions. Once the hydrogen has been exhausted, fusion of helium nuclei can occur. If the star is sufficiently massive, fusion of heavier and heavier nuclei can occur once the helium is used up. Consider a fusion reaction involving two nuclei with the same value of A. For this reaction to be exothermic, which of the following values of A are impossible? (a) 12 (b) 20 (c) 28 (d) 64

EXAMPLE 45.2 Energy Released in Fusion

Find the total energy released in the fusion reactions in the proton–proton cycle.

SOLUTION

Conceptualize The net nuclear result of the proton–proton cycle is to fuse four protons to form an alpha particle. Study the reactions above for the proton–proton cycle to be sure you understand how four protons become an alpha particle.

Categorize We use concepts discussed in this section, so we categorize this example as a substitution problem.

Find the initial mass of four protons using atomic masses from Table 44.2:

$$4(1.007\ 825\ \text{u}) = 4.031\ 300\ \text{u}$$

Find the change in mass of the system as this value minus the mass of the resultant alpha particle:

$$4.031\ 300\ \text{u} - 4.002\ 603\ \text{u} = 0.028\ 697\ \text{u}$$

Convert this mass change into energy units:

$$E = 0.028\ 697\ \text{u} \times 931.494\ \text{MeV/u} = 26.7\ \text{MeV}$$

This energy is shared among the alpha particle and other particles such as positrons, gamma rays, and neutrinos.

Terrestrial Fusion Reactions

The enormous amount of energy released in fusion reactions suggests the possibility of harnessing this energy for useful purposes. A great deal of effort is currently under way to develop a sustained and controllable thermonuclear reactor, a fusion power reactor. Controlled fusion is often called the ultimate energy source because of the availability of its fuel source: water. For example, if deuterium were used as the fuel, 0.12 g of it could be extracted from 1 gal of water at a cost of about four cents. This amount of deuterium would release approximately 10^{10} J if all nuclei underwent fusion. By comparison, 1 gal of gasoline releases approximately 10^8 J upon burning and costs far more than four cents.

An additional advantage of fusion reactors is that comparatively few radioactive by-products are formed. For the proton–proton cycle, for instance, the end product is safe, nonradioactive helium. Unfortunately, a thermonuclear reactor that can deliver a net power output spread over a reasonable time interval is not yet a reality, and many difficulties must be resolved before a successful device is constructed.

The Sun's energy is based in part on a set of reactions in which hydrogen is converted to helium. The proton–proton interaction is not suitable for use in a fusion reactor, however, because the event requires very high temperatures and densities. The process works in the Sun only because of the extremely high density of protons in the Sun's interior.

The reactions that appear most promising for a fusion power reactor involve deuterium ($^2_1$H) and tritium ($^3_1$H):

$$\begin{aligned} {}^2_1\text{H} + {}^2_1\text{H} &\rightarrow {}^3_2\text{He} + {}^1_0\text{n} \qquad Q = 3.27\text{ MeV} \\ {}^2_1\text{H} + {}^2_1\text{H} &\rightarrow {}^3_1\text{H} + {}^1_1\text{H} \qquad Q = 4.03\text{ MeV} \\ {}^2_1\text{H} + {}^3_1\text{H} &\rightarrow {}^4_2\text{He} + {}^1_0\text{n} \qquad Q = 17.59\text{ MeV} \end{aligned} \tag{45.4}$$

As noted earlier, deuterium is available in almost unlimited quantities from our lakes and oceans and is very inexpensive to extract. Tritium, however, is radioactive ($T_{1/2}$ = 12.3 yr) and undergoes beta decay to ^{3}He. For this reason, tritium does not occur naturally to any great extent and must be artificially produced.

One major problem in obtaining energy from nuclear fusion is that the Coulomb repulsive force between two nuclei, which carry positive charges, must be overcome before they can fuse. Figure 45.7 is a graph of potential energy as a function of the separation distance between two deuterons (deuterium nuclei, each having charge $+e$). The potential energy is positive in the region $r > R$, where the Coulomb repulsive force dominates ($R \approx 1$ fm), and negative in the region $r < R$, where the nuclear force dominates. The fundamental problem then is to give the two nuclei enough kinetic energy to overcome this repulsive force. This requirement can be accomplished by raising the fuel to extremely high temperatures (to approximately 10^8 K, far greater than the interior temperature of the Sun). At these high temperatures, the atoms are ionized and the system consists of a collection of electrons and nuclei, commonly referred to as a *plasma*.

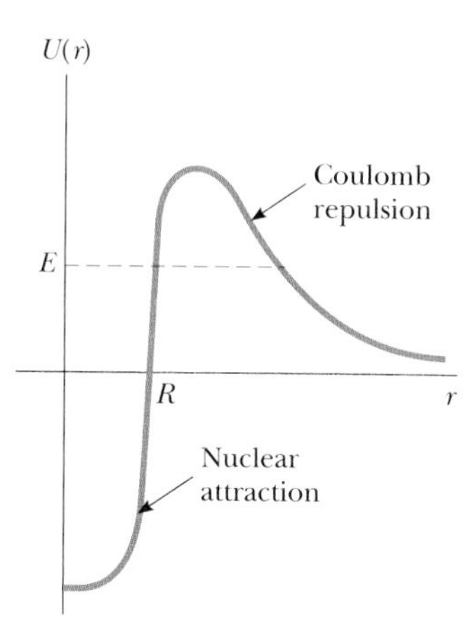

Figure 45.7 Potential energy as a function of separation distance between two deuterons. The Coulomb repulsive force is dominant at long range, and the nuclear force is dominant at short range, where R is on the order of 1 fm. If we neglect tunneling, the two deuterons require an energy E greater than the height of the barrier to undergo fusion.

EXAMPLE 45.3 The Fusion of Two Deuterons

For the nuclear force to overcome the repulsive Coulomb force, the separation distance between two deuterons must be approximately 1.0×10^{-14} m.

(A) Calculate the height of the potential barrier due to the repulsive force.

SOLUTION

Conceptualize Imagine moving two deuterons toward each other. As they move closer together, the Coulomb repulsion force becomes stronger. Work must be done on the system to push against this force, and this work appears in the system of two deuterons as electric potential energy.

Categorize We categorize this problem as one involving the electric potential energy of a system of two charged particles.

Analyze Evaluate the potential energy associated with two charges separated by a distance r (Eq. 25.13) for two deuterons:

$$U = k_e \frac{q_1 q_2}{r} = k_e \frac{(+e)^2}{r} = (8.99 \times 10^9\ \text{N}\cdot\text{m}^2/\text{C}^2)\frac{(1.60 \times 10^{-19}\ \text{C})^2}{1.0 \times 10^{-14}\ \text{m}}$$

$$= 2.3 \times 10^{-14}\ \text{J} = 0.14\ \text{MeV}$$

(B) Estimate the temperature required for a deuteron to overcome the potential barrier, assuming an energy of $\frac{3}{2}k_B T$ per deuteron (where k_B is Boltzmann's constant).

SOLUTION

Because the total Coulomb energy of the pair is 0.14 MeV, the Coulomb energy per deuteron is equal to 0.07 MeV = 1.1×10^{-14} J.

Set this energy equal to the average energy per deuteron:

$$\tfrac{3}{2}k_B T = 1.1 \times 10^{-14}\ \text{J}$$

Solve for T:

$$T = \frac{2(1.1 \times 10^{-14}\ \text{J})}{3(1.38 \times 10^{-23}\ \text{J/K})} = 5.6 \times 10^8\ \text{K}$$

(C) Find the energy released in the deuterium–deuterium reaction

$$^2_1\text{H} + ^2_1\text{H} \rightarrow ^3_1\text{H} + ^1_1\text{H}$$

SOLUTION

The mass of a single deuterium atom is equal to 2.014 102 u. Therefore, the total mass of the system before the reaction is 4.028 204 u.

Find the sum of the masses after the reaction: $3.016\,049\text{ u} + 1.007\,825\text{ u} = 4.023\,874\text{ u}$

Find the change in mass and convert to energy units: $4.028\,204\text{ u} - 4.023\,874\text{ u}$

$= 0.004\,33\text{ u}$

$= 0.004\,33\text{ u} \times 931.494\text{ MeV/u} = \boxed{4.03\text{ MeV}}$

Finalize The calculated temperature in part (B) is too high because the particles in the plasma have a Maxwellian speed distribution (Section 21.5) and therefore some fusion reactions are caused by particles in the high-energy tail of this distribution. Furthermore, even those particles that do not have enough energy to overcome the barrier have some probability of tunneling through. When these effects are taken into account, a temperature of "only" 4×10^8 K appears adequate to fuse two deuterons in a plasma. In part (C), notice that the energy value is consistent with that already given in Equation 45.4.

What If? Suppose the tritium resulting from the reaction in part (C) reacts with another deuterium in the reaction

$$^2_1\text{H} + ^3_1\text{H} \rightarrow ^4_2\text{He} + ^1_0\text{n}$$

How much energy is released in the sequence of two reactions?

Answer The overall effect of the sequence of two reactions is that three deuterium nuclei have combined to form a helium nucleus, a hydrogen nucleus, and a neutron. The initial mass is $3(2.014\,102\text{ u}) = 6.042\,306\text{ u}$. After the reaction, the sum of the masses is $4.002\,603\text{ u} + 1.007\,825\text{ u} + 1.008\,665 = 6.019\,093\text{ u}$. The excess mass is equal to 0.023 213 u, equivalent to an energy of 21.6 MeV. Notice that this value is the sum of the Q values for the second and third reactions in Equation 45.4.

The temperature at which the power generation rate in any fusion reaction exceeds the loss rate is called the **critical ignition temperature** T_{ignit}. This temperature for the deuterium–deuterium (D–D) reaction is 4×10^8 K. From the relationship $E \approx \frac{3}{2}k_BT$, the ignition temperature is equivalent to approximately 52 keV. The critical ignition temperature for the deuterium–tritium (D–T) reaction is approximately 4.5×10^7 K, or only 6 keV. A plot of the power $\mathcal{P}_{\text{gen}}$ generated by fusion versus temperature for the two reactions is shown in Figure 45.8. The straight green line represents the power $\mathcal{P}_{\text{lost}}$ lost via the radiation mechanism known as bremsstrahlung (Section 42.8). In this principal mechanism of energy loss, radiation (primarily x-rays) is emitted as the result of electron–ion collisions within the plasma. The intersections of the $\mathcal{P}_{\text{lost}}$ line with the $\mathcal{P}_{\text{gen}}$ curves give the critical ignition temperatures.

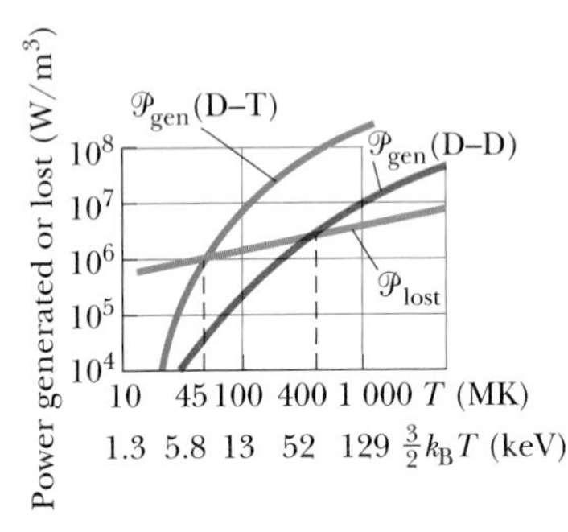

Figure 45.8 Power generated versus temperature for deuterium–deuterium (D–D) and deuterium–tritium (D–T) fusion. The green line represents power lost as a function of temperature. When the generation rate exceeds the loss rate, ignition takes place.

In addition to the high-temperature requirements, two other critical parameters determine whether or not a thermonuclear reactor is successful: the **ion density** n and **confinement time** τ, which is the time interval during which energy injected into the plasma remains within the plasma. British physicist J. D. Lawson has shown that both the ion density and confinement time must be large enough to ensure that more fusion energy is released than the amount required to raise the temperature of the plasma. For a given value of n, the probability of fusion between two particles increases as τ increases. For a given value of τ, the collision rate between nuclei increases as n increases. The product $n\tau$ is referred to as the **Lawson number** of a reaction. A graph of the value of $n\tau$ necessary to achieve a net energy output for the D–T and D–D reactions at different temperatures is shown in Figure 45.9. In particular, **Lawson's criterion** states that a net energy output is possible for values of $n\tau$ that meet the following conditions:

Lawson's criterion

$$n\tau \geq 10^{14}\ \text{s/cm}^3 \qquad \text{(D–T)} \tag{45.5}$$

$$n\tau \geq 10^{16}\ \text{s/cm}^3 \qquad \text{(D–D)}$$

These values represent the minima of the curves in Figure 45.9.

Lawson's criterion was arrived at by comparing the energy required to raise the temperature of a given plasma with the energy generated by the fusion process.[2] The energy E_{in} required to raise the temperature of the plasma is proportional to the ion density n, which we can express as $E_{\text{in}} = C_1 n$, where C_1 is some constant. The energy generated by the fusion process is proportional to $n^2\tau$, or $E_{\text{gen}} = C_2 n^2 \tau$. This dependence may be understood by realizing that the fusion energy released is proportional to both the rate at which interacting ions collide ($\propto n^2$) and the confinement time τ. Net energy is produced when $E_{\text{gen}} > E_{\text{in}}$. When the constants C_1 and C_2 are calculated for different reactions, the condition that $E_{\text{gen}} \geq E_{\text{in}}$ leads to Lawson's criterion.

Current efforts are aimed at meeting Lawson's criterion at temperatures exceeding T_{ignit}. Although the minimum required plasma densities have been achieved, the problem of confinement time is more difficult. The two basic techniques under investigation for solving this problem are magnetic confinement and inertial confinement.

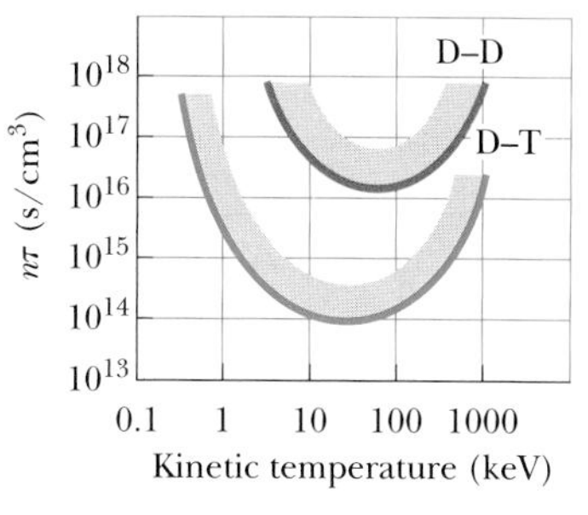

Figure 45.9 The Lawson number $n\tau$ at which net energy output is possible versus temperature for the D–T and D–D fusion reactions. The regions above the colored curves represent favorable conditions for fusion.

Magnetic Confinement

Many fusion-related plasma experiments use **magnetic confinement** to contain the plasma. A toroidal device called a **tokamak,** first developed in Russia, is shown in Figure 45.10a. A combination of two magnetic fields is used to confine and stabilize the plasma: (1) a strong toroidal field produced by the current in the toroidal windings surrounding a doughnut-shaped vacuum chamber and (2) a weaker "poloidal" field produced by the toroidal current. In addition to confining the

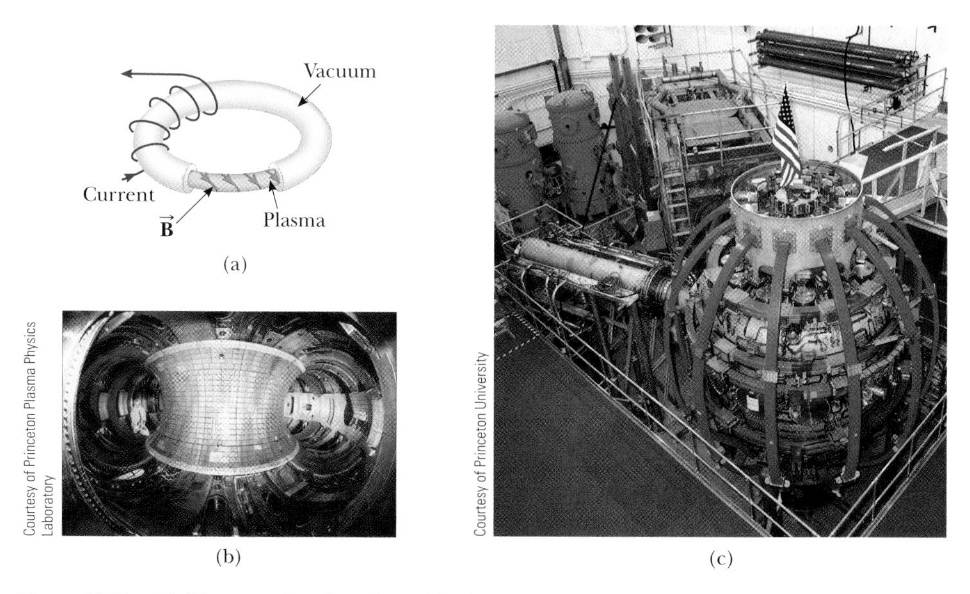

Figure 45.10 (a) Diagram of a tokamak used in the magnetic confinement scheme. (b) Interior view of the closed Tokamak Fusion Test Reactor (TFTR) vacuum vessel at the Princeton Plasma Physics Laboratory. (c) The National Spherical Torus Experiment (NSTX) that began operation in March 1999.

[2] Lawson's criterion neglects the energy needed to set up the strong magnetic field used to confine the hot plasma in a magnetic confinement approach. This energy is expected to be about 20 times greater than the energy required to raise the temperature of the plasma. It is therefore necessary either to have a magnetic energy recovery system or to use superconducting magnets.

plasma, the toroidal current is used to raise its temperature. The resultant helical magnetic field lines spiral around the plasma and keep it from touching the walls of the vacuum chamber. (If the plasma touches the walls, its temperature is reduced and heavy impurities sputtered from the walls "poison" it, leading to large power losses.)

One major breakthrough in magnetic confinement in the 1980s was in the area of auxiliary energy input to reach ignition temperatures. Experiments have shown that injecting a beam of energetic neutral particles into the plasma is a very efficient method of raising it to ignition temperatures. Radio-frequency energy input will probably be needed for reactor-size plasmas.

When it was in operation from 1982 to 1997, the Tokamak Fusion Test Reactor (TFTR, Fig. 45.10b) at Princeton University reported central ion temperatures of 510 million degrees Celsius, more than 30 times greater than the temperature at the center of the Sun. The $n\tau$ values in the TFTR for the D–T reaction were well above 10^{13} s/cm^3 and close to the value required by Lawson's criterion. In 1991, reaction rates of 6×10^{17} D–T fusions per second were reached in the JET tokamak at Abington, England.

One of the new generation of fusion experiments is the National Spherical Torus Experiment (NSTX) at the Princeton Plasma Physics Laboratory and shown in Figure 45.10c. This reactor was brought on line in February 1999 and has been running fusion experiments since then. Rather than the doughnut-shaped plasma of a tokamak, the NSTX produces a spherical plasma that has a hole through its center. The major advantage of the spherical configuration is its ability to confine the plasma at a higher pressure in a given magnetic field. This approach could lead to development of smaller, more economical fusion reactors.

An international collaborative effort involving the United States, the European Union, Japan, China, Korea, India, and Russia is currently under way to build a fusion reactor called ITER. This acronym stands for International Thermonuclear Experimental Reactor, although recently the emphasis has shifted to interpreting "iter" in terms of its Latin meaning, "the way." One reason proposed for this change is to avoid public misunderstanding and negative connotations toward the word *thermonuclear.* This facility will address the remaining technological and scientific issues concerning the feasibility of fusion power. The design is completed, and Cadarache, France, was chosen in June 2005 as the reactor site. Construction will require about 10 years, with fusion operation projected to begin in 2016. If the planned device works as expected, the Lawson number for ITER will be about six times greater than the current record holder, the JT-60 U tokamak in Japan. ITER is expected to produce 1.5 GW of power, and the energy content of the alpha particles inside the reactor will be so intense that they will sustain the fusion reaction, allowing the auxiliary energy sources to be turned off once the reaction is initiated.

EXAMPLE 45.4 Inside a Fusion Reactor

In 1998, the JT-60U tokamak in Japan operated with a D–T plasma density of 4.8×10^{13} cm^{-3} at a temperature (in energy units) of 24.1 keV. It confined this plasma inside a magnetic field for 1.1 s.

(A) Do these data meet Lawson's criterion?

SOLUTION

Conceptualize With the help of the third of Equations 45.4, imagine many such reactions occurring in a plasma of high temperature and high density.

Categorize We use the concept of the Lawson number discussed in this section, so we categorize this example as a substitution problem.

Evaluate the Lawson number for the JT-60U:

$$n\tau = (4.8 \times 10^{13}\ \text{cm}^{-3})(1.1\ \text{s}) = 5.3 \times 10^{13}\ \text{s/cm}^3$$

This value is close to meeting Lawson's criterion of 10^{14} s/cm^3 for a D–T plasma given in Equation 45.5. In fact, scientists recorded a power gain of 1.25, indicating that the reactor operated slightly past the break-even point and produced more energy than it required to maintain the plasma.

(B) How does the plasma density compare with the density of atoms in an ideal gas when the gas is under standard conditions ($T = 0°\text{C}$ and $P = 1$ atm)?

SOLUTION

Find the density of atoms in a sample of ideal gas by evaluating N_A/V_{mol}, where N_A is Avogadro's number and V_{mol} is the molar volume of an ideal gas under standard conditions, 2.24×10^{-2} m^3/mol:

$$\frac{N_A}{V_{\text{mol}}} = \frac{6.02 \times 10^{23} \text{ atoms/mol}}{2.24 \times 10^{-2} \text{ m}^3\text{/mol}} = 2.7 \times 10^{25} \text{ atoms/m}^3$$
$$= 2.7 \times 10^{19} \text{ atoms/cm}^3$$

This value is more than 500 000 times greater than the plasma density in the reactor.

Inertial Confinement

The second technique for confining a plasma, called **inertial confinement**, makes use of a D–T target that has a very high particle density. In this scheme, the confinement time is very short (typically 10^{-11} to 10^{-9} s), and, because of their own inertia, the particles do not have a chance to move appreciably from their initial positions. Therefore, Lawson's criterion can be satisfied by combining a high particle density with a short confinement time.

Laser fusion is the most common form of inertial confinement. A small D–T pellet, approximately 1 mm in diameter, is struck simultaneously by several focused, high-intensity laser beams, resulting in a large pulse of input energy that causes the surface of the fuel pellet to evaporate (Fig. 45.11). The escaping particles exert a third-law reaction force on the core of the pellet, resulting in a strong, inwardly moving compressive shock wave. This shock wave increases the pressure and density of the core and produces a corresponding increase in temperature. When the temperature of the core reaches ignition temperature, fusion reactions occur.

One of the leading laser fusion laboratories in the United States is the Omega facility at the University of Rochester in New York. This facility focuses 24 laser beams on the target. Currently under construction at the Lawrence Livermore National Laboratory in Livermore, California, is the National Ignition Facility. The research apparatus there will include 192 laser beams that can be focused on a deuterium–tritium pellet. Construction is expected to be final in 2009, with fusion ignition tests planned in 2010.

Figure 45.11 In inertial confinement, a D–T fuel pellet fuses when struck by several high-intensity laser beams simultaneously.

Fusion Reactor Design

In the D–T fusion reaction

$${}^2_1\text{H} + {}^3_1\text{H} \rightarrow {}^4_2\text{He} + {}^1_0\text{n} \qquad Q = 17.59 \text{ MeV}$$

the alpha particle carries 20% of the energy and the neutron carries 80%, or approximately 14 MeV. A diagram of the deuterium–tritium fusion reaction is shown in Active Figure 45.12. Because the alpha particles are charged, they are primarily absorbed by the plasma, causing the plasma's temperature to increase. In contrast, the 14-MeV neutrons, being electrically neutral, pass through the plasma and are absorbed by a surrounding blanket material, where their large kinetic energy is extracted and used to generate electric power.

One scheme is to use molten lithium metal as the neutron-absorbing material and to circulate the lithium in a closed heat-exchange loop, thereby producing steam and driving turbines as in a conventional power plant. Figure 45.13 (page 1342) shows a diagram of such a reactor. It is estimated that a blanket of lithium

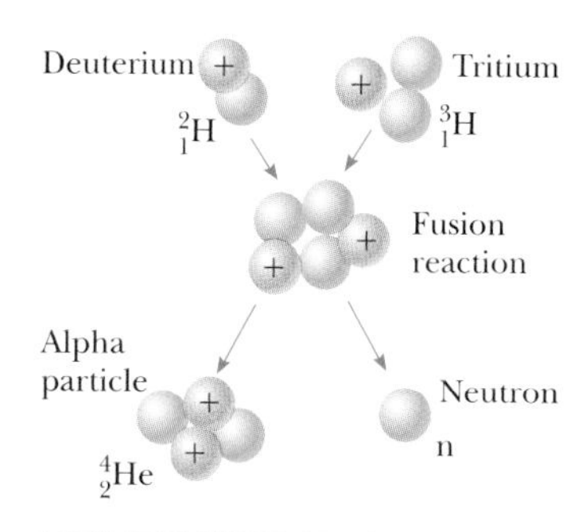

ACTIVE FIGURE 45.12

Deuterium–tritium fusion. Eighty percent of the energy released is in the 14-MeV neutron.

Sign in at www.thomsonedu.com and go to ThomsonNOW to observe several fusion reactions and measure the energy released.

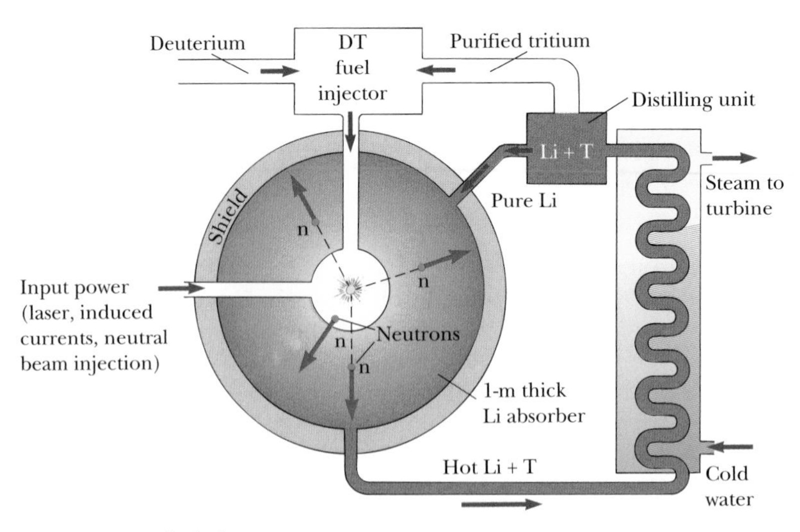

Figure 45.13 Diagram of a fusion reactor.

approximately 1 m thick will capture nearly 100% of the neutrons from the fusion of a small D–T pellet.

The capture of neutrons by lithium is described by the reaction

$$^{1}_{0}\text{n} + {}^{6}_{3}\text{Li} \rightarrow {}^{3}_{1}\text{H} + {}^{4}_{2}\text{He}$$

where the kinetic energies of the charged tritium $^{3}_{1}\text{H}$ and alpha particle are converted to internal energy in the molten lithium. An extra advantage of using lithium as the energy-transfer medium is that the tritium produced can be separated from the lithium and returned as fuel to the reactor.

Advantages and Problems of Fusion

If fusion power can ever be harnessed, it will offer several advantages over fission-generated power: (1) low cost and abundance of fuel (deuterium), (2) impossibility of runaway accidents, and (3) decreased radiation hazard. Some of the anticipated problems and disadvantages include (1) scarcity of lithium, (2) limited supply of helium, which is needed for cooling the superconducting magnets used to produce strong confining fields, and (3) structural damage and induced radioactivity caused by neutron bombardment. If such problems and the engineering design factors can be resolved, nuclear fusion may become a feasible source of energy by the middle of the twenty-first century.

45.5 Radiation Damage

In Chapter 34, we learned that electromagnetic radiation is all around us in the form of radio waves, microwaves, light waves, and so on. In this section, we describe forms of radiation that can cause severe damage as they pass through matter, such as radiation resulting from radioactive processes and radiation in the form of energetic particles such as neutrons and protons.

The degree and type of damage depend on several factors, including the type and energy of the radiation and the properties of the matter. The metals used in nuclear reactor structures can be severely weakened by high fluxes of energetic neutrons because these high fluxes often lead to metal fatigue. The damage in such situations is in the form of atomic displacements, often resulting in major alterations in the properties of the material.

Radiation damage in biological organisms is primarily due to ionization effects in cells. A cell's normal operation may be disrupted when highly reactive ions are formed as the result of ionizing radiation. For example, hydrogen and the hydroxyl radical OH^- produced from water molecules can induce chemical reactions that may break bonds in proteins and other vital molecules. Furthermore, the ionizing radiation may affect vital molecules directly by removing electrons from their structure. Large doses of radiation are especially dangerous because damage to a great number of molecules in a cell may cause the cell to die. Although the death of a single cell is usually not a problem, the death of many cells may result in irreversible damage to the organism. Cells that divide rapidly, such as those of the digestive tract, reproductive organs, and hair follicles, are especially susceptible. In addition, cells that survive the radiation may become defective. These defective cells can produce more defective cells and can lead to cancer.

In biological systems, it is common to separate radiation damage into two categories: somatic damage and genetic damage. *Somatic damage* is that associated with any body cell except the reproductive cells. Somatic damage can lead to cancer or can seriously alter the characteristics of specific organisms. *Genetic damage* affects only reproductive cells. Damage to the genes in reproductive cells can lead to defective offspring. It is important to be aware of the effect of diagnostic treatments, such as x-rays and other forms of radiation exposure, and to balance the significant benefits of treatment with the damaging effects.

Damage caused by radiation also depends on the radiation's penetrating power. Alpha particles cause extensive damage, but penetrate only to a shallow depth in a material due to the strong interaction with other charged particles. Neutrons do not interact via the electric force and hence penetrate deeper, causing significant damage. Gamma rays are high-energy photons that can cause severe damage, but often pass through matter without interaction.

Several units have been used historically to quantify the amount, or dose, of any radiation that interacts with a substance.

The **roentgen** (R) is that amount of ionizing radiation that produces an electric charge of 3.33×10^{-10} C in 1 cm^3 of air under standard conditions.

Equivalently, the roentgen is that amount of radiation that increases the energy of 1 kg of air by 8.76×10^{-3} J.

For most applications, the roentgen has been replaced by the rad (an acronym for *radiation absorbed dose*):

One **rad** is that amount of radiation that increases the energy of 1 kg of absorbing material by 1×10^{-2} J.

Although the rad is a perfectly good physical unit, it is not the best unit for measuring the degree of biological damage produced by radiation because damage depends not only on the dose but also on the type of the radiation. For example, a given dose of alpha particles causes about ten times more biological damage than an equal dose of x-rays. The **RBE** (relative biological effectiveness) factor for a given type of radiation is **the number of rads of x-radiation or gamma radiation that produces the same biological damage as 1 rad of the radiation being used.** The RBE factors for different types of radiation are given in Table 45.1 (page 1344). The values are only approximate because they vary with particle energy and with the form of the damage. The RBE factor should be considered only a first-approximation guide to the actual effects of radiation.

Finally, the **rem** (radiation equivalent in man) is the product of the dose in rad and the RBE factor:

$$\text{Dose in rem} \equiv \text{dose in rad} \times \text{RBE} \quad (45.6)$$

◀ Radiation dose in rem

TABLE 45.1

RBE Factors for Several Types of Radiation

Radiation	RBE Factor
X-rays and gamma rays	1.0
Beta particles	1.0–1.7
Alpha particles	10–20
Thermal neutrons	4–5
Fast neutrons and protons	10
Heavy ions	20

Note: RBE = relative biological effectiveness.

According to this definition, 1 rem of any two types of radiation produces the same amount of biological damage. Table 45.1 shows that a dose of 1 rad of fast neutrons represents an effective dose of 10 rem, but 1 rad of gamma radiation is equivalent to a dose of only 1 rem.

Low-level radiation from natural sources such as cosmic rays and radioactive rocks and soil delivers to each of us a dose of approximately 0.13 rem/yr. This radiation, called *background radiation,* varies with geography, with the main factors being altitude (exposure to cosmic rays) and geology (radon gas released by some rock formations, deposits of naturally radioactive minerals).

The upper limit of radiation dose rate recommended by the U.S. government (apart from background radiation) is approximately 0.5 rem/yr. Many occupations involve much higher radiation exposures, so an upper limit of 5 rem/yr has been set for combined whole-body exposure. Higher upper limits are permissible for certain parts of the body, such as the hands and the forearms. A dose of 400 to 500 rem results in a mortality rate of approximately 50% (which means that half the people exposed to this radiation level die). The most dangerous form of exposure for most people is either ingestion or inhalation of radioactive isotopes, especially isotopes of those elements the body retains and concentrates, such as ^{90}Sr.

This discussion has focused on measurements of radiation dosage in units such as rads and rems because these units are still widely used. They have, however, been formally replaced with new SI units. The rad has been replaced with the *gray* (Gy), equal to 100 rad, and the rem has been replaced with the *sievert* (Sv), equal to 100 rem. Table 45.2 summarizes the older and the current SI units of radiation dosage.

45.6 Radiation Detectors

Particles passing through matter interact with the matter in several ways. The particle can, for example, ionize atoms, scatter from atoms, or be absorbed by atoms. Radiation detectors exploit these interactions to allow a measurement of the particle's energy, momentum, or charge and sometimes the very existence of the parti-

TABLE 45.2

Units for Radiation Dosage

Quantity	SI Unit	Symbol	Relation to Other SI units	Older Unit	Conversion
Absorbed dose	gray	Gy	= 1 J/kg	rad	1 Gy = 100 rad
Dose equivalent	sievert	Sv	= 1 J/kg	rem	1 Sv = 100 rem

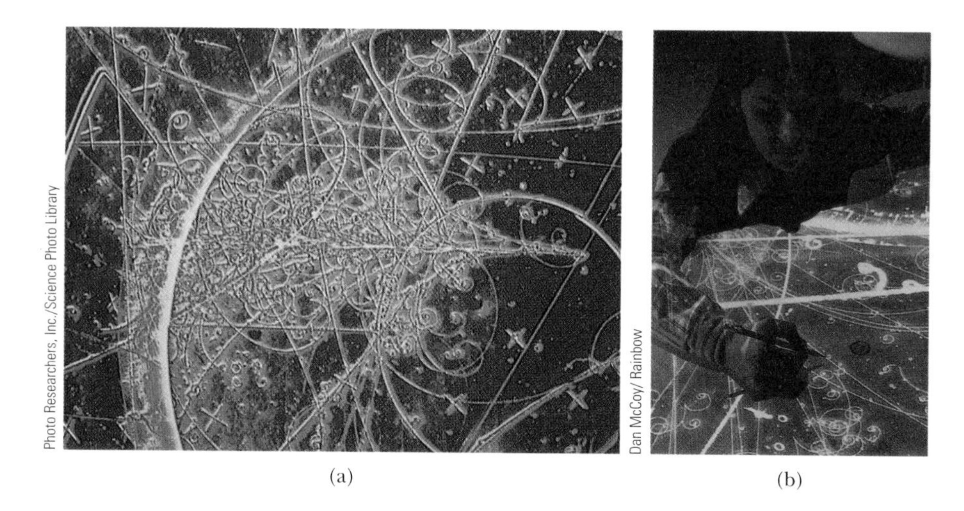

Figure 45.14 (a) Artificially colored bubble-chamber photograph showing tracks of particles that have passed through the chamber. (b) This research scientist is studying a photograph of particle tracks made in a bubble chamber at Fermilab. The curved tracks are produced by charged particles moving through the chamber in the presence of an applied magnetic field. Negatively charged particles deflect in one direction, and positively charged particles deflect in the opposite direction.

cle if it is otherwise difficult to detect. Various devices have been developed for detecting radiation. These devices are used for a variety of purposes, including medical diagnoses, radioactive dating measurements, measuring background radiation, and measuring the mass, energy, and momentum of particles created in high-energy nuclear reactions.

In the early part of the 20th century, detectors were much simpler than those used today. We discuss three of these early detectors first. A **photographic emulsion** is the simplest example of a detector. A charged particle ionizes the atoms in an emulsion layer. The particle's path corresponds to a family of points at which chemical changes have occurred in the emulsion. When the emulsion is developed, the particle's track becomes visible. A **cloud chamber** contains a gas that has been supercooled to slightly below its usual condensation point. An energetic particle passing through ionizes the gas along the particle's path. The ions serve as centers for condensation of the supercooled gas. The particle's track can be seen with the naked eye and can be photographed. A magnetic field can be applied to determine the charges of the particles as well as their momentum and energy. A device called a **bubble chamber** uses a liquid (usually liquid hydrogen) maintained near its boiling point. Ions produced by incoming charged particles leave bubble tracks, which can be photographed (Fig. 45.14). Because the density of the detecting medium in a bubble chamber is much higher than the density of the gas in a cloud chamber, the bubble chamber has a much higher sensitivity.

More contemporary detectors involve more sophisticated processes. In an **ion chamber** (Fig. 45.15), electron–ion pairs are generated as radiation passes through a gas and produces an electrical signal. Two plates in the chamber are connected to a voltage supply and thereby maintained at different electric potentials. The positive plate attracts the electrons, and the negative plate attracts positive ions, causing a current pulse that is proportional to the number of electron–ion pairs produced when a particle passes through the chamber. When an ion chamber is

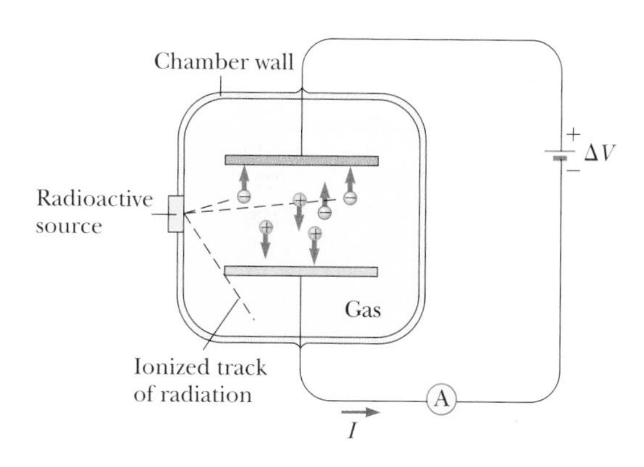

Figure 45.15 Simplified diagram of an ion chamber. The radioactive source creates electrons and positive ions that are collected by the charged plates. The current set up in the external circuit is proportional to a radioactive particle's kinetic energy if the particle stops in the chamber.

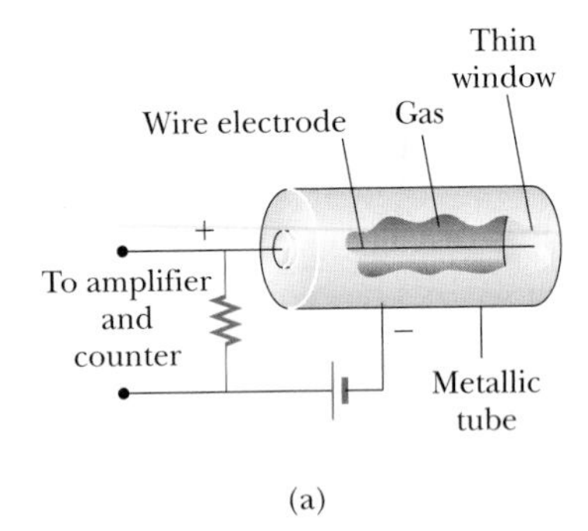

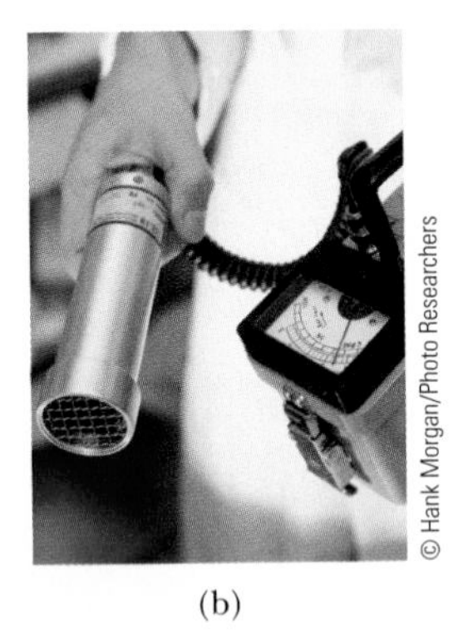

Figure 45.16 (a) Diagram of a Geiger counter. The voltage between the wire electrode and the metallic tube is usually approximately 1 000 V. (b) A scientist uses a Geiger counter to make a measurement.

used both to detect the presence of a particle and to measure its energy, it is called a **proportional counter.**

The **Geiger counter** (Fig. 45.16) is perhaps the most common form of ion chamber used to detect radioactivity. It can be considered the prototype of all counters that use the ionization of a medium as the basic detection process. A Geiger counter consists of a thin wire electrode aligned along the central axis of a cylindrical metallic tube filled with a gas at low pressure. The wire is maintained at a high positive electric potential (approximately 10^3 V) relative to the tube. When a high-energy particle resulting, for example, from a radioactive decay enters the tube through a thin window at one end, some of the gas atoms are ionized. The electrons removed from these atoms are attracted toward the wire electrode, and, in the process, they ionize other atoms in their path. This sequential ionization results in an *avalanche* of electrons that produces a current pulse. After the pulse has been amplified, it can either be used to trigger an electronic counter or be delivered to a loudspeaker that clicks each time a particle is detected. Although a Geiger counter easily detects the presence of a particle, the energy lost by the particle in the counter is *not* proportional to the current pulse produced. Therefore, a Geiger counter cannot be used to measure the energy of a particle.

A **semiconductor-diode detector** is essentially a reverse-bias p–n junction. Recall from Section 43.7 that a p–n junction passes current readily when forward-biased and prohibits a current when reverse-biased. As an energetic particle passes through the junction, electrons are excited into the conduction band and holes are formed in the valence band. The internal electric field sweeps the electrons toward the positive (n) side of the junction and the holes toward the negative (p) side. This movement of electrons and holes creates a pulse of current that is measured with an electronic counter. In a typical device, the duration of the pulse is 10^{-8} s.

A **scintillation counter** usually uses a solid or liquid material whose atoms are easily excited by radiation. The excited atoms then emit photons when they return to their ground state. Common materials used as scintillators are transparent crystals of sodium iodide and certain plastics. If the scintillator material is attached to a photomultiplier tube (Section 40.2), the photons emitted by the scintillator can be detected and an electrical signal produced.

Both the scintillator and the semiconductor-diode detector are much more sensitive than a Geiger counter mainly because of the higher density of the detecting medium. Both measure the total energy deposited in the detector, which can be very useful in particle identification. In addition, if the particle stops in the detector, both instruments can be used to measure the total particle energy.

Track detectors are devices used to view the tracks of charged particles directly. High-energy particles produced in particle accelerators may have energies ranging from 10^9 to 10^{12} eV. Therefore, they often cannot be stopped and cannot have their energy measured with the detectors already mentioned. Instead, the energy and momentum of these energetic particles are found from the curvature of their path in a magnetic field of known magnitude and direction.

A **spark chamber** is a counting device that consists of an array of conducting parallel plates and is capable of recording a three-dimensional track record. Even-numbered plates are grounded, and odd-numbered plates are maintained at a high electric potential (approximately 10 kV). The spaces between the plates contain an inert gas at atmospheric pressure. When a charged particle passes through the chamber, gas atoms are ionized, resulting in a current surge and visible sparks along the particle path. These sparks may be photographed or electronically detected and the data sent to a computer for path reconstruction and determination of particle mass, momentum, and energy.

Newer versions of the spark chamber have been developed. A **drift chamber** has thousands of high-voltage wires arrayed through the space of the detector, which is filled with gas. The result is an array of thousands of proportional counters. When a charged particle passes through the detector, it ionizes gas molecules and the ejected electrons drift toward the high-voltage wires, creating an electrical signal upon arrival. A computer detects the signals and reconstructs the path through the detector. A large-volume, sophisticated drift chamber that has provided significant results in studying particles formed in collisions of atoms is the Solenoidal Tracker at RHIC (STAR). (The acronym RHIC stands for Relativistic Heavy Ion Collider, a facility at Brookhaven National Laboratory that began operation in 2000.) This type of drift chamber is called a **time projection chamber.** A photograph of the STAR detector is shown in Figure 45.17.

Courtesy of Brookhaven National Laboratory/ RHIC-STAR

Figure 45.17 The STAR detector at the Relativistic Heavy Ion Collider at Brookhaven National Laboratory.

45.7 Uses of Radiation

Nuclear physics applications are extremely widespread in manufacturing, medicine, and biology. In this section, we present a few of these applications and the underlying theories supporting them.

Tracing

Radioactive tracers are used to track chemicals participating in various reactions. One of the most valuable uses of radioactive tracers is in medicine. For example, iodine, a nutrient needed by the human body, is obtained largely through the intake of iodized salt and seafood. To evaluate the performance of the thyroid, the patient drinks a very small amount of radioactive sodium iodide containing ^{131}I, an artificially produced isotope of iodine (the natural, nonradioactive isotope is ^{127}I). The amount of iodine in the thyroid gland is determined as a function of time by measuring the radiation intensity at the neck area. How much of the isotope ^{131}I remains in the thyroid is a measure of how well that gland is functioning.

A second medical application is indicated in Figure 45.18. A solution containing radioactive sodium is injected into a vein in the leg, and the time at which the radioisotope arrives at another part of the body is detected with a radiation counter. The elapsed time is a good indication of the presence or absence of constrictions in the circulatory system.

Tracers are also useful in agricultural research. Suppose the best method of fertilizing a plant is to be determined. A certain element in a fertilizer, such as nitrogen, can be *tagged* (identified) with one of its radioactive isotopes. The fertilizer is then sprayed on one group of plants, sprinkled on the ground for a second group, and raked into the soil for a third. A Geiger counter is then used to track the nitrogen through each of the three groups.

Tracing techniques are as wide ranging as human ingenuity can devise. Today, applications range from checking how teeth absorb fluoride to monitoring how cleansers contaminate food-processing equipment to studying deterioration inside an automobile engine. In this last case, a radioactive material is used in the manufacture of the car's piston rings and the oil is checked for radioactivity to determine the amount of wear on the rings.

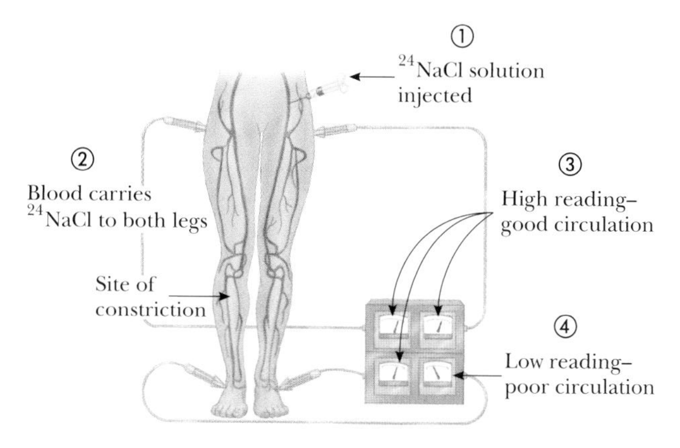

Figure 45.18 A tracer technique for determining the condition of the human circulatory system.

Materials Analysis

For centuries, a standard method of identifying the elements in a sample of material has been chemical analysis, which involves determining how the material reacts with various chemicals. A second method is spectral analysis, which works because each element, when excited, emits its own characteristic set of electromagnetic wavelengths. These methods are now supplemented by a third technique, **neutron activation analysis.** A disadvantage of both chemical and spectral methods is that a fairly large sample of the material must be destroyed for the analysis. In addition, extremely small quantities of an element may go undetected by either method. Neutron activation analysis has an advantage over chemical analysis and spectral analysis in both respects.

When a material is irradiated with neutrons, nuclei in the material absorb the neutrons and are changed to different isotopes, most of which are radioactive. For example, ^{65}Cu absorbs a neutron to become ^{66}Cu, which undergoes beta decay:

$$^{1}_{0}\text{n} + {}^{65}_{29}\text{Cu} \rightarrow {}^{66}_{29}\text{Cu} \rightarrow {}^{66}_{30}\text{Zn} + \text{e}^{-} + \overline{\nu}$$

The presence of the copper can be deduced because it is known that ^{66}Cu has a half-life of 5.1 min and decays with the emission of beta particles having maximum energies of 2.63 and 1.59 MeV. Also emitted in the decay of ^{66}Cu is a 1.04-MeV gamma ray. By examining the radiation emitted by a substance after it has been exposed to neutron irradiation, one can detect extremely small amounts of an element in that substance.

Neutron activation analysis is used routinely in a number of industries. In commercial aviation, for example, it is used to check airline luggage for hidden explosives. One nonroutine use is of historical interest. Napoleon died on the island of St. Helena in 1821, supposedly of natural causes. Over the years, suspicion has existed that his death was not all that natural. After his death, his head was shaved and locks of his hair were sold as souvenirs. In 1961, the amount of arsenic in a sample of this hair was measured by neutron activation analysis, and an unusually large quantity of arsenic was found. (Activation analysis is so sensitive that very small pieces of a single hair could be analyzed.) Results showed that the arsenic was fed to him irregularly. In fact, the arsenic concentration pattern corresponded to the fluctuations in the severity of Napoleon's illness as determined from historical records.

Art historians use neutron activation analysis to detect forgeries. The pigments used in paints have changed throughout history, and old and new pigments react differently to neutron activation. The method can even reveal hidden works of art behind existing paintings because an older, hidden layer of paint reacts differently than the surface layer to neutron activation.

Radiation Therapy

Radiation causes much damage to rapidly dividing cells. Therefore, it is useful in cancer treatment because tumor cells divide extremely rapidly. Several mechanisms can be used to deliver radiation to a tumor. In some cases, a narrow beam of x-rays or radiation from a source such as ^{60}Co is used as shown in Figure 45.19. In other situations, thin radioactive needles called *seeds* are implanted in the cancerous tissue. The radioactive isotope ^{131}I is used to treat cancer of the thyroid.

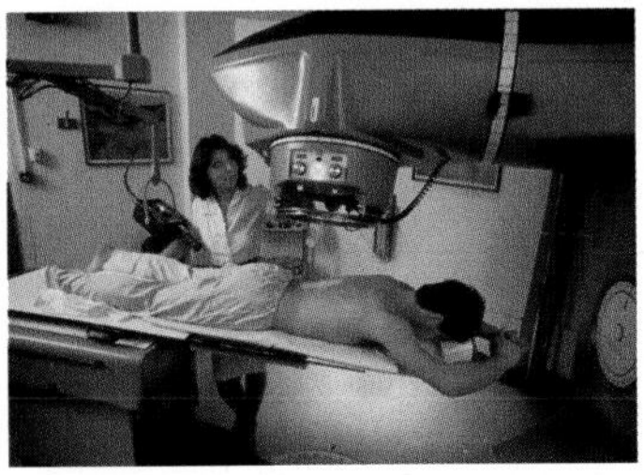

Martin Dohrn/Science Photo Library/ Photo Researchers

Figure 45.19 This large machine is being set to deliver a dose of radiation from ^{60}Co in an effort to destroy a cancerous tumor. Cancer cells are especially susceptible to this type of therapy because they tend to divide more often than cells of healthy tissue nearby.

Food Preservation

Radiation is finding increasing use as a means of preserving food because exposure to high levels of radiation can destroy or incapacitate bacteria and mold spores (Fig. 45.20). Techniques include exposing foods to gamma rays, high-energy electron beams, and x-rays. Food preserved by such exposure can be placed in a sealed container (to keep out new spoiling agents) and stored for long periods of time.

Figure 45.20 The strawberries on the left are untreated and have become moldy. The unspoiled strawberries on the right have been irradiated. The radiation has killed or incapacitated the mold spores that have spoiled the strawberries on the left.

There is little or no evidence of adverse effect on the taste or nutritional value of food from irradiation. The safety of irradiated foods has been endorsed by the World Health Organization, the Centers for Disease Control and Prevention, the U.S. Department of Agriculture, and the Food and Drug Administration.

Summary

ThomsonNOW™ Sign in at **www.thomsonedu.com** and go to ThomsonNOW to take a practice test for this chapter.

CONCEPTS AND PRINCIPLES

The probability that neutrons are captured as they move through matter generally increases with decreasing neutron energy. A **thermal neutron** is a slow-moving neutron that has a high probability of being captured by a nucleus in a **neutron capture event:**

$$ {}_0^1\text{n} + {}_Z^A\text{X} \rightarrow {}^{A+1}_{Z}\text{X}^* \rightarrow {}^{A+1}_{Z}\text{X} + \gamma \qquad \textbf{(45.1)} $$

where ${}^{A+1}_{Z}\text{X}^*$ is an excited intermediate nucleus that rapidly emits a photon.

Nuclear fission occurs when a very heavy nucleus, such as ^{235}U, splits into two smaller **fission fragments.** Thermal neutrons can create fission in ^{235}U:

$$ {}_0^1\text{n} + {}^{235}_{92}\text{U} \rightarrow {}^{236}_{92}\text{U}^* \rightarrow \text{X} + \text{Y} + \text{neutrons} \qquad \textbf{(45.2)} $$

where ^{236}U* is an intermediate excited state and X and Y are the fission fragments. On average, 2.5 neutrons are released per fission event. The fragments then undergo a series of beta and gamma decays to various stable isotopes. The energy released per fission event is approximately 200 MeV.

The **reproduction constant** K is the average number of neutrons released from each fission event that cause another event. In a fission reactor, it is necessary to maintain $K \approx 1$. The value of K is affected by such factors as reactor geometry, mean neutron energy, and probability of neutron capture.

In **nuclear fusion,** two light nuclei fuse to form a heavier nucleus and release energy. The major obstacle in obtaining useful energy from fusion is the large Coulomb repulsive force between the charged nuclei at small separation distances. The temperature required to produce fusion is on the order of 10^8 K, and at this temperature, all matter occurs as a plasma.

In a fusion reactor, the plasma temperature must reach the **critical ignition temperature,** the temperature at which the power generated by the fusion reactions exceeds the power lost in the system. The most promising fusion reaction is the D–T reaction, which has a critical ignition temperature of approximately 4.5×10^7 K. Two critical parameters in fusion reactor design are **ion density** n and **confinement time** τ, the time interval during which the interacting particles must be maintained at $T > T_{\text{ignit}}$. **Lawson's criterion** states that for the D–T reaction, $n\tau \geq 10^{14}$ s/cm^3.

Questions

☐ denotes answer available in *Student Solutions Manual/Study Guide;* **O** denotes objective question

1. Why is water a better shield against neutrons than lead or steel?

2. If a nucleus captures a slow-moving neutron, the product is left in a highly excited state, with an energy approximately 8 MeV above the ground state. Explain the source of the excitation energy.

3. O If the moderator were suddenly removed from a nuclear reactor in an electric generating station, what is the most likely consequence? (a) The reactor would go supercritical, and a runaway reaction would occur. (b) The nuclear reaction would proceed in the same way, but the reactor would overheat. (c) The reactor would become subcritical, and the reaction would die out. (d) No change would occur in the reactor's operation.

4. Discuss the advantages and disadvantages of fission reactors from the point of view of safety, pollution, and resources. Make a comparison with power generated from the burning of fossil fuels.

5. O On August 6, 1945, the United States dropped a nuclear bomb on Hiroshima. The Americans called it "little boy," and the Japanese called it the "original child bomb." The energy released was approximately 5×10^{13} J, equivalent to that from 12 000 metric tons of TNT. What is the order of magnitude of the mass converted into energy in this explosion? (a) 1 mg (b) 1 g (c) 1 kg (d) 1 000 kg (e) 10 000 metric tons

6. O A certain $^{235}_{92}$U nucleus absorbs a neutron and fissions into the products $^{137}_{53}$I and $^{96}_{39}$Y together with how many neutrons? (a) 0 (b) 1 (c) 2 (d) 3 (e) 4 (f) an indeterminate number

7. Why would a fusion reactor produce less radioactive waste than a fission reactor?

8. Lawson's criterion states that the product of ion density and confinement time must exceed a certain number before a break-even fusion reaction can occur. Why should these two parameters determine the outcome?

9. Discuss the similarities and differences between fusion and fission.

10. Discuss the advantages and disadvantages of fusion power from the viewpoint of safety, pollution, and resources.

11. O You may use Figure 44.5, the curve of binding energy, to answer this question. Three nuclear reactions take place, each involving 108 nucleons: (1) eighteen ^{6}Li nuclei fuse in pairs to form nine ^{12}C nuclei, (2) four nuclei each with 27 nucleons fuse in pairs to form two nuclei with 54 nucleons, and (3) one nucleus with 108 nucleons fissions to form two nuclei with 54 nucleons. Rank these three reactions according to the reaction energy from the largest positive Q value (representing energy output) to the largest negative value (representing energy input). Also include $Q = 0$ in your ranking to make clear which of the reactions put out energy and which absorb energy. Note any cases of equality in your ranking.

12. O In Figure 45.14, the particle tracks in the bubble chamber are generally spirals rather than sections of circles. What is the primary reason for this shape? (a) The magnetic field is not perpendicular to the velocity of the particles. (b) The magnetic field is not uniform in space. (c) The forces on the particles increase with time. (d) The speeds of the particles decrease with time.

13. O Choose all correct answers. In the operation of a Geiger counter, the amplitude of the current pulse is (a) proportional to the kinetic energy of the particle producing the pulse, (b) proportional to the number of particles entering the tube to produce the pulse, (c) proportional to the RBE factor of the type of particle producing the pulse, or (d) independent of all these factors.

14. The design of a photomultiplier tube (Fig. 40.12) might suggest that any number of dynodes may be used to amplify a weak signal. What factors do you suppose would limit the amplification in this device?

15. O If an alpha particle and an electron have the same kinetic energy, which undergoes the greater deflection when passed through a magnetic field? (a) The alpha particle does. (b) The electron does. (c) They undergo the same deflection. (d) Neither is deflected.

16. O Working with radioactive materials at a laboratory over one year, (a) Tom received 1 rem of alpha radiation, (b) Karen received 1 rad of fast neutrons, (c) Paul received 1 rad of thermal neutrons as a whole-body dose, and (d) Ingrid received 1 rad of thermal neutrons to her hands only. Rank these four doses according to the likely amount of biological damage from the greatest to the least, noting any cases of equality.

17. *And swift, and swift past comprehension*
Turn round Earth's beauty and her might.
The heavens blaze in alternation
With deep and chill and rainy night.
In mighty currents foams the ocean
Up from the rocks' abyssal base,
With rock and sea torn into motion
In ever-swift celestial race.
Corrosive, choking smoke is spraying.
Above infernos, lava flies.
A perilous bridge, the land is swaying
Between them and the gaping skies.
And tempests bluster in a contest
From sea to land, from land to sea.
In rage they forge a chain around us
Of primal meaning, energy.
There flames a lightning disaster
Before the thunder, in its way.
But all Your servants honor, Master,
The gentle order of Your day.

Johann Wolfgang von Goethe wrote the song of the archangels in *Faust* half a century before the law of conservation of energy was recognized. Students often find it useful to list several "forms of energy," from kinetic to nuclear. Argue for or against the view that these lines of poetry make an obvious or oblique reference to every form of energy and energy transfer.

Problems

WebAssign The Problems from this chapter may be assigned online in WebAssign.

ThomsonNOW Sign in at **www.thomsonedu.com** and go to ThomsonNOW to assess your understanding of this chapter's topics with additional quizzing and conceptual questions.

1, 2, 3 denotes straightforward, intermediate, challenging; □ denotes full solution available in *Student Solutions Manual/Study Guide;* ▲ denotes coached solution with hints available at **www.thomsonedu.com;** denotes developing symbolic reasoning; ● denotes asking for qualitative reasoning; denotes computer useful in solving problem

Section 45.2 Nuclear Fission

Note: Problem 47 in Chapter 25 and Problems 17 and 40 in Chapter 44 can be assigned with this chapter.

1. Burning one metric ton (1 000 kg) of coal can yield an energy of 3.30×10^{10} J. Fission of one nucleus of uranium-235 yields an average of approximately 208 MeV. What mass of uranium produces the same energy as one ton of coal?

2. Find the energy released in the fission reaction

$$^{1}_{0}\text{n} + ^{235}_{92}\text{U} \rightarrow ^{98}_{40}\text{Zr} + ^{135}_{52}\text{Te} + 3(^{1}_{0}\text{n})$$

The atomic masses of the fission products are $^{98}_{40}$Zr, 97.912 7 u; and $^{135}_{52}$Te, 134.916 5 u.

3. Strontium-90 is a particularly dangerous fission product of ^{235}U because it is radioactive and it substitutes for calcium in bones. What other direct fission products would accompany it in the neutron-induced fission of ^{235}U? *Note:* This reaction may release two, three, or four free neutrons.

4. (a) The following fission reaction is typical of those occurring in a nuclear electric generating station:

$$^{1}_{0}\text{n} + ^{235}_{92}\text{U} \rightarrow ^{141}_{56}\text{Ba} + ^{92}_{36}\text{Kr} + 3(^{1}_{0}\text{n})$$

Find the energy released. The masses of the products are

$$M(^{141}_{56}\text{Ba}) = 140.914\,4 \text{ u} \quad M(^{92}_{36}\text{Kr}) = 91.926\,2 \text{ u}$$

(b) What fraction of the initial mass of the system is transformed?

5. ▲ List the nuclear reactions required to produce ^{233}U from ^{232}Th under fast neutron bombardment.

6. ● A reaction that has been considered as a source of energy is the absorption of a proton by a boron-11 nucleus to produce three alpha particles:

$$^{1}_{1}\text{H} + ^{11}_{5}\text{B} \rightarrow 3(^{4}_{2}\text{He})$$

This reaction is an attractive possibility because boron is easily obtained from the Earth's crust. A disadvantage is that the protons and boron nuclei must have large kinetic energies for the reaction to take place. This requirement contrasts to the initiation of uranium fission by slow neutrons. (a) How much energy is released in each reaction? (b) Why must the reactant particles have high kinetic energies?

7. **Review problem.** Suppose enriched uranium containing 3.40% of the fissionable isotope $^{235}_{92}$U is used as fuel for a ship. The water exerts an average friction force of magnitude 1.00×10^5 N on the ship. How far can the ship travel per kilogram of fuel? Assume the energy released per fission event is 208 MeV and the ship's engine has an efficiency of 20.0%.

8. A typical nuclear fission power plant produces approximately 1.00 GW of electrical power. Assume the plant has an overall efficiency of 40.0% and each fission produces 200 MeV of energy. Calculate the mass of ^{235}U consumed each day.

Section 45.3 Nuclear Reactors

9. ▲ It has been estimated that on the order of 10^9 tons of natural uranium is available at concentrations exceeding 100 parts per million, of which 0.7% is the fissionable isotope ^{235}U. Assume all the world's energy use (7×10^{12} J/s) were supplied by ^{235}U fission in conventional nuclear reactors, releasing 208 MeV for each reaction. How long would the supply last? The estimate of uranium supply is taken from K. S. Deffeyes and I. D. MacGregor, "World Uranium Resources," *Scientific American* **242**(1): 66, 1980.

10. ● To minimize neutron leakage from a reactor, the surface area-to-volume ratio should be a minimum. For a given volume V, calculate this ratio for (a) a sphere, (b) a cube, and (c) a parallelepiped of dimensions $a \times a \times 2a$. (d) Which of these shapes would have minimum leakage? Which would have maximum leakage? Explain your answers.

11. If the reproduction constant is 1.000 25 for a chain reaction in a fission reactor and the average time interval between successive fissions is 1.20 ms, by what factor does the reaction rate increase in one minute?

12. A large nuclear power reactor produces approximately 3 000 MW of power in its core. Three months after a reactor is shut down, the core power from radioactive byproducts is 10.0 MW. Assuming each emission delivers 1.00 MeV of energy to the power, find the activity in becquerels three months after the reactor is shut down.

13. ● A particle cannot generally be localized to distances much smaller than its de Broglie wavelength. This fact can be taken to mean that a slow neutron appears to be larger to a target particle than does a fast neutron in the sense that the slow neutron has probabilities of being found over a larger volume of space. For a thermal neutron at room temperature of 300 K, find (a) the linear momentum and (b) the de Broglie wavelength. State how this effective size compares with both nuclear and atomic dimensions.

14. The probability of a nuclear reaction increases dramatically when the incident particle is given energy above the "Coulomb barrier," which is the electric potential energy of the two nuclei when their surfaces barely touch. Compute the Coulomb barrier for the absorption of an alpha particle by a gold nucleus.

Section 45.4 Nuclear Fusion

15. (a) Consider a fusion generator built to create 3.00 GW of power. Determine the rate of fuel burning in grams per hour if the D–T reaction is used. (b) Do the same for the D–D reaction assuming the reaction products are split evenly between (n, ^{3}He) and (p, ^{3}H).

16. ● Two nuclei having atomic numbers Z_1 and Z_2 approach each other with a total energy E. (a) When they are far apart, they interact only by electric repulsion. If they approach to a distance of 1.00×10^{-14} m, the nuclear force suddenly takes over to make them fuse. Find the minimum value of E, in terms of Z_1 and Z_2, required to produce fusion. (b) State how E depends on the atomic numbers. (c) If $Z_1 + Z_2$ is to have a certain target value such as 60, would it be energetically favorable to take $Z_1 = 1$ and $Z_2 = 59$, or $Z_1 = Z_2 = 30$, or what? Explain your answer. (d) Evaluate from your expression the minimum energy for fusion for the D–D and D–T reactions (the first and third reactions in Eq. 45.4).

17. ● **Review problem.** Consider the deuterium–tritium fusion reaction with the tritium nucleus at rest:

$$^2_1\text{H} + {}^3_1\text{H} \rightarrow {}^4_2\text{He} + {}^1_0\text{n}$$

(a) Suppose the reactant nuclei will spontaneously fuse if their surfaces touch. From Equation 44.1, determine the required distance of closest approach between their centers. (b) What is the electric potential energy (in eV) at this distance? (c) Suppose the deuteron is fired straight at an originally stationary tritium nucleus with just enough energy to reach the required distance of closest approach. What is the common speed of the deuterium and tritium nuclei, in terms of the initial deuteron speed v_i, as they touch? *Note:* At this point, the two nuclei have a common velocity equal to the center-of-mass velocity. (d) Use energy methods to find the minimum initial deuteron energy required to achieve fusion. (e) Why does the fusion reaction actually occur at much lower deuteron energies than the energy calculated in part (d)?

18. Of all the hydrogen in the oceans, 0.030 0% of the mass is deuterium. The oceans have a volume of 317 million mi^3. (a) If nuclear fusion were controlled and all the deuterium in the oceans were fused to $^4_2$He, how many joules of energy would be released? (b) **What If?** World power consumption is approximately 7.00×10^{12} W. If consumption were 100 times greater, how many years would the energy calculated in part (a) last?

19. ▲ To understand why plasma containment is necessary, consider the rate at which an unconfined plasma would be lost. (a) Estimate the rms speed of deuterons in a plasma at 4.00×10^8 K. (b) **What If?** Estimate the order of magnitude of the time interval during which such a plasma would remain in a 10-cm cube if no steps were taken to contain it.

20. It has been suggested that fusion reactors are safe from explosion because the plasma never contains enough energy to do much damage. (a) In 1992, the TFTR reactor achieved an ion temperature of 4.0×10^8 K, an ion density of 2.0×10^{13} cm^{-3}, and a confinement time of 1.4 s. Calculate the amount of energy stored in the plasma of the TFTR reactor. (b) How many kilograms of water could be boiled away by this much energy? The plasma volume of the TFTR reactor is approximately 50 m^3.

21. **Review problem.** To confine a stable plasma, the magnetic energy density in the magnetic field (Eq. 32.14) must exceed the pressure $2nk_BT$ of the plasma by a factor of at least 10. In this problem, assume a confinement time $\tau = 1.00$ s. (a) Using Lawson's criterion, determine the ion density required for the D–T reaction. (b) From the ignition-temperature criterion, determine the required plasma pressure. (c) Determine the magnitude of the magnetic field required to contain the plasma.

22. One old prediction for the future was to have a fusion reactor supply energy to dissociate the molecules in garbage into separate atoms and then to ionize the atoms. This material could be put through a giant mass spectrometer so that trash would be a new source of isotopically pure elements, the mine of the future. Assuming an average atomic mass of 56 and an average charge of 26 (a high estimate, considering all the organic materials) at a beam current of 1.00 MA, how long would it take to process 1.00 metric ton of trash?

Section 45.5 Radiation Damage

23. A small building has become accidentally contaminated with radioactivity. The longest-lived material in the building is strontium-90. ($^{90}_{38}$Sr has an atomic mass 89.907 7 u, and its half-life is 29.1 yr. It is particularly dangerous because it substitutes for calcium in bones.) Assume the building initially contained 5.00 kg of this substance uniformly distributed throughout the building and the safe level is defined as less than 10.0 decays/min (which is small compared to background radiation). How long will the building be unsafe?

24. **Review problem.** A particular radioactive source produces 100 mrad of 2-MeV gamma rays per hour at a distance of 1.00 m from the source. (a) How long could a person stand at this distance before accumulating an intolerable dose of 1 rem? (b) **What If?** Assuming the radioactive source is a point source, at what distance would a person receive a dose of 10.0 mrad/h?

25. ● Assume an x-ray technician takes an average of eight x-rays per day and receives a dose of 5 rem/yr as a result. (a) Estimate the dose in rem per x-ray taken. (b) Explain how the technician's exposure compares with low-level background radiation.

26. *Lead shielding.* When gamma rays are incident on matter, the intensity of the gamma rays passing through the material varies with depth x as $I(x) = I_0e^{-\mu x}$, where I_0 is the intensity of the radiation at the surface of the material and μ is the absorption coefficient. For 0.400-MeV gamma rays in lead, the absorption coefficient is 1.59 cm^{-1}. (a) Determine the "half-thickness" for lead, that is, the thickness of lead that would absorb half the incident gamma rays. (b) What thickness reduces the radiation by a factor of 10^4?

27. ▲ A "clever" technician decides to warm some water for his coffee with an x-ray machine. If the machine produces 10.0 rad/s, over what time interval will the temperature of an insulated cup of water rise by 50.0°C?

28. Review problem. The danger to the body from a high dose of gamma rays is not due to the amount of energy absorbed; rather, it is due to the ionizing nature of the radiation. As an illustration, calculate the rise in body temperature that results if a "lethal" dose of 1 000 rad is absorbed strictly as internal energy. Take the specific heat of living tissue as 4 186 J/kg · °C.

29. Technetium-99 is used in certain medical diagnostic procedures. Assume 1.00×10^{-8} g of ^{99}Tc is injected into a 60.0-kg patient and half of the 0.140-MeV gamma rays are absorbed in the body. Determine the total radiation dose received by the patient.

30. To destroy a cancerous tumor, a dose of gamma radiation with a total energy of 2.12 J is to be delivered in 30.0 days from implanted sealed capsules containing palladium-103. Assume this isotope has a half-life of 17.0 d and emits gamma rays of energy 21.0 keV, which are entirely absorbed within the tumor. (a) Find the initial activity of the set of capsules. (b) Find the total mass of radioactive palladium that these "seeds" should contain.

31. Strontium-90 from the testing of nuclear bombs can still be found in the atmosphere. Each decay of ^{90}Sr releases 1.1 MeV of energy into the bones of a person who has had strontium replace his or her body's calcium. Assume a 70.0-kg person receives 1.00 μg of ^{90}Sr from contaminated milk. Calculate the absorbed dose rate (in joules per kilogram) in one year. Take the half-life of ^{90}Sr to be 29.1 yr.

Section 45.6 Radiation Detectors

32. Assume a photomultiplier tube (Fig. 40.12) has seven dynodes with potentials of 100, 200, 300, . . . , 700 V. The average energy required to free an electron from the dynode surface is 10.0 eV. Assume only one electron is incident and the tube functions with 100% efficiency. (a) How many electrons are freed at the first dynode? (b) How many electrons are collected at the last dynode? (c) What is the energy available to the counter for each electron?

33. In a Geiger tube, the voltage between the electrodes is typically 1.00 kV and the current pulse discharges a 5.00-pF capacitor. (a) What is the energy amplification of this device for a 0.500-MeV electron? (b) How many electrons participate in the avalanche caused by the single initial electron?

34. ● (a) Your grandmother recounts to you how, as young children, your father, aunts, and uncles made the screen door slam continually as they ran between the house and the back yard. The time interval between one slam and the next varied randomly, but the average slamming rate remained constant at 38.0/h from dawn to dusk every summer day. If the slamming rate suddenly dropped to zero, the children would have found a nest of baby field mice or gotten into some other mischief requiring adult intervention. Approximately how long after the last screen-door slam would a prudent and attentive parent wait before leaving her or his tasks to see about the children? Explain your reasoning. (b) A student wishes to measure the half-life of a radioactive substance using a small sample. Consecutive clicks of her Geiger counter are randomly spaced in time. The counter registers 372 counts during one 5.00-min interval and 337 counts during the next 5.00 min. The average background rate is 15 counts per minute. Find the most probable value for the half-life. (c) Estimate the uncertainty in the half-life determination. Explain your reasoning.

Section 45.7 Uses of Radiation

35. During the manufacture of a steel engine component, radioactive iron (^{59}Fe) is included in the total mass of 0.200 kg. The component is placed in a test engine when the activity due to this isotope is 20.0 μCi. After a 1 000-h test period, some of the lubricating oil is removed from the engine and found to contain enough ^{59}Fe to produce 800 disintegrations/min/L of oil. The total volume of oil in the engine is 6.50 L. Calculate the total mass worn from the engine component per hour of operation. The half-life of ^{59}Fe is 45.1 d.

36. You want to find out how many atoms of the isotope ^{65}Cu are in a small sample of material. You bombard the sample with neutrons to ensure that on the order of 1% of these copper nuclei absorb a neutron. After activation, you turn off the neutron flux and then use a highly efficient detector to monitor the gamma radiation that comes out of the sample. Assume half of the ^{66}Cu nuclei emit a 1.04-MeV gamma ray in their decay. (The other half of the activated nuclei decay directly to the ground state of ^{66}Ni.) If after 10 min (two half-lives) you have detected 10^4 MeV of photon energy at 1.04 MeV, (a) approximately how many ^{65}Cu atoms are in the sample? (b) Assume the sample contains natural copper. Refer to the isotopic abundances listed in Table 44.2 and estimate the total mass of copper in the sample.

37. *Neutron activation analysis* is a method for chemical analysis at the level of isotopes. When a sample is irradiated by neutrons, radioactive atoms are produced continuously and then decay according to their characteristic half-lives. (a) Assume one species of radioactive nuclei is produced at a constant rate R and its decay is described by the conventional radioactive decay law. Assuming irradiation begins at time $t = 0$, show that the number of radioactive atoms accumulated at time t is

$$N = \frac{R}{\lambda}(1 - e^{-\lambda t})$$

(b) What is the maximum number of radioactive atoms that can be produced?

38. *A thickness gauge.* When gamma rays are incident on matter, the intensity of the gamma rays passing through the material varies with depth x as $I(x) = I_0 e^{-\mu x}$, where I_0 is the intensity of the radiation at the surface of the material and μ is the absorption coefficient. For low-energy gamma rays in steel, take the absorption coefficient to be 0.720 mm^{-1}. (a) Determine the "half-thickness" for steel, that is, the thickness of steel that would absorb half the incident gamma rays. (b) In a steel mill, the thickness of sheet steel passing into a roller is measured by monitoring the intensity of gamma radiation reaching a detector below the rapidly moving metal from a small source immediately above the metal. If the thickness of the sheet changes from 0.800 mm to 0.700 mm, by what percentage does the gamma-ray intensity change?

Additional Problems

39. Carbon detonations are powerful nuclear reactions that temporarily tear apart the cores inside massive stars late in their lives. These blasts are produced by carbon fusion, which requires a temperature of approximately 6×10^8 K to overcome the strong Coulomb repulsion between carbon nuclei. (a) Estimate the repulsive energy barrier to fusion, using the temperature required for carbon fusion. (In other words, what is the average kinetic energy of a carbon nucleus at 6×10^8 K?) (b) Calculate the energy (in MeV) released in each of these "carbon-burning" reactions:

$$^{12}\text{C} + {}^{12}\text{C} \rightarrow {}^{20}\text{Ne} + {}^{4}\text{He}$$

$$^{12}\text{C} + {}^{12}\text{C} \rightarrow {}^{24}\text{Mg} + \gamma$$

(c) Calculate the energy in kilowatt-hours given off when 2.00 kg of carbon completely fuses according to the first reaction.

40. **Review problem.** Consider a nucleus at rest, which then spontaneously splits into two fragments of masses m_1 and m_2. Show that the fraction of the total kinetic energy carried by fragment m_1 is

$$\frac{K_1}{K_{\text{tot}}} = \frac{m_2}{m_1 + m_2}$$

and the fraction carried by m_2 is

$$\frac{K_2}{K_{\text{tot}}} = \frac{m_1}{m_1 + m_2}$$

assuming relativistic corrections can be ignored. *Note:* If the parent nucleus was moving before the decay, the fission products still share the kinetic energy as shown as long as all velocities are measured in the center-of-mass frame of reference, in which the total momentum of the system is zero.

41. A stationary $^{236}_{92}\text{U}$ nucleus fissions spontaneously into two primary fragments, $^{87}_{35}\text{Br}$ and $^{149}_{57}\text{La}$. (a) Calculate the disintegration energy. The required atomic masses are 86.920 711 u for $^{87}_{35}\text{Br}$, 148.934 370 u for $^{149}_{57}\text{La}$, and 236.045 562 u for $^{236}_{92}\text{U}$. (b) How is the disintegration energy split between the two primary fragments? You may use the result of Problem 40. (c) Calculate the speed of each fragment immediately after the fission.

42. ● Explain how the fractional energy loss in a typical ^{235}U fission reaction compares with the fractional energy loss in D–T fusion.

43. ● The half-life of tritium is 12.3 yr. If the TFTR fusion reactor contained 50.0 m^3 of tritium at a density equal to 2.00×10^{14} ions/cm^3, how many curies of tritium were in the plasma? State how this value compares with a fission inventory (the estimated supply of fissionable material) of 4×10^{10} Ci.

44. A fission reactor is hit by a missile, and 5.00×10^6 Ci of ^{90}Sr, with half-life 27.7 yr, evaporates into the air. The strontium falls out over an area of 10^4 km^2. After what time interval will the activity of the ^{90}Sr reach the agriculturally "safe" level of 2.00 μCi/m^2?

45. **Review problem.** A nuclear power plant operates by using the energy released in nuclear fission to convert 20°C water into 400°C steam. How much water could theoretically be converted to steam by the complete fissioning of 1.00 g of ^{235}U at 200 MeV/fission?

46. **Review problem.** A nuclear power plant operates by using the energy released in nuclear fission to convert liquid water at T_c into steam at T_h. How much water could theoretically be converted to steam by the complete fissioning of a mass m of ^{235}U at 200 MeV/fission?

47. ● Consider a 1.00-kg sample of natural uranium composed primarily of ^{238}U, a smaller amount (0.720% by mass) of ^{235}U, and a trace (0.005%) of ^{234}U, which has a half-life of 2.44×10^5 yr. (a) Find the activity in curies due to each of the isotopes. (b) What fraction of the total activity is due to each isotope? (c) Explain whether the activity of this sample is dangerous.

48. **Review problem.** The first nuclear bomb was a fissioning mass of plutonium-239 exploded in the Trinity test before dawn on July 16, 1945, at Alamogordo, New Mexico. Enrico Fermi was 14 km away, lying on the ground facing away from the bomb. After the whole sky had flashed with unbelievable brightness, Fermi stood up and began dropping bits of paper to the ground. They first fell at his feet in the calm and silent air. As the shock wave passed, about 40 s after the explosion, the paper then in flight jumped approximately 5 cm away from ground zero. (a) Assume the shock wave in air propagated equally in all directions without absorption. Find the change in volume of a sphere of radius 14 km as it expands by 5 cm. (b) Find the work $P\,\Delta V$ done by the air in this sphere on the next layer of air farther from the center. (c) Assume the shock wave carried on the order of one-tenth of the energy of the explosion. Make an order-of-magnitude estimate of the bomb yield. (d) One ton of exploding TNT releases 4.2 GJ of energy. What was the order of magnitude of the energy of the Trinity test in equivalent tons of TNT? The dawn revealed the mushroom cloud. Fermi's immediate knowledge of the bomb yield agreed with that determined days later by analysis of elaborate measurements.

49. ● Approximately 1 of every 3 300 water molecules contains one deuterium atom. (a) If all the deuterium nuclei in 1 L of water are fused in pairs according to the D–D fusion reaction $^2\text{H} + {}^2\text{H} \rightarrow {}^3\text{He} + \text{n} + 3.27$ MeV, how much energy in joules is liberated? (b) **What If?** Burning gasoline produces approximately 3.40×10^7 J/L. State how the energy obtainable from the fusion of the deuterium in 1 L of water compares with the energy liberated from the burning of 1 L of gasoline.

50. **Review problem.** A very slow neutron (with speed approximately equal to zero) can initiate the reaction

$$^1_0\text{n} + {}^{10}_{5}\text{B} \rightarrow {}^7_3\text{Li} + {}^4_2\text{He}$$

The alpha particle moves away with speed 9.25×10^6 m/s. Calculate the kinetic energy of the lithium nucleus. Use nonrelativistic equations.

51. A certain nuclear plant generates internal energy at a rate of 3.065 GW and transfers energy out of the plant by electrical transmission at a rate of 1.000 GW. Of the wasted energy, 3.0% is ejected to the atmosphere and the remainder is passed into a river. A state law requires that the river water be warmed by no more than 3.50°C when it is returned to the river. (a) Determine the amount of cooling water necessary (in kilograms per hour and cubic meters per hour) to cool the plant. (b) Assume fission

generates 7.80×10^{10} J/g of ^{235}U. Determine the rate of fuel burning (in kilograms per hour) of ^{235}U.

52. The alpha-emitter polonium-210 ($^{210}_{84}$Po) is used in a nuclear energy source on a spacecraft (Fig. P45.52). Determine the initial power output of the source. Assume it contains 0.155 kg of ^{210}Po and the efficiency for conversion of radioactive decay energy to energy transferred by electrical transmission is 1.00%.

Figure P45.52 The Pioneer 10 spacecraft leaves the Solar System. It carries radioactive power supplies at the ends of two booms. Solar panels would not work far from the Sun.

53. ● Natural uranium must be processed to produce uranium enriched in ^{235}U for bombs and power plants. The processing yields a large quantity of nearly pure ^{238}U as a by-product, called "depleted uranium." Because of its high mass density, ^{238}U is used in armor-piercing artillery shells. (a) Find the edge dimension of a 70.0-kg cube of ^{238}U ($\rho = 18.7 \times 10^3$ kg/m^3). (b) The isotope ^{238}U has a long half-life of 4.47×10^9 yr. As soon as one nucleus decays, a relatively rapid series of 14 steps begins that together constitutes the net reaction

$$^{238}_{92}\text{U} \rightarrow 8(^4_2\text{He}) + 6(^{\ 0}_{-1}\text{e}) + ^{206}_{82}\text{Pb} + 6\bar{\nu} + Q_{\text{net}}$$

Find the net decay energy. (Refer to Table 44.2.) (c) Argue that a radioactive sample with decay rate R and decay energy Q has power output $\mathcal{P} = QR$. (d) Consider an artillery shell with a jacket of 70.0 kg of ^{238}U. Find its power output due to the radioactivity of the uranium and its daughters. Assume the shell is old enough that the daughters have reached steady-state amounts. Express the power in joules per year. (e) **What If?** A 17-year-old soldier of mass 70.0 kg works in an arsenal where many such artillery shells are stored. Assume his radiation exposure is limited to 5.00 rem per year. Find the rate in joules per year at which he can absorb energy of radiation. Assume an average RBE factor of 1.10.

54. A 2.0-MeV neutron is emitted in a fission reactor. If it loses half its kinetic energy in each collision with a moderator atom, how many collisions does it undergo as it becomes a thermal neutron, with energy 0.039 eV?

55. ▲ Assume a deuteron and a triton are at rest when they fuse according to the reaction

$$^2_1\text{H} + ^3_1\text{H} \rightarrow ^4_2\text{He} + ^1_0\text{n} + 17.6\text{ MeV}$$

Determine the kinetic energy acquired by the neutron.

56. A sealed capsule containing the radiopharmaceutical phosphorus-32 ($^{32}_{15}$P), an e$^-$ (beta) emitter, is implanted into a patient's tumor. The average kinetic energy of the beta particles is 700 keV. The initial activity is 5.22 MBq. Determine the absorbed dose during a 10.0-day period. Assume the beta particles are completely absorbed in 100 g of tissue. *Suggestion:* Find the number of beta particles emitted.

57. ● (a) Calculate the energy (in kilowatt-hours) released if 1.00 kg of ^{239}Pu undergoes complete fission and the energy released per fission event is 200 MeV. (b) Calculate the energy (in electron volts) released in the deuterium–tritium fusion reaction

$$^2_1\text{H} + ^3_1\text{H} \rightarrow ^4_2\text{He} + ^1_0\text{n}$$

(c) Calculate the energy (in kilowatt-hours) released if 1.00 kg of deuterium undergoes fusion according to this reaction. (d) **What If?** Calculate the energy (in kilowatt-hours) released by the combustion of 1.00 kg of coal if each $C + O_2 \rightarrow CO_2$ reaction yields 4.20 eV. (e) List advantages and disadvantages of each of these methods of energy generation.

58. The Sun radiates energy at the rate of 3.85×10^{26} W. Suppose the net reaction

$$4(^1_1\text{H}) + 2(^{\ 0}_{-1}\text{e}) \rightarrow ^4_2\text{He} + 2\nu + \gamma$$

accounts for all the energy released. Calculate the number of protons fused per second.

59. Consider the two nuclear reactions

$$\text{(I)} \quad \text{A} + \text{B} \rightarrow \text{C} + \text{E}$$
$$\text{(II)} \quad \text{C} + \text{D} \rightarrow \text{F} + \text{G}$$

(a) Show that the net disintegration energy for these two reactions ($Q_{\text{net}} = Q_{\text{I}} + Q_{\text{II}}$) is identical to the disintegration energy for the net reaction

$$\text{A} + \text{B} + \text{D} \rightarrow \text{E} + \text{F} + \text{G}$$

(b) One chain of reactions in the proton–proton cycle in the Sun's core is

$$^1_1\text{H} + ^1_1\text{H} \rightarrow ^2_1\text{H} + ^0_1\text{e} + \nu$$
$$^0_1\text{e} + ^{\ 0}_{-1}\text{e} \rightarrow 2\gamma$$
$$^1_1\text{H} + ^2_1\text{H} \rightarrow ^3_2\text{He} + \gamma$$
$$^1_1\text{H} + ^3_2\text{He} \rightarrow ^4_2\text{He} + ^0_1\text{e} + \nu$$
$$^0_1\text{e} + ^{\ 0}_{-1}\text{e} \rightarrow 2\gamma$$

Based on part (a), what is Q_{net} for this sequence?

60. Suppose the target in a laser fusion reactor is a sphere of solid hydrogen that has a diameter of 1.50×10^{-4} m and a density of 0.200 g/cm^3. Assume half of the nuclei are ^{2}H and half are ^{3}H. (a) If 1.00% of a 200-kJ laser pulse is delivered to this sphere, what temperature does the sphere reach? (b) If all the hydrogen "burns" according to the D–T reaction, how many joules of energy are released?

61. ● In addition to the proton–proton cycle described in the chapter text, the carbon cycle, first proposed by Hans Bethe in 1939, is another cycle by which energy is released in stars as hydrogen is converted to helium. The

carbon cycle requires higher temperatures than the proton–proton cycle. The series of reactions is

$$^{12}C + {}^{1}H \rightarrow {}^{13}N + \gamma$$
$$^{13}N \rightarrow {}^{13}C + e^+ + \nu$$
$$e^+ + e^- \rightarrow 2\gamma$$
$$^{13}C + {}^{1}H \rightarrow {}^{14}N + \gamma$$
$$^{14}N + {}^{1}H \rightarrow {}^{15}O + \gamma$$
$$^{15}O \rightarrow {}^{15}N + e^+ + \nu$$
$$e^+ + e^- \rightarrow 2\gamma$$
$$^{15}N + {}^{1}H \rightarrow {}^{12}C + {}^{4}He$$

(a) Assuming the proton–proton cycle requires a temperature of 1.5×10^7 K, estimate by proportion the temperature required for the carbon cycle. (b) Calculate the Q value for each step in the carbon cycle and the overall energy released. (c) Do you think the energy carried off by the neutrinos is deposited in the star? Explain.

62. When photons pass through matter, the intensity I of the beam (measured in watts per square meter) decreases exponentially according to

$$I = I_0 e^{-\mu x}$$

where I is the intensity of the beam that just passed through a thickness x of material and I_0 is the intensity of the incident beam. The constant μ is known as the *linear absorption coefficient*, and its value depends on the absorbing material and the wavelength of the photon beam. This wavelength (or energy) dependence allows us to filter out unwanted wavelengths from a broad-spectrum x-ray beam. (a) Two x-ray beams of wavelengths λ_1 and λ_2 and equal incident intensities pass through the same metal plate. Show that the ratio of the emergent beam intensities is

$$\frac{I_2}{I_1} = e^{-(\mu_2 - \mu_1)x}$$

(b) Compute the ratio of intensities emerging from an aluminum plate 1.00 mm thick if the incident beam contains equal intensities of 50 pm and 100 pm x-rays. The values of μ for aluminum at these two wavelengths are $\mu_1 = 5.4\ \text{cm}^{-1}$ at 50 pm and $\mu_2 = 41.0\ \text{cm}^{-1}$ at 100 pm. (c) Repeat part (b) for an aluminum plate 10.0 mm thick.

63. *To build a bomb.* (a) At time $t = 0$, a sample of uranium is exposed to a neutron source that causes N_0 nuclei to undergo fission. The sample is in a supercritical state, with a reproduction constant $K > 1$. A chain reaction occurs that proliferates fission throughout the mass of uranium. The chain reaction can be thought of as a succession of *generations*. The N_0 fissions produced initially are the zeroth generation of fissions. From this generation, N_0K neutrons go off to produce fission of new uranium nuclei. The N_0K fissions that occur subsequently are the first generation of fissions, and from this generation, N_0K^2 neutrons go in search of uranium nuclei in which to cause fission. The subsequent N_0K^2 fissions are the second generation of fissions. This process can continue until all the uranium nuclei have fissioned. Show that the cumulative total of fissions N that have occurred up to and including the nth generation after the zeroth generation is given by

$$N = N_0\left(\frac{K^{n+1} - 1}{K - 1}\right)$$

(b) Consider a hypothetical uranium bomb made from 5.50 kg of isotopically pure ^{235}U. The chain reaction has a reproduction constant of 1.10 and starts with a zeroth generation of 1.00×10^{20} fissions. The average time interval between one fission generation and the next is 10.0 ns. How long after the zeroth generation does it take the uranium in this bomb to fission completely? (c) Assume the bulk modulus of uranium is 150 GPa. Find the speed of sound in uranium. You may ignore the density difference between ^{235}U and natural uranium. (d) Find the time interval required for a compressional wave to cross the radius of a 5.50-kg sphere of uranium. This time interval indicates how quickly the motion of explosion begins. (e) Fission must occur in a time interval that is short compared with that in part (d); otherwise, most of the uranium will disperse in small chunks without having fissioned. Can the bomb considered in part (b) release the explosive energy of all its uranium? If so, how much energy does it release in equivalent tons of TNT? Assume one ton of TNT releases 4.20 GJ and each uranium fission releases 200 MeV of energy.

Answers to Quick Quizzes

45.1 (b). According to Figure 44.4, the ratio N/Z increases with increasing Z. As a result, when a heavy nucleus fissions to two lighter nuclei, the lighter nuclei tend to have too many neutrons for the nucleus to be stable. Beta decay in which electrons are ejected decreases the number of neutrons and increases the number of protons so as to stabilize the nucleus.

45.2 (a) and (b). In both of these cases, the Z and A values balance on the two sides of the equations. In reaction (c), $Z_{\text{left}} = Z_{\text{right}}$, but $A_{\text{left}} \neq A_{\text{right}}$.

45.3 (a). To reduce the value of K, more neutrons need to be absorbed, so a larger volume of the control rods must be inside the reactor core.

45.4 (d). Figure 44.5 shows that the curve representing the binding energy per nucleon peaks at $A \approx 60$. Consequently, combining two nuclei with equal values of $A > 60$ results in an increase in mass, so a fusion reaction will not occur.

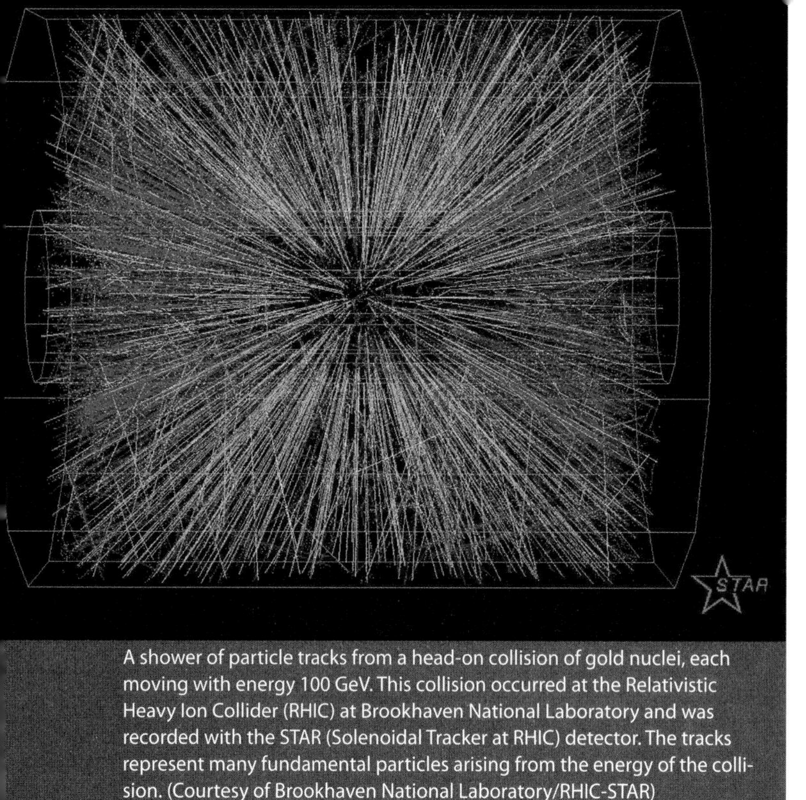

A shower of particle tracks from a head-on collision of gold nuclei, each moving with energy 100 GeV. This collision occurred at the Relativistic Heavy Ion Collider (RHIC) at Brookhaven National Laboratory and was recorded with the STAR (Solenoidal Tracker at RHIC) detector. The tracks represent many fundamental particles arising from the energy of the collision. (Courtesy of Brookhaven National Laboratory/RHIC-STAR)

46 Particle Physics and Cosmology

The word *atom* comes from the Greek *atomos*, which means "indivisible." The early Greeks believed that atoms were the indivisible constituents of matter; that is, they regarded them as elementary particles. After 1932, physicists viewed all matter as consisting of three constituent particles: electrons, protons, and neutrons. Beginning in the 1940s, many "new" particles were discovered in experiments involving high-energy collisions between known particles. The new particles are characteristically very unstable and have very short half-lives, ranging between 10^{-6} s and 10^{-23} s. So far, more than 300 of these particles have been catalogued.

Until the 1960s, physicists were bewildered by the great number and variety of subatomic particles that were being discovered. They wondered whether the particles had no systematic relationship connecting them or whether a pattern was emerging that would provide a better understanding of the elaborate structure in the subatomic world. That the neutron has a magnetic moment despite having zero electric charge (Section 44.8) suggests an underlying structure to the neutron. The periodic table explains how more than 100 elements can be formed from three types of particles (electrons, protons, and neutrons), which suggests there is, perhaps, a means of forming more than 300 subatomic particles from a small number of basic building blocks.

Recall Figure 1.2, which illustrated the various levels of structure in matter. We studied the atomic structure of matter in Chapter 42. In Chapter 44, we investigated the substructure of the atom by describing the structure of the nucleus. As mentioned in Section 1.2, the protons and neutrons in the nucleus, and a host of other exotic particles, are now known to be composed of six different varieties of particles called *quarks*. In this concluding chapter, we examine the current theory of elementary particles, in which all matter is constructed from only two families of particles, quarks and leptons. We also discuss how clarifications of such models might help scientists understand the birth and evolution of the Universe.

46.1 The Fundamental Forces in Nature

As noted in Section 5.1, all natural phenomena can be described by four fundamental forces acting between particles. In order of decreasing strength, they are the nuclear force, the electromagnetic force, the weak force, and the gravitational force.

The nuclear force discussed in Chapter 44 is an attractive force between nucleons. It has a very short range and is negligible for separation distances between nucleons greater than approximately 10^{-15} m (about the size of the nucleus). The electromagnetic force, which binds atoms and molecules together to form ordinary matter, has a strength of approximately 10^{-2} times that of the nuclear force. This long-range force decreases in magnitude as the inverse square of the separation between interacting particles. The weak force is a short-range force that tends to produce instability in certain nuclei. It is responsible for decay processes, and its strength is only about 10^{-5} times that of the nuclear force. Finally, the gravitational force is a long-range force that has a strength of only about 10^{-39} times that of the nuclear force. Although this familiar interaction is the force that holds the planets, stars, and galaxies together, its effect on elementary particles is negligible.

In Section 13.4, we discussed the difficulty early scientists had with the notion of the gravitational force acting at a distance, with no physical contact between the interacting objects. To resolve this difficulty, the concept of the gravitational field was introduced. Similarly, in Chapter 23, we introduced the electric field to describe the electric force acting between charged objects, and we followed that with a discussion of the magnetic field in Chapter 29. In modern physics, the nature of the interaction between particles is carried a step further. These interactions are described in terms of the exchange of entities called **field particles** or **exchange particles.** Field particles are also called **gauge bosons.**[1] The interacting particles continuously emit and absorb field particles. The emission of a field particle by one particle and its absorption by another manifests as a force between the two interacting particles. In the case of the electromagnetic interaction, for instance, the field particles are photons. In the language of modern physics, the electromagnetic force is said to be *mediated* by photons, and photons are the field particles of the electromagnetic field. Likewise, the nuclear force is mediated by field particles called *gluons.* The weak force is mediated by field particles called *W* and *Z bosons,* and the gravitational force is proposed to be mediated by field particles called *gravitons.* These interactions, their ranges, and their relative strengths are summarized in Table 46.1.

PAUL ADRIEN MAURICE DIRAC
British Physicist (1902–1984)
Dirac was instrumental in the understanding of antimatter and the unification of quantum mechanics and relativity. He made many contributions to the development of quantum physics and cosmology. In 1933, Dirac won a Nobel Prize in Physics.

46.2 Positrons and Other Antiparticles

In the 1920s, Paul Dirac developed a relativistic quantum-mechanical description of the electron that successfully explained the origin of the electron's spin and its magnetic moment. His theory had one major problem, however: its relativistic

[1] The word *bosons* suggests that the field particles have integral spin as discussed in Section 43.8. The word *gauge* comes from *gauge theory,* which is a sophisticated mathematical analysis that is beyond the scope of this book.

TABLE 46.1

Particle Interactions

Interaction	Relative Strength	Range of Force	Mediating Field Particle	Mass of Field Particle (GeV/c^2)
Nuclear	1	Short ($\approx$ 1 fm)	Gluon	0
Electromagnetic	10^{-2}	∞	Photon	0
Weak	10^{-5}	Short ($\approx 10^{-3}$ fm)	$W^{\pm}$, Z^0 bosons	80.4, 80.4, 91.2
Gravitational	10^{-39}	∞	Graviton	0

wave equation required solutions corresponding to negative energy states, and if negative energy states existed, an electron in a state of positive energy would be expected to make a rapid transition to one of these states, emitting a photon in the process.

Dirac circumvented this difficulty by postulating that all negative energy states are filled. The electrons occupying these negative energy states are collectively called the *Dirac sea.* Electrons in the Dirac sea are not directly observable because the Pauli exclusion principle does not allow them to react to external forces; there are no available states to which an electron can make a transition in response to an external force. Therefore, an electron in such a state acts as an isolated system unless an interaction with the environment is strong enough to excite the electron to a positive energy state. Such an excitation causes one of the negative energy states to be vacant as in Figure 46.1, leaving a hole in the sea of filled states. *The hole can react to external forces and is observable.* The hole reacts in a way similar to that of the electron except that it has a positive charge: it is the *antiparticle* to the electron.

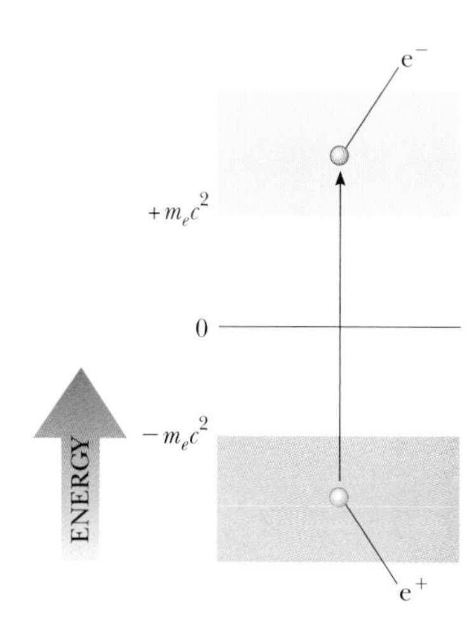

Figure 46.1 Dirac's model for the existence of antielectrons (positrons). The states lower in energy than $-m_ec^2$ are filled with electrons as indicated by the blue coloring. This set of filled states is called the Dirac sea. One of these electrons can make a transition out of its state only if it is provided with energy equal to or larger than $2m_ec^2$. This transition leaves a vacancy in the Dirac sea, which can behave as a particle identical to the electron except for its positive charge.

This theory strongly suggested that *an antiparticle exists for every particle,* not only for fermions such as electrons but also for bosons. That has subsequently been verified for *all* particles known today. The antiparticle for a charged particle has the same mass as the particle but opposite charge. For example, the electron's antiparticle (the *positron* mentioned in Section 44.4) has a rest energy of 0.511 MeV and a positive charge of $+1.60 \times 10^{-19}$ C.

Carl Anderson (1905–1991) observed the positron experimentally in 1932 and was awarded a Nobel Prize in Physics in 1936 for this achievement. Anderson discovered the positron while examining tracks created in a cloud chamber by electron-like particles of positive charge. (These early experiments used cosmic rays—mostly energetic protons passing through interstellar space—to initiate high-energy reactions on the order of several GeV.) To discriminate between positive and negative charges, Anderson placed the cloud chamber in a magnetic field, causing moving charges to follow curved paths. He noted that some of the electron-like tracks deflected in a direction corresponding to a positively charged particle.

PITFALL PREVENTION 46.1
Antiparticles

An antiparticle is not identified solely on the basis of opposite charge; even neutral particles have antiparticles, which are defined in terms of other properties, such as spin.

Since Anderson's discovery, positrons have been observed in a number of experiments. A common source of positrons is **pair production.** In this process, a gamma-ray photon with sufficiently high energy interacts with a nucleus and an electron–positron pair is created from the photon. (The presence of the nucleus allows the principle of conservation of momentum to be satisfied.) Because the total rest energy of the electron–positron pair is $2m_ec^2 = 1.02$ MeV (where m_e is the mass of the electron), the photon must have at least this much energy to create an electron–positron pair. The energy of a photon is converted to rest energy of the electron and positron in accordance with Einstein's relationship $E_R = mc^2$. If the gamma-ray photon has energy in excess of the rest energy of the electron–positron pair, the excess appears as kinetic energy of the two particles. Figure 46.2 (page 1360) shows early observations of tracks of electron–positron pairs in a bubble chamber created by 300-MeV gamma rays striking a lead sheet.

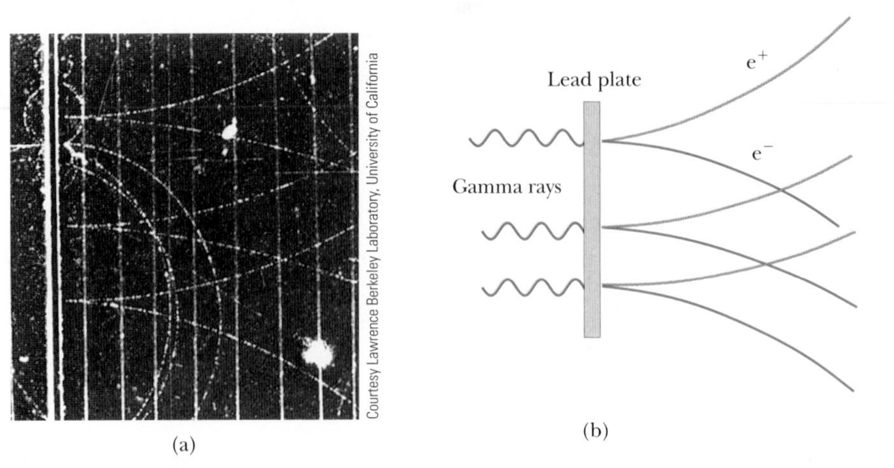

Figure 46.2 (a) Bubble-chamber tracks of electron–positron pairs produced by 300-MeV gamma rays striking a lead sheet from the left. (b) The pertinent pair-production events. The positrons deflect upward, and the electrons downward in an applied magnetic field.

Quick Quiz 46.1 Given the identification of the particles in Figure 46.2b, is the direction of the external magnetic field in Figure 46.2a (a) into the page, (b) out of the page, or (c) impossible to determine?

The reverse process can also occur. Under the proper conditions, an electron and a positron can annihilate each other to produce two gamma-ray photons that have a combined energy of at least 1.02 MeV:

$$e^- + e^+ \rightarrow 2\gamma$$

Because the initial momentum of the electron–positron system is approximately zero, the two gamma rays travel in opposite directions after the annihilation, satisfying the principle of conservation of momentum for the system.

Practically every known elementary particle has a distinct antiparticle. Among the exceptions are the photon and the neutral pion (π^0; see Section 46.3). Following the construction of high-energy accelerators in the 1950s, many other antiparticles were revealed. They included the antiproton, discovered by Emilio Segré (1905–1989) and Owen Chamberlain (1920–2006) in 1955, and the antineutron,[2] discovered shortly thereafter.

Electron–positron annihilation is used in the medical diagnostic technique called *positron-emission tomography* (PET). The patient is injected with a glucose solution containing a radioactive substance that decays by positron emission, and the material is carried throughout the body by the blood. A positron emitted during a decay event in one of the radioactive nuclei in the glucose solution annihilates with an electron in the surrounding tissue, resulting in two gamma-ray photons emitted in opposite directions. A gamma detector surrounding the patient pinpoints the source of the photons and, with the assistance of a computer, displays an image of the sites at which the glucose accumulates. (Glucose metabolizes rapidly in cancerous tumors and accumulates at those sites, providing a strong signal for a PET detector system.) The images from a PET scan can indicate a wide variety of disorders in the brain, including Alzheimer's disease (Fig. 46.3). In addition, because glucose metabolizes more rapidly in active areas of the brain, a PET scan can indicate areas of the brain involved in the activities in which the patient is engaging at the time of the scan, such as language use, music, and vision.

[2] Antiparticles of charged particles have the opposite charge. Antiparticles for uncharged particles, such as the neutron, are a little more difficult to describe. One basic process that can detect the existence of an antiparticle is pair annihilation. For example, a neutron and an antineutron can annihilate to form two gamma rays. Because the photon and the neutral pion do not have distinct antiparticles, pair annihilation is not observed with either of these particles.

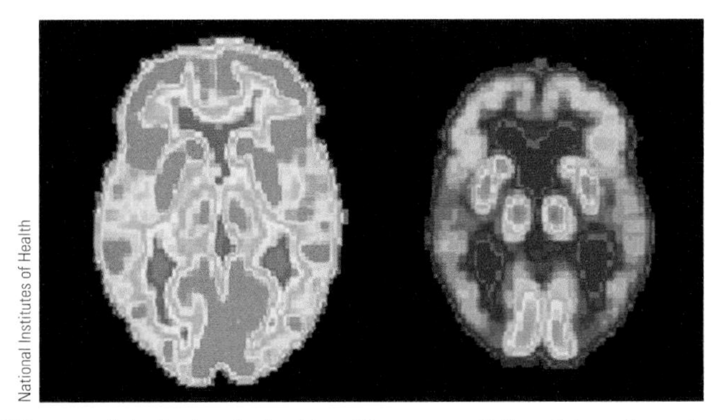
National Institutes of Health

Figure 46.3 PET scans of the brain of a healthy older person *(left)* and that of a patient suffering from Alzheimer's disease *(right)*. Lighter regions contain higher concentrations of radioactive glucose, indicating higher metabolism rates and therefore increased brain activity.

46.3 Mesons and the Beginning of Particle Physics

Physicists in the mid-1930s had a fairly simple view of the structure of matter. The building blocks were the proton, the electron, and the neutron. Three other particles were either known or postulated at the time: the photon, the neutrino, and the positron. Together these six particles were considered the fundamental constituents of matter. With this simple picture, however, no one was able to answer the following important question: the protons in any nucleus should strongly repel one another due to their charges of the same sign, so what is the nature of the force that holds the nucleus together? Scientists recognized that this mysterious force must be much stronger than anything encountered in nature up to that time. This force is the nuclear force discussed in Section 44.1 and examined in historical perspective in the following paragraphs.

The first theory to explain the nature of the nuclear force was proposed in 1935 by Japanese physicist Hideki Yukawa, an effort that earned him a Nobel Prize in Physics in 1949. To understand Yukawa's theory, recall the introduction of field particles in Section 46.1, which stated that each fundamental force is mediated by a field particle exchanged between the interacting particles. Yukawa used this idea to explain the nuclear force, proposing the existence of a new particle whose exchange between nucleons in the nucleus causes the nuclear force. He established that the range of the force is inversely proportional to the mass of this particle and predicted the mass to be approximately 200 times the mass of the electron. (Yukawa's predicted particle is *not* the gluon mentioned in Section 46.1, which is massless and is today considered to be the field particle for the nuclear force.) Because the new particle would have a mass between that of the electron and that of the proton, it was called a **meson** (from the Greek *meso,* "middle").

In efforts to substantiate Yukawa's predictions, physicists began experimental searches for the meson by studying cosmic rays entering the Earth's atmosphere. In 1937, Carl Anderson and his collaborators discovered a particle of mass 106 MeV/c^2, approximately 207 times the mass of the electron. This particle was thought to be Yukawa's meson. Subsequent experiments, however, showed that the particle interacted very weakly with matter and hence could not be the field particle for the nuclear force. That puzzling situation inspired several theoreticians to propose two mesons having slightly different masses equal to approximately 200 times that of the electron, one having been discovered by Anderson and the other, still undiscovered, predicted by Yukawa. This idea was confirmed in 1947 with the discovery of the **pi meson** (π), or simply **pion.** The particle discovered by Anderson in 1937, the one initially thought to be Yukawa's meson, is not really a meson. (We shall discuss the characteristics of mesons in Section 46.4.) Instead, it takes part in the weak and electromagnetic interactions only and is now called the **muon** (μ).

UPI/Corbis-Bettman

HIDEKI YUKAWA
Japanese Physicist (1907–1981)
Yukawa was awarded the Nobel Prize in Physics in 1949 for predicting the existence of mesons. This photograph of him at work was taken in 1950 in his office at Columbia University. Yukawa came to Columbia in 1949 after spending the early part of his career in Japan.

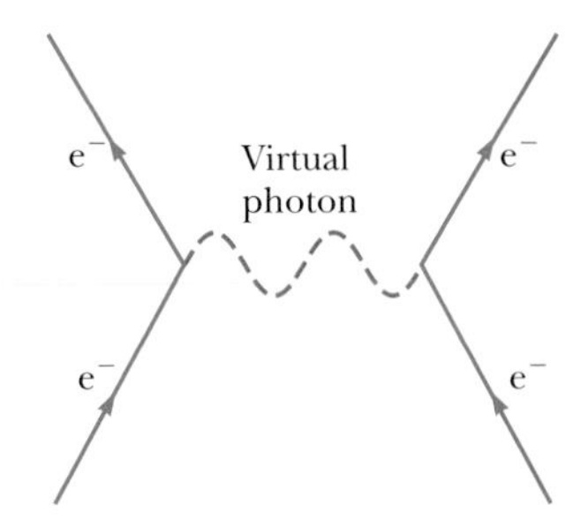

Figure 46.4 Feynman diagram representing a photon mediating the electromagnetic force between two electrons.

The pion comes in three varieties, corresponding to three charge states: π^+, π^-, and π^0. The π^+ and π^- particles (π^- is the antiparticle of π^+) each have a mass of 139.6 MeV/c^2, and the π^0 mass is 135.0 MeV/c^2. Two muons exist: μ^- and its antiparticle μ^+.

Pions and muons are very unstable particles. For example, the π^-, which has a mean lifetime of 2.6×10^{-8} s, decays to a muon and an antineutrino.[3] The muon, which has a mean lifetime of 2.2 μs, then decays to an electron, a neutrino, and an antineutrino:

$$\begin{aligned} \pi^- &\rightarrow \mu^- + \bar{\nu} \\ \mu^- &\rightarrow e^- + \nu + \bar{\nu} \end{aligned} \qquad (46.1)$$

For chargeless particles (as well as some charged particles, such as the proton), a bar over the symbol indicates an antiparticle, as for the neutrino in beta decay (see Section 44.5). Other antiparticles, such as e^+ and μ^+, use a different notation.

The interaction between two particles can be represented in a simple diagram called a **Feynman diagram,** developed by American physicist Richard P. Feynman. Figure 46.4 is such a diagram for the electromagnetic interaction between two electrons. A Feynman diagram is a qualitative graph of time on the vertical axis versus space on the horizontal axis. It is qualitative in the sense that the actual values of time and space are not important, but the overall appearance of the graph provides a pictorial representation of the process.

In the simple case of the electron–electron interaction in Figure 46.4, a photon (the field particle) mediates the electromagnetic force between the electrons. Notice that the entire interaction is represented in the diagram as occurring at a single point in time. Therefore, the paths of the electrons appear to undergo a discontinuous change in direction at the moment of interaction. The electron paths shown in Figure 46.4 are different from the *actual* paths, which would be curved due to the continuous exchange of large numbers of field particles.

RICHARD FEYNMAN
American Physicist (1918–1988)
Inspired by Dirac, Feynman developed quantum electrodynamics, the theory of the interaction of light and matter on a relativistic and quantum basis. In 1965, Feynman won the Nobel Prize in Physics. The prize was shared by Feynman, Julian Schwinger, and Sin Itiro Tomonaga. Early in Feynman's career, he was a leading member of the team developing the first nuclear weapon in the Manhattan Project. Toward the end of his career, he worked on the commission investigating the 1986 *Challenger* tragedy and demonstrated the effects of cold temperatures on the rubber O-rings used in the space shuttle.

In the electron–electron interaction, the photon, which transfers energy and momentum from one electron to the other, is called a *virtual photon* because it vanishes during the interaction without having been detected. In Chapter 40, we discussed that a photon has energy $E = hf$, where f is its frequency. Consequently, for a system of two electrons initially at rest, the system has energy $2m_ec^2$ before a virtual photon is released and energy $2m_ec^2 + hf$ after the virtual photon is released (plus any kinetic energy of the electron resulting from the emission of the photon). Is that a violation of the law of conservation of energy for an isolated system? No; this process does *not* violate the law of conservation of energy because the virtual photon has a very short lifetime Δt that makes the uncertainty in the energy $\Delta E \approx \hbar/2\,\Delta t$ of the system greater than the photon energy. Therefore, within the constraints of the uncertainty principle, the energy of the system is conserved.

Now consider a pion mediating the nuclear force between a proton and a neutron as in Yukawa's model (Fig. 46.5a). The rest energy E_R of a pion of mass m_π is given by Einstein's equation $E_R = m_\pi c^2$. To conserve energy, as with the photon in Figure 46.4, the uncertainty in the system energy must be greater than the rest energy of the pion: $\Delta E > E_R$. The existence of the pion would violate the law of conservation of energy if the particle existed for a time interval greater than $\Delta t \approx \hbar/2E_R$ (from the uncertainty principle), where E_R is the rest energy of the pion and Δt is the time interval required for the pion to transfer from one nucleon to the other. Therefore,

$$\Delta t \approx \frac{\hbar}{2E_R} = \frac{\hbar}{2m_\pi c^2} \quad \rightarrow \quad m_\pi c^2 = \frac{\hbar}{2\Delta t} \qquad (46.2)$$

[3] The antineutrino is another zero-charge particle for which the identification of the antiparticle is more difficult than that for a charged particle. Although the details are beyond the scope of this book, the neutrino and antineutrino can be differentiated by means of the relationship between the linear momentum and the spin angular momentum of the particles.

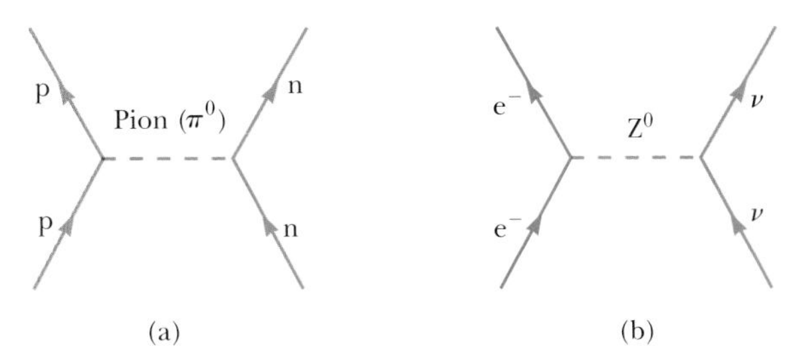

Figure 46.5 (a) Feynman diagram representing a proton and a neutron interacting via the nuclear force with a neutral pion mediating the force. (This model is *not* the current model for nucleon interaction.) (b) Feynman diagram for an electron and a neutrino interacting via the weak force, with a Z^0 boson mediating the force.

Because the pion cannot travel faster than the speed of light, the maximum distance it can travel in a time interval Δt is $d = c\,\Delta t$. Therefore,

$$m_\pi c^2 \approx \frac{\hbar}{2(d/c)} = \frac{\hbar c}{2d} \tag{46.3}$$

Table 46.1 shows that the range of the nuclear force is approximately 1×10^{-15} m. Using this value for d in Equation 46.3, we estimate the rest energy of the pion to be

$$m_\pi c^2 \approx \frac{(1.055 \times 10^{-34}\text{ J}\cdot\text{s})(3.00 \times 10^8\text{ m/s})}{2(1 \times 10^{-15}\text{ m})}$$

$$= 1.6 \times 10^{-11}\text{ J} \approx 100\text{ MeV}$$

Because this result is the same order of magnitude as the observed rest energies of the pions, we have some confidence in the field-particle model.

The concept just described is quite revolutionary. In effect, it says that a system of two nucleons can change into two nucleons plus a pion as long as it returns to its original state in a very short time interval. (Remember that this description is the older historical model, which assumes the pion is the field particle for the nuclear force; the gluon is the actual field particle in current models.) Physicists often say that a nucleon undergoes *fluctuations* as it emits and absorbs field particles. These fluctuations are a consequence of a combination of quantum mechanics (through the uncertainty principle) and special relativity (through Einstein's energy–mass relationship $E_R = mc^2$).

In this section, we discussed the field particles that were originally proposed to mediate the nuclear force (pions) and those that mediate the electromagnetic force (photons). The graviton, the field particle for the gravitational force, has yet to be observed. In 1983, $W^\pm$ and Z^0 particles, which mediate the weak force, were discovered by Italian physicist Carlo Rubbia (b. 1934) and his associates, using a proton–antiproton collider. Rubbia and Simon van der Meer (b. 1925), both at CERN,[4] shared the 1984 Nobel Prize in Physics for the discovery of the $W^\pm$ and Z^0 particles and the development of the proton–antiproton collider. Figure 46.5b shows a Feynman diagram for a weak interaction mediated by a Z^0 boson.

PITFALL PREVENTION 46.2

The Nuclear Force and the Strong Force

The nuclear force discussed in Chapter 44 was historically called the strong force. Once the quark theory (Section 46.8) was established, however, the phrase *strong force* was reserved for the force between quarks. We shall follow this convention: the strong force is between quarks or particles built from quarks, and the nuclear force is between nucleons in a nucleus. The nuclear force is a secondary result of the strong force as discussed in Section 46.9. It is sometimes called the *residual strong force*. Because of this historical development of the names for these forces, other books sometimes refer to the nuclear force as the strong force.

46.4 Classification of Particles

All particles other than field particles can be classified into two broad categories, *hadrons* and *leptons*. The criterion for separating these particles into categories is whether or not they interact via the strong force. The nuclear force between nucleons in a nucleus is a particular manifestation of the strong force, but we will use the term *strong force* to refer to any interaction between particles made up of

[4] CERN was originally the Conseil Européen pour la Recherche Nucléaire (European Organization for Nuclear Research); the name has been altered to European Laboratory for Particle Physics, but the CERN acronym has been retained.

TABLE 46.2

Some Particles and Their Properties

Category	Particle Name	Symbol	Anti-particle	Mass (MeV/c^2)	B	L_e	L_μ	L_τ	S	Lifetime(s)	Spin
Leptons	Electron	e^-	e^+	0.511	0	+1	0	0	0	Stable	$\frac{1}{2}$
	Electron–neutrino	ν_e	$\overline{\nu}_e$	$< 7\text{eV}/c^2$	0	+1	0	0	0	Stable	$\frac{1}{2}$
	Muon	μ^-	μ^+	105.7	0	0	+1	0	0	2.20×10^{-6}	$\frac{1}{2}$
	Muon–neutrino	ν_μ	$\overline{\nu}_\mu$	< 0.3	0	0	+1	0	0	Stable	$\frac{1}{2}$
	Tau	τ^-	τ^+	1 784	0	0	0	+1	0	$< 4 \times 10^{-13}$	$\frac{1}{2}$
	Tau–neutrino	ν_τ	$\overline{\nu}_\tau$	< 30	0	0	0	+1	0	Stable	$\frac{1}{2}$
Hadrons											
Mesons	Pion	π^+	π^-	139.6	0	0	0	0	0	2.60×10^{-8}	0
		π^0	Self	135.0	0	0	0	0	0	0.83×10^{-16}	0
	Kaon	K^+	K^-	493.7	0	0	0	0	+1	1.24×10^{-8}	0
		K^0_S	$\overline{K}^0_S$	497.7	0	0	0	0	+1	0.89×10^{-10}	0
		K^0_L	$\overline{K}^0_L$	497.7	0	0	0	0	+1	5.2×10^{-8}	0
	Eta	η	Self	548.8	0	0	0	0	0	$< 10^{-18}$	0
		η'	Self	958	0	0	0	0	0	2.2×10^{-21}	0
Baryons	Proton	p	$\overline{\text{p}}$	938.3	+1	0	0	0	0	Stable	$\frac{1}{2}$
	Neutron	n	$\overline{\text{n}}$	939.6	+1	0	0	0	0	614	$\frac{1}{2}$
	Lambda	Λ^0	$\overline{\Lambda}^0$	1 115.6	+1	0	0	0	−1	2.6×10^{-10}	$\frac{1}{2}$
	Sigma	Σ^+	$\overline{\Sigma}^-$	1 189.4	+1	0	0	0	−1	0.80×10^{-10}	$\frac{1}{2}$
		Σ^0	$\overline{\Sigma}^0$	1 192.5	+1	0	0	0	−1	6×10^{-20}	$\frac{1}{2}$
		Σ^-	$\overline{\Sigma}^+$	1 197.3	+1	0	0	0	−1	1.5×10^{-10}	$\frac{1}{2}$
	Delta	Δ^{++}	$\overline{\Delta}^{--}$	1 230	+1	0	0	0	0	6×10^{-24}	$\frac{3}{2}$
		Δ^+	$\overline{\Delta}^-$	1 231	+1	0	0	0	0	6×10^{-24}	$\frac{3}{2}$
		Δ^0	$\overline{\Delta}^0$	1 232	+1	0	0	0	0	6×10^{-24}	$\frac{3}{2}$
		Δ^-	$\overline{\Delta}^+$	1 234	+1	0	0	0	0	6×10^{-24}	$\frac{3}{2}$
	Xi	Ξ^0	$\overline{\Xi}^0$	1 315	+1	0	0	0	−2	2.9×10^{-10}	$\frac{1}{2}$
		Ξ^-	Ξ^+	1 321	+1	0	0	0	−2	1.64×10^{-10}	$\frac{1}{2}$
	Omega	Ω^-	Ω^+	1 672	+1	0	0	0	−3	0.82×10^{-10}	$\frac{3}{2}$

quarks. (For more detail on quarks and the strong force, see Section 46.8.) Table 46.2 provides a summary of the properties of hadrons and leptons.

Hadrons

Particles that interact through the strong force (as well as through the other fundamental forces) are called **hadrons.** The two classes of hadrons, *mesons* and *baryons,* are distinguished by their masses and spins.

Mesons all have zero or integer spin (0 or 1). As indicated in Section 46.3, the name comes from the expectation that Yukawa's proposed meson mass would lie between the masses of the electron and the proton. Several meson masses do lie in this range, although mesons having masses greater than that of the proton have been found to exist.

All mesons decay finally into electrons, positrons, neutrinos, and photons. The pions are the lightest known mesons and have masses of approximately 1.4×10^2 MeV/c^2, and all three pions—π^+, π^-, and π^0—have a spin of 0. (This spin-0 characteristic indicates that the particle discovered by Anderson in 1937, the muon, is not a meson. The muon has spin $\frac{1}{2}$ and belongs in the *lepton* classification, described below.)

Baryons, the second class of hadrons, have masses equal to or greater than the proton mass (the name *baryon* means "heavy" in Greek), and their spin is always a half-integer value ($\frac{1}{2}$ or $\frac{3}{2}$). Protons and neutrons are baryons, as are many other particles. With the exception of the proton, all baryons decay in such a way that the end products include a proton. For example, the baryon called the Ξ hyperon (Greek letter xi) decays to the Λ^0 baryon (Greek letter lambda) in approximately 10^{-10} s. The Λ^0 then decays to a proton and a π^- in approximately 3×10^{-10} s.

Today it is believed that hadrons are not elementary particles but instead are composed of more elementary units called quarks, per Section 46.8.

Leptons

Leptons (from the Greek *leptos,* meaning "small" or "light") are particles that do not interact by means of the strong force. All leptons have spin $\frac{1}{2}$. Unlike hadrons, which have size and structure, leptons appear to be truly elementary, meaning that they have no structure and are point-like.

Quite unlike the case with hadrons, the number of known leptons is small. Currently, scientists believe that only six leptons exist: the electron, the muon, the tau, and a neutrino associated with each: e^-, μ^-, τ^-, ν_e, ν_μ, and ν_τ. The tau lepton, discovered in 1975, has a mass about twice that of the proton. Direct experimental evidence for the neutrino associated with the tau was announced by the Fermi National Accelerator Laboratory (Fermilab) in July 2000. Each of the six leptons has an antiparticle.

Current studies indicate that neutrinos have a small but nonzero mass. If they do have mass, they cannot travel at the speed of light. In addition, because so many neutrinos exist, their combined mass may be sufficient to cause all the matter in the Universe to eventually collapse into a single point, which might then explode and create a completely new Universe! We shall discuss this possibility in more detail in Section 46.11.

46.5 Conservation Laws

The laws of conservation of energy, linear momentum, angular momentum, and electric charge provide us with a set of rules that all processes must follow. In Chapter 44, we learned that conservation laws are important for understanding why certain radioactive decays and nuclear reactions occur and others do not. In the study of elementary particles, a number of additional conservation laws are important. Although the two described here have no theoretical foundation, they are supported by abundant empirical evidence.

Baryon Number

Experimental results show that whenever a baryon is created in a decay or nuclear reaction, an antibaryon is also created. This scheme can be quantified by assigning every particle a quantum number, the **baryon number,** as follows: $B = +1$ for all baryons, $B = -1$ for all antibaryons, and $B = 0$ for all other particles. (See Table 46.2.) The **law of conservation of baryon number** states that **whenever a nuclear reaction or decay occurs, the sum of the baryon numbers before the process must equal the sum of the baryon numbers after the process.**

◀ Conservation of baryon number

If baryon number is conserved, the proton must be absolutely stable. For example, a decay of the proton to a positron and a neutral pion would satisfy conservation of energy, momentum, and electric charge. Such a decay has never been observed, however. The law of conservation of baryon number would be consistent with the absence of this decay because the proposed decay would involve the loss of a baryon. Based on experimental observations as pointed out in Example 46.2, all we can say at present is that protons have a half-life of at least 10^{33} years (the estimated age of the Universe is only 10^{10} years). Some recent theories, however, predict that the proton is unstable. According to this theory, baryon number is not absolutely conserved.

Quick Quiz 46.2 Consider the decays **(i)** $n \rightarrow \pi^+ + \pi^- + \mu^+ + \mu^-$ and **(ii)** $n \rightarrow p + \pi^-$. From the following choices, which conservation laws are violated by each decay? (a) energy (b) electric charge (c) baryon number (d) angular momentum (e) no conservation laws

EXAMPLE 46.1 Checking Baryon Numbers

Use the law of conservation of baryon number to determine whether each of the following reactions can occur:

(A) $p + n \rightarrow p + p + n + \overline{p}$

SOLUTION

Conceptualize Because the mass on the right is larger than the mass on the left, the initial particles must have sufficient kinetic energy that energy conservation is satisfied for this reaction.

Categorize We use a conservation law developed in this section, so we categorize this example as a substitution problem.

Evaluate the total baryon number for the left side of the reaction: $1 + 1 = 2$

Evaluate the total baryon number for the right side of the reaction: $1 + 1 + 1 + (-1) = 2$

Therefore, baryon number is conserved and the reaction can occur.

(B) $p + n \rightarrow p + p + \overline{p}$

SOLUTION

Evaluate the total baryon number for the left side of the reaction: $1 + 1 = 2$

Evaluate the total baryon number for the right side of the reaction: $1 + 1 + (-1) = 1$

Because baryon number is not conserved, the reaction cannot occur.

EXAMPLE 46.2 Detecting Proton Decay

Measurements taken at the Super Kamiokande neutrino detection facility (Fig. 46.6) indicate that the half-life of protons is at least 10^{33} yr.

(A) Estimate how long we would have to watch, on average, to see a proton in a glass of water decay.

SOLUTION

Conceptualize Imagine the number of protons in a glass of water. Although this number is huge, the probability of a single proton undergoing decay is small, so we would expect to wait for a long time interval before observing a decay.

Categorize Because a half-life is provided in the problem, we categorize this problem as one in which we can apply our statistical analysis techniques from Section 44.4.

Figure 46.6 (Example 46.2) This detector at the Super Kamiokande neutrino facility in Japan is used to study photons and neutrinos. It holds 50 000 metric tons of highly purified water and 13 000 photomultipliers. The photograph was taken while the detector was being filled. Technicians in a raft clean the photodetectors before they are submerged.

Analyze Let's estimate that a glass contains approximately 250 g of water.

Find the number of molecules of water in the glass:

$$\frac{(250\text{ g})(6.02 \times 10^{23}\text{ molecules/mol})}{18\text{ g/mol}} = 8.4 \times 10^{24}\text{ molecules}$$

Each water molecule contains one proton in each of its two hydrogen atoms plus eight protons in its oxygen atom, for a total of ten protons. Therefore, there are 8.4×10^{25} protons in the glass of water.

Find the decay constant (Section 44.4) from Equation 44.8:

$$(1)\quad \lambda = \frac{0.693}{T_{1/2}} = \frac{0.693}{10^{33}\text{ yr}} = 6.9 \times 10^{-34}\text{ yr}^{-1}$$

Find the activity of the protons from Equation 44.7:

$$(2)\quad R = \lambda N = (6.9 \times 10^{-34}\text{ yr}^{-1})(8.4 \times 10^{25})$$
$$= 5.8 \times 10^{-8}\text{ yr}^{-1}$$

Finalize Equation (1) represents the probability that *one* proton decays in one year. The probability that *any* proton in our glass of water decays in the one-year interval is given by Equation (2). Therefore, we must watch our glass of water for $1/R \approx$ 17 million years! That indeed is a long time interval, as expected.

(B) The Super Kamiokande neutrino facility contains 50 000 metric tons of water. Estimate the average time interval between detected proton decays in this much water if the half-life of a proton is 10^{33} yr.

SOLUTION

Analyze Find the ratio of the number of molecules in 50 000 metric tons of water to that in the glass of water in part (A), which is the same as the ratio of masses:

$$\frac{N_{\text{Kamiokande}}}{N_{\text{glass}}} = \frac{m_{\text{Kamiokande}}}{m_{\text{glass}}}$$
$$= \frac{50\,000\text{ metric ton}}{250\text{ g}}\left(\frac{1\,000\text{ kg}}{1\text{ metric ton}}\right)\left(\frac{1\,000\text{ g}}{1\text{ kg}}\right)$$
$$= 2.0 \times 10^{8}$$

Find the number of molecules in the Kamiokande facility:

$$N_{\text{Kamiokande}} = (2.0 \times 10^{8})N_{\text{glass}}$$
$$= (2.0 \times 10^{8})(8.4 \times 10^{24}\text{ molecules})$$
$$= 1.7 \times 10^{33}\text{ molecules}$$

Find the decay rate for protons in the facility:

$$R = (6.9 \times 10^{-34}\text{ yr}^{-1})(10\text{ protons/molecule})(1.7 \times 10^{33}\text{ molecules})$$
$$\approx 12\text{ yr}^{-1}$$

Finalize The average time interval between decays is about one-twelfth of a year, or approximately one month. That is much shorter than the time interval in part (A) due to the tremendous amount of water in the detector facility. Despite this rosy prediction of one proton decay per month, a proton decay has never been observed. This suggests that the half-life of the proton may be larger than 10^{33} years or that proton decay simply does not occur.

Lepton Number

◀ Conservation of electron lepton number

There are three conservation laws involving lepton numbers, one for each variety of lepton. The **law of conservation of electron lepton number** states that whenever a nuclear reaction or decay occurs, **the sum of the electron lepton numbers before the process must equal the sum of the electron lepton numbers after the process.**

The electron and the electron neutrino are assigned an electron lepton number $L_e = +1$, and the antileptons e^+ and $\bar{\nu}_e$ are assigned an electron lepton number $L_e = -1$. All other particles have $L_e = 0$. For example, consider the decay of the neutron:

$$n \rightarrow p + e^- + \bar{\nu}_e$$

Before the decay, the electron lepton number is $L_e = 0$; after the decay, it is $0 + 1 + (-1) = 0$. Therefore, electron lepton number is conserved. (Baryon number must also be conserved, of course, and it is: before the decay, $B = +1$, and after the decay, $B = +1 + 0 + 0 = +1$.)

Similarly, when a decay involves muons, the muon lepton number L_μ is conserved. The μ^- and the ν_μ are assigned a muon lepton number $L_\mu = +1$, and the antimuons μ^+ and $\bar{\nu}_\mu$ are assigned a muon lepton number $L_\mu = -1$. All other particles have $L_\mu = 0$.

Finally, tau lepton number L_τ is conserved with similar assignments made for the tau lepton, its neutrino, and their two antiparticles.

Quick Quiz 46.3 Consider the following decay: $\pi^0 \rightarrow \mu^- + e^+ + \nu_\mu$. What conservation laws are violated by this decay? (a) energy (b) angular momentum (c) electric charge (d) baryon number (e) electron lepton number (f) muon lepton number (g) tau lepton number (h) no conservation laws

Quick Quiz 46.4 Suppose a claim is made that the decay of the neutron is given by $n \rightarrow p + e^-$. What conservation laws are violated by this decay? (a) energy (b) angular momentum (c) electric charge (d) baryon number (e) electron lepton number (f) muon lepton number (g) tau lepton number (h) no conservation laws

EXAMPLE 46.3 Checking Lepton Numbers

Use the law of conservation of lepton numbers to determine whether each of the following decay schemes (A) and (B) can occur:

(A) $\mu^- \rightarrow e^- + \bar{\nu}_e + \nu_\mu$

SOLUTION

Conceptualize Because this decay involves a muon and an electron, L_μ and L_e must each be conserved separately if the decay is to occur.

Categorize We use a conservation law developed in this section, so we categorize this example as a substitution problem.

Evaluate the lepton numbers before the decay: $L_\mu = +1 \qquad L_e = 0$

Evaluate the total lepton numbers after the decay: $L_\mu = 0 + 0 + 1 = +1 \qquad L_e = +1 + (-1) + 0 = 0$

Therefore, both numbers are conserved and on this basis the decay is possible.

(B) $\pi^+ \rightarrow \mu^+ + \nu_\mu + \nu_e$

SOLUTION

Evaluate the lepton numbers before the decay: $L_\mu = 0 \qquad L_e = 0$

Evaluate the total lepton numbers after the decay: $L_\mu = -1 + 1 + 0 = 0 \qquad L_e = 0 + 0 + 1 = 1$

Therefore, the decay is not possible because electron lepton number is not conserved.

46.6 Strange Particles and Strangeness

Many particles discovered in the 1950s were produced by the interaction of pions with protons and neutrons in the atmosphere. A group of these—the kaon (K), lambda (Λ), and sigma (Σ) particles—exhibited unusual properties both as they were created and as they decayed; hence, they were called *strange particles.*

One unusual property of strange particles is that they are always produced in pairs. For example, when a pion collides with a proton, a highly probable result is the production of two neutral strange particles (Fig. 46.7):

$$\pi^- + \mathrm{p} \rightarrow \mathrm{K}^0 + \Lambda^0$$

The reaction $\pi^- + \mathrm{p} \rightarrow \mathrm{K}^0 + \mathrm{n}$, where only one of the final particles is strange, never occurs, however, even though no known conservation laws would be violated and even though the energy of the pion is sufficient to initiate the reaction.

The second peculiar feature of strange particles is that although they are produced in reactions involving the strong interaction at a high rate, they do not decay into particles that interact via the strong force at a high rate. Instead, they decay very slowly, which is characteristic of the weak interaction. Their half-lives are in the range 10^{-10} s to 10^{-8} s, whereas most other particles that interact via the strong force have much shorter lifetimes on the order of 10^{-23} s.

To explain these unusual properties of strange particles, a new quantum number S, called **strangeness,** was introduced, together with a conservation law. The strangeness numbers for some particles are given in Table 46.2. The production of strange particles in pairs is explained by assigning $S = +1$ to one of the particles, $S = -1$ to the other, and $S = 0$ to all nonstrange particles. The **law of conservation of strangeness** states that **in a nuclear reaction or decay that occurs via the**

◀ Conservation of strangeness

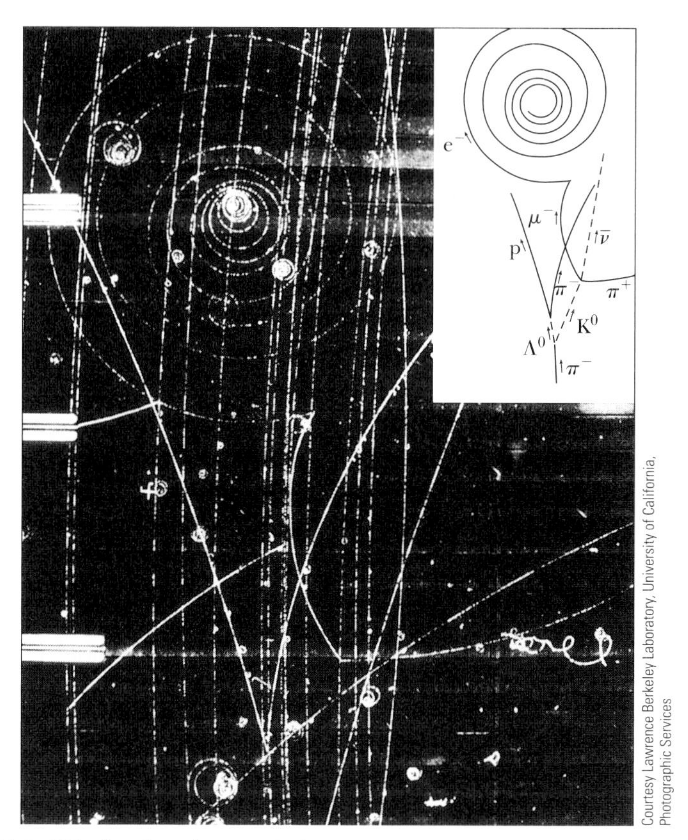

Figure 46.7 This bubble-chamber photograph shows many events, and the inset is a drawing of identified tracks. The strange particles Λ^0 and K^0 are formed at the bottom as a π^- particle interacts with a proton in the reaction $\pi^- + \mathrm{p} \rightarrow \Lambda^0 + \mathrm{K}^0$. (Notice that the neutral particles leave no tracks, as indicated by the dashed lines in the inset.) The Λ^0 then decays in the reaction $\Lambda^0 \rightarrow \pi^- + \mathrm{p}$ and the K^0 in the reaction $\mathrm{K}^0 \rightarrow \pi^+ + \mu^- + \bar{\nu}_\mu$.

strong force, strangeness is conserved; that is, the sum of the strangeness numbers before the process must equal the sum of the strangeness numbers after the process. In processes that occur via the weak interaction, strangeness may not be conserved.

The low decay rate of strange particles can be explained by assuming the strong and electromagnetic interactions obey the law of conservation of strangeness but the weak interaction does not. Because the decay of a strange particle involves the loss of one strange particle, it violates strangeness conservation and hence proceeds slowly via the weak interaction.

EXAMPLE 46.4 Is Strangeness Conserved?

(A) Use the law of strangeness conservation to determine whether the reaction $\pi^0 + \text{n} \rightarrow \text{K}^+ + \Sigma^-$ occurs.

SOLUTION

Conceptualize Because there are two strange particles on the right side of the reaction and none on the left, we see that we will need to investigate conservation of strangeness.

Categorize We use a conservation law developed in this section, so we categorize this example as a substitution problem.

Evaluate the strangeness for the left side of the reaction using Table 46.2: $S = 0 + 0 = 0$

Evaluate the strangeness for the right side of the reaction: $S = +1 - 1 = 0$

Therefore, strangeness is conserved and the reaction is allowed.

(B) Show that the reaction $\pi^- + \text{p} \rightarrow \pi^- + \Sigma^+$ does not conserve strangeness.

SOLUTION

Evaluate the strangeness for the left side of the reaction: $S = 0 + 0 = 0$

Evaluate the strangeness for the right side of the reaction: $S = 0 + (-1) = -1$

Therefore, strangeness is not conserved.

46.7 Finding Patterns in the Particles

One tool scientists use is the detection of patterns in data, patterns that contribute to our understanding of nature. One of the best examples of this tool's use is the development of the periodic table, which provides a fundamental understanding of the chemical behavior of the elements. As mentioned in the introduction, the periodic table explains how more than 100 elements can be formed from three particles, the electron, the proton, and the neutron. The table of nuclides, part of which is shown in Table 44.2, contains hundreds of nuclides, but all can be built from protons and neutrons.

The number of particles observed by particle physicists is also in the hundreds. Is it possible that a small number of entities exist from which all these can be built? Taking a hint from the success of the periodic table and the table of nuclides, let's explore the historical search for patterns among the particles.

Many classification schemes have been proposed for grouping particles into families. Consider, for instance, the baryons listed in Table 46.2 that have spins of $\frac{1}{2}$: p, n, Λ^0, Σ^+, Σ^0, Σ^-, Ξ^0, and Ξ^-. If we plot strangeness versus charge for these

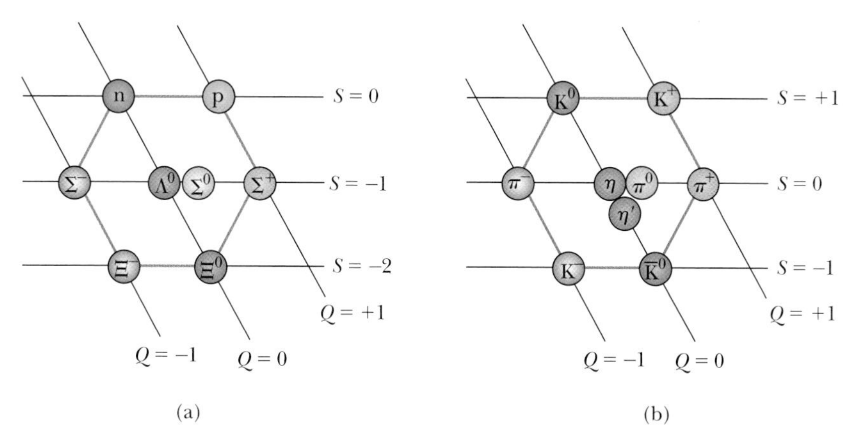

Figure 46.8 (a) The hexagonal eightfold-way pattern for the eight spin-$\frac{1}{2}$ baryons. This strangeness-versus-charge plot uses a sloping axis for charge number Q and a horizontal axis for strangeness S. (b) The eightfold-way pattern for the nine spin-zero mesons.

baryons using a sloping coordinate system as in Figure 46.8a, a fascinating pattern is observed: six of the baryons form a hexagon, and the remaining two are at the hexagon's center.

As a second example, consider the following nine spin-zero mesons listed in Table 46.2: π^+, π^0, π^-, K^+, K^0, K^-, η, η', and the antiparticle $\overline{K}^0$. Figure 46.8b is a plot of strangeness versus charge for this family. Again, a hexagonal pattern emerges. In this case, each particle on the perimeter of the hexagon lies opposite its antiparticle and the remaining three (which form their own antiparticles) are at the center of the hexagon. These and related symmetric patterns were developed independently in 1961 by Murray Gell-Mann and Yuval Ne'eman (1925–2006). Gell-Mann called the patterns the **eightfold way,** after the eightfold path to nirvana in Buddhism.

Groups of baryons and mesons can be displayed in many other symmetric patterns within the framework of the eightfold way. For example, the family of spin-$\frac{3}{2}$ baryons known in 1961 contains nine particles arranged in a pattern like that of the pins in a bowling alley as in Figure 46.9. (The particles Σ^{*+}, Σ^{*0}, Σ^{*-}, Ξ^{*0}, and Ξ^{*-} are excited states of the particles Σ^+, Σ^0, Σ^-, Ξ^0, and Ξ^-. In these higher-energy states, the spins of the three quarks (see Section 46.8) making up the particle are aligned so that the total spin of the particle is $\frac{3}{2}$.) When this pattern was proposed, an empty spot occurred in it (at the bottom position), corresponding to a particle that had never been observed. Gell-Mann predicted that the missing particle, which he called the omega minus (Ω^-), should have spin $\frac{3}{2}$, charge -1, strangeness -3, and rest energy of approximately 1 680 MeV. Shortly thereafter,

MURRAY GELL-MANN
American Physicist (b. 1929)
In 1969, Murray Gell-Mann was awarded the Nobel Prize in Physics for his theoretical studies dealing with subatomic particles.

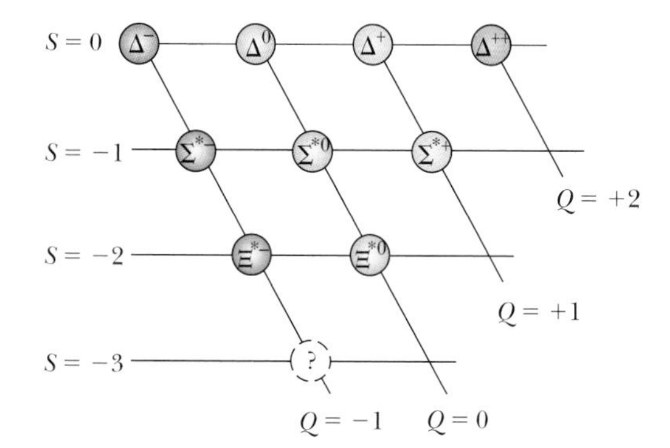

Figure 46.9 The pattern for the higher-mass, spin-$\frac{3}{2}$ baryons known at the time the pattern was proposed. The three Σ^* and two Ξ^* particles are excited states of the corresponding spin-$\frac{1}{2}$ particles in Figure 46.8. These excited states have higher mass and spin $\frac{3}{2}$. The absence of a particle in the bottom position was evidence of a new particle yet to be discovered, the Ω^-.

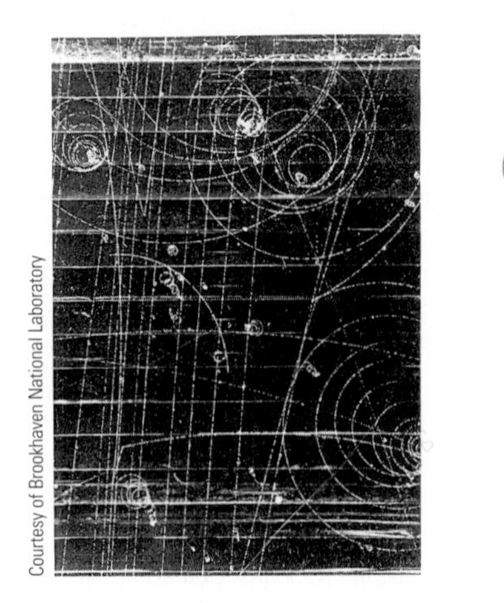

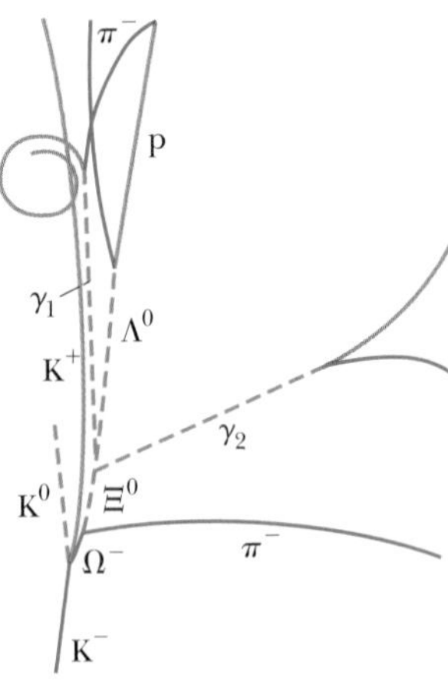

Figure 46.10 Discovery of the Ω^- particle. The photograph on the left shows the original bubble-chamber tracks. The drawing on the right isolates the tracks of the important events. The K^- particle at the bottom collides with a proton to produce the first detected Ω^- particle plus a K^0 and a K^+.

in 1964, scientists at the Brookhaven National Laboratory found the missing particle through careful analyses of bubble-chamber photographs (Fig. 46.10) and confirmed all its predicted properties.

The prediction of the missing particle in the eightfold way has much in common with the prediction of missing elements in the periodic table. Whenever a vacancy occurs in an organized pattern of information, experimentalists have a guide for their investigations.

46.8 Quarks

As mentioned earlier, leptons appear to be truly elementary particles because there are only a few types of them, and experiments indicate that they have no measurable size or internal structure. Hadrons, on the other hand, are complex particles having size and structure. The existence of the strangeness–charge patterns of the eightfold way suggests that hadrons have substructure. Furthermore, hundreds of types of hadrons exist and many decay into other hadrons.

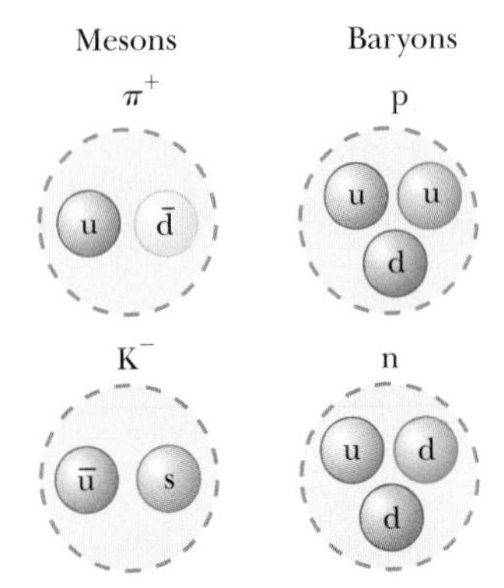

ACTIVE FIGURE 46.11

Quark composition of two mesons and two baryons.

Sign in at www.thomsonedu.com and go to ThomsonNOW to observe the quark composition for the mesons and baryons in Tables 46.4 and 46.5.

The Original Quark Model

In 1963, Gell-Mann and George Zweig (b. 1937) independently proposed a model for the substructure of hadrons. According to their model, all hadrons are composed of two or three elementary constituents called **quarks.** (Gell-Mann borrowed the word *quark* from the passage "Three quarks for Muster Mark" in James Joyce's *Finnegans Wake.* In Zweig's model, he called the constituents "aces.") The model has three types of quarks, designated by the symbols u, d, and s, that are given the arbitrary names **up, down,** and **strange.** The various types of quarks are called **flavors.** Active Figure 46.11 is a pictorial representation of the quark compositions of several hadrons.

An unusual property of quarks is that they carry a fractional electronic charge. The u, d, and s quarks have charges of $+2e/3$, $-e/3$, and $-e/3$, respectively, where e is the elementary charge 1.60×10^{-19} C. These and other properties of quarks and antiquarks are given in Table 46.3. Quarks have spin $\frac{1}{2}$, which means that all quarks are fermions, defined as any particle having half-integral spin, as pointed out in Section 43.8. As Table 46.3 shows, associated with each quark is an antiquark of opposite charge, baryon number, and strangeness.

TABLE 46.3

Properties of Quarks and Antiquarks

Quarks

Name	Symbol	Spin	Charge	Baryon Number	Strangeness	Charm	Bottomness	Topness
Up	u	$\frac{1}{2}$	$+\frac{2}{3}e$	$\frac{1}{3}$	0	0	0	0
Down	d	$\frac{1}{2}$	$-\frac{1}{3}e$	$\frac{1}{3}$	0	0	0	0
Strange	s	$\frac{1}{2}$	$-\frac{1}{3}e$	$\frac{1}{3}$	-1	0	0	0
Charmed	c	$\frac{1}{2}$	$+\frac{2}{3}e$	$\frac{1}{3}$	0	$+1$	0	0
Bottom	b	$\frac{1}{2}$	$-\frac{1}{3}e$	$\frac{1}{3}$	0	0	$+1$	0
Top	t	$\frac{1}{2}$	$+\frac{2}{3}e$	$\frac{1}{3}$	0	0	0	$+1$

Antiquarks

Name	Symbol	Spin	Charge	Baryon Number	Strangeness	Charm	Bottomness	Topness
Anti-up	$\bar{u}$	$\frac{1}{2}$	$-\frac{2}{3}e$	$-\frac{1}{3}$	0	0	0	0
Anti-down	$\bar{d}$	$\frac{1}{2}$	$+\frac{1}{3}e$	$-\frac{1}{3}$	0	0	0	0
Anti-strange	$\bar{s}$	$\frac{1}{2}$	$+\frac{1}{3}e$	$-\frac{1}{3}$	$+1$	0	0	0
Anti-charmed	$\bar{c}$	$\frac{1}{2}$	$-\frac{2}{3}e$	$-\frac{1}{3}$	0	-1	0	0
Anti-bottom	$\bar{b}$	$\frac{1}{2}$	$+\frac{1}{3}e$	$-\frac{1}{3}$	0	0	-1	0
Anti-top	$\bar{t}$	$\frac{1}{2}$	$-\frac{2}{3}e$	$-\frac{1}{3}$	0	0	0	-1

The compositions of all hadrons known when Gell-Mann and Zweig presented their model can be completely specified by three simple rules:

- A meson consists of one quark and one antiquark, giving it a baryon number of 0, as required.
- A baryon consists of three quarks.
- An antibaryon consists of three antiquarks.

The theory put forth by Gell-Mann and Zweig is referred to as the *original quark model.*

Quick Quiz 46.5 Using a coordinate system like that in Figure 46.8, draw an eightfold-way diagram for the three quarks in the original quark model.

Charm and Other Developments

Although the original quark model was highly successful in classifying particles into families, some discrepancies occurred between its predictions and certain experimental decay rates. Consequently, several physicists proposed a fourth quark flavor in 1967. They argued that if four types of leptons exist (as was thought at the time), there should also be four flavors of quarks because of an underlying symmetry in nature. The fourth quark, designated c, was assigned a property called **charm.** A *charmed* quark has charge $+2e/3$, just as the up quark does, but its charm distinguishes it from the other three quarks. This introduces a new quantum number C, representing charm. The new quark has charm $C = +1$, its antiquark has charm of $C = -1$, and all other quarks have $C = 0$. Charm, like strangeness, is conserved in strong and electromagnetic interactions but not in weak interactions.

Evidence that the charmed quark exists began to accumulate in 1974, when a heavy meson called the J/Ψ particle (or simply Ψ, Greek letter psi) was discovered

TABLE 46.4

Quark Composition of Mesons

		Antiquarks									
		$\bar{b}$		$\bar{c}$		$\bar{s}$		$\bar{d}$		$\bar{u}$	
Quarks	**b**	Υ	$(\bar{b}b)$	B_c^-	$(\bar{c}b)$	$\bar{B}_s^0$	$(\bar{s}b)$	$\bar{B}_d^0$	$(\bar{d}b)$	B^-	$(\bar{u}b)$
	c	B_c^+	$(\bar{b}c)$	J/Ψ	$(\bar{c}c)$	D_s^+	$(\bar{s}c)$	D^+	$(\bar{d}c)$	D^0	$(\bar{u}c)$
	s	B_s^0	$(\bar{b}s)$	D_s^-	$(\bar{c}s)$	η, η'	$(\bar{s}s)$	$\bar{K}^0$	$(\bar{d}s)$	K^-	$(\bar{u}s)$
	d	B_d^0	$(\bar{b}d)$	D^-	$(\bar{c}d)$	K^0	$(\bar{s}d)$	π^0, η, η'	$(\bar{d}d)$	π^-	$(\bar{u}d)$
	u	B^+	$(\bar{b}u)$	$\bar{D}^0$	$(\bar{c}u)$	K^+	$(\bar{s}u)$	π^+	$(\bar{d}u)$	π^0, η, η'	$(\bar{u}u)$

Note: The top quark does not form mesons because it decays too quickly.

independently by two groups, one led by Burton Richter (b. 1931) at the Stanford Linear Accelerator (SLAC), and the other led by Samuel Ting (b. 1936) at the Brookhaven National Laboratory. In 1976, Richter and Ting were awarded a Nobel Prize in Physics for this work. The J/Ψ particle does not fit into the three-quark model; instead, it has properties of a combination of the proposed charmed quark and its antiquark ($c\bar{c}$). It is much more massive than the other known mesons ($\sim$3 100 MeV/c^2), and its lifetime is much longer than the lifetimes of particles that interact via the strong force. Soon, related mesons were discovered, corresponding to such quark combinations as $\bar{c}d$ and $c\bar{d}$, all of which have great masses and long lifetimes. The existence of these new mesons provided firm evidence for the fourth quark flavor.

In 1975, researchers at Stanford University reported strong evidence for the tau (τ) lepton, mass 1 784 MeV/c^2. This fifth type of lepton led physicists to propose that more flavors of quarks might exist, on the basis of symmetry arguments similar to those leading to the proposal of the charmed quark. These proposals led to more elaborate quark models and the prediction of two new quarks, **top** (t) and **bottom** (b). (Some physicists prefer *truth* and *beauty*.) To distinguish these quarks from the others, quantum numbers called *topness* and *bottomness* (with allowed values +1, 0, −1) were assigned to all quarks and antiquarks (see Table 46.3). In 1977, researchers at the Fermi National Laboratory, under the direction of Leon Lederman (b. 1922), reported the discovery of a very massive new meson Υ (Greek letter upsilon), whose composition is considered to be $b\bar{b}$, providing evidence for the bottom quark. In March 1995, researchers at Fermilab announced the discovery of the top quark (supposedly the last of the quarks to be found), which has a mass of 173 GeV/c^2.

Table 46.4 lists the quark compositions of mesons formed from the up, down, strange, charmed, and bottom quarks. Table 46.5 shows the quark combinations for the baryons listed in Table 46.2. Notice that only two flavors of quarks, u and d, are contained in all hadrons encountered in ordinary matter (protons and neutrons).

Will the discoveries of elementary particles ever end? How many "building blocks" of matter actually exist? At present, physicists believe that the elementary particles in nature are six quarks and six leptons, together with their antiparticles, and the four field particles listed in Table 46.1. Table 46.6 lists the rest energies and charges of the quarks and leptons.

Despite extensive experimental effort, no isolated quark has ever been observed. Physicists now believe that at ordinary temperatures, quarks are permanently confined inside ordinary particles because of an exceptionally strong force that prevents them from escaping, called (appropriately) the **strong force**[5] (which we introduced at the beginning of Section 46.4 and will discuss further in Section 46.10). This force increases with separation distance, similar to the force exerted

TABLE 46.5

Quark Composition of Several Baryons

Particle	Quark Composition
p	uud
n	udd
Λ^0	uds
Σ^+	uus
Σ^0	uds
Σ^-	dds
Δ^{++}	uuu
Δ^+	uud
Δ^0	udd
Δ^-	ddd
Ξ^0	uss
Ξ^-	dss
Ω^-	sss

Note: Some baryons have the same quark composition, such as the p and the Δ^+ and the n and the Δ^0. In these cases, the Δ particles are considered to be excited states of the proton and neutron.

[5] As a reminder, the original meaning of the term *strong force* was the short-range attractive force between nucleons, which we have called the *nuclear force*. The nuclear force between nucleons is a secondary effect of the strong force between quarks.

TABLE 46.6

The Elementary Particles and Their Rest Energies and Charges

Particle	Rest Energy	Charge
Quarks		
u	360 MeV	$+\frac{2}{3}e$
d	360 MeV	$-\frac{1}{3}e$
s	540 MeV	$-\frac{1}{3}e$
c	1 500 MeV	$+\frac{2}{3}e$
b	5 GeV	$-\frac{1}{3}e$
t	173 GeV	$+\frac{2}{3}e$
Leptons		
e^-	511 keV	$-e$
μ^-	105.7 MeV	$-e$
τ^-	1 784 MeV	$-e$
ν_e	< 7 eV	0
ν_μ	< 0.3 MeV	0
ν_τ	< 30 MeV	0

by a stretched spring. Current efforts are under way to form a **quark–gluon plasma,** a state of matter in which the quarks are freed from neutrons and protons. In 2000, scientists at CERN announced evidence for a quark–gluon plasma formed by colliding lead nuclei. Experiments continue at CERN as well as at the Relativistic Heavy Ion Collider (RHIC) at Brookhaven to verify the production of a quark–gluon plasma.

Quick Quiz 46.6 Doubly charged baryons, such as the Δ^{++}, are known to exist. True or False: Doubly charged mesons also exist.

46.9 Multicolored Quarks

Shortly after the concept of quarks was proposed, scientists recognized that certain particles had quark compositions that violated the exclusion principle. In Section 42.7, we applied the exclusion principle to electrons in atoms. The principle is more general, however, and applies to all particles with half-integral spin ($\frac{1}{2}$, $\frac{3}{2}$, etc.), which are collectively called fermions. Because all quarks are fermions having spin $\frac{1}{2}$, they are expected to follow the exclusion principle. One example of a particle that appears to violate the exclusion principle is the Ω^- (sss) baryon, which contains three strange quarks having parallel spins, giving it a total spin of $\frac{3}{2}$. All three quarks have the same spin quantum number, in violation of the exclusion principle. Other examples of baryons made up of identical quarks having parallel spins are the Δ^{++} (uuu) and the Δ^- (ddd).

To resolve this problem, it was suggested that quarks possess an additional property called **color charge.** This property is similar in many respects to electric charge except that it occurs in six varieties rather than two. The colors assigned to quarks are red, green, and blue, and antiquarks have the colors antired, antigreen, and antiblue. Therefore, the colors red, green, and blue serve as the "quantum numbers" for the color of the quark. To satisfy the exclusion principle, the three quarks in any baryon must all have different colors. Look again at the quarks in the baryons in Active Figure 46.11 and notice the colors. The three colors "neutralize" to white. A quark and an antiquark in a meson must be of a color and the

PITFALL PREVENTION 46.3

Color Charge Is Not Really Color

The description of color for a quark has nothing to do with visual sensation from light. It is simply a convenient name for a property that is analogous to electric charge.

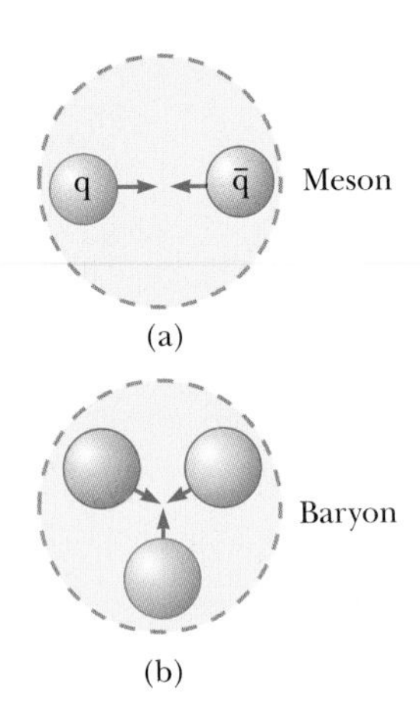

Figure 46.12 (a) A green quark is attracted to an antigreen quark. This forms a meson whose quark structure is $(q\bar{q})$. (b) Three quarks of different colors attract each other to form a baryon.

corresponding anticolor and will consequently neutralize to white, similar to the way electric charges + and − neutralize to zero net charge. (See the mesons in Active Fig. 46.11.) The apparent violation of the exclusion principle in the Ω^- baryon is removed because the three quarks in the particle have different colors.

The new property of color increases the number of quarks by a factor of 3 because each of the six quarks comes in three colors. Although the concept of color in the quark model was originally conceived to satisfy the exclusion principle, it also provided a better theory for explaining certain experimental results. For example, the modified theory correctly predicts the lifetime of the π^0 meson.

The theory of how quarks interact with each other is called **quantum chromodynamics,** or QCD, to parallel the name *quantum electrodynamics* (the theory of the electrical interaction between light and matter). In QCD, each quark is said to carry a color charge, in analogy to electric charge. The strong force between quarks is often called the **color force.** Therefore, the terms *strong force* and *color force* are used interchangeably.

In Section 46.1, we stated that the nuclear interaction between hadrons is mediated by massless field particles called **gluons.** As mentioned earlier, the nuclear force is actually a secondary effect of the strong force between quarks. The gluons are the mediators of the strong force. When a quark emits or absorbs a gluon, the quark's color may change. For example, a blue quark that emits a gluon may become a red quark and a red quark that absorbs this gluon becomes a blue quark.

The color force between quarks is analogous to the electric force between charges: particles with the same color repel, and those with opposite colors attract. Therefore, two green quarks repel each other, but a green quark is attracted to an antigreen quark. The attraction between quarks of opposite color to form a meson $(q\bar{q})$ is indicated in Figure 46.12a. Differently colored quarks also attract one another, although with less intensity than the oppositely colored quark and antiquark. For example, a cluster of red, blue, and green quarks all attract one another to form a baryon as in Figure 46.12b. Therefore, every baryon contains three quarks of three different colors.

Although the nuclear force between two colorless hadrons is negligible at large separations, the net strong force between their constituent quarks is not exactly zero at small separations. This residual strong force is the nuclear force that binds protons and neutrons to form nuclei. It is similar to the force between two electric dipoles. Each dipole is electrically neutral. An electric field surrounds the dipoles, however, because of the separation of the positive and negative charges (see Section 23.6). As a result, an electric interaction occurs between the dipoles that is weaker than the force between single charges. In Section 43.1, we explored how this interaction results in the Van der Waals force between neutral molecules.

According to QCD, a more basic explanation of the nuclear force can be given in terms of quarks and gluons. Figure 46.13a shows the nuclear interaction between a neutron and a proton by means of Yukawa's pion, in this case a π^-. This drawing differs from Figure 46.5a, in which the field particle is a π^0; there is no transfer of charge from one nucleon to the other in Figure 46.5a. In Figure 46.13a, the charged pion carries charge from one nucleon to the other, so the nucleons change identities, with the proton becoming a neutron and the neutron becoming a proton.

Let's look at the same interaction from the viewpoint of the quark model, shown in Figure 46.13b. In this Feynman diagram, the proton and neutron are represented by their quark constituents. Each quark in the neutron and proton is continuously emitting and absorbing gluons. The energy of a gluon can result in the creation of quark–antiquark pairs. This process is similar to the creation of electron–positron pairs in pair production, which we investigated in Section 46.2. When the neutron and proton approach to within 1 fm of each other, these gluons and quarks can be exchanged between the two nucleons, and such exchanges produce the nuclear force. Figure 46.13b depicts one possibility for the process shown in Figure 46.13a. A down quark in the neutron on the right emits a gluon.

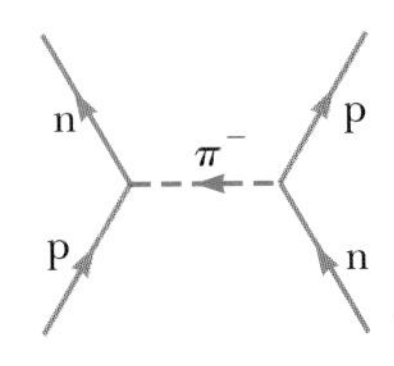

(a) Yukawa's pion model

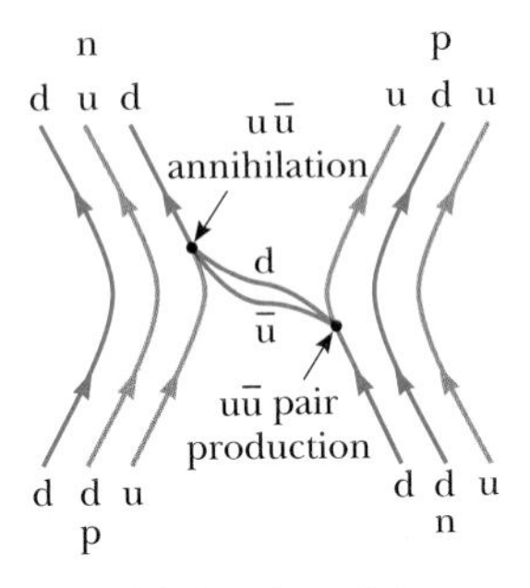

(b) Quark model

Figure 46.13 (a) A nuclear interaction between a proton and a neutron explained in terms of Yukawa's pion-exchange model. Because the pion carries charge, the proton and neutron switch identities. (b) The same interaction, explained in terms of quarks and gluons. The exchanged $\bar{u}d$ quark pair makes up a π^- meson.

The energy of the gluon is then transformed to create a u$\overline{\text{u}}$ pair. The u quark stays within the nucleon (which has now changed to a proton), and the recoiling d quark and the $\overline{\text{u}}$ antiquark are transmitted to the proton on the left side of the diagram. Here the $\overline{\text{u}}$ annihilates a u quark within the proton and the d is captured. The net effect is to change a u quark to a d quark, and the proton on the left has changed to a neutron.

As the d quark and $\overline{\text{u}}$ antiquark in Figure 46.13b transfer between the nucleons, the d and $\overline{\text{u}}$ exchange gluons with each other and can be considered to be bound to each other by means of the strong force. Looking back at Table 46.4, we see that this combination is a π^-, or Yukawa's field particle! Therefore, the quark model of interactions between nucleons is consistent with the pion-exchange model.

46.10 The Standard Model

Scientists now believe there are three classifications of truly elementary particles: leptons, quarks, and field particles. These three types of particles are further classified as either fermions or bosons. Quarks and leptons have spin $\frac{1}{2}$ and hence are fermions, whereas the field particles have integral spin of 1 or higher and are bosons.

Recall from Section 46.1 that the weak force is believed to be mediated by the W^+, W^-, and Z^0 bosons. These particles are said to have *weak charge,* just as quarks have color charge. Therefore, each elementary particle can have mass, electric charge, color charge, and weak charge. Of course, one or more of these could be zero.

In 1979, Sheldon Glashow (b. 1932), Abdus Salam (1926–1996), and Steven Weinberg (b. 1933) won a Nobel Prize in Physics for developing a theory that unifies the electromagnetic and weak interactions. This **electroweak theory** postulates that the weak and electromagnetic interactions have the same strength when the particles involved have very high energies. The two interactions are viewed as different manifestations of a single unifying electroweak interaction. The theory makes many concrete predictions, but perhaps the most spectacular is the prediction of the masses of the W and Z particles at approximately 82 GeV/c^2 and 93 GeV/c^2, respectively. These predictions are close to the masses in Table 46.1 determined by experiment.

The combination of the electroweak theory and QCD for the strong interaction is referred to in high-energy physics as the **Standard Model.** Although the details of the Standard Model are complex, its essential ingredients can be summarized with the help of Figure 46.14. (Although the Standard Model does not include the gravitational force at present, we include gravity in Figure 46.14 because physicists hope to eventually incorporate this force into a unified theory.) This diagram

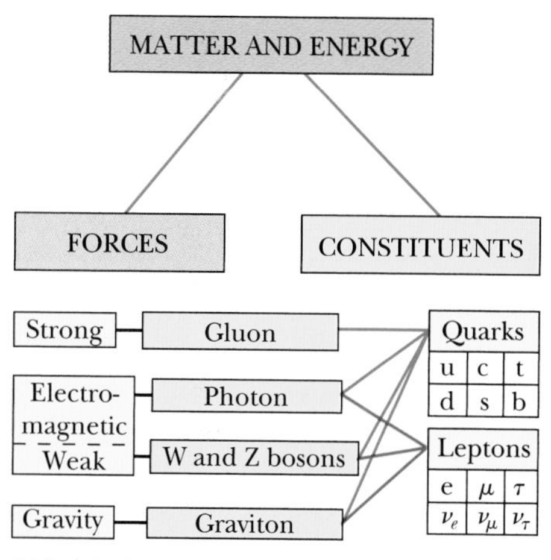

Figure 46.14 The Standard Model of particle physics.

Courtesy of CERN

Figure 46.15 A view from inside the Large Electron–Positron (LEP) Collider tunnel, which is 27 km in circumference.

shows that quarks participate in all the fundamental forces and that leptons participate in all except the strong force.

The Standard Model does not answer all questions. A major question still unanswered is why, of the two mediators of the electroweak interaction, the photon has no mass but the W and Z bosons do. Because of this mass difference, the electromagnetic and weak forces are quite distinct at low energies but become similar at very high energies, when the rest energy is negligible relative to the total energy. The behavior as one goes from high to low energies is called *symmetry breaking* because the forces are similar, or symmetric, at high energies but are very different at low energies. The nonzero rest energies of the W and Z bosons raise the question of the origin of particle masses. To resolve this problem, a hypothetical particle called the **Higgs boson,** which provides a mechanism for breaking the electroweak symmetry, has been proposed. The Standard Model modified to include the Higgs boson provides a logically consistent explanation of the massive nature of the W and Z bosons. Unfortunately, the Higgs boson has not yet been found, but physicists know that its rest energy should be less than 1 TeV. To determine whether the Higgs boson exists, two quarks each having at least 1 TeV of energy must collide. Calculations show that such a collision requires injecting 40 TeV of energy within the volume of a proton, however.

Because of the limited energy available in conventional accelerators using fixed targets, it is necessary to employ colliding-beam accelerators called **colliders.** The concept of colliders is straightforward. Particles that have equal masses and equal kinetic energies, traveling in opposite directions in an accelerator ring, collide head-on to produce the required reaction and form new particles. Because the total momentum of the interacting particles is zero, all their kinetic energy is available for the reaction. The Large Electron–Positron (LEP) Collider at CERN (Fig. 46.15) and the Stanford Linear Collider collide both electrons and positrons. The Super Proton Synchrotron at CERN accelerates protons and antiprotons to energies of 270 GeV. The world's highest-energy proton accelerator, the Tevatron at the Fermi National Laboratory in Illinois, produces protons at almost 1 000 GeV (1 TeV). The Relativistic Heavy Ion Collider at Brookhaven National Laboratory collides heavy ions to search for the quark–gluon plasma as discussed earlier. CERN expects a late 2007 completion date for the Large Hadron Collider, a proton–proton collider that will provide a center-of-mass energy of 14 TeV and enable exploration of Higgs-boson physics. The accelerator occupies the same 27-km circumference tunnel now housing the LEP Collider, and many countries will participate in the project.

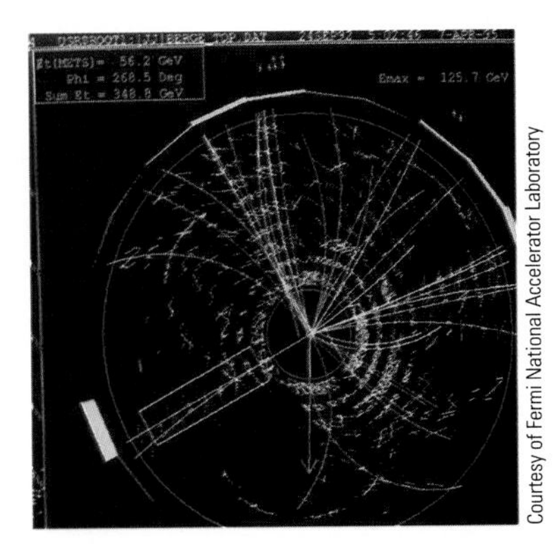

Courtesy of Fermi National Accelerator Laboratory

Figure 46.16 Computers at Fermilab create a pictorial representation such as this one of the paths of particles after a collision

In addition to increasing energies in modern accelerators, detection techniques have become increasingly sophisticated. We saw simple bubble-chamber photographs earlier in this chapter that required hours of analysis by hand. Figure 46.16 shows a modern detection display of particle tracks after a reaction; the tracks are analyzed rapidly by computer. The photograph at the beginning of this chapter shows a complex set of tracks from a collision of gold nuclei.

46.11 The Cosmic Connection

In this section, we describe one of the most fascinating theories in all science—the Big Bang theory of the creation of the Universe—and the experimental evidence that supports it. This theory of cosmology states that the Universe had a beginning and furthermore that the beginning was so cataclysmic that it is impossible to look back beyond it. According to this theory, the Universe erupted from an infinitely dense singularity between 10 and 20 billion years ago. The first few moments after the Big Bang saw such extremely high energy that it is believed that all four interactions of physics were unified and all matter was contained in a quark–gluon plasma.

The evolution of the four fundamental forces from the Big Bang to the present is shown in Figure 46.17. During the first 10^{-43} s (the ultrahot epoch, $T \sim 10^{32}$ K),

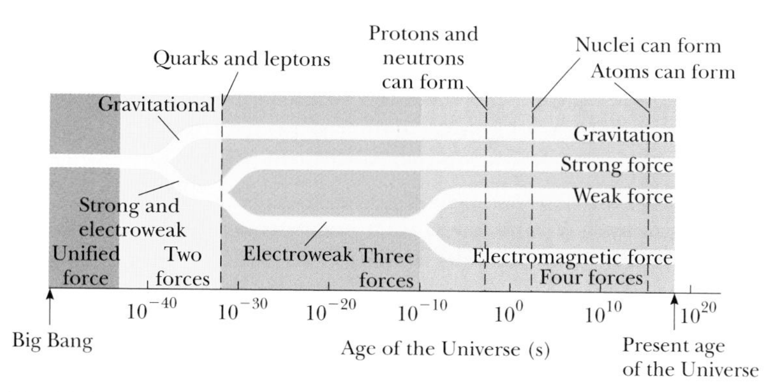

Figure 46.17 A brief history of the Universe from the Big Bang to the present. The four forces became distinguishable during the first nanosecond. Following that, all the quarks combined to form particles that interact via the nuclear force. The leptons, however, remained separate and to this day exist as individual, observable particles.

it is presumed the strong, electroweak, and gravitational forces were joined to form a completely unified force. In the first 10^{-35} s following the Big Bang (the hot epoch, $T \sim 10^{29}$ K), symmetry breaking occurred for gravity while the strong and electroweak forces remained unified. It was a period when particle energies were so great ($> 10^{16}$ GeV) that very massive particles as well as quarks, leptons, and their antiparticles existed. Then, after 10^{-35} s, the Universe rapidly expanded and cooled (the warm epoch, $T \sim 10^{29}$ to 10^{15} K) and the strong and electroweak forces parted company. As the Universe continued to cool, the electroweak force split into the weak force and the electromagnetic force approximately 10^{-10} s after the Big Bang.

After a few minutes, protons and neutrons condensed out of the plasma. For half an hour, the Universe underwent thermonuclear detonation, exploding as a hydrogen bomb and producing most of the helium nuclei that now exist. The Universe continued to expand, and its temperature dropped. Until about 700 000 years after the Big Bang, the Universe was dominated by radiation. Energetic radiation prevented matter from forming single hydrogen atoms because collisions would instantly ionize any atoms that happened to form. Photons experienced continuous Compton scattering from the vast numbers of free electrons, resulting in a Universe that was opaque to radiation. By the time the Universe was about 700 000 years old, it had expanded and cooled to approximately 3 000 K and protons could bind to electrons to form neutral hydrogen atoms. Because of the quantized energies of the atoms, far more wavelengths of radiation were not absorbed by atoms than were absorbed, and the Universe suddenly became transparent to photons. Radiation no longer dominated the Universe, and clumps of neutral matter steadily grew: first atoms, then molecules, gas clouds, stars, and finally galaxies.

Observation of Radiation from the Primordial Fireball

In 1965, Arno A. Penzias (b. 1933) and Robert W. Wilson (b. 1936) of Bell Laboratories were testing a sensitive microwave receiver and made an amazing discovery. A pesky signal producing a faint background hiss was interfering with their satellite communications experiments. The microwave horn that served as their receiving antenna is shown in Figure 46.18. Evicting a flock of pigeons from the 20-ft horn and cooling the microwave detector both failed to remove the signal.

The intensity of the detected signal remained unchanged as the antenna was pointed in different directions. That the radiation had equal strengths in all directions suggested that the entire Universe was the source of this radiation. Ultimately, it became clear that they were detecting microwave background radiation (at a wavelength of 7.35 cm), which represented the leftover "glow" from the Big Bang. Through a casual conversation, Penzias and Wilson discovered that a group

Figure 46.18 Robert W. Wilson *(left)* and Arno A. Penzias with the Bell Telephone Laboratories horn-reflector antenna.

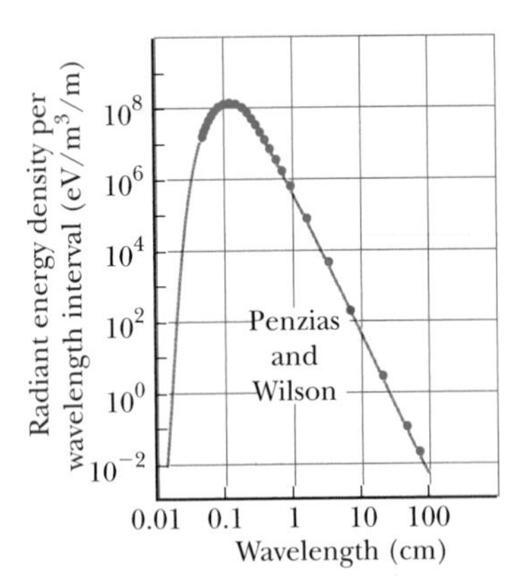

Figure 46.19 Theoretical blackbody (brown curve) and measured radiation spectra (blue points) of the Big Bang. Most of the data were collected from the Cosmic Background Explorer, or COBE, satellite. The datum of Penzias and Wilson is indicated in red.

at Princeton University had predicted the residual radiation from the Big Bang and were planning an experiment to attempt to confirm the theory. The excitement in the scientific community was high when Penzias and Wilson announced that they had already observed an excess microwave background compatible with a 3-K blackbody source, which was consistent with the predicted temperature of the Universe at this time after the Big Bang.

Because Penzias and Wilson made their measurements at a single wavelength, they did not completely confirm the radiation as 3-K blackbody radiation. Subsequent experiments by other groups added intensity data at different wavelengths as shown in Figure 46.19. The results confirm that the radiation is that of a black body at 2.7 K. This figure is perhaps the most clear-cut evidence for the Big Bang theory. The 1978 Nobel Prize in Physics was awarded to Penzias and Wilson for this most important discovery.

The discovery of the cosmic background radiation brought with it a problem, however: the radiation was too uniform. Scientists believed that slight fluctuations in this background had to occur to act as nucleation sites for the formation of the galaxies and other objects we now see in the sky. In 1989, NASA launched a satellite called COBE (KOH-bee), for Cosmic Background Explorer, to study this radiation in greater detail. In 1992, George Smoot (b. 1945) at the Lawrence Berkeley Laboratory found, on the basis of the data collected, that the background was not perfectly uniform but instead contained irregularities that corresponded to temperature variations of 0.000 3 K. Smoot won the 2006 Nobel Prize in Physics for this work. The Wilkinson Microwave Anisotropy Probe, launched in June 2001, exhibits data that allows observation of temperature differences in the cosmos in the microkelvin range.

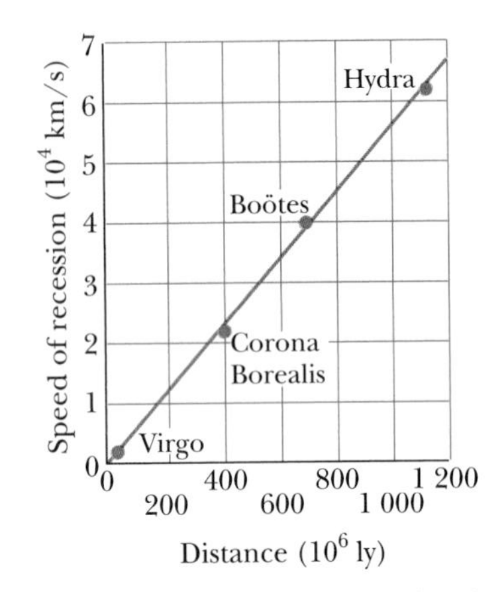

Figure 46.20 Hubble's law: a plot of speed of recession versus distance for four galaxies.

Other Evidence for an Expanding Universe

The Big Bang theory of cosmology predicts that the Universe is expanding. Most of the key discoveries supporting the theory of an expanding Universe were made in the 20th century. Vesto Melvin Slipher (1875–1969), an American astronomer, reported in 1912 that most nebulae are receding from the Earth at speeds up to several million miles per hour. Slipher was one of the first scientists to use Doppler shifts (see Section 17.4) in spectral lines to measure galaxy velocities.

In the late 1920s, Edwin P. Hubble (1889–1953) made the bold assertion that the whole Universe is expanding. From 1928 to 1936, until they reached the limits of the 100-inch telescope, Hubble and Milton Humason (1891–1972) worked at Mount Wilson in California to prove this assertion. The results of that work and of its continuation with the use of a 200-inch telescope in the 1940s showed that the speeds at which galaxies are receding from the Earth increase in direct proportion to their distance R from us (Fig. 46.20). This linear relationship, known as **Hubble's law,** may be written

Hubble's law ▶

$$v = HR \tag{46.4}$$

where H, called the **Hubble constant,** has the approximate value

$$H \approx 17 \times 10^{-3} \text{ m/s} \cdot \text{ly}$$

EXAMPLE 46.5 **Recession of a Quasar**

A quasar is an object that appears similar to a star and is very distant from the Earth. Its speed can be determined from Doppler-shift measurements in the light it emits. A certain quasar recedes from the Earth at a speed of $0.55c$. How far away is it?

SOLUTION

Conceptualize A common mental representation for the Hubble law is that of raisin bread cooking in an oven. Imagine yourself at the center of the loaf of bread. As the entire loaf of bread expands upon heating, raisins near you move slowly with respect to you. Raisins far away from you on the edge of the loaf move at a higher speed.

Categorize We use a concept developed in this section, so we categorize this example as a substitution problem.

Find the distance through Hubble's law:

$$R = \frac{v}{H} = \frac{(0.55)(3.00 \times 10^8 \text{ m/s})}{17 \times 10^{-3} \text{ m/s} \cdot \text{ly}} = 9.7 \times 10^9 \text{ ly}$$

What If? Suppose the quasar has moved at this speed ever since the Big Bang. With this assumption, estimate the age of the Universe.

Answer Let's approximate the distance from Earth to the quasar as the distance the quasar has moved from the singularity since the Big Bang. We can then find the time interval from the particle under constant speed model: $\Delta t = d/v = R/v = 1/H \approx 18$ billion years, which is in approximate agreement with other calculations.

Will the Universe Expand Forever?

In the 1950s and 1960s, Allan R. Sandage (b. 1926) used the 200-inch telescope at Mount Palomar to measure the speeds of galaxies at distances of up to 6 billion light-years away from the Earth. These measurements showed that these very distant galaxies were moving approximately 10 000 km/s faster than Hubble's law predicted. According to this result, the Universe must have been expanding more rapidly 1 billion years ago, and consequently we conclude from these data that the expansion rate is slowing.[6] Today, astronomers and physicists are trying to determine the rate of expansion. If the average mass density of the Universe is less than some critical value, the galaxies will slow in their outward rush but still escape to infinity. If the average density exceeds the critical value, the expansion will eventually stop and contraction will begin, possibly leading to a superdense state followed by another expansion. In this scenario, we have an oscillating Universe.

EXAMPLE 46.6 The Critical Density of the Universe

(A) Starting from energy conservation, derive an expression for the critical mass density of the Universe ρ_c in terms of the Hubble constant H and the universal gravitational constant G.

SOLUTION

Conceptualize Figure 46.21 shows a large section of the Universe, contained within a sphere of radius R. The total mass in this volume is M. A galaxy of mass $m \ll M$ that has a speed v at a distance R from the center of the sphere escapes to infinity (at which its speed approaches zero) if the sum of its kinetic energy and the gravitational potential energy of the system is zero.

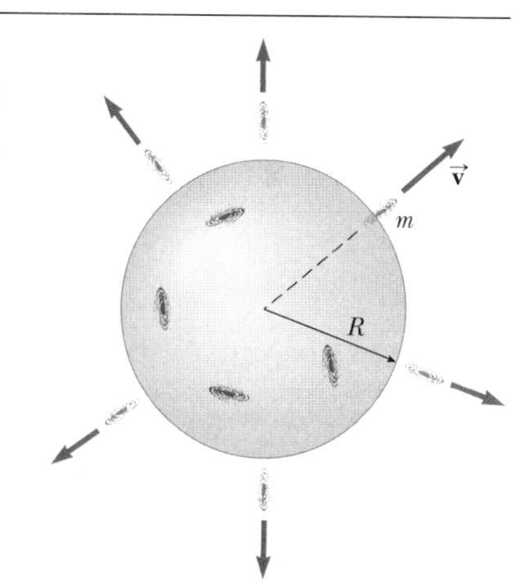

Figure 46.21 (Example 46.6) The galaxy marked with mass m is escaping from a large cluster of galaxies contained within a spherical volume of radius R. Only the mass within R slows the galaxy.

Categorize The Universe may be infinite in spatial extent, but Gauss's law for gravitation (Problem 61 in Chapter 24) implies that only the mass M inside the sphere contributes to the gravitational potential energy of the galaxy–sphere system. Therefore, we categorize this problem as one in which we apply Gauss's law for gravitation. We model the sphere in Figure 46.21 and the escaping galaxy as an isolated system.

Analyze Write an expression for the total mechanical energy of the system and set it equal to zero, representing the galaxy moving at the escape speed:

$$E_{\text{total}} = K + U = \tfrac{1}{2}mv^2 - \frac{GmM}{R} = 0$$

[6] The data at large distances have large observational uncertainties and may be systematically in error from effects such as abnormal brightness in the most distant visible clusters.

Substitute for the mass M contained within the sphere the product of the critical density and the volume of the sphere:

$$\tfrac{1}{2}mv^2 = \frac{Gm(\frac{4}{3}\pi R^3 \rho_c)}{R}$$

Solve for the critical density:

$$\rho_c = \frac{3v^2}{8\pi G R^2}$$

From Hubble's law, substitute for the ratio $v/R = H$:

$$(1) \quad \rho_c = \frac{3}{8\pi G}\left(\frac{v}{R}\right)^2 = \frac{3H^2}{8\pi G}$$

(B) Estimate a numerical value for the critical density in grams per cubic centimeter.

SOLUTION

In Equation (1), substitute numerical values for H and G:

$$\rho_c = \frac{3H^2}{8\pi G} = \frac{3(17 \times 10^{-3}\ \text{m/s}\cdot\text{ly})^2}{8\pi(6.67 \times 10^{-11}\ \text{N}\cdot\text{m}^2/\text{kg}^2)} = 5.17 \times 10^5\ \text{kg/m}\cdot(\text{ly})^2$$

Reconcile the units by converting light-years to meters:

$$\rho_c = 5.7 \times 10^5\ \text{kg/m}\cdot(\text{ly})^2\left(\frac{1\ \text{ly}}{9.46 \times 10^{15}\ \text{m}}\right)^2$$

$$= 6 \times 10^{-27}\ \text{kg/m}^3 = 6 \times 10^{-30}\ \text{g/cm}^3$$

Finalize Because the mass of a hydrogen atom is 1.67×10^{-24} g, this value of ρ_c corresponds to 3×10^{-6} hydrogen atoms per cubic centimeter or 3 atoms per cubic meter.

Missing Mass in the Universe?

The luminous matter in galaxies averages out to a Universe density of $5 \times 10^{-33}\ \text{g/cm}^3$. The radiation in the Universe has a mass equivalent of approximately 2% of the luminous matter. The total mass of all nonluminous matter (such as interstellar gas and black holes) may be estimated from the speeds of galaxies orbiting each other in a cluster. The higher the galaxy speeds, the more mass in the cluster. Measurements on the Coma cluster of galaxies indicate, surprisingly, that the amount of nonluminous matter is 20 to 30 times the amount of luminous matter present in stars and luminous gas clouds. Yet even this large, invisible component of *dark matter*, if extrapolated to the Universe as a whole, leaves the observed mass density a factor of 10 less than ρ_c calculated in Example 46.6. The deficit, called *missing mass*, has been the subject of intense theoretical and experimental work, with exotic particles such as axions, photinos, and superstring particles suggested as candidates for the missing mass. Some researchers have made the more mundane proposal that the missing mass is present in neutrinos. In fact, neutrinos are so abundant that a tiny neutrino rest energy on the order of only 20 eV would furnish the missing mass and "close" the Universe. Current experiments designed to measure the rest energy of the neutrino will have an effect on predictions for the future of the Universe.

Mysterious Energy in the Universe?

A surprising twist in the story of the Universe arose in 1998 with the observation of a class of supernovae that have a fixed absolute brightness. By combining the apparent brightness and the redshift of light from these explosions, their distance and speed of recession from the Earth can be determined. These observations led to the conclusion that the expansion of the Universe is not slowing down, but is accelerating! Observations by other groups also led to the same interpretation.

To explain this acceleration, physicists have proposed *dark energy*, which is energy possessed by the vacuum of space. In the early life of the Universe, gravity dominated over the dark energy. As the Universe expanded and the gravitational force between galaxies became smaller because of the great distances between them, the dark energy became more important. The dark energy results in an effective repulsive force that causes the expansion rate to increase.[7]

Although there is some degree of certainty about the beginning of the Universe, we are uncertain about how the story will end. Will the Universe keep on expanding forever, or will it someday collapse and then expand again, perhaps in an endless series of oscillations? Results and answers to these questions remain inconclusive, and the exciting controversy continues.

46.12 Problems and Perspectives

While particle physicists have been exploring the realm of the very small, cosmologists have been exploring cosmic history back to the first microsecond of the Big Bang. Observation of the events that occur when two particles collide in an accelerator is essential for reconstructing the early moments in cosmic history. For this reason, perhaps the key to understanding the early Universe is to first understand the world of elementary particles. Cosmologists and physicists now find that they have many common goals and are joining hands in an attempt to understand the physical world at its most fundamental level.

Our understanding of physics at short distances is far from complete. Particle physics is faced with many questions. Why does so little antimatter exist in the Universe? Is it possible to unify the strong and electroweak theories in a logical and consistent manner? Why do quarks and leptons form three similar but distinct families? Are muons the same as electrons apart from their difference in mass, or do they have other subtle differences that have not been detected? Why are some particles charged and others neutral? Why do quarks carry a fractional charge? What determines the masses of the elementary constituents of matter? Can isolated quarks exist?

An important and obvious question that remains is whether leptons and quarks have an underlying structure. If they do, we can envision an infinite number of deeper structure levels. If leptons and quarks are indeed the ultimate constituents of matter, however, scientists hope to construct a final theory of the structure of matter, just as Einstein dreamed of doing. This theory, whimsically called the Theory of Everything, is a combination of the Standard Model and a quantum theory of gravity.

String Theory: A New Perspective

Let's briefly discuss one current effort at answering some of these questions by proposing a new perspective on particles. While reading this book, you may recall starting off with the particle model and doing quite a bit of physics with it. In Part 2, we introduced the wave model, and there was more physics to be investigated via the properties of waves. We used a wave model for light in Part 5. Early in Part 6, however, we saw the need to return to the particle model for light. Furthermore, we found that material particles had wave-like characteristics. The quantum particle model discussed in Chapter 40 allowed us to build particles out of waves, suggesting that a wave is the fundamental entity. In this final chapter, however, we introduced elementary particles as the fundamental entities. It seems as if we cannot make up our mind! In this final section, we discuss a current research effort to build particles out of waves and vibrations.

[7] For an overview of dark energy, see S. Perlmutter, "Supernovae, Dark Energy, and the Accelerating Universe," *Physics Today* **56**(4): 53–60, April 2003.

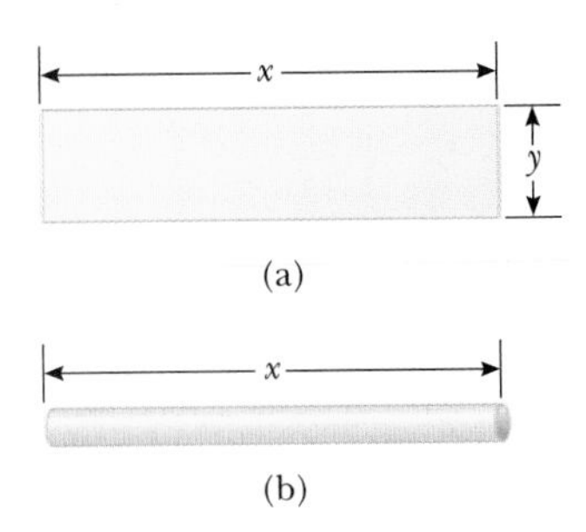

Figure 46.22 (a) A piece of paper is cut into a rectangular shape. As a rectangle, the shape has two dimensions. (b) The paper is rolled up into a soda straw. From far away, it appears to be one-dimensional. The curled-up second dimension is not visible when viewed from a distance that is large compared with the diameter of the straw.

String theory is an effort to unify the four fundamental forces by modeling all particles as various quantized vibrational modes of a single entity, an incredibly small string. The typical length of such a string is on the order of 10^{-35} m, called the **Planck length.** We have seen quantized modes before in the frequencies of vibrating guitar strings in Chapter 18 and the quantized energy levels of atoms in Chapter 42. In string theory, each quantized mode of vibration of the string corresponds to a different elementary particle in the Standard Model.

One complicating factor in string theory is that it requires space–time to have ten dimensions. Despite the theoretical and conceptual difficulties in dealing with ten dimensions, string theory holds promise in incorporating gravity with the other forces. Four of the ten dimensions—three space dimensions and one time dimension—are visible to us. The other six are said to be *compactified;* that is, the six dimensions are curled up so tightly that they are not visible in the macroscopic world.

As an analogy, consider a soda straw. You can build a soda straw by cutting a rectangular piece of paper (Fig. 46.22a), which clearly has two dimensions, and rolling it into a small tube (Fig. 46.22b). From far away, the soda straw looks like a one-dimensional straight line. The second dimension has been curled up and is not visible. String theory claims that six space–time dimensions are curled up in an analogous way, with the curling being on the size of the Planck length and impossible to see from our viewpoint.

Another complicating factor with string theory is that it is difficult for string theorists to guide experimentalists as to what to look for in an experiment. The Planck length is so small that direct experimentation on strings is impossible. Until the theory has been further developed, string theorists are restricted to applying the theory to known results and testing for consistency.

One of the predictions of string theory, called **supersymmetry,** or SUSY, suggests that every elementary particle has a superpartner that has not yet been observed. It is believed that supersymmetry is a broken symmetry (like the broken electroweak symmetry at low energies) and the masses of the superpartners are above our current capabilities of detection by accelerators. Some theorists claim that the mass of superpartners is the missing mass discussed in Section 46.11. Keeping with the whimsical trend in naming particles and their properties, superpartners are given names such as the *squark* (the superpartner to a quark), the *selectron* (electron), and the *gluinos* (gluon).

Other theorists are working on **M-theory,** which is an eleven-dimensional theory based on membranes rather than strings. In a way reminiscent of the correspondence principle, M-theory is claimed to reduce to string theory if one compactifies from eleven dimensions to ten dimensions.

The questions listed at the beginning of this section go on and on. Because of the rapid advances and new discoveries in the field of particle physics, many of these questions may be resolved in the next decade and other new questions may emerge.

Summary

ThomsonNOW™ Sign in at **www.thomsonedu.com** and go to ThomsonNOW to take a practice test for this chapter.

CONCEPTS AND PRINCIPLES

Before quark theory was developed, the four fundamental forces in nature were identified as nuclear, electromagnetic, weak, and gravitational. All the interactions in which these forces take part are mediated by **field particles.** The electromagnetic interaction is mediated by photons; the weak interaction is mediated by the $W^{\pm}$ and Z^0 bosons; the gravitational interaction is mediated by gravitons; and the nuclear interaction is mediated by gluons.

A charged particle and its **antiparticle** have the same mass but opposite charge, and other properties will have opposite values, such as lepton number and baryon number. It is possible to produce particle–antiparticle pairs in nuclear reactions if the available energy is greater than $2mc^2$, where m is the mass of the particle (or antiparticle).

Particles other than field particles are classified as hadrons or leptons. **Hadrons** interact via all four fundamental forces. They have size and structure and are not elementary particles. There are two types, **baryons** and **mesons.** Baryons, which generally are the most massive particles, have nonzero **baryon number** and a spin of $\frac{1}{2}$ or $\frac{3}{2}$. Mesons have baryon number zero and either zero or integral spin.

Leptons have no structure or size and are considered truly elementary. They interact only via the weak, gravitational, and electromagnetic forces. Six types of leptons exist: the electron e^-, the muon μ^-, and the tau τ^-, and their neutrinos ν_e, ν_μ, and ν_τ.

In all reactions and decays, quantities such as energy, linear momentum, angular momentum, electric charge, baryon number, and lepton number are strictly conserved. Certain particles have properties called **strangeness** and **charm.** These unusual properties are conserved in all decays and nuclear reactions except those that occur via the weak force.

Theorists in elementary particle physics have postulated that all hadrons are composed of smaller units known as **quarks,** and experimental evidence agrees with this model. Quarks have fractional electric charge and come in six **flavors:** up (u), down (d), strange (s), charmed (c), top (t), and bottom (b). Each baryon contains three quarks, and each meson contains one quark and one antiquark.

According to the theory of **quantum chromodynamics,** quarks have a property called **color;** the force between quarks is referred to as the **strong force** or the **color force.** The strong force is now considered to be a fundamental force. The nuclear force, which was originally considered to be fundamental, is now understood to be a secondary effect of the strong force due to gluon exchanges between hadrons.

The electromagnetic and weak forces are now considered to be manifestations of a single force called the **electroweak force.** The combination of quantum chromodynamics and the electroweak theory is called the **Standard Model.**

The background microwave radiation discovered by Penzias and Wilson strongly suggests that the Universe started with a Big Bang 10 to 20 billion years ago. The background radiation is equivalent to that of a black body at 3 K. Various astronomical measurements strongly suggest that the Universe is expanding. According to **Hubble's law,** distant galaxies are receding from the Earth at a speed $v = HR$, where H is the **Hubble constant,** $H \approx 17 \times 10^{-3}$ m/s · ly, and R is the distance from the Earth to the galaxy.

Questions

☐ denotes answer available in *Student Solutions Manual/Study Guide;* **O** denotes objective question

1. Name the four fundamental interactions and the field particle that mediates each.
2. **O** When an electron and a positron meet at low speed in empty space, they annihilate each other to produce two 0.511-MeV gamma rays. What law would be violated if they produced one gamma ray with an energy of 1.02 MeV? (a) conservation of energy (b) conservation of momentum (c) conservation of charge (d) conservation of baryon number (e) conservation of electron-lepton number (f) none of these answers
3. What are the differences between hadrons and leptons?
4. ☐ Describe the properties of baryons and mesons and the important differences between them.
5. **O** An isolated stationary muon decays into an electron, an electron antineutrino, and a muon neutrino. Is the total kinetic energy of these three particles (a) zero, (b) small, or (c) large compared to their rest energies, or (d) is any of these choices possible?
6. **O** The Ω^- particle is a baryon with spin $\frac{3}{2}$. The Ω^- particle has (a) three possible spin states in a magnetic field, (b) four possible spin states, (c) three times the charge of a spin-$\frac{1}{2}$ particle, (d) three times the mass of a spin-$\frac{1}{2}$ particle, or (e) none of these choices.
7. Kaons all decay into final states that contain no protons or neutrons. What is the baryon number for kaons?
8. ☐ The Ξ^0 particle decays by the weak interaction according to the decay mode $\Xi^0 \rightarrow \Lambda^0 + \pi^0$. Would you expect this decay to be fast or slow? Explain.
9. **O** What interactions affect protons in an atomic nucleus? Choose all correct answers. (a) the nuclear interaction (b) the weak interaction (c) the electromagnetic interaction (d) the gravitational interaction
10. Discuss the following conservation laws: energy, linear momentum, angular momentum, electric charge, baryon number, lepton number, and strangeness. Are all these laws based on fundamental properties of nature? Explain.
11. An antibaryon interacts with a meson. Can a baryon be produced in such an interaction? Explain.
12. Describe the essential features of the Standard Model of particle physics.
13. ☐ How many quarks are in each of the following: (a) a baryon, (b) an antibaryon, (c) a meson, (d) an antimeson? How do you explain that baryons have half-integral spins whereas mesons have spins of 0 or 1? *Note:* Quarks have spin $\frac{1}{2}$.
14. In the theory of quantum chromodynamics, quarks come in three colors. How would you justify the statement that "all baryons and mesons are colorless"?
15. The W and Z bosons were first produced at CERN in 1983 by causing a beam of protons and a beam of antiprotons to meet at high energy. Why was this discovery important?
16. **O** In one experiment, two balls of clay of the same mass travel with the same speed v toward each other. They collide head-on and come to rest. In a second experiment, two clay balls of the same mass are again used. One ball hangs at rest, suspended from the ceiling by a thread. The second ball is fired toward the first at speed v, to collide, stick to the first ball, and continue to move forward. Is the kinetic energy that is transformed into internal energy in the first experiment (a) one-fourth as much as in the second experiment, (b) one-half as much as in the second experiment, (c) the same as in the second experiment, (d) twice as much as in the second experiment, (e) four times as much as in the second experiment, or (f) none of these choices?
17. How did Edwin Hubble determine in 1928 that the Universe is expanding?
18. **O** Place the following events into the correct sequence from the earliest in the history of the Universe to the latest. (a) Neutral atoms form. (b) Protons and neutrons are no longer annihilated as fast as they form. (c) The Universe is a quark–gluon soup. (d) The Universe is like the core of a normal star today, forming helium by nuclear fusion. (e) The Universe is like the surface of a hot star today, consisting of a plasma of ionized atoms. (f) Polyatomic molecules form. (g) Solid materials form.
19. Neutral atoms did not exist until hundreds of thousands of years after the Big Bang. Why?
20. **O** Define the average density of the Solar System ρ_{SS} as the total mass of the Sun, planets, satellites, rings, asteroids, icy outliers, and comets, divided by the volume of a sphere around the Sun large enough to contain all these objects. The sphere extends about halfway to the nearest star, with a radius of approximately 2×10^{16} m, about two light-years. How does this average density of the Solar System compare with the critical density ρ_c required for the Universe to stop its Hubble's-law expansion? (a) ρ_{SS} is much greater than ρ_c. (b) ρ_{SS} is approximately or precisely equal to ρ_c. (c) ρ_{SS} is much less than ρ_c. (d) It is impossible to determine.
21. **Review question.** A girl and her grandmother grind corn while the woman tells the girl stories about what is most important. A boy keeps crows away from ripening corn while his grandfather sits in the shade and explains to him the Universe and his place in it. What the children do not understand this year they will better understand next year. Now you must take the part of the adults. State the most general, most fundamental, most universal truths you know. If you need to repeat someone else's ideas, get the best version of those ideas you can and state your source. If you do not understand something, make a plan to understand it better within the next year.

Problems

WebAssign The Problems from this chapter may be assigned online in WebAssign.

ThomsonNOW Sign in at **www.thomsonedu.com** and go to ThomsonNOW to assess your understanding of this chapter's topics with additional quizzing and conceptual questions.

1, 2, 3 denotes straightforward, intermediate, challenging; □ denotes full solution available in *Student Solutions Manual/Study Guide;* ▲ denotes coached solution with hints available at **www.thomsonedu.com;** denotes developing symbolic reasoning; ● denotes asking for qualitative reasoning; denotes computer useful in solving problem

Section 46.1 The Fundamental Forces in Nature

Section 46.2 Positrons and Other Antiparticles

1. A photon produces a proton–antiproton pair according to the reaction $\gamma \rightarrow p + \bar{p}$. What is the minimum possible frequency of the photon? What is its wavelength?

2. At some time in your life, you may find yourself in a hospital to have a PET, or positron-emission tomography, scan. In the procedure, a radioactive element that undergoes e^+ decay is introduced into your body. The equipment detects the gamma rays that result from pair annihilation when the emitted positron encounters an electron in your body's tissue. Suppose you receive an injection of glucose containing on the order of 10^{10} atoms of ^{14}O, with half-life 70.6 s. Assume the oxygen remaining after 5 min is uniformly distributed through 2 L of blood. What is then the order of magnitude of the oxygen atoms' activity in 1 cm^3 of the blood?

3. *Your two cents' worth.* Model a penny as 3.10 g of copper. Consider an anti-penny minted from 3.10 g of copper anti-atoms, each with 29 positrons in orbit around a nucleus comprising 29 antiprotons and 34 or 36 antineutrons. (a) Find the energy released if the two coins collide. (b) Find the value of this energy at the unit price of \$0.14/kWh, a representative retail rate for energy from the electric company.

4. Two photons are produced when a proton and antiproton annihilate each other. In the reference frame in which the center of mass of the proton–antiproton system is stationary, what are the minimum frequency and corresponding wavelength of each photon?

5. A photon with an energy $E_\gamma = 2.09$ GeV creates a proton–antiproton pair in which the proton has a kinetic energy of 95.0 MeV. What is the kinetic energy of the antiproton? *Note:* $m_pc^2 = 938.3$ MeV.

Section 46.3 Mesons and the Beginning of Particle Physics

6. Occasionally, high-energy muons collide with electrons and produce two neutrinos according to the reaction $\mu^+ + e^- \rightarrow 2\nu$. What kind of neutrinos are they?

7. ▲ One mediator of the weak interaction is the Z^0 boson, with mass 91 GeV/c^2. Use this information to find the order of magnitude of the range of the weak interaction.

8. ● (a) Prove that the exchange of a virtual particle of mass m can be associated with a force with a range given by

$$d = \frac{1\,240 \text{ eV} \cdot \text{nm}}{4\pi mc^2} = \frac{98.7 \text{ eV} \cdot \text{nm}}{mc^2}$$

(b) State the pattern of dependence of the range on the mass. (c) Of the interactions listed in Table 46.1, which are associated with field particles according to this rule and which are not? Explain your answer. (d) What is the range of the force that might be produced by the virtual exchange of a proton?

9. ▲ A neutral pion at rest decays into two photons according to $\pi^0 \rightarrow \gamma + \gamma$. Find the energy, momentum, and frequency of each photon.

10. When a high-energy proton or pion traveling near the speed of light collides with a nucleus, it travels an average distance of 3×10^{-15} m before interacting. From this information, find the order of magnitude of the time interval required for the strong interaction to occur.

11. ● A free neutron beta decays by creating a proton, an electron, and an antineutrino according to the reaction $n \rightarrow p + e^- + \bar{\nu}$. **What If?** Imagine that a free neutron were to decay by creating a proton and electron according to the reaction $n \rightarrow p + e^-$ and assume the neutron is initially at rest in the laboratory. (a) Determine the energy released in this reaction. (b) Energy and momentum are conserved in the reaction. Determine the speeds of the proton and electron after the reaction. (c) Is either of these particles moving at a relativistic speed? Explain.

Section 46.4 Classification of Particles

12. Identify the unknown particle on the left side of the reaction $? + p \rightarrow n + \mu^+$.

Section 46.5 Conservation Laws

13. Each of the following reactions is forbidden. Determine a conservation law that is violated for each reaction.
(a) $p + \bar{p} \rightarrow \mu^+ + e^-$
(b) $\pi^- + p \rightarrow p + \pi^+$
(c) $p + p \rightarrow p + \pi^+$
(d) $p + p \rightarrow p + p + n$
(e) $\gamma + p \rightarrow n + \pi^0$

14. ● (a) Show that baryon number and charge are conserved in the following reactions of a pion with a proton:

$$(1) \quad \pi^+ + p \rightarrow K^+ + \Sigma^+$$
$$(2) \quad \pi^+ + p \rightarrow \pi^+ + \Sigma^+$$

(b) The first reaction is observed, but the second never occurs. Explain.

15. ▲ The following reactions or decays involve one or more neutrinos. In each case, supply the missing neutrino (ν_e, ν_μ, or ν_τ) or antineutrino.
(a) $\pi^- \rightarrow \mu^- + ?$ (b) $K^+ \rightarrow \mu^+ + ?$
(c) $? + p \rightarrow n + e^+$ (d) $? + n \rightarrow p + e^-$
(e) $? + n \rightarrow p + \mu^-$ (f) $\mu^- \rightarrow e^- + ? + ?$

16. The first of the following two reactions can occur, but the second cannot. Explain.

$$\mathrm{K_S^0} \rightarrow \pi^+ + \pi^- \quad \text{(can occur)}$$

$$\Lambda^0 \rightarrow \pi^+ + \pi^- \quad \text{(cannot occur)}$$

17. ▲ Determine which of the following reactions can occur. For those that cannot occur, determine the conservation law (or laws) violated.
(a) $\mathrm{p} \rightarrow \pi^+ + \pi^0$ (b) $\mathrm{p} + \mathrm{p} \rightarrow \mathrm{p} + \mathrm{p} + \pi^0$
(c) $\mathrm{p} + \mathrm{p} \rightarrow \mathrm{p} + \pi^+$ (d) $\pi^+ \rightarrow \mu^+ + \nu_\mu$
(e) $\mathrm{n} \rightarrow \mathrm{p} + \mathrm{e}^- + \bar{\nu}_e$ (f) $\pi^+ \rightarrow \mu^+ + \mathrm{n}$

18. A $\mathrm{K_S^0}$ particle at rest decays into a π^+ and a π^-. What is the speed of each of the pions? The mass of the $\mathrm{K_S^0}$ is 497.7 MeV/c^2, and the mass of each π is 139.6 MeV/c^2.

19. (a) Show that the proton-decay reaction $\mathrm{p} \rightarrow \mathrm{e}^+ + \gamma$ cannot occur because it violates conservation of baryon number. (b) **What If?** Imagine that this reaction does occur, and the proton is initially at rest. Determine the energy and momentum of the positron and photon after the reaction. *Suggestion:* Recall that energy and momentum must be conserved in the reaction. (c) Determine the speed of the positron after the reaction.

20. Determine the type of neutrino or antineutrino involved in each of the following processes:
(a) $\pi^+ \rightarrow \pi^0 + \mathrm{e}^+ + ?$ (b) $? + \mathrm{p} \rightarrow \mu^- + \mathrm{p} + \pi^+$
(c) $\Lambda^0 \rightarrow \mathrm{p} + \mu^- + ?$ (d) $\tau^+ \rightarrow \mu^+ + ? + ?$

Section 46.6 Strange Particles and Strangeness

21. Determine whether or not strangeness is conserved in the following decays and reactions:
(a) $\Lambda^0 \rightarrow \mathrm{p} + \pi^-$ (b) $\pi^- + \mathrm{p} \rightarrow \Lambda^0 + \mathrm{K}^0$
(c) $\bar{\mathrm{p}} + \mathrm{p} \rightarrow \bar{\Lambda}^0 + \Lambda^0$ (d) $\pi^- + \mathrm{p} \rightarrow \pi^- + \Sigma^+$
(e) $\Xi^- \rightarrow \Lambda^0 + \pi^-$ (f) $\Xi^0 \rightarrow \mathrm{p} + \pi^-$

22. ● The neutral meson ρ^0 decays by the strong interaction into two pions:

$$\rho^0 \rightarrow \pi^+ + \pi^- \quad (T_{1/2} \sim 10^{-23}\ \mathrm{s})$$

The neutral kaon also decays into two pions:

$$\mathrm{K_S^0} \rightarrow \pi^+ + \pi^- \quad (T_{1/2} \sim 10^{-10}\ \mathrm{s})$$

How do you explain the difference in half-lives?

23. For each of the following forbidden decays, determine which conservation law is violated:
(a) $\mu^- \rightarrow \mathrm{e}^- + \gamma$ (b) $\mathrm{n} \rightarrow \mathrm{p} + \mathrm{e}^- + \nu_e$
(c) $\Lambda^0 \rightarrow \mathrm{p} + \pi^0$ (d) $\mathrm{p} \rightarrow \mathrm{e}^+ + \pi^0$
(e) $\Xi^0 \rightarrow \mathrm{n} + \pi^0$

24. Which of the following processes are allowed by the strong interaction, the electromagnetic interaction, the weak interaction, or no interaction at all?
(a) $\pi^- + \mathrm{p} \rightarrow 2\eta$ (b) $\mathrm{K}^- + \mathrm{n} \rightarrow \Lambda^0 + \pi^-$
(c) $\mathrm{K}^- \rightarrow \pi^- + \pi^0$ (d) $\Omega^- \rightarrow \Xi^- + \pi^0$
(e) $\eta \rightarrow 2\gamma$

25. Identify the conserved quantities in the following processes:
(a) $\Xi^- \rightarrow \Lambda^0 + \mu^- + \nu_\mu$ (b) $\mathrm{K_S^0} \rightarrow 2\pi^0$
(c) $\mathrm{K}^- + \mathrm{p} \rightarrow \Sigma^0 + \mathrm{n}$ (d) $\Sigma^0 \rightarrow \Lambda^0 + \gamma$
(e) $\mathrm{e}^+ + \mathrm{e}^- \rightarrow \mu^+ + \mu^-$ (f) $\bar{\mathrm{p}} + \mathrm{n} \rightarrow \bar{\Lambda}^0 + \Sigma^-$

26. Fill in the missing particle. Assume (a) occurs via the strong interaction and (b) and (c) involve the weak interaction.
(a) $\mathrm{K}^+ + \mathrm{p} \rightarrow ? + \mathrm{p}$ (b) $\Omega^- \rightarrow ? + \pi^-$
(c) $\mathrm{K}^+ \rightarrow ? + \mu^+ + \nu_\mu$

27. If a $\mathrm{K_S^0}$ meson at rest decays in 0.900×10^{-10} s, how far does a $\mathrm{K_S^0}$ meson travel if it is moving at $0.960c$?

28. The particle decay $\Sigma^+ \rightarrow \pi^+ + \mathrm{n}$ is observed in a bubble chamber. Figure P46.28 represents the curved tracks of the particles Σ^+ and π^+, and the invisible track of the neutron, in the presence of a uniform magnetic field of 1.15 T directed out of the page. The measured radii of curvature are 1.99 m for the Σ^+ particle and 0.580 m for the π^+ particle. (a) Find the momenta of the Σ^+ and the π^+ particles in units of MeV/c. (b) The angle between the momenta of the Σ^+ and the π^+ particles at the moment of decay is 64.5°. Find the momentum of the neutron. (c) Calculate the total energy of the π^+ particle and of the neutron from their known masses (m_π = 139.6 MeV/c^2, m_n = 939.6 MeV/c^2) and the relativistic energy–momentum relation. What is the total energy of the Σ^+ particle? (d) Calculate the mass and speed of the Σ^+ particle.

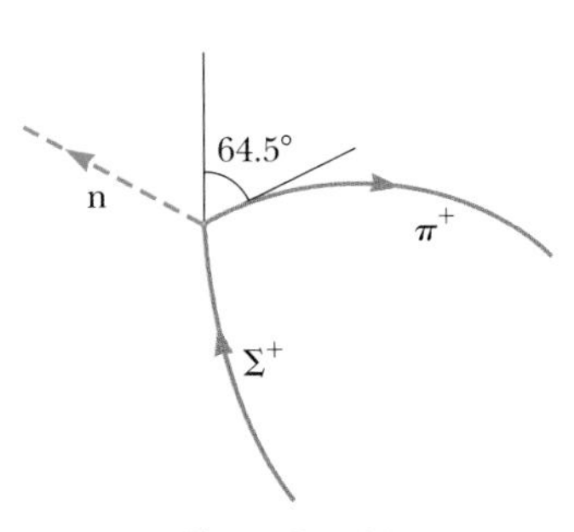

Figure P46.28

29. A particle of mass m_1 is fired at a stationary particle of mass m_2, and a reaction takes place in which new particles are created out of the incident kinetic energy. Taken together, the product particles have total mass m_3. The minimum kinetic energy the bombarding particle must have so as to induce the reaction is called the threshold energy. At this energy, the kinetic energy of the products is a minimum, so the fraction of the incident kinetic energy that is available to create new particles is a maximum. This condition is met when all the product particles have the same velocity and the particles have no kinetic energy of motion relative to one another. (a) By using conservation of relativistic energy and momentum, and the relativistic energy–momentum relation, show that the threshold energy is

$$K_{\min} = \frac{[m_3{}^2 - (m_1 + m_2)^2]c^2}{2m_2}$$

Calculate the threshold energy for each of the following reactions: (b) $\mathrm{p} + \mathrm{p} \rightarrow \mathrm{p} + \mathrm{p} + \mathrm{p} + \bar{\mathrm{p}}$ (One of the initial protons is at rest. Antiprotons are produced.) (c) $\pi^- + \mathrm{p} \rightarrow \mathrm{K}^0 + \Lambda^0$ (The proton is at rest. Strange particles are produced.) (d) $\mathrm{p} + \mathrm{p} \rightarrow \mathrm{p} + \mathrm{p} + \pi^0$ (One of the initial protons is at rest. Pions are produced.) (e) $\mathrm{p} + \bar{\mathrm{p}} \rightarrow \mathrm{Z}^0$ (One of the initial particles is at rest. Z^0 particles of mass 91.2 GeV/c^2 are produced.)

Section 46.7 Finding Patterns in the Particles

Section 46.8 Quarks

Section 46.9 Multicolored Quarks

Section 46.10 The Standard Model

Note: Problem 64 in Chapter 39 can be assigned with Section 46.10.

30. (a) Find the number of electrons and the number of each species of quarks in 1 L of water. (b) Make an order-of-magnitude estimate of the number of each kind of fundamental matter particle in your body. State your assumptions and the quantities you take as data.

31. The quark composition of the proton is uud and that of the neutron is udd. Show that in each case the charge, baryon number, and strangeness of the particle equal, respectively, the sums of these numbers for the quark constituents.

32. **What If?** Imagine that binding energies could be ignored. Find the masses of the u and d quarks from the masses of the proton and neutron.

33. The quark compositions of the K^0 and Λ^0 particles are $\bar{s}d$ and uds, respectively. Show that the charge, baryon number, and strangeness of these particles equal, respectively, the sums of these numbers for the quark constituents.

34. The reaction $\pi^- + p \rightarrow K^0 + \Lambda^0$ occurs with high probability, whereas the reaction $\pi^- + p \rightarrow K^0 + n$ never occurs. Analyze these reactions at the quark level. Show that the first reaction conserves the total number of each type of quark and the second reaction does not.

35. Analyze each of the following reactions in terms of constituent quarks:
(a) $\pi^- + p \rightarrow K^0 + \Lambda^0$
(b) $\pi^+ + p \rightarrow K^+ + \Sigma^+$
(c) $K^- + p \rightarrow K^+ + K^0 + \Omega^-$
(d) $p + p \rightarrow K^0 + p + \pi^+ + ?$
In the last reaction, identify the mystery particle.

36. A Σ^0 particle traveling through matter strikes a proton; then a Σ^+ and a gamma ray, as well as a third particle, emerge. Use the quark model of each to determine the identity of the third particle.

37. Identify the particles corresponding to the quark combinations (a) suu, (b) $\bar{u}d$, (c) $\bar{s}d$, and (d) ssd.

38. What is the electrical charge of the baryons with the quark compositions (a) $\bar{u}\bar{u}\bar{d}$ and (b) $\bar{u}\bar{d}\bar{d}$? What are these baryons called?

Section 46.11 The Cosmic Connection

Note: Problem 17 in Chapter 39 can be assigned with this section.

39. Imagine that all distances expand at a rate described by the Hubble constant of 17.0×10^{-3} m/s · ly. (a) At what rate would the 1.85-m height of a basketball player be increasing? (b) At what rate would the distance between the Earth and the Moon be increasing? In fact, gravitation and other forces prevent the Hubble's-law expansion from taking place except in systems larger than clusters of galaxies.

40. Review problem. Refer to Section 39.4. Prove that the Doppler shift in wavelength of electromagnetic waves is described by

$$\lambda' = \lambda\sqrt{\frac{1 + v/c}{1 - v/c}}$$

where λ' is the wavelength measured by an observer moving at speed v away from a source radiating waves of wavelength λ.

41. ▲ A distant quasar is moving away from the Earth at such high speed that the blue 434-nm H_γ line of hydrogen is observed at 510 nm, in the green portion of the spectrum (Fig. P46.41). (a) How fast is the quasar receding? You may use the result of Problem 40. (b) Edwin Hubble discovered that all objects outside the local group of galaxies are moving away from us, with speeds proportional to their distances. Hubble's law is expressed as $v = HR$, where Hubble's constant has the approximate value $H = 17 \times 10^{-3}$ m/s·ly. Determine the distance from the Earth to this quasar.

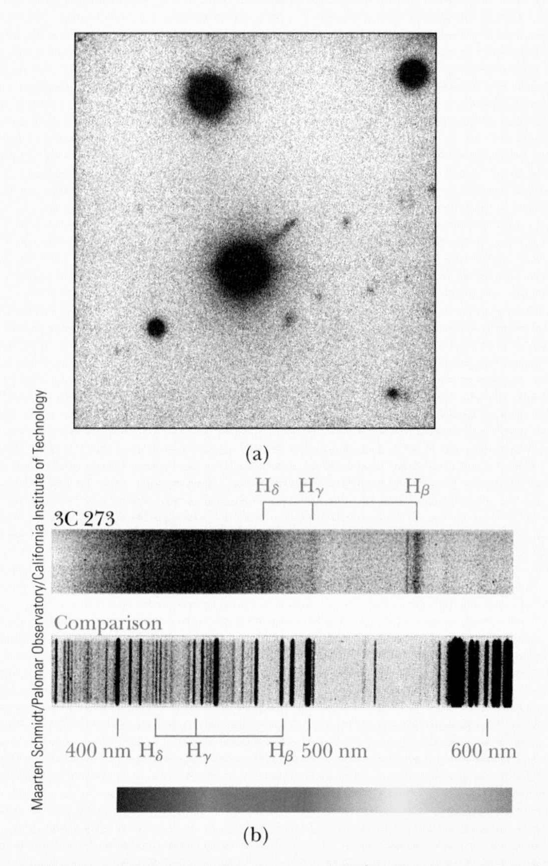

Figure P46.41 (a) Image of the quasar 3C273. (b) Spectrum of the quasar above a comparison spectrum emitted by stationary hydrogen and helium atoms. Both parts of the figure are printed as black-and-white photographic negatives to reveal detail.

42. The various spectral lines observed in the light from a distant quasar have longer wavelengths λ_n' than the wavelengths λ_n measured in light from a stationary source. Here n is an index taking different values for different spectral lines. The fractional change in wavelength toward

the red is the same for all spectral lines. That is, the redshift parameter Z defined by

$$Z = \frac{\lambda'_n - \lambda_n}{\lambda_n}$$

is common to all spectral lines for one object. In terms of Z, determine (a) the speed of recession of the quasar and (b) the distance from the Earth to this quasar. Use the result of Problem 40 and Hubble's law.

43. Using Hubble's law, find the wavelength of the 590-nm sodium line emitted from galaxies (a) 2.00×10^6 ly away from the Earth, (b) 2.00×10^8 ly away, and (c) 2.00×10^9 ly away. You may use the result of Problem 40.

44. ● The visible section of the Universe is a sphere centered on the bridge of your nose, with radius 13.7 billion light-years. (a) Explain why the visible Universe is getting larger, with its radius increasing by one light-year in every year. (b) Find the rate at which the volume of the visible section of the Universe is increasing.

45. ● Assume dark matter exists throughout space with a uniform density of 6.00×10^{-28} kg/m^3. (a) Find the amount of such dark matter inside a sphere centered on the Sun, having the Earth's orbit as its equator. (b) Explain whether the gravitational field of this dark matter would have a measurable effect on the Earth's revolution.

46. **Review problem.** The cosmic background radiation is blackbody radiation from a source at a temperature of 2.73 K. (a) Use Wien's law to determine the wavelength at which this radiation has its maximum intensity. (b) In what part of the electromagnetic spectrum is the peak of the distribution?

47. **Review problem.** Use Stefan's law to find the intensity of the cosmic background radiation emitted by the fireball of the Big Bang at a temperature of 2.73 K.

48. It is mostly your roommate's fault. Nosy astronomers have discovered enough junk and clutter in your dorm room to constitute the missing mass required to close the Universe. After observing your floor, closet, bed, and computer files, they extrapolate to slobs in other galaxies and calculate the average density of the observable Universe as $1.20\rho_c$. How many times larger will the Universe become before it begins to collapse? That is, by what factor will the distance between remote galaxies increase in the future?

49. The early Universe was dense with gamma-ray photons of energy $\sim k_B T$ and at such a high temperature that protons and antiprotons were created by the process $\gamma \rightarrow \text{p} + \bar{\text{p}}$ as rapidly as they annihilated each other. As the Universe cooled in adiabatic expansion, its temperature fell below a certain value and proton pair production became rare. At that time, slightly more protons than antiprotons existed, and essentially all the protons in the Universe today date from that time. (a) Estimate the order of magnitude of the temperature of the Universe when protons condensed out. (b) Estimate the order of magnitude of the temperature of the Universe when electrons condensed out.

50. If the average density of the Universe is small compared with the critical density, the expansion of the Universe described by Hubble's law proceeds with speeds that are nearly constant over time. (a) Prove that in this case the age of the Universe is given by the inverse of Hubble's constant. (b) Calculate $1/H$ and express it in years.

51. Assume the average density of the Universe is equal to the critical density. (a) Prove that the age of the Universe is given by $2/3H$. (b) Calculate $2/3H$ and express it in years.

52. Hubble's law can be stated in vector form as $\vec{\mathbf{v}} = H\vec{\mathbf{R}}$. Outside the local group of galaxies, all objects are moving away from us with velocities proportional to their positions relative to us. In this form, it sounds as if our location in the Universe is specially privileged. Prove that Hubble's law is equally true for an observer elsewhere in the Universe. Proceed as follows. Assume we are at the origin of coordinates, one galaxy cluster is at location $\vec{\mathbf{R}}_1$ and has velocity $\vec{\mathbf{v}}_1 = H\vec{\mathbf{R}}_1$ relative to us, and another galaxy cluster has position vector $\vec{\mathbf{R}}_2$ and velocity $\vec{\mathbf{v}}_2 = H\vec{\mathbf{R}}_2$. Suppose the speeds are nonrelativistic. Consider the frame of reference of an observer in the first of these galaxy clusters. Show that our velocity relative to her, together with the position vector of our galaxy cluster from hers, satisfies Hubble's law. Show that the position and velocity of cluster 2 relative to cluster 1 satisfy Hubble's law.

Section 46.12 Problems and Perspectives

53. ● Classical general relativity views the structure of space–time as deterministic and well defined down to arbitrarily small distances. On the other hand, quantum general relativity forbids distances smaller than the Planck length given by $L = (\hbar G/c^3)^{1/2}$. (a) Calculate the value of the Planck length. The quantum limitation suggests that after the Big Bang, when all the presently observable section of the Universe was contained within a point-like singularity, nothing could be observed until that singularity grew larger than the Planck length. Because the size of the singularity grew at the speed of light, we can infer that no observations were possible during the time interval required for light to travel the Planck length. (b) Calculate this time interval, known as the Planck time T, and state how it compares with the ultrahot epoch mentioned in the text. (c) Does this reasoning suggest we may never know what happened between the time $t = 0$ and the time $t = T$? Explain.

Additional Problems

54. **Review problem.** Supernova Shelton 1987A, located approximately 170 000 ly from the Earth, is estimated to have emitted a burst of neutrinos carrying energy $\sim 10^{46}$ J (Fig. P46.54). Suppose the average neutrino energy was 6 MeV and your mother's body presented cross-sectional area 5 000 cm^2. To an order of magnitude, how many of these neutrinos passed through her?

55. The most recent naked-eye supernova was Supernova Shelton 1987A (Fig. P46.54). It was 170 000 ly away in the next galaxy to ours, the Large Magellanic Cloud. Approximately 3 h before its optical brightening was noticed, two neutrino detection experiments simultaneously registered the first neutrinos from an identified source other than the Sun. The Irvine–Michigan–Brookhaven experiment in a salt mine in Ohio registered 8 neutrinos over a 6-s period, and the Kamiokande II experiment in a zinc mine in Japan counted 11 neutrinos in 13 s. (Because the supernova is far south in the sky, these neutrinos entered

Figure P46.54 (Problems 54 and 55) The giant star Sanduleak $-69°$ 202 in the "before" picture became Supernova Shelton 1987A in the "after" picture.

the detectors from below. They passed through the Earth before they were by chance absorbed by nuclei in the detectors.) The neutrino energies were between approximately 8 MeV and 40 MeV. If neutrinos have no mass, neutrinos of all energies should travel together at the speed of light, and the data are consistent with this possibility. The arrival times could vary simply because neutrinos were created at different moments as the core of the star collapsed into a neutron star. If neutrinos have nonzero mass, lower-energy neutrinos should move comparatively slowly. The data are consistent with a 10-MeV neutrino requiring at most approximately 10 s more than a photon would require to travel from the supernova to us. Find the upper limit that this observation sets on the mass of a neutrino. Other evidence sets an even tighter limit.

56. Name at least one conservation law that prevents each of the following reactions: (a) $\pi^- + p \rightarrow \Sigma^+ + \pi^0$ (b) $\mu^- \rightarrow \pi^- + \nu_e$ (c) $p \rightarrow \pi^+ + \pi^+ + \pi^-$

57. ▲ The energy flux carried by neutrinos from the Sun is estimated to be on the order of 0.4 W/m^2 at the Earth's surface. Estimate the fractional mass loss of the Sun over 10^9 yr due to the emission of neutrinos. The mass of the Sun is 2×10^{30} kg. The Earth–Sun distance is 1.5×10^{11} m.

58. Two protons approach each other head-on, each with 70.4 MeV of kinetic energy, and engage in a reaction in which a proton and positive pion emerge at rest. What third particle, obviously uncharged and therefore difficult to detect, must have been created?

59. A rocket engine for space travel using photon drive and matter–antimatter annihilation has been suggested. Suppose the fuel for a short-duration burn consists of N protons and N antiprotons, each with mass m. (a) Assume all the fuel is annihilated to produce photons. When the photons are ejected from the rocket, what momentum can be imparted to it? (b) **What If?** If half of the protons and antiprotons annihilate each other and the energy released is used to eject the remaining particles, what momentum could be given to the rocket? (c) Which scheme results in the greater change in speed for the rocket?

60. A gamma-ray photon strikes a stationary electron. Determine the minimum gamma-ray energy to make the following reaction occur:

$$\gamma + e^- \rightarrow e^- + e^- + e^+$$

61. Determine the kinetic energies of the proton and pion resulting from the decay of a Λ^0 at rest:

$$\Lambda^0 \rightarrow p + \pi^-$$

62. Two protons approach each other with velocities of equal magnitude in opposite directions. What is the minimum kinetic energy of each proton if they are to produce a π^+ meson at rest in the following reaction?

$$p + p \rightarrow p + n + \pi^+$$

63. A Σ^0 particle at rest decays according to $\Sigma^0 \rightarrow \Lambda^0 + \gamma$. Find the gamma-ray energy.

64. An unstable particle, initially at rest, decays into a proton (rest energy 938.3 MeV) and a negative pion (rest energy 139.6 MeV). A uniform magnetic field of 0.250 T exists perpendicular to the velocities of the created particles. The radius of curvature of each track is found to be 1.33 m. What is the mass of the original unstable particle?

65. A π-meson at rest decays according to $\pi^- \rightarrow \mu^- + \bar{\nu}_\mu$. What is the energy carried off by the neutrino? Assume the neutrino has no mass and moves off with the speed of light. Take $m_\pi c^2 = 139.6$ MeV and $m_\mu c^2 = 105.7$ MeV.

66. **Review problem.** Use the Boltzmann distribution function e^{-E/k_BT} to calculate the temperature at which 1.00% of a population of photons has energy greater than 1.00 eV. The energy required to excite an atom is on the order of 1 eV. Therefore, as the temperature of the Universe fell below the value you calculate, neutral atoms could form from plasma and the Universe became transparent. The cosmic background radiation represents our vastly red-shifted view of the opaque fireball of the Big Bang as it was at this time and temperature. The fireball surrounds us; we are embers.

67. What processes are described by the Feynman diagrams in Figure P46.67? What is the exchanged particle in each process?

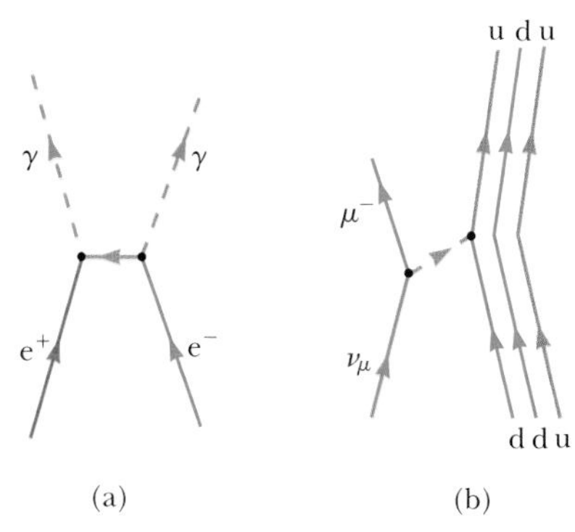

Figure P46.67

68. Identify the mediators for the two interactions described in the Feynman diagrams shown in Figure P46.68.

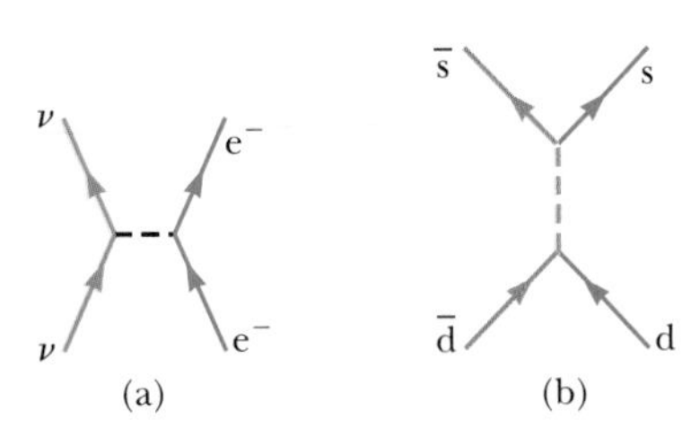

Figure P46.68

69. The cosmic rays of highest energy are mostly protons, accelerated by unknown sources. Their spectrum shows a cutoff at an energy on the order of 10^{20} eV. Above that energy, a proton interacts with a photon of cosmic microwave background radiation to produce mesons, for example according to $p + \gamma \rightarrow p + \pi^0$. Demonstrate this fact by taking the following steps. (a) Find the minimum photon energy required to produce this reaction in the reference frame where the total momentum of the photon–proton system is zero. The reaction was observed experimentally in the 1950s with photons of a few hundred MeV. (b) Use Wien's displacement law to find the wavelength of a photon at the peak of the blackbody spectrum of the primordial microwave background radiation, with a temperature of 2.73 K. (c) Find the energy of this photon. (d) Consider the reaction in part (a) in a moving reference frame so that the photon is the same as that in part (c). Calculate the energy of the proton in this frame, which represents the Earth reference frame.

Answers to Quick Quizzes

46.1 (a). The right-hand rule for the positive particle tells you that into the page is the direction that leads to a force directed toward the path's center of curvature.

46.2 **(i),** (c), (d). There is a baryon, the neutron, on the left of the reaction, but no baryon on the right. Therefore, baryon number is not conserved. The neutron has spin $\frac{1}{2}$. On the right side of the reaction, the pions each have integral spin and the combination of two muons must also have integral spin. Therefore, the total spin of the particles on the right-hand side must be integral and angular momentum is not conserved. **(ii),** (a). The sum of the proton and pion masses is larger than the mass of the neutron, so energy conservation is violated.

46.3 (b), (e), (f). The pion on the left has integral spin, whereas the three spin-$\frac{1}{2}$ leptons on the right must result in a total spin that is half-integral. Therefore, angular momentum is not conserved. Electron lepton number is zero on the left and -1 on the right. There are no muons on the left, but a muon and its neutrino on the right (both with $L_\mu = +1$). Therefore, muon lepton number is not conserved.

46.4 (b), (e). There is one spin-$\frac{1}{2}$ particle on the left and two on the right, so angular momentum is not conserved. There are no leptons on the left and one electron on the right, so electron lepton number is not conserved.

46.5 The diagram is

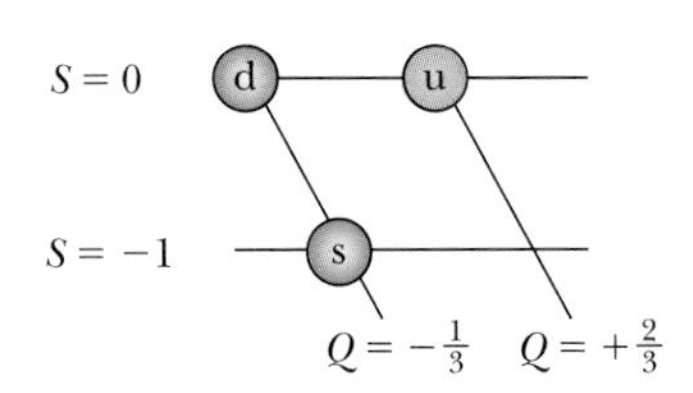

46.6 False. Because the charges on quarks are $+2e/3$ and $-e/3$, the maximum possible charge of a combination of a quark and an antiquark is $\pm e$.

The Meaning of Success

To earn the respect of intelligent people and to win the affection of children;

To appreciate the beauty in nature and all that surrounds us;

To seek out and nurture the best in others;

To give the gift of yourself to others without the slightest thought of return, for it is in giving that we receive;

To have accomplished a task, whether it be saving a lost soul, healing a sick child, writing a book, or risking your life for a friend;

To have celebrated and laughed with great joy and enthusiasm and sung with exultation;

To have hope even in times of despair, for as long as you have hope, you have life;

To love and be loved;

To be understood and to understand;

To know that even one life has breathed easier because you have lived;

This is the meaning of success.

—Ralph Waldo Emerson

Modified by Ray Serway, December 1989

TABLE A.1

Conversion Factors

Length

	m	cm	km	in.	ft	mi
1 meter	1	10^2	10^{-3}	39.37	3.281	6.214×10^{-4}
1 centimeter	10^{-2}	1	10^{-5}	0.393 7	3.281×10^{-2}	6.214×10^{-6}
1 kilometer	10^3	10^5	1	3.937×10^4	3.281×10^3	0.621 4
1 inch	2.540×10^{-2}	2.540	2.540×10^{-5}	1	8.333×10^{-2}	1.578×10^{-5}
1 foot	0.304 8	30.48	3.048×10^{-4}	12	1	1.894×10^{-4}
1 mile	1 609	1.609×10^5	1.609	6.336×10^4	5 280	1

Mass

	kg	g	slug	u
1 kilogram	1	10^3	6.852×10^{-2}	6.024×10^{26}
1 gram	10^{-3}	1	6.852×10^{-5}	6.024×10^{23}
1 slug	14.59	1.459×10^4	1	8.789×10^{27}
1 atomic mass unit	1.660×10^{-27}	1.660×10^{-24}	1.137×10^{-28}	1

Note: 1 metric ton = 1 000 kg.

Time

	s	min	h	day	yr
1 second	1	1.667×10^{-2}	2.778×10^{-4}	1.157×10^{-5}	3.169×10^{-8}
1 minute	60	1	1.667×10^{-2}	6.994×10^{-4}	1.901×10^{-6}
1 hour	3 600	60	1	4.167×10^{-2}	1.141×10^{-4}
1 day	8.640×10^4	1 440	24	1	2.738×10^{-5}
1 year	3.156×10^7	5.259×10^5	8.766×10^3	365.2	1

Speed

	m/s	cm/s	ft/s	mi/h
1 meter per second	1	10^2	3.281	2.237
1 centimeter per second	10^{-2}	1	3.281×10^{-2}	2.237×10^{-2}
1 foot per second	0.304 8	30.48	1	0.681 8
1 mile per hour	0.447 0	44.70	1.467	1

Note: 1 mi/min = 60 mi/h = 88 ft/s.

Force

	N	lb
1 newton	1	0.224 8
1 pound	4.448	1

(Continued)

TABLE A.1

Conversion Factors (*Continued*)

Energy, Energy Transfer

	J	ft · lb	eV
1 joule	1	0.737 6	6.242×10^{18}
1 foot-pound	1.356	1	8.464×10^{18}
1 electron volt	1.602×10^{-19}	1.182×10^{-19}	1
1 calorie	4.186	3.087	2.613×10^{19}
1 British thermal unit	1.055×10^{3}	7.779×10^{2}	6.585×10^{21}
1 kilowatt-hour	3.600×10^{6}	2.655×10^{6}	2.247×10^{25}

	cal	Btu	kWh
1 joule	0.238 9	9.481×10^{-4}	2.778×10^{-7}
1 foot-pound	0.323 9	1.285×10^{-3}	3.766×10^{-7}
1 electron volt	3.827×10^{-20}	1.519×10^{-22}	4.450×10^{-26}
1 calorie	1	3.968×10^{-3}	1.163×10^{-6}
1 British thermal unit	2.520×10^{2}	1	2.930×10^{-4}
1 kilowatt-hour	8.601×10^{5}	3.413×10^{2}	1

Pressure

	Pa	atm
1 pascal	1	9.869×10^{-6}
1 atmosphere	1.013×10^{5}	1
1 centimeter mercury[a]	1.333×10^{3}	1.316×10^{-2}
1 pound per square inch	6.895×10^{3}	6.805×10^{-2}
1 pound per square foot	47.88	4.725×10^{-4}

	cm Hg	lb/in.^{2}	lb/ft^2
1 pascal	7.501×10^{-4}	1.450×10^{-4}	2.089×10^{-2}
1 atmosphere	76	14.70	2.116×10^{3}
1 centimeter mercury[a]	1	0.194 3	27.85
1 pound per square inch	5.171	1	144
1 pound per square foot	3.591×10^{-2}	6.944×10^{-3}	1

[a]At 0°C and at a location where the free-fall acceleration has its "standard" value, 9.806 65 m/s^2.

TABLE A.2

Symbols, Dimensions, and Units of Physical Quantities

Quantity	Common Symbol	Unit[a]	Dimensions[b]	Unit in Terms of Base SI Units
Acceleration	$\vec{\mathbf{a}}$	m/s^2	L/T^2	m/s^2
Amount of substance	n	MOLE		mol
Angle	θ, ϕ	radian (rad)	1	
Angular acceleration	$\vec{\boldsymbol{\alpha}}$	rad/s^2	T^{-2}	s^{-2}
Angular frequency	ω	rad/s	T^{-1}	s^{-1}
Angular momentum	$\vec{\mathbf{L}}$	kg · m^2/s	ML^2/T	kg · m^2/s
Angular velocity	$\vec{\boldsymbol{\omega}}$	rad/s	T^{-1}	s^{-1}
Area	A	m^2	L^2	m^2
Atomic number	Z			
Capacitance	C	farad (F)	Q^{2}T^2/ML^2	A^2 · s^4/kg · m^2
Charge	q, Q, e	coulomb (C)	Q	A · s

(*Continued*)

TABLE A.2

Symbols, Dimensions, and Units of Physical Quantities *(Continued)*

Quantity	Common Symbol	Unit	Dimensions	Unit in Terms of Base SI Units
Charge density				
Line	λ	C/m	Q/L	A · s/m
Surface	σ	C/m^2	Q/L^2	A · s/m^2
Volume	ρ	C/m^3	Q/L^3	A · s/m^3
Conductivity	σ	1/Ω · m	Q^{2}T/ML^3	A^2 · s^3/kg · m^3
Current	I	AMPERE	Q/T	A
Current density	J	A/m^2	Q/TL^2	A/m^2
Density	ρ	kg/m^3	M/L^3	kg/m^3
Dielectric constant	κ			
Electric dipole moment	$\vec{\mathbf{p}}$	C · m	QL	A · s · m
Electric field	$\vec{\mathbf{E}}$	V/m	ML/QT^2	kg · m/A · s^3
Electric flux	Φ_E	V · m	ML^3/QT^2	kg · m^3/A · s^3
Electromotive force	$\mathcal{E}$	volt (V)	ML^2/QT^2	kg · m^2/A · s^3
Energy	E, U, K	joule (J)	ML^2/T^2	kg · m^2/s^2
Entropy	S	J/K	ML^2/T^{2}K	kg · m^2/s^2 · K
Force	$\vec{\mathbf{F}}$	newton (N)	ML/T^2	kg · m/s^2
Frequency	f	hertz (Hz)	T^{-1}	s^{-1}
Heat	Q	joule (J)	ML^2/T^2	kg · m^2/s^2
Inductance	L	henry (H)	ML^2/Q^2	kg · m^2/A^2 · s^2
Length	ℓ, L	METER	L	m
Displacement	$\Delta x, \Delta\vec{\mathbf{r}}$			
Distance	d, h			
Position	$x, y, z, \vec{\mathbf{r}}$			
Magnetic dipole moment	$\vec{\boldsymbol{\mu}}$	N · m/T	QL^2/T	A · m^2
Magnetic field	$\vec{\mathbf{B}}$	tesla (T) (= Wb/m^2)	M/QT	kg/A · s^2
Magnetic flux	Φ_B	weber (Wb)	ML^2/QT	kg · m^2/A · s^2
Mass	m, M	KILOGRAM	M	kg
Molar specific heat	C	J/mol · K		kg · m^2/s^2 · mol · K
Moment of inertia	I	kg · m^2	ML^2	kg · m^2
Momentum	$\vec{\mathbf{p}}$	kg · m/s	ML/T	kg · m/s
Period	T	s	T	s
Permeability of free space	μ_0	N/A^2 (= H/m)	ML/Q^2	kg · m/A^2 · s^2
Permittivity of free space	ϵ_0	C^2/N · m^2 (= F/m)	Q^{2}T^2/ML^3	A^2 · s^4/kg · m^3
Potential	V	volt (V) (= J/C)	ML^2/QT^2	kg · m^2/A · s^3
Power	$\mathcal{P}$	watt (W) (= J/s)	ML^2/T^3	kg · m^2/s^3
Pressure	P	pascal (Pa) (= N/m^2)	M/LT^2	kg/m · s^2
Resistance	R	ohm (Ω) (= V/A)	ML^2/Q^{2}T	kg · m^2/A^2 · s^3
Specific heat	c	J/kg · K	L^2/T^{2}K	m^2/s^2 · K
Speed	v	m/s	L/T	m/s
Temperature	T	KELVIN	K	K
Time	t	SECOND	T	s
Torque	$\vec{\boldsymbol{\tau}}$	N · m	ML^2/T^2	kg · m^2/s^2
Velocity	$\vec{\mathbf{v}}$	m/s	L/T	m/s
Volume	V	m^3	L^3	m^3
Wavelength	λ	m	L	m
Work	W	joule (J) (= N · m)	ML^2/T^2	kg · m^2/s^2

[a]The base SI units are given in uppercase letters.

[b]The symbols M, L, T, K, and Q denote mass, length, time, temperature, and charge, respectively.

This appendix in mathematics is intended as a brief review of operations and methods. Early in this course, you should be totally familiar with basic algebraic techniques, analytic geometry, and trigonometry. The sections on differential and integral calculus are more detailed and are intended for students who have difficulty applying calculus concepts to physical situations.

B.1 Scientific Notation

Many quantities used by scientists often have very large or very small values. The speed of light, for example, is about 300 000 000 m/s, and the ink required to make the dot over an *i* in this textbook has a mass of about 0.000 000 001 kg. Obviously, it is very cumbersome to read, write, and keep track of such numbers. We avoid this problem by using a method incorporating powers of the number 10:

$$10^0 = 1$$

$$10^1 = 10$$

$$10^2 = 10 \times 10 = 100$$

$$10^3 = 10 \times 10 \times 10 = 1\,000$$

$$10^4 = 10 \times 10 \times 10 \times 10 = 10\,000$$

$$10^5 = 10 \times 10 \times 10 \times 10 \times 10 = 100\,000$$

and so on. The number of zeros corresponds to the power to which ten is raised, called the **exponent** of ten. For example, the speed of light, 300 000 000 m/s, can be expressed as 3.00×10^8 m/s.

In this method, some representative numbers smaller than unity are the following:

$$10^{-1} = \frac{1}{10} = 0.1$$

$$10^{-2} = \frac{1}{10 \times 10} = 0.01$$

$$10^{-3} = \frac{1}{10 \times 10 \times 10} = 0.001$$

$$10^{-4} = \frac{1}{10 \times 10 \times 10 \times 10} = 0.000\,1$$

$$10^{-5} = \frac{1}{10 \times 10 \times 10 \times 10 \times 10} = 0.000\,01$$

In these cases, the number of places the decimal point is to the left of the digit 1 equals the value of the (negative) exponent. Numbers expressed as some power of ten multiplied by another number between one and ten are said to be in **scientific notation.** For example, the scientific notation for 5 943 000 000 is 5.943×10^9 and that for 0.000 083 2 is 8.32×10^{-5}.

When numbers expressed in scientific notation are being multiplied, the following general rule is very useful:

$$10^n \times 10^m = 10^{n+m} \qquad \textbf{(B.1)}$$

where n and m can be *any* numbers (not necessarily integers). For example, $10^2 \times 10^5 = 10^7$. The rule also applies if one of the exponents is negative: $10^3 \times 10^{-8} = 10^{-5}$.

When dividing numbers expressed in scientific notation, note that

$$\frac{10^n}{10^m} = 10^n \times 10^{-m} = 10^{n-m} \quad \textbf{(B.2)}$$

Exercises

With help from the preceding rules, verify the answers to the following equations:

1. $86\,400 = 8.64 \times 10^4$
2. $9\,816\,762.5 = 9.816\,762\,5 \times 10^6$
3. $0.000\,000\,039\,8 = 3.98 \times 10^{-8}$
4. $(4.0 \times 10^8)(9.0 \times 10^9) = 3.6 \times 10^{18}$
5. $(3.0 \times 10^7)(6.0 \times 10^{-12}) = 1.8 \times 10^{-4}$
6. $\dfrac{75 \times 10^{-11}}{5.0 \times 10^{-3}} = 1.5 \times 10^{-7}$
7. $\dfrac{(3 \times 10^6)(8 \times 10^{-2})}{(2 \times 10^{17})(6 \times 10^5)} = 2 \times 10^{-18}$

B.2 Algebra

Some Basic Rules

When algebraic operations are performed, the laws of arithmetic apply. Symbols such as x, y, and z are usually used to represent unspecified quantities, called the **unknowns.**

First, consider the equation

$$8x = 32$$

If we wish to solve for x, we can divide (or multiply) each side of the equation by the same factor without destroying the equality. In this case, if we divide both sides by 8, we have

$$\frac{8x}{8} = \frac{32}{8}$$

$$x = 4$$

Next consider the equation

$$x + 2 = 8$$

In this type of expression, we can add or subtract the same quantity from each side. If we subtract 2 from each side, we have

$$x + 2 - 2 = 8 - 2$$

$$x = 6$$

In general, if $x + a = b$, then $x = b - a$.

Now consider the equation

$$\frac{x}{5} = 9$$

If we multiply each side by 5, we are left with x on the left by itself and 45 on the right:

$$\left(\frac{x}{5}\right)(5) = 9 \times 5$$

$$x = 45$$

In all cases, *whatever operation is performed on the left side of the equality must also be performed on the right side.*

The following rules for multiplying, dividing, adding, and subtracting fractions should be recalled, where a, b, c, and d are four numbers:

	Rule	Example
Multiplying	$\left(\frac{a}{b}\right)\left(\frac{c}{d}\right) = \frac{ac}{bd}$	$\left(\frac{2}{3}\right)\left(\frac{4}{5}\right) = \frac{8}{15}$
Dividing	$\frac{(a/b)}{(c/d)} = \frac{ad}{bc}$	$\frac{2/3}{4/5} = \frac{(2)(5)}{(4)(3)} = \frac{10}{12}$
Adding	$\frac{a}{b} \pm \frac{c}{d} = \frac{ad \pm bc}{bd}$	$\frac{2}{3} - \frac{4}{5} = \frac{(2)(5) - (4)(3)}{(3)(5)} = -\frac{2}{15}$

Exercises

In the following exercises, solve for x.

	Answers
1. $a = \frac{1}{1+x}$	$x = \frac{1-a}{a}$
2. $3x - 5 = 13$	$x = 6$
3. $ax - 5 = bx + 2$	$x = \frac{7}{a-b}$
4. $\frac{5}{2x+6} = \frac{3}{4x+8}$	$x = -\frac{11}{7}$

Powers

When powers of a given quantity x are multiplied, the following rule applies:

$$x^n x^m = x^{n+m} \tag{B.3}$$

For example, $x^2x^4 = x^{2+4} = x^6$.

When dividing the powers of a given quantity, the rule is

$$\frac{x^n}{x^m} = x^{n-m} \tag{B.4}$$

For example, $x^8/x^2 = x^{8-2} = x^6$.

A power that is a fraction, such as $\frac{1}{3}$, corresponds to a root as follows:

$$x^{1/n} = \sqrt[n]{x} \tag{B.5}$$

For example, $4^{1/3} = \sqrt[3]{4} = 1.587\,4$. (A scientific calculator is useful for such calculations.)

Finally, any quantity x^n raised to the mth power is

$$(x^n)^m = x^{nm} \tag{B.6}$$

Table B.1 summarizes the rules of exponents.

TABLE B.1

Rules of Exponents
$x^0 = 1$
$x^1 = x$
$x^n x^m = x^{n+m}$
$x^n/x^m = x^{n-m}$
$x^{1/n} = \sqrt[n]{x}$
$(x^n)^m = x^{nm}$

Exercises

Verify the following equations:

1. $3^2 \times 3^3 = 243$
2. $x^5x^{-8} = x^{-3}$
3. $x^{10}/x^{-5} = x^{15}$
4. $5^{1/3} = 1.709\,976$ (Use your calculator.)
5. $60^{1/4} = 2.783\,158$ (Use your calculator.)
6. $(x^4)^3 = x^{12}$

Factoring

Some useful formulas for factoring an equation are the following:

$$ax + ay + az = a(x + y + z) \quad \text{common factor}$$

$$a^2 + 2ab + b^2 = (a + b)^2 \quad \text{perfect square}$$

$$a^2 - b^2 = (a + b)(a - b) \quad \text{differences of squares}$$

Quadratic Equations

The general form of a quadratic equation is

$$ax^2 + bx + c = 0 \tag{B.7}$$

where x is the unknown quantity and a, b, and c are numerical factors referred to as **coefficients** of the equation. This equation has two roots, given by

$$x = \frac{-b \pm \sqrt{b^2 - 4ac}}{2a} \tag{B.8}$$

If $b^2 \geq 4ac$, the roots are real.

EXAMPLE B.1

The equation $x^2 + 5x + 4 = 0$ has the following roots corresponding to the two signs of the square-root term:

$$x = \frac{-5 \pm \sqrt{5^2 - (4)(1)(4)}}{2(1)} = \frac{-5 \pm \sqrt{9}}{2} = \frac{-5 \pm 3}{2}$$

$$x_+ = \frac{-5 + 3}{2} = -1 \qquad x_- = \frac{-5 - 3}{2} = -4$$

where x_+ refers to the root corresponding to the positive sign and x_- refers to the root corresponding to the negative sign.

Exercises

Solve the following quadratic equations:

	Answers	
1. $x^2 + 2x - 3 = 0$	$x_+ = 1$	$x_- = -3$
2. $2x^2 - 5x + 2 = 0$	$x_+ = 2$	$x_- = \frac{1}{2}$
3. $2x^2 - 4x - 9 = 0$	$x_+ = 1 + \sqrt{22}/2$	$x_- = 1 - \sqrt{22}/2$

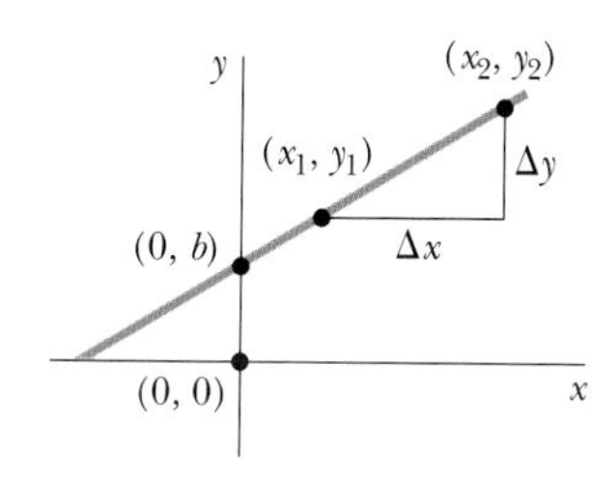

Figure B.1 A straight line graphed on an xy coordinate system. The slope of the line is the ratio of Δy to Δx.

Linear Equations

A linear equation has the general form

$$y = mx + b \tag{B.9}$$

where m and b are constants. This equation is referred to as linear because the graph of y versus x is a straight line as shown in Figure B.1. The constant b, called the ***y*-intercept,** represents the value of y at which the straight line intersects the y axis. The constant m is equal to the **slope** of the straight line. If any two points on the straight line are specified by the coordinates (x_1, y_1) and (x_2, y_2) as in Figure B.1, the slope of the straight line can be expressed as

$$\text{Slope} = \frac{y_2 - y_1}{x_2 - x_1} = \frac{\Delta y}{\Delta x} \tag{B.10}$$

Note that m and b can have either positive or negative values. If $m > 0$, the straight line has a *positive* slope as in Figure B.1. If $m < 0$, the straight line has a *negative* slope. In Figure B.1, both m and b are positive. Three other possible situations are shown in Figure B.2.

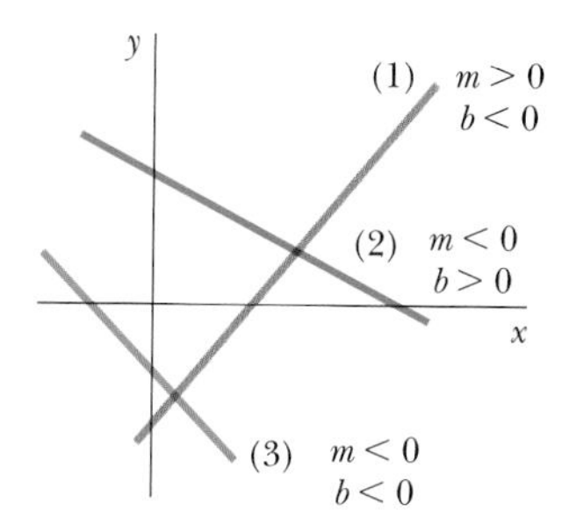

Figure B.2 The brown line has a positive slope and a negative y-intercept. The blue line has a negative slope and a positive y-intercept. The green line has a negative slope and a negative y-intercept.

Exercises

1. Draw graphs of the following straight lines: (a) $y = 5x + 3$ (b) $y = -2x + 4$ (c) $y = -3x - 6$
2. Find the slopes of the straight lines described in Exercise 1.

Answers (a) 5 (b) -2 (c) -3

3. Find the slopes of the straight lines that pass through the following sets of points: (a) $(0, -4)$ and $(4, 2)$ (b) $(0, 0)$ and $(2, -5)$ (c) $(-5, 2)$ and $(4, -2)$

Answers (a) $\frac{3}{2}$ (b) $-\frac{5}{2}$ (c) $-\frac{4}{9}$

Solving Simultaneous Linear Equations

Consider the equation $3x + 5y = 15$, which has two unknowns, x and y. Such an equation does not have a unique solution. For example, $(x = 0,\ y = 3)$, $(x = 5,\ y = 0)$, and $(x = 2,\ y = \frac{9}{5})$ are all solutions to this equation.

If a problem has two unknowns, a unique solution is possible only if we have *two* equations. In general, if a problem has n unknowns, its solution requires n equations. To solve two simultaneous equations involving two unknowns, x and y, we solve one of the equations for x in terms of y and substitute this expression into the other equation.

EXAMPLE B.2

Solve the two simultaneous equations

$$(1) \quad 5x + y = -8$$

$$(2) \quad 2x - 2y = 4$$

Solution From Equation (2), $x = y + 2$. Substitution of this equation into Equation (1) gives

$$5(y + 2) + y = -8$$

$$6y = -18$$

$$y = -3$$

$$x = y + 2 = -1$$

Alternative Solution Multiply each term in Equation (1) by the factor 2 and add the result to Equation (2):

$$10x + 2y = -16$$

$$2x - 2y = 4$$

$$12x = -12$$

$$x = -1$$

$$y = x - 2 = -3$$

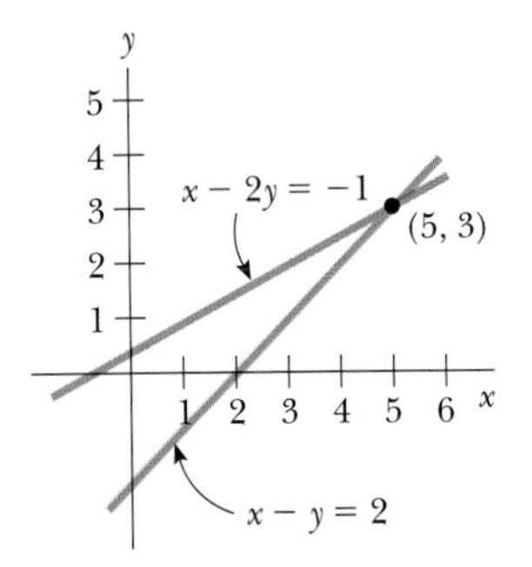

Figure B.3 A graphical solution for two linear equations.

Two linear equations containing two unknowns can also be solved by a graphical method. If the straight lines corresponding to the two equations are plotted in a conventional coordinate system, the intersection of the two lines represents the solution. For example, consider the two equations

$$x - y = 2$$

$$x - 2y = -1$$

These equations are plotted in Figure B.3. The intersection of the two lines has the coordinates $x = 5$ and $y = 3$, which represents the solution to the equations. You should check this solution by the analytical technique discussed earlier.

Exercises

Solve the following pairs of simultaneous equations involving two unknowns:

		Answers
1.	$x + y = 8$ $x - y = 2$	$x = 5, y = 3$
2.	$98 - T = 10a$ $T - 49 = 5a$	$T = 65, a = 3.27$
3.	$6x + 2y = 6$ $8x - 4y = 28$	$x = 2, y = -3$

Logarithms

Suppose a quantity x is expressed as a power of some quantity a:

$$x = a^y \tag{B.11}$$

The number a is called the **base** number. The **logarithm** of x with respect to the base a is equal to the exponent to which the base must be raised to satisfy the expression $x = a^y$:

$$y = \log_a x \tag{B.12}$$

Conversely, the **antilogarithm** of y is the number x:

$$x = \text{antilog}_a\ y \tag{B.13}$$

In practice, the two bases most often used are base 10, called the *common* logarithm base, and base $e = 2.718\ 282$, called Euler's constant or the *natural* logarithm base. When common logarithms are used,

$$y = \log_{10} x \quad (\text{or } x = 10^y) \tag{B.14}$$

When natural logarithms are used,

$$y = \ln x \quad (\text{or } x = e^y) \tag{B.15}$$

For example, $\log_{10} 52 = 1.716$, so $\text{antilog}_{10} 1.716 = 10^{1.716} = 52$. Likewise, $\ln 52 = 3.951$, so $\text{antiln}\ 3.951 = e^{3.951} = 52$.

In general, note you can convert between base 10 and base e with the equality

$$\ln x = (2.302\ 585) \log_{10} x \tag{B.16}$$

Finally, some useful properties of logarithms are the following:

$$\left.\begin{aligned} \log(ab) &= \log a + \log b \\ \log(a/b) &= \log a - \log b \\ \log(a^n) &= n \log a \end{aligned}\right\} \text{ any base}$$

$$\ln e = 1$$

$$\ln e^a = a$$

$$\ln\left(\frac{1}{a}\right) = -\ln a$$

B.3 Geometry

The **distance** d between two points having coordinates (x_1, y_1) and (x_2, y_2) is

$$d = \sqrt{(x_2 - x_1)^2 + (y_2 - y_1)^2} \tag{B.17}$$

Figure B.4 The angles are equal because their sides are perpendicular.

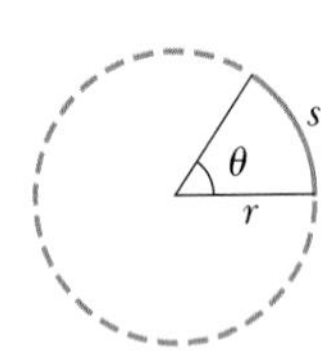

Figure B.5 The angle θ in radians is the ratio of the arc length s to the radius r of the circle.

TABLE B.2

Useful Information for Geometry

Shape	Area or Volume	Shape	Area or Volume
Rectangle (ℓ, w)	Area $= \ell w$	Sphere (r)	Surface area $= 4\pi r^2$ Volume $= \dfrac{4\pi r^3}{3}$
Circle (r)	Area $= \pi r^2$ Circumference $= 2\pi r$	Cylinder (r, ℓ)	Lateral surface area $= 2\pi r\ell$ Volume $= \pi r^2\ell$
Triangle (b, h)	Area $= \frac{1}{2}bh$	Rectangular box (ℓ, w, h)	Surface area $= 2(\ell h + \ell w + hw)$ Volume $= \ell wh$

Two angles are equal if their sides are perpendicular, right side to right side and left side to left side. For example, the two angles marked θ in Figure B.4 are the same because of the perpendicularity of the sides of the angles. To distinguish the left and right sides of an angle, imagine standing at the angle's apex and facing into the angle.

Radian measure: The arc length s of a circular arc (Fig. B.5) is proportional to the radius r for a fixed value of θ (in radians):

$$s = r\theta \qquad \theta = \frac{s}{r} \tag{B.18}$$

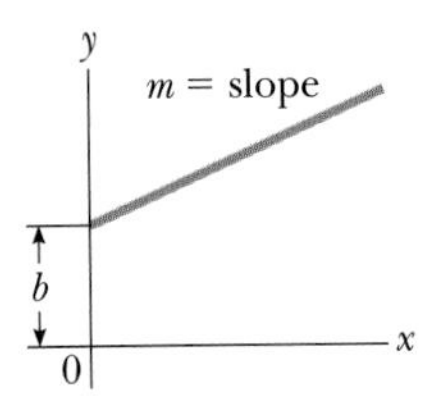

Figure B.6 A straight line with a slope of m and a y-intercept of b.

Table B.2 gives the **areas** and **volumes** for several geometric shapes used throughout this text.

The equation of a **straight line** (Fig. B.6) is

$$y = mx + b \tag{B.19}$$

where b is the y-intercept and m is the slope of the line.

The equation of a **circle** of radius R centered at the origin is

$$x^2 + y^2 = R^2 \tag{B.20}$$

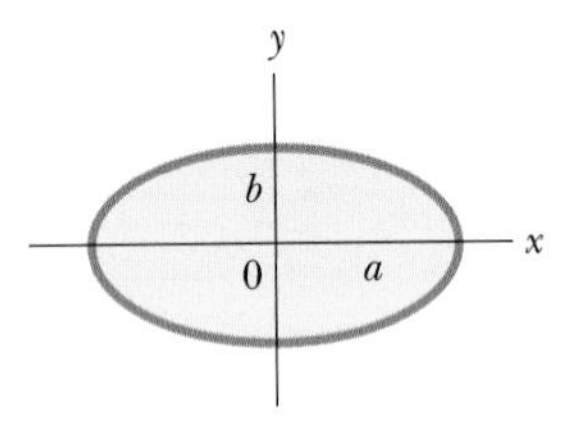

Figure B.7 An ellipse with semimajor axis a and semiminor axis b.

The equation of an **ellipse** having the origin at its center (Fig. B.7) is

$$\frac{x^2}{a^2} + \frac{y^2}{b^2} = 1 \tag{B.21}$$

where a is the length of the semimajor axis (the longer one) and b is the length of the semiminor axis (the shorter one).

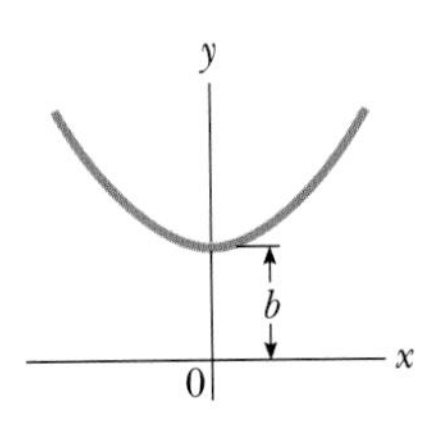

Figure B.8 A parabola with its vertex at $y = b$.

The equation of a **parabola** the vertex of which is at $y = b$ (Fig. B.8) is

$$y = ax^2 + b \tag{B.22}$$

The equation of a **rectangular hyperbola** (Fig. B.9) is

$$xy = \text{constant} \tag{B.23}$$

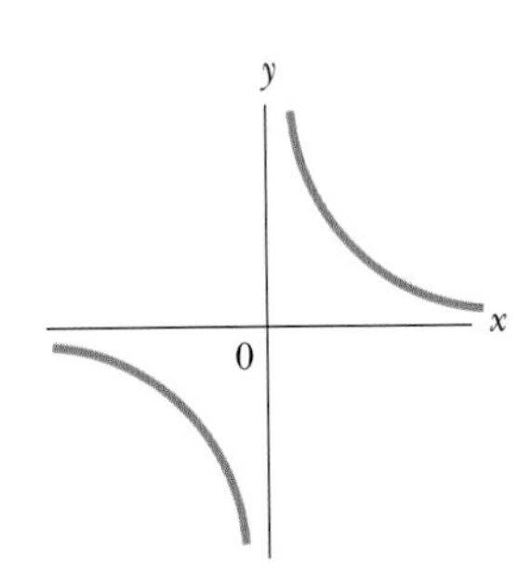

Figure B.9 A hyperbola.

B.4 Trigonometry

That portion of mathematics based on the special properties of the right triangle is called trigonometry. By definition, a right triangle is a triangle containing a 90° angle. Consider the right triangle shown in Figure B.10, where side a is opposite the angle θ, side b is adjacent to the angle θ, and side c is the hypotenuse of the triangle. The three basic trigonometric functions defined by such a triangle are the sine (sin), cosine (cos), and tangent (tan). In terms of the angle θ, these functions are defined as follows:

$$\sin\theta = \frac{\text{side opposite } \theta}{\text{hypotenuse}} = \frac{a}{c} \qquad \textbf{(B.24)}$$

$$\cos\theta = \frac{\text{side adjacent to } \theta}{\text{hypotenuse}} = \frac{b}{c} \qquad \textbf{(B.25)}$$

$$\tan\theta = \frac{\text{side opposite } \theta}{\text{side adjacent to } \theta} = \frac{a}{b} \qquad \textbf{(B.26)}$$

a = opposite side
b = adjacent side
c = hypotenuse

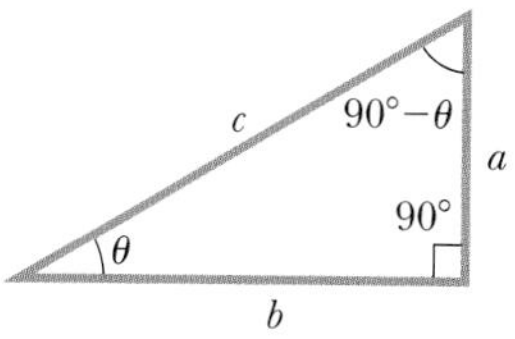

Figure B.10 A right triangle, used to define the basic functions of trigonometry.

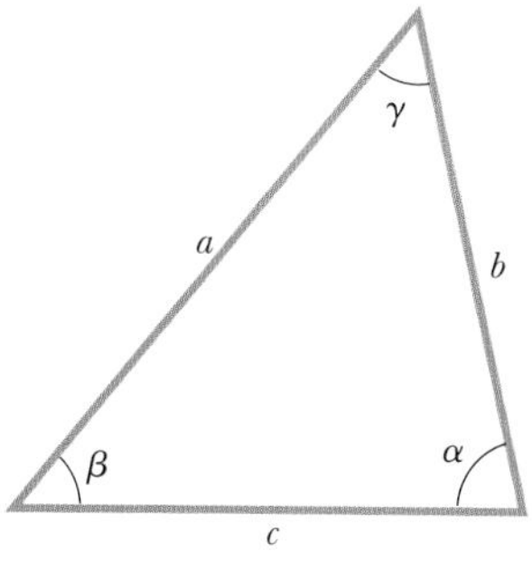

The Pythagorean theorem provides the following relationship among the sides of a right triangle:

$$c^2 = a^2 + b^2 \qquad \textbf{(B.27)}$$

From the preceding definitions and the Pythagorean theorem, it follows that

$$\sin^2\theta + \cos^2\theta = 1$$

$$\tan\theta = \frac{\sin\theta}{\cos\theta}$$

The cosecant, secant, and cotangent functions are defined by

$$\csc\theta = \frac{1}{\sin\theta} \qquad \sec\theta = \frac{1}{\cos\theta} \qquad \cot\theta = \frac{1}{\tan\theta}$$

The following relationships are derived directly from the right triangle shown in Figure B.10:

$$\sin\theta = \cos(90^\circ - \theta)$$
$$\cos\theta = \sin(90^\circ - \theta)$$
$$\cot\theta = \tan(90^\circ - \theta)$$

Some properties of trigonometric functions are the following:

$$\sin(-\theta) = -\sin\theta$$
$$\cos(-\theta) = \cos\theta$$
$$\tan(-\theta) = -\tan\theta$$

The following relationships apply to *any* triangle as shown in Figure B.11:

$$\alpha + \beta + \gamma = 180^\circ$$

Law of cosines
$$\begin{cases} a^2 = b^2 + c^2 - 2bc\cos\alpha \\ b^2 = a^2 + c^2 - 2ac\cos\beta \\ c^2 = a^2 + b^2 - 2ab\cos\gamma \end{cases}$$

Law of sines
$$\frac{a}{\sin\alpha} = \frac{b}{\sin\beta} = \frac{c}{\sin\gamma}$$

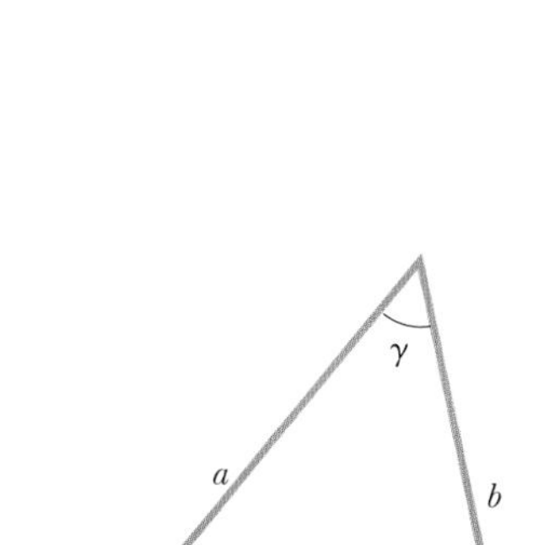

Figure B.11 An arbitrary, nonright triangle.

Table B.3 (page A-12) lists a number of useful trigonometric identities.

EXAMPLE B.3

Consider the right triangle in Figure B.12 in which $a = 2.00$, $b = 5.00$, and c is unknown. From the Pythagorean theorem, we have

$$c^2 = a^2 + b^2 = 2.00^2 + 5.00^2 = 4.00 + 25.0 = 29.0$$

$$c = \sqrt{29.0} = 5.39$$

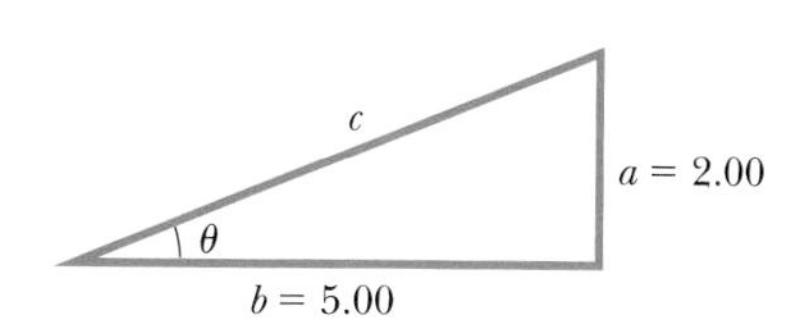

Figure B.12 (Example B.3)

To find the angle θ, note that

$$\tan\theta = \frac{a}{b} = \frac{2.00}{5.00} = 0.400$$

Using a calculator, we find that

$$\theta = \tan^{-1}(0.400) = 21.8°$$

where $\tan^{-1}(0.400)$ is the notation for "angle whose tangent is 0.400," sometimes written as arctan (0.400).

TABLE B.3

Some Trigonometric Identities

$\sin^2\theta + \cos^2\theta = 1$	$\csc^2\theta = 1 + \cot^2\theta$
$\sec^2\theta = 1 + \tan^2\theta$	$\sin^2\frac{\theta}{2} = \frac{1}{2}(1 - \cos\theta)$
$\sin 2\theta = 2\sin\theta\cos\theta$	$\cos^2\frac{\theta}{2} = \frac{1}{2}(1 + \cos\theta)$
$\cos 2\theta = \cos^2\theta - \sin^2\theta$	$1 - \cos\theta = 2\sin^2\frac{\theta}{2}$
$\tan 2\theta = \frac{2\tan\theta}{1 - \tan^2\theta}$	$\tan\frac{\theta}{2} = \sqrt{\frac{1 - \cos\theta}{1 + \cos\theta}}$

$\sin(A \pm B) = \sin A\cos B \pm \cos A\sin B$

$\cos(A \pm B) = \cos A\cos B \mp \sin A\sin B$

$\sin A \pm \sin B = 2\sin\left[\frac{1}{2}(A \pm B)\right]\cos\left[\frac{1}{2}(A \mp B)\right]$

$\cos A + \cos B = 2\cos\left[\frac{1}{2}(A + B)\right]\cos\left[\frac{1}{2}(A - B)\right]$

$\cos A - \cos B = 2\sin\left[\frac{1}{2}(A + B)\right]\sin\left[\frac{1}{2}(B - A)\right]$

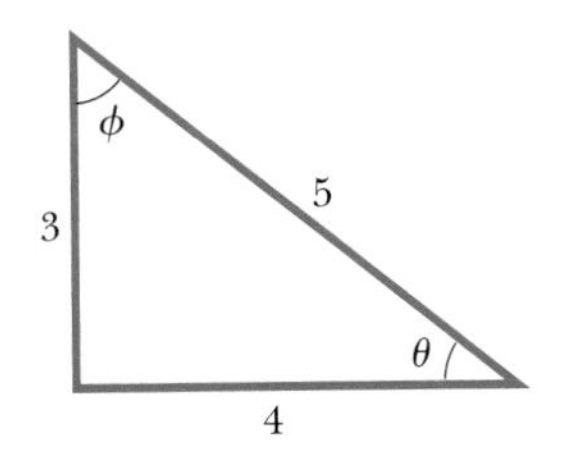

Figure B.13 (Exercise 1)

Exercises

1. In Figure B.13, identify (a) the side opposite θ (b) the side adjacent to ϕ and then find (c) $\cos\theta$, (d) $\sin\phi$, and (e) $\tan\phi$.

Answers (a) 3 (b) 3 (c) $\frac{4}{5}$ (d) $\frac{4}{5}$ (e) $\frac{4}{3}$

2. In a certain right triangle, the two sides that are perpendicular to each other are 5.00 m and 7.00 m long. What is the length of the third side?

Answer 8.60 m

3. A right triangle has a hypotenuse of length 3.0 m, and one of its angles is 30°. (a) What is the length of the side opposite the 30° angle? (b) What is the side adjacent to the 30° angle?

Answers (a) 1.5 m (b) 2.6 m

B.5 Series Expansions

$$(a + b)^n = a^n + \frac{n}{1!}a^{n-1}b + \frac{n(n-1)}{2!}a^{n-2}b^2 + \cdots$$

$$(1 + x)^n = 1 + nx + \frac{n(n-1)}{2!}x^2 + \cdots$$

$$e^x = 1 + x + \frac{x^2}{2!} + \frac{x^3}{3!} + \cdots$$

$$\ln(1 \pm x) = \pm x - \tfrac{1}{2}x^2 \pm \tfrac{1}{3}x^3 - \cdots$$

$$\left.\begin{aligned} \sin x &= x - \frac{x^3}{3!} + \frac{x^5}{5!} - \cdots \\ \cos x &= 1 - \frac{x^2}{2!} + \frac{x^4}{4!} - \cdots \\ \tan x &= x + \frac{x^3}{3} + \frac{2x^5}{15} + \cdots \quad |x| < \frac{\pi}{2} \end{aligned}\right\} x \text{ in radians}$$

For $x << 1$, the following approximations can be used:[1]

$$(1 + x)^n \approx 1 + nx \qquad \sin x \approx x$$

$$e^x \approx 1 + x \qquad \cos x \approx 1$$

$$\ln(1 \pm x) \approx \pm x \qquad \tan x \approx x$$

B.6 Differential Calculus

In various branches of science, it is sometimes necessary to use the basic tools of calculus, invented by Newton, to describe physical phenomena. The use of calculus is fundamental in the treatment of various problems in Newtonian mechanics, electricity, and magnetism. In this section, we simply state some basic properties and "rules of thumb" that should be a useful review to the student.

First, a **function** must be specified that relates one variable to another (e.g., a coordinate as a function of time). Suppose one of the variables is called y (the dependent variable), and the other x (the independent variable). We might have a function relationship such as

$$y(x) = ax^3 + bx^2 + cx + d$$

If a, b, c, and d are specified constants, y can be calculated for any value of x. We usually deal with continuous functions, that is, those for which y varies "smoothly" with x.

The **derivative** of y with respect to x is defined as the limit as Δx approaches zero of the slopes of chords drawn between two points on the y versus x curve. Mathematically, we write this definition as

$$\frac{dy}{dx} = \lim_{\Delta x \to 0} \frac{\Delta y}{\Delta x} = \lim_{\Delta x \to 0} \frac{y(x + \Delta x) - y(x)}{\Delta x} \tag{B.28}$$

where Δy and Δx are defined as $\Delta x = x_2 - x_1$ and $\Delta y = y_2 - y_1$ (Fig. B.14). Note that dy/dx *does not* mean dy divided by dx, but rather is simply a notation of the limiting process of the derivative as defined by Equation B.28.

A useful expression to remember when $y(x) = ax^n$, where a is a *constant* and n is *any* positive or negative number (integer or fraction), is

$$\frac{dy}{dx} = nax^{n-1} \tag{B.29}$$

If $y(x)$ is a polynomial or algebraic function of x, we apply Equation B.29 to *each* term in the polynomial and take $d[\text{constant}]/dx = 0$. In Examples B.4 through B.7, we evaluate the derivatives of several functions.

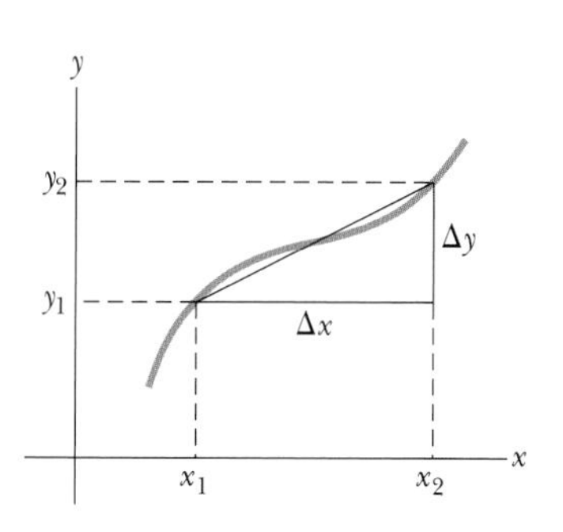

Figure B.14 The lengths Δx and Δy are used to define the derivative of this function at a point.

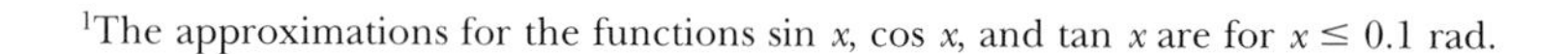

[1]The approximations for the functions sin x, cos x, and tan x are for $x \leq 0.1$ rad.

TABLE B.4

Derivative for Several Functions

$$\frac{d}{dx}(a) = 0$$

$$\frac{d}{dx}(ax^n) = nax^{n-1}$$

$$\frac{d}{dx}(e^{ax}) = ae^{ax}$$

$$\frac{d}{dx}(\sin ax) = a\cos ax$$

$$\frac{d}{dx}(\cos ax) = -a\sin ax$$

$$\frac{d}{dx}(\tan ax) = a\sec^2 ax$$

$$\frac{d}{dx}(\cot ax) = -a\csc^2 ax$$

$$\frac{d}{dx}(\sec x) = \tan x \sec x$$

$$\frac{d}{dx}(\csc x) = -\cot x \csc x$$

$$\frac{d}{dx}(\ln ax) = \frac{1}{x}$$

$$\frac{d}{dx}(\sin^{-1} ax) = \frac{a}{\sqrt{1 - a^2x^2}}$$

$$\frac{d}{dx}(\cos^{-1} ax) = \frac{-a}{\sqrt{1 - a^2x^2}}$$

$$\frac{d}{dx}(\tan^{-1} ax) = \frac{a}{1 + a^2x^2}$$

Note: The symbols a and n represent constants.

Special Properties of the Derivative

A. Derivative of the product of two functions If a function $f(x)$ is given by the product of two functions—say, $g(x)$ and $h(x)$—the derivative of $f(x)$ is defined as

$$\frac{d}{dx}f(x) = \frac{d}{dx}[g(x)h(x)] = g\frac{dh}{dx} + h\frac{dg}{dx} \tag{B.30}$$

B. Derivative of the sum of two functions If a function $f(x)$ is equal to the sum of two functions, the derivative of the sum is equal to the sum of the derivatives:

$$\frac{d}{dx}f(x) = \frac{d}{dx}[g(x) + h(x)] = \frac{dg}{dx} + \frac{dh}{dx} \tag{B.31}$$

C. Chain rule of differential calculus If $y = f(x)$ and $x = g(z)$, then dy/dz can be written as the product of two derivatives:

$$\frac{dy}{dz} = \frac{dy}{dx}\frac{dx}{dz} \tag{B.32}$$

D. The second derivative The second derivative of y with respect to x is defined as the derivative of the function dy/dx (the derivative of the derivative). It is usually written as

$$\frac{d^2y}{dx^2} = \frac{d}{dx}\left(\frac{dy}{dx}\right) \tag{B.33}$$

Some of the more commonly used derivatives of functions are listed in Table B.4.

EXAMPLE B.4

Suppose $y(x)$ (that is, y as a function of x) is given by

$$y(x) = ax^3 + bx + c$$

where a and b are constants. It follows that

$$y(x + \Delta x) = a(x + \Delta x)^3 + b(x + \Delta x) + c$$
$$= a(x^3 + 3x^2\,\Delta x + 3x\,\Delta x^2 + \Delta x^3) + b(x + \Delta x) + c$$

so

$$\Delta y = y(x + \Delta x) - y(x) = a(3x^2\,\Delta x + 3x\,\Delta x^2 + \Delta x^3) + b\,\Delta x$$

Substituting this into Equation B.28 gives

$$\frac{dy}{dx} = \lim_{\Delta x \to 0}\frac{\Delta y}{\Delta x} = \lim_{\Delta x \to 0}[3ax^2 + 3ax\,\Delta x + a\,\Delta x^2] + b$$

$$\frac{dy}{dx} = 3ax^2 + b$$

EXAMPLE B.5

Find the derivative of

$$y(x) = 8x^5 + 4x^3 + 2x + 7$$

Solution Applying Equation B.29 to each term independently and remembering that d/dx (constant) = 0, we have

$$\frac{dy}{dx} = 8(5)x^4 + 4(3)x^2 + 2(1)x^0 + 0$$

$$\frac{dy}{dx} = 40x^4 + 12x^2 + 2$$

EXAMPLE B.6

Find the derivative of $y(x) = x^3/(x+1)^2$ with respect to x.

Solution We can rewrite this function as $y(x) = x^3(x+1)^{-2}$ and apply Equation B.30:

$$\frac{dy}{dx} = (x+1)^{-2}\frac{d}{dx}(x^3) + x^3\frac{d}{dx}(x+1)^{-2}$$

$$= (x+1)^{-2}3x^2 + x^3(-2)(x+1)^{-3}$$

$$\frac{dy}{dx} = \frac{3x^2}{(x+1)^2} - \frac{2x^3}{(x+1)^3}$$

EXAMPLE B.7

A useful formula that follows from Equation B.30 is the derivative of the quotient of two functions. Show that

$$\frac{d}{dx}\left[\frac{g(x)}{h(x)}\right] = \frac{h\frac{dg}{dx} - g\frac{dh}{dx}}{h^2}$$

Solution We can write the quotient as gh^{-1} and then apply Equations B.29 and B.30:

$$\frac{d}{dx}\left(\frac{g}{h}\right) = \frac{d}{dx}(gh^{-1}) = g\frac{d}{dx}(h^{-1}) + h^{-1}\frac{d}{dx}(g)$$

$$= -gh^{-2}\frac{dh}{dx} + h^{-1}\frac{dg}{dx}$$

$$= \frac{h\frac{dg}{dx} - g\frac{dh}{dx}}{h^2}$$

B.7 Integral Calculus

We think of integration as the inverse of differentiation. As an example, consider the expression

$$f(x) = \frac{dy}{dx} = 3ax^2 + b \tag{B.34}$$

which was the result of differentiating the function

$$y(x) = ax^3 + bx + c$$

in Example B.4. We can write Equation B.34 as $dy = f(x)\,dx = (3ax^2 + b)\,dx$ and obtain $y(x)$ by "summing" over all values of x. Mathematically, we write this inverse operation as

$$y(x) = \int f(x)\,dx$$

For the function $f(x)$ given by Equation B.34, we have

$$y(x) = \int (3ax^2 + b)\,dx = ax^3 + bx + c$$

where c is a constant of the integration. This type of integral is called an *indefinite integral* because its value depends on the choice of c.

A general **indefinite integral** $I(x)$ is defined as

$$I(x) = \int f(x)\,dx \tag{B.35}$$

where $f(x)$ is called the *integrand* and $f(x) = dI(x)/dx$.

For a *general continuous* function $f(x)$, the integral can be described as the area under the curve bounded by $f(x)$ and the x axis, between two specified values of x, say, x_1 and x_2, as in Figure B.15.

The area of the blue element in Figure B.15 is approximately $f(x_i)\,\Delta x_i$. If we sum all these area elements between x_1 and x_2 and take the limit of this sum as $\Delta x_i \rightarrow 0$, we obtain the *true* area under the curve bounded by $f(x)$ and the x axis, between the limits x_1 and x_2:

$$\text{Area} = \lim_{\Delta x_i \rightarrow 0} \sum_i f(x_i)\Delta x_i = \int_{x_1}^{x_2} f(x)\,dx \tag{B.36}$$

Integrals of the type defined by Equation B.36 are called **definite integrals.**

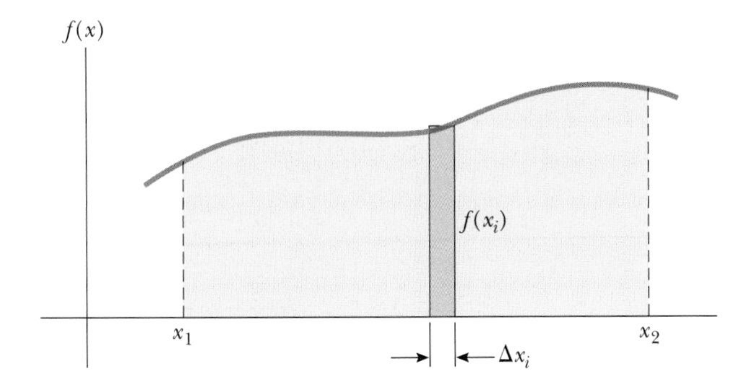

Figure B.15 The definite integral of a function is the area under the curve of the function between the limits x_1 and x_2.

One common integral that arises in practical situations has the form

$$\int x^n\,dx = \frac{x^{n+1}}{n+1} + c \quad (n \neq -1) \tag{B.37}$$

This result is obvious, being that differentiation of the right-hand side with respect to x gives $f(x) = x^n$ directly. If the limits of the integration are known, this integral becomes a *definite integral* and is written

$$\int_{x_1}^{x_2} x^n\,dx = \left.\frac{x^{n+1}}{n+1}\right|_{x_1}^{x_2} = \frac{x_2^{\,n+1} - x_1^{\,n+1}}{n+1} \quad (n \neq -1) \tag{B.38}$$

EXAMPLES

1. $\displaystyle\int_0^a x^2\,dx = \left.\frac{x^3}{3}\right]_0^a = \frac{a^3}{3}$

2. $\displaystyle\int_0^b x^{3/2}dx = \left.\frac{x^{5/2}}{5/2}\right]_0^b = \tfrac{2}{5}b^{5/2}$

3. $\displaystyle\int_3^5 x\,dx = \left.\frac{x^2}{2}\right]_3^5 = \frac{5^2 - 3^2}{2} = 8$

Partial Integration

Sometimes it is useful to apply the method of *partial integration* (also called "integrating by parts") to evaluate certain integrals. This method uses the property

$$\int u\,dv = uv - \int v\,du \tag{B.39}$$

where u and v are *carefully* chosen so as to reduce a complex integral to a simpler one. In many cases, several reductions have to be made. Consider the function

$$I(x) = \int x^2 e^x\,dx$$

which can be evaluated by integrating by parts twice. First, if we choose $u = x^2$, $v = e^x$, we obtain

$$\int x^2 e^x\,dx = \int x^2\,d(e^x) = x^2 e^x - 2\int e^x x\,dx + c_1$$

Now, in the second term, choose $u = x$, $v = e^x$, which gives

$$\int x^2 e^x\,dx = x^2 e^x - 2x\,e^x + 2\int e^x\,dx + c_1$$

or

$$\int x^2 e^x\,dx = x^2 e^x - 2xe^x + 2e^x + c_2$$

TABLE B.5

Some Indefinite Integrals (An arbitrary constant should be added to each of these integrals.)

$$\int x^n\,dx = \frac{x^{n+1}}{n+1} \quad \text{(provided } n \neq 1\text{)}$$

$$\int \frac{dx}{x} = \int x^{-1}\,dx = \ln x$$

$$\int \frac{dx}{a+bx} = \frac{1}{b}\ln(a+bx)$$

$$\int \frac{x\,dx}{a+bx} = \frac{x}{b} - \frac{a}{b^2}\ln(a+bx)$$

$$\int \frac{dx}{x(x+a)} = -\frac{1}{a}\ln\frac{x+a}{x}$$

$$\int \frac{dx}{(a+bx)^2} = -\frac{1}{b(a+bx)}$$

$$\int \frac{dx}{a^2+x^2} = \frac{1}{a}\tan^{-1}\frac{x}{a}$$

$$\int \frac{dx}{a^2-x^2} = \frac{1}{2a}\ln\frac{a+x}{a-x} \quad (a^2-x^2>0)$$

$$\int \frac{dx}{x^2-a^2} = \frac{1}{2a}\ln\frac{x-a}{x+a} \quad (x^2-a^2>0)$$

$$\int \frac{x\,dx}{a^2 \pm x^2} = \pm\tfrac{1}{2}\ln(a^2 \pm x^2)$$

$$\int \frac{dx}{\sqrt{a^2-x^2}} = \sin^{-1}\frac{x}{a} = -\cos^{-1}\frac{x}{a} \quad (a^2-x^2>0)$$

$$\int \frac{dx}{\sqrt{x^2+a^2}} = \ln(x+\sqrt{x^2 \pm a^2})$$

$$\int \frac{x\,dx}{\sqrt{a^2-x^2}} = -\sqrt{a^2-x^2}$$

$$\int \frac{x\,dx}{\sqrt{x^2 \pm a^2}} = \sqrt{x^2 \pm a^2}$$

$$\int \sqrt{a^2-x^2}\,dx = \tfrac{1}{2}\left(x\sqrt{a^2-x^2} + a^2\sin^{-1}\frac{x}{a}\right)$$

$$\int x\sqrt{a^2-x^2}\,dx = -\tfrac{1}{3}(a^2-x^2)^{3/2}$$

$$\int \sqrt{x^2 \pm a^2}\,dx = \tfrac{1}{2}\left[x\sqrt{x^2 \pm a^2} \pm a^2\ln(x+\sqrt{x^2 \pm a^2})\right]$$

$$\int x(\sqrt{x^2 \pm a^2})\,dx = \tfrac{1}{3}(x^2 \pm a^2)^{3/2}$$

$$\int e^{ax}\,dx = \frac{1}{a}e^{ax}$$

$$\int \ln ax\,dx = (x\ln ax) - x$$

$$\int xe^{ax}\,dx = \frac{e^{ax}}{a^2}(ax-1)$$

$$\int \frac{dx}{a+be^{cx}} = \frac{x}{a} - \frac{1}{ac}\ln(a+be^{cx})$$

$$\int \sin ax\,dx = \frac{1}{a}\cos ax$$

$$\int \cos ax\,dx = \frac{1}{a}\sin ax$$

$$\int \tan ax\,dx = -\frac{1}{a}\ln(\cos ax) = \frac{1}{a}\ln(\sec ax)$$

$$\int \cot ax\,dx = \frac{1}{a}\ln(\sin ax)$$

$$\int \sec ax\,dx = \frac{1}{a}\ln(\sec ax + \tan ax) = \frac{1}{a}\ln\left[\tan\left(\frac{ax}{2}+\frac{\pi}{4}\right)\right]$$

$$\int \csc ax\,dx = \frac{1}{a}\ln(\csc ax - \cot ax) = \frac{1}{a}\ln\left(\tan\frac{ax}{2}\right).$$

$$\int \sin^2 ax\,dx = \frac{x}{2} + \frac{\sin 2ax}{4a}$$

$$\int \cos^2 ax\,dx = \frac{x}{2} + \frac{\sin 2ax}{4a}$$

$$\int \frac{dx}{\sin^2 ax} = -\frac{1}{a}\cot ax$$

$$\int \frac{dx}{\cos^2 ax} = \frac{1}{a}\tan ax$$

$$\int \tan^2 ax\,dx = \frac{1}{a}(\tan ax) - x$$

$$\int \cot^2 ax\,dx = -\frac{1}{a}(\cot ax) - x$$

$$\int \sin^{-1} ax\,dx = x(\sin^{-1} ax) + \frac{\sqrt{1-a^2x^2}}{a}$$

$$\int \cos^{-1} ax\,dx = x(\cos^{-1} ax) - \frac{\sqrt{1-a^2x^2}}{a}$$

$$\int \frac{dx}{(x^2+a^2)^{3/2}} = \frac{x}{a^2\sqrt{x^2+a^2}}$$

$$\int \frac{x\,dx}{(x^2+a^2)^{3/2}} = -\frac{1}{\sqrt{x^2+a^2}}$$

TABLE B.6

Gauss's Probability Integral and Other Definite Integrals

$$\int_0^\infty x^n e^{-ax}\, dx = \frac{n!}{a^{n+1}}$$

$$I_0 = \int_0^\infty e^{-ax^2} dx = \frac{1}{2}\sqrt{\frac{\pi}{a}} \qquad \text{(Gauss's probability integral)}$$

$$I_1 = \int_0^\infty x e^{-ax^2}\, dx = \frac{1}{2a}$$

$$I_2 = \int_0^\infty x^2 e^{-ax^2} dx = -\frac{dI_0}{da} = \frac{1}{4}\sqrt{\frac{\pi}{a^3}}$$

$$I_3 = \int_0^\infty x^3 e^{-ax^2}\, dx = -\frac{dI_1}{da} = \frac{1}{2a^2}$$

$$I_4 = \int_0^\infty x^4 e^{-ax^2} dx = \frac{d^2 I_0}{da^2} = \frac{3}{8}\sqrt{\frac{\pi}{a^5}}$$

$$I_5 = \int_0^\infty x^5 e^{-ax^2}\, dx = \frac{d^2 I_1}{da^2} = \frac{1}{a^3}$$

$$\vdots$$

$$I_{2n} = (-1)^n \frac{d^n}{da^n} I_0$$

$$I_{2n+1} = (-1)^n \frac{d^n}{da^n} I_1$$

The Perfect Differential

Another useful method to remember is that of the *perfect differential,* in which we look for a change of variable such that the differential of the function is the differential of the independent variable appearing in the integrand. For example, consider the integral

$$I(x) = \int \cos^2 x \sin x\ dx$$

This integral becomes easy to evaluate if we rewrite the differential as $d(\cos x) = -\sin x\, dx$. The integral then becomes

$$\int \cos^2 x \sin x\ dx = -\int \cos^2 x\, d(\cos x)$$

If we now change variables, letting $y = \cos x$, we obtain

$$\int \cos^2 x \sin x\ dx = -\int y^2\, dy = -\frac{y^3}{3} + c = -\frac{\cos^3 x}{3} + c$$

Table B.5 (page A-18) lists some useful indefinite integrals. Table B.6 gives Gauss's probability integral and other definite integrals. A more complete list can be found in various handbooks, such as *The Handbook of Chemistry and Physics* (Boca Raton, FL: CRC Press, published annually).

B.8 Propagation of Uncertainty

In laboratory experiments, a common activity is to take measurements that act as raw data. These measurements are of several types—length, time interval, temperature, voltage, and so on—and are taken by a variety of instruments. Regardless of the measurement and the quality of the instrumentation, **there is always uncertainty associated with a physical measurement.** This uncertainty is a combination of that associated with the instrument and that related to the system being measured. An example of the former is the inability to exactly determine the position of a length measurement between the lines on a meterstick. An example of uncertainty related to the system being measured is the variation of temperature within a sample of water so that a single temperature for the sample is difficult to determine.

Uncertainties can be expressed in two ways. **Absolute uncertainty** refers to an uncertainty expressed in the same units as the measurement. Therefore, the length of a computer disk label might be expressed as (5.5 ± 0.1) cm. The uncertainty of ± 0.1 cm by itself is not descriptive enough for some purposes, however. This uncertainty is large if the measurement is 1.0 cm, but it is small if the measurement is 100 m. To give a more descriptive account of the uncertainty, **fractional uncertainty** or **percent uncertainty** is used. In this type of description, the uncertainty is divided by the actual measurement. Therefore, the length of the computer disk label could be expressed as

$$\ell = 5.5 \text{ cm} \pm \frac{0.1 \text{ cm}}{5.5 \text{ cm}} = 5.5 \text{ cm} \pm 0.018 \quad \text{(fractional uncertainty)}$$

or as

$$\ell = 5.5 \text{ cm} \pm 1.8\% \quad \text{(percent uncertainty)}$$

When combining measurements in a calculation, the percent uncertainty in the final result is generally larger than the uncertainty in the individual measurements. This is called **propagation of uncertainty** and is one of the challenges of experimental physics.

Some simple rules can provide a reasonable estimate of the uncertainty in a calculated result:

Multiplication and division: When measurements with uncertainties are multiplied or divided, add the *percent uncertainties* to obtain the percent uncertainty in the result.

Example: The Area of a Rectangular Plate

$$A = \ell w = (5.5 \text{ cm} \pm 1.8\%) \times (6.4 \text{ cm} \pm 1.6\%) = 35 \text{ cm}^2 \pm 3.4\%$$

$$= (35 \pm 1) \text{ cm}^2$$

Addition and subtraction: When measurements with uncertainties are added or subtracted, add the *absolute uncertainties* to obtain the absolute uncertainty in the result.

Example: A Change in Temperature

$$\Delta T = T_2 - T_1 = (99.2 \pm 1.5)°\text{C} - (27.6 \pm 1.5)°\text{C} = (71.6 \pm 3.0)°\text{C}$$

$$= 71.6°\text{C} \pm 4.2\%$$

Powers: If a measurement is taken to a power, the percent uncertainty is multiplied by that power to obtain the percent uncertainty in the result.

Example: The Volume of a Sphere

$$V = \tfrac{4}{3}\pi r^3 = \tfrac{4}{3}\pi(6.20 \text{ cm} \pm 2.0\%)^3 = 998 \text{ cm}^3 \pm 6.0\%$$

$$= (998 \pm 60) \text{ cm}^3$$

For complicated calculations, many uncertainties are added together, which can cause the uncertainty in the final result to be undesirably large. Experiments should be designed such that calculations are as simple as possible.

Notice that uncertainties in a calculation always add. As a result, an experiment involving a subtraction should be avoided if possible, especially if the measurements being subtracted are close together. The result of such a calculation is a small difference in the measurements and uncertainties that add together. It is possible that the uncertainty in the result could be larger than the result itself!

Symbol — Ca 20 — Atomic number; Atomic mass† — 40.078; $4s^2$ — Electron configuration

Group I	Group II	Transition elements						
H 1 1.007 9 $1s$								
Li 3 6.941 $2s^1$	Be 4 9.0122 $2s^2$							
Na 11 22.990 $3s^1$	Mg 12 24.305 $3s^2$							
K 19 39.098 $4s^1$	Ca 20 40.078 $4s^2$	Sc 21 44.956 $3d^14s^2$	Ti 22 47.867 $3d^24s^2$	V 23 50.942 $3d^34s^2$	Cr 24 51.996 $3d^54s^1$	Mn 25 54.938 $3d^54s^2$	Fe 26 55.845 $3d^64s^2$	Co 27 58.933 $3d^74s^2$
Rb 37 85.468 $5s^1$	Sr 38 87.62 $5s^2$	Y 39 88.906 $4d^15s^2$	Zr 40 91.224 $4d^25s^2$	Nb 41 92.906 $4d^45s^1$	Mo 42 95.94 $4d^55s^1$	Tc 43 (98) $4d^55s^2$	Ru 44 101.07 $4d^75s^1$	Rh 45 102.91 $4d^85s^1$
Cs 55 132.91 $6s^1$	Ba 56 137.33 $6s^2$	57–71*	Hf 72 178.49 $5d^26s^2$	Ta 73 180.95 $5d^36s^2$	W 74 183.84 $5d^46s^2$	Re 75 186.21 $5d^56s^2$	Os 76 190.23 $5d^66s^2$	Ir 77 192.2 $5d^76s^2$
Fr 87 (223) $7s^1$	Ra 88 (226) $7s^2$	89–103**	Rf 104 (261) $6d^27s^2$	Db 105 (262) $6d^37s^2$	Sg 106 (266)	Bh 107 (264)	Hs 108 (277)	Mt 109 (268)

*Lanthanide series	La 57 138.91 $5d^16s^2$	Ce 58 140.12 $5d^14f^16s^2$	Pr 59 140.91 $4f^36s^2$	Nd 60 144.24 $4f^46s^2$	Pm 61 (145) $4f^56s^2$	Sm 62 150.36 $4f^66s^2$
**Actinide series	Ac 89 (227) $6d^17s^2$	Th 90 232.04 $6d^27s^2$	Pa 91 231.04 $5f^26d^17s^2$	U 92 238.03 $5f^36d^17s^2$	Np 93 (237) $5f^46d^17s^2$	Pu 94 (244) $5f^66d^07s^2$

Note: Atomic mass values given are averaged over isotopes in the percentages in which they exist in nature.
†For an unstable element, mass number of the most stable known isotope is given in parentheses.
††Elements 112 and 114 have not yet been named.
†††For a description of the atomic data, visit *physics.nist.gov/PhysRefData/Elements/per_text.html*

			Group III	Group IV	Group V	Group VI	Group VII	Group 0
							H 1 1.007 9 $1s^1$	**He** 2 4.002 6 $1s^2$
			B 5 10.811 $2p^1$	**C** 6 12.011 $2p^2$	**N** 7 14.007 $2p^3$	**O** 8 15.999 $2p^4$	**F** 9 18.998 $2p^5$	**Ne** 10 20.180 $2p^6$
			Al 13 26.982 $3p^1$	**Si** 14 28.086 $3p^2$	**P** 15 30.974 $3p^3$	**S** 16 32.066 $3p^4$	**Cl** 17 35.453 $3p^5$	**Ar** 18 39.948 $3p^6$
Ni 28 58.693 $3d^8 4s^2$	**Cu** 29 63.546 $3d^{10} 4s^1$	**Zn** 30 65.41 $3d^{10} 4s^2$	**Ga** 31 69.723 $4p^1$	**Ge** 32 72.64 $4p^2$	**As** 33 74.922 $4p^3$	**Se** 34 78.96 $4p^4$	**Br** 35 79.904 $4p^5$	**Kr** 36 83.80 $4p^6$
Pd 46 106.42 $4d^{10}$	**Ag** 47 107.87 $4d^{10} 5s^1$	**Cd** 48 112.41 $4d^{10} 5s^2$	**In** 49 114.82 $5p^1$	**Sn** 50 118.71 $5p^2$	**Sb** 51 121.76 $5p^3$	**Te** 52 127.60 $5p^4$	**I** 53 126.90 $5p^5$	**Xe** 54 131.29 $5p^6$
Pt 78 195.08 $5d^9 6s^1$	**Au** 79 196.97 $5d^{10} 6s^1$	**Hg** 80 200.59 $5d^{10} 6s^2$	**Tl** 81 204.38 $6p^1$	**Pb** 82 207.2 $6p^2$	**Bi** 83 208.98 $6p^3$	**Po** 84 (209) $6p^4$	**At** 85 (210) $6p^5$	**Rn** 86 (222) $6p^6$
Ds 110 (271)	**Rg** 111 (272)	112†† (285)		114†† (289)				

Eu 63 151.96 $4f^7 6s^2$	**Gd** 64 157.25 $4f^7 5d^1 6s^2$	**Tb** 65 158.93 $4f^8 5d^1 6s^2$	**Dy** 66 162.50 $4f^{10} 6s^2$	**Ho** 67 164.93 $4f^{11} 6s^2$	**Er** 68 167.26 $4f^{12} 6s^2$	**Tm** 69 168.93 $4f^{13} 6s^2$	**Yb** 70 173.04 $4f^{14} 6s^2$	**Lu** 71 174.97 $4f^{14} 5d^1 6s^2$
Am 95 (243) $5f^7 7s^2$	**Cm** 96 (247) $5f^7 6d^1 7s^2$	**Bk** 97 (247) $5f^8 6d^1 7s^2$	**Cf** 98 (251) $5f^{10} 7s^2$	**Es** 99 (252) $5f^{11} 7s^2$	**Fm** 100 (257) $5f^{12} 7s^2$	**Md** 101 (258) $5f^{13} 7s^2$	**No** 102 (259) $5f^{14} 7s^2$	**Lr** 103 (262) $6d^1 5f^{14} 7s^2$

TABLE D.1

SI Units

Base Quantity	SI Base Unit Name	SI Base Unit Symbol
Length	meter	m
Mass	kilogram	kg
Time	second	s
Electric current	ampere	A
Temperature	kelvin	K
Amount of substance	mole	mol
Luminous intensity	candela	cd

TABLE D.2

Some Derived SI Units

Quantity	Name	Symbol	Expression in Terms of Base Units	Expression in Terms of Other SI Units
Plane angle	radian	rad	m/m	
Frequency	hertz	Hz	s^{-1}	
Force	newton	N	$kg \cdot m/s^2$	J/m
Pressure	pascal	Pa	$kg/m \cdot s^2$	N/m^2
Energy	joule	J	$kg \cdot m^2/s^2$	$N \cdot m$
Power	watt	W	$kg \cdot m^2/s^3$	J/s
Electric charge	coulomb	C	$A \cdot s$	
Electric potential	volt	V	$kg \cdot m^2/A \cdot s^3$	W/A
Capacitance	farad	F	$A^2 \cdot s^4/kg \cdot m^2$	C/V
Electric resistance	ohm	Ω	$kg \cdot m^2/A^2 \cdot s^3$	V/A
Magnetic flux	weber	Wb	$kg \cdot m^2/A \cdot s^2$	$V \cdot s$
Magnetic field	tesla	T	$kg/A \cdot s^2$	
Inductance	henry	H	$kg \cdot m^2/A^2 \cdot s^2$	$T \cdot m^2/A$

CHAPTER 1

1. 5.52×10^3 kg/m^3, between the density of aluminum and iron and greater than the densities of typical surface rocks
3. 23.0 kg
5. 7.69 cm
7. (b) only
9. The units of G are m^3/kg · s^2.
11. 1.39×10^3 m^2
13. Not with the pages from Volume 1, but yes with the pages from the full version. Each page has area 0.059 m^2. The room has wall area 37 m^2, requiring 630 sheets, which would be counted as 1 260 pages.
15. 11.4×10^3 kg/m^3
17. (a) 250 yr (b) 3.09×10^4 times
19. 1.00×10^{10} lb
21. 151 μm
23. 2.86 cm
25. $\sim 10^6$ balls
27. $\sim 10^2$ kg; $\sim 10^3$ kg
29. $\sim 10^2$ tuners
31. (a) 3 (b) 4 (c) 3 (d) 2
33. (a) 797 (b) 1.1 (c) 17.66
35. 8.80%
37. 9
39. 63
41. 108° and 288°
43. 48.6 kg
45. (a) smaller by nine times (b) Δt is inversely proportional to d^2. (c) Plot Δt on the vertical axis and $1/d^2$ on the horizontal axis. (d) $4QL/[k\pi(T_h - T_c)]$
47. (a) $m = 346$ g $-$ $(14.5$ g/cm$^3)a^3$ (b) $a = 0$ (c) 346 g (d) yes (e) $a = 2.60$ cm (f) 90.6 g (g) yes (h) 218 g (i) No; 218 g is not equal to 314 g. (j) Parts (b), (c), and (d) describe a uniform solid sphere with $\rho = 4.70$ g/cm^3 as a approaches zero. Parts (e), (f), and (g) describe a uniform liquid drop with $\rho = 1.23$ g/cm^3 as a approaches 2.60 cm. The function $m(a)$ is not a linear function, so a halfway between 0 and 2.60 cm does not give a value for m halfway between the minimum and maximum values. The graph of m versus a starts at $a = 0$ with a horizontal tangent. Then it curves down more and more steeply as a increases. The liquid drop of radius 1.30 cm has only one eighth the volume of the whole sphere, so its presence brings down the mass by only a small amount, from 346 g to 314 g. (k) The answer would not change as long as the wall of the shell is unbroken.
49. 5.0 m
51. $0.579t$ ft^3/s + $(1.19 \times 10^{-9})t^2$ ft^3/s^2
53. 3.41 m
55. 0.449%
57. (a) 0.529 cm/s (b) 11.5 cm/s
59. 1×10^{10} gal/yr

CHAPTER 2

1. (a) 5 m/s (b) 1.2 m/s (c) -2.5 m/s (d) -3.3 m/s (e) 0
3. (a) 3.75 m/s (b) 0
5. (a) -2.4 m/s (b) -3.8 m/s (c) 4.0 s
7. (a) and (c)

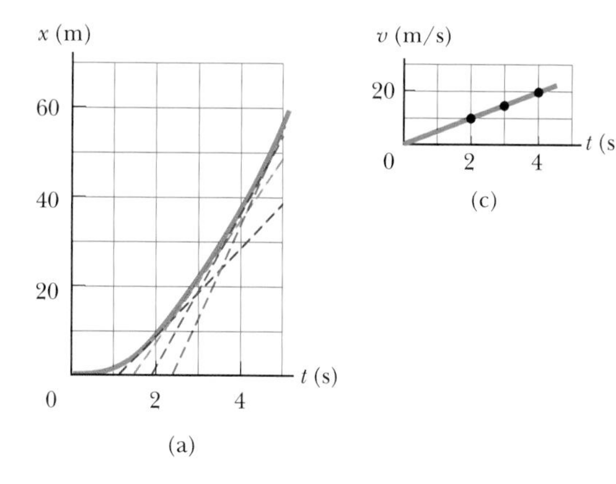

(b) $v_{t=5.0\text{ s}} = 23$ m/s, $v_{t=4.0\text{ s}} = 18$ m/s, $v_{t=3.0\text{ s}} = 14$ m/s, $v_{t=2.0\text{ s}} = 9.0$ m/s (c) 4.6 m/s^2 (d) 0
9. 5.00 m
11. (a) 20.0 m/s, 5.00 m/s (b) 262 m
13. (a) 2.00 m (b) -3.00 m/s (c) -2.00 m/s^2
15. (a) 13.0 m/s (b) 10.0 m/s, 16.0 m/s (c) 6.00 m/s^2 (d) 6.00 m/s^2
17.

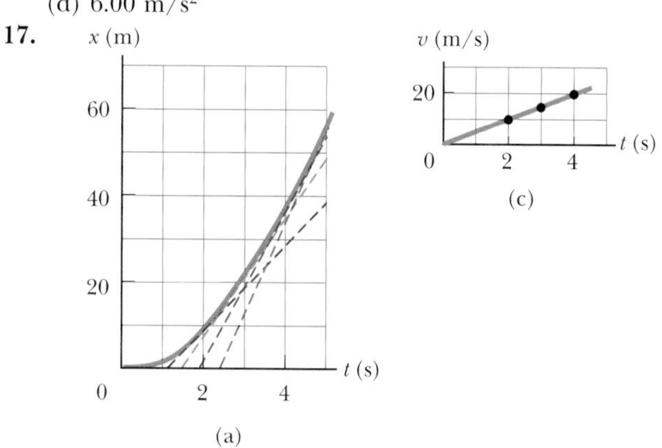

19. (a) 9.00 m/s (b) 5.00 m/s (c) 3.00 m/s (d) -3.00 m/s (e) 17.0 m/s (f) The graph of velocity versus time is a straight line passing through 13 m/s at 10:05 a.m. and sloping downward, decreasing by 4 m/s for each second thereafter. (g) If and only if we know the object's velocity at one instant of time, knowing its acceleration tells us its velocity at every other moment, as long as the acceleration is constant.
21. -16.0 cm/s^2
23. (a) 20.0 s (b) It cannot; it would need a longer runway.
25. 3.10 m/s
27. (a) -202 m/s^2 (b) 198 m
29. (a) 4.98×10^{-9} s (b) 1.20×10^{15} m/s^2
31. (a) False unless the acceleration is zero. We define constant acceleration to mean that the velocity is changing steadily in time. Then the velocity cannot be changing steadily in space. (b) True. Because the velocity is changing steadily in time, the velocity halfway through an interval is equal to the average of its initial and final values.
33. (a) 3.45 s (b) 10.0 ft
35. (a) 19.7 cm/s (b) 4.70 cm/s^2 (c) The time interval required for the speed to change between Ⓐ and Ⓑ is

sufficient to find the acceleration, more directly than we could find it from the distance between the points.

37. We ignore air resistance. We assume the worker's flight time, "a mile," and "a dollar" were measured to three-digit precision. We have interpreted "up in the sky" as referring to free-fall time, not to the launch and landing times. Therefore, the wage was \$99.3/h.

39. (a) 10.0 m/s up (b) 4.68 m/s down

41. (a) 29.4 m/s (b) 44.1 m

43. (a) 7.82 m (b) 0.782 s

45. 38.2 m

47. (a) $a_x(t) = a_{xi} + Jt$, $v_x(t) = v_{xi} + a_{xi}t + \frac{1}{2}Jt^2$, $x(t) = x_i + v_{xi}t + \frac{1}{2}a_{xi}t^2 + \frac{1}{6}Jt^3$

49. (a) 0 (b) 6.0 m/s^2 (c) -3.6 m/s^2 (d) 6 s and 18 s (e) 18 s (f) 84 m (g) 204 m

51. (a) 41.0 s (b) 1.73 km (c) -184 m/s

53. (a) 5.43 m/s^2 and 3.83 m/s^2 (b) 10.9 m/s and 11.5 m/s (c) Maggie by 2.62 m

55. 155 s, 129 s

57. (a) 3.00 s (b) -15.3 m/s (c) 31.4 m/s down and 34.8 m/s down

59. (a) 5.46 s (b) 73.0 m (c) $v_{\text{Stan}} = 22.6$ m/s, $v_{\text{Kathy}} = 26.7$ m/s

61. (a) yes, to two significant digits (b) 0.742 s (c) Yes; the braking distance is proportional to the square of the original speed. (d) -19.7 ft/s^2 = -6.01 m/s^2

63. $0.577v$

CHAPTER 3

1. $(-2.75, -4.76)$ m

3. (a) 2.24 m (b) 2.24 m at 26.6°

5. (a) r, $180° - \theta$ (b) $2r$, $180° + \theta$ (c) $3r$, $-\theta$

7. 70.0 m

9. (a) 10.0 m (b) 15.7 m (c) 0

11. (a) 5.2 m at 60° (b) 3.0 m at 330° (c) 3.0 m at 150° (d) 5.2 m at 300°

13. approximately 420 ft at $-3°$

15. 47.2 units at 122°

17. Yes. The speed of the camper should be 28.3 m/s or greater.

19. (a) $(-11.1\hat{\mathbf{i}} + 6.40\hat{\mathbf{j}})$ m (b) $(1.65\hat{\mathbf{i}} + 2.86\hat{\mathbf{j}})$ cm (c) $(-18.0\hat{\mathbf{i}} - 12.6\hat{\mathbf{j}})$ in.

21. 358 m at 2.00° S of E

23. 196 cm at 345°

25. (a) $2.00\hat{\mathbf{i}} - 6.00\hat{\mathbf{j}}$ (b) $4.00\hat{\mathbf{i}} + 2.00\hat{\mathbf{j}}$ (c) 6.32 (d) 4.47 (e) 288°, 26.6°

27. 9.48 m at 166°

29. 4.64 m at 78.6° N of E

31. (a) 185 N at 77.8° from the $+x$ axis (b) $(-39.3\hat{\mathbf{i}} - 181\hat{\mathbf{j}})$ N

33. $|\vec{\mathbf{B}}| = 7.81$, $\theta_x = 59.2°$, $\theta_y = 39.8°$, $\theta_z = 67.4°$

35. (a) 5.92 m is the magnitude of $(5.00\hat{\mathbf{i}} - 1.00\hat{\mathbf{j}} - 3.00\hat{\mathbf{k}})$ m. (b) 19.0 m is the magnitude of $(4.00\hat{\mathbf{i}} - 11.0\hat{\mathbf{j}} - 15.0\hat{\mathbf{k}})$ m.

37. (a) $8.00\hat{\mathbf{i}} + 12.0\hat{\mathbf{j}} - 4.00\hat{\mathbf{k}}$ (b) $2.00\hat{\mathbf{i}} + 3.00\hat{\mathbf{j}} - 1.00\hat{\mathbf{k}}$ (c) $-24.0\hat{\mathbf{i}} - 36.0\hat{\mathbf{j}} + 12.0\hat{\mathbf{k}}$

39. (a) $(3.12\hat{\mathbf{i}} + 5.02\hat{\mathbf{j}} - 2.20\hat{\mathbf{k}})$ km (b) 6.31 km

41. (a) $-3.00\hat{\mathbf{i}} + 2.00\hat{\mathbf{j}}$ (b) 3.61 at 146° (c) $3.00\hat{\mathbf{i}} - 6.00\hat{\mathbf{j}}$

43. (a) $49.5\hat{\mathbf{i}} + 27.1\hat{\mathbf{j}}$ (b) 56.4 units at 28.7°

45. (a) $[(5 + 11f)\hat{\mathbf{i}} + (3 + 9f)\hat{\mathbf{j}}]$ m (b) $(5\hat{\mathbf{i}} + 3\hat{\mathbf{j}})$ m is reasonable because it is the starting point. (c) $(16\hat{\mathbf{i}} + 12\hat{\mathbf{j}})$ m is reasonable because it is the endpoint.

47. 1.15°

49. 2.29 km

51. (a) 7.17 km (b) 6.15 km

53. 390 mi/h at 7.37° N of E

55. $(0.456\hat{\mathbf{i}} - 0.708\hat{\mathbf{j}})$ m

57. 240 m at 237°

59. (a) (10.0 m, 16.0 m) (b) You will arrive at the treasure if you take the trees in any order. The directions take you to the average position of the trees.

61. 106°

CHAPTER 4

1. (a) 4.87 km at 209° from E (b) 23.3 m/s (c) 13.5 m/s at 209°

3. 2.50 m/s

5. (a) $(0.800\hat{\mathbf{i}} - 0.300\hat{\mathbf{j}})$ m/s^2 (b) 339° (c) $(360\hat{\mathbf{i}} - 72.7\hat{\mathbf{j}})$ m, $-15.2°$

7. (a) $\vec{\mathbf{v}} = 5\hat{\mathbf{i}} + 4t^{3/2}\hat{\mathbf{j}}$ (b) $\vec{\mathbf{r}} = 5t\hat{\mathbf{i}} + 1.6t^{5/2}\hat{\mathbf{j}}$

9. (a) $3.34\hat{\mathbf{i}}$ m/s (b) $-50.9°$

11. $(7.23 \times 10^3$ m, 1.68×10^3 m$)$

13. 53.1°

15. (a) 22.6 m (b) 52.3 m (c) 1.18 s

17. (a) The ball clears by 0.889 m. (b) while descending

19. (a) 18.1 m/s (b) 1.13 m (c) 2.79 m

21. 9.91 m/s

23. $\tan^{-1}[(2gh)^{1/2}/v]$

25. 377 m/s^2

27. (a) 6.00 rev/s (b) 1.52 km/s^2 (c) 1.28 km/s^2

29. 1.48 m/s^2 inward and 29.9° backward

31. (a) 13.0 m/s^2 (b) 5.70 m/s (c) 7.50 m/s^2

33. (a) 57.7 km/h at 60.0° W of vertical (b) 28.9 km/h downward

35. 2.02×10^3 s; 21.0% longer

37. $t_{\text{Alan}} = \dfrac{2L/c}{1 - v^2/c^2}$, $t_{\text{Beth}} = \dfrac{2L/c}{\sqrt{1 - v^2/c^2}}$. Beth returns first.

39. 15.3 m

41. 27.7° E of N

43. (a) 9.80 m/s^2 down (b) 3.72 m

45. (a) 41.7 m/s (b) 3.81 s (c) $(34.1\hat{\mathbf{i}} - 13.4\hat{\mathbf{j}})$ m/s; 36.7 m/s

47. (a) 25.0 m/s^2; 9.80 m/s^2

(b)

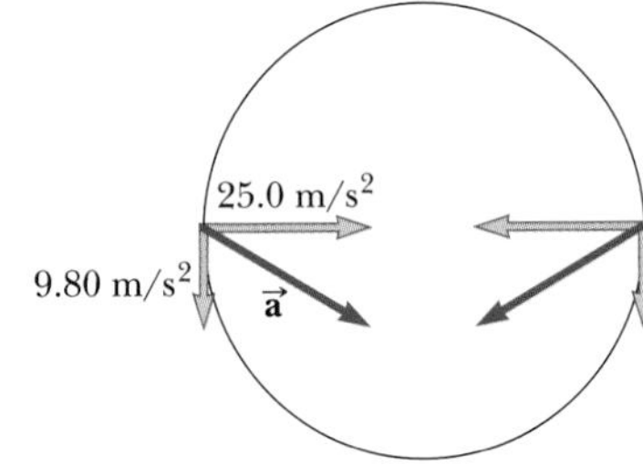

(c) 26.8 m/s^2 inward at 21.4° below the horizontal

49. (a)

t (s)	0	1	2	3	4	5
r (m)	0	45.7	82.0	109	127	136

t (s)	6	7	8	9	10
r (m)	138	133	124	117	120

(b) The vector $\vec{\mathbf{v}}$ tells how $\vec{\mathbf{r}}$ is changing. If $\vec{\mathbf{v}}$ at a particular point has a component along $\vec{\mathbf{r}}$, then $\vec{\mathbf{r}}$ will be increasing in magnitude (if $\vec{\mathbf{v}}$ is at an angle less than 90° from $\vec{\mathbf{r}}$) or decreasing (if the angle between $\vec{\mathbf{v}}$ and $\vec{\mathbf{r}}$ is more than 90°). To be at a maximum, the distance from the origin must be momentarily staying constant, and the only way that can happen is if the angle between velocity and position is a right angle. Then $\vec{\mathbf{r}}$ will be changing in direction at that point, but not in magnitude. (c) The requirement for perpendicularity can be defined as equality between the tangent of the angle between $\vec{\mathbf{v}}$ and the x direction and the tangent of the angle between $\vec{\mathbf{r}}$ and the y direction. In symbols, this equality can be written $(9.8t - 49)/12 = 12t/(49t - 4.9t^2)$, which has the solution $t = 5.70$ s, giving, in turn, $r = 138$ m. Alternatively, we can require $dr^2/dt = 0 = (d/dt)[(12t)^2 + (49t - 4.9t^2)^2]$, which results in the same equation with the same solution.

51. (a) 26.6° (b) 0.949
53. (a) 6.80 km (b) 3.00 km vertically above the impact point (c) 66.2°
55. (a) 46.5 m/s (b) −77.6° (c) 6.34 s
57. (a) 20.0 m/s, 5.00 s (b) $(16.0\hat{\mathbf{i}} - 27.1\hat{\mathbf{j}})$ m/s (c) 6.53 s (d) $24.5\hat{\mathbf{i}}$ m
59. (a) 43.2 m (b) $(9.66\hat{\mathbf{i}} - 25.5\hat{\mathbf{j}})$ m/s. Air resistance would ordinarily make the jump distance smaller and the final horizontal and vertical velocity components both somewhat smaller. When the skilled jumper makes his body into an airfoil, he deflects downward the air through which he passes so that it deflects him upward, giving him more time in the air and a longer jump.
61. Safe distances are less than 270 m or greater than 3.48×10^3 m from the western shore.

CHAPTER 5

1. $(6.00\hat{\mathbf{i}} + 15.0\hat{\mathbf{j}})$ N; 16.2 N
3. (a) $(2.50\hat{\mathbf{i}} + 5.00\hat{\mathbf{j}})$ N (b) 5.59 N
5. (a) 3.64×10^{-18} N (b) 8.93×10^{-30} N is 408 billion times smaller
7. 2.55 N for an 88.7-kg person
9. (a) 5.00 m/s² at 36.9° (b) 6.08 m/s² at 25.3°
11. (a) $\sim 10^{-22}$ m/s² (b) $\sim 10^{-23}$ m
13. (a) 15.0 lb up (b) 5.00 lb up (c) 0
15. (a) 3.43 kN (b) 0.967 m/s horizontally forward

17.

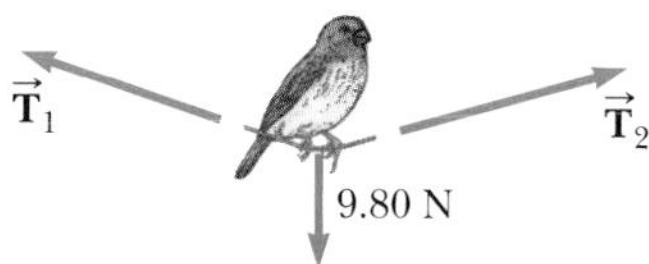

613 N
19. (a) $P \cos 40° - n = 0$ and $P \sin 40° - 220$ N $= 0$; $P = 342$ N and $n = 262$ N (b) $P - n \cos 40° - (220 \text{ N}) \sin 40° = 0$ and $n \sin 40 - (220 \text{ N}) \cos 40° = 0$; $n = 262$ N and $P = 342$ N (c) The results agree. The methods have a similar level of difficulty. Each involves one equation in one unknown and one equation in two unknowns. If we are interested in finding n without finding P, method (b) is simpler.
23. (a) 49.0 N (b) 49.0 N (c) 98.0 N (d) 24.5 N
25. 8.66 N east
27. (a) 646 N up (b) 646 N up (c) 627 N up (d) 589 N up
29. 3.73 m
31. (a) $F_x > 19.6$ N (b) $F_x \le -78.4$ N
(c)

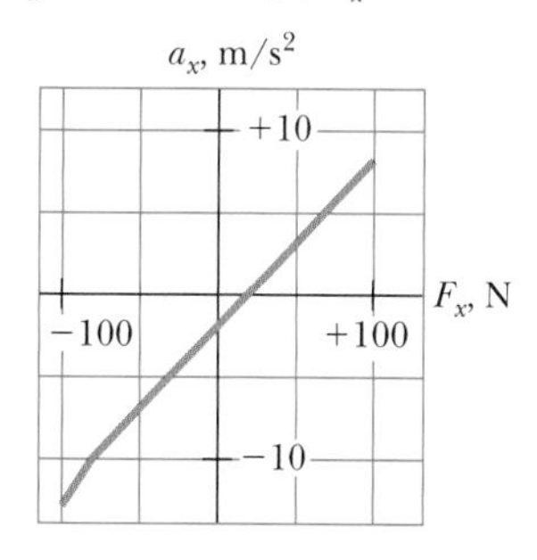

33. (a) 706 N (b) 814 N (c) 706 N (d) 648 N
35. (a) 256 m (b) 42.7 m
37. (a) no (b) 16.9 N backwards + 37.2 N upward = 40.9 N upward and backward at 65.6° with the horizontal
39. (a) 1.78 m/s² (b) 0.368 (c) 9.37 N (d) 2.67 m/s
41. 37.8 N
43. (a)

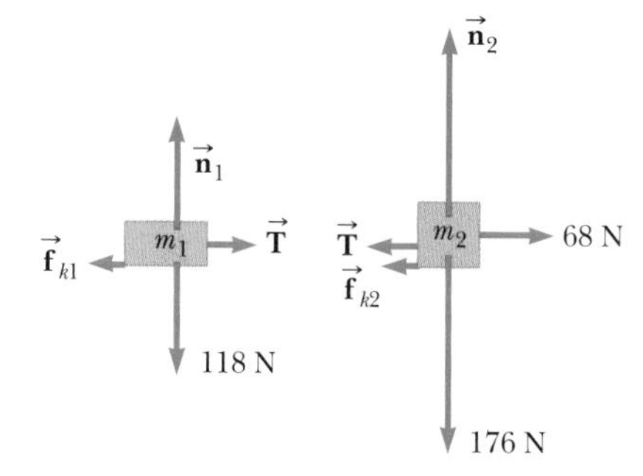

(b) 27.2 N, 1.29 m/s²
45. (a) $a = 0$ if $P < 8.11$ N; $a = -3.33$ m/s² + (1.41/kg) P to the right if $P > 8.11$ N (b) $a = 0$; 3.99 N horizontally backward (c) 10.8 m/s² to the right; 3.45 N to the left (d) The acceleration is zero for all values of P less than 8.11 N. When P passes this threshold, the acceleration jumps to its minimum nonzero value of 8.14 m/s². From there it increases linearly with P toward arbitrarily high values.
47. 72.0 N
49. (a) 2.94 m/s² forward (b) 2.45 m/s² forward (c) 1.19 m/s² up the incline (d) 0.711 m/s² up the incline (e) 16.7° (f) The mass makes no difference. Mathematically, the mass divides out in determinations of acceleration. If several packages of dishes were placed in the truck, they would all slide together, whether they were tied to one another or not.
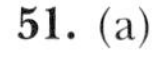
51. (a)

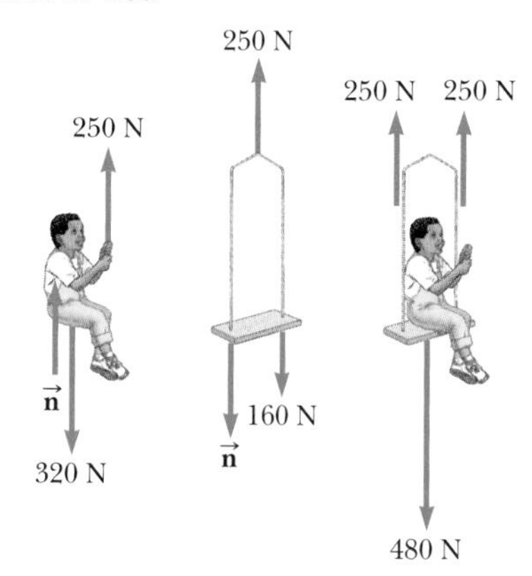

(b) 0.408 m/s² (c) 83.3 N
53. (a) 3.00 s (b) 20.1 m (c) $(18.0\hat{\mathbf{i}} - 9.00\hat{\mathbf{j}})$ m

55. (a) $a = 12\text{ N}/(4\text{ kg} + m_1)$ forward (b) $12\text{ N}/(1 + m_1/4\text{ kg})$ forward (c) 2.50 m/s² forward and 10.0 N forward (d) The force approaches zero (e) The force approaches 12.0 N (f) The tension in a cord of negligible mass is constant along its length.
57. (a) $Mg/2$, $Mg/2$, $Mg/2$, $3Mg/2$, Mg (b) $Mg/2$
59. (a) Both are equal respectively. (b) 1.61×10^4 N (c) 2.95×10^4 N (d) 0 N; 3.51 m/s upward. The first 3.50 m/s of the speed of 3.51 m/s needs no dynamic cause; the motion of the cable continues on its own, as described by the law of "inertia" or "pigheadedness." The increase from 3.50 m/s to 3.51 m/s must be caused by some total upward force on the section of cable. Because its mass is very small compared to a thousand kilograms, however, the force is very small compared to 1.61×10^4 N, the nearly uniform tension of this section of cable.
61. (b)

$\boldsymbol{\theta}$	0	15°	30°	45°	60°
P (N)	40.0	46.4	60.1	94.3	260

63. (a) The net force on the cushion is in a fixed direction, downward and forward making angle $\tan^{-1}(F/mg)$ with the vertical. Starting from rest, it will move along this line with (b) increasing speed. Its velocity changes in magnitude. (c) 1.63 m (d) It will move along a parabola. The axis of the parabola is parallel to the dashed line in the problem figure. If the cushion is thrown in a direction above the dashed line, its path will be concave downward, making its velocity become more and more nearly parallel to the dashed line over time. If the cushion is thrown down more steeply, its path will be concave upward, again making its velocity turn toward the fixed direction of its acceleration.
65. (a) 19.3° (b) 4.21 N
67. $(M + m_1 + m_2)(m_2 g/m_1)$
69. (a) 30.7° (b) 0.843 N
71. (a) $T_1 = \dfrac{2mg}{\sin\theta_1}$, $T_2 = \dfrac{mg}{\sin\theta_2} = \dfrac{mg}{\sin\left[\tan^{-1}\left(\frac{1}{2}\tan\theta_1\right)\right]}$,

$$T_3 = \frac{2\,mg}{\tan\theta_1}$$

(b) $\theta_2 = \tan^{-1}\left(\dfrac{\tan\theta_1}{2}\right)$

CHAPTER 6

1. Any speed up to 8.08 m/s
3. (a) 8.32×10^{-8} N toward the nucleus (b) 9.13×10^{22} m/s² inward
5. (a) static friction (b) 0.085 0
7. 2.14 rev/min
9. $v \leq 14.3$ m/s
11. (a) 108 N (b) 56.2 N
13. (a) 4.81 m/s (b) 700 N up
15. No. Tarzan needs a vine of tensile strength 1.38 kN.
17. 3.13 m/s
19. (a) 3.60 m/s² (b) zero (c) An observer in the car (a noninertial frame) claims an 18.0-N force toward the left and an 18.0-N force toward the right. An inertial observer (outside the car) claims only an 18.0-N force toward the right.
21. (a) 17.0° (b) 5.12 N
23. (a) 491 N (b) 50.1 kg (c) 2.00 m/s
25. 0.092 8°
27. (a) 32.7 s⁻¹ (b) 9.80 m/s² down (c) 4.90 m/s² down
29. 3.01 N up
31. (a) 1.47 N·s/m (b) 2.04×10^{-3} s (c) 2.94×10^{-2} N
33. (a) 78.3 m/s (b) 11.1 s (c) 121 m
35. (a) $x = k^{-1}\ln(1 + kv_0 t)$ (b) $v = v_0 e^{-kx}$
37. (a) 0.034 7 s⁻¹ (b) 2.50 m/s (c) $a = -cv$
39. $v = v_0 e^{-bt/m}$

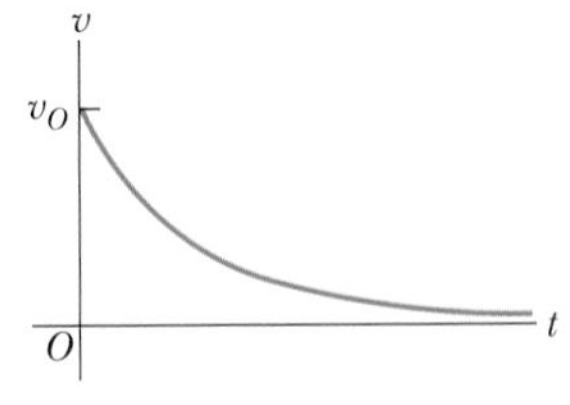

In this model, the object keeps moving forever. It travels a finite distance in an infinite time interval.
41. (a) 106 N up the incline (b) 0.396
43. (a) 11.5 kN (b) 14.1 m/s
45. (a) 0.016 2 kg/m (b) $\frac{1}{2}D\rho A$ (c) 0.778 (d) 1.5% (e) For stacked coffee filters falling in air at terminal speed, the graph of air resistance force as a function of the square of speed demonstrates that the force is proportional to the speed squared, within the experimental uncertainty estimated as 2%. This proportionality agrees with the theoretical model of air resistance at high speeds. The drag coefficient of a coffee filter is $D = 0.78 \pm 2\%$.
47. $g(\cos\phi\tan\theta - \sin\phi)$
49. (b) 732 N down at the equator and 735 N down at the poles
51. (a) The only horizontal force on the car is the force of friction, with a maximum value determined by the surface roughness (described by the coefficient of static friction) and the normal force (here equal to the gravitational force on the car). (b) 34.3 m (c) 68.6 m (d) Braking is better. You should not turn the wheel. If you used any of the available friction force to change the direction of the car, it would be unavailable to slow the car and the stopping distance would be longer. (e) The conclusion is true in general. The radius of the curve you can barely make is twice your minimum stopping distance.
53. (a) 5.19 m/s
(b) $T = 555$ N

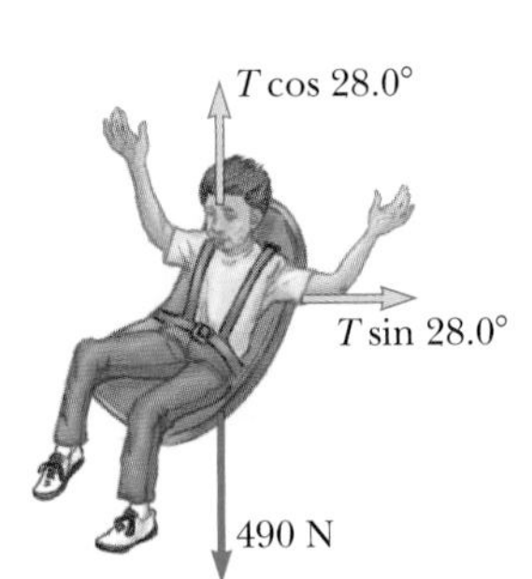

55. (b) 2.54 s; 23.6 rev/min (c) The gravitational and friction forces remain constant. The normal force increases. The person remains in motion with the wall. (d) The gravitational force remains constant. The normal and friction forces decrease. The person slides relative to the wall and downward into the pit.
57. (a) $v_{\min} = \sqrt{\dfrac{Rg(\tan\theta - \mu_s)}{1 + \mu_s\tan\theta}}$, $v_{\max} = \sqrt{\dfrac{Rg(\tan\theta + \mu_s)}{1 - \mu_s\tan\theta}}$

(b) $\mu_s = \tan\theta$ (c) 8.57 m/s $\leq v \leq$ 16.6 m/s
59. (a) 0.013 2 m/s (b) 1.03 m/s (c) 6.87 m/s
61. 12.8 N

CHAPTER 7

1. (a) 31.9 J (b) 0 (c) 0 (d) 31.9 J
3. −4.70 kJ
7. (a) 16.0 J (b) 36.9°
9. (a) 11.3° (b) 156° (c) 82.3°
11. $\vec{\mathbf{A}}$ = 7.05 m at 28.4°
13. (a) 24.0 J (b) −3.00 J (c) 21.0 J
15. (a) 7.50 J (b) 15.0 J (c) 7.50 J (d) 30.0 J
17. (a) 0.938 cm (b) 1.25 J
19. 7.37 N/m
21. 0.299 m/s
23. (a) 0.020 4 m (b) 720 N/m
25. (b) mgR
27. (a) 0.600 J (b) −0.600 J (c) 1.50 J
29. (a) 1.20 J (b) 5.00 m/s (c) 6.30 J
31. (a) 60.0 J (b) 60.0 J
33. 878 kN up
35. (a) 4.56 kJ (b) 6.34 kN (c) 422 km/s² (d) 6.34 kN (e) The forces are the same. The two theories agree.
37. (a) 259 kJ, 0, −259 kJ (b) 0, −259 kJ, −259 kJ
39. (a) −196 J (b) −196 J (c) −196 J. The force is conservative.
41. (a) 125 J (b) 50.0 J (c) 66.7 J (d) The force is nonconservative. The results differ.
43. (a) 40.0 J (b) −40.0 J (c) 62.5 J
45. (A/r^2) away from the other particle
47. (a) + at Ⓑ, − at Ⓓ, 0 at Ⓐ, Ⓒ, and Ⓔ (b) Ⓒ stable; Ⓐ and Ⓔ unstable
(c)

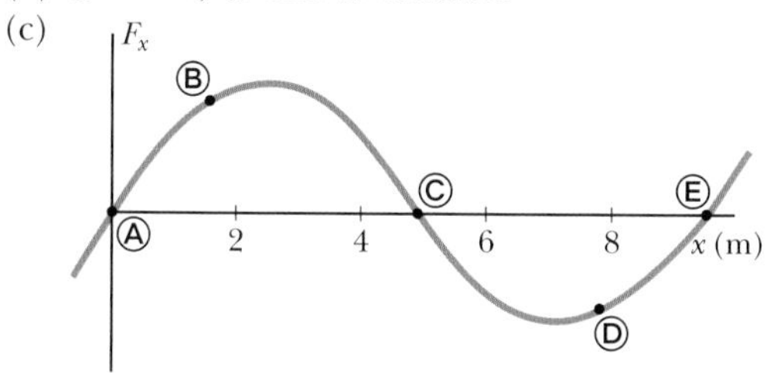

49. (c) Equilibrium at $x = 0$

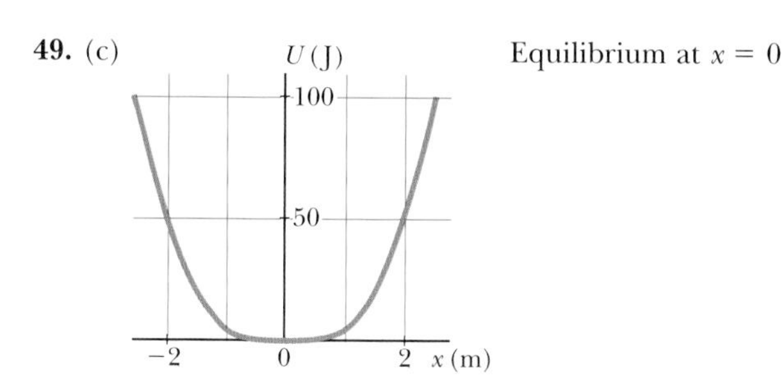

(d) 0.823 m/s
51. 90.0 J
53. (a) $x = (3.62\ m)/(4.30 - 23.4m)$ where x is in meters and m is in kilograms (b) 0.095 1 m (c) 0.492 m (d) 6.85 m (e) The situation is impossible. (f) The extension is directly proportional to m when m is only a few grams. Then it grows faster and faster, diverging to infinity for $m = 0.184$ kg.
55. $U(x) = 1 + 4e^{-2x}$. The force must be conservative because the work the force does on the object on which it acts depends only on the original and final positions of the object, not on the path between them.
57. 1.68 m/s
59. 0.799 J

CHAPTER 8

1. (a) $\Delta E_{int} = Q + T_{ET} + T_{ER}$ (b) $\Delta K + \Delta U + \Delta E_{int} = W + Q + T_{MW} + T_{MT}$ (c) $\Delta U = Q + T_{MT}$ (d) $0 = Q + T_{MT} + T_{ET} + T_{ER}$
3. (a) $v = (3gR)^{1/2}$ (b) 0.098 0 N down
5. 10.2 m
7. (a) 4.43 m/s (b) 5.00 m
9. 5.49 m/s
11. (a) 25.8 m (b) 27.1 m/s²
13. (a) 650 J (b) 588 J (c) 0 (d) 0 (e) 62.0 J (f) 1.76 m/s
15. (a) −168 J (b) 184 J (c) 500 J (d) 148 J (e) 5.65 m/s
17. 2.04 m
19. 3.74 m/s
21. (a) −160 J (b) 73.5 J (c) 28.8 N (d) 0.679
23. (a) 1.40 m/s (b) 4.60 cm after release (c) 1.79 m/s
25. (a) 0.381 m (b) 0.143 m (c) 0.371 m
27. (a) $a_x = -\mu_k gx/L$ (b) $v = (\mu_k gL)^{1/2}$
29. 875 W
31. ~ 10^4 W
33. $46.2
35. (a) 10.2 kW (b)10.6 kW (c) 5.82 MJ
37. (a) 11.1 m/s (b) 19.6 m/s² upward (c) 2.23×10^3 N upward (d) 1.01×10^3 J (e) 5.14 m/s (f) 1.35 m (g) 1.39 s
39. (a) $(2 + 24t^2 + 72t^4)$ J (b) $12t$ m/s²; $48t$ N (c) $(48t + 288t^3)$ W (d) 1 250 J
41. (a) 1.38×10^4 J (b) 3.02×10^4 W
43. (a) 4.12 m (b) 3.35 m
45. (a) 2.17 kW (b) 58.6 kW
47. (a) $x = -4.0$ mm (b) −1.0 cm
49. 33.4 kW
51. (a) 0.225 J (b) $\Delta E_{mech} = -0.363$ J (c) No. The normal force changes in a complicated way.
53. (a) 100 J (b) 0.410 m (c) 2.84 m/s (d) −9.80 mm (e) 2.85 m/s
55. 0.328
57. 1.24 m/s
59. (a) 0.400 m (b) 4.10 m/s (c) The block stays on the track.
61. $2m$
65. (a) 14.1 m/s (b) −7.90 kJ (c) 800 N (d) 771 N (e) 1.57 kN up

CHAPTER 9

1. (a) $(9.00\hat{\mathbf{i}} - 12.0\hat{\mathbf{j}})$ kg · m/s (b) 15.0 kg · m/s at 307°
3. ~ 10^{-23} m/s
5. (b) $p = \sqrt{2mK}$
7. (a) 13.5 N · s (b) 9.00 kN (c) 18.0 kN
9. 260 N normal to the wall
11. (a) $12.0\hat{\mathbf{i}}$ N · s (b) $4.80\hat{\mathbf{i}}$ m/s (c) $2.80\hat{\mathbf{i}}$ m/s (d) $2.40\hat{\mathbf{i}}$ N
13. (b) small (d) large (e) no difference
15. 301 m/s
17. (a) 2.50 m/s (b) 37.5 kJ (c) Each process is the time-reversal of the other. The same momentum conservation equation describes both.
19. 0.556 m
21. (a) $\vec{\mathbf{v}}_g = 1.15\hat{\mathbf{i}}$ m/s (b) $\vec{\mathbf{v}}_p = -0.346\hat{\mathbf{i}}$ m/s
23. (a) 0.284 (b) 115 fJ and 45.4 fJ

25. 91.2 m/s
27. 2.50 m/s at $-60.0°$
29. $v_{orange} = 3.99$ m/s, $v_{yellow} = 3.01$ m/s
31. $(3.00\hat{\mathbf{i}} - 1.20\hat{\mathbf{j}})$ m/s
33. (a) $(-9.33\hat{\mathbf{i}} - 8.33\hat{\mathbf{j}})$ Mm/s (b) 439 fJ
35. $\vec{\mathbf{r}}_{CM} = (0\hat{\mathbf{i}} + 1.00\hat{\mathbf{j}})$ m
37. $\vec{\mathbf{r}}_{CM} = (11.7\hat{\mathbf{i}} + 13.3\hat{\mathbf{j}})$ cm
39. (a) 15.9 g (b) 0.153 m
41. (a) $(1.40\hat{\mathbf{i}} + 2.40\hat{\mathbf{j}})$ m/s (b) $(7.00\hat{\mathbf{i}} + 12.0\hat{\mathbf{j}})$ kg · m/s
43. 0.700 m
45. (a) Yes. $18.0\hat{\mathbf{i}}$ kg · m/s. (b) No. The floor does zero work. (c) Yes. We could say that the final momentum of the cart came from the floor or from the Earth through the floor. (d) No. The kinetic energy came from the original gravitational energy of the elevated load, in amount 27.0 J. (e) Yes. The acceleration is caused by the static friction force exerted by the floor that prevents the caterpillar tracks from slipping backward.
47. (b) 2.06 m/s (c) Yes. The bumper continues to exert a force to the left until the particle has swung down to its lowest point.
49. (a) 3.75 kg · m/s² to the right (b) 3.75 N to the right (c) 3.75 N (d) 2.81 J (e) 1.41 J (f) Friction between sand and belt converts half of the input work into extra internal energy.
51. (a) 39.0 MN (b) 3.20 m/s² up
53. (a) 442 metric tons (b) 19.2 metric tons. This amount is much less than the value suggested. Mathematically, the logarithm in the rocket propulsion equation is not a linear function. Physically, a higher exhaust speed has an extra-large cumulative effect on the rocket frame's final speed, by counting again and again in the speed the frame attains second after second during its burn.
55. 240 s
57. $\left(\dfrac{M+m}{m}\right)\sqrt{\dfrac{gd^2}{2h}}$
59. (a) 0; inelastic
(b) $(-0.250\hat{\mathbf{i}} + 0.750\hat{\mathbf{j}} - 2.00\hat{\mathbf{k}})$ m/s; perfectly inelastic
(c) either $a = -6.74$ with $\vec{\mathbf{v}} = -0.419\,\hat{\mathbf{k}}$ m/s or $a = 2.74$ with $\vec{\mathbf{v}} = -3.58\,\hat{\mathbf{k}}$ m/s
61. (a) $m/M = 0.403$ (b) no changes; no difference
63. (b) 0.042 9 (c) 1.00 (d) Energy is an entirely different thing from momentum. A comparison: When children eat their soup, they do not eat the tablecloth. Another comparison: When a photographer's single-use flashbulb flashes, a magnesium filament oxidizes. Chemical energy disappears. (Internal energy appears and light carries some energy away.) The measured mass of the flashbulb is the same before and after. It can be the same despite the 100% energy conversion because energy and mass are totally different things in classical physics. In the ballistic pendulum, conversion of energy from mechanical into internal does not upset conservation of mass or conservation of momentum.
65. (a) $-0.256\hat{\mathbf{i}}$ m/s and $0.128\hat{\mathbf{i}}$ m/s
(b) $-0.064\,2\hat{\mathbf{i}}$ m/s and 0 (c) 0 and 0
67. (a) 100 m/s (b) 374 J
69. $(3Mgx/L)\hat{\mathbf{j}}$

CHAPTER 10

1. (a) 5.00 rad, 10.0 rad/s, 4.00 rad/s² (b) 53.0 rad, 22.0 rad/s, 4.00 rad/s²
3. (a) 4.00 rad/s² (b) 18.0 rad
5. (a) 5.24 s (b) 27.4 rad
7. (a) 7.27×10^{-5} rad/s (b) 2.57×10^4 s = 428 min
9. 50.0 rev
11. $\sim 10^7$ rev
13. (a) 8.00 rad/s (b) 8.00 m/s, $a_r = -64.0$ m/s², $a_t = 4.00$ m/s² (c) 9.00 rad
15. (a) $(-2.73\hat{\mathbf{i}} + 1.24\hat{\mathbf{j}})$ m (b) in the second quadrant, at 156° (c) $(-1.85\hat{\mathbf{i}} - 4.10\hat{\mathbf{j}})$ m/s (d) toward the third quadrant, at 246°

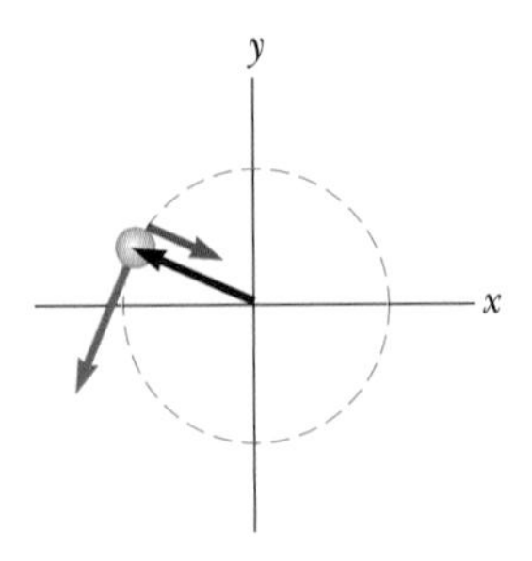

(e) $(6.15\hat{\mathbf{i}} - 2.78\hat{\mathbf{j}})$ m/s² (f) $(24.6\hat{\mathbf{i}} - 11.1\hat{\mathbf{j}})$ N
17. (a) 126 rad/s (b) 3.77 m/s (c) 1.26 km/s² (d) 20.1 m
19. 0.572
21. (a) 143 kg · m² (b) 2.57 kJ
25. (a) 24.5 m/s (b) no; no; no; no; yes
27. 1.28 kg · m²
29. $\sim 10^0$ kg · m²
33. -3.55 N · m
35. (a) 24.0 N · m (b) 0.035 6 rad/s² (c) 1.07 m/s²
37. (a) 0.309 m/s² (b) 7.67 N and 9.22 N
39. 21.5 N
41. 24.5 km
43. 149 rad/s
45. (a) 1.59 m/s (b) 53.1 rad/s
47. (a) 11.4 N, 7.57 m/s², 9.53 m/s down (b) 9.53 m/s
51. (a) $2(Rg/3)^{1/2}$ (b) $4(Rg/3)^{1/2}$ (c) $(Rg)^{1/2}$
53. (a) 500 J (b) 250 J (c) 750 J
55. (a) $\frac{2}{3}g \sin\theta$ for the disk, larger than $\frac{1}{2}g \sin\theta$ for the hoop (b) $\frac{1}{3}\tan\theta$
57. 1.21×10^{-4} kg · m²; height is unnecessary
59. $\frac{1}{3}\ell$
61. (a) 4.00 J (b) 1.60 s (c) yes
63. (a) $\omega = 3F\ell/b$ (b) $\alpha = 3F\ell/mL^2$ (c) and (d) Both larger. A component of the thrust force, exerted by the water about to spray from the ends of the arms, causes a forward torque on the rotor. Notice also that the rotor with bent arms has a slightly smaller moment of inertia than it would if the same metal tubes were straight.
65. (a) $(3g/L)^{1/2}$ (b) $3g/2L$ (c) $-\frac{3}{2}g\hat{\mathbf{i}} - \frac{3}{4}g\hat{\mathbf{j}}$
(d) $-\frac{3}{2}Mg\hat{\mathbf{i}} + \frac{1}{4}Mg\hat{\mathbf{j}}$
67. -0.322 rad/s²
71. (a) 118 N and 156 N (b) 1.17 kg · m²
73. (a) $\alpha = -0.176$ rad/s² (b) 1.29 rev (c) 9.26 rev
75. (a) $\omega(2h^3/g)^{1/2}$ (b) 0.011 6 m (c) Yes; the deflection is only 0.02% of the original height.
79. (a) $2.70R$ (b) $\Sigma F_x = -20mg/7$, $\Sigma F_y = -5mg/7$
81. (a) $(3gh/4)^{1/2}$ (b) $(3gh/4)^{1/2}$
83. (c) $(8Fd/3M)^{1/2}$
85. to the left

CHAPTER 11

1. $-7.00\hat{\mathbf{i}} + 16.0\hat{\mathbf{j}} - 10.0\hat{\mathbf{k}}$
3. (a) $-17.0\hat{\mathbf{k}}$ (b) 70.6°
5. 0.343 N · m horizontally north
7. 45.0°
9. $F_3 = F_1 + F_2$; no
11. $17.5\hat{\mathbf{k}}$ kg · m²/s
13. $(60.0\hat{\mathbf{k}})$ kg · m²/s
15. $mvR[\cos(vt/R) + 1]\hat{\mathbf{k}}$
17. (a) zero (b) $(-mv_i^3 \sin^2\theta \cos\theta/2g)\hat{\mathbf{k}}$ (c) $(-2mv_i^3 \sin^2\theta \cos\theta/g)\hat{\mathbf{k}}$ (d) The downward gravitational force exerts a torque in the $-z$ direction.
19. (a) $-m\ell gt \cos\theta\hat{\mathbf{k}}$ (b) The planet exerts a gravitational torque on the ball. (c) $-mg\ell \cos\theta\hat{\mathbf{k}}$
23. (a) 0.360 kg · m²/s (b) 0.540 kg · m²/s
25. (a) 0.433 kg · m²/s (b) 1.73 kg · m²/s
27. (a) 1.57×10^8 kg · m²/s (b) 6.26×10^3 s = 1.74 h
29. (a) $\omega_f = \omega_i I_1/(I_1 + I_2)$ (b) $I_1/(I_1 + I_2)$
31. (a) 11.1 rad/s counterclockwise (b) No. 507 J is transformed into internal energy. (c) No. The turntable bearing promptly imparts impulse 44.9 kg · m/s north into the turntable-clay system and thereafter keeps changing the system momentum.
33. 7.14 rev/min
35. (a) Mechanical energy is not conserved; some chemical energy is converted into mechanical energy. Momentum is not conserved. The turntable bearing exerts an external northward force on the axle. Angular momentum is conserved. (b) 0.360 rad/s counterclockwise (c) 99.9 J
37. (a) $mv\ell$ down (b) $M/(M + m)$
39. (a) $\omega = 2mv_i d/[M + 2m]R^2$ (b) No; some mechanical energy changes into internal energy. (c) Momentum is not conserved. The axle exerts a backward force on the cylinder.
41. $\sim 10^{-13}$ rad/s
43. 5.45×10^{22} N · m
45. (a) $1.67\hat{\mathbf{i}}$ m/s (b) 0.033 5 = 3.35% (c) $1.67\hat{\mathbf{i}}$ m/s (d) 15.8 rad/s (e) 1.00 = 100%
47. (a) $7md^2/3$ (b) $mgd\hat{\mathbf{k}}$ (c) $3g/7d$ counterclockwise (d) $2g/7$ upward (e) mgd (f) $\sqrt{6g/7d}$ (g) $m\sqrt{14gd^3/3}$ (h) $\sqrt{2gd/21}$
49. 0.910 km/s
51. (a) $v_i r_i/r$ (b) $T = (mv_i^2 r_i^2)r^{-3}$ (c) $\frac{1}{2}mv_i^2(r_i^2/r^2 - 1)$ (d) 4.50 m/s, 10.1 N, 0.450 J
53. (a) 3 750 kg · m²/s (b) 1.88 kJ (c) 3 750 kg · m²/s (d) 10.0 m/s (e) 7.50 kJ (f) 5.62 kJ
55. (a) $2mv_0$ (b) $2v_0/3$ (c) $4m\ell v_0/3$ (d) $4v_0/9\ell$ (e) mv_0^2 (f) $26mv_0^2/27$ (g) No horizontal forces act on the bola from outside after release, so the horizontal momentum stays constant. Its center of mass moves steadily with the horizontal velocity it had at release. No torques about its axis of rotation act on the bola, so its spin angular momentum stays constant. Internal forces cannot affect momentum conservation and angular momentum conservation, but they can affect mechanical energy. Energy $mv_0^2/27$ changes from mechanical energy into internal energy as the bola takes its stable configuration.
57. An increase of 0.550 s. It is not a significant change.

CHAPTER 12

1. $[(m_1 + m_b)d + m_1\ell/2]/m_2$
3. (3.85 cm, 6.85 cm)
5. (−1.50 m, −1.50 m)
7. (2.54 m, 4.75 m)
9. 177 kg
11. (a) $f_s = 268$ N, $n = 1\,300$ N (b) 0.324
13. 2.94 kN on each rear wheel and 4.41 kN on each front wheel
15. (a) 29.9 N (b) 22.2 N
17. (a) 1.73 rad/s² (b) 1.56 rad/s (c) $(-4.72\hat{\mathbf{i}} + 6.62\hat{\mathbf{j}})$ kN (d) $38.9\hat{\mathbf{j}}$ kN
19. 2.82 m
21. 88.2 N and 58.8 N
23. 4.90 mm
25. 23.8 μm
27. (a) 3.14×10^4 N (b) 6.28×10^4 N
29. 1.65×10^8 N/m²
31. 0.860 mm
33. $n_A = 5.98 \times 10^5$ N, $n_B = 4.80 \times 10^5$ N
35. 9.00 ft
37. (a)

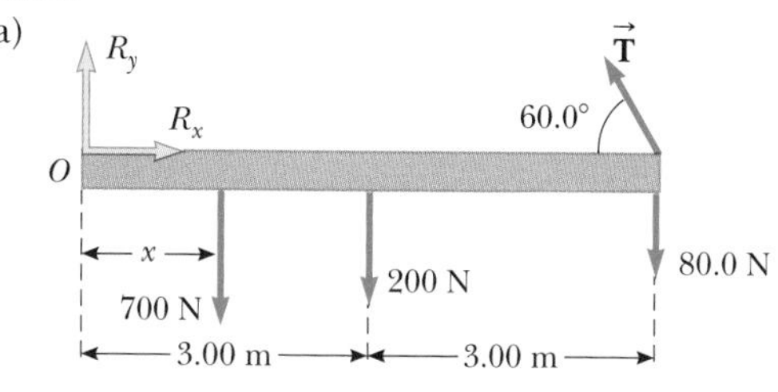

(b) $T = 343$ N, $R_x = 171$ N to the right, $R_y = 683$ N up (c) 5.13 m
39. (a) $T = F_g(L + d)/[\sin\theta\,(2L + d)]$ (b) $R_x = F_g(L + d)\cot\theta/(2L + d)$, $R_y = F_g L/(2L + d)$
41. $\vec{\mathbf{F}}_A = (-6.47 \times 10^5\hat{\mathbf{i}} + 1.27 \times 10^5\hat{\mathbf{j}})$ N, $\vec{\mathbf{F}}_B = 6.47 \times 10^5\hat{\mathbf{i}}$ N
43. 5.08 kN, $R_x = 4.77$ kN, $R_y = 8.26$ kN
45. (a) 20.1 cm to the left of the front edge; $\mu_k = 0.571$ (b) 0.501 m
47. (a) $M = (m/2)(2\mu_s \sin\theta - \cos\theta)(\cos\theta - \mu_s \sin\theta)^{-1}$ (b) $R = (m + M)g(1 + \mu_s^2)^{1/2}$ $F = g[M^2 + \mu_s^2(m + M)^2]^{1/2}$
49. (b) *AB* compression 732 N, *AC* tension 634 N, *BC* compression 897 N
51. (a) 133 N (b) $n_A = 429$ N and $n_B = 257$ N (c) $R_x = 133$ N and $R_y = -257$ N
55. 1.09 m
57. (a) 4 500 N (b) 4.50×10^6 N/m² (c) The board will break.
59. (a) $P_y = (F_g/L)(d - ah/g)$ (b) 0.306 m (c) $(-306\hat{\mathbf{i}} + 553\hat{\mathbf{j}})$ N

CHAPTER 13

1. $\sim 10^{-7}$ N toward you
3. (a) 2.50×10^{-5} N toward the 500-kg object (b) between the objects and 0.245 m from the 500-kg object
5. $(-100\hat{\mathbf{i}} + 59.3\hat{\mathbf{j}})$ pN
7. 7.41×10^{-10} N
9. 0.613 m/s² toward the Earth
11. $\rho_{\text{Moon}}/\rho_{\text{Earth}} = \frac{2}{3}$
13. 1.26×10^{32} kg
15. 1.90×10^{27} kg
17. 8.92×10^7 m
19. After 3.93 yr, Mercury would be farther from the Sun than Pluto.

21. $\vec{\mathbf{g}} = \frac{Gm}{\ell^2}\left(\frac{1}{2} + \sqrt{2}\right)$ toward the opposite corner

23. (a) $\vec{\mathbf{g}} = 2MGr(r^2 + a^2)^{-3/2}$ toward the center of mass (b) At $r = 0$, the fields of the two objects are equal in magnitude and opposite in direction, to add to zero. (d) When r is much greater than a, the fact that the two masses are separate is unimportant. They create a total field like that of a single object of mass $2M$.

25. (a) 1.84×10^9 kg/m^3 (b) 3.27×10^6 m/s^2 (c) -2.08×10^{13} J

27. (a) -1.67×10^{-14} J (b) Each object will slowly accelerate toward the center of the triangle, where the three will simultaneously collide.

29. (b) 340 s

31. 1.66×10^4 m/s

35. (a) 5.30×10^3 s (b) 7.79 km/s (c) 6.43×10^9 J

37. (b) 1.00×10^7 m (c) 1.00×10^4 m/s

39. (a) 0.980 (b) 127 yr (c) -2.13×10^{17} J

43. (b) $2[Gm^3(1/2r - 1/R)]^{1/2}$

45. (a) -7.04×10^4 J (b) -1.57×10^5 J (c) 13.2 m/s

47. 7.79×10^{14} kg

49. $\omega = 0.057\,2$ rad/s or 1 rev in 110 s

51. (a) $m_2(2G/d)^{1/2}(m_1 + m_2)^{-1/2}$ and $m_1(2G/d)^{1/2}(m_1 + m_2)^{-1/2}$; relative speed $(2G/d)^{1/2}(m_1 + m_2)^{1/2}$ (b) 1.07×10^{32} J and 2.67×10^{31} J

53. (a) 200 Myr (b) $\sim 10^{41}$ kg; $\sim 10^{11}$ stars

55. $(GM_E/4R_E)^{1/2}$

59. $(800 + 1.73 \times 10^{-4})\hat{\mathbf{i}}$ m/s and $(800 - 1.73 \times 10^{-4})\hat{\mathbf{i}}$ m/s

61. 18.2 ms

CHAPTER 14

1. 0.111 kg

3. 6.24 MPa

5. 1.62 m

7. 7.74×10^{-3} m^2

9. 271 kN horizontally backward

11. 5.88×10^6 N down; 196 kN outward; 588 kN outward

13. 0.722 mm

15. 10.5 m; no because some alcohol and water evaporate

17. 98.6 kPa

19. (a) 1.57 Pa, 1.55×10^{-2} atm, 11.8 mm Hg (b) The fluid level in the tap should rise. (c) blockage of flow of the cerebrospinal fluid

21. 0.258 N down

23. (a) $1.017\,9 \times 10^3$ N down, $1.029\,7 \times 10^3$ N up (b) 86.2 N (c) By either method of evaluation, the buoyant force is 11.8 N up.

25. (a) 1.20×10^3 N/s (b) 0

27. (a) 7.00 cm (b) 2.80 kg

31. 1 430 m^3

33. 1 250 kg/m^3 and 500 kg/m^3

35. (a) 17.7 m/s (b) 1.73 mm

37. 31.6 m/s

39. 0.247 cm

41. (a) 2.28 N toward Holland (b) 1.74×10^6 s

43. (a) 1 atm + 15.0 MPa (b) 2.95 m/s (c) 4.34 kPa

45. 2.51×10^{-3} m^3/s

47. (a) 4.43 m/s (b) The siphon can be no higher than 10.3 m.

49. 12.6 m/s

51. 1.91 m

55. 0.604 m

57. If the helicopter could create the air it expels downward, the mass flow rate of the air would have to be at least 233 kg/s. In reality, the rotor takes in air from above, which is moving over a larger area with lower speed, and blows it downward at higher speed. The amount of this air has to be at least a few times larger than 233 kg every second.

61. 17.3 N and 31.7 N

63. 90.04%

65. 758 Pa

67. 4.43 m/s

69. (a) 1.25 cm (b) 13.8 m/s

71. (c) 1.70 m^2

CHAPTER 15

1. (a) The motion repeats precisely. (b) 1.81 s (c) No, the force is not in the form of Hooke's law

3. (a) 1.50 Hz, 0.667 s (b) 4.00 m (c) π rad (d) 2.83 m

5. (b) 18.8 cm/s, 0.333 s (c) 178 cm/s^2, 0.500 s (d) 12.0 cm

7. 40.9 N/m

9. 18.8 m/s, 7.11 km/s^2

11. (a) 40.0 cm/s, 160 cm/s^2 (b) 32.0 cm/s, -96.0 cm/s^2 (c) 0.232 s

13. 0.628 m/s

15. 2.23 m/s

17. (a) 28.0 mJ (b) 1.02 m/s (c) 12.2 mJ (d) 15.8 mJ

19. 2.60 cm and -2.60 cm

21. (a) at 0.218 s and at 1.09 s (b) 0.014 6 W

23. (b) 0.628 s

25. Assuming simple harmonic motion, (a) 0.820 m/s, (b) 2.57 rad/s^2, and (c) 0.641 N. More precisely, (a) 0.817 m/s, (b) 2.54 rad/s^2, and (c) 0.634 N. The answers agree to two digits. The answers computed from conservation of energy and from Newton's second law are more precisely correct. With this amplitude, the motion of the pendulum is approximately simple harmonic.

29. 0.944 kg · m^2

33. (a) 5.00×10^{-7} kg · m^2 (b) 3.16×10^{-4} N · m/rad

35. 1.00×10^{-3} s^{-1}

37. (a) 7.00 Hz (b) 2.00% (c) 10.6 s

39. (a) 1.00 s (b) 5.09 cm

41. 318 N

43. 1.74 Hz

45. (a) 2.09 s (b) 0.477 Hz (c) 36.0 cm/s (d) $(0.064\,8\text{ m}^2/\text{s}^2)\,m$ (e) $(9.00/\text{s}^2)\,m$ (f) Period, frequency, and maximum speed are all independent of mass in this situation. The energy and the force constant are directly proportional to mass.

47. (a) $2Mg$, $Mg(1 + y/L)$ (b) $T = (4\pi/3)(2L/g)^{1/2}$, 2.68 s

49. 6.62 cm

51. 9.19×10^{13} Hz

53. (a)

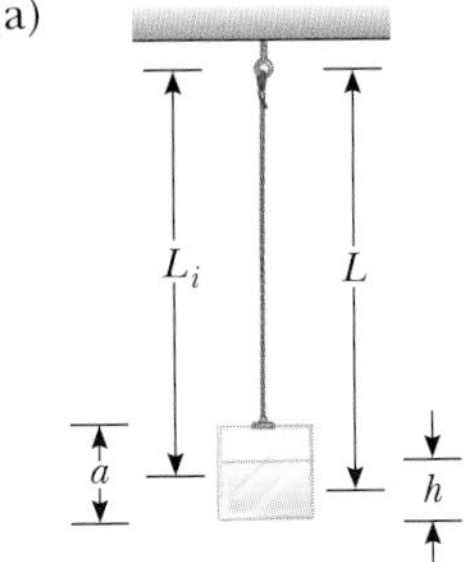

(b) $\frac{dT}{dt} = \frac{\pi \, dM/dt}{2\rho a^2 g^{1/2}[L_i + (dM/dt)t/2\rho a^2]^{1/2}}$

(c) $T = 2\pi g^{-1/2}\left[L_i + \left(\frac{dM}{dt}\right)\left(\frac{t}{2\rho a^2}\right)\right]^{1/2}$

55. $f = (2\pi L)^{-1}\left(gL + \frac{kh^2}{M}\right)^{1/2}$

57. (b) 1.23 Hz

59. (a) 3.00 s (b) 14.3 J (c) 25.5°

61. If the cyclist goes over washboard bumps at one certain speed, they can excite a resonance vibration of the bike, so large in amplitude as to make the rider lose control. $\sim 10^1$ m

69. (b) after 42.2 minutes

CHAPTER 16

1. $y = 6\,[(x - 4.5t)^2 + 3]^{-1}$

3. (a) the P wave (b) 665 s

5. (a) $(3.33\hat{\mathbf{i}})$ m/s (b) −5.48 cm (c) 0.667 m, 5.00 Hz (d) 11.0 m/s

7. 0.319 m

9. 2.00 cm, 2.98 m, 0.576 Hz, 1.72 m/s

11. (a) 31.4 rad/s (b) 1.57 rad/m (c) $y = (0.120 \text{ m}) \sin(1.57x - 31.4t)$ where x is in meters and t is in seconds (d) 3.77 m/s (e) 118 m/s²

13. (a) 0.250 m (b) 40.0 rad/s (c) 0.300 rad/m (d) 20.9 m (e) 133 m/s (f) $+x$

15. (a) $y = (8.00 \text{ cm}) \sin(7.85x + 6\pi t)$ (b) $y = (8.00 \text{ cm}) \sin(7.85x + 6\pi t - 0.785)$

17. (a) −1.51 m/s, 0 (b) 16.0 m, 0.500 s, 32.0 m/s

19. (a) 0.500 Hz, 3.14 rad/s (b) 3.14 rad/m (c) $(0.100 \text{ m}) \sin(3.14\, x/\text{m} - 3.14\, t/\text{s})$ (d) $(0.100 \text{ m}) \sin(-3.14\, t/\text{s})$ (e) $(0.100 \text{ m}) \sin(4.71 \text{ rad} - 3.14\, t/\text{s})$ (f) 0.314 m/s

21. 80.0 N

23. 520 m/s

25. 1.64 m/s²

27. 13.5 N

29. 185 m/s

31. 0.329 s

35. 55.1 Hz

37. (a) 62.5 m/s (b) 7.85 m (c) 7.96 Hz (d) 21.1 W

39. $\sqrt{2}\,\mathcal{P}_0$

41. (a) $A = 40$ (b) $A = 7.00$, $B = 0$, $C = 3.00$. One can take the dot product of the given equation with each one of $\hat{\mathbf{i}}, \hat{\mathbf{j}}$, and $\hat{\mathbf{k}}$. (c) $A = 0$, $B = 7.00$ mm, $C = 3.00$/m, $D = 4.00$/s, $E = 2.00$. Consider the average value of both sides of the given equation to find A. Then consider the maximum value of both sides to find B. You can evaluate the partial derivative of both sides of the given equation with respect to x and separately with respect to t to obtain equations yielding C and D upon chosen substitutions for x and t. Then substitute $x = 0$ and $t = 0$ to obtain E.

45. ~ 1 min

47. 0.456 m/s

49. (a) 39.2 N (b) 0.892 m (c) 83.6 m/s

51. (a) The energy a wave crest carries is constant in the absence of absorption. Then the rate at which energy moves beyond a fixed distance from the source, which is the power of the wave, is constant. The power is proportional to the square of the amplitude and to the wave speed. The speed decreases as the wave moves into shallower water near shore, so the amplitude must increase. (b) 8.31 m (c) As the water depth goes to zero, our model would predict zero speed and infinite amplitude. The amplitude must be finite as the wave comes ashore. As the speed decreases, the wavelength also decreases. When it becomes comparable to the water depth, or smaller, the expression $v = \sqrt{gd}$ no longer applies.

53. (a) $\mathcal{P} = (0.050\,0 \text{ kg/s})v_{y,\max}^2$ (b) The power is proportional to the square of the maximum element speed. (c) $(7.5 \times 10^{-4} \text{ kg})v_{y,\max}^2 = \frac{1}{2}m_3 v_{y,\max}^2$ (d) $(0.300 \text{ kg})v_{y,\max}^2$

55. 0.084 3 rad

59. (a) $(0.707)2(L/g)^{1/2}$ (b) $L/4$

61. 3.86×10^{-4}

63. (a) $\frac{\mu\omega^3}{2k}A_0^{\,2}e^{-2bx}$ (b) $\frac{\mu\omega^3}{2k}A_0^{\,2}$ (c) e^{-2bx}

65. (a) $\mu_0 + (\mu_L - \mu_0)x/L$

CHAPTER 17

1. 5.56 km. As long as the speed of light is much greater than the speed of sound, its actual value does not matter.

3. 0.196 s

5. 7.82 m

7. (a) 826 m (b) 1.47 s

9. (a) 0.625 mm (b) 1.50 mm to 75.0 μm

11. (a) 2.00 μm, 40.0 cm, 54.6 m/s (b) −0.433 μm (c) 1.72 mm/s

13. $\Delta P = (0.200 \text{ N/m}^2) \sin(62.8x/\text{m} - 2.16 \times 10^4 t/\text{s})$

15. 5.81 m

17. 66.0 dB

19. (a) 3.75 W/m² (b) 0.600 W/m²

21. (a) 2.34 m and 0.390 m (b) 0.161 N/m² for both notes (c) 4.25×10^{-7} m and 7.09×10^{-8} m (d) The wavelengths and displacement amplitudes would be larger by a factor of 1.09. The answer to part (b) would be unchanged.

23. (a) 1.32×10^{-4} W/m² (b) 81.2 dB

25. (a) 0.691 m (b) 691 km

27. 65.6 dB

29. (a) 30.0 m (b) 9.49×10^5 m

31. (a) 332 J (b) 46.4 dB

33. (a) 3.04 kHz (b) 2.08 kHz (c) 2.62 kHz, 2.40 kHz

35. 26.4 m/s

37. 19.3 m

39. (a) 56.3 s (b) 56.6 km farther along

41. 2.82×10^8 m/s

43. It is unreasonable, implying a sound level of 123 dB. Nearly all the missing mechanical energy becomes internal energy in the latch.

45. (a) f is a few hundred hertz. $\lambda \sim 1$ m, duration ~ 0.1 s. (b) Yes. The frequency can be close to 1 000 Hz. If the person clapping his or her hands is at the base of the pyramid, the echo can drop somewhat in frequency and in loudness as sound returns, with the later cycles coming from the smaller and more distant upper risers. The sound could imitate some particular bird and could in fact be a recording of the call.

49. (a) 0.515/min (b) 0.614/min

51. (a) 55.8 m/s (b) 2 500 Hz

53. 1 204.2 Hz

55. (a) 0.642 W (b) 0.004 28 = 0.428%

57. (a) The sound through the metal arrives first. (b) (365 m/s) Δt (c) 46.3 m (d) The answer becomes

$$\ell = \frac{\Delta t}{\dfrac{1}{331\text{ m/s}} - \dfrac{1}{v_r}}$$

where v_r is the speed of sound in the rod. As v_r goes to infinity, the travel time in the rod becomes negligible. The answer approaches (331 m/s) Δt, which is the distance the sound travels in air during the delay time.
59. (a) 0.948° (b) 4.40°
61. 1.34×10^4 N
63. (a) 6.45 (b) 0

CHAPTER 18

1. (a) −1.65 cm (b) −6.02 cm (c) 1.15 cm
3. (a) $+x$, $-x$ (b) 0.750 s (c) 1.00 m
5. (a) 9.24 m (b) 600 Hz
7. (a) 2 (b) 9.28 m and 1.99 m
9. (a) 156° (b) 0.058 4 cm
11. 15.7 m, 31.8 Hz, 500 m/s
13. At 0.089 1 m, 0.303 m, 0.518 m, 0.732 m, 0.947 m, 1.16 m from one speaker
15. (a) 4.24 cm (b) 6.00 cm (c) 6.00 cm (d) 0.500 cm, 1.50 cm, 2.50 cm
17. 0.786 Hz, 1.57 Hz, 2.36 Hz, 3.14 Hz
19. (a) 350 Hz (b) 400 kg
21. (a) 163 N (b) 660 Hz
23. $\dfrac{Mg}{4Lf^2 \tan\theta}$
25. (a) 3 loops (b) 16.7 Hz (c) 1 loop
27. (a) 3.66 m/s (b) 0.200 Hz
29. (a) 0.357 m (b) 0.715 m
31. 0.656 m and 1.64 m
33. n(206 Hz) for $n = 1$ to 9 and n(84.5 Hz) for $n = 2$ to 23
35. 50.0 Hz, 1.70 m
37. (a) 350 m/s (b) 1.14 m
39. (21.5 ± 0.1) m. The data suggest 0.6-Hz uncertainty in the frequency measurements, which is only a little more than 1%.
41. (a) 1.59 kHz (b) odd-numbered harmonics (c) 1.11 kHz
43. 5.64 beats/s
45. (a) 1.99 beats/s (b) 3.38 m/s
47. The second harmonic of E is close to the third harmonic of A, and the fourth harmonic of C# is close to the fifth harmonic of A.
49. (a) The yo-yo's downward speed is $dL/dt = (0.8\text{ m/s}^2)(1.2\text{ s}) = 0.960$ m/s. The instantaneous wavelength of the fundamental string wave is given by $d_{NN} = \lambda/2 = L$, so $\lambda = 2L$ and $d\lambda/dt = 2\,dL/dt = 2(0.96\text{ m/s}) = 1.92$ m/s. (b) For the second harmonic, the wavelength is equal to the length of the string. Then the rate of change of wavelength is equal to $dL/dt = 0.960$ m/s, half as much as for the first harmonic. (c) A yo-yo of different mass will hold the string under different tension to make each string wave vibrate with a different frequency, but the geometrical argument given in parts (a) and (b) still applies to the wavelength. The answers are unchanged: $d\lambda_1/dt = 1.92$ m/s and $d\lambda_2/dt = 0.960$ m/s.
51. (a) 34.8 m/s (b) 0.977 m
53. 3.85 m/s away from the station or 3.77 m/s toward the station
55. (a) 59.9 Hz (b) 20.0 cm
57. (a) $\frac{1}{2}$ (b) $[n/(n+1)]^2T$ (c) $\frac{9}{16}$
59. $y_1 + y_2 = 11.2 \sin(2.00x - 10.0t + 63.4°)$
61. (a) 78.9 N (b) 211 Hz

CHAPTER 19

1. (a) −274°C (b) 1.27 atm (c) 1.74 atm
3. (a) −320°F (b) 77.3 K
5. 3.27 cm
7. (a) 0.176 mm (b) 8.78 μm (c) 0.093 0 cm³
9. (a) −179°C is attainable. (b) −376°C is below 0 K and unattainable.
11. (a) 99.8 mL (b) about 6% of the volume change of the acetone
13. (a) 99.4 cm³ (b) 0.943 cm
15. 5 336 images
17. (a) 400 kPa (b) 449 kPa
19. 1.50×10^{29} molecules
21. 472 K
23. (a) 41.6 mol (b) 1.20 kg, nearly in agreement with the tabulated density
25. (a) 1.17 g (b) 11.5 mN (c) 1.01 kN (d) The molecules must be moving very fast.
27. 4.39 kg
29. (a) 7.13 m (b) The open end of the tube should be at the bottom after the bird surfaces so that the water can drain out. There is no other requirement. Air does not tend to bubble out of a narrow tube.
31. (a) 94.97 cm (b) 95.03 cm
33. 3.55 cm
35. It falls by 0.094 3 Hz.
37. (a) Expansion makes density drop. (b) $5 \times 10^{-5}(°\text{C})^{-1}$
39. (a) $h = nRT/(mg + P_0A)$ (b) 0.661 m
41. We assume $\alpha\,\Delta T$ is much less than 1.
43. Yes, as long as the coefficients of expansion remain constant. The lengths L_C and L_S at 0°C need to satisfy $17L_C = 11L_S$. Then the steel rod must be longer. With $L_S - L_C = 5.00$ cm, the only possibility is $L_S = 14.2$ cm and $L_C = 9.17$ cm.
45. (a) 0.340% (b) 0.480%
47. 2.74 m
49. (b) 1.33 kg/m³
53. No. Steel would need to be 2.30 times stronger.
55. (a) $L_f = L_i e^{\alpha\Delta T}$ (b) 2.00×10^{-4}%; 59.4%
57. (a) 6.17×10^{-3} kg/m (b) 632 N (c) 580 N; 192 Hz
59. 4.54 m

CHAPTER 20

1. (10.0 + 0.117)°C
3. 0.234 kJ/kg · °C
5. 1.78×10^4 kg
7. 29.6°C
9. (a) 0.435 cal/g · °C (b) We cannot make a definite identification. The material might be an unknown alloy or a material not listed in the table. It might be beryllium.
11. 23.6°C
13. 1.22×10^5 J
15. 0.294 g
17. 0.414 kg
19. (a) 0°C (b) 114 g
21. −1.18 MJ

23. −466 J

25. (a) $-4P_iV_i$ (b) It is proportional to the square of the volume, according to $T = (P_i/nRV_i)V^2$.

27. $Q = -720$ J

29.

	Q	W	ΔE_{int}
BC	−	0	−
CA	−	+	−
AB	+	−	+

31. (a) 7.50 kJ (b) 900 K

33. −3.10 kJ, 37.6 kJ

35. (a) 0.041 0 m³ (b) +5.48 kJ (c) −5.48 kJ

37. 10.0 kW

39. 51.2°C

41. 74.8 kJ

43. (a) 0.964 kg or more (b) The test samples and the inner surface of the insulation can be prewarmed to 37.0°C as the box is assembled. Then nothing changes in temperature during the test period, and the masses of the test samples and insulation make no difference.

45. 3.49×10^3 K

47. Intensity is defined as power per area perpendicular to the direction of energy flow. The direction of sunlight is along the line from the Sun to the object. The perpendicular area is the projected flat, circular area enclosed by the *terminator,* the line that separates day and night on the object. The object radiates infrared light outward in all directions. The area perpendicular to this energy flow is its spherical surface area. The steady-state surface temperature is 279 K = 6°C. We find this temperature to be chilly, well below comfortable room temperatures.

49. 2.27 km

51. (a) 16.8 L (b) 0.351 L/s

53. $c = \mathcal{P}/\rho R \Delta T$

55. 5.87×10^4°C

57. 5.31 h

59. 1.44 kg

61. 38.6 m³/d

63. 9.32 kW

65. (a) The equation $dT/dr = \mathcal{P}/4\pi kr^2$ represents the law of thermal conduction, incorporating the definition of thermal conductivity, applied to a spherical surface within the shell. The rate of energy transfer $\mathcal{P}$ must be the same for all radii so that each bit of material stays at a temperature that is constant in time. (b) We separate the variables T and r in the thermal conduction equation and integrate the equation between points on the interior and exterior surfaces. (c) 18.5 W (d) With $\mathcal{P}$ now known, we separate the variables again and integrate between a point on the interior surface and any point within the shell. (e) $T = 5°\text{C} + 184 \text{ cm} \cdot °\text{C}\,[1/(3 \text{ cm}) - 1/r]$ (f) 29.5°C

CHAPTER 21

1. (a) 4.00 u = 6.64×10^{-24} g (b) 55.9 u = 9.28×10^{-23} g (c) 207 u = 3.44×10^{-22} g

3. 0.943 N, 1.57 Pa

5. 3.21×10^{12} molecules

7. 3.32 mol

9. (a) 3.54×10^{23} atoms (b) 6.07×10^{-21} J (c) 1.35 km/s

11. (a) 8.76×10^{-21} J for both (b) 1.62 km/s for helium and 514 m/s for argon

13. (a) 3.46 kJ (b) 2.45 kJ (c) −1.01 kJ

15. Between 10^{-2}°C and 10^{-3}°C

17. 13.5PV

19. (a) 1.39 atm (b) 366 K, 253 K (c) 0, −4.66 kJ, −4.66 kJ

21. 227 K

23. (a)

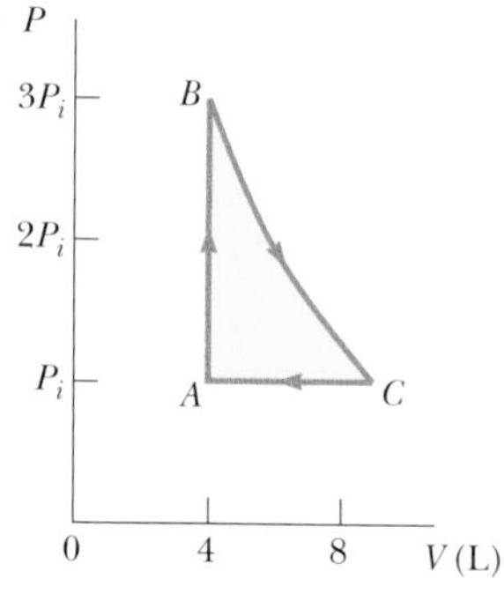

(b) 8.77 L (c) 900 K (d) 300 K (e) −336 J

25. (a) 28.0 kJ (b) 46.0 kJ (c) isothermal process: $P_f = 10.0$ atm; adiabatic process: $P_f = 25.1$ atm

27. (a) 9.95 cal/K, 13.9 cal/K (b) 13.9 cal/K, 17.9 cal/K

29. Sulfur dioxide is the gas in Table 21.2 with the greatest molecular mass. If the effective spring constants for various chemical bonds are comparable, SO_2 can then be expected to have low frequencies of atomic vibration. Vibration can be excited at lower temperature for sulfur dioxide than for the other gases. Some vibration may be going on at 300 K.

31. (a) 6.80 m/s (b) 7.41 m/s (c) 7.00 m/s

35. (a) 2.37×10^4 K (b) 1.06×10^3 K

37. (b) 0.278

39. (a) 100 kPa, 66.5 L, 400 K, 5.82 kJ, 7.48 kJ, −1.66 kJ
(b) 133 kPa, 49.9 L, 400 K, 5.82 kJ, 5.82 kJ, 0
(c) 120 kPa, 41.6 L, 300 K, 0, −909 J, +909 J
(d) 120 kPa, 43.3 L, 312 K, 722 J, 0, +722 J

41. (b) 447 J/kg·°C agrees with the tabulated value within 0.3%. (c) 127 J/kg·°C agrees with the tabulated value within 2%.

43. (b) The expressions are equal because $PV = nRT$ and $\gamma = (C_V + R)/C_V = 1 + R/C_V$ give $R = (\gamma - 1)C_V$, so $PV = n(\gamma - 1)C_VT$ and $PV/(\gamma - 1) = nC_VT$

45. 510 K and 290 K

47. 0.623

49. (a) Pressure increases as volume decreases. (d) 0.500 atm⁻¹, 0.300 atm⁻¹

51. (a) 7.27×10^{-20} J (b) 2.20 km/s (c) 3 510 K. The evaporating molecules are exceptional, at the high-speed tail of the distribution of molecular speeds. The average speed of molecules in the liquid and in the vapor is appropriate only to room temperature.

53. (a) 0.514 m³ (b) 2.06 m³ (c) 2.38×10^3 K (d) −480 kJ (e) 2.28 MJ

55. 1.09×10^{-3}, 2.69×10^{-2}, 0.529, 1.00, 0.199, 1.01×10^{-41}, $1.25 \times 10^{-1\,082}$

59. (a) 0.203 mol (b) $T_B = T_C = 900$ K, $V_C = 15.0$ L

(c, d)	P, atm	V, L	T, K	E_{int}, kJ
A	1.00	5.00	300	0.760
B	3.00	5.00	900	2.28
C	1.00	15.0	900	2.28
A	1.00	5.00	300	0.760

(e) Lock the piston in place and put the cylinder into an oven at 900 K. Keep the gas in the oven while gradually letting the gas expand to lift a load on the piston as far as it can. Move the cylinder from the oven back to the 300-K room and let the gas cool and contract.

(f, g)	Q, kJ	W, kJ	ΔE_{int}, kJ
AB	1.52	0	1.52
BC	1.67	−1.67	0
CA	−2.53	+1.01	−1.52
$ABCA$	0.656	−0.656	0

61. (b) 1.60×10^4 K

CHAPTER 22

1. (a) 6.94% (b) 335 J
3. (a) 10.7 kJ (b) 0.533 s
5. 55.4%
7. 77.8 W
9. (a) 67.2% (b) 58.8 kW
11. The actual efficiency of 0.069 8 is less than four-tenths of the Carnot efficiency of 0.177.
13. (a) 741 J (b) 459 J
15. (a) 564 K (b) 212 kW (c) 47.5%
17. (b) $1 - T_c/T_h$, the same as for a single reversible engine (c) $(T_c + T_h)/2$ (d) $(T_hT_c)^{1/2}$
19. 9.00
23. 72.2 J
25. 23.1 mW
27. (a) 244 kPa (b) 192 J
29. (a) 51.2% (b) 36.2%
33. 195 J/K
35. 1.02 kJ/K
37. $\sim 10^0$ W/K from metabolism; much more if you are using high-power electric appliances or an automobile
39. 5.76 J/K; the temperature is constant if the gas is ideal.
41. (a) 1 (b) 6
43. (a)

Result	Number of ways to draw
All R	1
2 R, 1 G	3
1 R, 2 G	3
All G	1

(b)

Result	Number of ways to draw
All R	1
4R, 1G	5
3R, 2G	10
2R, 3G	10
1R, 4G	5
All G	1

45. (a) 214 J, 64.3 J (b) −35.7 J, −35.7 J. The net effect would be the transport of energy by heat from the cold to the hot reservoir without expenditure of external work. (c) 333 J, 233 J (d) 83.3 J, 83.3 J, 0. The net effect would be converting energy, taken in by heat, entirely into energy output by work in a cyclic process. (e) −0.111 J/K. The entropy of the Universe would have decreased.
47. (a) 5.00 kW (b) 763 W
49. (a) $2nRT_i \ln 2$ (b) 0.273
51. 5.97×10^4 kg/s
53. (a) 8.48 kW (b) 1.52 kW (c) 1.09×10^4 J/K (d) The COP drops by 20.0%.
55. (a) $10.5nRT_i$ (b) $8.50nRT_i$ (c) 0.190 (d) This efficiency is much less than the 0.833 for a Carnot engine operating between the temperatures used here.
57. (a) $nC_P \ln 3$ (b) Both ask for the change in entropy between the same two states of the same system. Entropy is a state variable. The change in entropy does not depend on path, but only on original and final states.
61. (a) 20.0°C (c) $\Delta S = +4.88$ J/K (d) The mixing is irreversible. It is clear that warm water and cool water do not come unmixed, and the entropy change is positive.

CHAPTER 23

1. (a) +160 zC, 1.01 u (b) +160 zC, 23.0 u (c) −160 zC, 35.5 u (d) +320 zC, 40.1 u (e) −480 zC, 14.0 u (f) +640 zC, 14.0 u (g) +1.12 aC, 14.0 u (h) −160 zC, 18.0 u
3. The force is $\sim 10^{26}$ N.
5. (a) 1.59 nN away from the other (b) 1.24×10^{36} times larger (c) 8.61×10^{-11} C/kg
7. 0.872 N at 330°
9. (a) 2.16×10^{-5} N toward the other (b) 8.99×10^{-7} N away from the other
11. (a) 82.2 nN toward the other particle (b) 2.19 Mm/s
13. (a) 55.8 pN/C down (b) 102 nN/C up
15. The field at the origin can be to the right if the unknown charge is $-9Q$, or the field can be to the left if and only if the unknown charge is $+27Q$.
17. (a) $5.91k_eq/a^2$ at 58.8° (b) $5.91k_eq^2/a^2$ at 58.8°
19. (a) $k_eQx\hat{\mathbf{i}}/(R^2 + x^2)^{3/2}$ (b) As long as the charge is symmetrically placed, the number of charges does not matter. A continuous ring corresponds to n becoming larger without limit.
21. 1.59×10^6 N/C toward the rod
23. (a) $6.64\hat{\mathbf{i}}$ MN/C (b) $24.1\hat{\mathbf{i}}$ MN/C (c) $6.40\hat{\mathbf{i}}$ MN/C (d) $0.664\hat{\mathbf{i}}$ MN/C, taking the axis of the ring as the x axis
25. (a) 93.6 MN/C; the near-field approximation is 104 MN/C, about 11% high. (b) 0.516 MN/C; the charged-particle approximation is 0.519 MN/C, about 0.6% high.
27. $-21.6\hat{\mathbf{i}}$ MN/C
31. (a) 86.4 pC for each (b) 324 pC, 459 pC, 459 pC, 432 pC (c) 57.6 pC, 106 pC, 154 pC, 96.0 pC
33.

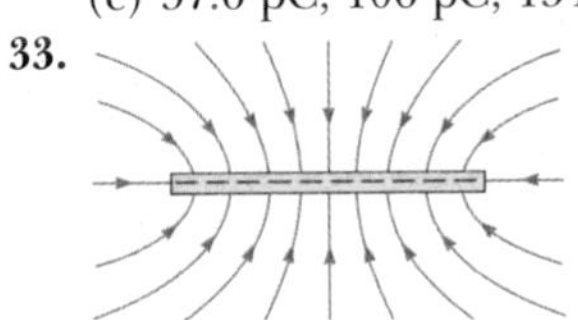

35. (a)

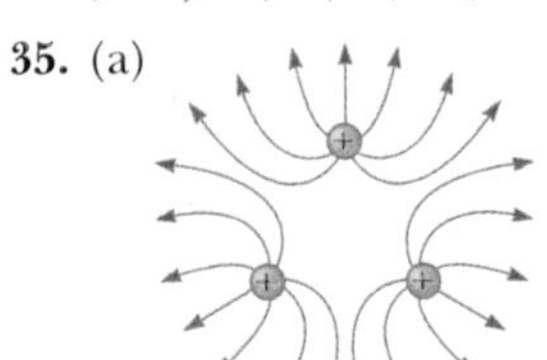

The field is zero at the center of the triangle. (b) $1.73\, k_eq\hat{\mathbf{j}}/a^2$
37. (a) 61.3 Gm/s² (b) 19.5 μs (c) 11.7 m (d) 1.20 fJ
39. K/ed in the direction of motion
41. (a) 111 ns (b) 5.68 mm (c) $(450\hat{\mathbf{i}} + 102\hat{\mathbf{j}})$ km/s

43. (a) 21.8 μm (b) 2.43 cm
45. (a) 10.9 nC (b) 5.44 mN
47. 40.9 N at 263°
49. $Q = 2L\sqrt{\dfrac{k(L - L_i)}{k_e}}$
53. $-707\hat{\mathbf{j}}$ mN
55. (a) $\theta_1 = \theta_2$
57. (a) 0.307 s (b) Yes. Ignoring gravity makes a difference of 2.28%.
59. (a) $\vec{\mathbf{F}} = 1.90(k_e q^2/s^2)(\hat{\mathbf{i}} + \hat{\mathbf{j}} + \hat{\mathbf{k}})$ (b) $\vec{\mathbf{F}} = 3.29(k_e q^2/s^2)$ in the direction away from the diagonally opposite vertex
65. $\dfrac{k_e \lambda_0}{2x_0}(-\hat{\mathbf{i}})$

CHAPTER 24

1. 4.14 MN/C
3. (a) aA (b) bA (c) 0
5. 1.87 kN · m²/C
7. (a) −6.89 MN · m²/C (b) The number of lines entering exceeds the number leaving by 2.91 times or more.
9. $-Q/\epsilon_0$ for S_1; 0 for S_2; $-2Q/\epsilon_0$ for S_3; 0 for S_4
11. (a) $+Q/2\epsilon_0$ (b) $-Q/2\epsilon_0$
13. −18.8 kN · m²/C
15. 0 if $R \leq d$; $(2\lambda/\epsilon_0)\sqrt{R^2 - d^2}$ if $R > d$
17. (a) 3.20 MN · m²/C (b) 19.2 MN · m²/C (c) The answer to part (a) could change, but the answer to part (b) would stay the same.
19. 2.33×10^{21} N/C
21. 508 kN/C up
23. −2.48 μC/m²
25. 5.94×10^5 m/s
27. $\vec{\mathbf{E}} = \rho r/2\epsilon_0$ away from the axis
29. (a) 0 (b) 7.19 MN/C away from the center
31. (a) 51.4 kN/C outward (b) 646 N · m²/C
33. (a) 0 (b) 5 400 N/C outward (c) 540 N/C outward
35. (a) +708 nC/m² and −708 nC/m² (b) +177 nC and −177 nC
37. 2.00 N
39. (a) $-\lambda, +3\lambda$ (b) $3\lambda/2\pi\epsilon_0 r$ radially outward
41. (a) 80.0 nC/m² on each face (b) $9.04\hat{\mathbf{k}}$ kN/C (c) $-9.04\hat{\mathbf{k}}$ kN/C
43. (b) $Q/2\epsilon_0$ (c) Q/ϵ_0
45. (a) The charge on the exterior surface is −55.7 nC distributed uniformly. (b) The charge on the interior surface is +55.7 nC. It might have any distribution. (c) The charge within the shell is −55.7 nC. It might have any distribution.
47. (a) $\rho r/3\epsilon_0$, $Q/4\pi\epsilon_0 r^2$, 0, $Q/4\pi\epsilon_0 r^2$, all radially outward (b) $-Q/4\pi b^2$ and $+Q/4\pi c^2$
49. $\theta = \tan^{-1}[qQ/(2\pi\epsilon_0 dmv^2)]$
51. (a) σ/ϵ_0 away from both plates (b) 0 (c) σ/ϵ_0 away from both plates
53. $\sigma/2\epsilon_0$ radially outward
57. $\vec{\mathbf{E}} = a/2\epsilon_0$ radially outward
61. (b) $\vec{\mathbf{g}} = GM_E r/R_E^3$ radially inward
63. (a) −4.00 nC (b) +9.56 nC (c) +4.00 nC and +5.56 nC
65. (a) If the volume charge density is nonzero, the field cannot be uniform in magnitude. (b) The field must be uniform in magnitude along any line in the direction of the field. The field magnitude can vary between points in a plane perpendicular to the field lines.

CHAPTER 25

1. (a) 152 km/s (b) 6.49 Mm/s
3. 1.67 MN/C
5. 38.9 V; the origin
7. (a) $2QE/k$ (b) QE/k (c) $2\pi\sqrt{m/k}$ (d) $2(QE - \mu_k mg)/k$
9. (a) 0.400 m/s (b) It is the same. Each bit of the rod feels a force of the same size as before.
11. (a) 1.44×10^{-7} V (b) -7.19×10^{-8} V (c) -1.44×10^{-7} V, $+7.19 \times 10^{-8}$ V
13. (a) 6.00 m (b) −2.00 μC
15. −11.0 MV
17. 8.95 J
21. (a) no point at a finite distance from the particles (b) $2k_e q/a$
23. (a) 10.8 m/s and 1.55 m/s (b) Greater. The conducting spheres will polarize each other, with most of the positive charge of one and of the negative charge of the other on their inside faces. Immediately before they collide, their centers of charge will be closer than their geometric centers, so they will have less electric potential energy and more kinetic energy.
25. $5k_e q^2/9d$
27. $\left[(1 + \sqrt{\tfrac{1}{8}})\dfrac{k_e q^2}{mL}\right]^{1/2}$
29. (a) 10.0 V, −11.0 V, −32.0 V (b) 7.00 N/C in the $+x$ direction
31. $\vec{\mathbf{E}} = (-5 + 6xy)\hat{\mathbf{i}} + (3x^2 - 2z^2)\hat{\mathbf{j}} - 4yz\hat{\mathbf{k}}$; 7.07 N/C
33. $E_y = \dfrac{k_e Q}{y\sqrt{\ell^2 + y^2}}$
35. (a) C/m² (b) $k_e\alpha[L - d\ln(1 + L/d)]$
37. −1.51 MV
39. (a) 0, 1.67 MV (b) 5.84 MN/C away, 1.17 MV (c) 11.9 MN/C away, 1.67 MV
41. (a) 248 nC/m² (b) 496 nC/m²
43. (a) 450 kV (b) 7.51 μC
45. (a) 1.42 mm (b) 9.20 kV/m
47. 253 MeV
49. (a) −27.2 eV (b) −6.80 eV (c) 0
51. (a) Yes. The inverse proportionality of potential to radius is sufficient to show that $200R = 150(R + 10\text{ cm})$, so $R = 30.0$ cm. Then $Q = 6.67$ nC. (b) Almost but not quite. Two possibilities exist: $R = 29.1$ cm with $Q = 6.79$ nC and $R = 3.44$ cm with $Q = 804$ pC.
53. 4.00 nC at (−1.00 m, 0) and −5.01 nC at (0, 2.00 m)
55. $k_e Q^2/2R$
57. $V_2 - V_1 = (-\lambda/2\pi\epsilon_0)\ln(r_2/r_1)$
61. (b) $E_r = 2k_e p\cos\theta/r^3$; $E_\theta = k_e p\sin\theta/r^3$; yes; no (c) $V = k_e py(x^2 + y^2)^{-3/2}$; $\vec{\mathbf{E}} = 3k_e pxy(x^2 + y^2)^{-5/2}\hat{\mathbf{i}} + k_e p(2y^2 - x^2)(x^2 + y^2)^{-5/2}\hat{\mathbf{j}}$
63. $V = \pi k_e C\left[R\sqrt{x^2 + R^2} + x^2\ln\left(\dfrac{x}{R + \sqrt{x^2 + R^2}}\right)\right]$
65. (a) 488 V (b) 78.1 aJ (c) 306 km/s (d) 390 Gm/s² toward the negative plate (e) 651 aN toward the negative plate (f) 4.07 kN/C
67. Outside the sphere, $E_x = 3E_0 a^3 xz(x^2 + y^2 + z^2)^{-5/2}$, $E_y = 3E_0 a^3 yz(x^2 + y^2 + z^2)^{-5/2}$, and $E_z = E_0 + E_0 a^3(2z^2 - x^2 - y^2)(x^2 + y^2 + z^2)^{-5/2}$. Inside the sphere, $E_x = E_y = E_z = 0$.

CHAPTER 26

1. (a) 48.0 μC (b) 6.00 μC
3. (a) 1.33 μC/m^2 (b) 13.3 pF
5. (a) 11.1 kV/m toward the negative plate (b) 98.3 nC/m^2 (c) 3.74 pF (d) 74.7 pC
7. 4.42 μm
9. (a) 2.68 nF (b) 3.02 kV
11. (a) 15.6 pF (b) 256 kV
13. (a) 3.53 μF (b) 6.35 V and 2.65 V (c) 31.8 μC on each
15. 6.00 pF and 3.00 pF
17. (a) 5.96 μF (b) 89.5 μC on 20 μF, 63.2 μC on 6 μF, 26.3 μC on 15 μF and on 3 μF
19. 120 μC; 80.0 μC and 40.0 μC
21. ten
23. 6.04 μF
25. 12.9 μF
27. (a) 216 μJ (b) 54.0 μJ
31. (a) 1.50 μC (b) 1.83 kV
35. 9.79 kg
37. (a) 81.3 pF (b) 2.40 kV
39. 1.04 m
41. 22.5 V
43. (b) -8.78×10^6 N/C·m; $-5.53 \times 10^{-2}\hat{\mathbf{i}}$ N
45. 19.0 kV
47. (a) 11.2 pF (b) 134 pC (c) 16.7 pF (d) 66.9 pC
49. (a) 40.0 μJ (b) 500 V
51. 0.188 m^2
55. Gasoline has 194 times the specific energy content of the battery and 727 000 times that of the capacitor.
57. (a) $Q_0^2 d(\ell - x)/(2\ell^3\epsilon_0)$ (b) $Q_0^2 d/(2\ell^3\epsilon_0)$ to the right (c) $Q_0^2/(2\ell^4\epsilon_0)$ (d) $Q_0^2/(2\ell^4\epsilon_0)$; they are precisely the same.
59. 4.29 μF
61. (a) The additional energy comes from work done by the electric field in the wires as it forces more charge onto the already-charged plates. (b) The charge increases according to $Q/Q_0 = \kappa$.
63. 750 μC on C_1 and 250 μC on C_2
65. $\frac{4}{3}C$

CHAPTER 27

1. 7.50×10^{15} electrons
3. (a) $0.632I_0\tau$ (b) $0.999\ 95I_0\tau$ (c) $I_0\tau$
5. (a) 17.0 A (b) 85.0 kA/m^2
7. (a) 2.55 A/m^2 (b) 5.31×10^{10} m^{-3} (c) 1.20×10^{10} s
9. (a) 221 nm (b) No. The deuterons are so far apart that one does not produce a significant potential at the location of the next.
11. 6.43 A
13. (a) 1.82 m (b) 280 μm
15. $6.00 \times 10^{-15}/\Omega\cdot$m
17. 0.180 V/m
19. (a) 31.5 n$\Omega\cdot$m (b) 6.35 MA/m^2 (c) 49.9 mA (d) 659 μm/s (e) 0.400 V
21. 0.125
23. 5.00 A, 24.0 Ω
25. 5.49 Ω
27. 36.1%
29. (a) 3.17 m (b) 340 W
31. (a) 0.660 kWh (b) $0.039 6
33. $0.232
35. $0.269/day
37. (a) 184 W (b) 461°C
39. ~ $1
41. Any diameter d and length ℓ related by $d^2 = (4.77 \times 10^{-8}\text{ m})\ell$, such as length 0.900 m and diameter 0.207 mm. Yes.
45. Experimental resistivity = 1.47 $\mu\Omega\cdot$m ±4%, in agreement with 1.50 $\mu\Omega\cdot$m
47. (a) 8.00 V/m in the x direction (b) 0.637 Ω (c) 6.28 A (d) 200 MA/m^2 in the x direction
49. (a) 667 A (b) 50.0 km
51.

Material	$\alpha' = \alpha/(1 - 20\alpha)$
Silver	4.1×10^{-3}/°C
Copper	4.2×10^{-3}/°C
Gold	3.6×10^{-3}/°C
Aluminum	4.2×10^{-3}/°C
Tungsten	4.9×10^{-3}/°C
Iron	5.6×10^{-3}/°C
Platinum	4.25×10^{-3}/°C
Lead	4.2×10^{-3}/°C
Nichrome	0.4×10^{-3}/°C
Carbon	-0.5×10^{-3}/°C
Germanium	-24×10^{-3}/°C
Silicon	-30×10^{-3}/°C

53. It is exact. The resistance can be written $R = \rho L^2/V$ and the stretched length as $L = L_i(1 + \delta)$. Then the result follows directly.
55. (b) Charge is conducted by current in the direction of decreasing potential. Energy is conducted by heat in the direction of decreasing temperature.
59. Coat the surfaces of entry and exit with a material of much higher conductivity than the bulk material of the object. The electric potential will be essentially uniform over each of these electrodes. Current will be distributed over the whole area where each electrode is in contact with the resistive object.
61. (a) $\dfrac{\epsilon_0\ell}{2d}(\ell + 2x + \kappa\ell - 2\kappa x)$ (b) $\dfrac{\epsilon_0\ell v\,\Delta V(\kappa - 1)}{d}$ clockwise
63. 2.71 MΩ
65. 2 020°C

CHAPTER 28

1. (a) 6.73 Ω (b) 1.97 Ω
3. (a) 12.4 V (b) 9.65 V
5. (a) 17.1 Ω (b) 1.99 A for 4 Ω and 9 Ω, 1.17 A for 7 Ω, 0.818 A for 10 Ω
7. (a) 227 mA (b) 5.68 V
9. (a) 75.0 V (b) 25.0 W, 6.25 W, and 6.25 W; 37.5 W
11. R_1 = 1.00 kΩ, R_2 = 2.00 kΩ, R_3 = 3.00 kΩ
13. It decreases. Closing the switch opens a new path with resistance of only 20 Ω. R = 14.0 Ω
15. 14.2 W to 2 Ω, 28.4 W to 4 Ω, 1.33 W to 3 Ω, 4.00 W to 1 Ω
17. 846 mA down in the 8-Ω resistor; 462 mA down in the middle branch; 1.31 A up in the right-hand branch
19. (a) −222 J and 1.88 kJ (b) 687 J, 128 J, 25.6 J, 616 J, 205 J (c) 1.66 kJ of chemical energy is transformed into internal energy.

21. 0.395 A and 1.50 V
23. 1.00 A up in 200 Ω, 4.00 A up in 70 Ω, 3.00 A up in 80 Ω, 8.00 A down in 20 Ω, 200 V
25. (a) 909 mA (b) $-1.82\text{ V} = V_b - V_a$
27. (a) 5.00 s (b) 150 μC (c) 4.06 μA
29. (a) −61.6 mA (b) 0.235 μC (c) 1.96 A
31. (a) 6.00 V (b) 8.29 μs
33. 0.302 Ω
35. 16.6 kΩ
37. (a) 12.5 A, 6.25 A, 8.33 A (b) No. Together they would require 27.1 A.
39. (a) 1.02 A down (b) 0.364 A down (c) 1.38 A up (d) 0 (e) 66.0 μC
41. 2.22 h
43. *a* is 4.00 V higher
45. 87.3 %
47. 6.00 Ω, 3.00 Ω
49. (a) $I_1 = \dfrac{IR_2}{R_1 + R_2}$, $I_2 = \dfrac{IR_1}{R_1 + R_2}$
51. (a) $R \leq 1\,050\ \Omega$ (b) $R \geq 10.0\ \Omega$
53. (a) 9.93 μC (b) 33.7 nA (c) 334 nW (d) 337 nW
55. (a) 40.0 W (b) 80.0 V, 40.0 V, 40.0 V
57. (a) 9.30 V, 2.51 Ω (b) 186 V and 3.70 A (c) 1.09 A (d) 143 W (e) 0.162 Ω (f) 3.00 mW (g) 2.21 W (h) The power output of the emf depends on the resistance connected to it. A question about "the rest of the power" is not meaningful when it compares circuits with different currents. The net emf produces more current in the circuit where the copper wire is used. The net emf delivers more power when the copper wire is used, 687 W rather than 203 W without the wire. Nearly all this power results in extra internal energy in the internal resistance of the batteries, which rapidly rise to a high temperature. The circuit with the copper wire is unsafe because the batteries overheat. The circuit without the copper wire is unsafe because it delivers an electric shock to the experimenter.
61. (a) 0 in 3 kΩ and 333 μA in 12 kΩ and 15 kΩ (b) 50.0 μC (c) $(278\ \mu\text{A})e^{-t/180\text{ ms}}$ (d) 290 ms
63. (a) $R_x = R_2 - R_1/4$ (b) $R_x = 2.75\ \Omega$. The station is inadequately grounded.
65. (a) $2\Delta t/3$ (b) $3\Delta t$

CHAPTER 29

1. (a) up (b) toward you, out of the plane of the paper (c) no deflection (d) into the plane of the paper
3. $(-20.9\hat{\mathbf{j}})$ mT
5. 48.9° or 131°
7. 2.34 aN
9. (a) 49.6 aN south (b) 1.29 km
11. $r_\alpha = r_d = \sqrt{2}r_p$
13. (a) 5.00 cm (b) 8.78×10^6 m/s
15. 7.88 pT
17. 244 kV/m
19. 0.278 m
21. (a) 4.31×10^7 rad/s (b) 51.7 Mm/s
23. 70.1 mT
25. 0.245 T east
27. (a) 4.73 N (b) 5.46 N (c) 4.73 N
29. 1.07 m/s
31. $2\pi rIB \sin\theta$ up
33. 2.98 μN west
35. 9.98 N · m clockwise as seen looking down from above
37. (a) Minimum: pointing north at 48.0° below the horizontal; maximum: pointing south at 48.0° above the horizontal. (b) 1.07 μJ
39. The magnetic moment cannot go to infinity. Its maximum value is 5.37 mA · m² for a single-turn circle. Smaller by 21% and by 40% are the magnetic moments for the single-turn square and triangle. Circular coils with several turns have magnetic moments inversely proportional to the number of turns, approaching zero as the number of turns goes to infinity.
41. 43.1 μT
43. (a) The electric current experiences a magnetic force. (c) no, no, no
45. 12.5 km. It will not hit the Earth, but it will perform a hairpin turn and go back parallel to its original direction.
47. (a) -8.00×10^{-21} kg · m/s (b) 8.90°
49. (a) $(3.52\hat{\mathbf{i}} - 1.60\hat{\mathbf{j}})$ aN (b) 24.4°
51. 128 mT north at an angle of 78.7° below the horizontal
53. 0.588 T
55. 0.713 A counterclockwise as seen from above
57. 2.75 Mrad/s
59. 3.70×10^{-24} N · m
61. (a) 1.33 m/s (b) Positive ions moving toward you in magnetic field to the right feel upward magnetic force and migrate upward in the blood vessel. Negative ions moving toward you feel downward magnetic force and accumulate at the bottom of this section of vessel. Therefore, both species can participate in the generation of the emf.
63. (a) $v = qBh/m$. If its speed is slightly less than the critical value, the particle moves in a semicircle of radius h and leaves the field with velocity $-v\hat{\mathbf{j}}$. If its speed is incrementally greater, the particle moves in a quarter circle of the same radius and moves along the boundary outside the field with velocity $v\hat{\mathbf{i}}$. (b) The particle moves in a smaller semicircle of radius mv/qB, attaining final velocity $-v\hat{\mathbf{j}}$. (c) The particle moves in a circular arc of radius $r = mv/qB$, leaving the field with velocity $v\sin\theta\hat{\mathbf{i}} + v\cos\theta\hat{\mathbf{j}}$, where $\theta = \sin^{-1}(h/r)$.
65. (a) For small angular displacements, the torque on the dipole is equal to a negative constant times the displacement.
(b) $f = \dfrac{1}{2\pi}\sqrt{\dfrac{\mu B}{I}}$
(c) The equilibrium orientation of the needle shows the direction of the field. In a stronger field, the frequency is higher. The frequency is easy to measure precisely over a wide range of values. 2.04 mT.

CHAPTER 30

1. 12.5 T
3. (a) 28.3 μT into the paper (b) 24.7 μT into the paper
5. $\dfrac{\mu_0 I}{4\pi x}$ into the paper
7. (a) $2I_1$ out of the page (b) $6I_1$ into the page
9. (a) along the line ($y = -0.420$ m, $z = 0$) (b) $(-34.7\hat{\mathbf{j}})$ mN (c) $(17.3\hat{\mathbf{j}})$ kN/C
11. at *A*, 53.3 μT toward the bottom of the page; at *B*, 20.0 μT toward the bottom of the page; at *C*, zero.
13. (a) $4.5\dfrac{\mu_0 I}{\pi L}$

(b) Stronger. Each of the two sides meeting at the nearby vertex contributes more than twice as much to the net field at the new point.

15. $(-13.0\hat{\mathbf{j}})\ \mu\text{T}$

17. $(-27.0\hat{\mathbf{i}})\ \mu\text{N}$

19. parallel to the wires and 0.167 m below the upper wire

21. (a) opposite directions (b) 67.8 A (c) Smaller. A smaller gravitational force would be pulling down on the wires, therefore tending to reduce the angle.

23. 20.0 μT toward the bottom of the page

25. at a, 200 μT toward the top of the page; at b, 133 μT toward the bottom of the page

27. (a) 6.34 mN/m inward (b) Greater. The magnetic field increases toward the outside of the bundle, where more net current lies inside a particular radius. The larger field exerts a stronger force on the strand we choose to monitor.

29. (a) 0

(b) $\dfrac{\mu_0 I}{2\pi R}$ tangent to the wall in a counterclockwise sense

(c) $\dfrac{\mu_0 I^2}{(2\pi R)^2}$ inward

31. (a) $\mu_0 b r_1{}^2/3$ (b) $\mu_0 b R^3/3r_2$

35. 31.8 mA

37. 226 μN away from the center of the loop, 0

39. (a) 3.13 mWb (b) 0

41. (a) 7.40 μWb (b) 2.27 μWb

43. 2.02

45. (a) 8.63×10^{45} electrons (b) 4.01×10^{20} kg

47. $\dfrac{\mu_0 I}{2\pi w} \ln\left(1 + \dfrac{w}{b}\right)\hat{\mathbf{k}}$

49. $(-12.0\hat{\mathbf{k}})$ mN

51. 143 pT

57. (a) 2.46 N upward (b) The magnetic field at the center of the loop or on its axis is much weaker than the magnetic field immediately outside the wire. The wire has negligible curvature on the scale of 1 mm, so we model the lower loop as a long, straight wire to find the field it creates at the location of the upper wire. (c) 107 m/s^2 upward

59. (a) 274 μT (b) $(-274\hat{\mathbf{j}})\ \mu\text{T}$ (c) $(1.15\hat{\mathbf{i}})$ mN (d) $(0.384\hat{\mathbf{i}})$ m/s^2 (e) acceleration is constant (f) $(0.999\hat{\mathbf{i}})$ m/s

61. $\dfrac{\mu_0 I_1 I_2 L}{\pi R}$ to the right

65. $\frac{1}{3}\rho\mu_0\omega R^2$

67. (a) $\dfrac{\mu_0 I(2r^2 - a^2)}{\pi r(4r^2 - a^2)}$ to the left

(b) $\dfrac{\mu_0 I(2r^2 + a^2)}{\pi r(4r^2 + a^2)}$ toward the top of the page

CHAPTER 31

1. (a) 101 μV tending to produce clockwise current as seen from above (b) It is twice as large in magnitude and in the opposite sense.

3. 9.82 mV

5. (b) 3.79 mV (c) 28.0 mV

7. 160 A

9. (a) 1.60 A counterclockwise (b) 20.1 μT (c) left

11. $-(14.2\text{ mV})\cos(120t)$

13. 283 μA upward

15. $(68.2\text{ mV})e^{-1.6t}$, tending to produce counterclockwise current

17. 272 m

19. 13.3 mA counterclockwise in the lower loop and clockwise in the upper loop

21. (a) 1.18 mV. The wingtip on the pilot's left is positive. (b) no change (c) No. If we try to connect the wings into a circuit with the lightbulb, we run an extra insulated wire along the wing. In a uniform field, the total emf generated in the one-turn coil is zero.

23. (a) 3.00 N to the right (b) 6.00 W

25. 24.1 V with the outer contact positive

27. 2.83 mV

29. (a) $F = N^2B^2w^2v/R$ to the left (b) 0 (c) $F = N^2B^2w^2v/R$ to the left

31. 145 μA upward in the picture

33. 1.80 mN/C upward and to the left, perpendicular to $\vec{r}_1$

35. (a) 7.54 kV (b) The plane of the loop is parallel to $\vec{\mathbf{B}}$.

37. $(28.6\text{ mV})\sin(4\pi t)$

39. (a) 110 V (b) 8.53 W (c) 1.22 kW

41. Both are correct. The current in the magnet creates an upward magnetic field, so the N and S poles on the solenoid core are shown correctly. On the rail in front of the brake, the upward magnetic flux increases as the coil approaches, so a current is induced here to create downward magnetic field. This current is clockwise, so the S pole on the rail is shown correctly. On the rail behind the brake, the upward magnetic flux is decreasing. The induced current in the rail will produce upward magnetic field by being counterclockwise  as the picture correctly shows.

43. (b) Larger R makes current smaller, so the loop must travel faster to maintain equality of magnetic force and weight. (c) The magnetic force is proportional to the product of field and current, while the current is itself proportional to field. If B becomes two times smaller, the speed must become four times larger to compensate.

45. $-(7.22\text{ mV})\cos(2\pi\ 523t/\text{s})$

47.

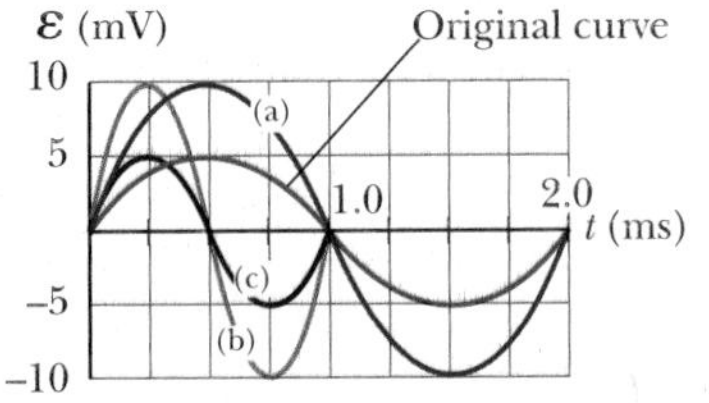

ℰ (mV)
Original curve
10
5
−5
−10
(a)
(b)
(c)
1.0
2.0
t (ms)

(a) Doubling N doubles amplitude. (b) Doubling ω doubles the amplitude and halves the period. (c) Doubling ω and halving N leaves the amplitude the same and cuts the period in half.

49. (a) 3.50 A up in 2 Ω, and 1.40 A up in 5 Ω (b) 34.3 W (c) 4.29 N

51. ~10^{-4} V, by reversing a 20-turn coil of diameter 3 cm in 0.1 s in a field of 10^{-3} T

53. 1.20 μC

55. (a) 0.900 A from b toward a (b) 0.108 N (c) b (d) No. Instead of decreasing downward magnetic flux to induce clockwise current, the new loop will see increasing downward flux to cause counterclockwise current, but the current in the resistor is still from b to a.

57. (a) $C\pi a^2K$ (b) the upper plate (c) The changing magnetic field within the loop induces an electric field around the circumference, which pushes on charged particles in the wire.

59. (a) 36.0 V (b) 600 mWb/s (c) 35.9 V (d) 4.32 N · m

63. 6.00 A

67. $(-87.1\text{ mV})\cos(200\pi t + \phi)$

CHAPTER 32

1. 100 V

3. $-(18.8\text{ V})\cos(377t)$

5. -0.421 A/s

7. (a) 188 μT (b) 33.3 nT · m^2 (c) 0.375 mH (d) B and Φ_B are proportional to current; L is independent of current

9. $\mathcal{E}_0/k^2L$

11. (a) 0.139 s (b) 0.461 s

13. (a) 2.00 ms (b) 0.176 A (c) 1.50 A (d) 3.22 ms

15. (a) 0.800 (b) 0

17. (a) 6.67 A/s (b) 0.332 A/s

19. (a) 1.00 kΩ (b) 3.00 ms

21. (a) 5.66 ms (b) 1.22 A (c) 58.1 ms

23. 2.44 μJ

25. 44.2 nJ/m^3 for the $\vec{\mathbf{E}}$ field and 995 μJ/m^3 for the $\vec{\mathbf{B}}$ field

27. (a) 66.0 W (b) 45.0 W (c) 21.0 W (d) At all instants after the connection is made, the battery power is equal to the sum of the power delivered to the resistor and the power delivered to the magnetic field. Immediately after $t = 0$, the resistor power is nearly zero and nearly all the battery power is going into the magnetic field. Long after the connection is made, the magnetic field is absorbing no more power and the battery power is going into the resistor.

29. $\dfrac{2\pi B_0^2R^3}{\mu_0} = 2.70 \times 10^{18}$ J

31. 1.00 V

33. (a) 18.0 mH (b) 34.3 mH (c) -9.00 mV

35. $M = \dfrac{N_1N_2\pi\mu_0R_1^2R_2^2}{2(x^2 + R_1^2)^{3/2}}$

37. 400 mA

39. 281 mH

41. 608 pF

43. (a) 6.03 J (b) 0.529 J (c) 6.56 J

45. (a) 4.47 krad/s (b) 4.36 krad/s (c) 2.53%

47. (a) $0.693(2L/R)$ (b) $0.347(2L/R)$

49. (a) -20.0 mV (b) $-(10.0\text{ MV/s}^2)t^2$ (c) 63.2 μs

51. $(Q/2N)(3L/C)^{1/2}$

53. (a) Immediately after the circuit is connected, the potential difference across the resistor is zero and the emf across the coil is 24.0 V. (b) After several seconds, the potential difference across the resistor is 24.0 V and that across the coil is 0. (c) The two voltages are equal to each other, both being 12.0 V, only once, at 0.578 ms after the circuit is connected. (d) As the current decays, the potential difference across the resistor is always equal to the emf across the coil.

55.

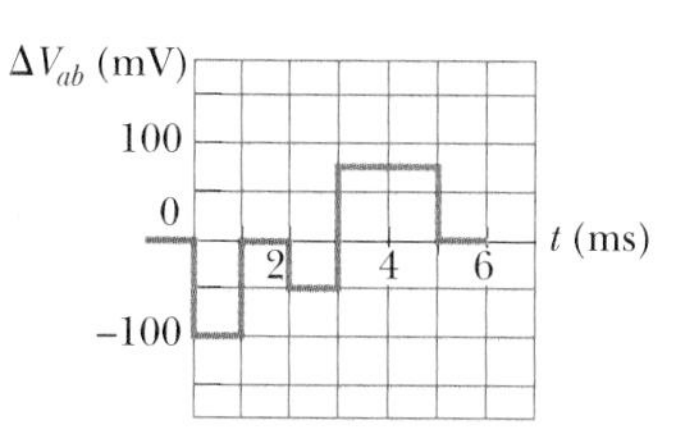

57. (b) 91.2 μH (c) 90.9 μH is only 0.3% smaller

61. (a) 72.0 V; b
(b)

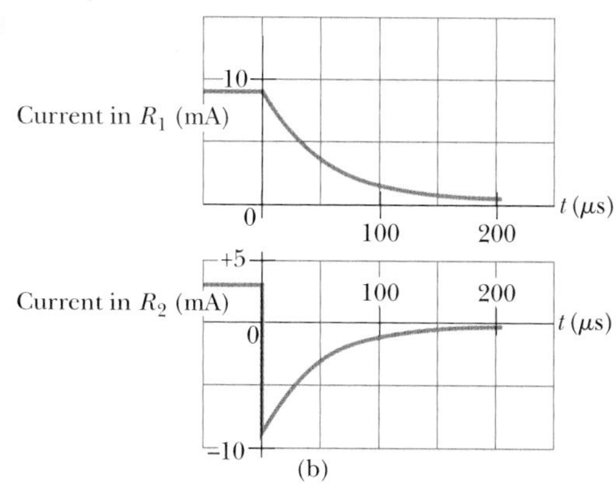

(c) 75.2 μs

63. 300 Ω

65. (a) 62.5 GJ (b) 2 000 N

67. (a) 2.93 mT up (b) 3.42 Pa (c) The supercurrents must be clockwise to produce a downward magnetic field that cancels the upward field of the current in the windings. (d) The field of the windings is upward and radially outward around the top of the solenoid. It exerts a force radially inward and upward on each bit of the clockwise supercurrent. The total force on the supercurrents in the bar is upward. (e) 1.30 mN

CHAPTER 33

1. $\Delta v(t) = (283\text{ V})\sin(628t)$

3. 2.95 A, 70.7 V

5. 14.6 Hz

7. (a) 42.4 mH (b) 942 rad/s

9. 5.60 A

11. 0.450 Wb

13. (a) 141 mA (b) 235 mA

15. 100 mA

17. (a) 194 V (b) current leads by 49.9°

19. (a) 78.5 Ω (b) 1.59 kΩ (c) 1.52 kΩ (d) 138 mA (e) $-84.3°$

21. (a) 17.4° (b) The voltage leads the current.

23. 1.88 V

25.

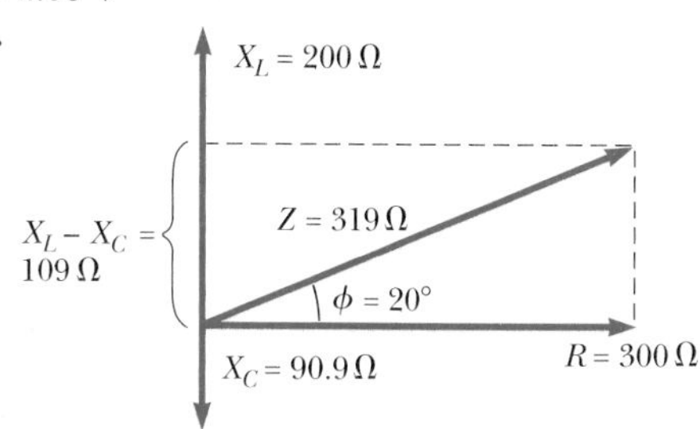

27. 8.00 W

29. (a) 16.0 Ω (b) $-12.0\ \Omega$

31. (a) 39.5 V · m/ΔV (b) The diameter is inversely proportional to the potential difference. (c) 26.3 mm (d) 13.2 kV

33. $11(\Delta V_{rms})^2/14R$
35. 1.82 pF
37. 242 mJ
39. 0.591 and 0.987; the circuit in Problem 21
41. 687 V
43. (a) 29.0 kW (b) 5.80×10^{-3} (c) If the generator were limited to 4 500 V, no more than 17.5 kW could be delivered to the load, never 5 000 kW.
45. (b) 0; 1 (c) $f_h = (10.88RC)^{-1}$
47. (a) 613 μF (b) 0.756
49. (a) 580 μH and 54.6 μF (b) 1 (c) 894 Hz (d) Δv_{out} leads Δv_{in} by 60.0° at 200 Hz. Δv_{out} and Δv_{in} are in phase at 894 Hz. Δv_{out} lags Δv_{in} by 60.0° at 4 000 Hz. (e) 1.56 W, 6.25 W, 1.56 W (f) 0.408
51. (a) X_C could be 53.8 Ω or it could be 1.35 kΩ. (b) X_C must be 53.8 Ω. (c) X_C must be 1.43 kΩ.
53. 56.7 W
55. Tension T and separation d must be related by $T = (274 \text{ N/m}^2)d^2$. One possibility is $T = 10.9$ N and $d = 0.200$ m.
57. (a) 225 mA (b) 450 mA
59. (a) 1.25 A (b) The current lags the voltage by 46.7°.
61. (a) 200 mA; voltage leads by 36.8° (b) 40.0 V; $\phi = 0°$ (c) 20.0 V; $\phi = -90.0°$ (d) 50.0 V; $\phi = +90.0°$
63. (b) 31.6
67. (a)

f (Hz)	X_L (Ω)	X_C (Ω)	Z (Ω)
300	283	12 600	12 300
600	565	6 280	5 720
800	754	4 710	3 960
1 000	942	3 770	2 830
1 500	1 410	2 510	1 100
2 000	1 880	1 880	40.0
3 000	2 830	1 260	1 570
4 000	3 770	942	2 830
6 000	5 650	628	5 020
10 000	9 420	377	9 040

(b)

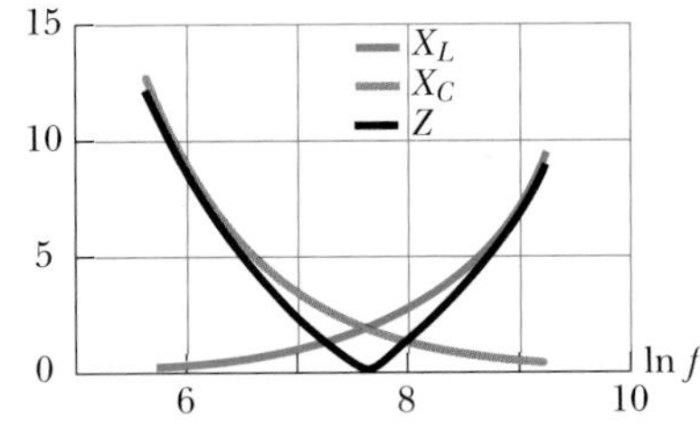

69. (a) and (b) 19.7 cm at 35.0°. The answers are identical. (c) 9.36 cm at 169°

CHAPTER 34

1. (a) 11.3 GV · m/s (b) 0.100 A
3. 1.85 aT up
5. $(-2.87\hat{\mathbf{j}} + 5.75\hat{\mathbf{k}})$ Gm/s^2
7. (a) the year 2.69×10^3 (b) 499 s (c) 2.56 s (d) 0.133 s (e) 33.3 μs
9. (a) 6.00 MHz (b) $(-73.3\hat{\mathbf{k}})$ nT (c) $\vec{\mathbf{B}} = [(-73.3\hat{\mathbf{k}}) \text{ nT}] \cos(0.126x - 3.77 \times 10^7 t)$
11. (a) 0.333 μT (b) 0.628 μm (c) 477 THz
13. 75.0 MHz
15. 3.33 μJ/m^3
17. 307 μW/m^2
19. 3.33×10^3 m^2
21. (a) 332 kW/m^2 radially inward (b) 1.88 kV/m and 222 μT
23. (a) $\vec{\mathbf{E}} \cdot \vec{\mathbf{B}} = 0$ (b) $(11.5\hat{\mathbf{i}} - 28.6\hat{\mathbf{j}})$ W/m^2
25. (a) 2.33 mT (b) 650 MW/m^2 (c) 510 W
27. (a) 88.8 nW/m^2 (b) 11.3 MW
29. 83.3 nPa
31. (a) 1.90 kN/C (b) 50.0 pJ (c) 1.67×10^{-19} kg · m/s
33. (a) 590 W/m^2 (b) 2.10×10^{16} W (c) 70.1 MN (d) The gravitational force is ~ 10^{13} times stronger and in the opposite direction. (e) On the Earth, the Sun's gravitational force is also ~ 10^{13} times stronger than the light-pressure force and in the opposite direction.
35. (a) 134 m (b) 46.9 m
37. (a) away along the perpendicular bisector of the line segment joining the antennas (b) along the extensions of the line segment joining the antennas
39. (a) $\vec{\mathbf{E}} = \frac{1}{2}\mu_0 c J_{max}[\cos(kx - \omega t)]\hat{\mathbf{j}}$ (b) $\vec{\mathbf{S}} = \frac{1}{4}\mu_0 c J_{max}^2[\cos^2(kx - \omega t)]\hat{\mathbf{i}}$ (c) $I = \dfrac{\mu_0 c J_{max}^2}{8}$ (d) 3.48 A/m
41. (a) 6.00 pm (b) 7.50 cm
43. (a) 4.17 m to 4.55 m (b) 3.41 m to 3.66 m (c) 1.61 m to 1.67 m
45. 1.00 Mm = 621 mi; not very practical
47. (a) 3.85×10^{26} W (b) 1.02 kV/m and 3.39 μT
49. (a)

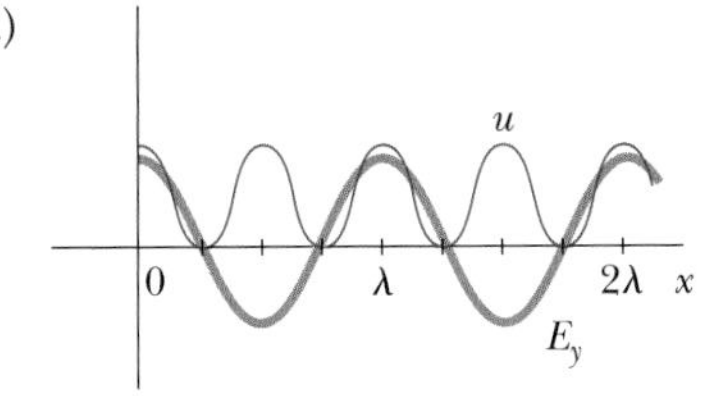

(b), (c) $u_E = u_B = \frac{1}{2}\epsilon_0 E_{max}^2 \cos^2(kx)$
(d) $u = \epsilon_0 E_{max}^2 \cos^2(kx)$ (e) $E_\lambda = \frac{1}{2}A\lambda\epsilon_0 E_{max}^2$
(f) $I = \frac{1}{2}c\epsilon_0 E_{max}^2 = \frac{1}{2}\sqrt{\dfrac{\epsilon_0}{\mu_0}}\,E_{max}^2$. This result agrees with $I = \dfrac{E_{max}^2}{2\mu_0 c}$ in Equation 34.24.
51. (a) 6.67×10^{-16} T (b) 5.31×10^{-17} W/m^2 (c) 1.67×10^{-14} W (d) 5.56×10^{-23} N
53. 95.1 mV/m
55. (a) $B_{max} = 583$ nT, $k = 419$ rad/m, $\omega = 126$ Grad/s; $\vec{\mathbf{B}}$ vibrates in xz plane (b) $\vec{\mathbf{S}}_{avg} = (40.6\hat{\mathbf{i}})$ W/m^2 (c) 271 nPa (d) $(406\hat{\mathbf{i}})$ nm/s^2
57. (a) 22.6 h (b) 30.6 s
59. (a) 8.32×10^7 W/m^2 (b) 1.05 kW
61. (b) 17.6 Tm/s^2, 1.75×10^{-27} W (c) 1.80×10^{-24} W
63. (a) $2\pi^2 r^2 f B_{max} \cos\theta$, where θ is the angle between the magnetic field and the normal to the loop (b) The loop

should be in the vertical plane containing the line of sight to the transmitter.

65. (a) 388 K (b) 363 K

CHAPTER 35

1. 300 Mm/s. The sizes of the objects need to be accounted for; otherwise, the answer would be too large by 2%.
3. 114 rad/s
5. (a) 1.94 m (b) 50.0° above the horizontal
7. 23.3°
9. 25.5°, 442 nm
11. 19.5° above the horizon
13. 22.5°
15. (a) 181 Mm/s (b) 225 Mm/s (c) 136 Mm/s
17. 30.0° and 19.5° at entry; 19.5° and 30.0° at exit
19. 3.88 mm
21. 30.4° and 22.3°
23. (a) yes, if the angle of incidence is 58.9° (b) No. Both the reduction in speed and the bending toward the normal reduce the component of velocity parallel to the interface. This component cannot remain constant unless the angle of incidence is 0°.
25. 86.8°
27. 27.9°
29. (b) 37.2° (c) 37.3° (d) 37.3°
31. 4.61°
33. (a) 24.4° (b) 37.0° (c) 49.8°
35. 1.000 08
37. (a) $dn/(n-1)$ (b) yes; yes; yes (c) 350 μm
39. Skylight incident from above travels down the plastic. If the index of refraction of the plastic is greater than 1.41, the rays close in direction to the vertical are totally reflected from the side walls of the slab and from both facets at the bottom of the plastic, where it is not immersed in gasoline. This light returns up inside the plastic and makes it look bright. Unless the index of refraction of the plastic is unrealistically large (greater than about 2.1), total internal reflection is frustrated where the plastic is immersed in gasoline. There the downward-propagating light passes from the plastic out into the gasoline. Little light is reflected up, and the gauge looks dark.
41. Scattered light leaving the photograph in all forward horizontal directions in air is gathered by refraction into a fan in the water of half-angle 48.6°. At larger angles, you see things on the other side of the globe, reflected by total internal reflection at the back surface of the cylinder.
43. 77.5°
45. 2.27 m
47. 62.2%
49. 82 reflections
51. 27.5°
53. (a) Total internal reflection occurs for all values of θ, or the maximum angle is 90°. (b) 30.3° (c) Total internal reflection never occurs as the light moves from lower-index polystyrene into higher-index carbon disulfide.
55. 2.36 cm
57. $\theta = \sin^{-1}\left[\dfrac{L}{R^2}\left(\sqrt{n^2R^2 - L^2} - \sqrt{R^2 - L^2}\right)\right]$
61. (a) nR_1 (b) R_2
63. (a) 1.20 (b) 3.40 ns

CHAPTER 36

1. $\sim 10^{-9}$ s younger
3. 35.0 in.
5. (a) $-(p_1 + h)$ (b) virtual (c) upright (d) 1.00 (e) no
7. (a) −12.0 cm; 0.400 (b) −15.0 cm; 0.250 (c) upright
9. (a) q = 45.0 cm; M = −0.500
(b) q = −60.0 cm; M = 3.00
(c) Image (a) is real, inverted, and diminished. Image (b) is virtual, upright, and enlarged.

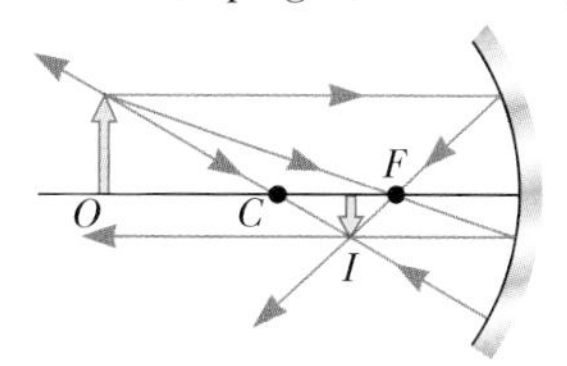

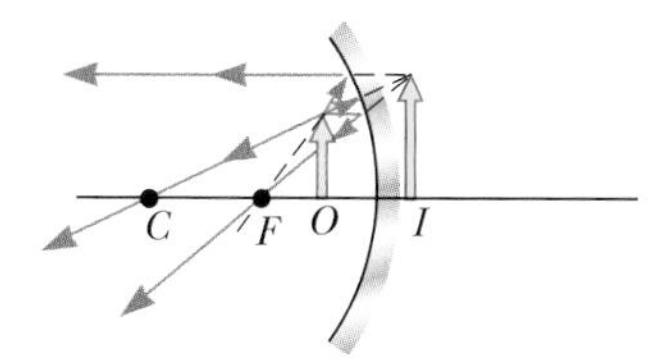

11. (a) 2.22 cm (b) 10.0
13. (a) 160 mm (b) R = −267 mm
15. (a) convex (b) at the 30.0 cm mark (c) −20.0 cm
17. (a) a concave mirror with radius of curvature 2.08 m
(b) 1.25 m from the object
19. (a) The image starts 60.0 cm above the mirror and moves up faster and faster, running out to an infinite distance above the mirror. At that moment, the image rays are parallel and the image is equally well described as infinitely far below the mirror. From there, the image moves up, slowing down as it moves, to reach the mirror vertex. (b) at 0.639 s and 0.782 s
21. 38.2 cm below the top surface of the ice
23. 8.57 cm
25. (a) inside the tank, 24.9 cm behind the front wall; virtual, right side up, enlarged (b) inside the tank, 93.9 cm behind the front wall; virtual, right side up, enlarged (c) 1.10 and 1.39 (d) 9.92 cm and 12.5 cm (e) The plastic has uniform thickness, so the surfaces of entry and exit for any particular ray are very nearly parallel. The ray is slightly displaced, but it would not be changed in direction by going through the plastic wall with air on both sides. Only the difference between the air and water is responsible for the refraction of the light.
27. (a) 16.4 cm (b) 16.4 cm
29. (a) q = 40.0 cm, real and inverted, actual size M = −1.00 (b) $q = \infty$, $M = \infty$, no image is formed (c) q = −20.0 cm, upright, virtual, enlarged M = +2.00
31. 2.84 cm
33. (a) −12.3 cm, to the left of the lens (b) 0.615
(c)

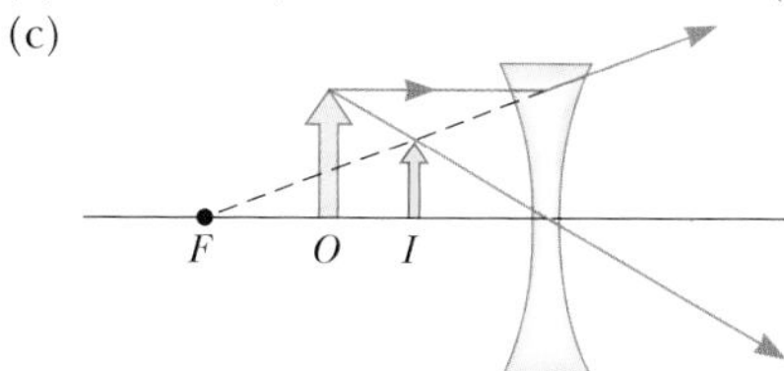

35. (a) 5.36 cm (b) −18.8 cm (c) virtual, right side up, enlarged (d) A magnifying glass with focal length 7.50 cm is used to form an image of a stamp, enlarged 3.50 times. Find the object distance. Locate and describe the image.

37. (a) $p = \frac{d}{2} \pm \sqrt{\frac{d^2}{4} - fd}$
(b) Both images are real and inverted. One is enlarged, the other diminished.
39. 2.18 mm away from the film plane
41. 21.3 cm
43. −4.00 diopters, a diverging lens
45. (a) at 4.17 cm (b) 6.00
47. (a) −800 (b) image is inverted
49. 3.38 min
51. −40.0 cm
53. −25.0 cm
55. $x' = (1\,024 \text{ cm} - 58x) \text{ cm}/(6 \text{ cm} - x)$. The image starts at the position $x_i' = 171$ cm and moves in the positive x direction, faster and faster, until it is out at infinity when the object is at the position $x = 6$ cm. At this instant, the rays from the top of the object are parallel as they leave the lens. Their intersection point can be described as at $x' = \infty$ to the right or equally well as at $x' = -\infty$ on the left. From $x' = -\infty$, the image continues moving to the right, now slowing down. It reaches, for example, −280 cm when the object is at 8 cm and −55 cm when the object is finally at 12 cm. The image has traveled always to the right, to infinity and beyond.
57. Align the lenses on the same axis and 9.00 cm apart. Let the light pass first through the diverging lens and then through the converging lens. The diameter increases by a factor of 1.75.
59. 0.107 m to the right of the vertex of the hemispherical face
61. 8.00 cm. Ray diagram:

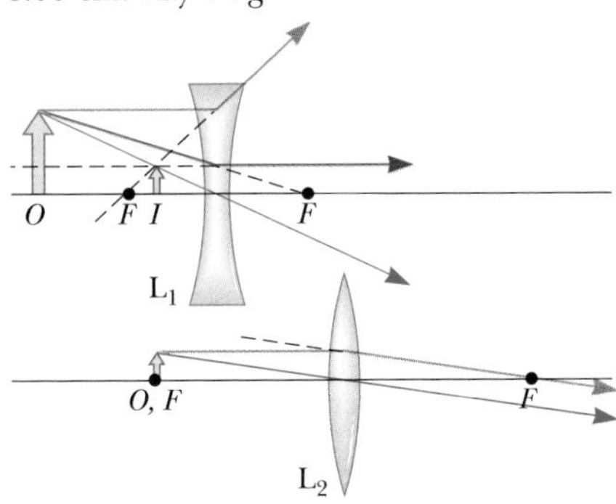

63. 1.50 m in front of the mirror; 1.40 cm (inverted)
65. (a) 30.0 cm and 120 cm (b) 24.0 cm (c) real, inverted, diminished with $M = -0.250$
67. (a) 20.0 cm to the right of the second lens, −6.00
(b) inverted (c) 6.67 cm to the right of the second lens, −2.00, inverted

CHAPTER 37

1. 1.58 cm
3. (a) 55.7 m (b) 124 m
5. 1.54 mm
7. (a) 2.62 mm (b) 2.62 mm
9. 36.2 cm
11. (a) 10.0 m (b) 516 m (c) Only the runway centerline is a maximum for the interference patterns for both frequencies. If the frequencies were related by a ratio of small integers k/ℓ, the plane could by mistake fly along the kth side maximum of one signal, where it coincides with the ℓth side maximum of the other. The plane cannot make a sharp turn at the end of the runway, so it would then be headed in the wrong direction to land.
13. (a) 13.2 rad (b) 6.28 rad (c) 0.012 7 degree
(d) 0.059 7 degree
15. (a) 1.93 μm (b) 3.00λ (c) It corresponds to a maximum. The path difference is an integer multiple of the wavelength.
17. 48.0 μm
19. $E_1 + E_2 = 10.0 \sin(100\pi t + 0.927)$
21. (a) 7.95 rad (b) 0.453
23. (a) green (b) violet
25. 512 nm
27. 96.2 nm
29. (a) 238 nm (b) The wavelength increases because of thermal expansion of the filter material. (c) 328 nm
31. 4.35 μm
33. 1.20 mm
35. 39.6 μm
37. 1.62 cm
39. 1.25 m
41. (a) ~ 10^{-3} degree (b) ~ 10^{11} Hz, microwave
43. 2.52 cm
45. 20.0×10^{-6} °C^{-1}
47. 3.58°
49. 1.62 km
51. 421 nm
55. (b) 266 nm
57. $y' = (n - 1)tL/d$
59. (a) 70.6 m (b) 136 m
61. (a) 14.7 μm (b) 1.53 cm (c) −16.0 m
63. (a) 4.86 cm from the top (b) 78.9 nm and 128 nm
(c) 2.63×10^{-6} rad
65. 0.505 mm
67. 0.498 mm

CHAPTER 38

1. 4.22 mm
3. 0.230 mm
5. three maxima, at 0° and near 46° on both sides
7. 0.016 2
9.

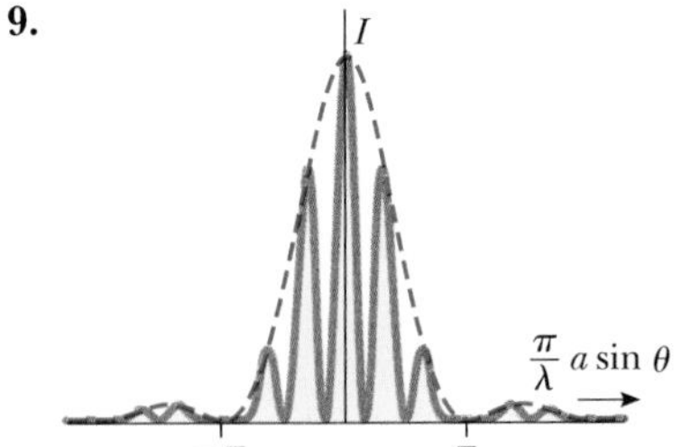

11. 1.00 mrad
13. 3.09 m
15. 13.1 m
17. Neither. It can resolve no objects closer than several centimeters apart.
19. 7.35°
21. 5.91° in first order, 13.2° in second order, 26.5° in third order
23. (a) 478.7 nm, 647.6 nm, and 696.6 nm (b) 20.51°, 28.30°, and 30.66°
25. three, at 0° and at 45.2° to the right and left.

27. (a) five orders (b) ten orders in the short-wavelength region

29. 2

31. 14.4°

33. The crystal cannot produce diffracted beams of visible light. Bragg's law cannot be satisfied for a wavelength much larger than the distance between atomic planes in the crystal.

35. (a) 54.7° (b) 63.4° (c) 71.6°

37. 60.5°

39. (b) For light confined to a plane, yes.
$$\left|\tan^{-1}\left(\frac{n_3}{n_2}\right) - \tan^{-1}\left(\frac{n_1}{n_2}\right)\right|$$

41. (a) 0.875 (b) 0.789 (c) 0.670 (d) You can get more and more of the incident light through the stack of ideal filters, approaching 50%, by reducing the angle between the axes of each one and the next.

43. (a) 6 (b) 7.50°

45. (a) 0.045 0 (b) 0.016 2

47. 632.8 nm

49. (a) 25.6° (b) 19.0°

51. 545 nm

53. (a) 3.53×10^3 grooves/cm (c) 11 maxima

55. 4.58 μm $< d <$ 5.23 μm

57. 15.4

59. (a) 41.8° (b) 0.593 (c) 0.262 m

61. (b) 3.77 nm/cm

63. (b) 15.3 μm

65. ϕ = 1.391 557 4 after 17 steps or fewer

67. a = 99.5 μm ± 1%

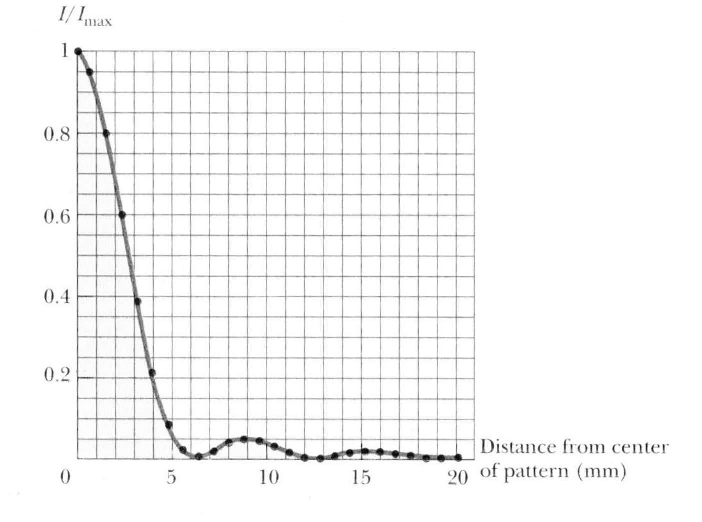

CHAPTER 39

5. $0.866c$

7. 1.54 ns

9. $0.800c$

11. (a) 39.2 μs (b) It was accurate to one digit. Cooper aged 1.78 μs less in each single orbit.

13. (a) 20.0 m (b) 19.0 m (c) $0.312c$

15. 11.3 kHz

17. (b) $0.050\,4c$

19. (a) $0.943c$ (b) 2.55 km

21. B occurred 4.44×10^{-7} s before A

23. $0.960c$

25. (a) 2.73×10^{-24} kg · m/s (b) 1.58×10^{-22} kg · m/s (c) 5.64×10^{-22} kg · m/s

27. 4.50×10^{-14}

29. $0.285c$

31. (a) 5.37×10^{-11} J (b) 1.33×10^{-9} J

33. 1.63×10^3 MeV/c

35. (a) 938 MeV (b) 3.00 GeV (c) 2.07 GeV

39. (a) $0.979c$ (b) $0.065\,2c$ (c) $0.914c$ = 274 Mm/s (d) $0.999\,999\,97c$; $0.948c$; $0.052\,3c$ = 15.7 Mm/s

41. 4.08 MeV and 29.6 MeV

43. smaller by 3.18×10^{-12} kg, which is too small a fraction of 9 g to be measured

45. 4.28×10^9 kg/s

47. (a) 26.6 Mm (b) 3.87 km/s (c) -8.34×10^{-11} (d) 5.29×10^{-10} (e) $+4.46 \times 10^{-10}$

49. (a) a few hundred seconds (b) ~ 10^8 km

51. (a) $u = c\left(\dfrac{2H + H^2}{1 + 2H + H^2}\right)^{1/2}$, where $H = K/mc^2$

(b) u goes to 0 as K goes to 0. (c) u approaches c as K increases without limit.

(d) $a = \dfrac{\mathcal{P}}{mcH^{1/2}(2 + H)^{1/2}(1 + H)^2}$

(e) $a = \dfrac{\mathcal{P}}{mc\,(2H)^{1/2}} = \dfrac{\mathcal{P}}{(2mK)^{1/2}}$, in agreement with the nonrelativistic case.

(f) a approaches $\mathcal{P}/mcH^3 = \mathcal{P}m^2c^5/K^3$ (g) As energy is steadily imparted to the particle, the particle's acceleration decreases. It decreases steeply, proportionally to $1/K^3$ at high energy. In this way, the particle's speed cannot reach or surpass a certain upper limit, which is the speed of light in vacuum.

53. 0.712%

55. (a) 76.0 min (b) 52.1 min

57. (a) $0.946c$ (b) 0.160 ly (c) 0.114 yr (d) 7.50×10^{22} J

59. yes, with 18.8 m to spare

61. (b) For u small compared with c, the relativistic expression agrees with the classical expression. As u approaches c, the acceleration approaches zero, so the object can never reach or surpass the speed of light.
(c) Perform the operation $\int(1 - u^2/c^2)^{-3/2}du = (qE/m)\int dt$ to obtain $u = qEct(m^2c^2 + q^2E^2t^2)^{-1/2}$ and then $\int dx = \int qEct(m^2c^2 + q^2E^2t^2)^{-1/2}dt$ to obtain $x = (c/qE)[(m^2c^2 + q^2E^2t^2)^{1/2} - mc]$

63. (a) $M = \dfrac{2m\sqrt{4 - u^2/c^2}}{3\sqrt{1 - u^2/c^2}}$

(b) $M = 4m/3$. The result agrees with the arithmetic sum of the masses of the two colliding particles.

65. (a) The refugees conclude that Tau Ceti exploded 16.0 yr before the Sun. (b) A stationary observer at the midpoint concludes that Tau Ceti and the Sun exploded simultaneously.

67. 1.82×10^{-3} eV

CHAPTER 40

1. 5.18×10^3 K

3. approximately 5 200 K. A firefly cannot be at this temperature, so its light cannot be blackbody radiation.

5. 1.30×10^{15}/s

7. (a) 5.78×10^3 K (b) 501 nm

9. (a) 2.57 eV (b) 12.8 μeV (c) 191 neV (d) 484 nm (visible), 9.68 cm, and 6.52 m (radio waves)
11. 2.27×10^{30} photons/s
15. (a) 1.90 eV (b) 0.216 V
17. (a) We find the energy of a photon with wavelength 400 nm and check whether it exceeds the work function. Of these metals, only lithium shows the photoelectric effect. (b) 0.808 eV
19. 8.41 pC
21. 1.78 eV, 9.47×10^{-28} kg · m/s
23. 70.0°
25. (a) 43.0° (b) 602 keV, 3.21×10^{-22} kg · m/s (c) 278 keV, 3.21×10^{-22} kg · m/s
27. (a) 2.88 pm (b) 101°
29. It is sufficient because Compton's equation and the conservation of vector momentum give three independent equations in the unknowns λ', λ_0, and u. Wavelength is 3.82 pm
31. (a) 0.667 (b) 0.001 09
33. (a) 14.0 kV/m, 46.8 μT (b) 4.19 nN (c) 10.2 g
35. (a) 0.174 nm (b) 5.37 pm or 5.49 pm ignoring relativistic correction
37. (a) ~ 100 MeV or more (b) ~ −1 MeV. No. With kinetic energy much larger than the magnitude of its negative electric potential energy, the electron would immediately escape.
39. (b) No. $\lambda^{-2} + \lambda_C^{-2}$ cannot be equal to λ^{-2}
41. (a) 14.9 keV or, ignoring relativistic correction, 15.1 keV (b) 124 keV
43. (a) 3.91×10^4 (b) 20.0 GeV/c = 1.07×10^{-17} kg · m/s (c) 6.22×10^{-17} m, small compared with the size of the nucleus. The scattering of the electrons can give information about the particles forming the nucleus.
47. (a) 993 nm (b) 4.96 mm (c) If its detection forms part of an interference pattern, the neutron must have passed through both slits. If we test to see which slit a particular neutron passes through, the neutron will not form part of the interference pattern.
49. Within 1.16 mm for the electron, 5.28×10^{-32} m for the bullet
51. 3.79×10^{28} m, 190 times the diameter of the observable Universe
53. (b) 519 am
55. (a)

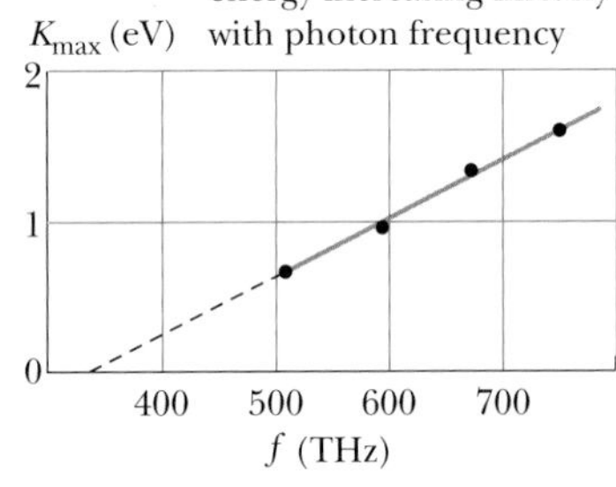

(b) 6.4×10^{-34} J · s ± 8% (c) 1.4 eV
57. $\dfrac{hc}{\lambda} - \dfrac{e^2B^2R^2}{2m_e}$
63. 0.143 nm. The wavelength is comparable to the distance between atoms in a crystal, so diffraction can be observed.
67. 2.81×10^{-8}

CHAPTER 41

1. (a) 126 pm (b) 5.27×10^{-24} kg · m/s (c) 95.5 eV
3. (a) 0.434 nm (b) 6.00 eV
5. (a)

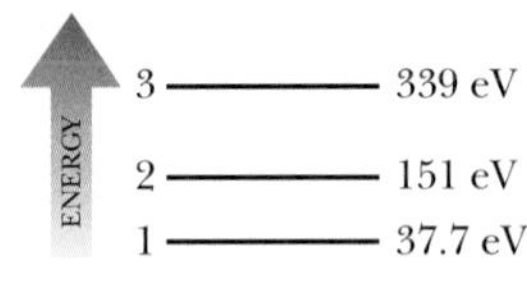

(b) 2.20 nm, 2.75 nm, 4.12 nm, 4.71 nm, 6.60 nm, 11.0 nm
7. 0.793 nm
9. 6.16 MeV, 202 fm, a gamma ray
11. 0.513 MeV, 2.05 MeV, 4.62 MeV. They do; the MeV is the natural unit for energy radiated by an atomic nucleus.
13. (a) $\Delta p \approx \hbar/2L$ (b) $E \approx \hbar^2/8mL^2$. This estimate is too low by $4\pi^2 \approx 40$ times. It correctly displays the pattern of dependence of the energy on the mass and on the length of the well.
17. At $L/4$ and at $3L/4$. We look for $\sin(2\pi x/L)$ taking on its extreme values 1 and −1 so that the squared wave function is as large as it can be.
19. (a) 0.196 (b) 0.609 (c) The classical probability, $\frac{1}{3}$, is very different.
23. (a) $E = \hbar^2/mL^2$

(b) $\displaystyle\int_{-L}^{L} A^2(1 - x^2/L^2)^2 dx = 1$ gives $A = \left(\dfrac{15}{16L}\right)^{1/2}$.

(c) $\frac{47}{81} = 0.580$
27. (a)

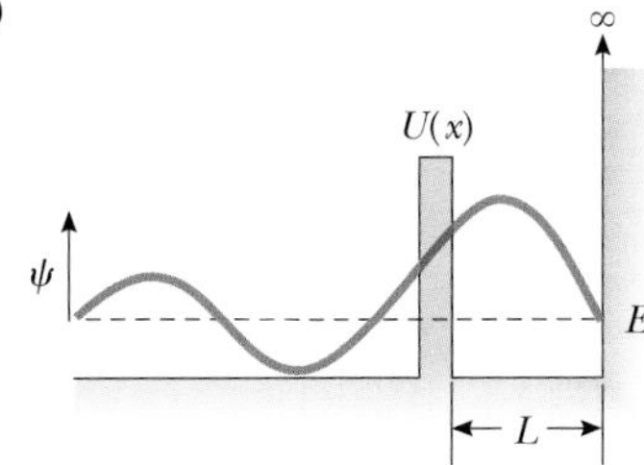

(b) $2L$
29. By 0.959 nm, to 1.91 nm
31. (a) 0.010 3 (b) 0.990
33. 1.35
37. 600 nm
39. (b) The acceleration is equal to a negative constant times the excursion from equilibrium. The frequency is $\dfrac{1}{2\pi}\sqrt{\dfrac{k}{\mu}}$.
41. $\sim 10^{-10^{30}}$
43. (a)

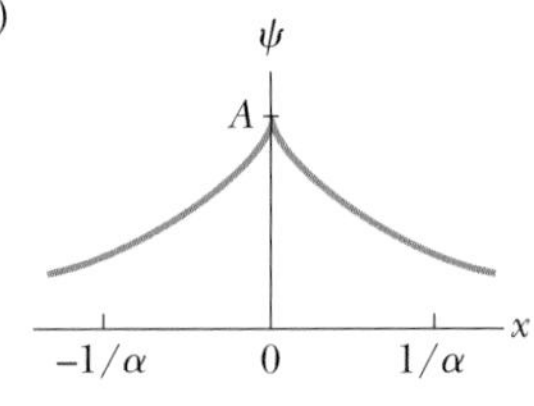

(b)

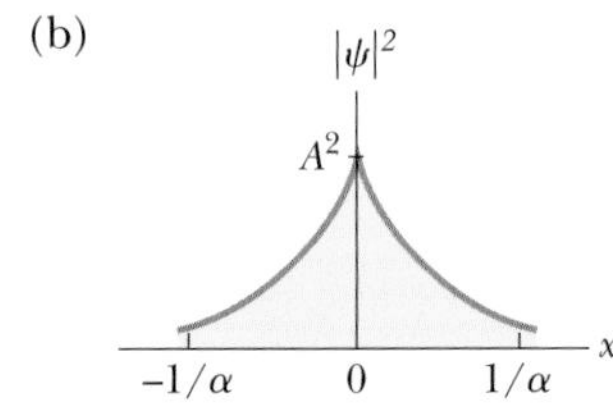

(c) The wave function is continuous. It shows localization by approaching zero as $x \to \pm\infty$. It is everywhere finite and can be normalized. (d) $A = \sqrt{\alpha}$ (e) 0.632

45. (b) For the transition from level 2 to level 1, $\lambda = 1.38\ \mu$m infrared. For level 4 to level 1, $\lambda = 276$ nm ultraviolet. For level 3 to level 2, $\lambda = 827$ nm infrared. For level 4 to level 2, $\lambda = 344$ nm near ultraviolet. For level 4 to level 3, $\lambda = 590$ nm yellow-orange visible.

47. 0.029 4

49. (a) 434 THz (b) 691 nm (c) 165 peV or more

53. (a)

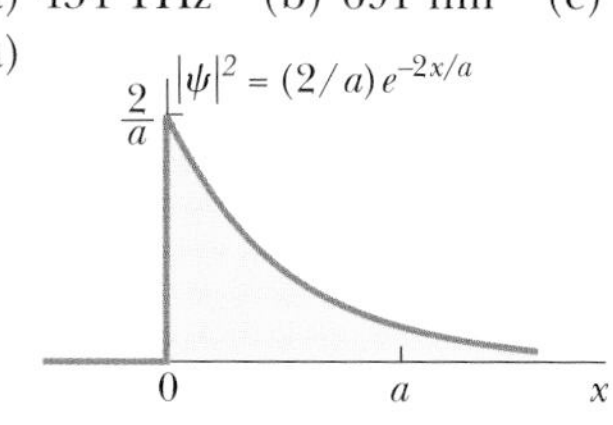

(b) 0 (c) 0.865

55. (a) $-7k_ee^2/3d$ (b) $\hbar^2/36m_ed^2$ (c) 49.9 pm (d) The lithium interatomic spacing of 280 pm is 5.62 times larger. Therefore, it is of the same order of magnitude as the interatomic spacing $2d$ here.

57. (a) $A = \left(\dfrac{2}{17L}\right)^{1/2}$ (b) $|A|^2 + |B|^2 = 1/a$

59. 2.25

CHAPTER 42

1. (a) 5 (b) no; no

3. (b) 0.846 ns

5. (a) 1.89 eV, 656 nm (b) 3.40 eV, 365 nm

7. (a) 0.212 nm (b) 9.95×10^{-25} kg · m/s (c) 2.11×10^{-34} kg · m²/s (d) 3.40 eV (e) −6.80 eV (f) −3.40 eV

9. (a) $E_n = -54.4\text{ eV}/n^2$ for $n = 1, 2, 3, \ldots$

(b) 54.4 eV

11. $r_n = (0.106\text{ nm})n^2$, $E_n = -6.80\text{ eV}/n^2$, for $n = 1, 2, 3, \ldots$

13. (a) 1.31 μm (b) 164 nm

15. (a) $\Delta p \geq \hbar/2r$ (b) Choosing $p \approx \hbar/r$, we find that $E = K + U = \hbar^2/2m_er^2 - k_ee^2/r$. (c) $r = \hbar^2/m_e k_e e^2 = a_0$ and $E = -13.6$ eV, in agreement with the Bohr theory

17. (b) 0.497

19. It does, with $E = -k_ee^2/2a_0$.

21. (a)

n	ℓ	m_ℓ	m_s
3	2	2	$\frac{1}{2}$
3	2	2	$-\frac{1}{2}$
3	2	1	$\frac{1}{2}$
3	2	1	$-\frac{1}{2}$
3	2	0	$\frac{1}{2}$
3	2	0	$-\frac{1}{2}$
3	2	−1	$\frac{1}{2}$
3	2	−1	$-\frac{1}{2}$
3	2	−2	$\frac{1}{2}$
3	2	−2	$-\frac{1}{2}$

(b)

n	ℓ	m_ℓ	m_s
3	1	1	$\frac{1}{2}$
3	1	1	$-\frac{1}{2}$
3	1	0	$\frac{1}{2}$
3	1	0	$-\frac{1}{2}$
3	1	−1	$\frac{1}{2}$
3	1	−1	$-\frac{1}{2}$

23. (a) Find the orbital quantum number of an electron in a state in which it has orbital angular momentum equal to 4.714×10^{-34} J · s. (b) orbital quantum number = 4

25. (a) 2 (b) 8 (c) 18 (d) 32 (e) 50

27. (a) 3.99×10^{17} kg/m³ (b) 81.7 am (c) 1.77 Tm/s (d) It is $5.91 \times 10^3 c$, which is huge compared with the speed of light and impossible.

29. $n = 3$; $\ell = 2$; $m_\ell = -2, -1, 0, 1$, or 2; $s = 1$; $m_s = -1, 0$, or 1, for a total of 15 states

31. The 4*s* subshell is filled first. We would expect [Ar]$3d^44s^2$ to have lower energy, but [Ar]$3d^54s^1$ has more unpaired spins and lower energy according to Hund's rule. It is the ground-state configuration of chromium.

33. aluminum

35. (a) 1*s*, 2*s*, 2*p*, 3*s*, 3*p*, 4*s*, 3*d*, 4*p*, 5*s*, 4*d*, 5*p*, 6*s*, 4*f*, 5*d*, 6*p*, 7*s* (b) Element 15 should have valence +5 or −3, and it does. Element 47 should have valence −1, but it has valence +1. Element 86 should be inert, and it is.

37. 18.4 T

39. 1.4 and 1.0. When the outermost electron in sodium is promoted from the 3*s* state into a 3*p* state, its wave function still overlaps somewhat with the ten electrons below it. It therefore sees the $+11e$ nuclear charge not fully screened and on average moves in an electric field like that created by a particle with charge $+11e - 9.6e = 1.4e$. When this valence electron is lifted farther to a 3*p* state, it is essentially entirely outside the cloud of ten electrons below it and moves in the field of a net charge $+11e - 10e = 1e$.

41. 0.072 5 nm

43. iron

45. 28.2 THz, 10.6 μm, infrared

47. 3.49×10^{16} photons

49. (a) 217 nm (b) 93.1 nm

51. (a) 609 μeV (b) 6.9 μeV (c) 147 GHz, 2.04 mm

53. The classical frequency is $4\pi^2 m_e k_e^2 e^4/h^3 n^3$.

55. (a) −8.16 eV, −2.04 eV, −0.902 eV, −0.508 eV, −0.325 eV (b) 1 090 nm, 811 nm, 724 nm, 609 nm (d) The spectrum could be that of hydrogen, Doppler-shifted by motion away from us at speed $0.471c$.

57. (a) 1.57×10^{14} m$^{-3/2}$ (b) 2.47×10^{28} m^{-3} (c) 8.69×10^{8} m^{-1}

59. (a) 4.20 mm (b) 1.05×10^{19} photons (c) 8.82×10^{16}/mm^3

61. $3h^2/4mL^2$

65. 5.39 keV

67. 0.125

69. 9.79 GHz

CHAPTER 43

1. (a) 921 pN toward the other ion (b) −2.88 eV

3. (a) $(2A/B)^{1/6}$ (b) $B^2/4A$ (c) 74.2 pm, 4.46 eV

5. ~ 10 K

7. (a) 40.0 μeV, 9.66 GHz (b) If r is 10% too small, then f is 20% too large.

9. 5.63 Trad/s

13. 2.72×10^{-47} kg · m^2

15. 0.358 eV

17. (18.4 μeV)$J(J + 1)$, where J = 0, 1, 2, 3, . . .

19. 2.9×10^{-47} kg · m^2

21. only 64.1 THz

23. −7.84 eV

27. An average atom contributes 0.981 electron to the conduction band.

29. (a) 1.57 Mm/s (b) It is larger by ten orders of magnitude.

31. (a) 4.23 eV (b) 3.27×10^4 K

33. 5.28 eV

37. (a) 1.10 (b) 1.47×10^{-25}. It is vastly smaller. Very few states well above the Fermi energy are occupied at room temperature.

39. (a) 275 THz (b) 1.09 μm

41. The gap should be less than or equal to 1.24 eV. Silicon's energy gap of 1.14 eV means that it can absorb the energy of nearly all the photons in sunlight and is an appropriate material for a solar energy collector.

43. 226 nm

45. (a) 59.5 mV (b) −59.5 mV

47. 4.19 mA

49. 203 A to produce a magnetic field in the direction of the original field

51. (a)

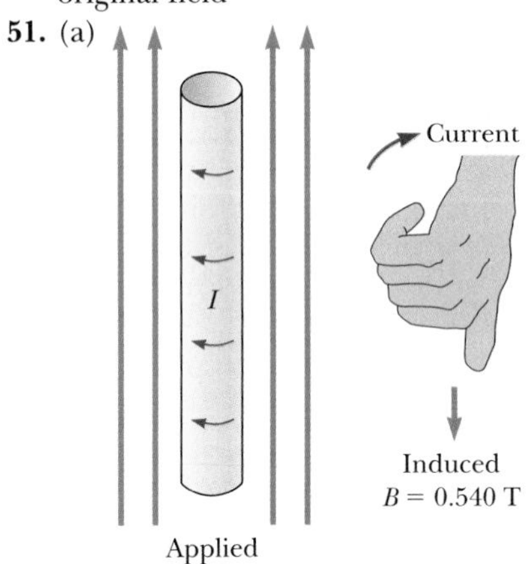

(b) 10.7 kA

53. (a) 6.15×10^{13} Hz (b) 1.59×10^{-46} kg · m^2 (c) 4.79 μm or 4.96 μm

55. 7

59. (a) 0.350 nm (b) −7.02 eV (c) $-1.20\hat{\mathbf{i}}$ nN

61. (a) r_0 (b) B (c) $(a/\pi)[B/2\mu]^{1/2}$ (d) $B - (ha/\pi)[B/8\mu]^{1/2}$

CHAPTER 44

1. ~ 10^{28}; ~ 10^{28}; ~ 10^{28}

3. (a) 2.81×10^{-5} (b) 1.38×10^{-14}

5. (a) 455 fm (b) 6.04 Mm/s

7. 16.0 km

9. (a) 1.11 MeV/nucleon (b) 7.07 MeV/nucleon (c) 8.79 MeV/nucleon (d) 7.57 MeV/nucleon

11. (a) $^{139}_{55}$Cs (b) $^{139}_{57}$La (c) $^{139}_{55}$Cs

13. greater for $^{15}_{7}$N by 3.54 MeV

15. 7.93 MeV

17. 200 MeV

19. 1.16 ks

21. (a) 1.55×10^{-5}/s, 12.4 h (b) 2.39×10^{13} atoms (c) 1.88 mCi

23. 9.47×10^{9} nuclei

25. 1.41

27. 2.66 d

29. 4.27 MeV

31. (a) $N_d = 10^6(1 - e^{-0.069\,3t})$, where t is in hours (b) The number of daughter nuclei starts from zero at $t = 0$. It first increases most rapidly, at 6.93×10^4/h, and then more and more slowly. Its rate of change approaches zero while the number asymptotically approaches 1.00×10^6 as t increases without limit.

33. (a) $e^- + p \rightarrow n + \nu$ (b) $^{15}_{8}$O atom → $^{15}_{7}$N atom + ν (c) 2.75 MeV

35.

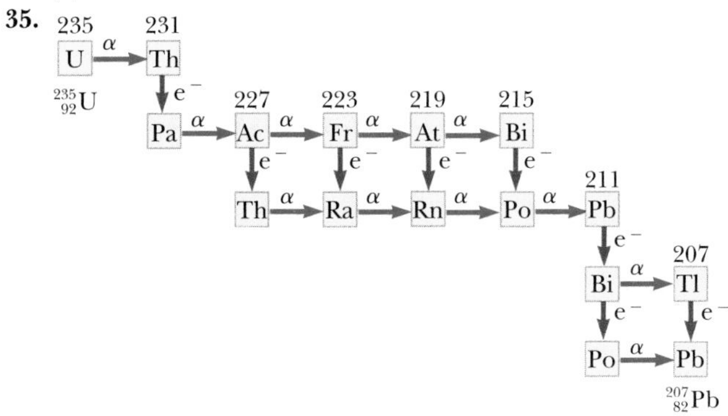

37. (a) 148 Bq/m^3 (b) 7.05×10^7 atoms/m^3 (c) 2.17×10^{-17}

39. (a) $^{21}_{10}$Ne (b) $^{144}_{54}$Xe (c) $^{0}_{1}$e$^+$ + ν

41. $^{197}_{79}$Au + $^{1}_{0}$n → $^{198}_{80}$Hg + $^{0}_{-1}$e$^-$ + $\bar{\nu}$ (b) 7.89 MeV

43. 8.005 3 u; 10.013 5 u

45. (a) 29.2 MHz (b) 42.6 MHz (c) 2.13 kHz

47. (a) 5.70 MeV (b) 3.27 MeV, exothermic

49. (b) 1.53 MeV

51. Find the reaction energy (Q value) of the reaction $^{10}_{5}$B + $^{4}_{2}$He → $^{13}_{6}$C + $^{1}_{1}$H. Solving, Q = 4.06 MeV is the energy released by the reaction as it is converted from rest energy into other forms.

53. (a) The process cannot occur because energy input would be required. (b) can occur (c) K_e between 0 and 156 keV

55. (a) conservation of energy (b) Electric potential energy of the parent nucleus can supply the required energy. (c) 1.20 MeV

57. (a) 61.8 Bq/L (b) 40.3 d

59. 5.94 Gyr
61. 2.20 μeV
63. 0.400%
65. (a) 3.91 Gyr (b) It could be no older. We must make some assumption about the original quantity of radioactive material. In part (a), we assumed the rock originally contained no strontium.

CHAPTER 45

1. 0.387 g
3. ${}^{144}_{54}$Xe, ${}^{143}_{54}$Xe, and ${}^{142}_{54}$Xe
5. ${}^{1}_{0}$n + ^{232}Th $\rightarrow$ ^{233}Th $\rightarrow$ ^{233}Pa + e$^-$ + $\overline{\nu}$, ^{233}Pa $\rightarrow$ ^{233}U + e$^-$ + $\overline{\nu}$
7. 5.80 Mm
9. approximately 3 000 yr
11. 2.68×10^5
13. (a) 4.55×10^{-24} kg $\cdot$ m/s (b) 0.146 nm. This size is the same order of magnitude as an atom's outer electron cloud and is vastly larger than a nucleus.
15. (a) 31.9 g/h (b) 122 g/h
17. (a) 3.24 fm (b) 444 keV (c) $\frac{2}{5}v_i$ (d) 740 keV (e) possibly by tunneling
19. (a) 2.22 Mm/s (b) $\sim 10^{-7}$ s
21. (a) 10^{14} cm^{-3} (b) 1.24×10^5 J/m^3 (c) 1.77 T
23. 1.66×10^3 yr
25. (a) 2.5 mrem per x-ray image (b) The technician's occupational exposure is high compared to background radiation; it is 38 times 0.13 rem/yr.
27. 2.09×10^6 s
29. 1.14 rad
31. 3.96×10^{-4} J/kg
33. (a) 3.12×10^7 (b) 3.12×10^{10} electrons
35. 4.45×10^{-8} kg/h
37. (b) R/λ
39. (a) 8×10^4 eV (b) 4.62 MeV and 13.9 MeV (c) 1.03×10^7 kWh
41. (a) 177 MeV (b) $K_{Br} = 112$ MeV, $K_{La} = 65.4$ MeV (c) $v_{Br} = 15.8$ Mm/s, $v_{La} = 9.20$ Mm/s
43. 482 Ci, less than the fission inventory by on the order of a hundred million times
45. 2.56×10^4 kg
47. (a) 333 μCi, 15.5 μCi, 312 μCi (b) 50.4%, 2.35%, 47.3% (c) It is potentially dangerous, notably if the material is inhaled as a powder. With precautions to minimize human contact, microcurie sources can be routinely used in laboratories.
49. (a) 2.65 GJ (b) The fusion energy is 78.0 times larger.
51. (a) 4.91×10^8 kg/h = 4.91×10^5 m^3/h (b) 0.141 kg/h
53. (a) 15.5 cm (b) 51.7 MeV (c) The number of decays per second is the decay rate R, and the energy released in each decay is Q. Then the energy released per unit time interval is $\mathcal{P} = QR$. (d) 227 kJ/yr (e) 3.18 J/yr
55. 14.0 MeV or, ignoring relativistic correction, 14.1 MeV
57. (a) 2.24×10^7 kWh (b) 17.6 MeV (c) 2.34×10^8 kWh (d) 9.36 kWh (e) Coal is cheap at this moment in human history. We hope that safety and waste disposal problems can be solved so that nuclear energy can be affordable before scarcity drives up the price of fossil fuels.
59. (b) 26.7 MeV
61. (a) 5×10^7 K (b) 1.94 MeV, 1.20 MeV, 1.02 MeV, 7.55 MeV, 7.30 MeV, 1.73 MeV, 1.02 MeV, 4.97 MeV, 26.7 MeV (c) Most of the neutrinos leave the star directly after their creation, without interacting with any other particles.
63. (b) 1.00 μs (c) 2.83 km/s (d) 14.6 μs (e) yes; 107 kilotons of TNT

CHAPTER 46

1. 453 ZHz; 662 am
3. (a) 558 TJ (b) $\$2.17 \times 10^7$
5. 118 MeV
7. $\sim 10^{-18}$ m
9. 67.5 MeV, 67.5 MeV/c, 16.3 ZHz
11. (a) 0.782 MeV (b) $v_e = 0.919c$, $v_p = 380$ km/s (c) The electron is moving faster than one-tenth the speed of light, and the proton is not. Relativistic kinetic energy and momentum equations need be used only for the electron.
13. (a) muon lepton number L_μ and electron lepton number L_e (b) charge (c) baryon number (d) baryon number (e) charge
15. (a) $\overline{\nu}_\mu$ (b) ν_μ (c) $\overline{\nu}_e$ (d) ν_e (e) ν_μ (f) $\overline{\nu}_e + \nu_\mu$
17. Reactions (a), (c), and (f) violate baryon number conservation. Reactions (b), (d), and (e) can occur. Reaction (f) violates muon lepton number conservation.
19. (b) $E_e = E_\gamma = 469$ MeV, $p_e = p_\gamma = 469$ MeV/c (c) 0.999 999 4c
21. Reactions (b) and (c) conserve strangeness. Reactions (a), (d), (e), and (f) violate strangeness conservation.
23. (a) electron and muon lepton numbers (b) electron lepton number (c) charge and strangeness (d) baryon number (e) strangeness
25. (a) baryon number, charge, L_e and L_τ (b) baryon number, charge, L_e, L_μ, and L_τ (c) strangeness, charge, L_e, L_μ, and L_τ (d) baryon number, strangeness, charge, L_e, L_μ, and L_τ (e) baryon number, strangeness, charge, L_e, L_μ, and L_τ (f) baryon number, strangeness, charge, L_e, L_μ, and L_τ
27. 9.26 cm
29. (b) 5.63 GeV (c) 768 MeV (d) 280 MeV (e) 4.43 TeV
35. (a) The reaction $\overline{\text{u}}$d + uud $\rightarrow$ $\overline{\text{s}}$d + uds has a total of 1 u, 2 d, and 0 s quarks originally and finally. (b) The reaction $\overline{\text{d}}$u + uud $\rightarrow$ $\overline{\text{s}}$u + uus has a net of 3 u, 0 d, and 0 s before and after. (c) The reaction $\overline{\text{u}}$s + uud $\rightarrow$ $\overline{\text{s}}$u + $\overline{\text{s}}$d + sss shows conservation at 1 u, 1 d, and 1 s quark. (d) The process uud + uud $\rightarrow$ $\overline{\text{s}}$d + uud + $\overline{\text{d}}$u + uds nets 4 u, 2 d, and 0 s quarks initially and finally; the mystery particle is a Λ^0 or a Σ^0.
37. (a) Σ^+ (b) π^- (c) K^0 (d) Ξ^-
39. (a) 3.32 am/s (b) 690 pm/s
41. (a) 0.160c (b) 2.82×10^9 ly
43. (a) 590.07 nm (b) 597 nm (c) 661 nm
45. (a) 8.41×10^6 kg (b) No. It is only the fraction 4.23×10^{-24} of the mass of the Sun.
47. 3.15 μW/m^2
49. (a) $\sim 10^{13}$ K (b) $\sim 10^{10}$ K
51. (b) 11.8 Gyr
53. (a) 1.61×10^{-35} m (b) 5.38×10^{-44} s, of the same order of magnitude as the ultrahot epoch (c) Yes. The opaque fireball of the Big Bang, measured as the cosmic microwave background radiation, prevents us from receiving visible light from things before the Universe was a few hundred thousand years old. Walls of more profound fire hide all information from still earlier times.

55. 19 eV/c^2
57. one part in 50 000 000
59. (a) $2Nmc$ (b) $3^{1/2}Nmc$ (c) method (a)
61. 5.35 MeV and 32.3 MeV
63. 74.4 MeV
65. 29.8 MeV
67. Diagram (a): electron–positron annihilation; e^-. Diagram (b): a neutrino collides with a neutron, producing a proton and a muon; W^+
69. (a) 127 MeV (b) 1.06 mm (c) 1.17 meV (d) 5.81×10^{19} eV

Index

Locator note: **boldface** indicates a definition; *italics* indicates a figure; *t* indicates a table.